Watching Ultrafast Molecular Motions with 2D IR Chemical Exchange Spectroscopy

Selected Works of M D Fayer

IISc CENTENARY LECTURE SERIES

Series Editor: Anurag Kumar *(Indian Institute of Science, India)*

Description:

World Scientific Publishing Company Singapore and Indian Institute of Science (IISc), Bangalore, will co-publish a series of prestigious lectures delivered during IISc's centenary year (2008-09), and a series of textbooks and monographs, by prominent scientists and engineers from IISc and other institutions.

This pioneering collaboration will contribute significantly in disseminating current Indian scientific advancement worldwide. In addition, the collaboration also proposes to bring the best scientific ideas and thoughts across the world in areas of priority to India through specially designed India editions.

The "IISc Centenary Lecture Series" will comprise lectures by designated Centenary Professors – eminent teachers and researchers from all over the world.

Published

Vol. 1   Trends in Chemistry of Materials
Selected Research Papers of C N R Rao
by C N R Rao

Vol. 2   The Foundations of the Digital Wireless World
Selected Works of A J Viterbi
by A J Viterbi

Vol. 3   Science and Sustainable Food Security
Selected Papers of M S Swaminathan
by M S Swaminathan

Vol. 4   Watching Ultrafast Molecular Motions with 2D IR Chemical Exchange Spectroscopy
Selected Works of M D Fayer
by M D Fayer

Vol. 5   Tailoring Surfaces
Modifying Surface Composition and Structure for Applications in Tribology,
Biology and Catalysis
by N D Spencer

IISc Centenary Lecture Series

Watching Ultrafast Molecular Motions with 2D IR Chemical Exchange Spectroscopy

Selected Works of M D Fayer

Michael D Fayer

Stanford University, USA

NEW JERSEY · LONDON · SINGAPORE · BEIJING · SHANGHAI · HONG KONG · TAIPEI · CHENNAI

Published by

World Scientific Publishing Co. Pte. Ltd.

5 Toh Tuck Link, Singapore 596224

USA office: 27 Warren Street, Suite 401-402, Hackensack, NJ 07601

UK office: 57 Shelton Street, Covent Garden, London WC2H 9HE

British Library Cataloguing-in-Publication Data
A catalogue record for this book is available from the British Library.

IISc Centenary Lecture Series — Vol. 4
**WATCHING ULTRAFAST MOLECULAR MOTIONS WITH 2D IR CHEMICAL
EXCHANGE SPECTROSCOPY**
Selected Works of M D Fayer

Copyright © 2011 by World Scientific Publishing Co. Pte. Ltd.

ISBN-13 978-981-4355-62-9
ISBN-10 981-4355-62-3

Printed by FuIsland Offset Printing (S) Pte Ltd. Singapore

Contents

<u>Part I</u>

Watching Ultrafast Molecular Motions in Liquids, Molecules, and Proteins — Ultrafast 2D IR Echo Chemical Exchange Spectroscopy

Part I

Watching Ultrafast Molecular Motions in Liquids, Molecules, and Proteins — Ultrafast 2D IR Echo Chemical Exchange Spectroscopy

I. Introduction

This monograph is a written rendition of the Indian Institute of Science Centenary Lecture
I presented on July 16, 2008 in Bangalore, India, as part of the celebration of the hundredth
anniversary of the founding of the Institute. In the lecture and in this monograph, I address
the problem of how to experimentally examine fast molecular processes that are occurring
under thermal equilibrium conditions without changing them. A vast amount of chemistry
occurs at or near room temperature in the ground electronic state of molecules. From
synthetic organic chemistry to biological chemistry, chemical processes occur on fast time
scales. Although the rate of an overall process may be slow, key events on the molecular level
are fast. For example, in a second order reaction, the rate of the reaction can be controlled
by the diffusion of the species into contact, but the actual reactive event occurs on a very
fast time scale. The binding of a substrate to an enzyme can be very slow, but the structural
changes of the enzyme or the reaction of the substrate once the substrate finds its way into
the binding site can be very fast. A new approach to this important problem is presented
— ultrafast two dimensional infrared vibrational echo chemical exchange spectroscopy.

Why do we want to study fast molecular dynamics? On Earth, an enormous fraction
of molecular processes in chemistry, biology, and materials science occur at or near room
temperature. These processes are driven by the thermal energy contained in the systems.
They occur in the ground electronic state rather than through promotion to a high energy
state by the absorption of light. While there are important chemical systems that are driven
by the absorption of light, for example, the initial step in photosynthesis, most chemistry
occurs through the dynamics induced by ambient thermal energy.

When I look at nature, I divide it into three distance scales. There is the very large,
10^{20} to 10^{25} meters. These are the distance scales of galaxies and the entire universe. There
is interesting physics to be studied on these large distance scales, the motions of galaxies,
cities of galaxies, and the expansion of the universe. These are fundamental problems of
great complexity. However, on Earth on the time scale of human life, the issues involving
great distances with the associated billion year time scales have little impact on daily life.
Then there is the very small, distances less than 10^{-15} meters. This is the distance scale of
subatomic particle physics. Again there are incredibly complex problems associated with
the properties and interactions of subatomic particles that can answer questions such as the
nature of the big bang. Again, events that occur on subatomic distance scales have little
impact on humans and the Earthly environment we inhabit.

The third distance scale is the nanometer, 10^{-9} meters (nm). This is the size of molecules
and atoms. A typical molecular size is about a nanometer. An atom is a few tenths of a
nanometer. Even proteins, which are very large molecules that can be 50 nm in diameter,

are composed of amino acid subunits that are about a nanometer. The nanometer is the distance scale of chemical processes that are responsible for everything that we see around us every day. Molecular and atomic processes underlie biology and life as well as geology. They are responsible for the production of pharmaceuticals and the production of polymers that are the materials of synthetic cloth and plastics. Therefore, in everyday life, it is the distance scale of the nanometer that is of primary importance.

But why should we be interested in fast molecular dynamics? Molecules are small. Therefore, the intrinsic time scale for molecular motions is very fast, on the order of picoseconds (ps), 10^{-12} seconds. Fundamental steps in molecular processes occur on very fast time scales. Slow processes, which are frequently observed in chemistry and biology, occur through sequences of very fast steps. Furthermore, as mentioned above, most chemistry and biology are driven thermally, that is by the heat that is present in systems at normal temperatures. Therefore, understanding fast molecular dynamics under thermal equilibrium conditions is central to understanding the nature of the world around us.

Over the past thirty years a great deal has been learned about ultrafast chemical processes using visible and ultraviolet femtosecond experimental techniques to study molecules in their electronic excited states. Figure 1 is a schematic representation of an experiment to study ultrafast dynamics of electronically excited molecules. An ultrashort pulse, $\sim$100 fs (femtosecond, 10^{-15} second) is tuned to an electronic absorption of the molecule under study. Following the pump pulse, which promotes the molecule to the electronic excited state, a probe pulse is brought into the sample with a variable delay. The probe pulse and detection electronics records the electronic excited state spectrum of the molecule as a function of time. From the change in the spectrum, details of molecular dynamics can be inferred.

As an example, a system that has been studied extensively using ultrafast electronic excited state spectroscopy is the stilbene trans-cis isomerization. Figure 2 is based on the work of D. C. Todd and G. R. Fleming published in the Journal of Chemical Physics in 1993.[1] The trans form of the stilbene molecule is shown in the lower left portion of the figure. In the ground electronic state, stilbene has a carbon–carbon double bond that prevents rotation about the bond. This is indicated in the potential energy diagram by

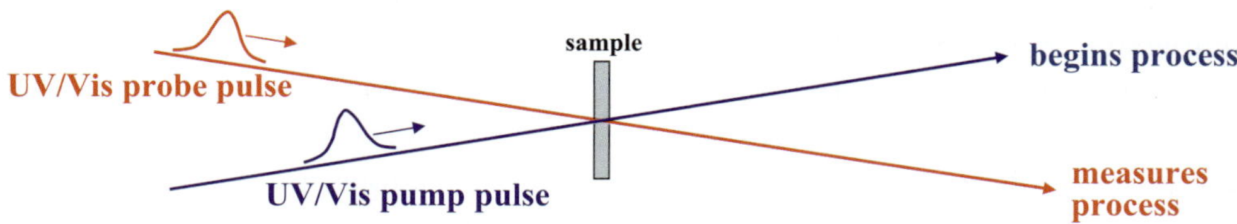

Pump pulse – promotes molecule to electronic excited state.
Probe pulse – measures change in spectrum as time increases.

Fig. 1. Schematic illustration of an electronic excited state transient absorption experiment. The pump pulse promotes molecules to an electronic excited state. The variably time delayed probe pulse measures the dynamics of the electronically excited state system.

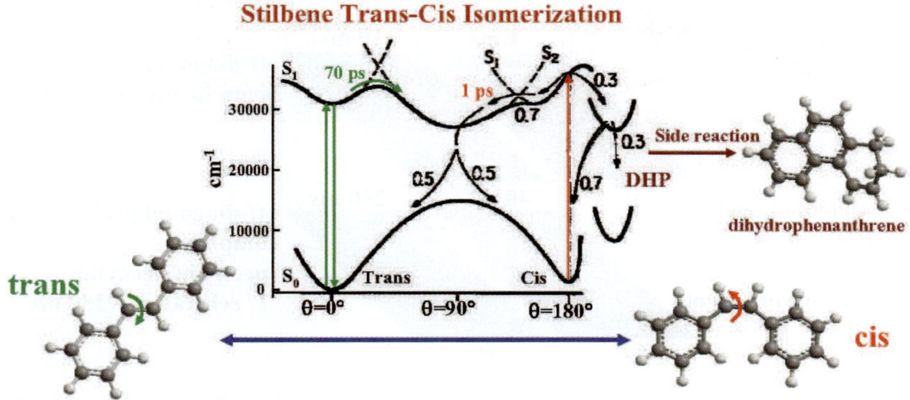

Fig. 2. The molecules and the ground and electronic excited state potential surface diagram for the transient absorption experiment measuring the excited state transformation of trans stilbene to cis stilbene.

the very large barrier for isomerization (rotation around the carbon–carbon bond) to form cis-stilbene, which is shown in the lower right portion of the figure. Therefore, isomerization will not occur in the ground electronic state under thermal equilibrium conditions. However excitation to the electronic excited state with UV light changes the electronic structure of the molecule. Effectively the double bond no longer exists in the excited state, and the molecule can undergo isomerization from trans to cis. When the molecules relax to the ground state, some of them will be trapped in the cis configuration. These can be pumped again to the electronic excited state, and some of them will undergo cis to trans isomerization. The times for these processes are shown in the figure. Placing molecules in an electronic excited state can also cause them to undergo irreversible chemistry. Such a photochemical reaction is shown in the upper right portion of Figure 2. The high energy state of the molecules when they are electronically excited causes some of the stilbene to convert to dihydrophenanthrene.

The type of electronic excited state dynamics experiment briefly illustrated by stilbene isomerization has taught us a great deal about molecular dynamics, but not under thermal equilibrium conditions in the ground electronic state. Nuclear magnetic resonance (NMR) particularly two dimensional (2D) NMR can measure dynamics of molecular systems in the ground electronic state under thermal equilibrium conditions. However, 2D NMR can only measure processes on very slow time scales, on the order of a millisecond (ms). 2D NMR pulse sequences can measure chemical exchange in equilibrium systems. Chemical exchange occurs when chemical species are interconverting, but there is no net change in the number of each species. The pulse sequence labels the species at time, $t = 0$, and then can track the species as they interconvert. Figure 3 is a modified version of a figure published by C. L. Perrin and T. J. Dwyer in Chemical Reviews in 1990.[2] It shows the interconversion of the five species that are obtained when the two liquids, $SnCl_4$ and $SnBr_4$, are combined in equal proportions. There are five molecules that are constantly changing one to another, $SnCl_4$ $\leftrightarrow$ $SnCl_3Br$ $\leftrightarrow$ $SnCl_2Br_2$ $\leftrightarrow$ $SnCl_1Br_3$ $\leftrightarrow$ $SnBr_4$. The upper left hand spectrum at 1 ms

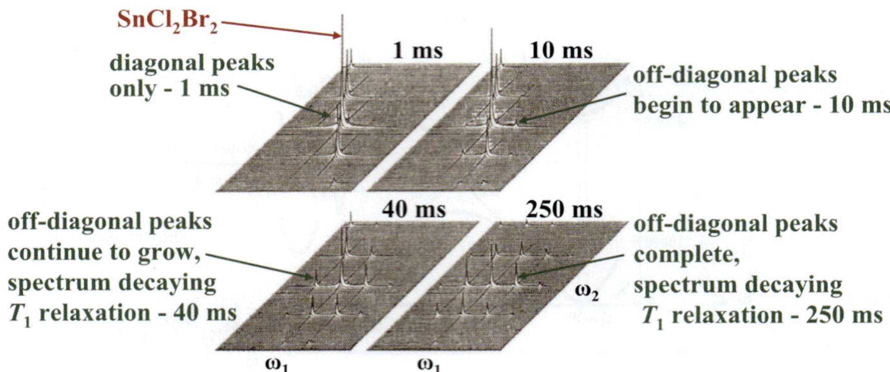

Fig. 3. An illustration of data from a 2D NMR chemical exchange experiment. There are five molecules that are constantly changing one to another, $SnCl_4 \leftrightarrow SnCl_3Br \leftrightarrow SnCl_2Br_2 \leftrightarrow SnCl_1Br_3 \leftrightarrow SnBr_4$. As time proceeds on the millisecond time scale, interconversion of the species causes off-diagonal peaks in the 2D NMR spectrum to grow in.

shows peaks only on the diagonal. There is one peak for each type of molecule. By 10 ms, off-diagonal peaks are beginning to grow in. At 40 ms the off-diagonal peaks are quite prominent. As will be discussed in detail below in the context of the ultrafast 2D infrared (IR) vibrational echo chemical exchange spectroscopy, the growth of the off-diagonal peaks with time is directly related to the time dependence of the exchange of one species into another. Because of the lifetime (T_1) of the spin states prepared by the initial pulses in the pulse sequence, the peaks decay with time. By 250 ms (bottom right spectrum), the off-diagonal peaks have grown to their maximum extent but all of the peaks have shrunk because of the lifetime of the excited spin states. Detailed analysis of the growth of the off-diagonal peaks gives the rates of interconversion of one species into another.

Ultrafast electronic excited state spectroscopy can measure fast chemical dynamics but not under ground state thermal equilibrium conditions. 2D NMR can measure chemical dynamics occurring at thermal equilibrium, but only slow processes. Below the new spectroscopic approach of ultrafast 2D IR vibrational echo chemical exchange spectroscopy will be discussed. It makes possible the measurement of ultrafast molecular dynamics under thermal equilibrium conditions. The method is in many respects analogous to 2D NMR, but it uses vibrations rather than spins to probe molecular dynamics, and it operates on ultrafast time scales. The 2D IR vibrational echo experiments utilize new technology. To set the stage it is useful to compare the histories of the technologies associated with 2D NMR vs. 2D IR vibrational echoes.

The first demonstration of radio took place in approximately 1890. By 1945, radio wave technology was highly developed. It was possible to make short high powered pulses of radio waves and very sensitive radio frequency detectors existed. The first NMR experiment was performed in 1946. These initial experiments involved continuous radio wave irradiation of the sample and scanning the magnetic field. The first pulsed coherent spectroscopy experiment, the NMR spin echo, was performed only four years later in 1950. The spin echo was the first experiment that used pulses to irradiate the sample and detected the

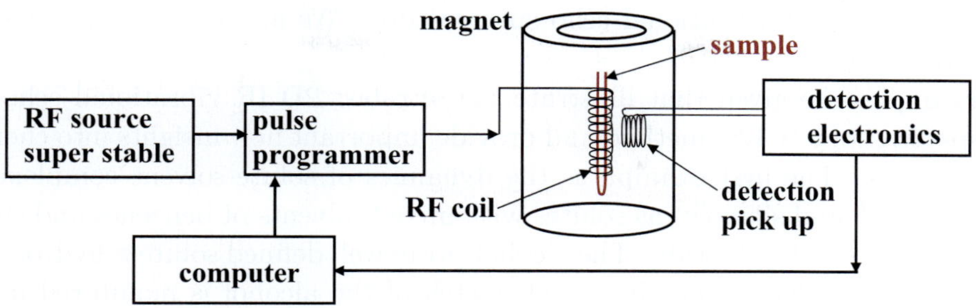

Fig. 4. A schematic of an NMR instrument.

time dependent signal that emerged following the pulses. All modern NMR pulse sequences including 2D NMR are based on extensions of the spin echo experiment. The first 2D NMR experiment was performed in 1976, and the first magnetic resonance imaging experiment was performed in approximately 1977. Today, thirty years later, there have been tremendous advances in all areas of NMR that have been made possible by continued improvements in the underlying technology.

Figure 4 shows a schematic diagram of an NMR machine. Today, we have super stable RF sources, pulse program electronics to make any sequence of pulses with variable pulse timings, lengths, and phases. There exist highly sensitive low noise detectors and very sophisticated software for data processing. One hundred and twenty years after the advent of radio frequency technology and more than sixty years after the first NMR experiment, NMR is very well developed and mature spectroscopy.

In contrast, the technology for 2D IR vibrational echo spectroscopy is very new. The first laser, which produced visible light, was invented in 1960. The first IR laser, was a gas CO_2 laser in 1964. It generated essentially fixed wavelength light with very long pulses. The first picosecond laser, which produced visible light, was developed in 1970. The first IR picosecond pulses were produced using nonlinear crystals in 1974. While the first simple IR absorption experiment was conducted in ~1700 by Newton, it was not until 1993 that my research group performed the first IR ultrafast coherent experiment, the ultrafast 1D vibrational echo. To perform these first experiments on liquids, glasses, and proteins required the use of a free electron laser (FEL) to generate the necessary short IR pulses. The FEL, located in the Stanford Physics Department, is approximately two football fields long, and required a crew of about 10 people to run it. Although very complex and difficult to use, the FEL made the first experiments possible. The time between the first NMR experiment and the spin echo was four years. The time between the first IR experiment and the first ultrafast vibrational echo experiment was approximately three hundred years. The modern era of ultrafast IR spectroscopy did not really begin until ~2000 when IR femtosecond table top sources were developed. These sources led to the first 2D IR vibrational echo experiments in ~2003. While we are in the ~120th year of RF technology, the ~60th year of NMR, and the ~30th year of 2D NMR, we are only in the ~10th year of ultrafast IR technology necessary to perform 2D IR vibrational echoes and the $\sim$5th year of 2D IR vibrational echo spectroscopy. We are at the beginning of a new era of spectroscopy.

Three examples are given that illustrate the ultrafast 2D IR vibrational echo chemical exchange spectroscopy (CES) method and provide important new insights into chemical and biological systems. The first example is the dynamics of solute-solvent complexes formed between phenols[3–6] and silanols[7] as solutes with mixed solvents of benzenes and substituted benzenes and carbon tetrachloride. The alcohols form well defined solute π hydrogen bonded complexes with the solvent. The hydroxyl stretch of the alcohol is monitored in the CES experiment.

The second example is rotational gauche-trans isomerization around a carbon–carbon single bond of a substituted ethane.[8] During the course of isomerization a molecule exchanges between relatively stable conformations by passing through unstable configurations. Rotational isomerization is a major factor in the dynamics, reactivity, and biological activity of a multiplicity of molecular structures. Techniques such as NMR can only be applied at low temperature to systems with very high barriers to slow the dynamics to the millisecond time scale.[9] CES directly measures the isomerization time for a low barrier system at room temperature and permits a determination of the ethane isomerization time.

The final example is the dynamics of an elementary structural change in a protein. Proteins can fold into a number of well defined structural substates. Under thermal equilibrium conditions, a protein undergoes transitions from one structural substate to another by passing over relatively low barriers. Proteins undergo large conformational changes on long time scales, milliseconds to seconds, such as those that occur following substrate binding to an enzyme. However, these slow transformations occur through a vast number of fast elementary conformational steps. CES was used to make the first measurement of the time for an elementary conformational step.[10]

II. The Ultrafast 2D IR Vibrational Echo Chemical Exchange Spectroscopy Method

As discussed above, the challenge is to directly observe fast time scale thermal equilibrium chemical events without changing the systems equilibrium behavior. Ultrafast two dimensional infrared (2D IR) vibrational echo chemical exchange spectroscopy is meeting this challenge.[3,4,8,10–13] 2D IR vibrational echo chemical exchange spectroscopy is akin to a 2D NMR chemical exchange experiment[14] except that it can operate on picosecond time scales, and it directly probes the structural degrees of freedom through the time evolution of the 2D IR vibrational echo spectrum.

Consider two molecular species that differ structurally, e.g., two distinct molecular structural isomers. A vibrational mode has a frequency that is different for the two species. The molecular vibrational oscillators of each subset are excited simultaneously by the first pulse in the 2D IR vibrational echo pulse sequence. The later pulses in the pulse sequence generate observable signals that are sensitive to interconversion between species even if the aggregate populations of the two distinct species (the observable in linear absorption spectroscopy) do not change. As discussed in detail below, ultrafast 2D IR vibrational echo CES makes it possible to directly measure the rate of interconversion of chemical species under thermal equilibrium conditions.

Details of the methods used for these 2D IR vibrational echo CES experiments have been described previously.[4,12,15,16] Figure 5 is a schematic representation of the experiment and the pulse sequence. There are three excitation pulses, $\sim$60 fs in duration produced with a Ti:Sapphire regenerative amplifier pumped optical parametric amplifier (OPA) (see below). The pulse sequence is shown at the bottom of the figure. The OPA is tuned to the frequency of the vibrational mode under investigation. The IR pulses have sufficient bandwidth to span the spectral region of interest. The times between pulses 1 and 2 and between pulses 2 and 3 are called τ, and T_w, respectively. The vibrational echo signal radiates from the sample at a time $\leq \tau$ after the third pulse in a unique direction. The vibrational echo signals are recorded by scanning τ at fixed T_w. The signal is spatially and temporally overlapped with a local oscillator for heterodyne detection, and the combined pulse is dispersed by a monochromator onto an IR array detector. Heterodyne detection provides both amplitude and phase information. If the vibrational echo is detected as it emerges from the sample, there is no phase information. The experiment produces a 2D spectrum in the frequency domain. The measurements are made in the time domain. To go from the frequency domain to the time domain requires two Fourier transforms. To perform a Fourier transform requires

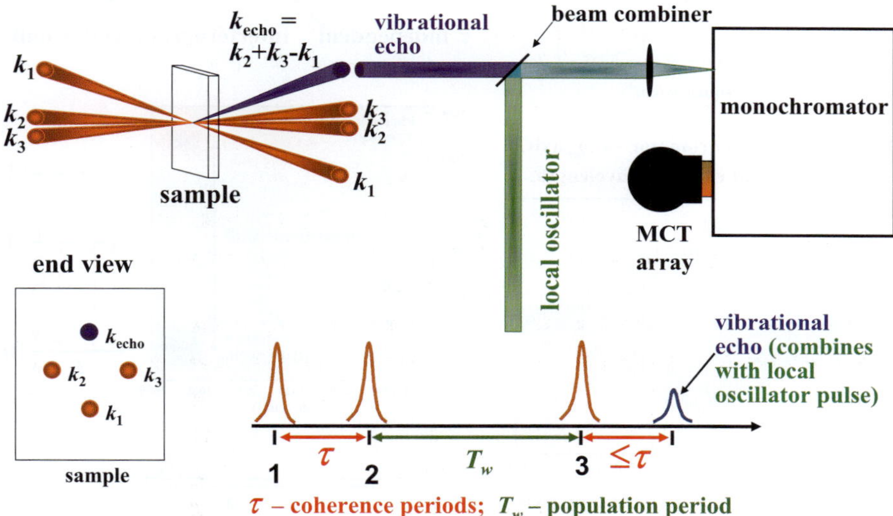

Fig. 5. A schematic representation of the ultrafast 2D IR vibrational echo experiment and the pulse sequence. There are three excitation pulses. The pulse sequence is shown at the bottom of the figure. The times between pulses 1 and 2 and between pulses 2 and 3 are called τ, and T_w, respectively. The vibrational echo signal radiates from the sample in a unique direction at a time $\leq \tau$ after the third pulse. The vibrational echo signals are recorded by scanning τ at fixed T_w. The signal is spatially and temporally overlapped with a local oscillator for heterodyne detection, and the combined pulse is dispersed by a monochromator onto an IR array detector. Heterodyne detection provides both amplitude and phase information.

both amplitude and phase information. The local oscillator is another IR fs pulse. When it overlaps with the vibrational echo wave packet, interference occurs. As shown below, when the time τ is scanned, the vibrational echo pulses moves in time across the fixed local oscillator pulse. The result is a temporal interferogram that provides the necessary phase information.

As shown in Figure 5, the combined vibrational echo/local oscillator pulse is passed through a spectrometer and recorded by an IR array detector. The array detector allows us to measure 32 wavelengths simultaneously, reducing the data acquisition time by a factor of 32. Taking the spectrum of the heterodyne-detected vibrational echo signal performs one of the two Fourier transforms and gives the ω_m axis (m for monochromator) in the 2D IR spectra. When τ is scanned, a temporal interferogram is obtained at each ω_m. The temporal interferograms are numerically Fourier transformed to give the other axis, the ω_τ axis. 2D IR spectra are obtained for a range of T_ws.

Figure 6 shows a very simple example of the nature of the data. The upper left portion of the figure shows the spectrum of the hydroxyl stretching mode of phenol-OD. The H in the OH hydroxyl group has been replaced by a D. The IR absorption spectrum shows a single

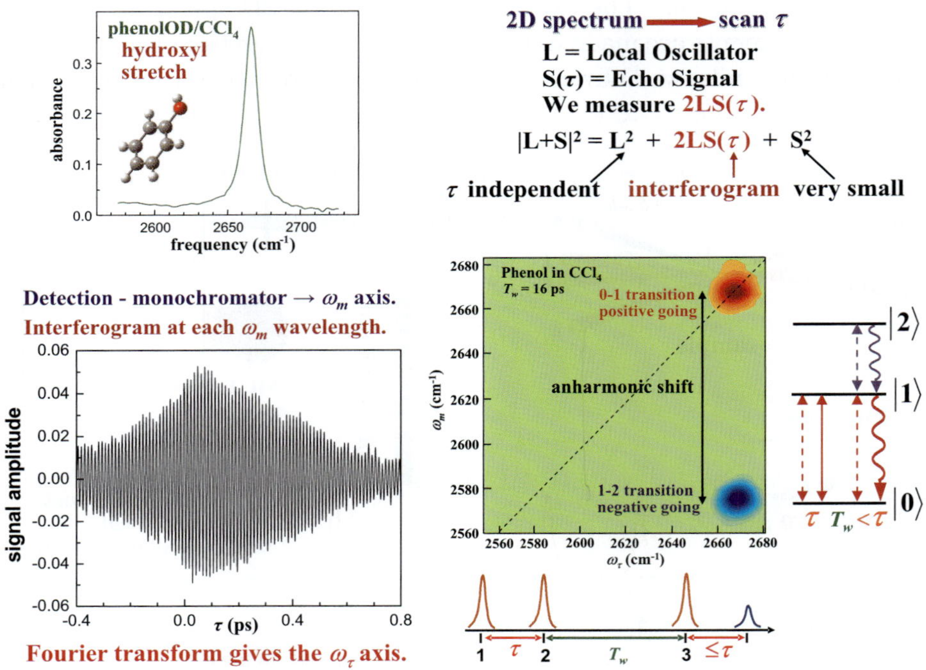

Fig. 6. Illustration of a 2D IR vibrational echo experiment for the hydroxyl stretch of phenol-OD. The upper left shows the spectrum of the OD stretch and the structure of the molecule. The lower left shows an interferogram recorded as τ is scanned at a single frequency ω_m. The lower right shows the 2D IR vibrational echo spectrum. The peak on the diagonal (positive going, red) is from vibrational echo emission at the 0-1 transition frequency. The peak off-diagonal (negative going, blue) is from vibrational echo emission at the 1-2 transition frequency. To the right of the spectrum is an energy level diagram showing the quantum pathways that give rise to the signals (see text).

peak. The structure of the molecule is shown as an inset. As outlined to the right of the spectrum, when τ is scanned, the time of the output of the vibrational echo electric field, which is the signal (S), changes relative to the fixed in time local oscillator electric field (L). L is much larger than S. The detector measures the absolute value squared of the sum of the electric fields. This gives the three terms. L^2 is constant in time. S^2 is very small and not detected relative to the much larger cross term, $2LS(\tau)$. As τ is scanned, the temporal interferogram is created. A portion of an interferogram is shown in the bottom left side of the figure. There is one such interferogram for each monochromator frequency, ω_m. ω_m is the vertical axis of the 2D IR spectrum shown in the lower right portion of the figure. This axis is obtained from the frequency measurements performed by the monochromator. The horizontal axis, the ω_τ axis, is obtained by numerical Fourier transformation of an interferogram like that shown in the figure. There is one interferogram at each ω_m where there is signal.

The 2D IR vibrational spectrum at the time $T_w = 16$ ps has two peaks. The peak on the diagonal (the dashed line) is positive going (red). The peak off-diagonal is negative going (blue). The ω_τ axis is the frequency of the first interaction of the molecules with the radiation field (first pulse). This first interaction is represented by the dashed arrow on the left side of the energy level diagram that is to the right of the spectrum in Figure 6. The first pulse makes a coherent superposition state of the vibrational ground state (0) and the first vibrationally excited state, (1). The second pulse, represented by the solid arrow in the energy level diagram transfers the vibrational coherence to populations in the 0 and 1 levels. Phase information is stored in the populations. The third pulse between 0 and 1 (dashed arrow between 0 and 1) again produces a coherent superposition state (a coherence) which gives rise to the vibrational echo emission (wavy arrow) at the frequency the molecules interact with the third pulse. The ω_m axis is the axis of the vibrational echo emission. The 0-1 vibrational peak is on the diagonal because the first interaction of the molecules with the first radiation field (ω_τ axis) is at the same frequency as the last interaction and vibrational echo emission (ω_m axis).

The off-diagonal 1-2 peak in the spectrum is shifted along the ω_m axis by the vibrational anharmonicity (unequal spacing of the energy levels as shown in the energy level diagram). The first two interactions are the same as discussed above. They produce population in the 1 level. Because the bandwidth of the laser is large, it is greater than the anharmonic shift of the levels. So the third interaction can produce a coherence between the 1 and 2 levels, which is represented by the dashed arrow connecting the 1 and 2 levels. The 1-2 coherence gives rise to vibrational echo emission at the 1-2 transition frequency. Because the first interaction (ω_τ axis) is at the 0-1 transition frequency and the third interaction and vibrational echo emission (ω_m axis) is at the shifted 1-2 transition frequency, the 1-2 peak appears off-diagonal. For what follows, the difference between diagonal and off-diagonal peaks is very important. *If the first and last interactions are at the same frequency, a peak will be on the diagonal. If the first and last interactions are at different frequencies, a peak will be off-diagonal.*

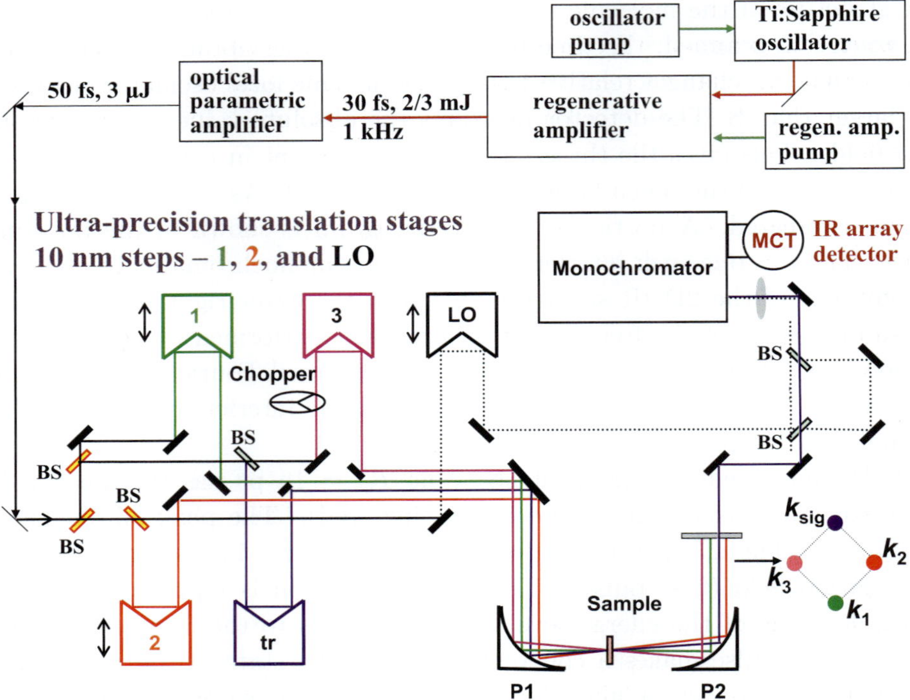

Fig. 7. A schematic of the 2D IR vibrational echo instrument. The upper portion of the figure shows the laser equipment. The lower portion of the figure shows the optical setup.

Figure 7 is a simplified schematic of the equipment used to perform the 2D IR vibrational echo experiments. The upper portion of the figure shows the laser equipment. A CW diode pumped neodymium vanadate laser is used to pump a Ti:Sapphire oscillator, which seeds a Ti:Sapphire regenerative amplifier. The regen is pumped with 7 watts by a 1 kHz diode pumped Nd:YLF Q-switched laser. The output of the regen at 800 nm is 2/3 mJ, 30 fs pulses. The 800 nm pulses drive a multi-stage OPA, which produces 3 μJ pulses with 50 fs duration. These are extremely short pulses. At a wavelength of 4 μm, the pulses are only ∼4 cycles of light. This number of cycles of light would be the equivalent of an ∼6 fs pulse with a wavelength in the middle of the visible spectrum. The lower portion of Figure 7 shows the optical setup that does the vibrational echo experiment. The input IR pulse from the OPA is split into 5 beams using beam splitters (BS). Three of them are the three excitation pulses, 1, 2, and 3 in the pulse sequence shown at the bottom of Figue 5. A fourth pulse is the local oscillator (see Figure 5). The fifth pulse is called the tracer. It is aligned along the path that the echo will emerge from the sample (see Figure 5) and is used only for alignment purposes. It is blocked during the experiments. Pulses 1, 2, and the LO are passed down ultra-precision translation stages labeled 1, 2, and LO. These stages can take 10 nm steps and are exceedingly reproducible in their position. The ultra-precision is necessary because the vibrational echo is a type of IR holography. The interferogram shown in Figure 6 must be accurate and reproducible. The translation stages labeled 3 and tr (tracer) are manual translation stages. Once these timings are set, they do not change. The three excitation beams are focused into the sample using an off-axis parabolic reflector (P1). After passing through the sample the three excitation beams and the vibrational echo beam are collimated by a second off-axis parabolic reflector (P2). The beams other than the vibrational echo are blocked. The LO is split into two beams. One is combined with the vibrational echo. The combined pulses pass through the monochromator and are detected by the upper stripe of the dual stripe IR 32 element array detector. The other portion of the LO passes through the monochromator and is detected by the lower stripe of the array. The signal from the lower stripe is used to normalize the signal on the upper stripe for the IR spectrum and shot-to-shot intensity fluctuations. The signals on the array detector are read out after each laser shot by a computer for processing into the 2D IR vibrational echo spectrum.

A 2D IR vibrational echo chemical exchange experiment works in the following manner. The first laser pulse "labels" the initial structures of the species by establishing their initial frequencies, the ω_τ axis. The second pulse ends the first time period τ and starts the reaction time period T_w during which the labeled species undergo chemical exchange, i.e., one type of species interconverts into the other. The third pulse ends the population period of length T_w and begins a third period of length $\leq \tau$, which ends with the emission of the vibrational echo pulse of frequency ω_m. The vibrational echo signal reads out information about the final structures of all labeled species by their frequencies, ω_m. During the period T_w between pulses 2 and 3, chemical exchange occurs. The exchange causes new off-diagonal peaks to grow in as T_w is increased. The growth of the off-diagonal peaks in the 2D IR spectra with increasing T_w is used to obtain the chemical exchange time constants.

Figure 8 illustrates the influence of chemical exchange between two species A and B that are in chemical equilibrium. A's are constantly turning into B's, and B's are constantly turning into A's, but the net numbers of A's and B's are not changing. The upper third of the figure shows what happens at very short time, a time short compared to any chemical exchange, that is interconversion between A's and B's. In the 0-1 portion of the spectrum there will be two peaks on the diagonal, one for species A at ω_A and one for species B at ω_B. The middle portion of the figures shows what happens at a long time, a time long compared to the time for chemical exchange. The energy level diagrams are like the one in Figure 8 except there are two species and two frequencies. During the T_w period between pulses 2 and 3, A's turn into B's and B's turn into A's. When A's turn into B's, the first interaction is at ω_A but the last interaction is at ω_B. Therefore, an off-diagonal peak will be created in the lower right portion of the schematic spectrum. When B's turn into A's, the first interaction is at ω_B but the last interaction is at ω_A. Therefore, an off-diagonal peak will be created in the upper left portion of the schematic spectrum. The important point is that chemical exchange will cause off-diagonal peaks to grow in as T_w goes from a short time to a long time. The rate of growth of the off-diagonal peaks can be used to directly determine the rate of interconversion of A's and B's under equilibrium conditions. This is analogous to the 2D NMR example shown in Figure 3. The bottom portion of Figure 8 shows the nature of the spectrum at long time when chemical exchange has occurred and

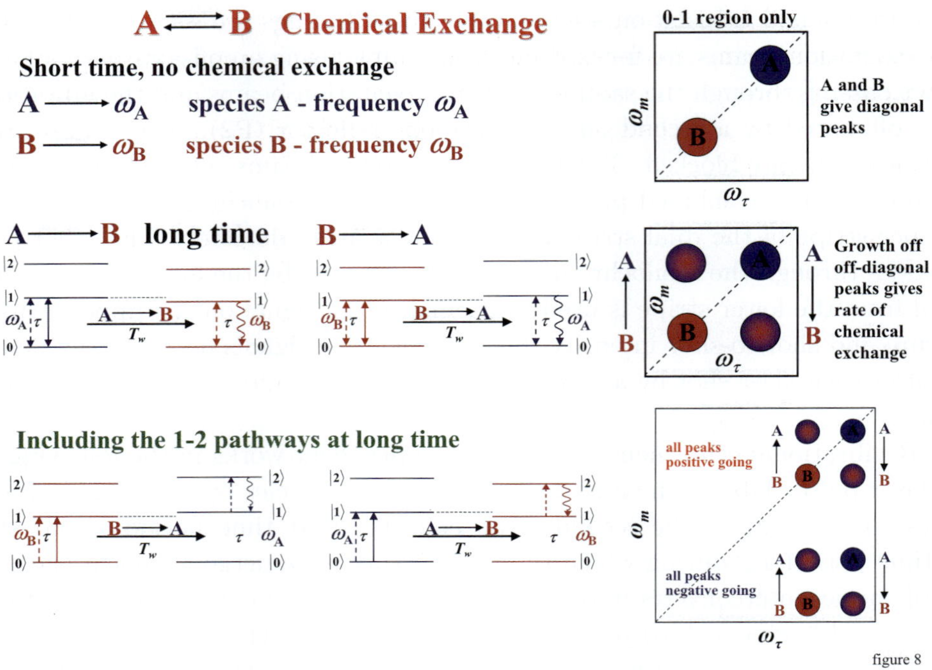

Fig. 8. Illustration of the influence of chemical exchange between two species A and B that are in chemical equilibrium. Upper portion: very short time. No chemical exchange has occurred. There are two peaks on the diagonal, one for each species. Middle portion: sufficient time for chemical exchange to occur. Off-diagonal peaks have grown in. The quantum pathways are shown. Bottom portion: the 0-1 and 1-2 portions of the spectrum are shown at long time. New peaks have grown in in both regions of the spectrum.

the 1-2 portion of the spectrum is included. Chemical exchange causes peaks to form in both the 0-1 and 1-2 regions of the spectrum.

Also needed for the data analysis are the vibrational population relaxation time constants and orientational relaxation time constants, which are measured independently with the polarization selective IR pump-probe experiments.[4, 17] The enthalpies of formation of the solute-solvent complexes discussed below were determined by measuring the temperature dependence of the equilibrium constants with FT-IR.[5, 13] Expression and purification of the mutant sperm whale Mb L29I discussed below were performed as described previously.[18] The CO forms of mutant Mb was prepared according to published protocols.[19]

III. Chemical Exchange Spectroscopy Studies of Structural Dynamics

A. *Solute-Solvent Complexes*

Solvents play a substantial role in practical chemistry by influencing the reactivity of dissolved solutes.[20] Specific intermolecular interactions, such as hydrogen bonding, can lead to structurally characterizable solute-solvent complexes that are constantly forming and dissociating on very short time scales under thermal equilibrium conditions.[21] Dynamics of these transient species can play an important role in the physical and chemical properties of a solute-solvent system by affecting reaction rates, reaction mechanisms, and product ratios.[20] For the majority of organic and other types of non-aqueous solutions, the solute-solvent complexes form and dissociate on sub-nanosecond timescales[3,5,7] that cannot be measured by NMR and other methods.

Figure 9 illustrates some widely used descriptions of a solute in a solvent. The upper portion of the figure shows a solute in a solvent that is treated as a homogenous continuum. The properties of the solvent are described by the solvent's dielectric constant. A more accurate description of a solute in a solvent is shown in the lower portion of the figure. Both the solute and the solvent are described as molecular. They are treated as spheres with some form of potential to account for the intermolecular interactions between the solute molecule with the solvent molecules, and the solvent molecules with each other. The interactions give rise to solvent shells around the solute and a radial distribution function,

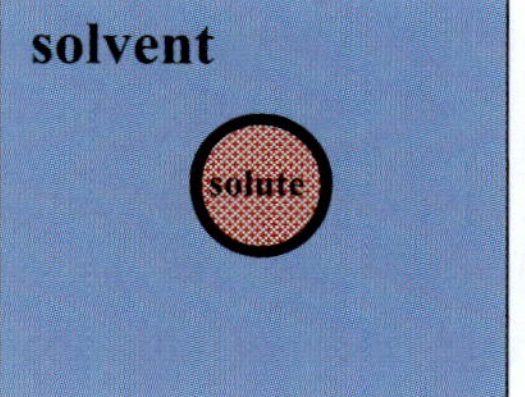

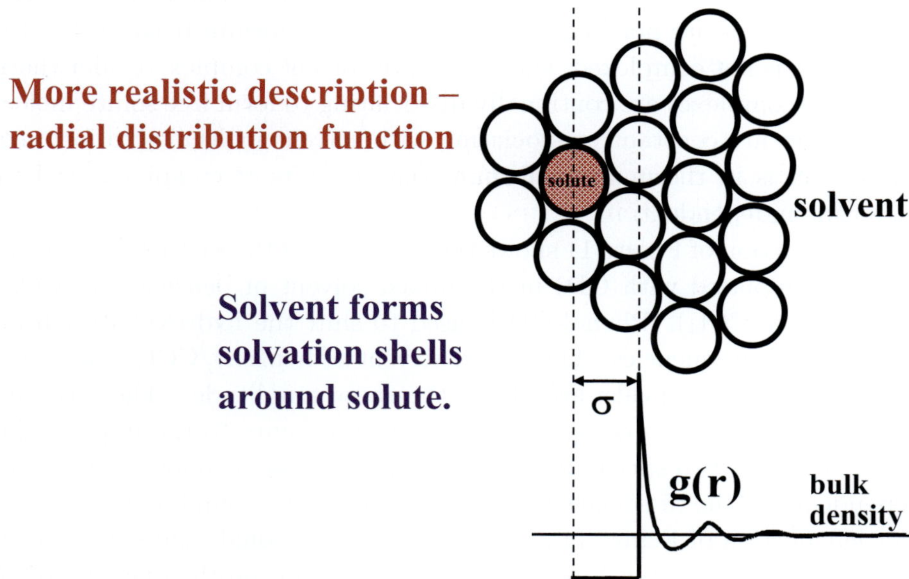

Fig. 9. Models for a solute in a solvent. Upper portion: the solvent is treated as a homogeneous continuum. Lower portion: the solute and solvent are discreet and described in terms of a radial distribution function, $g(r)$.

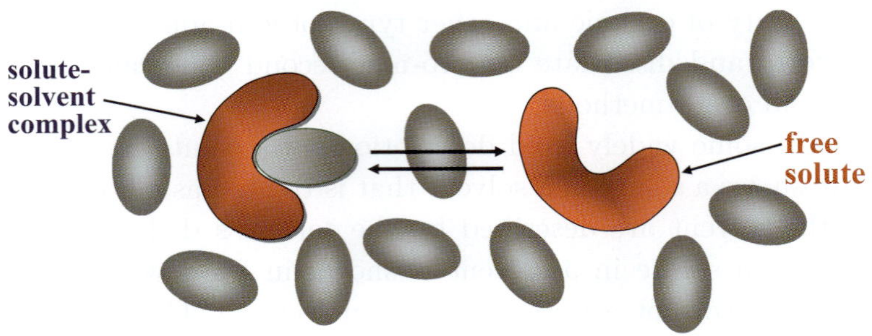

Fig. 10. A schematic representation of an anisotropic solute in an anisotropic solvent. Specific solute-solvent interactions can lead to the formation of well defined solute-solvent complexes. The complexed form of the solute is in equilibrium with the free solute (solute that is not complexed).

$g(r)$, which gives the probability of finding a solvent molecule some distance from the solute. However, this description does not account for the shapes of molecules and the anisotropic intermolecular interactions between a solute and a solvent molecule.

Figure 10 is an illustration of an anisotropic solute molecule and a solute-solvent complex. The anisotropic intermolecular interactions can lead to the formation of a well defined complex. The interactions are weak, comparable to the thermal energy in the system. Therefore, the complexes are not long lived. There is equilibrium between the free solute, that is a solute that is not complexed, and the solute-solvent complex. Under thermal equilibrium conditions, complexes are continually dissociating to form free solute molecules, and free solute molecules are constantly associating with solvent molecules to form complexes. Because the system is in thermal equilibrium, the numbers of complexes and free solute molecules are time independent.

The upper left portion of Figure 11 shows the spectrum of the hydroxyl stretch of phenol-OD (OH hydroxyl replaced with OD) in the mixed solvent of benzene and CCl_4 (29 mol % benzene, 71 mol % CCl_4). Phenol-OD is used to shift the hydroxyl stretch absorption below the C-H stretch frequencies. The mixed solvent of benzene/CCl_4 is used to shift the equilibrium so that the two peaks are about the same amplitude. The structure of free phenol and the phenol-benzene complex is shown in the figure to the upper right.[3,6] The spectrum shows that the two species coexist, but the spectrum cannot yield information on the time dependence of the dissociation and formation of the complexes.

The lower left portion of Figure 11 shows a 2D IR vibrational echo spectrum taken at a T_w short compared to the chemical exchange. The two bands on the diagonal (dashed line) arise from the two peaks in the absorption spectrum. As discussed above, the red bands are positive going and correspond to vibrational echo emission at the 0-1 vibrational transition frequency. The two blue bands (negative going) below the diagonal arise from vibrational

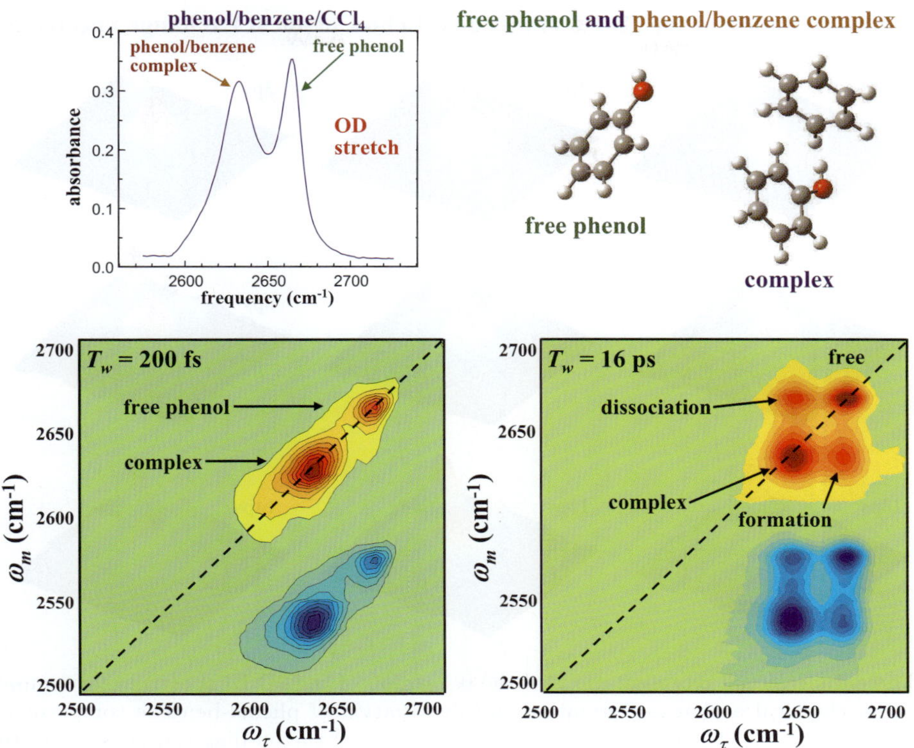

Fig. 11. The spectrum of the hydroxyl (OD) stretch of phenol-OD when it is free and when it has formed a complex with benzene. The structures of free phenol and the phenol-benzene complex are shown. Lower left portion: the 2D IR vibrational echo spectrum at short T_w. No exchange has occurred. Lower right portion: the 2D IR vibrational echo spectrum at $T_w = 16$ ps. Substantial chemical exchange has occurred (complexes have dissociated to become free phenol, and free phenol has formed complexes). New peaks have grown in.

echo emission at the 1-2 transition frequency. These peaks are shifted along the ω_m axis by the vibrational anharmonicity of the OD hydroxyl stretching potential. The ω_τ axis is the frequency of the first interaction with the radiation field (first pulse). The bottom right side of Figure 11 shows the 2D IR spectrum at a time that is sufficiently long for a substantial amount of chemical exchange to have occurred. Off-diagonal peaks have grown in. Consider the peak labeled dissociation in the 0-1 portion of the spectrum. At the time of the first pulse, species that are complexes have frequency $\omega_\tau = 2630$ cm^{-1}. Complexes that dissociate during the interval $T_w = 14$ ps become free and give rise to vibrational echo emission at frequency $\omega_m = 2665$ cm^{-1}. Therefore, the peak will be off-diagonal and to the upper left of the diagonal. This peak corresponds to dissociation. The peak labeled formation arises in the same manner. Free species form complexes and give rise to the other off-diagonal peak. The peaks on the diagonal correspond to species that have not changed their character. The important point is that the time dependent growth of the off-diagonal peaks yields direct information on the rate of chemical exchange. This actual data is completely analogous to the schematic illustration of chemical exchange shown in Fig. 8.

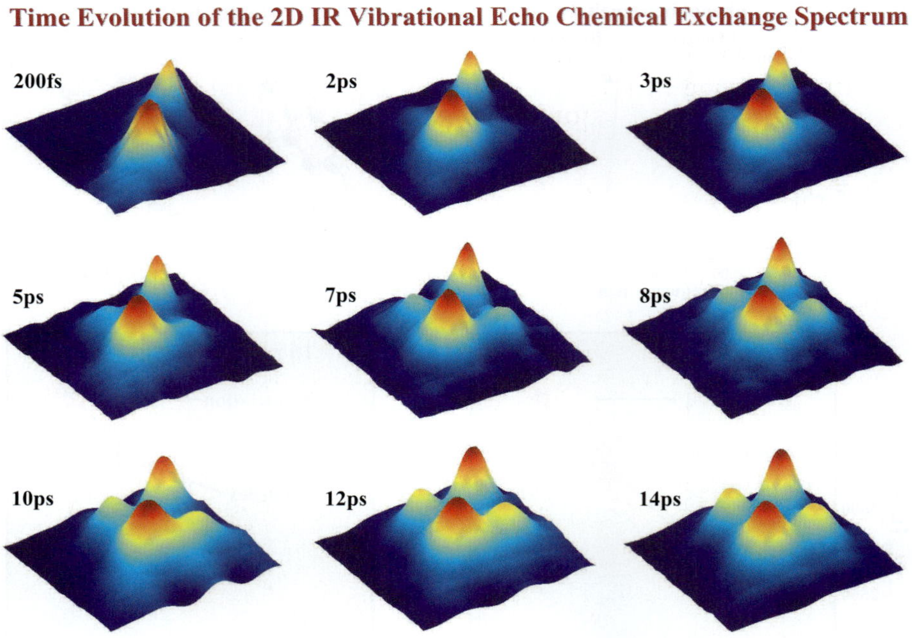

Time Evolution of the 2D IR Vibrational Echo Chemical Exchange Spectrum

200fs   2ps   3ps

5ps   7ps   8ps

10ps   12ps   14ps

Fig. 12. Three dimensional representations of the 0-1 region of the 2D IR spectrum for a number of T_ws. As T_w increases, chemical exchange (formation and dissociation of phenol-benzene complexes) causes off-diagonal peaks to grow in. The time dependence of the growth of the off-diagonal peaks gives the chemical exchange rate.

Figure 12 shows three dimensional representations of the 0-1 region of the 2D IR spectrum for a number of T_ws. As T_w increases, off-diagonal peaks grow in. At 200 fs, no off-diagonal peaks are visible. By 5 ps, they are becoming apparent, and at 14 ps the off-diagonal peaks are very prominent. From simple inspection, it is clear that the chemical exchange time is greater than a picosecond and on the order of a number of picoseconds. To analyze the data in detail, it is necessary to take into account the other dynamics that influence 2D IR vibrational echo CES.[3,12] The vibrational excited states decay to the ground state with lifetimes, T_1. The free and complex species undergo orientational relaxation, with orientational relaxation time constants, τ_r. The T_1 causes decay of all of the peaks to zero. The τ_r causes all of the peaks to be reduced in amplitude, but not decay to zero. The chemical exchange causes the diagonal peaks to decay and the off-diagonal peaks to grow. Another contribution is spectral diffusion, which is caused by thermally induced molecular motions that produce fluctuations in the transition frequency of a species.[22–27] Spectral diffusion changes the shape and amplitude of the peaks, but it does not change the peak volumes.[3,12]

In fitting the data, all of the necessary input parameters except for the chemical exchange rate can be measured independently. The T_1s and τ_rs are measured with IR polarization and wavelength selective pump-probe spectroscopy. The time dependent peak volumes are used rather than the peak amplitudes to eliminate the contribution from spectral diffusion.[3,12] In addition, it is necessary to know the equilibrium constant and the ratio of the complex and free OD stretch transition dipoles. These are obtained with IR absorption spectroscopy.[3,5] Therefore, the time dependence the CES data can be fit with a single adjustable parameter, the complex dissociation time, τ_d, which is the inverse of the dissociation rate constant. The single parameter τ_d can be used because the system is in equilibrium,[3] and therefore, the rate of formation is equal to the rate of dissociation.

Figure 13 shows the results of fitting the CES data for phenol-OD in the mixed benzene-CCl_4 solvent. The T_w dependent data in Figure 13 (symbols) are for the 0-1 portion of the 2D IR spectra. There are four peaks, two diagonal and two off-diagonal. Because the system's thermal equilibrium is not perturbed by excitation of the hydroxyl stretch,[3] the off-diagonal peaks grow in at the same rate. The curves through the data points are obtained from *the fit using the single adjustable parameter,* τ_d. Clearly, the fit is very good. The fit yields the phenol-benzene complex dissociation time, $\tau_d = 10$ ps.[3,5]

Thirteen solute-solvent complexes have been studied with 2D IR vibrational echo CES. Eight of them have phenol or substituted phenols as the solutes,[5] and the other five have triethylsilanol-OD (TES) as the solute.[7] All have solvents of substituted benzenes-CCl_4 mixed solvents except for one which is TES in acetonitrile-CCl_4. The enthalpies of formation of the complexes, ΔH^0, range from -0.6 kcal/mol to -3.3 kcal/mol. Figure 14 shows data for 5 of the phenol systems at the same waiting time, $T_w = 7$ ps. From left to right, ΔH^0

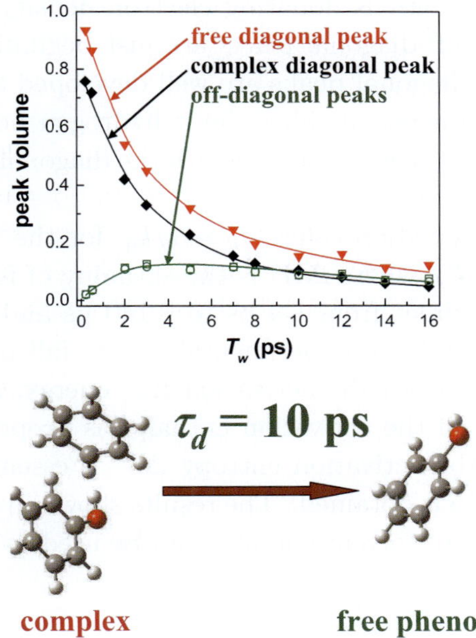

Fig. 13. The results of fitting the chemical exchange spectroscopy data for phenol-OD in the mixed benzene-CCl_4 solvent. The T_w dependent data (symbols) are for the 0-1 portion of the 2D IR spectra. There are four peaks, two diagonal and two off-diagonal. The curves through the data points are obtained from *the fit using the single adjustable parameter,* τ_d, the dissociation time of the phenol-benzene complex. The fit yields the phenol-benzene complex dissociation time, $\tau_d = 10$ ps.

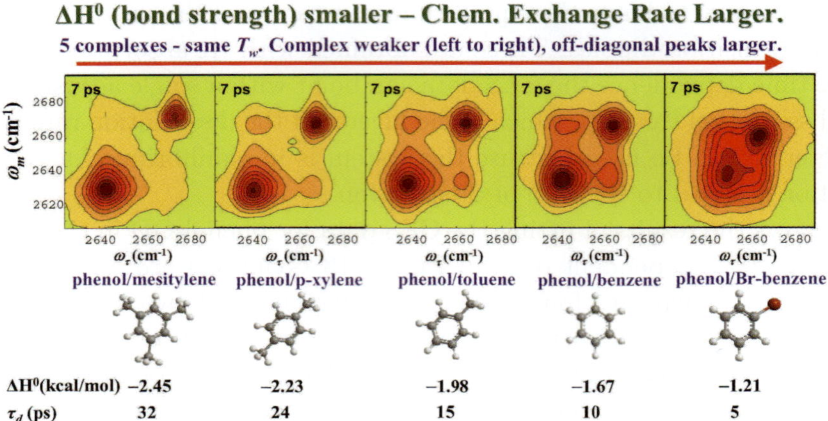

ΔH⁰(kcal/mol)	−2.45	−2.23	−1.98	−1.67	−1.21
τ_d (ps)	32	24	15	10	5

Fig. 14. Data for 5 of the phenol systems at the same waiting time, $T_w = 7$ ps. From left to right, ΔH^0 (the enthalpy of formation of the complex) becomes smaller and τ_d becomes shorter. This is evident by the size of the off-diagonal peaks. For mesitylene the off-diagonal peaks are just beginning to appear at 7 ps. For toluene, the off-diagonal peaks are well developed at 7 ps. For bromobenzene the off-diagonal peaks are so large at 7 ps that they have merged with the diagonal peaks to give the "square" shape. The structures of the solvent molecules that form the complexes with phenol are shown, and the ΔH^0s and dissociation times, τ_d, are given.

becomes smaller and τ_d becomes shorter. This is evident by the size of the off-diagonal peaks. For mesitylene (three methyls donating electron density to the ring making the π hydrogen bond stronger) the off-diagonal peaks are just beginning to appear at 7 ps. For toluene (one methyl), the off-diagonal peaks are well developed at 7 ps. For bromobenzene (the bromo is electron withdrawing, making the π hydrogen bond weak) the off-diagonal peaks are so large at 7 ps that they have merged with the diagonal peaks to give the "square" shape.

Figure 15 displays the dissociation times, $\tau_d = 1/k_d$, for the 13 complexes.[5,7] The data are plotted vs. $\exp(-\Delta H^0/RT)$, where ΔH^0 is the enthalpy of formation of the complexes. Over a range of dissociation times from ~4 ps to ~140 ps and ΔH^0 values ranging from −0.6 kcal/mol to −3.3 kcal/mol the experimental points fall on a line. Transition state theory[28] states that k_d depends on the activation free energy, ΔG^*, not on the enthalpy of formation, ΔH^0. However, if the activation enthalpy is proportional to the dissociation enthalpy, $\Delta H^* \propto \Delta H^0$, and the activation entropy ΔS^* is essentially a constant, then the behavior displayed in Figure 15 is obtained. The results shown in Figure 15 indicate that the enthalpy of formation of a solute-solvent complex can be used as a guide for its dissociation time.

B. *Orientational Isomerization about a Carbon–Carbon Single Bond*

Ethane and its derivatives are textbook examples of molecules that undergo orientational isomerization around a carbon–carbon single bond.[29] Ethane isomerization is illustrated in the upper portion of Figure 16. In ethane the transition from one staggered state to

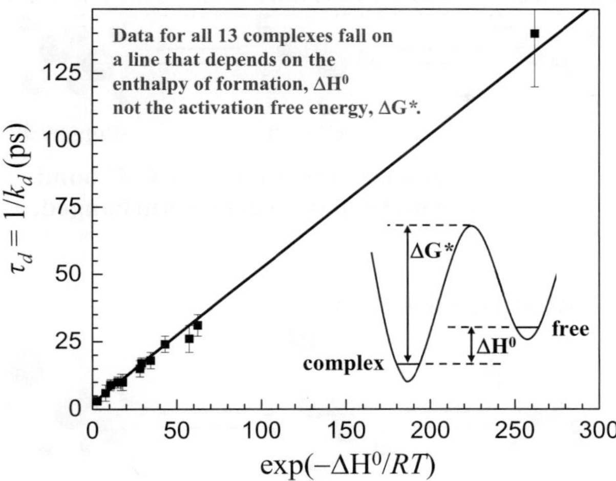

Fig. 15. The dissociation times, $\tau_d = 1/k_d$, for 13 complexes vs. $\exp(-\Delta H^0/RT)$, where ΔH^0 is the enthalpy of formation of the complexes. Over a range of dissociation times from ~4 ps to ~140 ps and ΔH^0 values ranging from -0.6 kcal/mol to -3.3 kcal/mol the experimental points fall on a line. The inset is a schematic of the energy surface.

another leaves ethane structurally identical. Therefore, there is no change in the frequency of a vibration needed to perform CES. As shown in the bottom portion of Figure 16, a 1, 2-di-substituted ethane has two distinct staggered conformations, gauche and trans, that have distinguishing characteristics because of the relative positions of the two substituents.[29]

The trans-gauche isomerization of 1, 2-disubstituted ethane derivatives, e.g. n-butane, is one of the simplest cases of a first-order chemical reaction. This type of isomerization has served as a basic model for modern chemical reaction kinetic theory and molecular dynamics (MD) simulation studies in condensed phases.[30–35] In spite of extensive theoretical investigation, until recently no corresponding kinetic experiments had been performed to test the results.[8] The experimental difficulty was due to the low rotational energy barrier of the n-butane (~3.3 kcal/mol),and other simple 1, 2-disubstituted ethane derivatives.[36] According to theoretical studies,[30–35] the isomerization time scale ($1/k$, k is the rate constant) is in the range of 10 to 100 ps in room temperature liquids.

Ultrafast 2D IR vibrational echo CES was used to measure the gauche-trans isomerization about the carbon–carbon single bond of 1-fluoro-2-isocyanato-ethane (FICE).[8] FICE was studied in CCl_4 at room temperature. The isocyanate group (N=C=O) was used as the vibrational probe. The structure of FICE in the gauche, eclipsed, and trans conformations, and the spectrum of the isocyanate antisymmetric stretching mode in the gauche and trans conformations are shown in Figure 17. The structures and their assignment to the peaks in the spectrum were obtained from electronic structure calculations using density functional theory[37] (DFT) at the B3LYP level and $6-31+G(d,p)$ basis set for the isolated molecules. The calculation also gave the barrier height for the eclipsed transition state as 3.3 kcal/mole.[8]

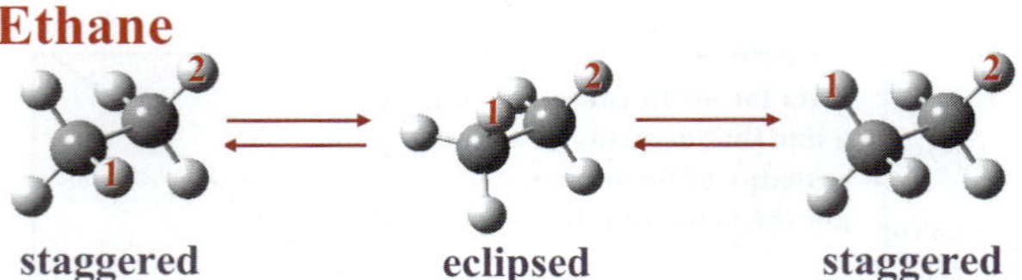

**Ethane isomerization about the C-C bond
leaves the molecular structure unchanged.**

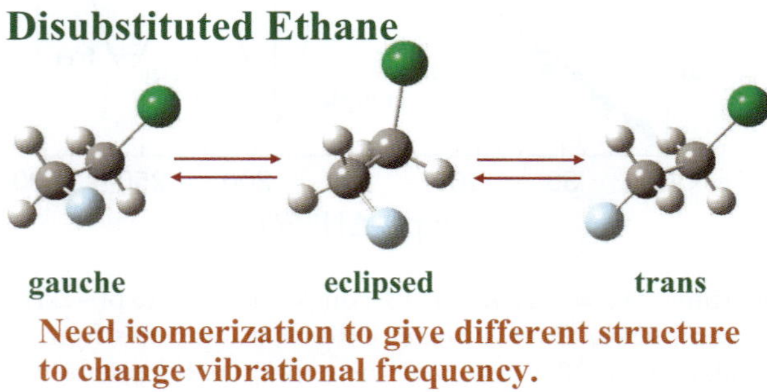

**Need isomerization to give different structure
to change vibrational frequency.**

Fig. 16. Ethane isomerization is illustrated in the upper portion. The bottom portion shows that a 1, 2-di-substituted ethane has two distinct staggered conformations, gauche and trans, that have distinguishing characteristics because of the relative positions of the two substituents.

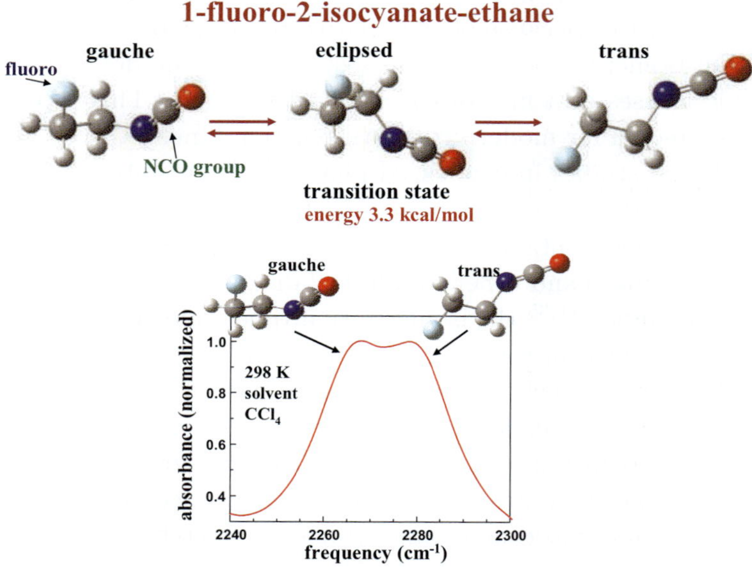

Fig. 17. The gauche-trans isomerization about the carbon–carbon single bond of 1-fluoro-2-isocyanato-ethane (FICE). FICE was studied in CCl_4 at room temperature. The isocyanate group (N=C=O) was used as the vibrational probe. The spectrum of the isocyanate antisymmetric stretching mode in the gauche and trans conformations are shown.

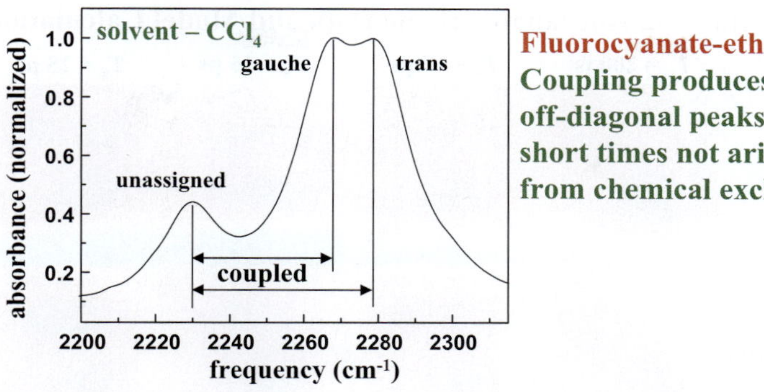

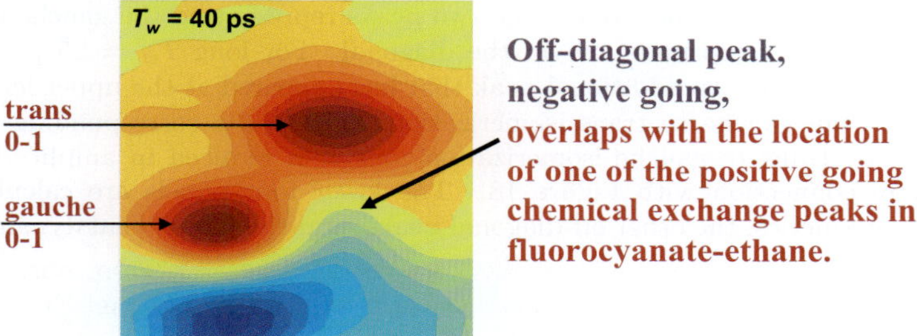

Fig. 18. In addition to the two absorption peaks shown in Figure 17, there is another smaller unassigned combination band or overtone absorption at 2230 cm^{-1}. This small peak produces additional off-diagonal peaks in the spectrum. The 2D IR vibrational echo spectra of 1-bromo-2-isocyanato-ethane is shown in the lower portion of the figure. The bromo group is so bulky that no gauche-trans isomerization occurs on the experimental time scale of ∼50 ps. The experiments on the bromo compound show that there is a negative going off-diagonal band that overlaps the off-diagonal position where one of the positive going chemical exchange peaks will grow in due to isomerization in the fluoro compound. The presence of this additional negative going peak is included in the data analysis.

As can be seen in Figure 18, in addition to the two peaks shown in Figure 17, there is another smaller unassigned combination band or overtone absorption at 2230 cm^{-1}.[8] This small peak is coupled by anharmonic terms in the molecular potential to the antisymmetric isocyanate stretching mode. The coupling produces additional off-diagonal peaks in the spectrum.[8,38] The nature of these additional off-diagonal peaks was investigated in detail by observing the 2D IR vibrational echo spectra of 1-bromo-2-isocyanato-ethane.[8] The 2D IR spectrum of the bromo compound is shown in the bottom portion of Figure 18. The bromo group is so bulky, that it raises the barrier height to the point were no gauche-trans isomerization occurs on the experimental time scale of ∼50 ps. As can be seen in the 2D IR spectrum in Figure 18, the results of the experiments on the bromo compound show that there is a negative going off-diagonal band that overlaps the off-diagonal position where one of the positive going chemical exchange peaks will grow in due to isomerization in the fluoro compound. The presence of this additional negative going peak is included in the data analysis.

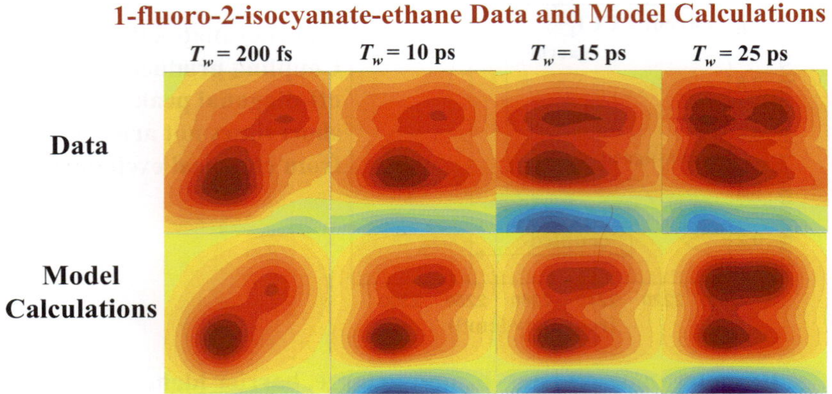

Fig. 19. 2D IR spectra of FICE in a CCl_4 solution at room temperature at four T_ws. The 200 fs panel: short T_w, negligible isomerization has occurred. The two peaks representing the gauche ($\omega_m = 2265$ cm^{-1}) and trans ($\omega_m = 2280$ cm^{-1}) conformers are on the diagonal. For long $T_w = 25$ ps, isomerization has proceeded to a substantial degree. An additional peak that has appeared at the upper left ($\omega_t = 2265$ cm^{-1} and $\omega_m = 2280$ cm^{-1}) from gauche to trans isomerization. There is a corresponding peak to the lower right that is generated by trans to gauche isomerization, but it is reduced in amplitude by the negative going peak discussed in connection with Figure 18. The lower four panels are calculated spectra that include the isomerization kinetics, the other off-diagonal peaks, and the vibrational dynamics other than the isomerization.

The top panels of Figure 19 are 2D IR spectra of FICE in a CCl_4 solution at room temperature taken at four T_ws. The 200 fs panel corresponds to a short T_w at which negligible isomerization has occurred. The two peaks representing the gauche ($\omega_m = 2265$ cm^{-1}) and trans ($\omega_m = 2280$ cm^{-1}) conformers are on the diagonal. For a long time ($T_w = 25$ ps), isomerization has proceeded to a substantial degree. The obvious change is the additional peak that has appeared at the upper left of the $T_w = 25$ ps panel ($\omega_\tau = 2265$ cm^{-1} and $\omega_m = 2280$ cm^{-1}). This peak arises from gauche to trans isomerization. There is a corresponding peak to the lower right that is generated by trans to gauche isomerization, but it is reduced in amplitude by the negative going peak discussed briefly above.[8,38] The lower four panels are calculated spectra that include the isomerization kinetics, the other off-diagonal peaks, and the vibrational dynamics other than the isomerization. Although the system is more complicated than the solute-solvent complex chemical exchange discussed above, the data can be reproduced well by the calculations.[8]

Figure 20 shows the T_w dependent data for the diagonal and off-diagonal peaks. As in the solute-solvent complex experiments, all of the necessary input parameters are known except for the isomerization time, $\tau_{\mathrm{iso}} = 1/k_{GT} = 1/k_{TG}$.[8] In the analysis, the gauche to trans rate constant was taken to be equal to the trans to gauche rate constant. Within experimental error, the difference in the two rate constants could not be discerned. The solid curves through the data are the result of the fit with the single adjustable parameter, τ_{iso}. The isomerization time is $\tau_{\mathrm{iso}} = 43 \pm 10$ ps. The error bars arise from the uncertainty in the parameters that go into the calculations. This is the first determination of isomerization about a carbon–carbon single bond in a room temperature liquid for a system with a typically low barrier height.

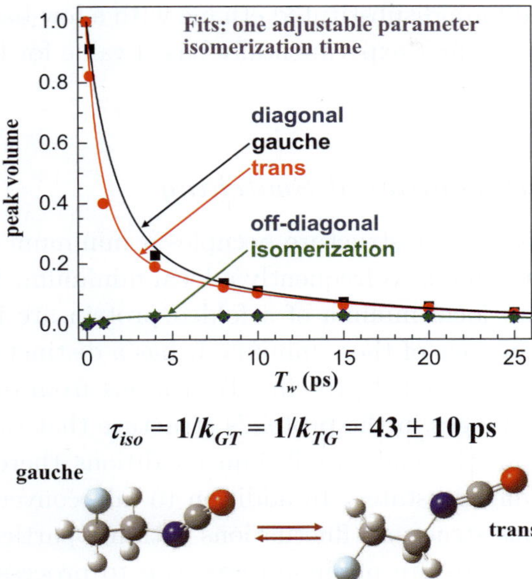

Fig. 20. T_w dependent data for the diagonal and off-diagonal peaks. The solid curves through the data are the result of the fit with the single adjustable parameter, τ_{iso}. The isomerization time is $\tau_{\text{iso}} = 43 \pm 10$ ps.

Based on the experimental results for the 1-fluoro-2-isocyanato-ethane, it is possible to calculate approximately the gauche-trans isomerization rate of n-butane and the rotational isomerization rate of ethane under the same conditions used in this study (CCl_4 solution at room temperature, 297 K) using transition state theory.[39] It was assumed that the prefactors are the same because the transition states and the barrier heights are quite similar for the three systems.[8] The barrier heights were calculated with DFT calculations on all the systems using the same method (the B3LYP level and $6-31+G(d,p)$ basis set). With the zero point energy correction, the trans to gauche isomerization of n-butane has a barrier of 3.3 kcal/mol. The barrier for ethane was calculated to be 2.5 kcal/mol,[8] which is smaller than the result of more extensive calculations, 2.9 kcal/mol.[40,41] The 2.5 kcal/mol value was used for ethane so that all of the barriers were obtained with the same method, which should result in some cancellation of errors.

Using the calculated barriers for FICE and for n-butane gave 43 ps time constant for the n-butane trans to gauche isomerization time constant ($1/k_{TG}$). Rosenberg, Berne, and Chandler reported a 43 ps time constant for this process (in CCl_4 at 300 K) from MD simulations.[33] Other MD simulations gave isomerization rates in liquid n-butane at slightly lower temperatures: 52 ps (292 K),[34] 57 ps (292 K),[32] 50 ps (273 K)[32] and 61 ps (< 292 K).[35] All of these values are quite close to the value based on the experimental measurements on FICE. This comparison is the first experimental confirmation of the MD simulations. In the same manner, the isomerization time constant for ethane is found to be $\sim$12 ps. This value can be improved by better electronic structure calculations on FICE and calculations for both FICE and ethane that include the CCl_4 in determining the barriers. Thus, this first experimental determination of the time constant for the orientational isomerization about a carbon–carbon single bond for a disubstituted ethane with a low barrier in solution at room temperature also provides the first experimentally based value for the ethane isomerization time constant.

C. *Fast Protein Conformational Switching*

A folded protein with a particular structure occupies a minimum on its free energy landscape.[42–44] However, the minimum is frequently a local minimum. Other minima of similar energy can also exist. Two local minima of a folded protein are illustrated in Figure 21. When a protein occupies any one of these minima, it has a distinct structure. The different structures are substates of the folded protein. Transitions from one minimum to another correspond to dynamical changes in the protein's structure that take the protein from one substate to another. Under thermal equilibrium conditions there will be continual conformational switching among substates. In addition to interconversion between substates, proteins undergo continuous structural fluctuations within a particular substate minimum. Such fluctuations within a substate minimum give rise to processes such as small ligand "diffusion" through a protein to an active site.[45]

The ability of proteins to undergo conformational switching is central to protein function. When an enzyme binds a substrate, the protein conformation will change.[46] On the path of protein folding, a protein will sample many conformations as it progresses toward the native folded structure.[47] Proteins can undergo large global conformational changes, which occur on long time scales, milliseconds to seconds. However, these large slow conformational changes, such as those that occur following substrate binding to an enzyme, involve a vast number of more local elementary conformational steps. The experimental determination for the time scale of elementary conformational steps is a long standing problem that has

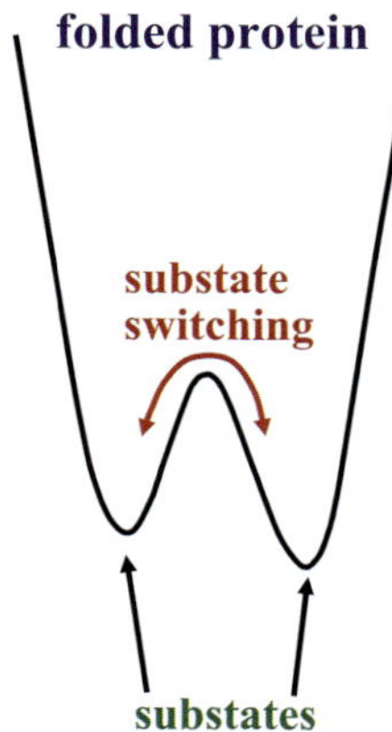

Fig. 21. A schematic diagram of two local minima of a folded protein. When a protein occupies a minima, it has a distinct structure. The different structures are substates of the folded protein. Transitions from one minimum to another correspond to dynamical changes in the protein's structure that take the protein from one substate to another.

now been successfully addressed using ultrafast two dimensional infrared vibrational echo chemical exchange spectroscopy.

The problem of multiple substates has been studied extensively for the protein myoglobin with the ligand CO bound at the active site (Mb-CO).[48–50] The infrared spectrum of the heme-ligated CO stretching mode of Mb shows two major absorption bands, denoted A_1 (1945 cm^{-1}), and A_3 (1932 cm^{-1}).[51] Mb-CO interconverts between these two conformational substates under thermal equilibrium conditions. The distal histidine, His64, plays a prominent role in determining the conformational substates of Mb. Here a study of a Mb mutant, L29I (leucine replaced by an isoleucine) is discussed.[10] The structure and the CO absorption spectrum of L29I-CO are shown in Figure 22. The replacement of leucine with isoleucine makes the A_1 and A_3 CO bands almost equal in amplitude. Changes in the configuration of the E helix (see Figure 22) cause the distal histidine's imidazole side group to move relative to the CO[51,52] (see lower right portion of Figure 22). The lower frequency of A_3 compared to A_1 reflects a closer proximity of the protonated epsilon nitrogen of the imidazole side group to the CO in A_3.[19,52,53] Each A substate exhibits a distinct ligand binding rate.[43,48] Therefore, the peaks in the FT-IR spectrum of Mb-CO and L29I-CO reflect functionally distinct conformational substates.

Figure 23 displays 2D-IR spectra of CO bound to L29I at several T_ws. The red bands are positive going and correspond to the 0-1 vibrational transitions. The blue bands (negative going) are from the 1-2 transition. For $T_w = 0.5$ ps, only the two diagonal peaks are

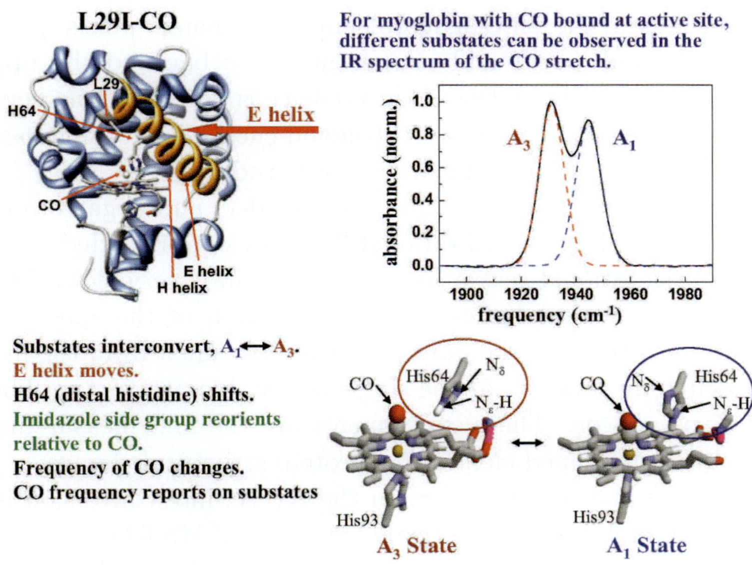

Fig. 22. The structure and the CO absorption spectrum of the myoglobin mutant L29I-CO (replacement of leucine with isoleucine). The A_1 and A_3 peaks in the CO spectrum correspond to distinct conformational substates. Changes in the configuration of the E helix cause the distal histidine's imidazole side group to move relative to the CO (see lower right portion of the figure). The lower frequency of A_3 in the absorption spectrum compared to A_1 reflects a closer proximity of the protonated epsilon nitrogen of the imidazole side group to the CO in A_3.

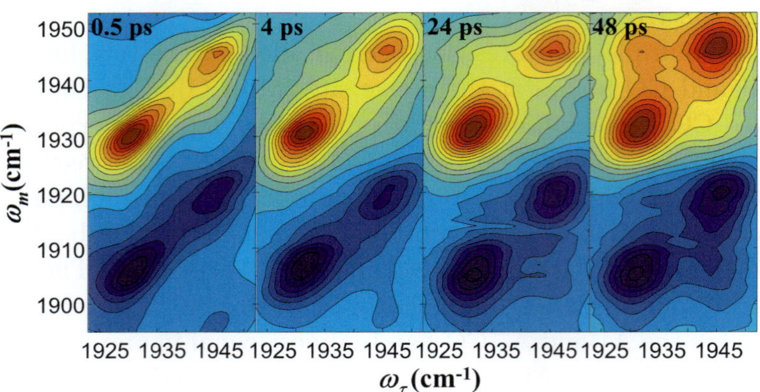

Fig. 23. The 2D-IR vibrational echo spectra of CO bound to L29I at several T_ws. The red bands (positive) correspond to the 0-1 vibrational transitions. The blue bands (negative) are from the 1-2 transition. $T_w =$ 0.5 ps: only the two diagonal peaks are observed, corresponding to the A_1 and A_3 bands in the FT-IR spectrum shown in Figure 22. As T_w increases, the off-diagonal chemical exchange peaks grow in. By $T_w = 48$ ps the off-diagonal bands are readily apparent. The band to the upper left in the $T_w = 48$ ps panel is strong. Because the anharmonicity is not large, the negative going 1-2 diagonal band partially overlaps the positive going off-diagonal chemical exchange peak to the lower right of the two 0-1 diagonal peaks, reducing its amplitude.

observed. These correspond to the A_1 and A_3 bands in the FT-IR spectrum shown in Figure 22. As T_w increases, the off-diagonal chemical exchange peaks grow in. By $T_w = 48$ ps the off-diagonal bands are readily apparent. The band to the upper left in the $T_w = 48$ ps panel is strong. Because the anharmonicity is not large, the negative going 1-2 band partially overlaps the positive going off-diagonal chemical exchange peak to the lower right of the two 0-1 diagonal peaks, reducing its amplitude.

The peak volumes were fit, and both the positive (0-1) and negative (1-2) peaks were included in the analysis.[10] Only the vibrational lifetimes were included in the kinetic data analysis because the orientational relaxation of the protein is very slow. The experimental diagonal and off-diagonal peak volumes for the 0-1 region of the spectra are plotted in Figure 24. The solid lines through the data points are obtained from the fitting procedure with the substate switching time, τ_s, as the single adjustable parameter. The results of the fitting yield $\tau_s = 47$ ps $\pm$ 8 ps. These experiments are the first measurement of the time dependence of a single well defined elementary protein structural change.

That significant structural change occurs in the A_1–A_3 interconversion is supported by X-ray experiments. The high resolution crystal structure of Mb-CO that contains two conformations has enabled modeling of the structure of A_1 and A_3 substates.[52,54] Although the distal histidine plays a critical role in determining the substates of Mb, structural comparison between the A_1 and A_3 substates shows that the A_3 substate contains an additional cavity, Xe3, and another transient cavity found in simulations.[54] Xe is used as a probe to identify the locations of cavities in proteins.[55,56] In Mb crystals, four Xe atoms (Xe1,Xe2, Xe3, and Xe4) occupy cavities, which may be involved in gas ligand migration.[54,55] The

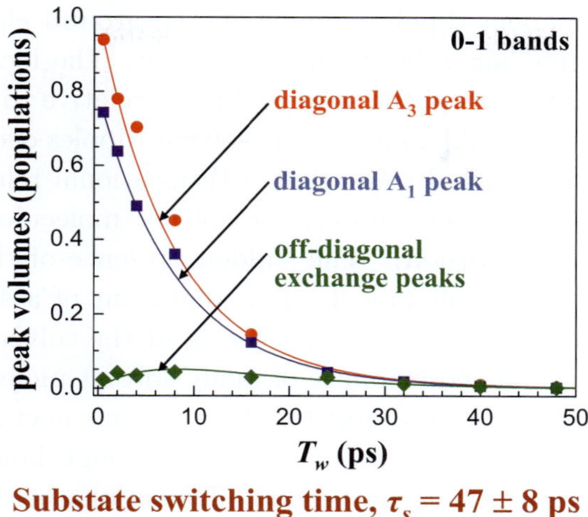

Substate switching time, $\tau_s = 47 \pm 8$ ps

Fig. 24. The experimental diagonal and off-diagonal peak volumes (symbols) for the 0-1 region of the spectra are plotted. The solid lines through the data points are obtained from the fitting procedure with the substate switching time, τ_s, as the single adjustable parameter. The results of the fitting yield $\tau_s = 47$ ps $\pm$ 8 ps.

Xe3 site, which is near the surface and far from the iron atom, involves Trp7 and is located between helices E and H[55] (see Figure 22). The existence of Xe3 in the A_3 substate but not in the A1 substate demonstrates that the difference in the substates is significantly more than the rotation of the imidazole side group of the distal histidine.

The fast time constant, 47 ps, for interconversion is in line with the structural difference between the A_1 and A_3 substates. Incoherent quasielastic neutron scattering experiments on native bovine α-lactalbumin observed collective motions on the tens of picosecond to 100 ps time scale.[57] The correlation length for such fluctuations was reported to be 18 Å.[57] The switching time observed here falls into this time range, which is consistent with a reconfiguration of the E helix in the interconversion between the A_1 and A_3 substates.

NMR techniques probe protein motions in the microsecond, millisecond, and longer timescales.[58] Conformational switching processes studied by NMR involve large structural changes that occur, for example, in enzyme catalytic processes.[59] Such structural changes require the reconfiguration of many amino acids, helices, and various protein structures. To place the time dependent measurements presented here of substate switching, a single elementary structural change, in a broader context it is useful to discuss the kinetics of large scale structural changes in terms of the ideas of linear response. Linear response theory describes how thermal fluctuations and elementary steps bring a system into a new configuration.

The concept of linear response comes from the fluctuation dissipation theorem.[60] It states that a system in thermodynamic equilibrium has a response to a small perturbation with a time dependence determined by the system's equilibrium fluctuations. The time dependent fluorescence Stokes shift experiment is a well studied example of relaxation to a new equilibrium structure following a perturbation.[61] In the experiment, a solute molecule with a small or zero permanent dipole moment in the ground electronic state is excited to the first excited state that has a large dipole moment. The local solvent structure will evolve to produce a net alignment of the solvent dipoles relative to the newly created large solute dipole. The orientational alignment of the solvent dipoles occurs in the same manner as thermal equilibrium orientational relaxation. (Translational relaxation will also occur.) Prior to the generation of the solute dipole, the solvent molecules are executing angular random walks (orientational diffusion). The sudden presence of the solute dipole slightly biases the random walks toward alignment. The alignment of a single solvent dipole can be viewed as an elementary step in the restructuring of the solvent. The solvation of the solute dipole requires the combined response of many solvent molecules.

Figure 21 displays a schematic of a portion of a protein's energy landscape having two substates. Each substate minimum is actually a broad rough landscape with many local minima separated by barriers of varying heights. Transitions between these local minima are responsible for protein structural fluctuations about a specific structure associated with a particular substate minimum.[3,42,44] These fluctuations can vary in time scales from sub-picosecond to tens and hundreds of picoseconds to much longer.[42–44] The fluctuations range from the fastest motions involving just a few atoms to low frequency acoustic type modes of the entire protein. The distinction between structural fluctuations that come from making transitions among the local minima on the landscape of a particular substate and transitions between substates is in some sense operational. A substate has a structural aspect that is qualitatively distinct. Fluctuations make interconversions among substates possible by sampling the energy landscape in a particular substate potential well (see Figure 21) and bringing the protein to the transition state between substates.

When a protein experiences a perturbation such as substrate binding or a temperature jump that induces folding or unfolding, in accord with linear response, it will respond to the perturbation by fluctuation driven elementary steps (substate changes) that can occur in the absence of the perturbation. However, the sampling of the substates will be biased by the perturbation, and the protein will relax to a new structure. The protein is always executing a multidimensional walk among substates. The walk will be skewed by the perturbation. The skewing can result from the shifting of substate minima and barriers. The relaxation to the new structure can be slow because it requires many elementary steps. A slow response to a perturbation results from structural fluctuation sampling that produces elementary steps, which in turn combine to produce major restructuring.

Using ultrafast 2D-IR vibrational echo chemical exchange spectroscopy we have measured the time dependence of a single elementary step, the A_1–A_3 substate switching. The time constant for the substate switching is 47 ps. The A_1–A_3 substate switching time will be an important target for future MD simulations because the barrier crossing will need to be simulated accurately. Simulations will determine if the dynamics of real systems undergoing an elementary structural change can be reproduced, and they will provide details of the structural changes that accompany the A_1–A_3 interconversion.

III. Concluding Remarks

Ultrafast 2D IR vibrational echo chemical exchange spectroscopy was described and illustrated with three examples, solute-solvent complex dynamics, isomerization about a carbon–carbon single bond, and protein conformational switching. The method measures dynamics in the ground electronic state under thermal equilibrium conditions. The method can be applied to a wide variety of problems. However, there are three conditions necessary to apply the technique. (1) There must be at least one IR active mode that has a distinct frequency for each species undergoing exchange; (2) the concentrations of the species must be high enough for detection; and (3) the exchange rate must fall within the experimental window set by the lifetime of the vibration used as the probe. The vibrational mode used as the probe need not be directly involved in the exchange process. It is sufficient that the exchange causes the mode frequency to change.

In the CES experiments presented above, only two species underwent interconversion. It is possible to observe more than two distinct species undergoing interconversion as illustrated by the 2D NMR experiments shown in Figure 3. A closely related type of ultrafast 2D IR vibrational echo experiment is the measurement of spectral diffusion. Consider a sample such as water in which the hydroxyl stretch is observed in the 2D IR vibrational echo experiments.[24,25] The hydroxyl stretch vibrational frequency is determined by the strength of hydrogen bonding and the number of hydrogen bonds. Stronger (shorter) hydrogen bonds and more hydrogen bonds shift the stretch frequency to lower frequency (the red). Weaker (longer) hydrogen bonds and fewer hydrogen bonds shift the frequency to higher frequency (the blue). The water hydroxyl stretch absorption spectrum is very broad, which reflects the broad continuous distribution of hydrogen bonding configurations in water.

In the CES experiments discussed above, the structural dynamics are manifested by the growth of off-diagonal peaks. In a system like water, there are so many configurations that distinct diagonal and off-diagonal peaks are not observed. Rather, the structural dynamics are manifested by a change in shape of the 2D IR vibrational echo spectrum. For short T_ws, the spectrum will be elongated along the diagonal. As T_w increases, the shape will change and the spectrum will become symmetrical at long T_w. Analysis of the time dependence of the shape change provides detailed information on the structural dynamics of systems with a vast number of interconverting structures.

Ultrafast 2D IR vibrational echo chemical exchange spectroscopy and the 2D IR vibrational echo method in general represent the opening of a new era in spectroscopy. The laser equipment that is necessary to perform the experiments is now commercially available and improving rapidly. Soon it will be close to turn key operation. The complex optical system shown schematically in Figure 7 can be assembled from all commercially available parts. Recently, we have made major advances which make the critical timing alignments self-correcting through computer controlled measurements and readjustments. It is reasonable to expect that in the future complete instruments will be available for purchase in the same way that complex NMR instruments can be purchased. It is important to remember that sophisticated commercial NMR machines are generally run by a dedicated operator with an advanced degree in the field. The need for an NMR technical assistant does not prevent NMR from being generally accessible to a wide variety of users in all fields of science. Commercial ultrafast 2D IR vibrational echo instruments will most likely also involve an expert operator. Commercial 2D IR vibrational echo instruments will open the field to a wide range of users.

In the early days of NMR, it was impossible to anticipate that some day an entire human being could be placed inside the magnet of an NMR machine to perform magnetic resonance imaging. It is difficult to predict the path and impact of ultrafast 2D IR vibrational echo spectroscopy on scientific research. The field is very young, and applications and advances are coming rapidly. In the future, ultrafast 2D IR vibrational echo methods will make major contributions to our understanding of nature.

Acknowledgments

I would like to thank the Indian Institute of Science, Bangalore, India, and particularly Director P. Balaram, for inviting me to present the Centenary Lecture in celebration of the 100th anniversary of the founding of the Institute. I would also like to thank Junrong Zheng, Kyungwon Kwak, Haruto Ishikawa, Jean K. Chung and Seongheun Kim who performed the experimental and theoretical work described in this monograph. I also thank the Air Force Office of Scientific Research (F49620-01-1-0018), the National Science Foundation (DMR 0652232), and the National Institutes of Health (2 R01 GM-061137-05) for support of this work.

I would like to acknowledge the following publishers for granting copyright permission to republish the listed journal articles.

Science Magazine

"Ultrafast Dynamics of Solute-Solvent Complexation Observed at Thermal Equilibrium in Real Time," Junrong Zheng, Kyungwon Kwak, John Asbury, Xin Chen, I. Piletic and M. D. Fayer, *Science* **309**, 1338–1343 (2005).

"Ultrafast Carbon-Carbon Single Bond Rotational Isomerization in Room Temperature Solution," Junrong Zheng, Kyungwon Kwak, Jia Xie and M. D. Fayer, *Science* **313**, 1951–1955 (2006).

American Chemical Society

"Vibrational Echo Studies of Myoglobin-CO", C. W. Rella, K. D. Rector, Alfred Kwok, Jeffrey R. Hill, H. A. Schwettman, Dana D. Dlott and M. D. Fayer, *J. Phys. Chem.* **100**, 15620–15629 (1996).

"A Dynamical Transition in the Protein Myoglobin Observed by Infrared Vibrational Echo Experiments," K. D. Rector, J. R. Engholm, C. W. Rella, J. R. Hill, D. D. Dlott and M. D. Fayer, *J. Phys. Chem. A* **103**, 2381–2387 (1999).

"Myoglobin-CO Substate Structures and Dynamics: Multidimensional Vibrational Echoes and Molecular Dynamics Simulations," Kusai A. Merchant, W. G. Noid, Ryo Akiyama, Ilya Finkelstein, Alexei Goun, Brian L. McClain, Roger F. Loring and M. D. Fayer, *J. Am. Chem. Soc.* **125**, 13804–13818 (2003).

"Formation and Dissociation of Intra-intermolecular Hydrogen Bonded Solute-Solvent Complexes: Chemical Exchange 2D IR Vibrational Echo Spectroscopy," Junrong Zheng, Kyungwon Kwak, Xin Chen and M. D. Fayer, *J. Am. Chem. Soc.* **128**, 2977–2987 (2006).

"Ultrafast 2D IR Vibrational Echo Spectroscopy," Junrong Zheng, Kyungwon Kwak and M. D. Fayer, *Acc. of Chem. Res.* **40**, 75–83 (2007).

"Ultrafast 2D IR Vibrational Echo Chemical Exchange Experiments and Theory," Kyungwon Kwak, Junrong Zheng, Hu Cang and M. D. Fayer, *J. Phys. Chem. B.* **110**, 19998–20013 (2006).

"Testing the Core/Shell Model of Nanoconfined Water in Reverse Micelles Using Linear and Nonlinear IR Spectroscopy," Ivan R. Piletic, David E. Moilanen, D. B. Spry and Nancy E. Levinger, M. D. Fayer, *J. Phys. Chem. A* **110**, 4985–4999 (2006).

"Hydrogen Bond Lifetimes and Energetics for Solute-Solvent Complexes Studied with 2D-IR Vibrational Echo Spectroscopy," Junrong Zheng and M. D. Fayer, *J. Am. Chem. Soc.* **129**, 4328–4335 (2007).

"Solute-Solvent Complex Kinetics and Thermodynamics Probe by 2D-IR Vibrational Echo Chemical Exchange Spectroscopy," Junrong Zheng and M. D. Fayer, *J. Phys. Chem. B* **112**, 10221–10227 (2008).

"Infrared Vibrational Photon Echo Experiments in Liquids and Glasses," A. Tokmakoff D. Zimdars, R. S. Urdahl, R. S. Francis, A. S. Kwok and M. D. Fayer, *J. Phys. Chem.* **99**, 13310–13320 (1995).

"Native and Unfolded Cytochrome C — Comparison of Dynamics using 2D-IR Vibrational Echo Spectroscopy," Seongheun Kim, Jean K. Chung, Kyungwon Kwak, Sarah E. J. Bowman, Kara L. Bren, Biman Bagchi and M. D. Fayer, *J. Phys. Chem. B* **112**, 10054–10063 (2008).

The American Institute of Physics
"Homogeneous Vibrational Dynamics and Inhomogeneous Broadening in Glass-Forming Liquids: Infrared Photon Echo Experiments from Room Temperature to 10 K," A. Tokmakoff and M. D. Fayer, *J. Chem. Phys.* **103**, 2810–2826 (1995).

"Vibrational Anharmonicity and Multilevel Vibrational Dephasing from Vibrational Echo Beats," K. D. Rector, A. S. Kwok, C. Ferrante, A. Tokmakoff, C. W. Rella and M. D. Fayer, *J. Chem. Phys.* **106**, 10027–10036 (1997).

"Dynamics of Water Probed with Vibrational Echo Correlation Spectroscopy," John B. Asbury, Tobias Steinel, Kyungwon Kwak, S. A. Corcelli, C. P. Lawrence, J. L. Skinner and M. D. Fayer, *J. Chem. Phys.* **121**, 12431–12446 (2004).

"Phenol-Benzene Complexation Dynamics: Quantum Chemistry Calculation, MD Simulations, and 2D IR Spectroscopy," Kijeong Kwac, Chewook Lee, Yousung Jung, Jaebeom Han, Kyungwon Kwak, Junrong Zheng, M. D. Fayer and Minhaeng Cho, *J. Chem. Phys.* **125**, 244508 (2006).

"Vibrational Photon Echoes in a Liquid and Glass: Room Temperature to 10 K," A. Tokmakoff, D. Zimdars, B. Sauter, R. S. Francis, A. S. Kwok and M. D. Fayer, *J. Chem. Phys.* **101**, 1741–1744 (1994).

"Vibrational Dephasing Mechanisms in Liquids and Glasses: Vibrational Echo Experiments," K. D. Rector and M. D. Fayer, *J. Chem. Phys.* **108**, 1794–1803 (1998).

"Frequency-Frequency Correlation Functions and Apodization in 2D-IR Vibrational Echo Spectroscopy, a New Approach," Kyungwon Kwak, Sungnam Park, Ilya J. Finkelstein and M. D. Fayer, *J. Chem. Phys.* **127**, 124503 (2007).

"Taking Apart 2D-IR Vibrational Echo Spectra: More Information and Elimination of Distortions," Kyungwon Kwak, Daniel E. Rosenfeld and M. D. Fayer, *J. Chem. Phys.* **128**, 204505 (2008).

The American Physical Society
"Taking Apart 2D-IR Vibrational Echo Spectra: More Information and Elimination of Distortions," Kyungwon Kwak, Daniel E. Rosenfeld and M. D. Fayer, *J. Chem. Phys.* **128**, 204505 (2008).

"Vibrational Echo Studies of Protein Dynamics," C. W. Rella, A. S. Kwok, K. D. Rector, Jeffrey R. Hill, H. A. Schwettman, Dana D. Dlott and M. D. Fayer, *Phys. Rev. Lett.* **77**, 1648–1651 (1996).

"Two-Dimensional Time-Frequency Ultrafast Infrared Vibrational Echo Spectroscopy," K. A. Merchant, David E. Thompson and M. D. Fayer, *Phys. Rev. Lett.* **86**, 3899–3902 (2001).

"Hydrogen Bond Dynamics Probed with Ultrafast Infrared Heterodyne Detected Multidimensional Vibrational Stimulated Echoes," John B. Asbury, Tobias Steinel, C. Stromberg, K. J. Gaffney, I. R. Piletic, Alexi Goun and M. D. Fayer, *Phys. Rev. Lett.* **91**, 237402 (2003).

"Dynamics of Water Confined on a Nanometer Length Scale in Reverse Micelles: Ultrafast Infrared Vibrational Echo Spectroscopy," Howe-Siang Tan, Ivan R. Piletic, Ruth E. Riter, Nancy E. Levinger and M. D. Fayer, *Phys. Rev. Lett.* **94**, 057405(4) (2005).

Proceedings of the National Academy of Sciences of the United States of America
"Substrate Binding and Protein Conformational Dynamics Measured via 2D-IR Vibrational Echo Spectroscopy," Ilya J. Finkelstein, Haruto Ishikawa, Seongheun Kim, Aaron M. Massari and M. D. Fayer, *Proc. Nat. Acad. Sci.* **104**, 2637–2642, (2007).

"Hydrogen Bond Dynamics in Aqueous NaBr Solutions," Sungnam Park and M. D. Fayer, *Proc. Nat. Acad. Sci. USA* **104**, 16731–16738 (2007).

"Neuroglobin Dynamics Observed with Ultrafast 2D-IR Vibrational Echo Spectroscopy," Haruto Ishikawa, Ilya J. Finkelstein, Seongheun Kim, Kyungwon Kwak, Jean K. Chung, Keisuke Wakasugi, Aaron M. Massari and M. D. Fayer, *Proc. Nat. Acad. Sci. USA* **104**, 16116–16121 (2007).

"Direct Observation of Fast Protein Conformational Switching," Haruto Ishikawa, Kyung-won Kwak, Jean K. Chung, Seongheun Kim and M. D. Fayer, *Proc. Nat. Acad. Sci. USA* **105**, 8619–8624 (2008).

Royal Society of Chemistry

"Probing Dynamics of Complex Molecular Systems with Ultrafast 2D IR Vibrational Echo Spectroscopy," Ilya J. Finkelstein, Junrong Zheng, Haruto Ishikawa, Seongheun Kim, Kyungwon Kwak and M. D. Fayer, *Phys. Chem. Chem. Phys.* **9**, 1533–1549, (2007).

Laser Physics

"Ultrafast 2D-IR Vibrational Echo Spectroscopy: A Probe of Molecular Dynamics," Sung-nam Park, Kyungwon Kwak and M. D. Fayer, *Laser Phys. Lett.* **4**, 704–718 (2007).

References

1. Todd, D. C.; Fleming, G. R. *J. Chem. Phys.* **1993**, *98*, 269.
2. Perrin, C. L.; Dwyer, T. J. *Chem. Rev.* **1990**, *90*, 935.
3. Zheng, J.; Kwak, K.; Asbury, J. B.; Chen, X.; Piletic, I.; Fayer, M. D. *Science* **2005**, *309*, 1338.
4. Zheng, J.; Kwak, K.; Chen, X.; Asbury, J. B.; Fayer, M. D. *J. Am. Chem. Soc* **2006**, *128*, 2977.
5. Zheng, J.; Fayer, M. D. *J. Am. Chem. Soc.* **2007**, *129*, 4328.
6. Kwac, K.; Lee, C.; Jung, Y.; Han, J.; Kwak, K.; Zheng, J.; Fayer, M. D.; Cho, M. *J. Chem. Phys.* **2006**, *125*, 244508.
7. Zheng, J.; Fayer, M. D. *J. Phys. Chem. B* **2008**, submitted.
8. Zheng, J.; Kwac, K.; Xie, J.; Fayer, M. D. *Science* **2006**, *313*, 1951.
9. Jackman, L. M.; Cotton, F. A. *Dynamic Nuclear Magnetic Resonance Spectroscopy*; Academic Press: New Yoyk, 1975.
10. Ishikawa, H.; Kwak, K.; Chung, J. K.; Kim, S.; Fayer, M. D. *Proc. Natl. Acad. Sci. of the USA* **2008**, *105*, 8619.
11. Kim, Y. S.; Hochstrasser, R. M. *Proc. Natl. Acad. Sci.* **2005**, *102*, 11185.
12. Kwak, K.; Zheng, J.; Cang, H.; Fayer, M. D. *J. Phys. Chem. B* **2006**, *110*, 19998.
13. Zheng, J.; Kwak, K.; Fayer, M. D. *Acc. of Chem. Res.* **2007**, *40*, 75.
14. Jeener, J.; Meier, B. H.; Bachmann, P.; Ernst, R. R. *J. Chem. Phys.* **1979**, *71*, 4546.
15. Asbury, J. B.; Steinel, T.; Fayer, M. D. *J. Luminescence* **2004**, *107*, 271.
16. Park, S.; Kwak, K.; Fayer, M. D. *Laser Phys. Lett.* **2007**, *4*, 704.
17. Tan, H.-S.; Piletic, I. R.; Fayer, M. D. *J. Chem. Phys.* **2005**, *122*, 174501(9).
18. Varadarajan, R.; Szabo, A.; Boxer, S. G. *Proc. Natl. Acad. Sci. USA* **1985**, *82*, 5681.
19. Merchant, K. A.; Noid, W. G.; Akiyama, R.; Finkelstein, I.; Goun, A.; McClain, B. L.; Loring, R. F.; Fayer, M. D. *J. Am. Chem. Soc.* **2003**, *125*, 13804.
20. Reichardt, C. *Solvents and Solvent Effects in Organic Chemistry*; Wiley-VCH, 2003.
21. Vinogradov, S. N.; Linnell, R. H. *Hydrogen Bonding*; Van Nostrand Reinhold Company: New York, 1971.

22. Mukamel, S. *Ann. Rev. Phys. Chem.* **2000**, *51*, 691.

23. Mukamel, S. *Principles of Nonlinear Optical Spectroscopy*; Oxford University Press: New York, 1995.

24. Asbury, J. B.; Steinel, T.; Stromberg, C.; Corcelli, S. A.; Lawrence, C. P.; Skinner, J. L.; Fayer, M. D. *J. Phys.Chem. A* **2004**, *108*, 1107.

25. Asbury, J. B.; Steinel, T.; Kwak, K.; Corcelli, S.; Lawrence, C. P.; Skinner, J. L.; Fayer, M. D. *J. Chem. Phys.* **2004**, *121*, 12431.

26. Kwak, K.; Rosenfeld, D. E.; Fayer, M. D. *J. Chem. Phys.* **2008**, accepted.

27. Kwak, K.; Park, S.; Finkelstein, I. J.; Fayer, M. D. *J. Chem. Phys.* **2007**, *127*, 124503.

28. Chang, R. *Physical Chemistry for the Chemical and Biological Sciences*; University Science Books: Sausalito, 2000.

29. March, J. *Advanced Organic Chemistry*, 3rd ed.; John Wiley & Sons: New Yoyk, 1985.

30. Chandler, D. *J. Chem. Phys.* **1978**, *68*, 2959.

31. Weber, T. A. *J. Chem. Phys.* **1978**, *69*, 2347.

32. Brown, D.; Clarke, J. H. R. *J. Chem. Phys.* **1990**, *92*, 3062.

33. Rosenberg, R. O.; Berne, B. J.; Chandler, D. *Chem. Phys. Lett.* **1980**, *75*, 162.

34. Edberg, R.; Evans, D. J.; Morris, G. P. *J. Chem. Phys.* **1987**, *87*, 5700.

35. Ramirez, J.; Laso, M. *J. Chem. Phys.* **2001**, *115*, 7285.

36. Streitwieser, A.; Taft, R. W. *Progress in Physical Organic Chemistry*; John Wiley & Sons: New York, 1968; Vol. 6.

37. Parr, R. G.; Yang, W. *Density Functional Theory of Atoms and Molecules*; Oxford University Press: New York, 1989.

38. Khalil, M.; Demirdoven, N.; Tokmakoff, A. *J. Chem. Phys.* **2004**, *121*, 362.

39. Levine, I. N. *Physical Chemistry*; McGraw-Hill Book Company: New York, 1978.

40. Pophristic, V.; Goodman, L. *Nature* **2001**, *411*, 565.

41. Bickelhaupt, F. M.; Baerends, E. J. *Angew. Chem., Int. Ed.* **2003**, *42*, 4183.

42. Frauenfelder, H.; Parak, F.; Young, R. D. *Ann. Rev. Biophys. Biophys. Chem.* **1988**, *17*, 471.

43. Frauenfelder, H.; Sligar, S. G.; Wolynes, P. G. *Science* **1991**, *254*, 1598.

44. Austin, R. H.; Beeson, K.; Eisenstein, L.; Frauenfelder, H.; Gunsalus, I. C.; Marshal, V. P. *Phys. Rev. Lett.* **1974**, *32*, 403.

45. Case, D. A.; Karplus, M. *J. Mol. Biol.* **1979**, *132*, 343.

46. Schnell, J. R.; Dyson, H. J.; Wright, P. E. *Annu. Rev. Biophys. Biomol. Struct.* **2004**, *33*, 119.

47. Oliveberg, M.; Wolynes, P. G. *Q. Rev. Biophys.* **2005**, *38*, 245.

48. Ansari, A.; Berendzen, J.; Braunstein, D.; Cowen, B. R.; Frauenfelder, H.; Hong, M. K.; Iben, I. E. T.; Johnson, J. B.; Ormos, P.; Sauke, T. B.; Scholl, R.; Schulte, A.; Steinbach, P. J.; Vittitow, J.; Young, R. D. *Biophys. Chem.* **1987**, *26*, 337.

49. Tian, W. D., Sage, J. T., Champion, P. M. *J. Mol. Bio.* **1993**, *233*, 155.

50. Muller, J. D.; McMahon, B. H.; Chen, E. Y. T.; Sligar, S. G.; Nienhaus, G. U. *Biophys. J.* **1999**, *77*, 1036.

51. Li, T. S.; Quillin, M. L.; Phillips, G. N., Jr.; Olson, J. S. *Biochemistry* **1994**, *33*, 1433.

52. Vojtechovsky, J.; Chu, K.; Berendzen, J.; Sweet, R. M.; Schlichting, I. *Biophys. J.* **1999**, *77*, 2153.

53. Johnson, J. B.; Lamb, D. C.; Frauenfelder, H.; Mller, J. D.; McMahon, B.; Nienhaus, G. U.; Young, R. D. *Biophys. J.* **1996**, *71*, 1563.

54. Teeter, M. M. *Protein Sci.* **2004**, *13*, 313.

55. Tilton, R. F., Jr.; Kuntz, I. D., Jr.; Petsko, G. A. *Biochemistry* **1984**, *23*, 2849.

56. Doukov, T. I.; Blasiak, L. C.; Seravalli, J.; Ragsdale, S. W.; Drennan, C. L. *Biochemistry* **2008**, *47*, 3474.

57. Bu, Z.; Neumann, D. A.; Lee, S. H.; Brown, C. M.; Engelman, D. M.; Han, C. C. *J. Mol. Bio.* **2000**, *301*, 525.

58. Palmer, A. G. r. *Curr. Opin. Struct. Biol.* **1997**, *7*, 732.

59. Boehr, D. D.; Dyson, H. J.; Wright, P. E. *Chem. Rev.* **2006**, *106*, 3055.

60. Kubo, R.; Toda, M.; Hashitusme, N. *Statistical Physics Ii: Nonequilibrium Statistical Mechanics*; Springer-Verlag: New York, 1985.

61. Fleming, G. R.; Mihnaeng, C. *Ann. Rev. Phys. Chem.* **1996**, *47*, 109.

Part II

Selected Publications of M. D. Fayer

The following contains selected publications on ultrafast vibrational echo experiments. The publications are divided into four sections.

Section 1. Ultrafast 2D IR Vibrational Echo Chemical Exchange Spectroscopy

Section 2. The First and Early Ultrafast Vibrational Echo Experiments on Liquids, Glasses, and Proteins

Section 3. Ultrafast 2D IR Vibrational Echo Spectral Diffusion Studies of Liquids and Proteins

Section 4. Recent Review Articles

Each section is preceded by a brief discussion of the topic discussed in the papers followed by a list of publications in the section.

Ultrafast 2D IR Vibrational Echo Chemical Exchange Spectroscopy

The following papers describe experiments conducted to date using ultrafast 2D IR vibrational echo chemical exchange spectroscopy. This method is an application of the 2D IR vibrational echo method to study a variety of fast chemical kinetic processes under thermal equilibrium conditions. In the papers presentd below, three different problems are discussed. They are the formation and dissociation of organic solute-solvent complexes in liquids, the isomerization of an organic molecule around a carbon–carbon single bond, and the interconversion of one protein structre to another. All of these, and many other future applications, are thermal equilibrium processes that occur in the ground electronic state. Such conditions are found in the vast majority of chemical and biological processes. The generic system involves two species, A and B that are constantly interconverting one to another. Because the species are in equilibrium, the concentrations of A and B do not change. Nonetheless, there are dynamics as A's become B's and B's become A's. The rate of A going to B is equal to the rate of B going to A, leaving the concentrations time independent. There can also be more than two species undergoing exchange. The problem is how to observe such equilibrium processes without disturbing the equilibrium nature of the system.

To use 2D IR vibrational echo chemical exchange spectroscopy it is necessary for the two species to have an IR active vibration that changes frequency when one species interconverts into the other. In the IR absorption spectrum there will be two peaks. The two peaks can be used to show that there are two species, but the absorption spectrum does not give information on the dynamical back and forth interchange of one species to the other. The 2D IR vibrational echo pulse sequence "labels" the species at time $t = 0$. The subsequent pulses in the sequence read out the identity of the labeled species and allow the chemical exchange to be observed. At a time short compared to the interconversion of the species, two peaks will be observed on the diagonal of the 2D IR spectrum. These correspond to the two peaks in the absorption spectrum. As the time between the second and third pulses in the sequence is increased, there is more time for the chemical exchange to occur. The chemical exchange is manifested by the growth of off-diagonal peaks in the spectrum. From the time dependence of the growth of the off idagonal peaks, the rate of interconversion among species can be directly determined.

As mentioned above, for the method to be applied, the interconversion of one species to another must produce a change in the vibrational frequency. The two peaks (or more peaks if there are more than two interconverting species) must be large enough to be observed. In addition, the time constant of interconversion that can be measured must be comparable to the vibrational lifetime of the vibration under observation. It is possible to take high quality 2D IR vibrational echo data out to three to five vibrational lifetimes. This sets the limit on how slow a process can be and still be observed.

The papers that follow go into details of how the experiments are conducted and how the data is anlayzed. The three types of systems studied provide important new insights into dynamical processes. The range of applications of this technique is rapidly increasing.

Section 1. Ultrafast 2D IR Vibrational Echo Chemical Exchange Spectroscopy

Ultrafast Dynamics of Solute-Solvent Complexation Observed at Thermal Equilibrium in Real Time

Junrong Zheng, Kyungwon Kwak, John Asbury,*
Xin Chen, Ivan R. Piletic, M. D. Fayer†

In general, the formation and dissociation of solute-solvent complexes have been too rapid to measure without disturbing the thermal equilibrium. We were able to do so with the use of two-dimensional infrared vibrational echo spectroscopy, an ultrafast vibrational analog of two-dimensional nuclear magnetic resonance spectroscopy. The equilibrium dynamics of phenol complexation to benzene in a benzene–carbon tetrachloride solvent mixture were measured in real time by the appearance of off-diagonal peaks in the two-dimensional vibrational echo spectrum of the phenol hydroxyl stretch. The dissociation time constant τ_d for the phenol-benzene complex was 8 picoseconds. Adding two electron-donating methyl groups to the benzene nearly tripled the value of τ_d and stabilized the complex, whereas bromobenzene, with an electron-withdrawing bromo group, formed a slightly weaker complex with a slightly lower τ_d. The spectroscopic method holds promise for studying a wide variety of other fast chemical exchange processes.

Solvents play an enormous role in practical chemistry by influencing the reactivity of dissolved substrates (*1*). In part, the influence stems from polarization effects or nonspecific solute-solvent interactions for which the solvent structure around a solute can be described in terms of an isotropic radial distribution function (*2*). However, specific intermolecular interactions, such as hydrogen bonding, can lead to structurally characterizable solute-solvent complexes that are constantly forming and dissociating under thermal equilibrium conditions on very short time scales (*3*). The dynamics of these transient species can play an important role in the physical and chemical properties of a solute-solvent system by affecting reaction rates, reaction mechanisms, and product ratios (*1*).

If the complexes persist for microseconds or longer, their thermal equilibrium dissociation and exchange dynamics can be studied by means of two-dimensional (2D) nuclear magnetic resonance (NMR) techniques (*4–6*). However, for the majority of organic and other types of nonaqueous solutions, in which a vast number of reactions take place, the solute-solvent complexes are bound by energies on the order of a few k_BT (where k_B is the Boltzmann constant and T is absolute temper-

ature; $k_BT \approx 0.6$ kcal/mol at room temperature) and therefore form and dissociate on subnanosecond time scales. As a result, NMR studies cannot distinguish the rapid exchange events and can offer only a dynamically averaged view of the system.

Ultrashort laser pulses have long offered a means of probing rapid dynamics. However, ultrafast absorption and fluorescence techniques have been hindered by the need to perturb the chemical properties of the system in order to study it. These techniques measure net changes in state populations. Thus, electronic excitation can be used to initiate a reaction on a femtosecond time scale by placing a molecule in a higher electronic state, with subsequent fast probing of new product formation, but the system is then no longer observed in its chemical equilibrium state. What is needed to probe equilibrium solvent-solute interactions is an ultrashort analog of 2D NMR spectroscopy. Here, we apply such an analog—ultrafast 2D infrared (2D IR) vibrational echo spectroscopy—to probe the dynamics of phenol complexation with several benzene-based solvents.

Coherent spectroscopy: From NMR to IR. Both the NMR and IR vibrational echo techniques involve pulse sequences that induce and then probe the coherent evolution of excitations (nuclear spins for NMR and vibrations for IR) of a molecular system. The molecules in a given environment (for example, free versus complexed) are induced to "oscillate" in spin states or vibrational states, all at the same time and with the same phase

by the first pulse in the sequence. The effect of the first pulse, along with the manipulation of the phase relationships among the vibrational oscillators by the following pulses in the sequence, is an important feature that 2D IR vibrational echo spectroscopy has in common with 2D NMR. The later pulses generate observable signals that are sensitive to change in environments of individual molecules during the experiment (e.g., from free to complexed solute or vice versa), even if there is no change of aggregate populations in the distinct environments (the observable in linear spectroscopy). The critical difference between the IR and NMR variants is that the IR pulse sequence acts on a time scale six orders of magnitude faster than the NMR sequence.

Vibrational echo–based 2D IR experiments have been applied to the study of the intramolecular coupling of vibrational modes (v), the structure of proteins, and the dynamics of hydrogen bonds by observing the positions of peaks or the change in the shape of peaks in the 2D spectrum (*7–9*). In the present study, intermolecular chemical exchange causes new peaks in the spectrum to appear and grow, yielding the rate of chemical exchange.

Phenol binds weakly to benzene. Intermolecular chemical exchange under thermal equilibrium conditions is ubiquitous in nature. It forms the basis for supramolecular chemistry (*10*), host-guest chemistry (*10*), chemical and biological recognition (*11*), and self-assembly (*10*). The dynamics of complexes involving noncovalent interactions with aromatic rings are pivotal to the protein-ligand recognition and concomitantly to drug design (*11*). The phenol-benzene complex cuts to the essence of interactions between protic and hydrophobic groups so widespread in proteins and surfactants. More fundamentally, the specific attraction between the polar OH group on phenol and the polarizable π-electron cloud on benzene is an intriguing intermediate case between dispersion forces and hydrogen bonding. The phenol-benzene complex appears to be mainly a van der Waals interaction (*12*), although it is also referred to as π-hydrogen bonding (*11, 13*) and has recently been found to be of biological importance. π-Hydrogen bonding can stabilize α helices in proteins and plays an important role in cellular and synaptic signal transmission (*13*).

The formation enthalpy for the gas-phase phenol-benzene complex is ~4 kcal/mol (*12*). The gas-phase structure has been determined by electronic structure calculations (*14*). Figure 1 shows two views of the structure of the phenol-benzene complex. In contrast to chemical intuition, the hydroxyl group does not point to the center of the benzene ring, but rather points between adjacent carbons. One of the ortho hydrogens on the phenol also points

Department of Chemistry, Stanford University, Stanford, CA 94305, USA.

*Present address: Department of Chemistry, Pennsylvania State University, University Park, PA 16802, USA.

†To whom correspondence should be addressed. E-mail: fayer@stanford.edu

44

between a pair of carbons that are on the opposite side of the benzene ring. Additional higher level electronic structure calculations performed as part of this study confirm this structure and show that the other complexes studied here have similar structures. Previous studies in liquids demonstrate the existence of the phenol-benzene complex and have measured its enthalpy of formation in CCl_4 (1.56 kcal/mol at room temperature) in temperature-dependent linear IR absorption experiments (15) (see below).

We used 2D IR vibrational echo spectroscopy to extract the binding kinetics of phenol-benzene complexes in solution. In addition, we have examined the effect of modifying benzene with electron-donating groups (*p*-xylene or *p*-dimethylbenzene) and an electron-withdrawing group (bromobenzene). Experimental specifics are described below for the benzene studies, with results from the corresponding substituted benzene studies given at the end for comparison.

For the purposes of our experimental study, phenol presents several advantages. The hydroxyl stretch is an isolated vibration with a strong absorption cross section and relatively long vibrational lifetime, so IR excitation is efficient and there is a sufficient time window to observe its dynamics. Moreover, the hydroxyl stretch frequency is sensitive to formation of the complex with benzene. By deuterating the hydroxyl group (OD rather than OH), we transferred the stretching mode away from aromatic C-H stretching bands of similar frequency. The OD stretching band of free phenol in pure CCl_4 is centered at 2665 cm^{-1} (Fig. 2, dotted curve). The spectrum of phenol in pure benzene shows

a red-shifted (lower energy) band in this region at 2631 cm^{-1} (Fig. 2, dashed curve), resulting from formation of the phenol-benzene complex. By diluting the benzene with CCl_4, we can shift the equilibrium toward more free phenol. At the concentrations used here in the mixed solution (phenol/benzene/CCl_4 molar ratio 2:40:100, respectively), the absorptions of free and complexed phenol are both prominent in the spectrum (Fig. 2, solid line). Integration of these bands, calibrated from the pure samples, yields the concentrations of complexed and free phenol, which in turn determine the equilibrium constant for complex formation (K_{eq} = [complex]/[phenol][benzene] = 0.26 at 298 K).

To measure the formation and dissociation kinetics of the complex in this solution, we apply three successive IR pulses with the same polarization, which induce the subsequent emission in a distinct direction of a time-delayed fourth pulse—the vibrational echo. The transform-limited pulses (50 fs, <4 cycles of light) are produced using a Ti:sapphire regeneratively amplified laser system coupled to an optical parametric oscillator, and they span sufficient bandwidth (300 cm^{-1} centered at ~4 μm, or 2500 cm^{-1}) to cover the $v = 0$ to $v = 1$ (hereafter denoted 0-1) and 1-2 transitions of the hydroxyl OD stretching modes in both free and complexed phenol. The echo pulse is detected, with frequency and phase resolution, by combining with a fifth (local oscillator) pulse, and the combined pulse is dispersed in a spectrograph. Data are thus obtained as a function of three variables: the emitted echo frequency ω_m (measured directly by the spectrograph), the variable time delay between the first and second pulses (τ), and the variable time delay between the second and third pulses (T_w, the variable "waiting" time). By numerical Fourier transform (FT), the τ scan data taken at each T_w are mapped to a second frequency variable ω_τ. The data are then plotted in three

dimensions, showing the amplitude as a function of both ω_τ and ω_m (which correspond to the ω_1 and ω_3 axes, respectively, in 2D NMR). The experimental setup is shown in fig. S1; further experimental details are given in (16).

Chemical exchange creates new off-diagonal peaks. Figure 3 displays the 2D IR vibrational echo spectra as contour plots at a very short T_w (200 fs, Fig. 3A) and a long T_w (14 ps, Fig. 3B) (1 fs = 10^{-3} ps = 10^{-15} s). The data have been normalized to the largest peak at each T_w. The red contours are positive-going (0-1 vibrational transition) and the blue contours are negative-going (1-2 vibrational transition). As discussed below, the 0-1 signal comes from two quantum pathways that are related to bleaching of the ground state and stimulated emission, both of which produce a vibrational echo pulse that is in phase with (and therefore adds to) the local oscillator pulse to produce a positive-going signal. The 1-2 signal arises because there is a new absorption that was not present before the first two excitation pulses. The 1-2 vibrational echo pulse is 180° out of phase with (and thus subtracts from) the local oscillator to produce

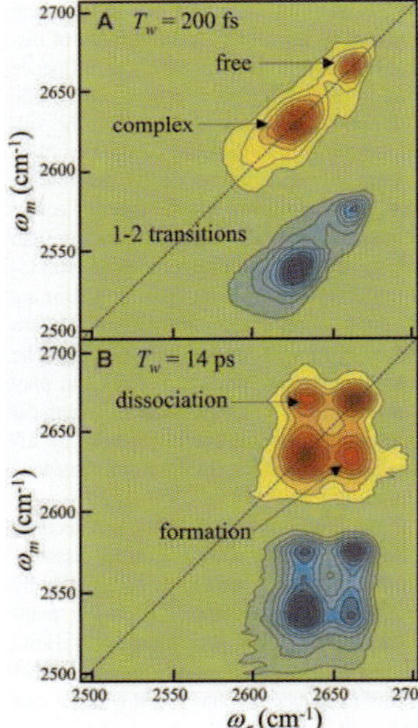

Fig. 3. 2D IR vibrational echo spectra of the OD stretch of phenol in the mixed benzene-CCl_4 solvent. Each contour represents a 10% change. (**A**) Data for T_w = 200 fs. The red contours on the diagonal (positive going) are the 0-1 transitions of the free phenol and the phenol-benzene complex. The blue contours off the diagonal (negative going) are the corresponding 1-2 transition peaks. (**B**) Data for T_w = 14 ps. Additional peaks have grown in because of chemical exchange, that is, the formation and dissociation of the phenol-benzene complex.

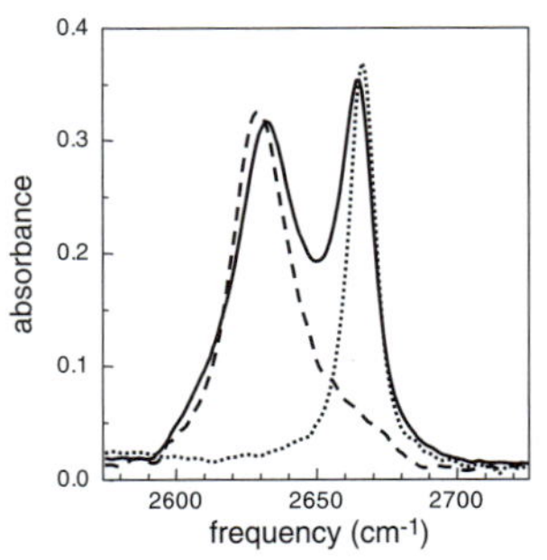

Fig. 2. FT-IR absorption spectra of the OD stretch of phenol-OD (hydroxyl H replaced with D) in CCl_4 (free phenol, dotted curve), phenol in benzene (benzene-phenol complex, dashed curve), and phenol in the mixed benzene-CCl_4 solvent (2:5 molar ratio), which displays absorptions of both free and complexed phenol (solid curve).

Fig. 1. Two views of the structure of the phenol-benzene complex calculated with DFT at B3LYP/ 6-31+ G (d,p) level in the gas phase.

a negative-going signal. At $T_w = 200$ fs, there are two peaks on the diagonal (0-1 transitions) and the corresponding 1-2 transition peaks are off-diagonal. There are no off-diagonal peaks in the 0-1 region because 200 fs is short relative to the exchange time; by contrast, $T_w = 14$ ps (Fig. 3B) is long relative to the exchange time, and additional peaks have grown in.

The 2D IR vibrational echo spectra in Fig. 3, A and B, can be understood in terms of time-dependent diagrammatic perturbation theory, which describes the nonlinear optical interactions with the molecular vibrations (17, 18). (See fig. S2 for detailed diagrams showing how all the peaks are generated.) The frequency at which the first pulse excites a mode is the mode frequency on the ω_τ axis (horizontal axis): 2631 cm^{-1} for the 0-1 transition in free phenol and 2665 cm^{-1} for the complex. The third pulse induces the vibrational echo pulse, which is emitted after a time delay at the precise frequency of the vibrational transition that interacted with that third pulse. The frequency of the vibrational echo emission is the frequency on the ω_m axis (the vertical axis). In Fig. 3A, the data are taken before chemical exchange. For the 0-1 vibrational transitions, the third pulse induces the vibrational echo emission at the same frequencies excited by the first pulse, so there are two peaks on the diagonal where $\omega_\tau = \omega_m$ (red peaks in Fig. 3A). If the frequency of vibrational echo emission (ω_m, third pulse frequency) is different from the frequency of initial excitation (ω_τ, first pulse frequency), peaks will appear off-diagonal. Again, in Fig. 3A, the blue peaks are off-diagonal by the vibrational anharmonicity (19) because the modes are initially excited at their 0-1 frequencies (ω_τ) but the third pulse causes vibrational echo emission at their 1-2 frequencies (ω_m). Even in the absence of chemical exchange, the peaks observed at very short T_w delays (Fig. 3A) will undergo evolution with increasing T_w, because of both spectral diffusion (17, 18, 20) (see below), which changes the shapes of the peaks, and vibrational lifetime decay and orientational relaxation, which cause the peaks to decay in amplitude.

The influence of chemical exchange on the 2D correlation spectrum can be easily understood in terms of the ideas presented above. If some complexed phenols are liberated during the T_w period, then the third pulse will cause the emission of the vibrational echo at the frequency of the free phenol OD stretch. The frequency of emission ω_m then differs from the excitation frequency ω_τ for these specific molecules. The result will be an off-diagonal peak that appears only if chemical exchange occurs. Because the free phenol absorbs at higher frequency than complexed phenol, this off-diagonal peak is shifted to higher frequency along the ω_m axis by the frequency difference (34 cm^{-1}) between the free and complexed modes. Conversely, if some free phenols

bind to benzene during the T_w period, then the third pulse will produce an off-diagonal peak for these (formerly) free phenols, which is shifted to lower frequency along the ω_m axis by the same amount. Identical considerations apply for both the 0-1 and 1-2 regions of the spectrum. This behavior is shown in Fig. 3B, wherein substantial chemical exchange has led to the generation of a block of four red peaks and a block of four blue peaks; the two new peaks in each block were not present at $T_w = 200$ fs. Clearly some complexes have dissociated and others have formed. The growth of the additional off-diagonal peaks with increasing T_w is directly related to the time dependence of the chemical exchange.

The phenol-benzene complex persists for 8 ps. Figure 4 displays the time evolution of the correlation spectrum in the 0-1 transition region as 3D representations. (More plots are shown in fig. S3.) The data have been normalized to the largest peak at each T_w. As in Fig. 3, at $T_w = 200$ fs there are only diagonal peaks. The peak in the foreground is the vibration echo of complexed phenol. At $T_w = 2$ ps, two changes are evident. First, the shapes of the diagonal peaks have altered. Whereas at $T_w = 200$ fs the peaks are elongated along the diagonal, by 2 ps they have become symmetrically rounded. The change in shape is caused by spectral diffusion. At short times, the vibrational transitions are inhomogeneously broadened. As time proceeds, the fluctuations in the solvent environment cause the transition frequency to sample all possible values, and each line becomes dynamically broadened. The change in shape provides information on the dynamic solute-solvent interactions, but here we focus our analysis exclusively on the chemical exchange. At 2 ps, the off-diagonal peaks are just becoming visible. By 5 ps, the off-diagonal peaks are clearly evident, and they continue to grow in amplitude, as shown in the 10-ps and 14-ps plots.

Simple inspection of the data reveals that the phenol-benzene complexes form and dissociate on a picosecond time scale. To obtain quantitative rates, we fit the data with the use of time-dependent diagrammatic perturbation theory and kinetic equations to describe the exchange dynamics. Because of spectral diffusion, the shapes of the peaks change. In the absence of all other dynamical processes, the change in shape preserves the volume of a peak, but the peak amplitude is reduced as the peak broadens along the ω_τ axis. Therefore, the integrated peak volumes are fit to obtain the population dynamics. There are three processes that contribute to the change in the peak volumes: the OD vibrational lifetime relaxation in free and complexed phenol (lifetimes $T_1^{\,f}$ and $T_1^{\,c}$), the orientational relaxation (time constants $\tau_r^{\,f}$ and $\tau_r^{\,c}$), and the chemical exchange (dissociation and formation rate constants k_d and k_f). The kinetic scheme is shown in Fig. 5A; more

details are described in fig. S4. The vibrational relaxation and orientational relaxation lead to diminishing intensities of all peaks with increasing time. Even complete orientational randomization during the T_w period does not cause the vibrational echo to decay to zero. Therefore, the chemical exchange time need not be short relative to the orientational relaxation time. The chemical exchange causes the diagonal peaks to diminish and the off-diagonal peaks to grow, as can be seen in Figs. 3 and 4. The dissociation rate (number per unit time) of the complex equals the formation rate if the system is in equilibrium (see below). The complex dissociation time constant τ_d is independent of concentration and is therefore used here ($\tau_d = 1/k_d$, where k_d is the dissociation rate constant; the dissociation rate is k_d[complex]).

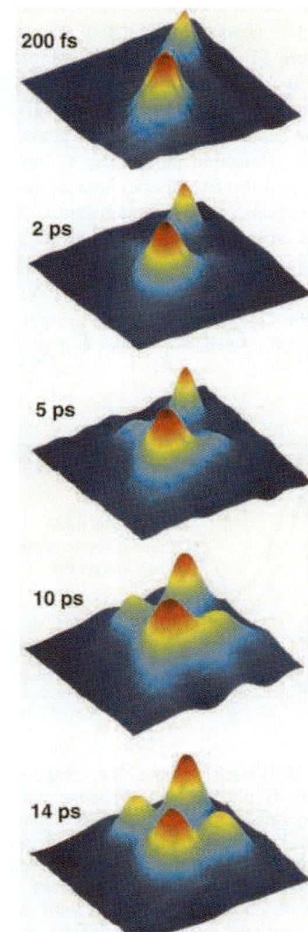

Fig. 4. The time dependence of the 2D IR vibrational echo spectrum in the 0-1 transition region. As T_w increases, the off-diagonal peaks grow in because of chemical exchange (formation and dissociation of the benzene-phenol complex). Between 200 fs and 2 ps, the peaks change shape because of spectral diffusion. Inspection of the data shows a picosecond time scale for the chemical exchange (growth of the off-diagonal peaks).

The data for the 0-1 transition region consist of four time-dependent components: the two diagonal peaks (complexed and free phenol) and the two off-diagonal peaks (dissociation and formation of complexes during the experiment). A similar treatment is applied for the 1-2 region below. All four peaks can be reproduced with the single fitting parameter τ_d. The input parameters used are $T_1^c = 10$ ps, $T_1^f = 12.5$ ps, $\tau_r^c = 3.4$ ps, $\tau_r^f = 2.9$ ps, and the ratio of the complexed and free phenol concentrations [complex]/[free] = 0.8. The lifetimes and the orientational relaxation times were measured with polarization-selective IR pump-probe experiments (21) on the OD stretch of the complex in pure benzene solvent and free phenol in pure CCl_4 solvent. The orientational relaxation time constants were corrected for the small viscosity difference between the pure solvents and mixed solvents. The concentration ratio was measured with linear absorption FT-IR spectroscopy.

Figure 5B shows the peak volume data for the 0-1 transition region of the spectrum as a function of T_w. The fits (solid curves) to the time dependence of all of the peaks using a single adjustable parameter, τ_d, are very good. From the fits, dissociation of the complex proceeds with a time constant $\tau_d = 8 \pm 2$ ps. The error bars obtained from the single-parameter fits are smaller than 2 ps. The cited error bars mainly arise from the uncertainties in all of the parameters in the calculation, rather than the quality of the fit.

Vibrational excitation does not perturb the equilibrium. An important aspect of 2D NMR is that the manipulation of spin states by a pulse sequence produces a negligible perturbation of the molecular system, and therefore the experiment does not move the system away from equilibrium. It is possible to determine whether vibrational excitation in the current system shifts the concentrations away from their equilibrium values or changes τ_d. Even if vibrational excitation does shift the system out of equilibrium, the dynamics for the equilibrium system can nonetheless be extracted. If the thermal equilibrium is not perturbed by vibrational excitation, both off-diagonal peaks (formation and dissociation of the complex) should have the same time dependence. In the 0-1 transition region, the integrated off-diagonal intensities (circles and squares in Fig. 5B) do show matching growth rates as a function of T_w, confirming an undisturbed equilibrium. In other words, the pulse sequence has no impact on the aggregate populations of free and bound phenol in solution. For each complex formed during the T_w period, another complex liberates a phenol.

It is possible that vibrational excitation changes both the dissociation rate and the complex formation rate. If both rates were changed to the same extent, the equilibrium concentra-

tions would be unchanged and the off-diagonal peaks would grow in at the same rate, as observed. It is possible to test whether vibrational excitation changes the dissociation time constant τ_d directly and conclusively. Of more importance, it is possible to see how the equilibrium exchange time constants can be extracted even if vibrational excitation does change the dynamics of dissociation and formation of complexes.

A series of qualitative energy level/kinetic diagrams (Fig. 6) clarifies how to do this and enables a more detailed understanding of the 2D vibrational echo measurements of chemical exchange. These diagrams are not rigorous quantum mechanical diagrams as used in diagrammatic perturbation theory, but their simplicity has important heuristic value. (The full set of double-sided Feynman diagrams pertaining to this problem is given in fig. S2.) The diagrams show how the off-diagonal peaks for the dissociation of complexed phenol to free phenol ($c \rightarrow f$) are generated in both the 0-1 (Fig. 6, A and B) and the 1-2 (Fig. 6C) portions of the 2D spectrum. These are the peaks at $\omega_\tau = 2631$ cm^{-1} and $\omega_m = 2665$ cm^{-1} (0-1), and at $\omega_\tau = 2631$ cm^{-1} and $\omega_m = 2575$ cm^{-1} (1-2) in Fig. 3B. The other off-diagonal peaks, representing complex formation, arise in the same manner.

The signal for the off-diagonal 0-1 portion $c \rightarrow f$ peak has two contributions illustrated in Fig. 6, A and B. In Fig. 6A, the arrow representing pulse 1 makes a coherent superposition state of the $v = 0$ and $v = 1$ levels of the phenol-benzene complexes at frequency

A

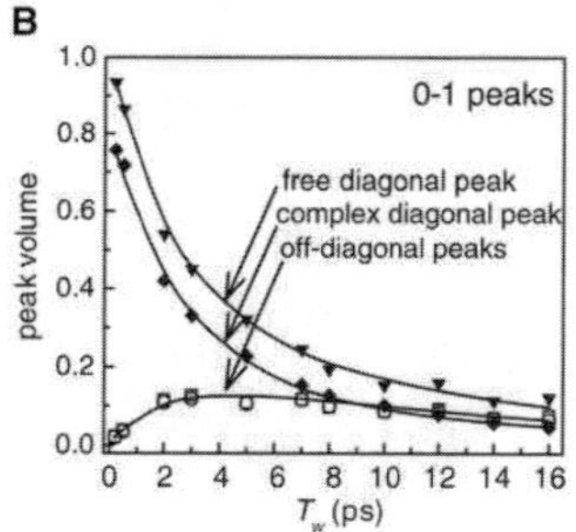

B

Fig. 5. (A) Kinetic scheme. The complexed (c) and free (f) phenols can exchange with dissociation rate constant k_d and formation rate constant k_f. The exchange process causes the off-diagonal peaks to grow in and the diagonal peaks to diminish. All of the peaks decay because of vibrational relaxation with lifetime T_1^i and rotational relaxation with time constant τ_r^i. (B) Peak volume data (scaled with extinction coefficient differences; see fig. S4) and fits to the data. There is one adjustable parameter to fit all of the data: the dissociation time constant, τ_d, of the phenol-benzene complex. Fits to the data for the four peaks in the 0-1 transition region give $\tau_d = 8$ ps. The two off-diagonal peaks (circles and squares) grow in at the same rate, showing that the thermal equilibrium is not perturbed by vibrational excitation to the first vibrationally excited state of the hydroxyl stretch.

Fig. 6. (A to C) Energy level/kinetic diagrams that show the pulse sequence pathways that give rise to the vibrational echo signal for the off-diagonal peaks created by the dissociation of the phenol benzene complex in the 0-1 and 1-2 portions of the 2D vibrational echo spectrum (Fig. 3B). (A) and (B) show the two pathways for the 0-1 portion of the spectrum. Each pathway contributes half of the 0-1 dissociation off-diagonal peak. In pathway (A), complexes are in the $v = 1$ excited state during the T_w period, and dissociation of complexes in the $v = 1$ state contributes to the off-diagonal peak. In pathway (B), complexes are in the $v = 0$ ground state during the T_w period, and dissociation of complexes in the

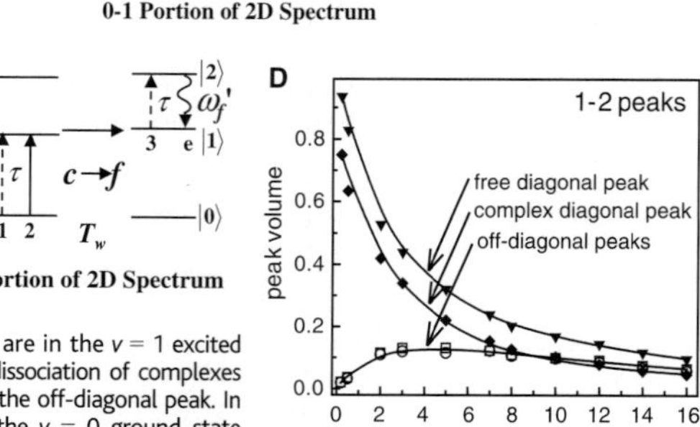

$v = 0$ state contributes to the off-diagonal peak. (C) shows the single pathway that contributes to the 0-1 dissociation off-diagonal peak in the 1-2 portion of the spectrum. In pathway (C), complexes are in the $v = 1$ excited state during the T_w period. Only complexes that dissociate while in the excited state contribute to the dissociation off-diagonal peak in the 1-2 portion of the spectrum. The difference between the two contributions to the signal for the 0-1 portion and the single contribution for the 1-2 portion permit the extraction of the equilibrium dissociation time even if vibrational excitation changes the dissociation and formation rate constants from their equilibrium values (see text). (D) Data and calculated curves for the 1-2 portion of the spectrum (blue peaks in Fig. 3B) give the same τ_d value as the 0-1 portion, demonstrating that excitation of the hydroxyl stretch does not influence the dissociation rate constant.

D

ω_c. In all diagrams, a dashed arrow represents the creation of a coherent superposition state. A coherent superposition state is equivalent to an in-plane precessing magnetization that is generated by the first pulse in a 2D NMR pulse sequence. During the τ period between pulses 1 and 2, exchange does not contribute to the growth of the off-diagonal peak, because the exchange time τ_d is long relative to the inverse of the frequency difference between the c and f peaks. Therefore, jumps from c to f will produce an ensemble of superposition states in f with random phase, resulting in no contribution to the off-diagonal peak signal. After time τ, the second pulse generates a population in the excited $v = 1$ state. A solid arrow in all diagrams represents the generation of a population. Some of the molecules that are initially complexes have now been "labeled" in the $v = 1$ state. If some of these complexes dissociate ($c{\to}f$) during the period T_w, pulse 3 will produce a coherent superposition state (dashed arrow in Fig. 6A) oscillating at frequency ω_f. The vibrational echo pulse will be emitted (wavy arrow) at frequency ω_f. Therefore, the ω_τ frequency is ω_c, but the ω_m frequency is ω_f. The $c{\to}f$ off-diagonal peak is produced. The pathway in Fig. 6A for which the dissociation occurs with molecules labeled in $v = 1$ accounts for half of the $c{\to}f$ off-diagonal peak signal in the 0-1 portion of the spectrum.

The pathway shown in Fig. 6B accounts for the other half of the $c{\to}f$ off-diagonal peak signal. Again, pulse 1 produces a 0-1 coherent superposition state (dashed arrow), but pulse 2 produces a population (solid arrow) in the $v = 0$ state instead of in the $v = 1$ state, again labeling molecules that are initially complexes. This pathway involves molecules labeled in the $v = 0$ state during the T_w period. For those complexes that undergo dissociation during the T_w period, pulse 3 will produce a coherent superposition state (dashed arrow) at frequency ω_f, and the vibrational echo (wavy arrow) will be emitted at ω_f. As in the path shown in Fig. 6A, in Fig. 6B the ω_τ frequency is ω_c, but the ω_m frequency is ω_f, and this path contributes to the $c{\to}f$ off-diagonal peak. For both pathways, pulse 3 produces a coherent superposition state. In NMR, this coherent superposition state yields an in-plane precessing magnetization (a macroscopic oscillating magnetic dipole moment) that is detected with a pickup coil. In the vibrational experiment, the coherent superposition state in each pathway gives rise to a macroscopic oscillating electric dipole moment that emits light, the vibrational echo pulse (wavy arrows).

The important point to see from Fig. 6, A and B, is that half of the signal comes from molecules that are in their $v = 1$ state during the T_w period and half comes from molecules in the $v = 0$ state. If vibrational excitation influences the equilibrium dynamics, then the time evolution of the exchange off-diagonal peaks will be affected and τ_d can be altered from its equilibrium value. Whether τ_d is changed by vibrational excitation can be tested; if it is, then its equilibrium value can be recovered by examining the 1-2 portion of the 2D vibrational echo spectrum in addition to the 0-1 portion. Figure 6C shows the single diagram that contributes to the $c{\to}f$ off-diagonal peak in the 1-2 portion of the spectrum. In Fig. 6C, pulse 1 creates a 0-1 coherent superposition state (dashed arrow) and pulse 2 generates a population (solid arrow) of complexed phenols in the $v = 1$ state. There is only one path because it is necessary to have molecules in the $v = 1$ state so that pulse 3 can produce a coherent superposition state (dashed arrow) of $v = 1$ and $v = 2$, which is then followed by vibrational echo emission at the 1-2 frequency (blue peaks in Fig. 3B). If dissociation of complexed phenols occurs during the T_w period, the blue $c{\to}f$ off-diagonal peak is generated. Thus, the entire signal for this off-diagonal peak arises solely from complexes that were in the $v = 1$ state and underwent dissociation during the T_w period.

This property does not hold for the red $c{\to}f$ off-diagonal peak, for which half of the signal is generated by molecules that were in the $v = 1$ state and half by molecules in the $v = 0$ state during the T_w period. If excitation to the vibrationally excited $v = 1$ state changes the equilibrium dynamics, then the τ_d determined from the 0-1 portion of the spectrum will be different from the τ_d determined from the 1-2 portion of the spectrum. The τ_d value determined from the 0-1 portion of the spectrum is an average of its ground state and first vibrationally excited state value. Using the two values of τ_d determined from the 0-1 and 1-2 portions of the 2D vibrational echo spectrum, the value for the ground state (unperturbed thermal equilibrium) pathway can be easily obtained. If the two values are the same, then vibrational excitation did not have an experimentally measurable influence on the equilibrium dynamics of the system.

Figure 6D shows the data and calculated curves for the 1-2 region of the 2D vibrational echo spectrum. There are no adjustable parameters. The data are reproduced with the identical parameters used for the 0-1 region, including $\tau_d = 8$ ps. Variation of τ_d does not improve the agreement. Thus, within experimental uncertainty, for the phenol-benzene complex and the other systems studied here, exciting the hydroxyl stretch has no influence on the dissociation time and, as discussed above, vibrational excitation does not change the equilibrium constant.

Electron-rich benzene derivatives make stronger complexes. In addition to the phenol-benzene complex, the phenol-bromobenzene and phenol–p-xylene complexes were studied in the identical manner. The dissociation time for the phenol–p-xylene complex is considerably slower than for the phenol-benzene complex, with $\tau_d = 21 \pm 3$ ps. The dissociation time constant for the phenol-bromobenzene complex is $\tau_d = 6 \pm 3$ ps. This value is within experimental uncertainty of the phenol-benzene complex τ_d. However, the error bars are determined in large part by the errors of all of the parameters used in the calculations, rather than by the quality of the fits. A direct comparison for the phenol-benzene and phenol-bromobenzene complexes (at, for example, 7 ps) of the sizes of their diagonal peaks versus their respective dissociation-induced off-diagonal peaks clearly shows that the bromobenzene complex dissociates more rapidly.

To investigate the trend in the τ_d values, we measured the temperature dependences of the complexation equilibria for all three systems by IR absorption and then used these values to determine the bond enthalpies, ΔH^0, of the complexes. (Here, the standard solvent concentrations are defined to be the concentrations used in the experiments.) The ΔH^0 values extracted from van't Hoff plots were -1.21 kcal/mol for the phenol-bromobenzene complex, -1.67 kcal/mol for the phenol-benzene complex, and -2.23 kcal/mol for the phenol–p-xylene complex (fig. S5). Thus, as the bond enthalpy increases (stronger bond), the dissociation time also increases. If the free energy of activation for dissociation ($\Delta G^\ddagger$) scales with the bond enthalpies, we can qualitatively understand the trend in τ_d. The electron-withdrawing bromine weakens the complex by reducing the π-electron density of the ring, leading to a faster dissociation time. Conversely, in p-xylene, the electron-donating methyl groups increase the π-electron density of the ring, resulting in a stronger complex and correspondingly longer dissociation time.

The 2D IR vibrational echo technique is general. The method used here is general for measuring fast chemical exchange dynamics in the ground electronic state, and appears promising for studies of a wide range of molecular isomerizations and electron and proton transfer processes under thermal equilibrium conditions. There are three conditions necessary to apply this technique to study thermal equilibrium exchange phenomena: (i) There must be at least one IR active mode that has a distinct frequency for each species undergoing exchange, (ii) the concentrations of all the equilibrated species must be high enough for detection, and (iii) the exchange rate must be comparable to or shorter than the vibrational lifetime of the vibration that is being used as the probe. It is important to note that the vibrational mode that is used as the probe need not be directly involved in the exchange process, as is the hydroxyl stretch used in this study. It is sufficient that the exchange causes the mode frequency to change.

References and Notes

1. C. Reichardt, *Solvents and Solvent Effects in Organic Chemistry* (Wiley-VCH, Weinheim, Germany, ed. 3, 2003).
2. G. J. Throop, R. J. Bearman, *J. Chem. Phys.* **42**, 2408 (1965).
3. S. N. Vinogradov, R. H. Linnell, *Hydrogen Bonding* (Van Nostrand Reinhold, New York, 1971).
4. J. Jeener, B. H. Meier, P. Bachmann, R. R. Ernst, *J. Chem. Phys.* **71**, 4546 (1979).
5. B. H. Meier, R. R. Ernst, *J. Am. Chem. Soc.* **101**, 6441 (1979).
6. A. G. Palmer, *Chem. Rev.* **104**, 3623 (2004).
7. N. Demirdoven, M. Khalil, O. Golonzka, A. Tokmakoff, *J. Phys. Chem. A* **105**, 8030 (2001).
8. Y. Kim, R. M. Hochstrasser, *J. Phys. Chem. B* **109**, 6884 (2005).
9. J. B. Asbury *et al.*, *Phys. Rev. Lett.* **91**, 237402 (2003).
10. H.-J. Schneider, A. K. Yatsimirsky, *Principles and Methods in Supramolecular Chemistry* (Wiley, Chichester, UK, 2000).
11. E. A. Meyer, R. K. Castellano, F. Diederich, *Angew. Chem. Int. Ed. Engl.* **42**, 1210 (2003).
12. J. L. Knee, L. R. Khundkar, A. H. Zewail, *J. Chem. Phys.* **87**, 115 (1987).
13. M. F. Perutz, *Philos. Trans. R. Soc. London Ser. A* **345**, 105 (1993).
14. A. Fujii, T. Ebata, N. Mikami, *J. Phys. Chem. A* **106**, 8554 (2002).
15. G. C. Pimentel, A. L. McClellan, *Annu. Rev. Phys. Chem.* **22**, 347 (1971).
16. J. B. Asbury, T. Steinel, M. D. Fayer, *J. Lumin.* **107**, 271 (2004).
17. S. Mukamel, *Annu. Rev. Phys. Chem.* **51**, 691 (2000).
18. S. Mukamel, *Principles of Nonlinear Optical Spectroscopy* (Oxford Univ. Press, New York, 1995).
19. O. Golonzka, M. Khalil, N. Demirdoven, A. Tokmakoff, *Phys. Rev. Lett.* **86**, 2154 (2001).
20. J. B. Asbury *et al.*, *J. Chem. Phys.* **121**, 12431 (2004).
21. H.-S. Tan, I. R. Piletic, M. D. Fayer, *J. Chem. Phys.* **122**, 174501(9) (2005).
22. We thank J. I. Brauman for insightful discussions, J. Xie for his aid in chemical preparation, and H.-S. Tan for his aid in pump-probe measurements. Supported by grants from the Air Force Office of Scientific Research (F49620-01-1-0018) and NSF Division of Materials Research (DMR-0332692).

Supporting Online Material
www.sciencemag.org/cgi/content/full/1116213/DC1
Materials and Methods
Figs. S1 to S5

16 June 2005; accepted 26 July 2005
Published online 4 August 2005;
10.1126/science.1116213
Include this information when citing this paper.

Formation and Dissociation of Intra−Intermolecular Hydrogen-Bonded Solute−Solvent Complexes: Chemical Exchange Two-Dimensional Infrared Vibrational Echo Spectroscopy

Junrong Zheng, Kyungwon Kwak, Xin Chen, John B. Asbury,[†] and M. D. Fayer*

Contribution from the Department of Chemistry, Stanford University, Stanford, California 94305

Received October 17, 2005; E-mail: fayer@stanford.edu

Abstract: 2-Methoxyphenol (2MP) solutes form weak complexes with toluene solvent molecules. The complexes are unusual in that the 2MP hydroxyl has an intramolecular hydrogen bond and simultaneously forms an intermolecular hydrogen bond with toluene and other aromatic solvents. In the equilibrated solute−solvent solution, there exists approximately the same concentration of 2MP−toluene complex and free 2MP. The very fast formation and dissociation (chemical exchange) of this type of three-centered hydrogen bond complex were observed in real time under thermal equilibrium conditions with two-dimensional (2D) infrared vibrational echo spectroscopy. Chemical exchange is manifested in the 2D spectrum by the growth of off-diagonal peaks. Both the formation and dissociation can be characterized in terms of the dissociation time constant, which was determined to be 3 ps for the 2MP−toluene complex. The intra−intermolecular hydrogen bond formation is influenced by subtle details of the molecular structure. Although 2MP forms a complex with toluene, it is demonstrated that 2-ethoxyphenol (2EP) does not form complexes to any significant extent. Density functional calculations at the B3LYP/6-31+G(d,p) level suggest that steric effects caused by the extra methyl group in 2EP are responsible for the difference.

I. Introduction

Hydrogen bonding is ubiquitous in nature. It is involved in the most basic and important chemical and biological phenomena.[1−3] The strength of hydrogen bonds lies between van der Waals forces and covalent bonds. Although not a true chemical bond, a hydrogen bond is sufficiently strong and directional to become the driving force for molecules to assemble into delicate architectures in supramolecular chemistry, molecular recognition, and self-assembly. The strength of hydrogen bonds is in the range of energies that permits rapid association and dissociation under ambient conditions. Such rapid hydrogen bond dynamics are important in a wide variety of systems, such as the properties of water[4] and biological recognition.[3] Hydrogen bonding has been studied extensively in many contexts since the birth of the concept in the early 1900s.[2,3,5]

Hydrogen bonds can be separated into two categories: intermolecular hydrogen bonds in which the hydrogen bond donor and acceptor are in different molecules and intramolecular hydrogen bonds in which the donor and acceptor are in the same molecule.[5] In general, intramolecular hydrogen bonds form five-, six-, or seven-membered rings,[2] where the geometric restrictions make them relatively weak compared to intermolecular hydrogen bonds involving the same type of donor and acceptor. Because of geometrical constraints, intramolecular hydrogen bonds cannot form with the optimal geometry compared to the intermolecular ones, creating the possibility that the donor of an intramolecular hydrogen bond can form an additional intermolecular hydrogen bond with an appropriate acceptor located on another molecule. A special type of hydrogen bond will form under these conditions. This type of hydrogen bond involves three centers, one donor, and two acceptors. There is another type of three-centered hydrogen bond with one acceptor and two donors, which is common when the acceptor has two electron lone pairs; e.g., in a molecule such as water, the oxygen can be the acceptor for two hydrogen donors on two other water molecules. The "intra−intermolecular hydrogen bond" discussed above has a single donor with two acceptors. This type of three-centered hydrogen bond, which is also called "bifurcated",[6] plays important roles in many chemical and biological systems, e.g., chiral molecular recognition,[7] proteins,[8] RNA,[9] DNA,[10] and carbohydrates.[11] Kinetic studies of the formation and dissociation

† Permanent address: Department of Chemistry, Penn State University, University Park, PA 16802.

(1) Pimentel, G. C.; McClellan, A. L. *Annu. Rev. Phys. Chem.* **1971**, *22*, 347−385.
(2) Pimentel, G. C.; McClellan, A. L. *The Hydrogen Bond*; W. H. Freeman and Co.: San Francisco, 1960.
(3) Desiraju, G. R.; Steiner, T. *The Weak Hydrogen Bond*; Oxford: New York, 1999.
(4) Bertolini, D.; Cassettari, M.; Ferrario, M.; Grigolini, P.; Salvetti, G. *Adv. Chem. Phys.* **1985**, *62*, 277−320.
(5) Jeffrey, G. A. *An Introduction to Hydrogen Bonding*; Oxford University Press: New York, 1997.

(6) Steiner, T. *Angew. Chem., Int. Ed.* **2002**, *41*, 48−76.
(7) Kim, S.-G.; Kim, K.-H.; Kim, Y. K.; Shin, S. K.; Ahn, K. H. *J. Am. Chem. Soc.* **2003**, *125*, 13819−13824.
(8) Sundaralingam, M.; Sekharudu, Y. C. *Science* **1989**, *244*, 1333−1337.
(9) Auffinger, P.; Westhol, E. *J. Mol. Biol.* **1999**, *292*, 467−483.
(10) Nelson, H. C. M.; Finch, J. T.; Luisi, B. F.; Klug, A. *Nature* **1987**, *330*, 221−226.

10.1021/ja0570584 CCC: $33.50 © 2006 American Chemical Society

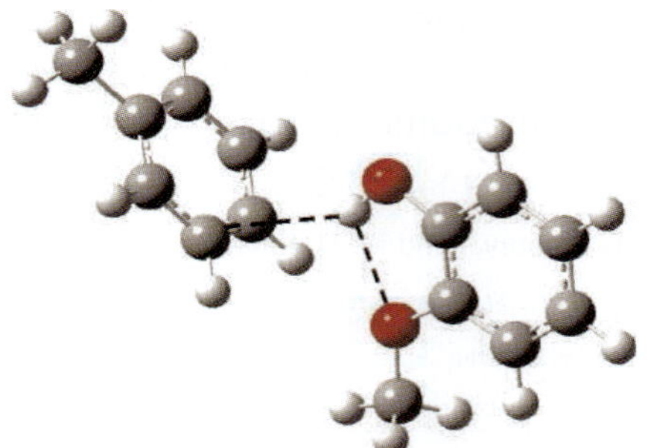

Figure 1. Structure of the intra—intermolecular hydrogen-bonded complex of 2-methoxyphenol and toluene in the isolated state calculated with DFT at the B3LYP/6-31+G(d,p) level.

of this type of hydrogen bond in liquid solutions are challenging because of its ultrafast time scale.

As described in detail below, 2-methoxyphenol (2MP) is an example of intra—intermolecular hydrogen bonding when it is dissolved in toluene or other aromatic solvents. 2MP forms an intramolecular hydrogen bond with the oxygen on the methoxyl group, the acceptor for the hydroxyl hydrogen donor. The formation enthalpy of the intramolecular hydrogen bond of 2MP in cyclohexane was determined to be -2.0 kcal/mol on the basis of the spectroscopic data.[12] When 2MP is dissolved in toluene, an intra—intermolecular hydrogen bond is formed, in which the π-electrons of the benzene ring act as a weak hydrogen bond acceptor. The formation enthalpy of the intermolecular part of this three-centered hydrogen bond was determined to be ~-0.6 kcal/mol (see below). The calculated structure of the 2MP—toluene complex is shown in Figure 1. The calculation was done with density functional theory (DFT) at the B3LYP/6-31+G-(d,p) level. The calculation is for the isolated molecules; that is, there is no solvent in the calculation. The distance between the hydrogen atom and the intramolecular acceptor oxygen is ~2.18 Å, whereas for the pure intramolecular hydrogen bond, the distance is ~2.11 Å. The distance change shows that the intramolecular interaction is weakened by the intermolecular interaction. The 2MP hydroxyl hydrogen atom points roughly to the center of one of the C—C bonds that is para to the methyl group of the toluene. The distance between the 2MP hydrogen atom and the toluene *meta*-carbon atom is ~2.74 Å, and the distance to the *para*-carbon is ~2.82 Å.

In the following, linear spectroscopic evidence will be presented for the existence of the complex shown in Figure 1 and for the formation enthalpy of the intra—intermolecular hydrogen bond. Then, ultrafast two-dimensional (2D) IR vibrational echo spectroscopy (2D IR VES) is used to measure the dynamics of chemical exchange between the 2MP—toluene complex and free 2MP in a solution of low concentration of 2MP and toluene as the solvent and to support the species assignments made by linear spectroscopy. The formation and dissociation of the three-centered hydrogen bond are directly monitored by observing the growth of off-diagonal peaks in the 2D IR vibrational echo spectrum.[13–15] The measurements provide the fast dynamics of solute—solvent complex formation

and dissociation under thermal equilibrium conditions, as has recently been demonstrated for a different type of solute—solvent complex system.[13,14] Another 2D IR technique, narrow bandwidth pump/broad bandwidth probe, has been used to study chemical exchange.[16] The 2D IR VES method is a Fourier transform spectroscopy like Fourier transform NMR and Fourier transform IR, which can frequently provide superior data when compared to non-Fourier transform methods. The observation of fast chemical exchange processes[14,15] is another advance in 2D IR VES. 2D IR VES has been developed and applied to a variety of problems over the last several years.[17–26]

II. Experimental Procedures

The 2D IR VES experimental setup is similar to that described previously.[27] Briefly, three successive IR pulses (~1 μJ/pulse) with the same polarization were applied to induce the subsequent emission in a distinct direction of a time-delayed 4th pulse, the vibrational echo. The transform-limited pulses (~50 fs, ~4 cycles of light) are produced using a Ti/Sapphire regeneratively amplified laser system pumping an optical parametric amplifier. The IR pulses span sufficient bandwidth (300 cm^{-1} centered at ~4 μm or 2500 cm^{-1}) to cover the $v = 0$ to $v = 1$ transition (denoted 0–1) and the $v = 1$ to $v = 2$ transition (denoted 1–2) of the hydroxyl OD stretching modes in both intramolecular and intra—intermolecular hydrogen-bonded 2MPOD. The OD hydroxyl stretch was studied rather than OH because the 2500 cm^{-1} frequency region associated with the OD stretch has less overlap with other modes. The vibrational echo pulse is detected with frequency and phase resolution by combining it with a 5th (local oscillator) pulse, and the combined pulses are dispersed in a spectrograph. The function of the local oscillator is to phase resolve and amplify the vibrational echo signal. Data are thus obtained as a function of three variables: the emitted vibrational echo frequencies ω_m and the variable time delays between the first and second pulses (τ) and the second and third pulses (T_w, the variable "waiting" time). By numerical Fourier transform (FT), the τ scan data taken at every ω_m are mapped to a second frequency variable ω_τ for each T_w. The data are then plotted in three dimensions, the amplitude as a function of both ω_τ and ω_m, which correspond to the ω_1 and ω_3 axes, respectively, in 2D NMR. The 2D vibrational echo spectra presented here are obtained with the dual scan technique[27–30]

(11) Taylor, R.; Kennard, O.; Versichel, W. *J. Am. Chem. Soc.* **1984**, *106*, 244–248.

(12) Carlson, G. L.; Fateley, W. G. *J. Phys. Chem.* **1973**, *77*, 1157–1163.

(13) Fayer, M. D.; Zheng, J.; Kwak, K.; Asbury, J. B. In *Molecular Dynamics/Theoretical Chemistry Meeting*, Monterey, California, May 22–24, 2005; Berman, M., Ed.

(14) Zheng, J.; Kwak, K.; Asbury, J. B.; Chen, X.; Piletic, I.; Fayer, M. D. *Science* **2005**, *309*, 1338–1343.

(15) Kim, Y. S.; Hochstrasser, R. M. *Proc. Natl. Acad. Sci. U.S.A.* **2005**, *102*, 11185–11190.

(16) Woutersen, S.; Mu, Y.; Stock, G.; Hamm, P. *Chem. Phys.* **2001**, *266*, 137–147.

(17) Zanni, M. T.; Hochstrasser, R. M. *Curr. Opin. Struct. Biol.* **2001**, *11*, 516–522.

(18) Golonzka, O.; Khalil, M.; Demirdoven, N.; Tokmakoff, A. *Phys. Rev. Lett.* **2001**, *86*, 2154–2157.

(19) Merchant, K. A.; Thompson, D. E.; Fayer, M. D. *Phys. Rev. Lett.* **2001**, *86*, 3899–3902.

(20) Volkov, V.; Schanz, R.; Hamm, P. *Opt. Lett.* **2005**, *30*, 2010–2012.

(21) Zanni, M. T.; Gnanakaran, S.; Stenger, J.; Hochstrasser, R. M. *J. Phys. Chem. B* **2001**, *105*, 6520–6535.

(22) Asbury, J. B.; Steinel, T.; Stromberg, C.; Gaffney, K. J.; Piletic, I. R.; Goun, A.; Fayer, M. D. *Phys. Rev. Lett.* **2003**, *91*, 237402.

(23) Steinel, T.; Asbury, J. B.; Corcelli, S. A.; Lawrence, C. P.; Skinner, J. L.; Fayer, M. D. *Chem. Phys. Lett.* **2004**, *386*, 295–300.

(24) Asbury, J. B.; Steinel, T.; Fayer, M. D. *Chem. Phys. Lett.* **2003**, *381*, 139–146.

(25) Zheng, J.; Kwak, K.; Steinel, T.; Asbury, J. B.; Chen, X.; Xie, J.; Fayer, M. D. *J. Chem. Phys.* **2005**, *123*, 164301.

(26) Khalil, M.; Demirdoven, N.; Tokmakoff, A. *J. Phys. Chem. A* **2003**, *107*, 5258–5279.

(27) Asbury, J. B.; Steinel, T.; Fayer, M. D. *J. Lumin.* **2004**, *107*, 217–286.

(28) Khalil, M.; Demirdoven, N.; Tokmakoff, A. *Phys. Rev. Lett.* **2003**, *90*, 047401 (047404).

(29) Asbury, J. B.; Steinel, T.; Stromberg, C.; Gaffney, K. J.; Piletic, I. R.; Fayer, M. D. *J. Chem. Phys.* **2003**, *119*, 12981–12997.

(30) Asbury, J. B.; Steinel, T.; Stromberg, C.; Gaffney, K. J.; Piletic, I. R.; Goun, A.; Fayer, M. D. *Chem. Phys. Lett.* **2003**, *374*, 362–371.

to reduce or eliminate dispersive contributions to the spectra. The phase correction procedure followed the method described previously.[27] The samples for the vibrational echo experiments are 3 wt % 2MPOD (OD = hydroxyl OH replaced by OD), *p*-methyl-*o*-methoxyphenolOD (pMMPOD), and 2-exothyphenolOD (2EPOD) in toluene between 3-mm thick CaF$_2$ windows with a 0.2-mm Teflon spacer.

The OD stretch vibrational lifetimes and rotational relaxation times for the samples were measured with the polarization selective IR pump—probe experiments. The polarization selective IR pump—probe experimental setup is similar to one described previously.[31] The laser source is the same as that used for the vibrational echo experiments. For the pump—probe experiments, the mid-IR pulse was spitted into two beams of intensity ratio 20:1. The beam with higher intensity served as the pump. The weaker one was the probe beam. The pump beam had horizontal polarization, and the probe beam polarization was 45° relative to the pump beam. The probe beam was passed through a spectrograph and detected by a 32-element MCT array detector. A polarizer was placed in front of the monochromator aligned to selectively measure the parallel or perpendicular polarized signal relative to the pump beam. Because the orientational relaxation is relatively fast compared to the vibrational lifetime, it was possible to tail match the signals for the two polarizations to eliminate possible errors introduced by differences in the amplitudes of the signals for the two polarizations. The method eliminates possible sources of error such as phase shifts caused by mirrors and different diffraction efficiencies of the grating for different polarizations.[32] The sample concentrations and cells are the same as those used in the vibrational echo experiments.

Because the intramolecular and intra—intermolecular hydrogen-bonded 2MPODs in toluene exchange on the same time scale as the OD stretch vibrational and rotational decays, the vibrational lifetimes and rational relaxation constants for the species cannot be obtained directly by performing pump—probe experiments on the 2MPOD/toluene solution. The vibrational lifetime for the intramolecular hydrogen-bonded 2MPOD in toluene was obtained by performing experiments on 2EPOD/toluene solution because 2EPOD does not form a complex with toluene, as discussed in detail below. It is assumed that the addition of one methyl will not make a significant change in the vibrational lifetime as it is well removed from the OD. The rotational relaxation time constant for the intramolecular hydrogen-bonded 2MP was obtained by correcting the measured value for the 2EPOD by the small volume difference between 2MPOD and 2EPOD. The rotational relaxation constant for the intra—intermolecular hydrogen-bonded 2MP in toluene was obtained by measuring that of *p*-methoxyphenolOD (pMPOD) in toluene solution. pMPOD forms a much stronger complex than 2MPOD because of the lack of the intramolecular hydrogen bond. In toluene, there are essentially only pMPOD—toluene complexes, and therefore, the complex orientational relaxation time can be measured. The OD stretch lifetime of the intra—intermolecular bonded 2MPOD was determined using the method described in Section III. B.

The formation enthalpy of the intermolecular part of the intra—intermolecular hydrogen bonding for 2MPOD in toluene was obtained by performing temperature-dependent FTIR measurements on the 3 wt % 2MPOD in toluene solution. The same sample cells as those used in the vibrational echo experiments were used. The equilibrium constant for the two species in the 2MPOD/toluene solution was obtained by comparing the areas of each peak obtained by fitting the FTIR spectrum to Gaussian line shapes and correcting for the difference in extinction coefficients. The extinction coefficient ratio for the intramolecular and intra—intermolecular hydrogen-bonded 2MPOD in toluene was obtained by comparing the 2D and one-dimensional (1D) IR results. The volume of each peak at a very short T_w in a 2D IR spectrum is proportional to the product of the concentration and the 4th power of the transition

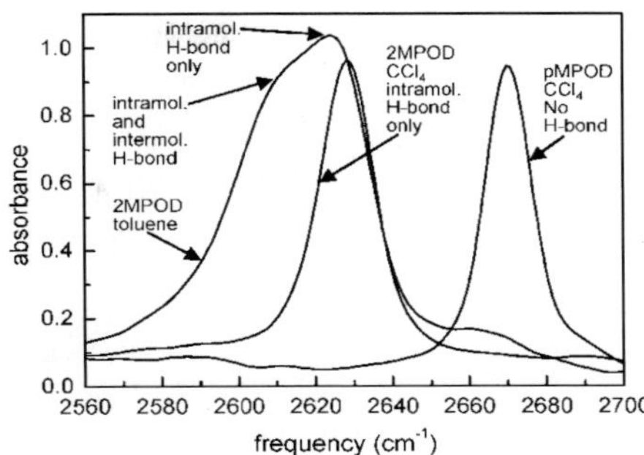

Figure 2. FTIR spectra of 0.25 wt % *p*-methoxyphenolOD in CCl$_4$, 1.3 wt % 2-methoxyphenolOD in CCl$_4$, and 2 wt % 2-methoxyphenolOD in toluene, showing the non-hydrogen-bonded, intramolecular hydrogen-bonded, and intra—intermolecular hydrogen-bonded OD stretch peaks, respectively.

dipole moment of the species.[33] The area of each peak in a 1D IR spectrum is proportional to the product of the concentration and the extinction coefficient (the square of the transition dipole moment) of each species.[34] Dividing the volume from the 2D IR spectrum at $T_w =$ 200 fs by the area from the 1D IR spectrum for the same species from the same sample determined the extinction coefficient ratio for the intra—intermolecular to the intramolecular hydrogen-bonded 2MPOD to be 2.0.

All chemicals were purchased from Aldrich. They were used as received. The deuterated hydroxyl hydrogen (OD) of the phenol derivatives was prepared by deuterium exchange with methanolOD. To obtain the deuterated phenol derivatives, 1 g of each undeuterated phenol derivative was dissolved in 10 g of methanolOD and stirred for 0.5 h. The solvent was then removed under vacuum. The procedure was repeated three times, and compounds with >90% deuteration of the hydroxyl hydrogen were obtained.

The electronic structure calculations were carried out using density functional theory,[35] as implemented in the Gaussian 98 program suite. The level and basis set used were Becke's three-parameter hybrid functional combined with the Lee—Yang—Parr correction functional, abbreviated as B3LYP and 6-31+G(d,p). All results reported here do not include the surrounding solvent and therefore are for the isolated molecules.

III. Results and Discussions

A. Linear Spectroscopy. The experimental evidence for the non-hydrogen-bonded, intramolecular hydrogen-bonded, and intra—intermolecular hydrogen-bonded species is shown in Figure 2. FTIR spectra in Figure 2 are for 0.25 wt % pMPOD in CCl$_4$, 1.3 wt % 2MPOD in CCl$_4$, and 2 wt % 2MPOD in toluene. The peak at 2670 cm^{-1} is for the free OD stretch. In CCl$_4$, phenol does not form an intermolecular complex when the concentration is low enough to avoid phenol oligomers.[1,14] A para-substituted phenol has chemical properties similar to those for an ortho-substituted one, so the non-hydrogen-bonded OD stretch of the pMPOD will have essentially the same frequency as 2MPOD would have if it did not form an intramolecular hydrogen bond. Therefore, the spectrum of pMPOD displays the OD stretch in the absence of both

(31) Tan, H.-S.; Piletic, I. R.; Fayer, M. D. *J. Chem. Phys.* **2005**, *122*, 174501 (174509).
(32) Tan, H.-S.; Piletic, I. R.; Fayer, M. D. *J. Opt. Soc. Am. B* **2005**, *22*, 2009—2017.
(33) Mukamel, S. *Principles of Nonlinear Optical Spectroscopy*; Oxford University Press: New York, 1995.
(34) Fayer, M. D. *Elements of Quantum Mechanics*; Oxford University Press: New York, 2001.
(35) Parr, R. G.; Yang, W. *Density Functional Theory of Atoms and Molecules*; Oxford University Press: New York, 1989.

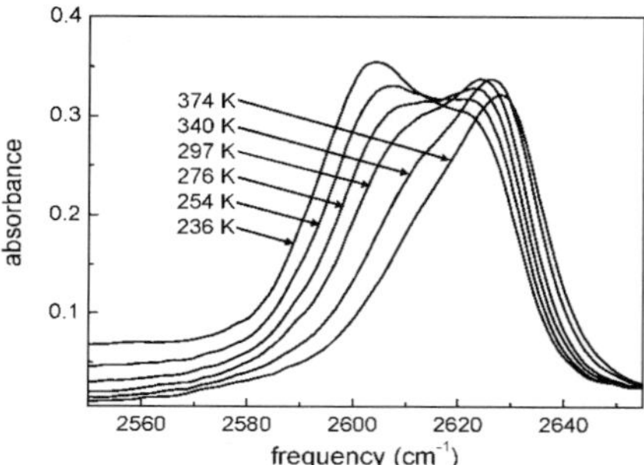

Figure 3. Temperature-dependent FTIR spectra of 3 wt % 2-methoxy-phenolOD in toluene from 236 to 374 K.

intermolecular and intramolecular hydrogen bonding. The 2MPOD spectrum in CCl$_4$ displays the spectrum of the intramolecular hydrogen-bonded OD stretch at 2629 cm^{-1}. The hydrogen bond produces a shift to lower frequency (red shift).[2] The spectrum for 2MPOD in toluene is considerably broader than the spectrum in CCl$_4$. The spectrum consists of one main peak and an approximately equal amplitude shoulder. The main peak has the same frequency as the peak in CCl$_4$. Therefore, we attribute it to the intramolecular hydrogen-bonded OD stretch in the absence of the intermolecular complex formation (free 2MPOD). The shoulder at around 2611 cm^{-1} arises from the intra—intermolecular hydrogen-bonded OD stretch, which is further red shifted because of the additional hydrogen bonding. Both the intramolecular hydrogen-bonded and intra—intermolecular hydrogen-bonded 2MP are prominent in the toluene solution because the formation Gibbs free energy of the intermolecular part of the intra—intermolecular hydrogen bond is very small (see below). It was found that the frequency change for the hydrogen-bonded hydroxyl stretch is linearly proportional to the formation enthalpy of the hydrogen bond for the same donor.[1] Our results are consistent with this observation. Additional FTIR experiments and the 2D IR VES data presented below confirm the identification of the main peak and the shoulder as the free 2MPOD and the 2MPOD—toluene complex, respectively.

Figure 3 is the temperature-dependent FTIR spectra of the 2MPOD in toluene from 236 to 374 K. It is clear that the lower-frequency peak becomes more pronounced when the temperature decreases. These results show that the lower-frequency peak is less favorable entropically, consistent with our assigning it to the OD stretch of the intra—intermolecular hydrogen-bonded 2MPOD. Lower entropy is consistent with a complex vs a free 2MPOD and a free solvent molecule. The shoulder at ~2610 cm^{-1} is the peak of the 2MPOD—toluene complex. The increasing amplitude of this peak with decreasing temperature shows that the equilibrium shifts to more complexes at lower temperatures.

Figure 4A and B are van't Hoff Plots[36] for 2MPOD in toluene plotted from 297 to 345 K and from 250 to 291 K (independent measurements). The enthalpy of formation of the 2MPOD/toluene complex in the toluene solution (or the intermolecular part of the intra—intermolecular hydrogen bonding) is −0.61 kcal/mol for both plots. (The concentration of the toluene in

the solution is taken to be the standard concentration.) The energy of formation of the 2MP—toluene complex from DFT calculations is −0.5 kcal/mol (with zero-point energy correction). The entropy of formation is −12.1 and −11.1 J/(mol K) from Figure 4A and B, respectively. The population ratio (also the equilibrium constant, assuming the concentration of the toluene in the solution to be the standard concentration) of the complex to free 2MPOD in toluene at 297 K (room temperature) is 0.66.

The data discussed above clearly show that an equilibrium exists between two species, which have been taken to be the three-centered hydrogen-bonded 2MPOD—toluene complex in equilibrium with free 2MPOD. However, it is reasonable to consider other possibilities. In CCl$_4$ and other nonaromatic solvents such as hexane, only a single peak is observed in the IR spectrum (see Figure 2). Therefore, the two species in equilibrium must be associated with a complex to the aromatic solvent. A possibility other than the proposed three-centered hydrogen-bonded complex in equilibrium with the free intramolecular hydrogen-bonded 2MPOD is a two-centered complex in which the intramolecular hydrogen bond is broken and a pure intermolecular hydrogen bond between the hydroxyl and toluene is formed. Figure 5 shows FTIR spectra of 2MPOD and pMPOD in toluene. pMPOD does not have an intramolecular hydrogen bond. The smaller peak for pMPOD at 2668 cm^{-1} is for the free OD stretch. The main peak at 2630 cm^{-1} belongs to the intermolecular hydrogen-bonded OD stretch (the hydrogen bond acceptor is the toluene π-electron system). Intermolecular hydrogen bonding produces a red shift.[2,14] The magnitude of the shift is consistent with that observed for the phenol—benzene complex.[14] In the 2MPOD spectrum, the peak at ~2625 cm^{-1} is the free 2MPOD hydroxyl stretch red shifted by the intramolecular hydrogen bond (the hydrogen bond acceptor is the electron pair of the oxygen of the methoxyl group; see Figure 2). Then, the additional red shifting of the peak at ~2610 cm^{-1} is produced by additional intermolecular hydrogen bonding to form the three-centered complex. This peak has a larger red shift (more hydrogen bonding) than either the intermolecular or intramolecular hydrogen bonds by themselves. It is interesting to note that the 2MPOD pure intramolecular hydrogen bond OD stretch is at 2629 cm^{-1} (Figure 2), which is only 1 cm^{-1} different from that of the pure intermolecular hydrogen bond of pMPOD with toluene. The intramolecular hydrogen bond is a weak hydrogen bond because of the unfavorable geometry.[2] The fact that the two peaks have virtually the same frequency shows that the intermolecular bond of the hydroxyl to the aromatic ring is about the same strength as the weak intramolecular hydrogen bond. The fact that the two types of hydrogen bonds have about the same strength is a play off influences. Normally, hydrogen-bonding interactions would be stronger for an oxygen acceptor than for the π-electrons of toluene as the acceptor. However, the geometric constraint of the intramolecular hydrogen bond weakens the hydrogen bond. The unfavorable geometry of the intramolecular hydrogen bond reduces the efficacy associated with the oxygen acceptor compared to the intermolecular π-electrons as the acceptor, resulting in intra- and intermolecular hydrogen bonds of about the same strength.

For 2MPOD/toluene, the peak at 2625 cm^{-1} is assigned to an intramolecular hydrogen bond only, and the shoulder at 2610

(36) Chang, R. *Physical Chemistry for the Chemical and Biological Sciences*; University Science Books: Sausalito, 2000.

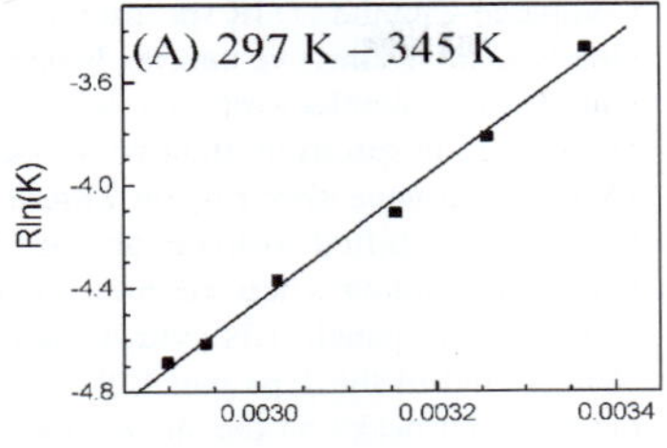
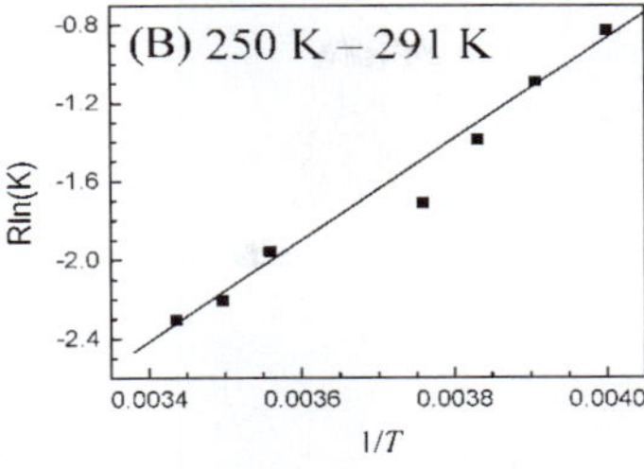

Figure 4. van't Hoff Plots derived from two sets of temperature-dependent FTIR measurements. The formation enthalpy for the intra−intermolecular hydrogen bond is found to be −0.6 kcal/mol from both measurements.

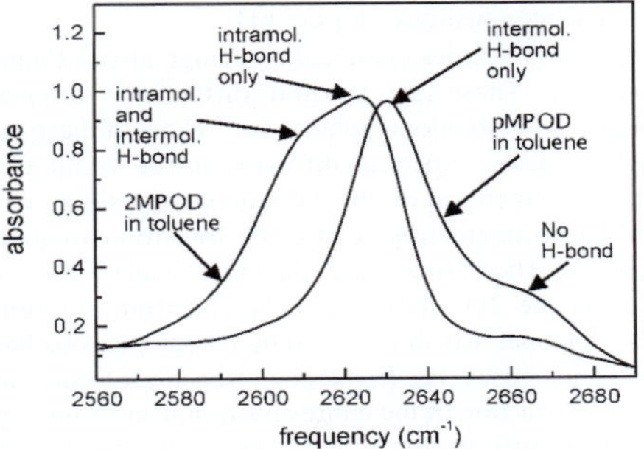

Figure 5. Hydroxyl stretch FTIR spectra of 2 wt % p-methoxyphenolOD (pMPOD) and 2-methoxyphenolOD (2MPOD) in toluene. The pMPOD spectrum shows two bands. The higher-frequency small shoulder is the free species, and the lower-frequency band is the intermolecular hydrogen-bonded complex. The 2MPOD spectrum shows two bands. The higher-frequency band is the free species with its intramolecular hydrogen bond, and the lower-frequency band is the intra−intermolecular hydrogen-bonded complex.

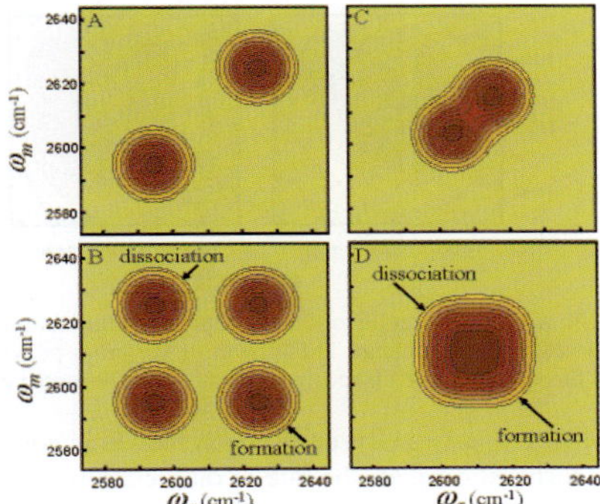

Figure 6. Schematic illustrations showing the peaks appearances at short and long T_w in 2D IR spectra for two peaks that are well resolved in the linear absorption spectrum (A before exchange and B after exchange) and for two overlapping peaks (C before exchange and D after exchange). The growth of the off-diagonal peaks will form a square rather than resolved off-diagonal peaks for two peaks that overlap substantially in the linear absorption spectrum.

cm^{-1} is assigned to the intra−intermolecular bond. The peak at 2625 cm^{-1} has almost the same position as the intermolecular hydrogen bond only peak for pMPOD/toluene (Figure 5) and the intramolecular hydrogen bond only peak for 2MPOD/CCl$_4$ (Figure 2). Another possibility is that the 2MPOD/toluene peak at 2625 cm^{-1} is the intermolecular hydrogen bond only, and the equilibrium with the intra−intermolecular bond (shoulder at 2610 cm^{-1}) involves making and breaking the intramolecular hydrogen bond rather than formation and dissociation of the solute−solvent complex. Although this seems implausible, it cannot in fact be ruled out by the linear spectroscopy alone. However, electronic structure calculations show that the activation energy to break the intramolecular hydrogen bond is >5 kcal/mol. This high barrier would produce slow exchange, estimated to be >100 ps. As shown below, the 2D IR results measure 3 ps, confirming the assignments given in Figure 5.

B. 2D IR Vibrational Echo Experiments and Analysis of the Dynamics. The dissociation enthalpy (negative sign of the formation enthalpy) of the intra−intermolecular hydrogen bonding was determined to be 0.6 kcal/mol, which is almost identical to the thermal energy at room temperature. Therefore, it might be anticipated that the 2MPOD−toluene complex dissociation time is fast. Recently, it has been demonstrated that 2D IR VES can be used to directly measure chemical exchange in real time under thermal equilibrium conditions.[13−15] In addition, 2D NMR has been used to measure chemical exchange on much slower time scales for some time,[37] although there are significant differences between 2D IR VES chemical exchange

measurement on ultrafast time scales and 2D NMR measurements. Here, a qualitative discussion will be presented.

First, consider the case in which the linear IR spectrum shows well-resolved peaks for two species, e.g., a hydrogen-bonded complex and a nonbonded one (the phenol/benzene−CCl$_4$ system is of this type; see ref 14). The peak for the complex is at lower frequency. At T_w's that are short compared to the exchange time, the 2D IR vibrational echo spectrum will consist of two peaks on the diagonal, reflecting the 0−1 transitions of the two vibrations. This is shown schematically in Figure 6A. Again, the peak for the complex is at a lower frequency along the ω_m axis. Two other peaks (not shown), corresponding to the 1−2 transitions, will appear off diagonal, shifted to lower frequency along the ω_m axis by their vibrational anharmonicities.[14,38] These peaks have been discussed in detail in the context of 2D IR VES chemical exchange[14] and will be returned to when needed later. At T_w's that are long compared to the chemical exchange time (inverse of the rate constant of formation and dissociation of the complex), off-diagonal peaks will have grown in, which is shown schematically in Figure 6B. The peak labeled dissociation appears as the complex dissociates, and the peak labeled formation appears as the complex forms. The rate of growth of the off-diagonal peaks provides the chemical exchange

(37) Ernst, R. R.; Bodenhausen, G.; Wokaun, A. *Nuclear Magnetic Resonance in One and Two Dimensions*; Oxford University Press: Oxford, 1987.
(38) Rector, K. D.; Kwok, A. S.; Ferrante, C.; Tokmakoff, A.; Rella, C. W.; Fayer, M. D. *J. Chem. Phys.* **1997**, *106*, 10027.

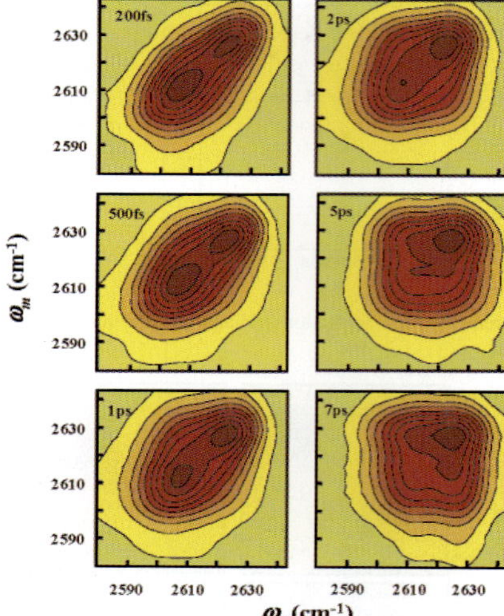

Figure 7. Shows the 0−1 region of the T_w dependent 2D vibrational echo spectrum for 2MPOD in toluene. The growth of the cross-peaks changes the shape of the spectrum, which becomes square after substantial exchange.

kinetics when properly analyzed. If the system is in equilibrium, the rate of formation equals the rate of dissociation, and the two off-diagonal peaks grow in together (see further discussion below).

For the situation in which the linear IR absorption of the complex appears as a shoulder on the low-frequency side of the free 2MPOD (see Figure 2), at a short T_w, the spectrum will appear as in Figure 6C. The peaks in the 2D spectrum appear on the diagonal but overlap in a manner related to the overlap in the linear spectrum. For T_w's that are long compared to the exchange time, off-diagonal peaks will have grown in. As with the diagonal peaks, the off-diagonal peaks will overlap the diagonal peaks, as illustrated schematically in Figure 6D. Although the off-diagonal peaks are not well resolved as they are in Figure 6B, they are clearly in evidence by the change in shape from the short time spectrum to an approximately square shape. It is straightforward to obtain the exchange rate by observing the growth of the off-diagonal peaks. As has been discussed in detail, whether the off-diagonal peaks are well resolved or not, to obtain the chemical exchange rate constant, it is necessary to fit all four peaks (diagonal and off-diagonal) in the 2D IR vibrational echo spectrum.

Figure 7 displays T_w-dependent 2D IR vibrational echo spectra of the 2MPOD−toluene system for the 0−1 transition region. For each panel, there are the equivalent peaks for the 1−2 transition region shifted to lower frequency along the ω_m axis by the 100 cm^{-1} anharmonicities of the transitions. These are not shown but were used in part of the data analysis as discussed below. Each contour represents a 10% change in amplitude of the peaks. The 200 fs panel corresponds to a short T_w at which negligible chemical exchange has occurred. In Figure 2, the complex appears as the low-energy shoulder on the 2MPOD in the toluene spectrum. In the 200 fs panel of Figure 7, the two peaks, the 2MPOD−toluene complex and free 2MPOD, are clearly visible on the diagonal. The complex and free peaks are the lower-frequency (2611 cm^{-1}) and higher-frequency (2629 cm^{-1}) peaks along the ω_m axis, respectively.

Compared with the 1D IR spectrum in Figure 2, the 2D spectrum clearly resolves the two species better. In the 1D spectrum, the complex is a shoulder on the low-energy side of the free peak, and in the 2D spectrum, the two peaks are distinct. As with the 2MPOD−toluene spectrum in Figure 2, the hydroxyl stretch transition is shifted to lower frequency because of increased hydrogen bonding when the three-centered complex is formed. In the 500 fs panel, it is evident that the shape has begun to change, and in the 1 ps panel, the effect of the growth of the off-diagonal peaks on the shape of the spectrum is very clear. In the 2 ps spectrum, the shape is almost the approximately square shape discussed in connection with Figure 6, and the 5 ps spectrum has obtained the approximately square shape illustrated schematically in Figure 6D.

Other factors besides chemical exchange also influence the 2D spectrum. These are spectral diffusion,[39] orientational relaxation,[40] and vibrational relaxation.[41] None of these produce off-diagonal peaks. Spectral diffusion is the result of time-dependent interactions of the vibrational transition with the solvent. These interactions cause the transition frequency to fluctuate. At short time, the diagonal peaks' line shapes measured in the 2D vibrational echo spectrum are elongated along the diagonal, which is caused by inhomogeneous broadening. As T_w increases, the frequency of each molecule samples an increasing fraction of the entire absorption spectrum (spectral diffusion). At sufficiently long T_w, the entire line is sampled, and the dynamic line width is equal to the absorption line width. In the 2D spectrum, complete spectral diffusion is manifested by a change in the 2D line shape from elongated along the diagonal to symmetrical about the diagonal.[33] This change can be seen most clearly by comparing the central contours of the free peak at 200 fs and 2 ps.

Spectral diffusion changes the shapes of the peaks but preserves their volumes.[14,33] Orientational relaxation and vibrational relaxation cause all peaks to decay in volume in contrast to chemical exchange, which causes the off-diagonal peaks to grow and the diagonal peaks to shrink. Orientational relaxation during the T_w period reduces the peak volumes but does not cause them to decay to zero.[14,42] Vibrational relaxation does cause the peaks to eventually decay to zero. The relaxation times are not the same for the two species, which can be seen in the 2D spectra. Each spectrum has been normalized to the largest peak at the associated T_w. By comparing the 200 fs and 2 ps data, particularly the central contour, it is clear that the diagonal peak for the complex has decayed more rapidly than the diagonal peak for free 2MPOD.

All of the dynamical phenomena discussed qualitatively above are combined in the quantitative analysis of the data. The data are analyzed to obtain the formation and dissociation rates of the three-centered hydrogen-bonded complex using a combination of time-dependent diagrammatic perturbation theory, which describes the nonlinear optical interactions with the molecular vibrations,[33,43] and kinetic equations.[14] A brief discussion is presented here. Figure 8A,B qualitatively illustrates the radiation

(39) Tokmakoff, A.; Urdahl, R. S.; Zimdars, D.; Francis, R. S.; Kwok, S.; Fayer, M. D. *J. Chem. Phys.* **1995**, *102*, 3919−3931.
(40) Berne, B. J.; Pecora, R. *Dynamic Light Scattering*; J. Wiley: New York, 1976.
(41) Egorov, S. A.; Everitt, K. F.; Skinner, J. L. *J. Phys. Chem. A* **1999**, *103*, 9494−9499.
(42) Berne, B. J.; Pecora, R. *Dynamic Light Scattering*: J. Wiley: New York, 1990.

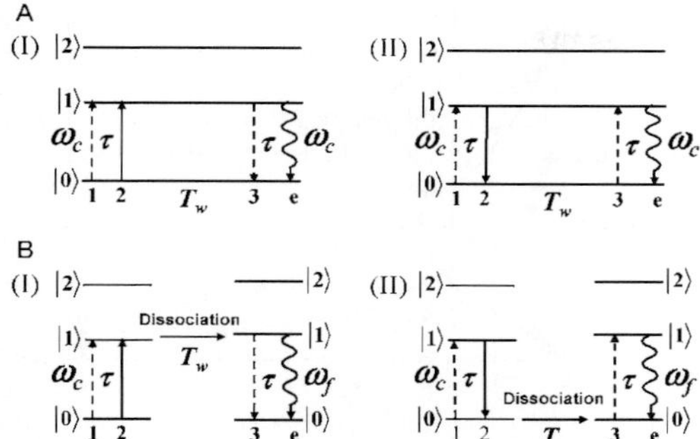

Figure 8. Energy level diagrams with radiation field interactions representing the pathways that give rise to the diagonal and off-diagonal peaks in Figure 7. See text.

field-matter interactions that give rise to the diagonal peak for the three-centered complex [$(\omega_\tau, \omega_m) = (2611$ cm, 2611 cm^{-1})] and the off-diagonal peak created by dissociation of the complex $(2611, 2629$ cm^{-1}) in the $0-1$ region of the spectrum, respectively. The frequency of the complex is $\omega_c = 2611$ cm^{-1}, and the frequency of free 2MPOD is $\omega_f = 2629$ cm^{-1}. The other peaks in the $0-1$ and $1-2$ regions are obtained in the analogous manner. The frequency at which the first radiation field interaction (first pulse) excites a mode is the mode frequency on the ω_τ axis (horizontal axis) in the 2D spectrum. The first interaction produces a coherence (coherent superposition state) between the 0 and 1 states. In Figure 8, coherences are represented by dashed arrows. The second interaction with the radiation field (second pulse) produces a population (solid arrows) either in the 1 state (I in Figure 8) or in the 0 state (II in Figure 8). The third interaction (third pulse) again produces a $0-1$ coherence followed by the vibrational echo emission (wavy arrow) at the same frequency as the coherence induced by the third pulse. The frequency of the vibrational echo emission is the frequency on the ω_m axis (the vertical axis). In Figure 8, the pathways I and II each contribute half of the signal.

First, consider Figure 8A for complexes that have not dissociated during the T_w period (or have dissociated and reformed a complex prior to the third pulse). Both pathways (I and II) begin at frequency ω_c and emit the vibrational echo at ω_c. Therefore, $\omega_\tau = \omega_m = \omega_c$, giving rise to the diagonal peak for the complex $(2611, 2611$ cm^{-1}). Now, consider Figure 8B for complexes that dissociate during the T_w period. The first interaction is at ω_c, but the last interaction and vibrational echo emission are at ω_f. Therefore, $\omega_\tau = \omega_c$ but $\omega_m = \omega_f$, giving rise to the off-diagonal peak for dissociation $(2611, 2629$ cm^{-1}). At short T_w (Figure 7, 200 fs; Figures 6A,C) only Figure 8A comes into play, and there are only peaks on the diagonal. At long T_w (Figure 7, 5 ps; Figure 6B,D) both 8A and B contribute to the signal, and there are diagonal and off-diagonal peaks.

The exchange during the τ period (time between pulses 1 and 2) is not considered here because for slow or moderate exchange rates it causes decay of the diagonal peaks but does not contribute to the off-diagonal peaks. If the exchange is much faster than the frequency difference between two peaks (18 cm^{-1} → 1.85 ps), motionally narrowing will result in a single peak in the spectrum.[44−46] In the systems considered here, the exchange rate is relatively slow.

(43) Mukamel, S. *Annu. Rev. Phys. Chem.* **2000**, *51*, 691−729.

Inspection of the data in Figure 7 shows that complexes form and dissociate on a few picoseconds time scale. Because of spectral diffusion, the shapes of the peaks change with T_w. In the absence of all other dynamical processes, the change in shape preserves the volume of a peak, but the peak amplitude is reduced as the peak broadens along the ω_τ axis. Therefore, the integrated peak volumes are fit to obtain the population dynamics. The volume of each peak as a function of T_w is obtained for each species by fitting the entire spectrum at each T_w with four 2D Gaussians. These fits can reproduce the spectrum at each T_w almost perfectly. The volumes are then scaled appropriately by the transition dipole moment ratio of the complexed and free species to obtain the population for each species. The populations are fit with the kinetic equations to yield the exchange rate constants. By using the peak volumes, spectral diffusion is accounted for.

As described above, there are three processes that contribute to the change in the peak volumes: the OD vibrational relaxation in free and complexed 2MPOD (lifetimes $T_f = 1/k_f$ and $T_c = 1/k_c$), the orientational relaxation (time constants $\tau_f = 1/(6D_f)$ and $\tau_c = 1/(6D_c)$; D_i is the orientational diffusion constant), and the chemical exchange dissociation and association rate constants k_d and k_a. The vibrational relaxation and orientational relaxation lead to diminishing intensities of all peaks with increasing T_w. Even complete orientational randomization during the T_w period does not cause the vibrational echo to decay to zero. Therefore, the chemical exchange time does not have to be short compared to the orientational relaxation time. The chemical exchange causes the diagonal peaks to diminish and the off-diagonal peaks to grow, as can be seen in Figure 7. On the basis of these considerations, a kinetic model was constructed.[14,47] The kinetic scheme is shown here for two peaks, the diagonal peak at 2611 and 2611 cm^{-1} (CC) and the off-diagonal peak at 2611 and 2629 cm^{-1} (CF) in Figure 7. CC represents the population of complexes at the end of the T_w period (the time of the third pulse) that were also complexes immediately after the second pulse. CF represents the population that were complexes immediately after the second pulse but are free 2MPODs at the end of the T_w period. Some complexes may have dissociated after the second pulse but reassociated by T_w. These are part of the CC population. It is also possible for a complex to dissociate, reassociate, and dissociate again during the T_w period. These would be part of the CF population and so forth. In this model, orientational relaxation is taken to be diffusive.[47] This is born by the fact that the measured time-dependent anisotropies decay exponentially. The model is illustrated schematically as:

$$\underset{\text{decay}}{\overset{\tau_c, T_c}{\longleftarrow}}\ \text{CC}\ \underset{k_a}{\overset{k_d}{\rightleftharpoons}}\ \text{CF}\ \underset{\text{decay}}{\overset{T_f, \tau_f}{\longrightarrow}}$$

The differential equations derived from this kinetic model are

$$\frac{d([CC(t)] \times f_c(t,\theta))}{dt} = -(k_c + k_d + D_c I^2) \times ([CC(t)] \times f_c(t,\theta)) + k_a \times ([CF(t)] \times f_f(t,\theta))$$

$$\frac{d([CF(t)] \times f_f(t,\theta))}{dt} = -(k_f + k_a + D_f I^2) \times ([CF(t)] \times f_f(t,\theta)) + k_d \times ([CC(t)] \times f_c(t,\theta)) \quad (1)$$

where the quantities in square brackets are the time-dependent concentrations, $f_i(t,\theta)$ is the angular distribution function for the complex (c) and the free species (f), Γ^2 is the spherical harmonic operator, D_c and D_f are the rotational diffusion constants, and k_c and k_f are the vibrational decay constants. k_a and k_d are the association and dissociation rate constants.[47] From these two equations, the equations for the time-dependent volumes of peak CC ($V_{CC}(t)$) and CF ($V_{CF}(t)$) are derived. The complete solutions of these equations are given in the Supporting Information. In the above equations, only k_a and k_d are unknown variables, and all the other parameters are experimentally measured. The dissociation rate (number per unit time) of the complex equals the formation rate if the system is in equilibrium. The complex dissociation time constant, τ_d (given in picoseconds below), is independent of concentration. Therefore, τ_d is reported and discussed. $\tau_d = 1/k_d$, where $k_d = k_a/0.66$ is the dissociation rate constant. The dissociation rate is k_d[complex]. Because the equilibrium constant is known, there is actually a single unknown parameter. The volumes are scaled appropriately by the transition dipole moment ratio for the two species. The volumes of the diagonal peaks, V_{CC} and V_{FF}, depend on the fourth power of their respective transition dipole moments, μ_c^4 and μ_f^4. The off-diagonal peaks, V_{CF} and V_{FC}, both depend on $\mu_c^2\mu_f^2$. μ_c and μ_f are the 0–1 transition dipole moments of the OD stretch for the complexed 2MPOD and free 2MPOD. It was determined experimentally that $\mu_c^2 = 2\mu_f^2$ by combining the 1D and 2D IR measurements as mentioned above. Thus, the scaled volume is the actual volume divided by 1 for the free peak, divided by 4 for the complexed peak, and divided by 2 for the cross-peaks. A set of equations equivalent to eq 1 applies for the free diagonal peak (FF) and the free associating to complex off-diagonal peak (FC).

The data for the 0–1 transition region consist of four time-dependent components: the two diagonal peaks (the three-centered complex and the free 2MPOD) and the two off-diagonal peaks (dissociation and formation of the complex during the T_w period). All four peaks can be reproduced with the single adjustable parameter, τ_d, by inputting the known constants and fitting the data with the kinetic equations. The input parameters used are $T_c = 5$ ps, $T_f = 6.2$ ps, $\tau_c = 6.6$ ps, $\tau_f = 4.0$ ps, and the ratio of the complexed and free 2MPOD concentrations, [complex]/[free] = 0.66. The intramolecular hydrogen-bonded (free) 2MPOD vibrational lifetime (T_f) was directly obtained by measuring that of 2EPOD in toluene, and the rotational relaxation time constant was measured for 2EPOD and scaled by the volume ratio of 2MP and 2EP ($\tau_f = 4.4 \times 0.9 = 4$ ps). For the intra–intermolecular hydrogen-bonded 2MPOD, the rotational relaxation time constant was directly obtained from measuring that of the pMPOD in toluene (they have virtually the same volumes). The vibrational lifetime of the 2MPOD–toluene complex was obtained by performing a pump–probe experiment at 2610 cm^{-1}, which is on the shoulder of the spectrum in Figure 2. A fit to the spectrum showed that this wavelength is virtually pure complex absorption. However, exchange will modify the measured lifetime. The observed pump–probe decay is 4.9 ps. The lifetime of the free 2MPOD

(44) Rollefso, R. J. *Phys. Rev. Lett.* **1972**, *29*, 410–412.
(45) Oxtoby, D. W. *J. Chem. Phys.* **1981**, *74*, 1503.
(46) Levinger, N. E.; Davis, P. H.; Behera, P.; Myers, D. J.; Stromberg, C.; Fayer, M. D. *J. Chem. Phys.* **2003**, *118*, 1312–11326.
(47) Cang, H. Thesis: Dynamics in Complex Liquids. Stanford University, Standford, CA, 2004.

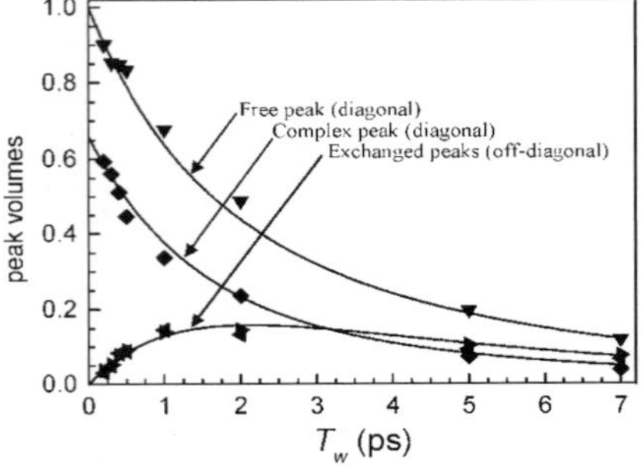

Figure 9. T_w dependent data (symbols) for 2MPOD showing the time dependence of the two diagonal and two off-diagonal peaks, in the 0–1 region of the 2D vibrational echo spectra as in Figure 7. The curves are from a fit to the data with one adjustable parameter, τ_d, using the kinetic model. Other parameters in the model are determined experimentally. $\tau_d = 3 \pm 0.6$ ps.

is 6.2 ps. Exchange will cause the observed pump–probe decay for the complex to be slower than the true lifetime. However, the exchange rate determined below is not highly sensitive to the lifetime. We found that variation of the value of 4.9 ps by ± 1 ps did not change the value of the exchange rate. Once the exchange rate was determined, we used its value, the lifetime of the free species, the observed pump–probe decay for the complex, and the appropriate pair of coupled differential equations to determine the true lifetime of the complex to be 4.6 ps. Because the vibrational and rotational parameters for the 2MPOD solution were not measured directly, we allowed each parameter to vary ± 1 ps in fitting the data and used the results to determine the error bars for τ_d.

Figure 9 shows the peak volume data for the 0–1 transition region of the spectrum as a function of T_w. The data are normalized to the largest peak at $T_w = 0$. The simultaneous fit (solid curves) to the time dependence of all of the peaks using *a single adjustable parameter*, τ_d, is very good. From the fits, the dissociation time for the complex is $\tau_d = 3.0 \pm 0.6$ ps.

In addition to 2MPOD, pMMPOD was also studied in toluene. The data have the same appearance as those for 2MPOD (Figure 7). Figure 10A displays the peak volume data for the 0–1 transition region of the spectrum. The data were fit using the same data processing procedure described above. The input parameters are [complex]/[free] = 0.8, extinction coefficient ratio (complex/free) = 1.55, $T_c = 5$ ps, $T_f = 6.2$ ps, $\tau_c = 7.3$ ps, and $\tau_f = 4.4$ ps. The solid curves in Figure 10A are the results of the fits and yield $\tau_d = 3.4 \pm 0.6$ ps. The addition of a methyl in the para position may have a small influence on the complex dissociation time, but the difference falls within the error bars of the measurements.

The question arises as to whether τ_d's and the related formation and dissociation rates of the two systems studied here are in fact the dynamics under thermal equilibrium conditions. It is possible that the excitation of the hydroxyl stretch shifts the equilibrium, and the dynamics shown in Figures 9 and 10A in part reflect the relaxation to the new equilibrium concentrations. Two tests can be performed to determine if vibrational excitation has influenced the measurement of equilibrium dynamics. First, if the system is in thermal equilibrium, the two off-diagonal peaks will grow at the same rate because the rates of dissociation and formation are the same at equilibrium.[14] As

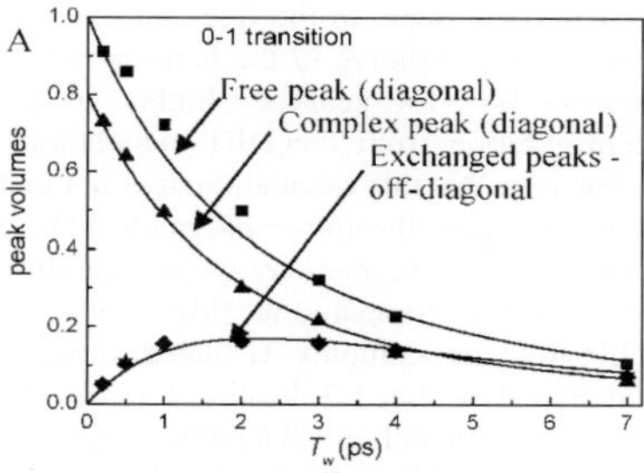
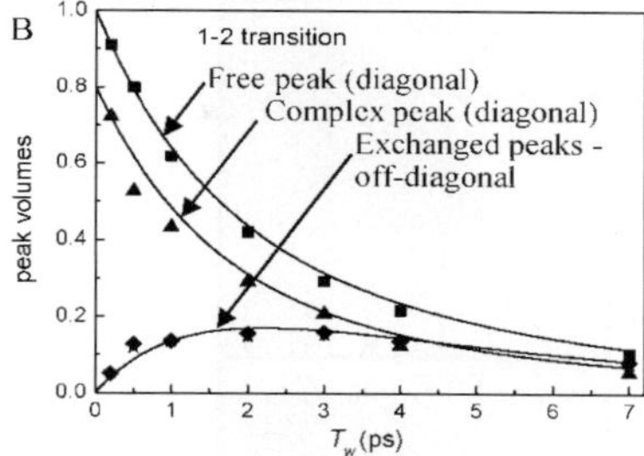

Figure 10. T_w dependent data (symbols) for pMMPOD in toluene showing the time dependence of the two diagonal and two off-diagonal peaks, in the $0-1$ region (A) and the $1-2$ region (B) of the 2D vibrational echo spectra. The curves in A are from a fit to the data with one adjustable parameter, τ_d, using the kinetic model. Other parameters in the model are determined experimentally. $\tau_d = 3.4 \pm 0.6$ ps. The curves in B are obtained without adjustable parameters using the same parameters as those in A. The agreement shows that the thermal equilibrium is not perturbed by vibrational excitation.

can be seen from Figures 9 and 10A, the growth of the formation and dissociation peaks (off-diagonal chemical exchange peaks) is identical within experimental error.

Another test involves comparing the $0-1$ data to the equivalent data for the $1-2$ transition. In Figure 8B, it can be seen that there are two pathways that contribute equally to the $0-1$ signal. For one pathway, the system is in the ground state ($v = 0$) during the T_w period (Figure 8B(I)). For the other pathway, the system is in the excited state ($v = 1$) during the T_w period (Figure 8B(II)). Half, 50%, of the signal arises from each of these pathways. For the $1-2$ portion of the 2D vibrational echo spectrum, there is one pathway that has the same first two interactions as those in Figure 8B(I). After the second interaction, the system is in the $v = 1$ state. The third interaction (third pulse) can produce a coherence between the $v = 1$ and $v = 2$ levels, and the vibrational echo is then emitted at the frequency of the $1-2$ transition, which is shifted to lower frequency by the vibrational anharmonicity (100 cm^{-1}). The portion of the 2D spectrum that occurs from vibrational echo emission at the $1-2$ frequency only involves a pathway in which the system is in the $v = 1$ state during the T_w period. Therefore, the $0-1$ portion of the spectrum involves 50% of the signal in which the dynamics might be influenced by vibrational excitation whereas the $1-2$ portion of the spectrum involves 100% of the signal that might be affected by vibrational excitation. If the measured dynamics are the same for these two portions of the spectrum, then vibrational excitation did not influence the thermal equilibrium chemical exchange rates.

2MPOD has an accidental degeneracy between the $v = 2$ level of the hydroxyl stretch and a combination band.[25] The accidental degeneracy results in additional overlapping peaks in the $1-2$ portion of the spectrum which makes it difficult to analyze the time dependence of the chemical exchange. The addition of the *para*-methyl group in pMMPOD shifts the energies of the states sufficiently to eliminate the accidental degeneracy and clean $1-2$ spectra results. Therefore, we performed the second test for perturbation of the thermal equilibrium dynamics on pMMPOD. Figure 10B shows the data for the $1-2$ portion of the 2D spectrum. The curves through the data are not fits. They are calculated with no adjustable parameters using the parameters including τ_d obtained for the $0-1$ transition. We find no difference in the kinetics for the $0-1$ and $1-2$ portions of the spectrum, within experimental error. Therefore, in these systems, vibrational excitation does not perturb the thermal equilibrium chemical exchange kinetics.

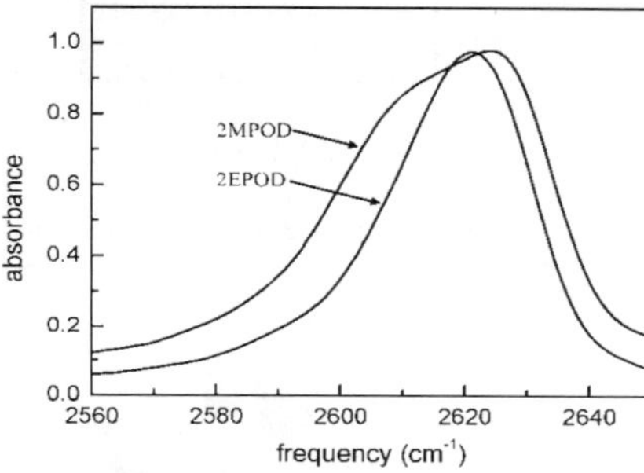

Figure 11. FTIR spectra of 2-methoxyphenolOD and 2-ethoxyphenolOD in toluene. The lack of a low-frequency shoulder in the 2EPOD spectrum shows there is little or no complex formation.

The same result was found for the phenol–benzene complex formation and dissociation.[14] It is important to point out that if vibrational excitation does perturb the chemical exchange kinetics of a system the ground-state thermal equilibrium kinetics can be recovered by using the observed kinetics from both the $0-1$ and $1-2$ portions of the spectrum in the analysis.

C. Structural Influences on the Intra–Intermolecular Hydrogen Bond. To explore some general chemical properties of the intra–intermolecular hydrogen bond, we varied the structures of both intramolecular and intermolecular acceptors and found that the three-centered hydrogen bond is very sensitive to small structural changes.

When one more methylene group is added to the methoxyl group of 2MPOD to become 2-ethoxyphenolOD (2EPOD), there is virtually no intermolecular hydrogen bonding between 2EPOD and toluene at room temperature, in contrast to 2MPOD. The difference between the two solutes' tendency to form complexes can be seen in the linear absorption spectrum shown in Figure 11. The 2EPOD spectrum has at most a very small shoulder on the low-frequency side in contrast to the 2MPOD spectrum. The shoulder in the 2MPOD spectrum arises from complex formation. 2D IR vibrational echo measurements confirmed the results from the linear IR spectra. Figure 12 shows 2D vibrational echo spectra for 2EPOD at $T_w = 200$ fs (A) and 2.5 ps (B). At 200 fs, there is a single peak elongated along the diagonal in contrast to the 200 fs panel of Figure 7. The elongation is caused by inhomogeneous broadening at short T_w. At 2.5 ps, the shape has changed substantially, becoming almost round, because of spectral diffusion. However, the characteristic approximately

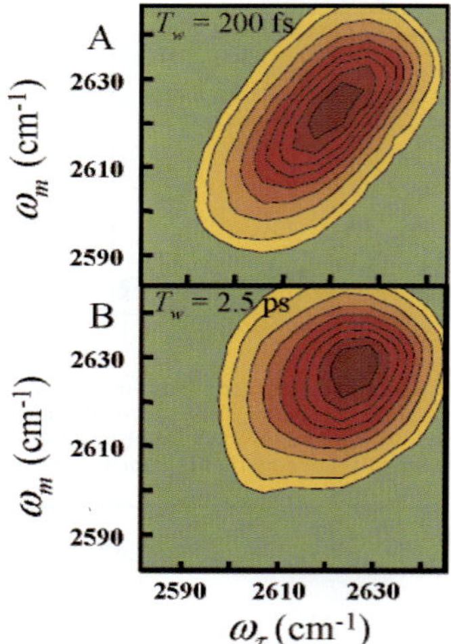

Figure 12. 2D IR spectra of 2-ethoxyphenolOD in toluene at (A) T_w = 200 fs and (B) T_w = 2.5 ps. A single peak appears in both spectra showing that no intra—intermolecular hydrogen-bonded complex is formed. The peak shape changes because of spectral diffusion.

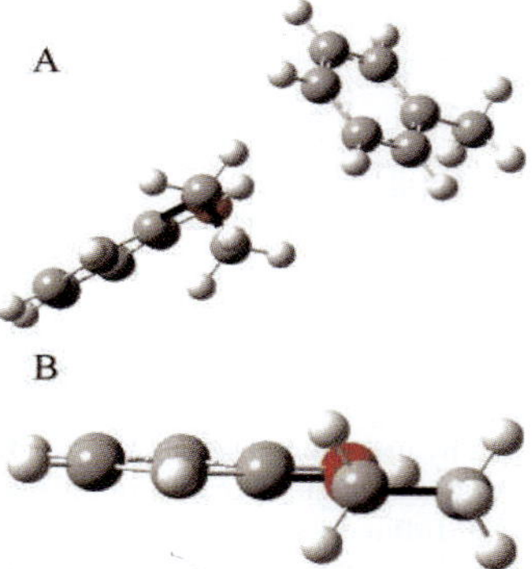

Figure 13. The structures of (A) the 2-ethoxyphenol/toluene complex and (B) 2-ethoxyphenol calculated with DFT at the B3LYP/6-31+G(d,p) level. See text.

square shape seen at the longer T_w's for 2MPOD (Figure 7) is absent. As discussed in conjunction with Figures 6 and 7, the square shape is indicative of the growth of off-diagonal peaks caused by chemical exchange between the complex and free species.

DFT structure calculations provide a possible explanation for the difference between 2EPOD and 2MPOD in their tendency to form a three-centered hydrogen bond with toluene. Figure 13A,B shows the optimized isolated molecule structures for the 2EP/toluene complex and the free 2EP molecule, respectively. To form the intra—intermolecular hydrogen bond with toluene, the end methyl group of 2EP has to be in the gauche position relative to the benzene ring. As can be seen in Figure 13A, the methyl group is below the plane of the benzene ring. If the ethyl substituent is in the anti configuration, the terminal methyl group prevents the toluene from obtaining the necessary position to form the intermolecular hydrogen bond. In contrast to the gauche configuration of 2EP necessary to form the three-

centered complex, in the free 2EP, the methyl group is in the anti position relative to the benzene ring. The gauche configuration is ~1.4 kcal/mol higher in energy than the anti configuration from the DFT calculations. It is important to emphasize that this calculation does not include toluene solvent. Thus, to form the intra—intermolecular hydrogen bond, 2EP must assume a higher energy configuration, which destabilizes the complex, compared to 2MP. The formation energy for the 2EP—toluene complex (isolated molecules, no solvent) is calculated to be 1.3 kcal/mol (with the zero-point energy correction), which is 1.8 kcal/mol higher than that of the 2MP/toluene complex (−0.5 kcal/mol from the equivalent calculation). On the basis of this energy difference, the Boltzmann distribution at room temperature gives the number of 2EP—toluene complexes to be <5% of the number of 2MP—toluene complexes. This is consistent with the lack of any obvious indication of 2EP—toluene complexes in either the absorption or 2D vibrational echo spectra.

The intra—intermolecular hydrogen bond is also formed between toluene and other ortho-substituted phenol derivatives, e.g., 2,6-dichlorophenol, 2-chlorophenol, and 2-bromophenol. In some sense, this intra—intermolecular hydrogen bond can also be called a π-hydrogen bond,[3] but it is different from the usual π-hydrogen bonds such as the bonding between phenol and benzene, in that the three-centered hydrogen bond is much more sensitive to steric effects on the benzene ring of the hydrogen bond acceptors. According to the literature[48,49] and our own results,[14] the binding energy between phenol and benzene (zero additional carbons on the ring), toluene (one additional carbon on the ring), p-xylene (two additional carbons on the ring), and mesitylene (three additional carbons on the ring) increases in the order of benzene < toluene < p-xylene < mesitylene because the π-electron density on the ring increases with increasing number of methyl substituents. The increase in the strength of the phenol complexes with an increased number of methyls on the acceptor occurs even though the steric effects that can inhibit the formation of the π-hydrogen bond increase with increased number of methyls. Therefore, in complexes with phenol, the π-electron density increase outweighs the adverse steric effects. The binding energies between 2MPOD and these aromatic solvents show the opposite trend, as demonstrated by the FTIR measurements displayed in Figure 14. In the linear IR spectra, the proportion of the shoulder at ~2611 cm^{-1} for the intra—intermolecular hydrogen-bonded OD stretch clearly decreases as the number of methyl groups on the solvent increases. The strengths of the complexes are benzene > toluene > p-xylene > mesitylene. The results imply that the adverse steric effects that occur with an increased number of methyl groups are more important than the increases in π-electron density in forming the three-centered hydrogen bond in this series of solvents.

IV. Concluding Remarks

In this paper, the formation and dissociation of organic solute—solvent complexes were investigated. By performing temperature-dependent linear IR absorption experiments and 2D IR nonlinear vibrational echo measurements, it was determined

(48) Arnett, E. M.; Joris, L.; Mitchell, E.; Murty, T. S. S. R.; Gorrie, T. M.; Schleyer, P. v. R. *J. Am. Chem. Soc.* **1970**, *92*, 2365−2377.
(49) Fuchs, R.; Peacock, L. A.; Stephenson, W. K. *Can. J. Chem.* **1982**, *60*, 1953−1958.

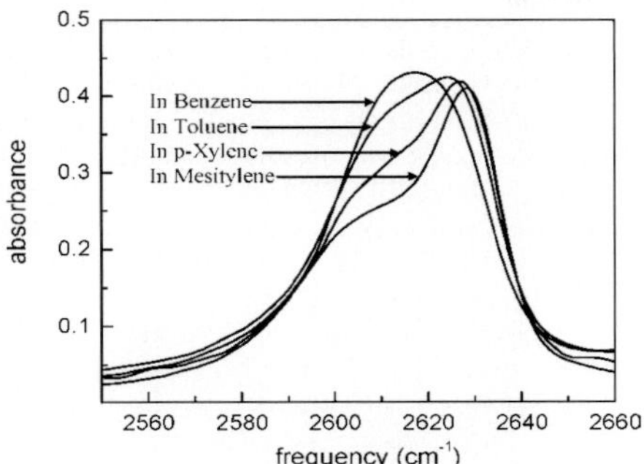

Figure 14. FTIR spectra of 2-methoxyphenolOD in benzene, toluene, *p*-xylene, and mesitylene. The intensity of the intra—intermolecular hydrogen-bonded complex peak decreases as more methyl groups are added to the benzene ring.

that the solute 2-methoxyphenol forms an unusual complex with the solvent toluene. The complex involves a three-centered hydrogen bond in which the 2MP hydroxyl hydrogen forms an intramolecular hydrogen bond with the methoxy oxygen and an intermolecular π-type hydrogen bond with the toluene. Both the thermodynamic and kinetic behaviors of the three-centered hydrogen bond were studied. The formation enthalpy of the 2MP—toluene complex was determined to be -0.6 ± 0.1 kcal/mol. 2D IR vibrational echo experiments showed that the formation and dissociation of the intra—intermolecular hydrogen-bonded complex happen on a picosecond time scale. The thermal equilibrium kinetics is characterized by the dissociation time, which was determined to be 3 ± 0.6 ps. It was experimentally demonstrated that the thermal equilibrium between the 2MPOD complex and free 2POD and intramolecular hydrogen-bonded 2MPOD and the exchange rates were not perturbed by the vibrational excitation on the hydroxyl stretch.

In addition to 2MP and toluene, several other related solutes and solvents were studied. It was found that the three-centered hydrogen bond is very sensitive to steric effects. The addition of a methylene group to the methoxyl group, that is, 2-ethoxyphenol (2EP), inhibits the formation of the intra—intermolecular hydrogen bond. DFT calculations show that the 2EP has to adopt a gauche conformation to be able to form the complex because complex formation is sterically prevented when 2EP is in the more stable anti conformation. The increased energy of the gauche conformation compared to the complex binding energy makes the 2EP—toluene complex improbable. It was also determined that the addition of methyl groups to the solvent molecules reduces the stability of the complex despite increased π-electron density in the solvent ring that would increase the strength of a conventional two-centered solute—solvent π-hydrogen bonding system, such as phenol—benzene.

Acknowledgment. We would like to thank Professor John I. Brauman for insightful discussions. This work was supported by a grant from AFOSR (F49620-01-1-0018) and from NSF (DMR-0332692).

Supporting Information Available: Analytical solutions for the kinetic model and 2D IR spectra including both 0—1 and 1—2 transitions. This material is available free of charge via the Internet at http://pubs.acs.org.

JA0570584

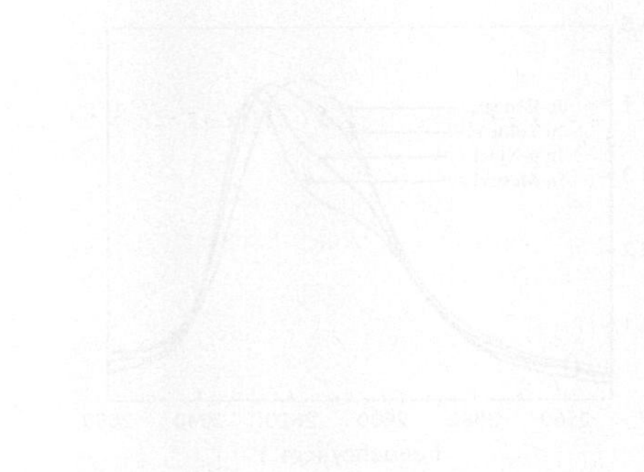

19998   *J. Phys. Chem. B* **2006,** *110,* 19998−20013

Ultrafast Two-Dimensional Infrared Vibrational Echo Chemical Exchange Experiments and Theory[†]

Kyungwon Kwak, Junrong Zheng, Hu Cang, and M. D. Fayer*

Department of Chemistry, Stanford University, Stanford, California 94305

Received: April 22, 2006; In Final Form: July 12, 2006

Ultrafast two-dimensional (2D) infrared vibrational echo experiments and theory are used to examine chemical exchange between solute−solvent complexes and the free solute for the solute phenol and three solvent complex partners, *p*-xylene, benzene, and bromobenzene, in mixed solvents of the partner and CCl_4. The experiments measure the time evolution of the 2D spectra of the hydroxyl (OD) stretching mode of the phenol. The time-dependent 2D spectra are analyzed using time-dependent diagrammatic perturbation theory with a model that includes the chemical exchange (formation and dissociation of the complexes), spectral diffusion of both the complex and the free phenol, orientational relaxation of the complexes and free phenol, and the vibrational lifetimes. The detailed calculations are able to reproduce the experimental results and demonstrate that a method employed previously that used a kinetic model for the volumes of the peaks is adequate to extract the exchange kinetics. The current analysis also yields the spectral diffusion (time evolution of the dynamic line widths) and shows that the spectral diffusion is significantly different for phenol complexes and free phenol.

I. Introduction

In this paper ultrafast two-dimensional (2D) infrared vibrational echo experiments that measure the formation and dissociation of solute−solvent complexes on a picosecond time scale under thermal equilibrium conditions are presented, and the theoretical underpinnings of such experiments are laid out in detail. Recent papers on solute−solvent chemical exchange experiments employed a kinetic scheme to analyze the data that included orientational relaxation rates, vibrational lifetimes, and an indirect accounting for spectral diffusion,[1,2] but the fundamental nature of the experiments, including the multiple light−matter interactions and how they relate to the experimental observables, was not treated.

Chemical exchange reactions on the picosecond time scale include intermolecular exchange, isomerization, proton transfer, and electron transfer. Intermolecular exchange, discussed here, is important in chemistry and biology. It forms the basis for supramolecular chemistry,[3] host−guest chemistry,[3] chemical and biological recognition,[4] and self-assembly.[3] Specific intermolecular interactions, such as hydrogen bonding, can lead to structurally unique solute−solvent complexes that are constantly forming and dissociating under thermal equilibrium conditions on very short time scales.[1,2,5] Dynamics of these transient species play a role in the physical and chemical properties of solute−solvent systems by affecting reaction rates, reaction mechanisms, and product ratios.[6] Although solute−solvent chemical exchange experiments will be discussed here, the theoretical description of the 2D IR vibrational echo chemical exchange experiment is applicable to all types of chemical exchange phenomena.

Until recently, chemical exchange on the picosecond time scale could not be measured under thermal equilibrium conditions. Two IR methods, 2D IR heterodyne-detected vibrational echoes[1,2,7,8] and two color IR pump−probe experiments,[9] can

be used. The 2D vibrational echo method explicated here has some advantages that generally accrue to Fourier transform spectroscopies, but more important, it does not have the time-bandwidth limitations of narrow band pump/broad band probe experiments.

There are methods that can measure chemical exchange for reactions slower than 10^{-10} s.[10] For reactions in the 10^{-12} s range, line shape analysis of linear-IR and Raman vibrational spectra has been used,[11] but line shape analysis is fraught with difficulties because of the multiple dynamic processes that can contribute to the line shape in addition to exchange.[12−14]

Two-dimensional NMR provides an excellent method for studying chemical exchange on time scales of microseconds or longer.[15−17] However, for solute−solvent complexes in organic and other types of nonaqueous solutions that are of interest here, the complexes are bound by energies on the order of a few RT ($RT \approx 0.6$ kcal/mol at room temperature, where R is the gas constant) and, therefore, form and dissociate on very rapidly. The 2D IR vibrational echo chemical exchange technique[1,2,7,8] is the ultrafast analogue of 2D NMR chemical exchange spectroscopy.[15−17] Both the 2D NMR and the 2D vibrational echo techniques involve pulse sequences that induce and then probe the coherent evolution of excitations (nuclear spins for NMR and vibrations for IR) of a molecular system. In the 2D vibrational echo experiment, a selected molecular vibration of molecules in a given environment (for example, a free solute versus complexed solute) is placed in a coherent superposition state by the first pulse in the sequence. The effect of the first pulse and the manipulation of the phase relationships among the excitations by the following pulses in the sequence is an important common feature of 2D vibrational echo spectroscopy and 2D NMR. The later pulses generate observable signals that are sensitive to chemical exchanges. The critical difference between the IR and NMR variants is that the IR pulse sequence acts on a time scale 6−9 orders of magnitude faster than the radio frequency pulse sequence in NMR.

[†] Part of the special issue "Charles B. Harris Festschrift".
* Author to whom correspondence should be addressed. E-mail: fayer@stanford.edu.

10.1021/jp0624808 CCC: $33.50   © 2006 American Chemical Society
Published on Web 08/30/2006

Vibrational-echo-based 2D IR experiments have been applied to a wide variety of problems, and both experiment and theory have been presented.[18−29] However, its application to chemical exchange is quite new,[1,2,7,8] and a full theoretical description with comparisons to experiments has not been given. Previously, data for the solute phenol complexed with benzene in a mixed solvent of benzene and CCl_4 was presented.[1] More recently, solute−solvent chemical exchange for the solute 2-methoxyphenol was published.[2] Here, the phenol/benzene data along with the phenol/*p*-xylene and phenol/bromobenzene data will be analyzed in detail using time-dependent diagrammatic perturbation theory[30] and a model that includes chemical exchange, spectral diffusion, orientational relaxation, and the vibrational lifetimes. The model is able to describe the data extremely well. One of the important results is that the detailed theory presented here confirms that the much simpler method of analysis used previously[1,2] is sufficient to extract the exchange kinetics. Application of the full theory also provides details of the spectral diffusion for the complexed and free phenol.

In the experiments presented below, a solute, phenol-OD (hydroxyl hydrogen replaced by a deuterium, which will be referred to as phenol for simplicity) exists in equilibrium as a complex with benzene (or *p*-xylene or bromobenzene) or as the free uncomplexed form. There are two peaks in the linear Fourier transform IR spectrum, one for the complex and one for the free species.[1] In the 2D IR vibrational echo spectrum at very short time (small interval between pulses 2 and 3 in the pulse sequence, see below) there are two peaks on the diagonal, one for the complex and one for free phenol. As time is increased, complexes form and dissociate under thermal equilibrium conditions. The chemical exchange produces off-diagonal peaks, which increase as time is increased. If no other processes occurred, then the growth in amplitude of the off-diagonal peaks would be simply related to the rate of the chemical exchange rate, that is the rate of complex formation and dissociation, which are equal for an equilibrium system.[1] However, spectral diffusion causes the peaks to change shape, and therefore, the peak heights cannot be used to analyze the data. In addition, orientational relaxation causes all peaks to decrease in amplitude. Because, in general, the orientational relaxation rates of the two species are not the same, the orientational relaxation with different rates for the species that are interconverting must be taken into account in the theoretical analysis. The species are also undergoing vibrational population relaxation (T_1), which decreases all peaks as time increases. Again, the lifetimes of the species are not the same, and the lifetimes with the species interconverting must be properly accounted for in the theoretical analysis. Another important feature of the problem is that the species, complexed and free phenol, do not have the same transition dipole matrix elements. All of these properties of the physical system are accounted for in the theoretical development presented below.

II. Experimental Procedures

A. Sample Preparation. Phenol forms weak complexes with benzene and its derivatives.[1,4,31,32] The polar hydroxyl group on phenol (OD in these experiments) and polarizable π-electron cloud of benzene lead to the attraction between these molecules. The resulting weak hydrogen bond that forms the complex shifts the OD stretching frequency to a lower frequency than that in the non-hydrogen-bonded free phenol. Phenol was deuterated by deuterium exchange with methanol-OD. In pure benzene solvent, only one OD stretch peak, corresponding to the complex, is observed, although there is a low-amplitude, high-

frequency tail that corresponds to free phenol.[1] To observe the formation and dissociation of the phenol−benzene complex, a mixed solvent of benzene and CCl_4 was used to shift the equilibrium to more of the free species. In pure CCl_4, only the free species is present. The free phenol has a higher-frequency hydroxyl stretch because of the lack of the hydrogen bond. Infrared spectra of free phenol in pure CCl_4, phenol−benzene complex in pure benzene, and phenol−benzene complex and free phenol in the benzene/CCl_4 mixed solvent have been published.[1] The spectra in the mixed solvents for the three systems studied here are shown in Figures 3a, 5a, and 7a. The mixed solvent creates a well-defined double potential minimum with similar populations along the exchange coordinate.

B. Optical System and Methodology. The ultrashort IR pulses ($\sim$50 fs) employed in the experiments were generated using a Ti:sapphire regeneratively amplified laser/optical parametric amplifier (OPA) system. The output of the regen is 40 fs transform-limited $^2/_3$ mJ pulses at a 1 kHz repetition rate. These are used to pump the short-pulse IR OPA. The output of the OPA is compressed to produce 50 fs transform-limited IR pulses in the actual sample cell using a purely nonresonant signal and frequency-resolved optical grating (FROG)[33] measurements. The pulses span sufficient bandwidth (300 cm^{-1} centered at 2500 cm^{-1}) to cover the $v = 0$ to $v = 1$ (hereafter denoted 0−1) and 1−2 transitions of the hydroxyl OD stretching modes in both free and complexed phenol.

The compressed IR output of the OPA is split into five beams. Three equal-intensity IR pulses impinge on the sample and stimulate the emission of the vibrational echo, a pulse that leaves the sample in a unique direction at a time following the third pulse. Another beam is made collinear with the vibrational echo path through the sample. It is used only for alignment and is blocked during the actual experiments. A fifth pulse, the local oscillator, does not pass through the sample and is overlapped with the vibrational echo pulse for heterodyne detection of the echo. The combined vibrational echo/local oscillator beam is directed into a spectrograph, and the spectrally dispersed signal is measured using a 32 element mercury cadmium telluride (MCT) array to record 32 frequencies simultaneously. The center frequency of the spectrograph is moved to record sets of 32 frequencies to span the entire spectrum.

The frequency- and phase-resolved, stimulated vibrational echo is measured as a function of one frequency variable from the spectrograph, ω_m, and two time variables, τ and T_w, the time between the first and second and the second and third IR pulses, respectively. ω_m provides one frequency axis of the 2D vibrational echo spectrum. By numerical Fourier transformation, the τ scan data are converted into the second frequency variable, providing the ω_τ axis. The resulting interferogram contains both the absorptive and the dispersive components of the vibrational echo signal.[34] To greatly reduce the dispersive contribution and obtain close to pure absorptive features, two sets of quantum pathways are measured independently by appropriate time ordering of the pulses in the experiment.[34] With pulses 1 and 2 at the time origin, pathway 1 or 2 is obtained by scanning pulse 1 or 2 to negative time, respectively. By addition of the Fourier transform of the combined interferograms from the two pathways, the dispersive component is substantially eliminated, which greatly narrows the features in the spectra. The 2D IR vibrational echo spectra are constructed by plotting the amplitude of the nominally absorptive part of the vibrational echo as a function of both ω_m and ω_τ. Additional experimental details, including procedures to ensure that phase relationships in the spectrum are proper, are discussed elsewhere.[35]

20000 *J. Phys. Chem. B, Vol. 110, No. 40, 2006*

Kwak et al.

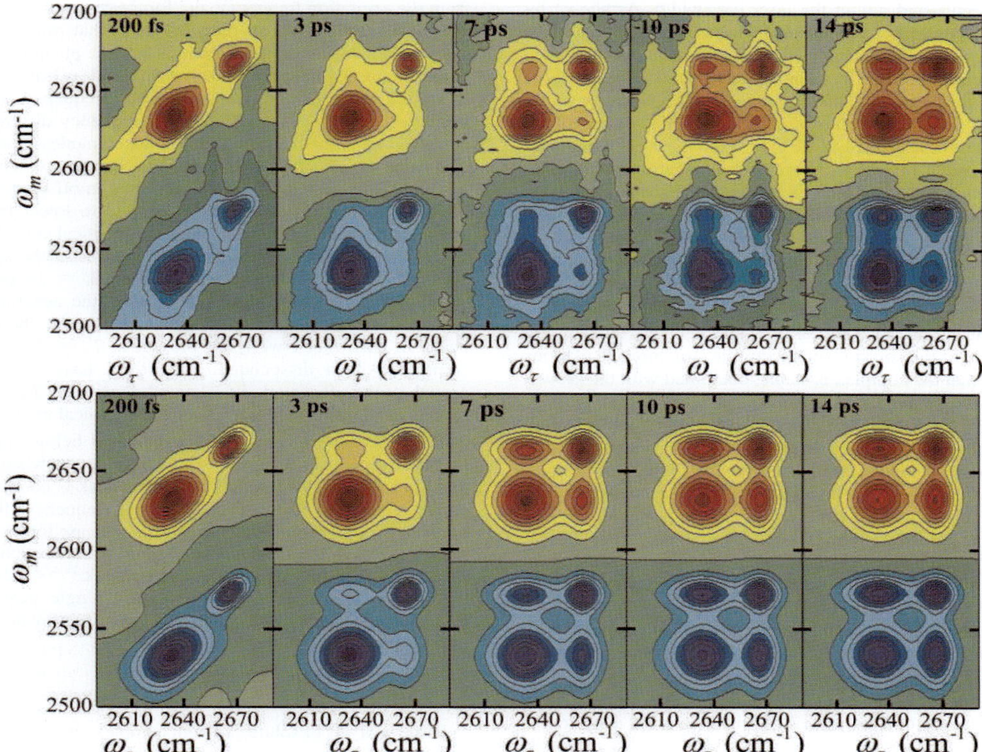

Figure 1. T_w-dependent 2D IR vibrational echo spectra (top) of the hydroxyl stretch of phenol-OD in benzene/CCl₄ mixed solvents (molar ratio of phenol/benzene/CCl₄ = 2:40:100). The red peaks (positive) are from the 0−1 vibration transitions, and the blue peaks (negative) are from the 1−2 vibrational transitions. At 200 fs, there are two peaks on the diagonal (red) and two peaks below these (blue) shifted by the anharmonicity. As T_w increases, additional peaks appear due to chemical exchange, that is, dissociation and formation of the phenol−benzene complex. The bottom portion displays response function calculations of the data as discussed in section V.

The dual-scan method can remove a substantial portion of the dispersive contributions to the 2D spectrum. In a 2D vibrational echo spectrum of two or more modes of a molecule with intramolecular coupling between them, there is an inherent imbalance of interaction pathways that results in substantial uncanceled dispersive contributions that are greater for the off-diagonal peaks than for the diagonal peaks.[28,34] However, there is no imbalance of interaction pathways in any of the peaks produced by the exchange process as shown by the Feynman pathways (see below). There are the same numbers of non-rephasing and rephrasing pathways for each peak. Therefore, the dual-scan method can, to a large extent, remove dispersive contributions to all peaks in the 2D spectrum if the off-diagonal peaks are produced solely by exchange. The phasing procedure using the projection theorem was applied,[35,36] and correctly phased diagonal and off-diagonal peaks were obtained.

III. Results

Figure 1 (top portion) displays the 2D IR vibrational echo spectra as contour plots (each contour is a 10% change) at various T_w points (from 200 fs to 14 ps) for the phenol−benzene complex system. (Two-dimensional data for the other systems are shown in Figures 4a and 6a.) The data have been normalized to the largest peak at each T_w. The red contours are positive going (0−1 vibrational transition), and the blue contours are negative going (1−2 vibrational transition). The 0−1 signal comes from two quantum pathways that are related to bleaching of the ground state and stimulated emission, both of which produce a vibrational echo pulse that is in phase with and therefore adds to the local oscillator pulse to produce a positive going signal. The 1−2 signal arises because there is a new absorption that was not present prior to the first two excitation pulses. The 1−2 vibrational echo pulse is 180° out of phase with the local oscillator and thus subtracts from the local oscillator to produce a negative going signal. At T_w = 200 fs, there are two peaks on the diagonal (0−1 transitions) and the corresponding 1−2 transition peaks off-diagonal. As T_w (waiting time) is increased, the off-diagonal peaks of the 0−1 transition and the corresponding 1−2 peaks appear.

There are a number of phenomena that can produce off-diagonal peaks. Two coupled modes on the same molecule will have the peaks for each mode on the diagonal and coherence transfer peaks off-diagonal.[28] In addition there will be negative going peaks off-diagonal below the coherence transfer peaks. These peaks are present at T_w = 0 and do not increase in amplitude. It is possible to have incoherent population relaxation between two modes on the same molecule that are coupled by anharmonic terms in the molecular potential. Such population relaxation will produce off-diagonal peaks that appear with increasing T_w.[37] However, the two modes have to be on the

J. Phys. Chem. B, Vol. 110, No. 40, 2006 **20001**

same molecule and within the laser bandwidth. As shown by the linear spectroscopy, the phenol system has a single OD stretch.[1] It has one frequency when it is free and another when it is a complex. The two frequencies correspond to the complex and free phenol peaks. Therefore, each species has a single mode, and population relaxation between modes cannot be responsible for the growth of the off-diagonal peaks. Vibrational Förster excitation transfer can transfer a vibrational excitation between modes on distinct molecules.[38,39] However, this is extremely short range because of very small vibrational transition dipoles and short vibrational lifetimes, with a range of 2−3 Å for hydroxyl stretches.[39] Given the low concentration of the complex and free phenol in the solution, Förster transfer is not a possibility.

The linear spectroscopy of the species, the temperature dependence of the spectra, and electronic structure calculations that show that the complex is stable, combined with the growth of the off-diagonal peaks, demonstrate that chemical exchange between complexed and free phenol is being observed.[1] Through the use of time-dependent diagrammatic perturbation theory, which describes nonlinear optical interactions with the molecular vibrations,[40,41] the chemical exchange experiments can be described quantitatively. It is necessary to include in the calculations not only chemical exchange but also orientational relaxation of each species, vibrational relaxation of each species, and the differences in the transition dipole moment matrix elements of the species. The detailed theory will be presented in the next section. First a brief description of the origin of the off-diagonal peaks will be given to set a qualitative stage for the detailed theoretical description.

The frequency at which the first pulse excites a mode is the mode frequency on the ω_τ axis (horizontal axis), 2631 cm^{-1} for the 0−1 transition in complex and 2665 cm^{-1} for the free phenol. The third pulse causes a mode to emit the time-delayed vibrational echo pulse at the frequency of the third interaction (third pulse) of the radiation field with the mode. The frequency of the vibrational echo emission is the frequency on the ω_m axis (the vertical axis). First consider $T_w = 200$ fs in Figure 1 in which the data are taken on a short time scale in comparison to the rate of chemical exchange. For the 0−1 vibrational transitions, the third pulse induces the vibrational echo emission at the same frequencies excited by the first pulse, so there are two peaks on the diagonal where $\omega_\tau = \omega_m$ (red peaks in Figure 1). If the frequency of vibrational echo emission (ω_m, third interaction frequency) is different from the frequency of initial excitation (ω_τ, first interaction frequency), then the peaks will appear off-diagonal. Again, for $T_w = 200$ fs in Figure 1, the blue peaks are off-diagonal by the vibrational anharmonicity because the modes are initially excited at their 0−1 frequencies (ω_τ), but the third pulse causes vibrational echo emission at their 1−2 frequencies (ω_m). Even in the absence of chemical exchange, the peaks observed at very short T_w delays undergo evolution with increasing T_w, because of spectral diffusion, which changes the shapes of the peaks, and vibrational lifetime decay and orientational relaxation, which cause the peaks to decay in amplitude.

The influence of chemical exchange on the 2D correlation spectrum can be understood qualitatively as follows. After the first two pulses in the vibrational echo sequence, if some of the complexed phenols dissociate during the T_w period, then the third pulse will cause the emission of the vibrational echo at the frequency of the free phenol OD stretch for these newly dissociated phenols. The frequency of emission, ω_m, then differs

from the excitation frequency, ω_τ, for these specific molecules. The result will be an off-diagonal peak that only appears if chemical exchange occurs. Because the free phenol absorbs at higher frequency than complexed phenol, this off-diagonal peak is shifted from the complexed phenol frequency to higher frequency along the ω_m axis by the frequency difference (34 cm^{-1}) between the free and the complexed modes. Conversely, if during the T_w period some free phenols associate with benzene, then the third pulse will produce an off-diagonal peak for these (formerly) free phenols, shifted to lower frequency along the ω_m axis by the same amount. Identical considerations apply for both the 0−1 and the 1−2 regions of the spectrum. This behavior is shown in Figure 1 at $T_w = 14$ ps, where substantial chemical exchange has led to the generation of a block of 4 red peaks and a block of 4 blue peaks; the two new peaks in each block were not present at $T_w = 200$ fs. Some complexes have dissociated, and others have formed. The growth of the off-diagonal peaks with increasing T_w is directly related to the time dependence of the chemical exchange.

The description given above and treated below applies for moderate or slow exchange. Fast exchange occurs when the time to jump to a new species and then jump back is comparable to or fast compared to the inverse of the frequency difference between the two peaks.[12,42,43] This in not the case for the systems studied here. In the fast exchange limit, the vibrational linear absorption spectrum will show a single peak.[14,42,43] The 2D IR vibrational echo spectrum will show a single peak on the diagonal; off-diagonal peaks will not appear to provide information on the rate of chemical exchange.

IV. Theory

In the analysis of solute−solvent complex chemical exchange data presented previously,[1,2] a kinetic model was used to fit the data and extract the exchange rate. The model did not explicitly deal with spectral diffusion but took it into account by fitting the time dependence of the peak volumes rather than the peak amplitudes. Orientational relaxation, vibrational lifetimes, and differences in transition dipoles were included. Here, diagrammatic perturbation theory methodology for this problem will be developed and applied. Spectral diffusion is included explicitly and analyzed. One of the important results in addition to obtaining the time dependence of the spectral diffusion is the demonstration that the much simpler kinetic model is accurate if only the chemical exchange is of interest.

In NMR, exchange effects have been studied extensively in both 1D and 2D spectroscopies.[15,17,42,43] Aspects of the NMR theoretical description of chemical exchange can be applied to the current problem so long as the important differences between the NMR problem and the vibrational problem are dealt with. The theoretical treatment includes the effects of chemical exchange during evolution and detection periods. The exchange during the first coherence period does not contribute to the growth of the off-diagonal peaks because the exchange time is long compared to the inverse frequency difference of the peaks (slow to moderate exchange limit). Then, during the first coherence period, jumps from the complex to free form or vice versa will produce ensembles of superposition states in both forms with random phases resulting in no contribution to the off-diagonal peaks. Such jumps do produce dynamic line broadening of the diagonal peaks, which is included in the treatment. Therefore, only exchange during the population period (time between pulses 2 and 3, T_w) contributes to the growth of the off-diagonal peaks.

20002 *J. Phys. Chem. B, Vol. 110, No. 40, 2006*

Spectral diffusion is slow enough that at least for the jumps that occur at short times the lines are inhomogeneously broadened. In the model presented here, it is assumed that there is no correlation in the frequency prior to a jump with the frequency following a jump between species. That is, the species prior to a jump has some position in its inhomogeneous line. After the jump, it can be anywhere in the new line only weighted by the line shape. This is a very reasonable assumption given the nature of the system in which complexes form and dissociate. The ramifications of the lack of correlation will be described below. It is possible to extend the formalism presented here to include frequency correlation, which has been observed in a different type of system.[44]

The processes other than chemical exchange can be divided into two groups. One is the spectral diffusion, which changes the shapes of peaks. The others are the finite lifetimes of excited states and orientational relaxation, both of which diminish the size of the peaks. In NMR, the coupled equations for exchange and population relaxation were derived.[45] The equation of motion of the density matrix for the spin system includes population relaxation via a Redfield-type matrix and exchange via a Kubo–Sack matrix.[46,47] To describe the vibrational problem, orientational relaxation must also be included. Vibration and rotation are generally separated in the theoretical description of vibrational spectroscopy.[48] This separation will be used here during the coherence periods. However, the dynamic partition model can avoid this separation of vibrational motion and rotation during the population period. As described above, the off-diagonal peaks are the result of an exchange process during the population period. Thus, it is important to properly treat various dynamic processes during the population period to accurately extract information about chemical exchange from 2D IR vibrational echo spectra.

Recently, theoretical descriptions of third-order response functions with exchange were published by two groups. Mukamel and co-workers used a Stochastic Liouville Equation[40] to simulate the vibrational echo spectrum of water with a four-site exchange model. Cho and co-workers derived analytical expressions for linear and third-order response functions using a two-species model with exchange and more importantly showed that the truncated cumulant expansion can be used in this two-species model.[41] In their response functions, the exchange effect was inserted using conditional probabilities that describe the probability of each pathway occurring at different T_w values. Both groups use molecular dynamics (MD) simulations to obtain information about exchange rates. Through the analysis of MD trajectories and using appropriate criteria for hydrogen-bonding configurations, species with different hydrogen bond structures were separated. From this separation, the exchange time and lifetime of each species were determined. In these theoretical treatments, the exchange time scale is much faster than vibrational and rotational relaxation time scales, so it was safely assumed that exchange had the dominant effect on the population dynamics; the other factors were not included.

However, in general chemical exchange, orientational relaxation and population relaxation can all occur on similar time scales. This is the situation for the experiments on organic solute–solvent chemical exchange analyzed below and presented previously.[1,2,7] As a result, a theoretical treatment that includes all aspects of the problem that influence the 2D spectrum is required. In the following, first the method for treating the orientational relaxation will be presented. Then, this

method will be used with the third-order response functions to obtain a model that describes the chemical exchange problem including spectral diffusion, orientational relaxation, and vibrational population decay. This modified response function approach is used to fit the experimental results with the exchange rate as a parameter as well as a form for the frequency–frequency correlation function (FFCF),[30] which yields a description of the spectral diffusion. Because the vibrational lifetimes and orientational relaxation rates are measured independently using IR pump–probe experiments, these are not adjustable parameters in the fits to the data. The exchange rate and the spectral diffusion have very different effects on the 2D vibrational echo spectra. Therefore, these can be determined with confidence independently from each other. Finally, it will be shown that the simple peak volume fitting approach is sufficient to determine the exchange rate if information on spectral diffusion is not desired.

A. Dynamic Partition Model. During the population period, molecules will experience three different processes that affect the population and finally determine the signal size. As explained above, three different processes include vibrational relaxation, orientational relaxation, and exchange between different species. All three relaxation processes are coupled in this treatment. The contribution of each relaxation process is schematically illustrated as

$$\overset{\tau_r^c,T_1^c}{\underset{\text{decay}}{\longleftarrow}}\ c\ \overset{k_{cf}}{\underset{k_{fc}}{\rightleftharpoons}}\ f\ \overset{\tau_r^f,T_1^f}{\underset{\text{decay}}{\longrightarrow}}$$

where $\tau_r^i = 1/6D_i$ and $T_1^i = 1/k_i$ are the orientational relaxation time constant (D_i is the orientational diffusion constant) and the vibrational lifetime of the ith species, respectively, and k_{cf} and k_{fc} are the complex dissociation (complex to free) and formation (free to complex) rate constants, respectively. In the treatment given below, what we refer to as effective populations are calculated. While the vibrational lifetimes and exchange can cause changes in the number of molecules of a given species, orientational relaxation can occur with no change in actual population. However, randomization of the transition dipole direction following interactions with the radiation fields reduces the signals for the diagonal and off-diagonal peaks. Thus, a population's contribution to the signal of a given peak is in some sense a vector quantity, including the magnitude and the projection of the transition dipole on the radiation field direction. These quantities are the effective populations that give rise to the diagonal and off-diagonal peaks.

Because the molecules (complex or free) have initial random orientations, signal size at a frequency ω_m (the frequency of the third interaction producing the coherence and echo emission at that frequency) is proportional to the orientational ensemble average of the population with the coherence frequency of each population

$$I_i(\omega_m = \omega_i) \propto \langle N_i(t) \cos^2 \theta \rangle_i \quad (i = \text{c,f}) \tag{1}$$

Here the orientational ensemble average is defined as

$$\langle \cos^2 \theta \rangle_i = \int_0^{\pi/2} \sin(\theta)\, d\theta\, f_i(\theta) \cos^2(\theta) \tag{2}$$

$f_i(\theta)$ is the angular distribution function for the complex ($i = $ c) and free ($i = $ f) forms.

　　　　

These effective populations (before the orientational ensemble average of the populations) will evolve as

$$\frac{d}{dt}\begin{pmatrix} N_f(t)f_f(\theta,t) \\ N_c(t)f_c(\theta,t) \end{pmatrix} =$$

$$\begin{pmatrix} -(k_f + k_{fc} + D_f l^2) & k_{cf} \\ k_{fc} & -(k_c + k_{cf} + D_c l^2) \end{pmatrix}\begin{pmatrix} N_f(t)f_f(\theta,t) \\ N_c(t)f_c(\theta,t) \end{pmatrix} \quad (3)$$

D_f and D_c are the orientational diffusion constants for the complex and free forms, which in general are not equal. l^2 is the spherical operator. Its eigenfunctions are the spherical harmonics, Y_m^l. As indicated by the use of diffusion constants, we make the reasonable assumption that the orientational relaxation is diffusive and described by[49]

$$\frac{\partial}{\partial t}f(\theta,t) = -Dl^2 f(\theta,t) \quad (4)$$

The formal solution of the coupled differential equations can be written in matrix form as

$$\begin{pmatrix} N_f(t)f_f(\theta,t) \\ N_c(t)f_c(\theta,t) \end{pmatrix} =$$

$$\exp\left[\begin{pmatrix} -(k_f + k_{fc} + D_f l^2) & k_{cf} \\ k_{fc} & -(k_c + k_{cf} + D_c l^2) \end{pmatrix}t\right] \times$$

$$\begin{pmatrix} N_f(0)f_f(\theta,0) \\ N_c(0)f_c(\theta,0) \end{pmatrix} \quad (5)$$

with the boundary conditions of the angular distribution function $f_i(\theta,0)$

$$f_i(\theta,0) = 3\cos^2(\theta)$$

$$f_i(\theta,\infty) = 1 \quad (6)$$

The boundary condition given in eq 6 needs to be discussed in some detail. This boundary condition is correct for pump–probe and fluorescence measurements because there is only one excitation pulse rather than two pulses with a time delay to produce a population in the vibrational echo experiments. Therefore, orientation relaxation should be considered during the coherence period (the period τ between the first and second pulses). The orientational relaxation part for the third-order experiment can be described using the probability evolution Green's function that satisfies the diffusion equation. The result can be expressed as the product of two first-order Legendre polynomials and one second-order Legendre polynomial. The former one describes the orientational relaxation during the two coherence periods and the latter one for population period. Through the use of this result, orientational relaxation for the vibrational echo experiment can be described as

$$R_{zzzz} = \frac{1}{9}C_1(t_1)\left(1 + \frac{4}{5}C_2(t_2)\right)C_1(t_3)$$

$$C_l(t_i) = \exp[-l(l+1)Dt_i] \quad (7)$$

Because of the orientational relaxation during first coherence period, the initial condition of $f_i(\theta,0)$ (eq 6) is not exactly

$3\cos^2\theta$. However, orientational relaxation during the first coherence period has a decay constant that is a factor of 3 smaller than the decay constant during the population period because of the difference in the coefficient $l(l+1)$. Furthermore, the coherence period is relatively short, ~ 1 ps, compared to the $C_1(t)$ relaxation time for these experiments (~ 9 ps). Therefore, we assume that the initial condition given in eq 6 is a reasonable approximation for the initial orientational state at the beginning of the population period. With this assumption, $f_i(\theta,0)$ can be expanded in terms of the spherical harmonics as

$$f_i(\theta,0) = 3\cos^2(\theta) = 2\sqrt{\frac{4\pi}{5}}Y_{20} + \sqrt{4\pi}Y_{00} \quad (8)$$

The orientational relaxation during the coherence periods is handled in the conventional manner as exponential decays that multiply the response function (see below). However, because chemical exchange produces the growth of the off-diagonal, it is necessary to explicitly account for jumps back and forth between the two species that are undergoing orientational relaxation at different rates.

Through the use of this initial condition, eq 5 can be solved

$$\begin{pmatrix} N_f(t)f_f(\theta,t) \\ N_c(t)f_c(\theta,t) \end{pmatrix} =$$

$$\exp\left[\begin{pmatrix} -(k_f + k_{fc} + 6D_f) & k_{cf} \\ k_{fc} & -(k_c + k_{cf} + 6D_c) \end{pmatrix}t\right] \times$$

$$\begin{pmatrix} N_f(0) \\ N_c(0) \end{pmatrix}2\sqrt{\frac{4\pi}{5}}Y_{20} +$$

$$\exp\left[\begin{pmatrix} -(k_f + k_{fc}) & k_{cf} \\ k_{fc} & -(k_c + k_{cf}) \end{pmatrix}t\right]\begin{pmatrix} N_f(0) \\ N_c(0) \end{pmatrix}\sqrt{4\pi}Y_{00} \quad (9)$$

When the orientational ensemble average is performed, the result is

$$\begin{pmatrix} \langle N_f(t)f_f(\theta,t)\rangle \\ \langle N_c(t)f_c(\theta,t)\rangle \end{pmatrix} =$$

$$\exp\left[\begin{pmatrix} -(k_f + k_{fc} + 6D_f) & k_{cf} \\ k_{fc} & -(k_c + k_{cf} + 6D_c) \end{pmatrix}t\right] \times$$

$$\begin{pmatrix} N_f(0) \\ N_c(0) \end{pmatrix}\frac{4}{15} +$$

$$\exp\left[\begin{pmatrix} -(k_f + k_{fc}) & k_{cf} \\ k_{fc} & -(k_c + k_{cf}) \end{pmatrix}t\right]\begin{pmatrix} N_f(0) \\ N_c(0) \end{pmatrix}\frac{1}{3} \quad (10)$$

Equation 10 can be solved analytically using the methods devised by Putzer.[50] The solutions to eq 10 yield both the diagonal and the off-diagonal solutions. The ensemble-averaged solutions for the diagonal peaks are labeled $N_{ff}(t)$ and $N_{cc}(t)$, indication that a species began and ended the pulse sequence in the same form. The off-diagonal peaks are labeled $N_{cf}(t)$ and $N_{fc}(t)$, indicating that a species began as a complex and ended free or vice versa, respectively

20004 *J. Phys. Chem. B, Vol. 110, No. 40, 2006*

Kwak et al.

$$N_{cc}(t) = \frac{4}{9}\, e^{-\alpha T_w}\{\cosh(\beta T_w) - \gamma\,\sinh(\beta T_w)\} +$$
$$\frac{5}{9}\, e^{-\phi T_w}\{\cosh(\varphi T_w) - \theta\,\sinh(\varphi T_w)\}$$

$$N_{ff}(t) = \frac{4}{9}\, e^{-\alpha T_w}\{\cosh(\beta T_w) + \gamma\,\sinh(\beta T_w)\} +$$
$$\frac{5}{9}\, e^{-\phi T_w}\{\cosh(\varphi T_w) + \theta\,\sinh(\varphi T_w)\}$$

$$N_{cf}(t) = N_{fc}(t) = \frac{4}{9}\frac{k_{cf}}{\beta}\, e^{-\alpha T_w}\sinh(\beta T_w) +$$
$$\frac{5}{9}\frac{k_{cf}}{\varphi}\, e^{-\phi T_w}\sinh(\varphi T_w) \quad (11)$$

The definitions of α, β, γ, ϕ, φ, and θ are given in the appendix.

The solutions of eq 10 for $\langle N_f(t)f_f(\theta,t)\rangle$ are composed of contributions from both $N_f(0)$ and $N_c(0)$. The former represents the effective number of oscillating dipoles with $\omega_m = \omega_f$ (vibrational echo emission) that were "frequency-labeled" during the first coherence period as $\omega_\tau = \omega_f$ (frequency of the first interaction). So the contribution from $N_f(0)$ describes the diagonal peak because $\omega_\tau = \omega_f$ and $\omega_m = \omega_f$; the frequency of the first interaction is the same as the frequency of the third interaction and, therefore, vibrational echo emission. The second contribution to $\langle N_f(t)f_f(\theta,t)\rangle$, from $N_c(0)$, involves complexes frequency-labeled by the first interaction with $\omega_\tau = \omega_c$ followed by the third interaction and echo emission at $\omega_m = \omega_f$. This is off-diagonal, corresponding to the dissociation of complexes because the initial interaction is at ω_c, but the final interaction and echo emission is at ω_f. The same considerations apply to $\langle N_c(t)f_c(\theta,t)\rangle$.

The contributions to the diagonal peaks $N_{cc}(t)$ and $N_{ff}(t)$ have two physically different origins. One contribution is from molecules staying in the initial state during the population period without exchange. This subensemble produces an echo signal that is identical to that which would arise from a system with a single species and, therefore, no possibility of chemical exchange. This subensemble can be represented using the response functions for a single oscillator. The other subensemble consists of molecules that undergo an even number of exchanges. For example, a molecule can start in the free form and end in the free form after two exchanges. As a result, such a molecule spends time in the complex form with vibrational lifetime and orientational relaxation rate of the complex form. This type of behavior is included in the kinetic equation and does not require additional treatment.

However, dephasing, described in terms of the response functions given below, is different for the two cases (no exchange versus exchange) even if the echo signals result in a peak at the same position in the 2D spectrum because the two cases involve different quantum pathways. These two cases are described using the Feynman diagrams shown in Figure 2. The contributions from the two cases can be separated using the solutions to the kinetic equations given in eq 11. Consider the effective population N_{ff}. The treatment is identical for N_{cc}. First, N_{ff} without exchange can be calculated by inserting exchange rate as 0, $N_{ff}(t;k_{cf} = 0)$. Then the effective population with exchange, $N_{ff}(t;k_{cf})$, which includes all molecules regardless of whether they happen to exchange or not, is obtained. The difference between these two effective populations gives the population that undergoes multiple exchanges

$$N_{ff}^{ex}(t) = N_{ff}(t;k_{cf} = 0) - N_{ff}(t;k_{cf}) \quad (12)$$

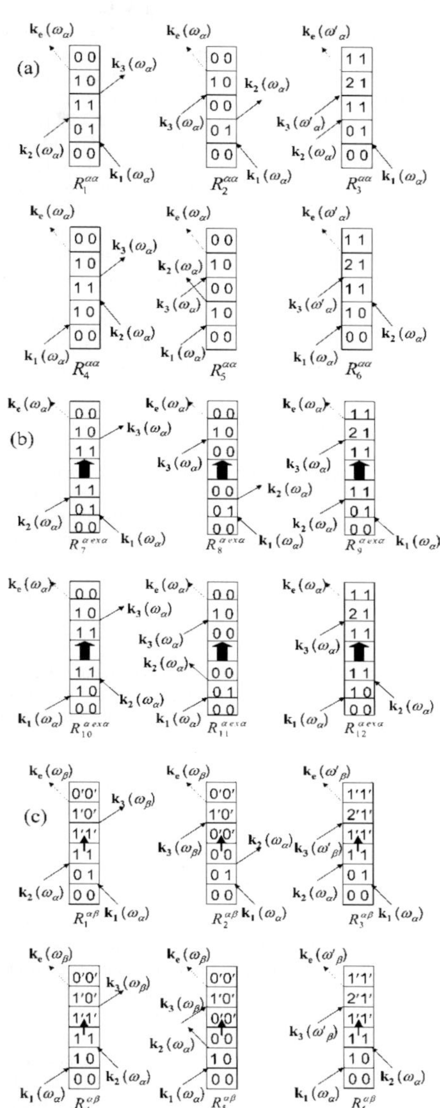

Figure 2. (a) Feynman diagrams corresponding to the first set of response functions. (b) Feynman diagrams corresponding to the second set of response functions. (c) Feynman diagrams corresponding to the third set of response functions.

As a result, the effective population that stays as a free molecule for the entire population period is

$$N_{ff}^{s}(t) = N_{ff}(t;k_c) - N_{ff}^{ex}(t) \quad (13)$$

The effective populations that undergo multiple exchanges, $N_{ff}^{ex}(t)$, and no exchanges, $N_{ff}^{s}(t)$, contribute differently to the total 2D IR spectrum, and these differences are included in the response functions derived in the following section.

B. Response Functions with Exchange. The following response functions were derived using a two-species model.

Kwac et al. obtained response functions for a two-species model.[41] However, the influence of orientational relaxation and lifetimes of the species were not included in the treatment. As discussed with the introduction of the dynamic partition model, for the 2D IR spectroscopic observables, the exchange process and the relaxation processes are coupled. To address this coupling during T_w, the relaxation functions $\Gamma(t_3,T_w,t_1)$ were modified to include exchange. A large number of parameters were included into the calculations. It was possible to measure many of the parameters independently, so that only the exchange rate and the FFCFs were varied to fit the data. The Feynman diagrams for nonlinear response functions are given in Figure 2.

Through the use of the two-species model, the linear response function, $R(t)$, can also be obtained, and they can be expressed as the sum of two terms, one for each species. This model can be generalized to more than two species.

$$R(t) = x_f R_f(t) + x_c R_c(t) \tag{14}$$

Here, x_f and x_c represent free and complex forms' relative populations in the ground state. The x_i can be determined from the analysis of IR spectra as described below. Through the use of the cumulant expansion, the linear response function for each species can be expressed as

$$R_\alpha(t) = |\mu_{0,\alpha}|^2 \, e^{-i\langle\omega_{0,\alpha}\rangle t} \exp\left[-g_{\alpha\alpha}(t) - \frac{1}{2T_1^\alpha}t\right] \tag{15}$$

where $\mu_{0,\alpha}$ is the transition dipole of the α form, $\langle\omega_{0,\alpha}\rangle$ is the ensemble average $0-1$ transition frequency of the α form, and T_1^α is the vibrational lifetime of the α form. The line shape function $g_{\alpha\alpha}(t)$ is defined as

$$g_{\alpha\alpha}(t) = \int_0^t d\tau_2 \int_0^{\tau_2} d\tau_1 \, \langle\delta\omega_{\alpha,0}(\tau_1)\delta\omega_{\alpha,0}(0)\rangle \tag{16}$$

where $\langle\delta\omega_{\alpha,0}(\tau_1)\delta\omega_{\alpha,0}(0)\rangle$ is the FFCF.

The analytical expressions for the nonlinear third-order response function use the standard approach and approximations, that is, the Franck–Condon approximation and the cumulant expansion, which apply well to the narrow Gaussian absorption bands.[51] The response functions are composed of two parts, the dephasing functions and the relaxation functions. The dephasing portion describes the time-dependent broadening of the dynamic lines. The relaxation function describes the time-dependent probability of an oscillator contributing to the signal following initial excitation. It can include both a lifetime and an orientational relaxation term as a multiplicative factor for the response function. However, as described above, this is an inadequate method for the problem with exchange. This approach can be used during the coherence periods, but the relaxation function during T_w must be replaced by the solution of the dynamic partition model given in eq 11. The dynamic partition model brings chemical exchange, orientational relaxation, and vibrational relaxation into the calculation in a proper manner that accounts for the differences in lifetimes and orientational relaxation rates of the two species.

For a single line or for two species with no exchange, the degree of frequency correlation between the first coherence period and the final coherence period is determined by the interaction between the oscillator and the bath. Such a correlation still applies to oscillators that do not undergo exchange in a system where exchange is occurring. In the treatment presented here, *we assume that chemical exchange destroys all frequency correlation*. That is, after exchange, an oscillator can assume any frequency in the spectral line with a probability that is only determined by the line shape. There is no memory of the oscillator's location in the spectral line of the species from which it originated. In the systems considered here, a complex and a free species are undergoing chemical exchange. For example, the phenol–benzene complex becomes free phenol in a benzene/CCl$_4$ mixed solvent. There is an abrupt change in the potential surface, and the surrounding solvent structure must also change. There is little likelihood of correlation. A lack of correlation may also apply to other types of exchange problems, such as isomerization or proton and electron transfer, because of the substantial changes in the nature of the species and the response of the solvent. Nonzero cross correlation is possible in, for example, a hydrogen-bonding system, where the changes in local structure may be small at least on some time scales.[25,44,52]

The assumption of no frequency correlation following chemical exchange means $\langle\delta\omega_\alpha(t)\delta\omega_\beta(0)\rangle = 0$, where α and β represent different species, that is, complex or free. As a result, the line shape of the off-diagonal peaks becomes the product of the linear line shape of the free and complex forms. The lack of frequency correlation following chemical exchange also affects the diagonal peaks because some fraction of the diagonal peaks' signals comes from species that have undergone an even number of exchanges during the T_w period. As a result, part of a diagonal peak's signal will consist of the free induction decay of that species. However, another portion of the signal from oscillators that have not undergone any exchange will have a contribution to the dynamic line shape determined by the coupling to the bath, which is expressed through the FFCF. All of the Feynman diagrams, including all pathways for diagonal peaks, are presented in Figure 2.

The third-order response function for each pathway can be derived analytically. The subscripts SE, GB, and TA in the relaxation functions $\Gamma(t_3,T_w,t_1)$ indicate stimulated emission, ground-state bleaching, and transient absorption, respectively. In the following, $\mu_{0,\alpha}$ and $\mu_{\alpha,2\alpha}$ are the transition dipole matrix elements for the $0-1$ and the $1-2$ vibrational transitions of the α species. $\Delta_{\alpha,\alpha}$ is the vibrational anharmonicity of the α species. $\omega_{0,\alpha}$ is the center frequency of the $0-1$ vibrational transition of the α species.

The first set of response functions (Figure 2a) are for the portion of the diagonal peaks that arise because a subensemble of a species undergoes no exchange, $C \rightarrow \{C\} \rightarrow C$. These can be expressed as

$$R_1^{\alpha\alpha}(t_3,T_w,t_1) = |\mu_{0,\alpha}|^4 \, e^{-i\langle\omega_{0,\alpha}\rangle(-t_1+t_3)}\Gamma_{SE}^{\alpha\alpha}(t_3,T_w,t_1) \times$$
$$\exp[-g_{\alpha\alpha}^*(t_1) + g_{\alpha\alpha}^*(T_w) - g_{\alpha\alpha}(t_3) - g_{\alpha\alpha}^*(t_1 + T_w) -$$
$$g_{\alpha\alpha}^*(T_w + t_3) + g_{\alpha\alpha}^*(t_1 + T_w + t_3)]$$

$$R_2^{\alpha\alpha}(t_3,T_w,t_1) = |\mu_{0,\alpha}|^4 \, e^{-i\langle\omega_{0,\alpha}\rangle(-t_1+t_3)}\Gamma_{GB}^{\alpha\alpha}(t_3,T_w,t_1) \times$$
$$\exp[-g_{\alpha\alpha}^*(t_1) + g_{\alpha\alpha}(T_w) - g_{\alpha\alpha}^*(t_3) - g_{\alpha\alpha}^*(t_1 + T_w) -$$
$$g_{\alpha\alpha}(T_w + t_3) + g_{\alpha\alpha}^*(t_1 + T_w + t_3)]$$

$$R_3^{\alpha\alpha}(t_3,T_w,t_1) =$$
$$-|\mu_{0,\alpha}|^2|\mu_{\alpha,2\alpha}|^2 \, e^{-i[\langle\omega_{0,\alpha}\rangle(-t_1+t_3)-\Delta_{\alpha,\alpha}t_3]}\Gamma_{TA}^{\alpha\alpha}(t_3,T_w,t_1) \times$$
$$\exp[-g_{\alpha\alpha}^*(t_1) + g_{\alpha\alpha}(T_w) - g_{\alpha\alpha}(t_3) - g_{\alpha\alpha}(t_1 + T_w) -$$
$$g_{\alpha\alpha}(T_w + t_3) + g_{\alpha\alpha}(t_1 + T_w + t_3)]$$

$$R_4^{\alpha\alpha}(t_3,T_w,t_1) = |\mu_{0,\alpha}|^4 \, e^{-i\langle\omega_{0,\alpha}\rangle(-t_1+t_3)}\Gamma_{SE}^{\alpha\alpha}(t_3,T_w,t_1) \times$$
$$\exp[-g_{\alpha\alpha}(t_1) - g_{\alpha\alpha}(T_w) - g_{\alpha\alpha}(t_3) + g_{\alpha\alpha}(t_1 + T_w) +$$
$$g_{\alpha\alpha}(T_w + t_3) - g_{\alpha\alpha}(t_1 + T_w + t_3)]$$

$$R_5^{\alpha\alpha}(t_3,T_w,t_1) = |\mu_{0,\alpha}|^4 \, e^{-i\langle\omega_{0,\alpha}\rangle(-t_1+t_3)}\Gamma_{GB}^{\alpha\alpha}(t_3,T_w,t_1) \times$$
$$\exp[-g_{\alpha\alpha}(t_1) - g_{\alpha\alpha}^*(T_w) - g_{\alpha\alpha}^*(t_3) + g_{\alpha\alpha}(t_1 + T_w) +$$
$$g_{\alpha\alpha}^*(T_w + t_3) - g_{\alpha\alpha}(t_1 + T_w + t_3)]$$

$$R_6^{\alpha\alpha}(t_3,T_w,t_1) =$$
$$-|\mu_{0,\alpha}|^2|\mu_{\alpha,2\alpha}|^2 \, e^{-i[\langle\omega_{0,\alpha}\rangle(-t_1+t_3)-\Delta_{\alpha,\alpha}t_3]}\Gamma_{TA}^{\alpha\alpha}(t_3,T_w,t_1) \times$$
$$\exp[- g_{\alpha\alpha}(t_1) - g_{\alpha\alpha}^*(T_w) - g_{\alpha\alpha}(t_3) + g_{\alpha\alpha}^*(t_1 + T_w) +$$
$$g_{\alpha\alpha}^*(T_w + t_3) - g_{\alpha\alpha}^*(t_1 + T_w + t_3)]$$

$$\Gamma_{SE}^{\alpha\alpha}(t_3,T_w,t_1) =$$
$$\exp\left[-2D_\alpha t_1 - 2D_\alpha t_3 - \frac{1}{2T_1}{}_\alpha t_1 - \frac{1}{2T_1}{}_\alpha t_3\right]N_{\alpha\alpha}^s(T_w)$$

$$\Gamma_{GB}^{\alpha\alpha}(t_3,T_w,t_1) =$$
$$\exp\left[-2D_\alpha t_1 - 2D_\alpha t_3 - \frac{1}{2T_1}{}_\alpha t_1 - \frac{1}{2T_1}{}_\alpha t_3\right]N_{\alpha\alpha}^s(T_w)$$

$$\Gamma_{TA}^{\alpha\alpha}(t_3,T_w,t_1) =$$
$$\exp\left[-2D_\alpha t_1 - 2D_\alpha t_3 - \frac{1}{2T_1}{}_\alpha t_1 - \frac{1}{T_1}{}_\alpha t_3\right]N_{\alpha\alpha}^s(T_w)$$

As discussed above, these response functions show the same behavior as a species without exchange.

The second set of response functions (Figure 2b) for the diagonal peaks describes the subensemble of a species that undergoes an even number of exchanges, C → {C → F → C} → C. As discussed above, these exchanges destroy all frequency correlation. So the final expressions for these pathways are

$$R_7^{\alpha\alpha}(t_3,T_w,t_1) = |\mu_{0,\alpha}|^4 \, e^{-i\langle\omega_{0,\alpha}\rangle(-t_1+t_3)}\Gamma_{SE}^{\alpha ex\alpha}(t_3,T_w,t_1) \times$$
$$\exp[-g_{\alpha\alpha}^*(t_1) - g_\alpha\alpha(t_3)]$$

$$R_8^{\alpha\alpha}(t_3,T_w,t_1) = |\mu_{0,\alpha}|^4 \, e^{-i\langle\omega_{0,\alpha}\rangle(-t_1+t_3)}\Gamma_{GB}^{\alpha ex\alpha}(t_3,T_w,t_1) \times$$
$$\exp[-g_{\alpha\alpha}^*(t_1) - g_{\alpha\alpha}^*(t_3)]$$

$$R_9^{\alpha\alpha}(t_3,T_w,t_1) =$$
$$-|\mu_{0,\alpha}|^2|\mu_{\alpha,2\alpha}|^2 \, e^{-i[\langle\omega_{0,\alpha}\rangle(-t_1+t_3)-\Delta_{\alpha,\alpha}t_3]}\Gamma_{TA}^{\alpha ex\alpha}(t_3,T_w,t_1) \times$$
$$\exp[-g_{\alpha\alpha}^*(t_1) - g_{\alpha\alpha}(t_3)]$$

$$R_{10}^{\alpha\alpha}(t_3,T_w,t_1) = |\mu_{0,\alpha}|^4 \, e^{-i\langle\omega_{0,\alpha}\rangle(-t_1+t_3)}\Gamma_{SE}^{\alpha ex\alpha}(t_3,T_w,t_1) \times$$
$$\exp[-g_\alpha\alpha(t_1) - g_{\alpha\alpha}^*(t_3)]$$

$$R_{11}^{\alpha\alpha}(t_3,T_w,t_1) = |\mu_{0,\alpha}|^4 \, e^{-i\langle\omega_{0,\alpha}\rangle(-t_1+t_3)}\Gamma_{GB}^{\alpha ex\alpha}(t_3,T_w,t_1) \times$$
$$\exp[-g_{\alpha\alpha}(t_1) - g_{\alpha\alpha}^*(t_3)]$$

$$R_{12}^{\alpha\alpha}(t_3,T_w,t_1) =$$
$$-|\mu_{0,\alpha}|^2|\mu_{\alpha,2\alpha}|^2 \, e^{-i\langle\omega_{0,\alpha}\rangle(-t_1+t_3)-\Delta_{\alpha,\alpha}t_3}\Gamma_{TA}^{\alpha ex\alpha}(t_3,T_w,t_1) \times$$
$$\exp[-g_{\alpha\alpha}(t_1) - g_{\alpha\alpha}(t_3)]$$

$$\Gamma_{SE}^{\alpha ex\alpha}(t_3,T_w,t_1) =$$
$$\exp\left[-2D_\alpha t_1 - 2D_\alpha t_3 - \frac{1}{2T_1}{}_\alpha t_1 - \frac{1}{2T_1}{}_\alpha t_3\right]N_{\alpha\alpha}^{ex}(T_w)$$

$$\Gamma_{GB}^{\alpha ex\alpha}(t_3,T_w,t_1) =$$
$$\exp\left[-2D_\alpha t_1 - 2D_\alpha t_3 - \frac{1}{2T_1}{}_\alpha t_1 - \frac{1}{2T_1}{}_\alpha t_3\right]N_{\alpha\alpha}^{ex}(T_w)$$

$$\Gamma_{TA}^{\alpha ex\alpha}(t_3,T_w,t_1) =$$
$$\exp\left[-2D_\alpha t_1 - 2D_\alpha t_3 - \frac{1}{2T_1}{}_\alpha t_1 - \frac{1}{T_1}{}_\alpha t_3\right]N_{\alpha\alpha}^{ex}(T_w)$$

The dephasing functions for these pathways show no dependence on T_w because of the assumption of no frequency correlation following exchange. So the contribution to the line shape from these pathways has no information for spectral diffusion dynamics. There is a dependence on T_w contained in the effective population term, $N_{\alpha\alpha}^{ex}(T_w)$. $N_{\alpha\alpha}^{ex}(T_w)$ determines the contribution of the fully broadened line shape to the dynamic line shape of the diagonal peaks. Without the separation into the first set of response functions and the second set of response functions for diagonal peaks, an observed fast broadening of the diagonal peaks caused by exchange could be misinterpreted as a result of spectral diffusion.

The third set of response functions (Figure 2c) for the off-diagonal peaks from a subensemble of a species that undergoes at least one exchange or any odd number of exchanges, C → {C → F} → F. All contributions to the off-diagonal peaks include at least one exchange, which destroys all frequency correlation. For this reason, the line shape of the off-diagonal peaks is a product of the linear line shapes of the two species.

$$R_1^{\alpha\beta}(t_3,T_w,t_1) = |\mu_{0,\alpha}|^2|\mu_{0,\beta}|^2 \, e^{i\langle\omega_{0,\alpha}\rangle t_1 - i\langle\omega_{0,\beta}\rangle t_3}\Gamma_{SE}^{\alpha\beta}(t_3,T_w,t_1) \times$$
$$\exp[-g_{\alpha\alpha}^*(t_1) - g_{\beta\beta}(t_3)]$$

$$R_2^{\alpha\beta}(t_3,T_w,t_1) = |\mu_{0,\alpha}|^2|\mu_{0,\beta}|^2 \, e^{i\langle\omega_{0,\alpha}\rangle t_1 - i\langle\omega_{0,\beta}\rangle t_3}\Gamma_{GB}^{\alpha\beta}(t_3,T_w,t_1) \times$$
$$\exp[-g_{\alpha\alpha}^*(t_1) - g_{\beta\beta}^*(t_3)]$$

$$R_3^{\alpha\beta}(t_3,T_w,t_1) =$$
$$-|\mu_{0,\alpha}|^2|\mu_{\beta,2\beta}|^2 \, e^{i\langle\omega_{0,\alpha}\rangle t_1 - i[\langle\omega_{0,\beta}\rangle-\Delta_{\beta,\beta}]t_3}\Gamma_{TA}^{\alpha\beta}(t_3,T_w,t_1) \times$$
$$\exp[-g_{\alpha\alpha}^*(t_1) - g_{\beta\beta}(t_3)]$$

$$R_4^{\alpha\beta}(t_3,T_w,t_1) = |\mu_{0,\alpha}|^2|\mu_{0,\beta}|^2 \, e^{-i\langle\omega_{0,\alpha}\rangle t_1 - i\langle\omega_{0,\beta}\rangle t_3}\Gamma_{SE}^{\alpha\beta}(t_3,T_w,t_1) \times$$
$$\exp[-g_{\alpha\alpha}(t_1) - g_{\beta\beta}(t_3)]$$

$$R_5^{\alpha\beta}(t_3,T_w,t_1) = |\mu_{0,\alpha}|^2|\mu_{0,\beta}|^2 \, e^{-i\langle\omega_{0,\alpha}\rangle t_1 - i\langle\omega_{0,\beta}\rangle t_3}\Gamma_{GB}^{\alpha\beta}(t_3,T_w,t_1) \times$$
$$\exp[-g_\alpha\alpha(t_1) - g_{\beta\beta}^*(t_3)]$$

$$R_6^{\alpha\beta}(t_3,T_w,t_1) =$$
$$-|\mu_{0,\alpha}|^2|\mu_{\beta,2\beta}|^2 \, e^{-i\langle\omega_{0,\alpha}\rangle t_1 - i[\langle\omega_{0,\beta}\rangle-\Delta_{\beta,\beta}]t_3}\Gamma_{TA}^{\alpha\beta}(t_3,T_w,t_1) \times$$
$$\exp[-g_{\alpha\alpha}(t_1) - g_{\beta\beta}(t_3)]$$

$$\Gamma_{SE}^{\alpha\beta}(t_3,T_w,t_1) =$$
$$\exp\left[-2D_\alpha t_1 - 2D_\beta t_3 - \frac{1}{2T_1}{}_\alpha t_1 - \frac{1}{2T_1}{}_\beta t_3\right]N_{\alpha\beta}(T_w)$$

$$\Gamma_{GB}^{\alpha\beta}(t_3,T_w,t_1) =$$
$$\exp\left[-2D_\alpha t_1 - 2D_\beta t_3 - \frac{1}{2T_1}{}_\alpha t_1 - \frac{1}{2T_1}{}_\beta t_3\right]N_{\alpha\beta}(T_w)$$

$$\Gamma_{TA}^{\alpha\beta}(t_3,T_w,t_1) =$$
$$\exp\left[-2D_\alpha t_1 - 2D_\beta t_3 - \frac{1}{2T_1}{}_\alpha t_1 - \frac{1}{T_1}{}_\beta t_3\right]N_{\alpha\beta}(T_w)$$

As discussed in the Experimental Section, rephasing (R) and nonrephasing (NR) signals are collected separately and added after Fourier transformation to eliminate a substantial portion of the dispersive contribution to the signal.[34] To emulate the experimental signal, calculation of the data mimicked he experimental procedure. The total rephrasing response function $R_R(t_1,T_w,t_3)$ and nonrephrasing response function $R_{NR}(t_1,T_w,t_3)$ are defined as

Ultrafast 2D IR Vibrational Echo Chemical Exchange

$$R_R(t_1,T_w,t_3) = \sum_{i=1}^{3} \sum_{\alpha,\beta} (R_i^{\alpha\alpha}(t_1,T_w,t_3) + R_{6+i}^{\alpha\alpha}(t_1,T_w,t_3) +$$
$$R_i^{\alpha\beta}(t_1,T_w,t_3)) \quad (17)$$

$$R_{NR}(t_1,T_w,t_3) = \sum_{i=4}^{6} \sum_{\alpha,\beta} (R_i^{\alpha}\alpha(t_1,T_w,t_3) + R_{6+i}^{\alpha\alpha}(t_1,T_w,t_3) +$$
$$R_i^{\alpha\beta}(t_1,T_w,t_3)) \quad (18)$$

The final 2D vibrational echo spectrum is

$$S_{2D}(\omega_\tau,\omega_m,T_w) \propto \{Re\}[\tilde{R}_R(\omega_\tau,\omega_m,T_w) + \tilde{R}_{NR}(\omega_\tau,\omega_m,T_w)] \quad (19)$$

where $\tilde{R}_R$ and $\tilde{R}_{NR}$ are defined as

$$\tilde{R}_R(\omega_\tau,\omega_m,T_w) = \int_0^\infty dt_1 \int_0^\infty dt_3$$
$$\exp(i\omega_m t_3 - i\omega_\tau t_1)R_R(t_1,T_w,t_3) \quad (20)$$

$$\tilde{R}_{NR}(\omega_\tau,\omega_m,T_w) = \int_0^\infty dt_1 \int_0^\infty dt_3$$
$$\exp(i\omega_m t_3 + i\omega_\tau t_1)R_{NR}(t_1,T_w,t_3) \quad (21)$$

V. Data Calculations Using the Response Functions with Exchange

The experimental 2D spectra are fit using the response functions with the exchange rate and FFCF as adjustable parameters. The entire region of 2D spectra including the 0−1 and 1−2 portions were calculated and compared to the experiments. Necessary input parameters, that is, the ratio of the transition dipole matrix elements for the complex and free form, the steady-state ratio of the complex and free populations (equilibrium constant), the vibrational lifetimes, and the orientational relaxation rates, were determined from linear-IR and pump−probe experiments.

For each of the three complex systems, phenol with benzene, p-xylene, and bromobenzene, the same procedures were employed. Each system contained phenol as the solute and one of the three complex partners as a mixed solvent with CCl_4. The methodology is discussed here for the phenol. All of the systems were treated in the same manner.

Pump−probe measurements were performed on three different samples, phenol in the mixed solvent (benzene/CCl_4), phenol in pure CCl_4, and phenol in pure benzene (Table 1). The pump−probe spectrum of phenol in benzene/CCl_4 is used for obtaining the properly "phased" 2D vibration echo spectra by employing the projection slice theorem.[35,36] The measurements of the pump−probe decays on phenol in the two pure solvents were used to obtained the vibrational lifetimes and orientational relaxation rates (Table 1). With benzene as the solvent, the equilibrium is shifted to virtually all complex, and the pump−probe experiment gives the vibrational lifetime of the complex. In CCl_4 there is no complex, and the lifetime of the free species is obtained. The mixed solvent may have a small effect on the lifetimes, but the calculation is not highly sensitive to small uncertainties in the lifetimes. The orientational relaxation obtained from polarized pump−probe experiments on the two pure solvents gives the orientational relaxation rates for the complex and the free form. These were corrected for the change in viscosity in going to the mixed solvent using the Debye−Stokes−Einstein equation, $\tau_r = V_{eff}\eta/k_BT$, where V_{eff} is the effective volume, η is the viscosity, k_B is Boltzmann's constant, and T is the temperature. The viscosity of each pure and mixed

TABLE 1: Vibrational Lifetimes (T_1) and Orientational Relaxation Time Constants (τ_r) of Phenol in Various Pure Solvents Measured with the Pump−Probe Experiment[a]

solvent	T_1 (ps)	τ_r (ps)
CCl_4	12.5	2.9
benzene	10	3.4
p-xylene	9.2	5.0
bromobenzene	10	3.1

[a] In CCl_4, phenol is not complexed. In the other three solvents, it is a complex with a solvent molecule. The τ_r values used in the analysis were corrected for the changes in viscosity from the pure solvents to the mixed solvents. See Table 2 for values used in calculations.

TABLE 2: Constants Used in the Response Function Calculations[a]

solvent	phenol species	T_1 (ps)	τ_r (ps)	μ_c/μ_f	[complex]/ [free]
p-xylene/CCl_4	free	12.5	2.9	1.6	0.83
	complex	9.2	4.3		
benzene/CCl_4	free	12.5	2.9	1.5	0.8
	complex	10	3.4		
bromobenzene/CCl_4	free	12.5	2.9	1.23	1.32
	complex	10	3.1		

[a] The orientational relaxation times were corrected from the measured values (Table 1) for the changes in viscosity in the mixed solvents.

solvent was measured at the experimental temperature (24 °C). The values used in the response function calculations are given in Table 2. These experimentally determined lifetimes and orientational relaxation rates were used without adjustment in the response function calculations. The transition dipole matrix elements for the two species were determined by measuring the absorption spectra in the two pure solvents for a known concentration of phenol. Once the ratio of the transition dipole matrix elements was known (Table 2), it was used to analyze the spectrum in the mixed solvent in which the complex peak and the free peak have approximately the same amplitudes (Figure 3a). The spectra were fit, and using the transition dipole matrix element ratio, the equilibrium constant and therefore the ratio of the concentrations of the two species were determined (Table 2).

To determine the exchange rate and the FFCF, the entire 2D vibrational echo spectrum was fit as follows. It is important to note that the exchange rate and the FFCF are relatively independent. The exchange rate determines the growth of the off-diagonal peaks and contributes to the decay of the diagonal peaks. The FFCF determines the time-dependent shape of the diagonal peaks. However, the rate of exchange also has an influence on the shape of the diagonal peaks, but the FFCF has no influence on the growth of the off-diagonal peaks. Various functional forms of the FFCF were tested, and it was determined that a biexponential function was sufficient to reproduce the data.

$$\langle \delta\omega(t)\delta\omega(0) \rangle = \Delta_0^{\,2} \exp(-t/\tau_0) + \Delta_1^{\,2} \exp(-t/\tau_1) \quad (22)$$

The biexponential FFCF includes a slow component (>1 ps) and a fast component (<1ps). These constraints were implemented in the fitting routine with the amplitudes and decay constants allowed to float. First, the 2D spectrum for each T_w point was fit separately. The parameters were iterated to minimize the residuals. Two-dimensional matrices that contain the intensity of each (ω_τ,ω_m) point from the experiment and the calculation are compared. A nonlinear multivariable fitting

20008 *J. Phys. Chem. B, Vol. 110, No. 40, 2006*

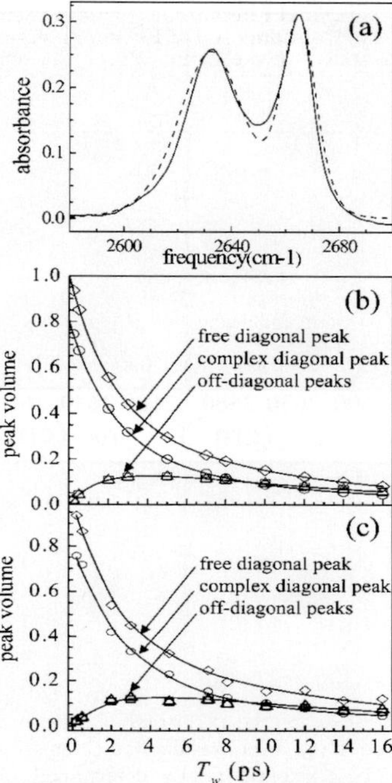

TABLE 3: Frequency–Frequency Correlation Function (FFCF) Parameters (Eq 22) and Complex Dissociation Times, $\tau_d = 1/k_{cf}$, for the Solute–Solvent Systems that Form the Complexes Phenol–p-Xylene, Phenol–Benzene, and Phenol–Bromobenzene[a]

complex		Δ_0 (rad/ps)	τ_0 (ps)	Δ_1 (rad/ps)	τ_1 (ps)	τ_d (ps)
phenol–p-xylene	free			1.1	2.3	21
	complex	3.0	0.10	1.7	2.4	
phenol–benzene	free			1.3	1.9	8
	complex	2.5	0.44	1.4	1.8	
phenol–bromobenzene	free			1.5	1.1	6
	complex	2.6	0.71	1.2	1.7	

[a] The FFCF parameters are for both the complex and the free phenol.

Figure 3. (a) Spectrum of the hydroxyl stretch of phenol-OD in the benzene/CCl$_4$ mixed solvent (solid curve) and the calculated spectrum (dashed curve). The high-frequency peak is the free phenol, and the low-frequency peak is the phenol–benzene complex. The calculated spectrum used the FFCF obtained by fitting the 2D vibrational echo spectra (Figure 1, bottom). (b) The points are peak volumes obtained from the response function calculation. The solid curves were obtained from fits using the peak volume method discussed in section VI. (c) The points are peak volumes obtained by fitting the 2D vibrational echo spectrum (Figure 1, top) using the peak volume method (section VI). The solid curves are fits to the peak volumes.

routine with a direct line search algorithm is used. To avoid false minima, many different initial conditions were used in the fitting program. As a check on the resulting parameters, the linear spectrum was calculated using the FFCF. This procedure was repeated for every data set with different T_w points. After optimized parameters for each T_w point were obtained, they were averaged to produce one parameter set. These averaged parameters were then used as the initial parameters for fitting the 2D spectra at all T_w values simultaneously. As a final test of the resulting parameters, the linear absorption line shape was calculated and compared to the data. The resulting complex dissociation times and FFCFs for the three systems studied are presented in the Table 3. The dissociation times are listed because this is the single parameter required to describe the chemical exchange process. The complex and the free form are in equilibrium. Therefore, the rate of complex dissociation is equal to the rate of complex formation, and the rate of complex dissociation can be characterized by the dissociation time.

In the top of Figure 1, data from a few of the T_w points are shown. The calculated 2D spectra for these points are shown in the bottom portion of the figure. Inspection of the two sets of figures shows that the calculation does a good job of reproducing the data. The response function calculations reproduce the experimental spectra including the growth of the

off-diagonal peaks and the tilt of the diagonal peaks at short T_w, which disappears as time progresses because of spectral diffusion. (An insufficient number of 2D spectra are shown at short times to see the progression of the spectral diffusion.)

Figure 3 provides additional insights as to the ability of the calculation to quantitatively reproduce the data. Figure 3a shows the absorption spectrum of the phenol-OD hydroxyl stretch along with the calculated spectrum obtained from the FFCF and eqs 15 and 16 with only a scaling factor as an adjustable parameter. Figure 3b shows calculations of the volumes of each of the four peaks (two diagonal and two off-diagonal) in the 0–1 region of the spectrum. The volumes were obtained by using the portions of the response functions that give rise to each peak individually and then integrating the resulting peak. These are the points in the figure. The calculated points for the two off-diagonal peaks fall on top of each other because the system is in thermal equilibrium. Therefore, the rate of complex formation is equal to the rate of complex dissociation. Tests showing that the systems are in thermal equilibrium have been presented.[1] The curves through the points were obtained previously[1] using the peak volume fitting method, which will be discussed in section VI. The agreement between the response function calculated points and the calculations that do not include spectral diffusion explicitly shows that the use of the more detailed response function method does not distort analysis of the chemical exchange dynamics.

Figures 4 and 5 display data and calculations for the phenol–p-xylene system, and Figures 6 and 7 display data and calculations for the phenol–bromobenzene system. The top portions of Figures 4 and 6 show the 2D spectra, and the bottom portions show the calculated 2D spectra. Figures 5a and 7a show the linear absorption spectra and the spectra calculated using the FFCF obtained from fitting the 2D spectra. Figures 5b and 7b show the response function method calculated peak volumes (points) and the curves obtained by using the peak volume fitting method (discussed in section VI). All of the procedures discussed in terms of the phenol–benzene system were applied in an identical manner to the other systems. In all cases the response function calculations do a good job of reproducing the 2D spectra and the linear spectra and agree with the less detailed peak volume method that gives only the chemical exchange dynamics.

The FFCF parameters obtained from the response function fits to the data are given in Table 3. For each solute–solvent system there are two species, the complex and the free phenol. For each system, the complex hydroxyl stretch dephasing is significantly different from the dephasing of the free phenol hydroxyl stretch. Although a biexponential form of the FFCF was used (eq 22), it was found that the fit for the data from the

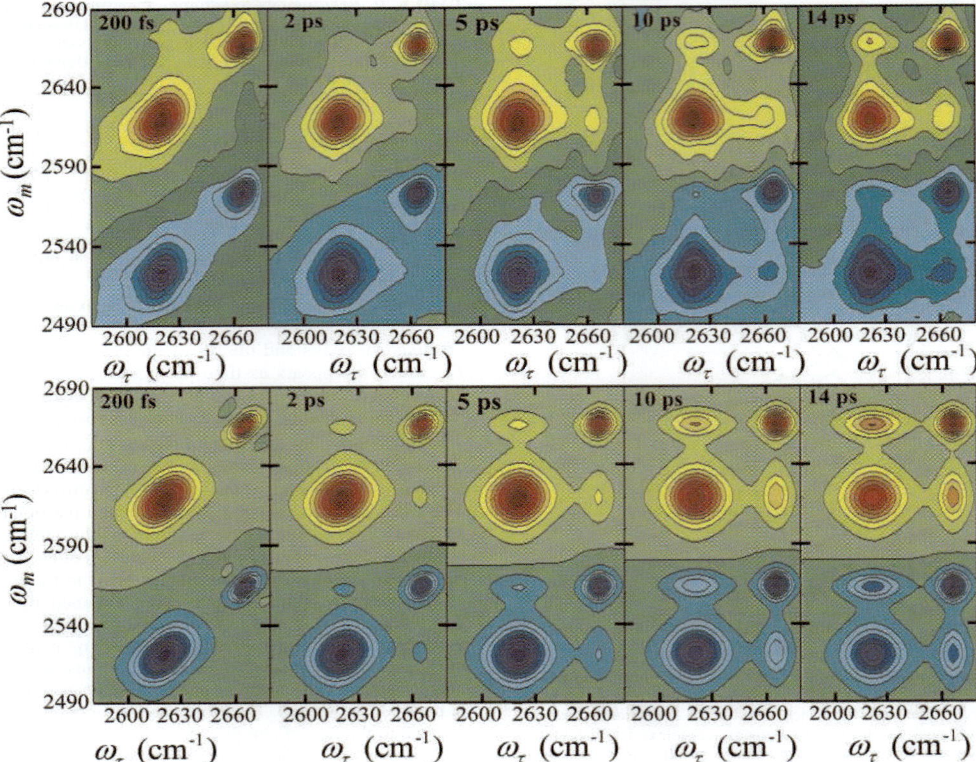

Figure 4. T_w-dependent 2D IR vibrational echo spectra (top) of the hydroxyl stretch of phenol-OD in *p*-xylene/CCl$_4$ mixed solvents (molar ratios of phenol/*p*-xylene/CCl$_4$ = 1:21:100). The red peaks (positive) are from the 0−1 vibration transitions, and the blue peaks (negative) are from the 1−2 vibrational transitions. At 200 fs, there are two peaks on the diagonal (red) and two peaks below these (blue) shifted by the anharmonicity. As T_w increases, additional peaks appear due to chemical exchange, that is, dissociation and formation of the phenol−*p*-xylene complex. The bottom portion displays response function calculations of the data.

free species converged to a single exponential. However, the data from the complex could not be fit well with a single-exponential FFCF. (It is important to note that these are single and biexponential FFCFs, both of which give rise to time-dependent observable broadening of the diagonal peaks along the ω_τ axis that is not exponential or biexponential.)

In each of the three systems, Δ_1 and τ_1 in Table 3 are similar for the complex and free forms. This similarity strongly suggests that this component of the FFCF arises from the effect of solvent fluctuation on the hydroxyl stretching frequency. These solvent fluctuations and their influence are not strongly dependent on complexation. However, only the complexed forms have the fast component, that is, Δ_0 and τ_0, which strongly suggests that this component is caused by fluctuation in the actual complex structure. The complex involves a weak π-hydrogen bond between the hydroxyl and the solvent aromatic ring. Electronic structure calculations have shown the structure of the phenol−benzene complex.[1] It would be expected that fluctuation of the complex structure would be a major source of vibrational dephasing.

The dissociation times listed in Table 3 decrease as the solute−solvent complex becomes weaker. In *p*-xylene, the methyl groups donate electron density to the benzene π-system, which results in a stronger complex than that with benzene. In bromobenzene, the bromo group withdraws electron density,

producing a weaker complex than that with benzene. These qualitative considerations are born out by measurements of the bond enthalpies, ΔH_0, of the complexes.[1] The ΔH_0 values extracted from van't Hoff plots were −1.21 kcal/mol for the phenol−bromobenzene complex, −1.67 kcal/mol for the phenol−benzene complex, and −2.23 kcal/mol for the phenol−*p*-xylene complex. Thus, as the bond enthalpy increases (stronger bond), the dissociation time τ_d also increases, because the free energy of activation for dissociation ($\Delta G^\ddagger$) would be expected to scale with the bond enthalpies.

It is also interesting to note that the fast component of the FFCF, τ_0, also appears to change with the change in the bond enthalpies. The Δ_0 values are very similar for the three complexes. This similarity means that the ranges of frequencies sampled because of fluctuations in the structures of the complexes are about the same. However, τ_0 becomes faster as the complex bond becomes stronger. The faster decay of this component of the FFCF with increasing bond strength might reflect a higher-frequency intermolecular quasi-vibration associated with the stronger complex.

VI. Comparison to the Peak Volume Only Fitting Method

Through the use of the response functions with the dynamic partition model, very good fits were obtained to the 2D spectra,

20010 *J. Phys. Chem. B, Vol. 110, No. 40, 2006*

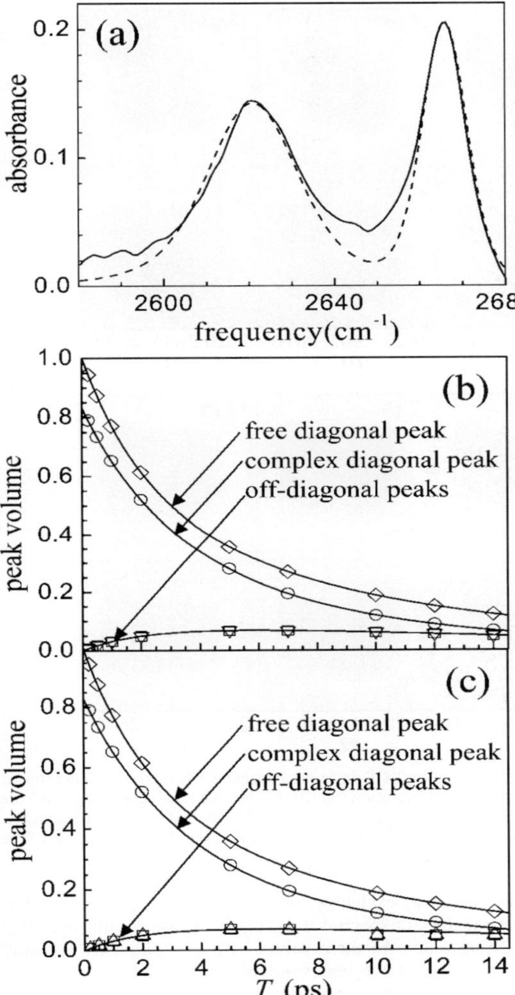

Figure 5. (a) Spectrum of the hydroxyl stretch of phenol-OD in the *p*-xylene/CCl$_4$ mixed solvent (solid curve) and the calculated spectrum (dashed curve). The high-frequency peak is the free phenol, and the low-frequency peak is the phenol$-p$-xylene complex. The calculated spectrum used the FFCF obtained by fitting the 2D vibrational echo spectra (Figure 4, bottom). (b) The points are peak volumes obtained from the response function calculation. The solid curves were obtained from fits using the peak volume method discussed in section VI. (c) The points are peak volumes obtained by fitting the 2D vibrational echo spectra (Figure 4, top) using the peak volume method (section VI). The solid curves are fits to the peak volumes.

and both the parameters for the FFCFs and the dissociation times were extracted from the data. Previously, a much simpler method was used to obtain the complex dissociation times for both phenol complexes[1] discussed here and complexes involving 2-methoxyphenol with several aromatic solvents.[2] This method provides no information on the spectral diffusion (FFCF) but uses the peak volumes and the dynamic partition model to extract the exchange kinetics. In the absence of any other process, spectral diffusion broadens the 2D peaks along the ω_τ

axis. However, the volumes of the peaks are preserved. When exchange, vibrational, and orientational relaxation are occurring in addition to spectral diffusion, the peak amplitudes cannot be used to determine the population kinetics because of the change in shapes of the peaks produced by spectral diffusion. It was proposed that if the peak volumes were fit as a function of $T_{\rm w}$, then the influence of spectral diffusion would be swept into the fit, and a detailed treatment, as presented here, was unnecessary to determine the exchange kinetics.[1] It is important to test this volume fitting approach because it is relatively simple to apply to obtain the exchange kinetics.

In the volume fitting method, the volume of each peak in the 2D IR spectrum was determined by fitting the entire 0$-$1 spectrum (or the full spectrum) composed of overlapping peaks. Each peak was approximated as a "tilted" two-dimensional Gaussian function

$$F(\phi,\omega_1,\omega_2,\sigma({\rm d}),\sigma({\rm a}),A) =$$

$$\sum_{i=1}^{4} A_i \frac{\exp(-((\cos\phi)(\omega_{\rm m} - \omega_j) + (\sin\phi)(\omega_\tau - \omega_j))^2)}{2(\sigma_i({\rm d}))^2} \times$$

$$\frac{\exp(-(-(\sin\phi)(\omega_{\rm m} - \omega_j) + (\cos\phi)(\omega_\tau - \omega_j))^2)}{2(\sigma_i({\rm a}))^2} \quad (23)$$

This function was used to reproduce the shape of each peak. Here, tilt angle (ϕ), line widths along the diagonal ($\sigma({\rm d})$) and antidiagonal ($\sigma({\rm a})$), and amplitude (A) are used as parameters for fitting the experimental data and obtaining the volume corresponding to each individual peak.

Each peak volume is corrected by the appropriate products of the transition dipole moments (μ_i^4 for the diagonal peaks and $\mu_i^2\mu_j^2$ for the off-diagonal peaks), and the largest volume peak at the shortest time is normalized to the corresponding solution of eq 11 at that time. All other peaks at any time have a volume relative to this normalized volume. Equations 11 for effective populations, $N_{\alpha\beta}(t)$, are fit to the volumes of the set of peaks at each $T_{\rm w}$. As with the response function calculations, the independently measured lifetimes, orientational relaxation rates, transition dipole moment ratio, and the equilibrium population ratio are used as fixed input parameters. The result is that there is a single adjustable parameter, the dissociation time, $\tau_{\rm d}$. All of the 2D spectra are fit simultaneously with this single parameter.

Figures 3c, 5c, and 7c show the results of using the peak volume fitting method. The identical complex dissociation times are obtained using the full response function calculations or the peak volume method. The curves in Figures 3b, 5b, and 7b are the same curves as those in Figures 3c, 5c, and 7c. In the parts b of the figures, the points were obtained from the response function calculations that fit the full 2D spectra including spectral diffusion. Therefore, if the chemical exchange kinetics are the sole interest, then these can be obtained without the complexity of using the response function approach but with the loss of information from the FFCFs. It is important to point out that to extract the chemical exchange kinetics with the peak volume fitting method quantitatively is still not simple. First, it is necessary to use the dynamic partition model (eqs 11) to account for orientational relaxation rates and vibrational lifetimes in addition to the chemical exchange. To reduce the input parameters, it is necessary to independently measure orientational relaxation rates, lifetimes, the transition dipole ratio, and the equilibrium population ratio. However, with these inputs

J. Phys. Chem. B, Vol. 110, No. 40, 2006 **20011**

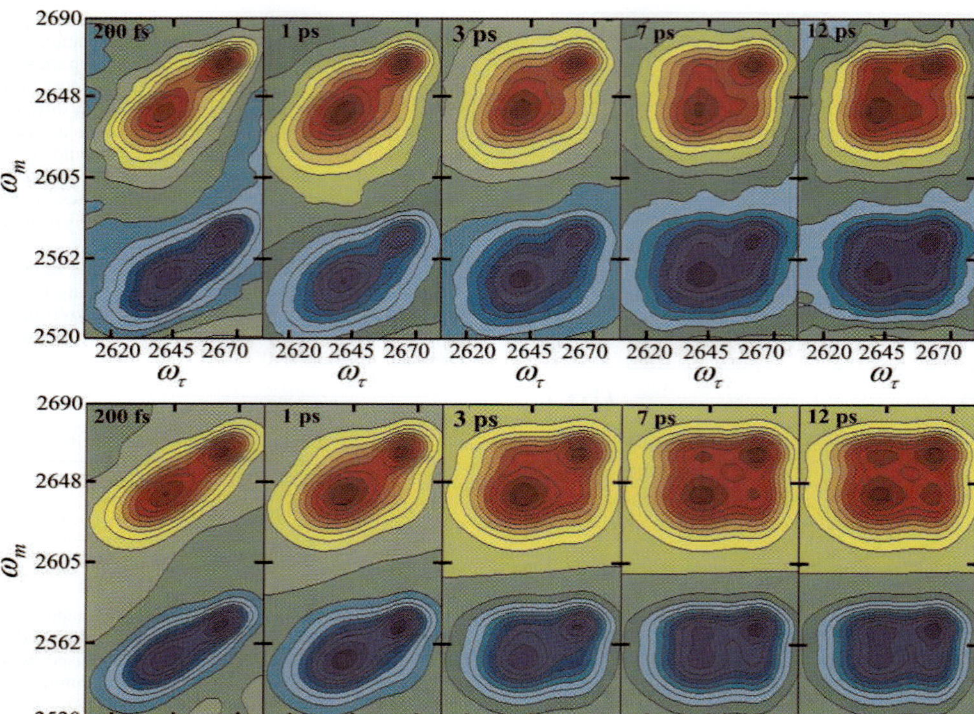

Figure 6. T_w-dependent 2D IR vibrational echo spectra (top) of the hydroxyl stretch of phenol-OD in bromobenzene/CCl$_4$ mixed solvents (molar ratio of phenol/bromobenzene/CCl$_4$ = 2:98:100). The red peaks (positive) are from the 0−1 vibration transitions, and the blue peaks (negative) are from the 1−2 vibrational transitions. At 200 fs, there are two peaks on the diagonal (red) and two peaks below these (blue) shifted by the anharmonicity. As T_w increases, additional peaks appear due to chemical exchange, that is, dissociation and formation of the phenol−bromobenzene complex. The bottom portion displays response function calculations of the data.

and the proper analysis, there is only one adjustable parameter, the dissociation time, τ_d.

One aspect of the volume fitting method is worth noting. The phenol−bromobenzene system has a linear spectrum in which the peaks are only slightly separated (Figure 7a). The result is that the diagonal and off-diagonal peaks in the 2D spectra have a great deal of overlap (Figure 6, top). The substantial overlap makes it tricky to extract the peak volumes in Figure 7c, particularly of the off-diagonal peaks. Nonetheless, there is sufficient accuracy to obtain the complex dissociation time with the same value as the full response function analysis. Note that the set of off-diagonal peak volumes extracted from the full response function calculations actually falls on the curve obtained by the peak volume fitting method (Figure 7b) better than points obtained by straight peak volume fitting (Figure 7c). Therefore, for systems such as bromobenzene, the full response function approach may give more accurate results in some instances.

VII. Concluding Remarks

Ultrafast 2D IR vibrational echo chemical exchange data were presented for the fast dissociation and formation of three organic solute−solvent complexes under thermal equilibrium conditions, and a detailed theoretical treatment of the 2D IR vibrational

echo chemical exchange observables was presented. The experimental 2D spectral data, taken on the phenol hydroxyl stretching mode for three solute−solvent complexes (phenol/benzene, phenol/p-xylene, and phenol/bromobenzene), were analyzed using the theory. The theory includes the important dynamical processes of orientational relaxation, vibrational lifetime, and spectral diffusion in addition to the chemical exchange itself. The orientational relaxation, vibrational relaxation, and exchange process are introduced through the dynamic partition model (eq 11), which gives the kinetic equations for the effective population as a function of time. The effective population includes the decrease in signal caused by orientational relaxation. The difficulty in handling orientational relaxation and vibrational relaxation is caused by the fact that the two species undergo these relaxation processes at different rates. A species begins relaxation with certain rates, converts to the other species, and continues to relax with different rates. It then can revert to the initial species and relax further with the original rates and so on.

The analytical results of the dynamic partition model were then used with a time-dependent diagrammatic perturbation theory treatment to obtain analytical expressions for the response functions with exchange, spectral diffusion, orientational relaxation, and vibrational relaxation. Quantum pathways for

20012 *J. Phys. Chem. B, Vol. 110, No. 40, 2006*

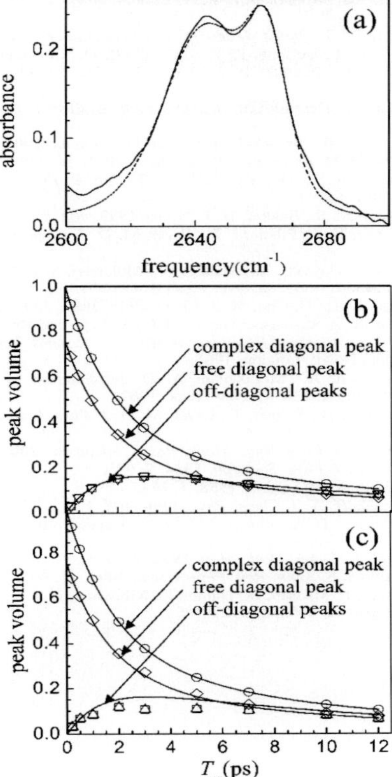

Figure 7. (a) Spectrum of the hydroxyl stretch of phenol-OD in the bromobenzene/CCl₄ mixed solvent (solid curve) and the calculated spectrum (dashed curve). The high-frequency peak is the free phenol, and the low-frequency peak is the phenol—bromobenzene complex. The calculated spectrum used the FFCF obtained by fitting the 2D vibrational echo spectra (Figure 6, bottom). (b) The points are peak volumes obtained from the response function calculations. The solid curves were obtained from fits using the peak volume method discussed in section VI. (c) The points are peak volumes obtained by fitting the 2D vibrational echo spectra (Figure 6, top) using the peak volume method (section VI). The solid curves are fits to the peak volumes.

diagonal peaks were divided into two classes, no exchange and multiple exchanges, because these two classes of pathways show different T_w-dependent line broadening. The line broadening of the portion of the diagonal peaks that undergo no exchange is determined by the spectral diffusion. However, the T_w-dependent line broadening from the pathways with multiple exchanges (an even number) is the result of the chemical exchange itself. To properly account for the T_w-dependent shape of the diagonal peaks and obtain the FFCF, it is necessary to separate the two contributions to the diagonal peak broadening.

Because a large number of the necessary input parameters were measured independently (Table 2), the only adjustable parameters in the calculations of the 2D spectra were the exchange rate and the FFCF. The calculations reproduced 2D IR vibrational echo spectra very well for all three species (Figures 1, 4, and 6), yielding exchange rates and the FFCF parameters of both the phenol complex and the free phenol for each of the three systems (Table 3). There is a marked difference in the spectral diffusion of the complex and the free species.

The free species reflects the influence of solvent fluctuations on the hydroxyl stretch frequency. For the complexes, there is an additional contribution from the relative motions of the phenol and its complex partner.

The results of the full response function calculations were compared to those of the previously employed peak volume fitting method. The peak volume method does not determine the spectral diffusion because it fits the time-dependent peak volumes without analyzing the change in shapes of the peaks. However, it can be used to extract the exchange rate from the 2D data but does not determine the FFCF. Comparisons to the full response function calculations show that the simpler to implement peak volume method is accurate and a reasonable approach to obtain the exchange rate information.

Acknowledgment. This research was supported by grants from the Air Force Office of Scientific Research (F49620-01-1-0018) and from the National Science Foundation (DMR-0332692).

Appendix

In eqs 11, the detailed solutions of dynamic partition model were given. Here, the various symbols used in the equation are defined.

$$\alpha = \frac{D_c + D_f + k_c + k_f + k_{cf} + k_{fc}}{2}$$

$$\beta = \frac{\sqrt{(D_c + D_f + k_c + k_f + k_{cf} + k_{fc})^2 - 4(D_cD_f + D_fk_c + D_fk_{cf} + D_ck_f + k_ck_f + k_{cf}k_f + D_ck_{fc} + k_ck_{fc})}}{2}$$

$$\gamma = \frac{(D_c + D_f + k_c + k_f + k_{cf} + k_{fc}) - 2(D_f + k_f + k_{fc})}{\sqrt{(D_c + D_f + k_c + k_f + k_{cf} + k_{fc})^2 - 4(D_cD_f + D_fk_c + D_fk_{cf} + D_ck_f + k_ck_f + k_{cf}k_f + D_ck_{fc} + k_ck_{fc})}}$$

$$\phi = \frac{k_c + k_{cf} + k_f + k_{fc}}{2}$$

$$\varphi = \frac{\sqrt{(k_c + k_{cf} + k_f + k_{fc})^2 - 4(k_ck_f + k_{cf}k_f + k_ck_{fc})}}{2}$$

$$\theta = \frac{(k_c + k_{cf} + k_f + k_{fc}) - 2(k_f + k_{fc})}{\sqrt{(k_c + k_{cf} + k_f + k_{fc})^2 - 4(k_ck_f + k_{cf}k_f + k_ck_{fc})}}$$

References and Notes

(1) Zheng, J.; Kwak, K.; Asbury, J. B.; Chen, X.; Piletic, I.; Fayer, M. D. *Science* **2005**, *309*, 1338.

(2) Zheng, J.; Kwak, K.; Chen, X.; Asbury, J. B.; Fayer, M. D. *J. Am. Chem. Soc* **2006**, *128*, 2977.

(3) Schneider, H.-J.; Yatsimirsky, A. K. *Principles and Methods in Supramolecular Chemistry*; John Wiley & Sons: New York, 2000.

(4) Meyer, E. A.; Castellano, R. K.; Diederich, F. *Angew. Chem., Int. Ed.* **2003**, *42*, 1210.

(5) Vinogradov, S. N.; Linnell, R. H. *Hydrogen Bonding*; Van Nostrand Reinhold: New York, 1971.

(6) Reichardt, C. *Solvents and Solvent Effects in Organic Chemistry*; Wiley-VCH: Weinheim, Germany, 2003.

(7) Fayer, M. D.; Zheng, J.; Kwak, K.; Asbury, J. B. Molecular Dynamics/Theoretical Chemistry Meeting, May, 2005, Monterey, California.

(8) Kim, Y. S.; Hochstrasser, R. M. *Proc. Natl. Acad. Sci. U.S.A.* **2005**, *102*, 11185.

(9) Woutersen, S.; Mu, Y.; Stock, G.; Hamm, P. *Chem. Phys.* **2001**, *266*, 137.

(10) Dunand, F. A.; Helm, L.; Merbach, A. E. *Adv. Inorg. Chem.* **2003**, *54*, 1.

(11) Strehlow, H. *Rapid Reactions in Solutions*; VCH Publishers: New York, 1992.

(12) Wood, K. A.; Strauss, H. L. *J. Phys. Chem.* **1990**, *94*, 5677.

(13) Wood, K. A.; Strauss, H. L. *Ber. Bunsen-Ges.* **1989**, *93*, 615.

(14) Levinger, N. E.; Davis, P. H.; Behera, P.; Myers, D. J.; Stromberg, C.; Fayer, M. D. *J. Chem. Phys.* **2003**, *118*, 1312.

(15) Jeener, J.; Meier, B. H.; Bachmann, P.; Ernst, R. R. *J. Chem. Phys.* **1979**, *71*, 4546.

(16) Meier, B. H.; Ernst, R. R. *J. Am. Chem. Soc.* **1979**, *101*, 6441.

(17) Palmer, A. G. *Chem. Rev.* **2004**, *104*, 3623.

(18) Demirdoven, N.; Khalil, M.; Golonzka, O.; Tokmakoff, A. *J. Phys. Chem. A* **2001**, *105*, 8030.

(19) Zanni, M. T.; Hochstrasser, R. M. *Curr. Opin. Struct. Biol.* **2001**, *11*, 516.

(20) Golonzka, O.; Khalil, M.; Demirdoven, N.; Tokmakoff, A. *Phys. Rev. Lett.* **2001**, *86*, 2154.

(21) Merchant, K. A.; Thompson, D. E.; Fayer, M. D. *Phys. Rev. Lett.* **2001**, *86*, 3899.

(22) Volkov, V.; Schanz, R.; Hamm, P. *Opt. Lett.* **2005**, *30*, 2010.

(23) Zanni, M. T.; Gnanakaran, S.; Stenger, J.; Hochstrasser, R. M. *J. Phys. Chem. B* **2001**, *105*, 6520.

(24) Asbury, J. B.; Steinel, T.; Stromberg, C.; Gaffney, K. J.; Piletic, I. R.; Goun, A.; Fayer, M. D. *Phys. Rev. Lett.* **2003**, *91*, 237402.

(25) Steinel, T.; Asbury, J. B.; Corcelli, S. A.; Lawrence, C. P.; Skinner, J. L.; Fayer, M. D. *Chem. Phys. Lett.* **2004**, *386*, 295.

(26) Asbury, J. B.; Steinel, T.; Fayer, M. D. *Chem. Phys. Lett.* **2003**, *381*, 139.

(27) Zheng, J.; Kwak, K.; Steinel, T.; Asbury, J. B.; Chen, X.; Xie, J.; Fayer, M. D. *J. Chem. Phys.* **2005**, *123*, 164301.

(28) Khalil, M.; Demirdoven, N.; Tokmakoff, A. *J. Phys. Chem. A.* **2003**, *107*, 5258.

(29) Kim, Y.; Hochstrasser, R. M. *J. Phys. Chem. B.* **2005**, *109*, 6884.

(30) Mukamel, S. *Principles of Nonlinear Optical Spectroscopy*; Oxford University Press: New York, 1995.

(31) Knee, J. L.; Khundkar, L. R.; Zewail, A. H. *J. Chem. Phys.* **1987**, *87*, 115.

(32) Perutz, M. F. *Philos. Trans. R. Soc. London, Ser. A* **1993**, 105.

(33) Trebino, R.; DeLong, K. W.; Fittinghoff, D. N.; Sweetser, J. N.; Krumbugel, M. A.; Richman, B. A.; Kane, D. J. *Rev. Sci. Instrum.* **1997**, *69*, 3277.

(34) Khalil, M.; Demirdoven, N.; Tokmakoff, A. *Phys. Rev. Lett.* **2003**, *90*, 047401.

(35) Asbury, J. B.; Steinel, T.; Fayer, M. D. *J. Lumin.* **2004**, *107*, 271.

(36) Jonas, D. M. *Annu. Rev. Phys. Chem.* **2003**, *54*, 425.

(37) Woutersen, S.; Mu, Y.; Stock, G.; Hamm, P. *Proc. Natl. Acad. Sci. U.S.A.* **2001**, *98*, 11254.

(38) Woutersen, S.; Bakker, H. J. *Nature* **1999**, *402*, 507.

(39) Gaffney, K. J.; Piletic, I. R.; Fayer, M. D. *J. Chem. Phys.* **2003**, *118*, 2270.

(40) Jasen, J.; Hayashi, T.; Zhuang, W.; Mukamel, S. *J. Chem. Phys.* **2005**, *123*, 114504.

(41) Kwac, K. L.; H. Cho, M. *J. Chem. Phys.* **2004**, *120*, 1477.

(42) Kubo, R. A Stochastic Theory of Line-Shape and Relaxation. In *Fluctuation, Relaxation and Resonance in Magnetic Systems*; Ter Haar, D., Ed.; Oliver and Boyd: London, 1961.

(43) Carrington, A.; McLachlan, A. D. *Introduction to Magnetic Resonance*; Harper and Row: New York, 1967.

(44) Asbury, J. B.; Steinel, T.; Fayer, M. D. *J. Phys. Chem. B.* **2004**, *108*, 6544.

(45) Bain, A. D. *Prog. Nucl. Magn. Reson. Spectrosc.* **2003**, *43*, 63.

(46) Kubo, R. *J. Phys. Soc. Jpn.* **1954**, *9*, 935.

(47) Sack, R. A. *Mol. Phys.* **1958**, *1*, 163.

(48) Tokmakoff, A. *J. Chem. Phys.* **1996**, *105*, 1.

(49) McQuarrie, D. A. *Statistical Mechnics*; Haper-Collins: New York, 1973.

(50) Putzer, E. J. *Am. Math. Mon.* **1966**, *73*, 2.

(51) Mukamel, S. *Annu. Rev. Phys. Chem.* **2000**, *51*, 691.

(52) Stenger, J.; Madsen, D.; Hamm, P.; Nibbering, E. T. J.; Elsaesser, T. *J. Phys. Chem. A.* **2002**, *106*, 2341.

Ultrafast Carbon-Carbon Single-Bond Rotational Isomerization in Room-Temperature Solution

Junrong Zheng, Kyungwon Kwak, Jia Xie, M. D. Fayer

Generally, rotational isomerization about the carbon-carbon single bond in simple ethane derivatives in room-temperature solution under thermal equilibrium conditions has been too fast to measure. We achieved this goal using two-dimensional infrared vibrational echo spectroscopy to observe isomerization between the gauche and trans conformations of an ethane derivative, 1-fluoro-2-isocyanato-ethane (**1**), in a CCl_4 solution at room temperature. The isomerization time constant is 43 picoseconds (ps, 10^{-12} s). Based on this value and on density functional theory calculations of the barrier heights of **1**, n-butane, and ethane, the time constants for n-butane and ethane internal rotation under the same conditions are $\sim$40 and $\sim$12 ps, respectively.

Many molecules can undergo rotational isomerization around one or more of their chemical bonds. During the course of isomerization, a molecule exchanges between relatively stable conformations by passing through unstable configurations. Rotational isomerization is a major factor in the dynamics, reactivity, and biological activity of a multiplicity of molecular structures. Ethane and its derivatives are textbook examples of molecules that undergo this type of isomerization. (*1*) In ethane, as one of the two methyl groups rotates 360° around the central carbon-carbon single bond, it will alternate three times between an unstable eclipsed conformation and the

Department of Chemistry, Stanford University, Stanford, CA 94305, USA.

preferred staggered conformation. The transition from one staggered state to another leaves ethane structurally identical. Therefore, the result of ethane isomerization cannot be observed through a change in chemical structure. In a 1,2-disubstituted ethane derivative, the molecule can undergo a similar isomerization. However, a 1,2-disubstituted ethane has two distinct staggered conformations, gauche and trans (anti), and two eclipsed conformations, anticlinal and synperiplanar, because of the distinguishing characteristics of the relative positions of the two substituents (Fig. 1A) (1). The isomerization is frequently referred to as "hindered internal rotation" (2). It has been the subject of intense theoretical and experimental study since Bischoff found 100 years ago that rotation about the C-C single bond in ethane is not completely free. (2)

The trans-gauche isomerization of 1,2-disubstituted ethane derivatives, such as n-butane, is one of the simplest cases of a first-order chemical reaction. This type of isomerization has served as a basic model for modern chemical reaction kinetic theory and molecular dynamics (MD) simulation studies in condensed phases of matter (3–8). In spite of extensive theoretical investigation, no corresponding kinetic experiments have been performed to test the results, partially because of the low rotational energy barrier of the n-butane (~3.4 kcal/mol) and of other simple 1,2-disubstituted ethane derivatives (9). According to theoretical studies (3–8), the isomerization time scale (1/k, where k is the rate constant) is 10 to 100 ps at room temperature in liquids.

The room-temperature time scale is much shorter than the microsecond and longer–time scale measurements that can be made with dynamic nuclear magnetic resonance (DNMR) spectroscopy, a widely used method for studying slow temperature-dependent isomerization kinetics (10). The picosecond kinetics at room temperature cannot be deduced from microsecond or millisecond dynamics at low temperature because the rate constant is not a simple function of temperature over a wide temperature range

(11). Thus, DNMR does not afford accurate estimates of isomerization rates at room temperature for molecules with small barriers, such as ethane, that rotate in tens of picoseconds. Other methods to study fast isomerization dynamics under thermal equilibrium, such as linear infrared (IR) and Raman line shape analysis (12, 13), are hampered by multiple contributing factors apart from isomerization (14–16).

Two-dimensional (2D) IR vibrational echo chemical exchange spectroscopy has recently proven useful for studying fast dynamical processes under thermal equilibrium conditions (17–19). We applied this method to study the ultrafast trans-gauche isomerization dynamics of a simple 1,2-disubstituted ethane derivative, 1-fluoro-2-isocyanato-ethane (1), at room temperature in liquid solution. The experiments were performed by observing the time dependence of the 2D spectrum of the isocyanate group's antisymmetric stretching mode.

Details of the methodology of 2D IR vibrational echo spectroscopy have been described previously (20). Very briefly, in a 2D IR vibrational echo experiment, three ultrashort IR pulses tuned to the frequency of the vibrational modes of interest are crossed in the sample. Because the pulses are very short, they have a broad bandwidth that makes it possible to simultaneously excite a number of vibrational modes. The first laser pulse "labels" the initial structures of the species in the sample by setting their initial frequency, ω_τ. The second pulse ends the first time period τ and starts clocking the reaction time period T_w during which the labeled species undergo isomerization and other population dynamics changes such as vibrational relaxation to the ground state and orientational relaxation of the entire molecule. The third pulse ends the population dynamics period of length T_w and begins a third period of length $\leq\tau$, which ends with the emission of the vibrational echo pulse of frequency ω_m. The vibrational echo signal reads out information about the final structures of all labeled species by their frequency ω_m.

There are two types of time periods in the experiment. The times between pulses 1 and 2 and between pulse 3 and the echo pulse are called coherence periods. During these periods, the vibrations are in coherent superpositions of two vibrational states. Fast vibrational oscillator frequency fluctuations induced by fast structural fluctuation of the system cause dynamic dephasing, which is one contribution to the line shapes in the conventional 1D absorption spectrum. During the period T_w between pulses 2 and 3, called the population period, a vibration is in a distinct eigenstate rather than a superposition state. Slow structural fluctuations of the system, termed spectral diffusion, contribute to the 2D line shapes. Other processes during the population period, particularly chemical exchange, also produce changes in the 2D spectrum. Chemical exchange occurs when two species in equilibrium interconvert without changing the overall number of either species. Isomerization back and forth between gauche and trans conformations of 1 is a type of chemical exchange. In other contexts, it has been demonstrated that chemical exchange causes new peaks to grow in as T_w is increased (17–19). In our experiments, the growth of off-diagonal peaks in the 2D vibrational echo spectrum of 1 with increasing T_w was used to extract the gauche-trans isomerization rate.

The calculated structures and potential energy of 1 as it undergoes rotational isomerization about the central carbon-carbon single bond (Fig. 1, A and B) were obtained using density functional theory (DFT) (21) at the B3LYP level and 6-31+G(d,p) basis set for isolated molecules. The energy values were corrected for the zero point energy. The gauche-trans isomerization has two possible transition states: the anticlinal conformer, where the F atom is eclipsed by the H atom; and the synperiplanar conformer, where the F atom is eclipsed by the N atom. Calculations show that the anticlinal conformer is the transition state because it has a much lower energy (3.3 kcal/mol) than the synperiplanar conformer (>7 cal/mol). From the cal-

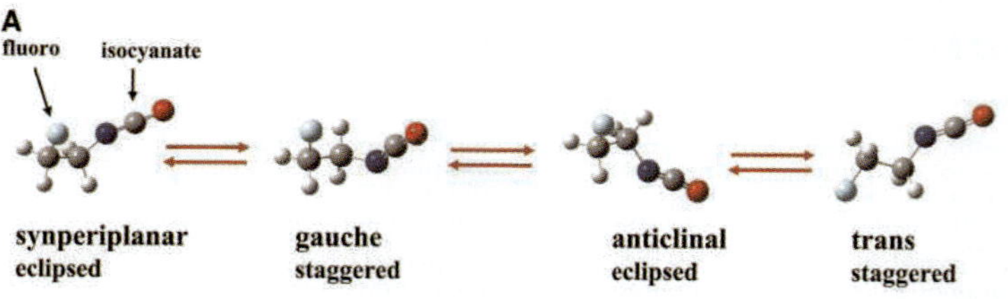

Fig. 1. (A) Calculated structures for two eclipsed conformations (anticlinal and synperiplanar) and two staggered conformations [gauche and trans (or anti)] of **1**. **(B)** Calculated energy of **1** undergoing isomerization around the central carbon-carbon single bond. The values are corrected for zero point energy and apply to isolated molecules (no solvent). The DFT calculations show that the C-C single bond rotational isomerization has an internal barrier (~3.3 kcal/mol). The CN motion contributes only slightly to the energy of the configurational change. Thus, the exchange rate between the gauche and trans conformers is essentially the C-C single-bond rotational isomerization rate. The calculations also demonstrate that during exchange between the gauche and trans conformers, the molecule passes through a transition state, the eclipsed conformation (anticlinal), that is similar to the one calculated for ethane isomerization.

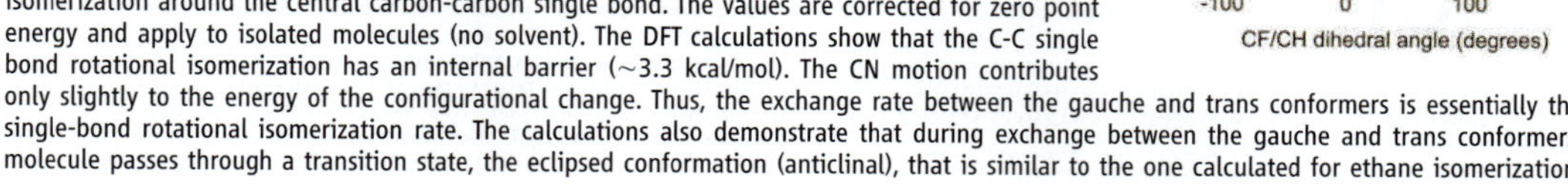

culations for the isolated molecule (no solvent), the gauche conformer is about 0.5 kcal/mol more stable than the trans conformer.

The gauche and trans conformers have different geometries and intramolecular interactions, resulting in different vibrational frequencies for the antisymmetric stretching mode of the isocyanate group (NCO) and different dipole moments. Calculations show that the trans conformer has an NCO stretching frequency $\sim$15 cm^{-1} higher than the gauche conformer. In the Fourier transform infrared (FTIR) spectroscopy spectrum of the NCO antisymmetric stretching mode of **1** in CCl$_4$ at room temperature, there are two peaks of similar intensity (Fig. 2). Based on the calculations, the peak at 2280 cm^{-1} is assigned to the trans conformer and the one at 2265 cm^{-1} to the gauche. The population ratio of trans/gauche (the equilibrium constant) is $\sim$1:1, obtained by analyzing both the 1D and 2D IR data (*19*) (fig. S1). The population ratio was experimentally determined to be temperature- and solvent polarity–dependent, demonstrating that two equilibrated species exist in the system. However, the linear IR spectrum cannot provide information about the isomerization kinetics.

Figure 3A displays six T_w-dependent 2D IR spectra of **1** in a CCl$_4$ solution at room temperature. The 0-fs panel corresponds to the shortest T_w, at which negligible isomerization has occurred. As discussed in detail previously (*17*), when no isomerization has occurred, the initial and final structures of each labeled species in the sample are unchanged. Therefore, the ω_τ and ω_m values of each peak are identical, and the peaks appear only on the diagonal. The two peaks representing the gauche ($\omega_m =$ 2265 cm^{-1}) and trans ($\omega_m =$ 2280 cm^{-1}) conformers are clearly visible on the diagonal. After a long reaction period ($T_w =$ 25 ps), isomerization has proceeded to a substantial degree. The obvious change is the additional peak that has appeared at the upper left ($\omega_\tau =$ 2265 cm^{-1} and $\omega_m =$ 2280 cm^{-1}). This peak arises from gauche-to-trans isomerization. There is a corresponding peak to the lower right that is

generated by trans-to-gauche isomerization, but it is somewhat negated by a negative-trending peak produced by population relaxation (*22*) between the antisymmetric isocyanate mode and another mode (fig. S1). The negative-trending (blue) peaks due to population relaxation are discussed further below and in the supporting material. They are included in the detailed fitting of the data. The diagonal peaks arise from molecules with the same initial and final structures; that is, molecules that either did not undergo isomerization during the time period T_w or else underwent multiple isomerization cycles that left them in their initial conformation at the end of the T_w period. The growth of the off-diagonal peaks with increasing T_w permits determination of the isomerization rate (chemical exchange rate), although it is necessary to analyze the growth and decay of all the peaks for accuracy (*19*).

During the T_w period, other factors besides chemical exchange also influence the 2D spectrum. These phenomena include spectral diffusion, orientational relaxation, and vibrational relaxation (*17*). Spectral diffusion changes the shape of each peak, and the orientational and vibrational relaxations cause all peaks to decrease in amplitude. Only chemical exchange produces growing off-diagonal positive-trending (red) peaks for the two distinct species. There is

an additional vibrational relaxation pathway, distinct from the regular vibrational lifetime decay, that stems from the coupling of the NCO antisymmetric stretch of both conformers to a vibrational mode of unassigned nature at $\sim$2230 cm^{-1} (fig. S1). The coupling induces fast back-and-forth population equilibration between the unassigned mode and the NCO antisymmetric stretch [equilibration time constants are 0.9 ps for the trans conformation and 1.9 ps for the gauche conformation (fig. S2)]. The equilibration via vibrational relaxation also produces additional negative (blue) peaks (*22*) just below each positive (red) peak. As shown at the very bottom of each panel in Fig. 3A, two blue peaks at low frequency along ω_m grow in with T_w. The off-diagonal red peak at $\omega_\tau =$ 2280 cm^{-1} and $\omega_m =$ 2265 cm^{-1} (lower right) is smaller than the upper left off-diagonal peak because of canceling overlap with a negative vibrational relaxation blue peak at approximately $\omega_\tau =$ 2280 cm^{-1} and $\omega_m =$ 2270 cm^{-1}.

Assignment of the growing positive (red) off-diagonal peaks to isomerization, with the one at the lower right offset by an overlapping negative-trending (blue) peak, can be confirmed by examining 1-bromo-2-isocyanato-ethane. The bromo group is so large that it generates substantial steric hindrance in the eclipsed form

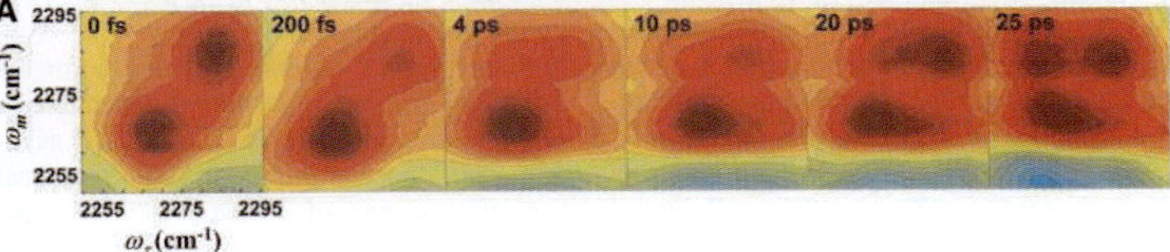

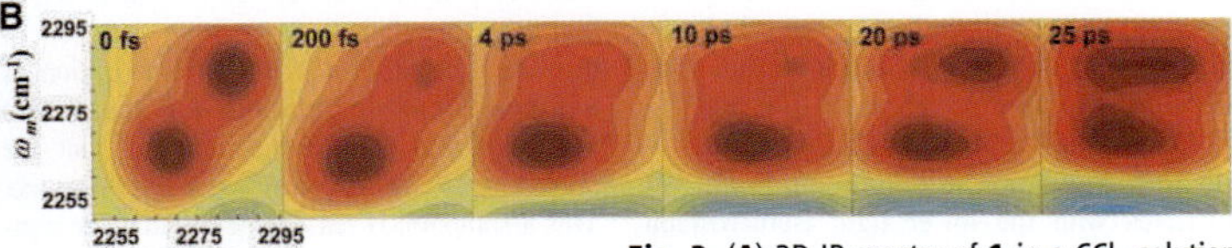

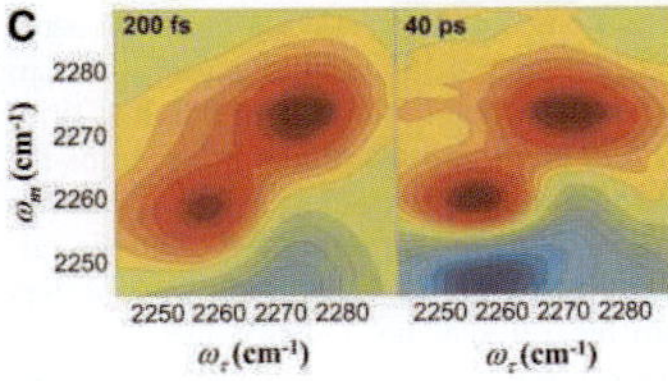

Fig. 3. (**A**) 2D IR spectra of **1** in a CCl$_4$ solution at room temperature. The data have been normalized to the largest peak at each T_w. Each contour is a 10% change. The red contours are positive-trending and the blue contours are negative-trending. At short times, only two diagonal red peaks appear, representing the gauche (lower frequency) and trans (higher frequency) conformers. At long times, isomerization causes two additional red peaks on the off-diagonal to grow in. The peak at the upper left is larger than the peak at the lower right, because the lower right peak overlaps with a negative-trending peak (Fig. 3C). (**B**) Calculated 2D IR spectra using the model and procedures discussed in the text and in the supporting online material. The agreement between the experiments and the calculated spectra is very good. (**C**) 2D IR spectra of 1-bromo-2-isocyanato-ethane in a CCl$_4$ solution at room temperature. The bulky bromo group prevents isomerization from occurring past $T_w =$ 40 ps. All other aspects of the system are the same as for **1**. The results show that the positive-trending (red) off-diagonal peaks do not grow in without isomerization. The negative trending peak to the lower right that interferes with the positive-trending isomerization peak in Fig. 3A is apparent.

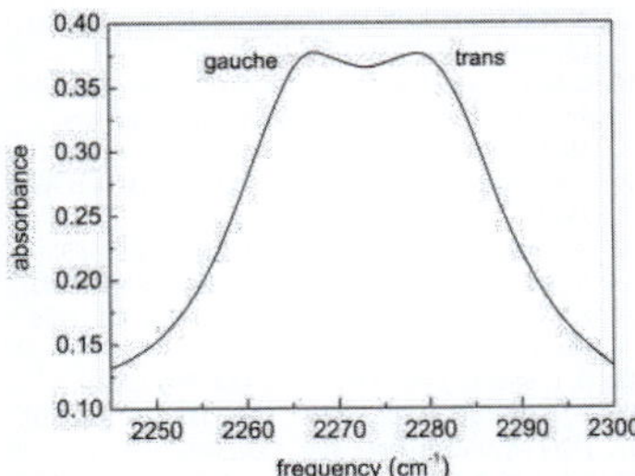

Fig. 2. FTIR spectrum of **1** in a CCl$_4$ solution at room temperature. Calculations assign the peak at $\sim$2280 cm^{-1} to the trans conformer and the peak at $\sim$2265 cm^{-1} to the gauche conformer.

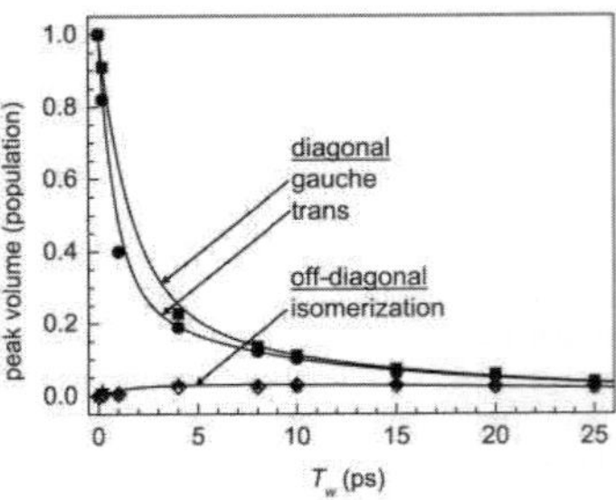

Fig. 4. Data (points) and calculated curves for the 2D spectra, some of which are shown in Fig. 3A. The points are the T_w-dependent peak volumes for the two diagonal and two off-diagonal peaks. The solid curves are all calculated with a single adjustable parameter, the isomerization time constant. The agreement is very good and yields an isomerization time constant of 43 ps. Input parameters of the model, all measured experimentally, are the equilibrium constant $K_{eq} = 1$; orientational relaxation time constants $\tau_T = 3.2$ ps (T, trans) and $\tau_G = 3.4$ ps (G, gauche); and vibrational relaxation time constants $T_T^{fast} = 0.91$ ps with normalized amplitude 0.63, $T_T^{slow} = 18.8$ ps with normalized amplitude 0.37, $T_G^{fast} = 1.95$ ps with normalized amplitude 0.6, and $T_G^{slow} = 18.4$ ps with normalized amplitude 0.4.

and greatly raises the barrier for isomerization. On the time scale of interest here, isomerization does not occur. All other aspects of the system are the same, including the vibrational population equilibration with the unassigned peak. Figure 3C displays 2D IR spectra for a 1-bromo-2-isocyanato-ethane/CCl$_4$ solution at room temperature at short and long T_w. At 40 ps, no positive-trending (red) off-diagonal peaks have appeared, indicating that the isomerization time constant is much greater than 100 ps. In the absence of the positive-trending off-diagonal isomerization peaks, the negative-trending population equilibration peak that interferes with the lower right isomerization peak in the spectrum of **1** is clearly visible.

To quantitatively model the 2D spectra and extract the kinetic parameters from the 2D IR data, a modified version of the kinetic model described previously was used (*17, 19*). The trans and gauche conformers undergo isomerization, which results in chemical exchange in the spectra. The amplitudes of the signals for each species decay because of orientational relaxation, fast vibrational relaxation (vibrational equilibration), and slower vibrational lifetime decay to the ground state (fig. S3). The kinetic model requires as inputs the orientational relaxation times and the fast (equilibration) and slow (decay to the ground state) vibrational relaxation times to model the data and obtain the isomerization rate constant. The orientational relaxation and vibrational relaxation time constants were measured with polarization-selective

pump-probe experiments (*19, 23*) (fig. S2). The equilibrium constant ($K_{eq} = 1$) and the vibrational transition dipole moment ratio (~ 1) were obtained by analyzing both the 1D IR spectrum and the 2D IR spectrum at $T_w = 0$ fs (fig. S1). (*19*) The time-dependent populations are provided by the peak volumes of the four red peaks in 2D IR spectra scaled with the transition dipole moment ratio. Therefore, only one unknown parameter, the isomerization rate constant $k_{TG} = k_{GT}$, was used in the fitting. The trans-to-gauche (TG) and gauche-to-trans (GT) rate constants are taken to be equal within experimental error because the equilibrium constant is 1.

Figure 3B shows calculated 2D spectra using the known input parameters and the results of fitting k_{TG}. Both the measured and calculated spectra are normalized by scaling to make the largest peak at each time equal to unity. Given the complexity of the system, the model calculations do a very good job of reproducing the time-dependent data. Of particular importance is the growth of the off-diagonal red peaks and the negative-trending peaks at the bottom of each panel. The data are well fit using the isomerization rate constant as the only adjustable parameter (Fig. 4). The off-diagonal peaks grow in at the same rate, consistent with $k_{TG} = k_{GT}$ within experimental error. The fits yield $1/k_{TG} = 1/k_{GT} = 43 \pm 10$ ps. The error bars arise from the uncertainty in the parameters that go into the calculations.

Based on the experimental results for 1-fluoro-2-isocyanato-ethane, it is possible to calculate approximately the gauche-trans isomerization rate of n-butane and the rotational isomerization rate of ethane under the same conditions used in this study (CCl$_4$ solution at room temperature, 297 K). We have analyzed n-butane because there is a large number of theoretical calculations for the isomerization of this molecule (*5–8*). Transition state theory (*11*) was employed, with the assumption that the prefactors for all of the systems are the same. This assumption is reasonable because the transition states and the barrier heights are quite similar for the three systems. We performed DFT calculations on all the systems using the same method [the B3LYP level and 6-31+ G(d,p) basis set] to obtain the barrier heights. With the zero point energy correction, the trans-to-gauche isomerization of n-butane has a barrier of 3.3 kcal/mol. The barrier for ethane is calculated to be 2.5 kcal/mol. This value differs from the 2.9 kcal/mol (*24, 25*) that has been obtained using more extensive electronic structure calculations. However, here we will employ the 2.5 kcal/mol value for comparison with 3.3 kcal/mol obtained for **1**. By calculating the two barriers with the same method, there should be some cancellation of errors.

Using the calculated barriers for **1** and for n-butane and the assumption that the prefactors are the same, we obtain a $\sim$40 ps time

constant for the n-butane trans-to-gauche isomerization time constant ($1/k_{TG}$). Twenty-six years ago, Rosenberg, Berne, and Chandler reported a 43-ps time constant for this process (in CCl$_4$ at 300 K) from MD simulations. (*6*) Other MD simulations gave isomerization rates in liquid n-butane at slightly lower temperatures: 52 ps (292 K) (*7*), 57 ps (292 K) (*5*), 50 ps (273 K) (*5*), and 61 ps (<292 K) (*8*). All of these values are reasonably close to the value obtained here based on the experimental measurements of **1**. In the same manner, the isomerization time constant for ethane is found to be $\sim$12 ps. This value can be improved by better electronic structure calculations on **1** and calculations for both **1** and ethane that include the CCl$_4$ in determining the barriers.

This 2D IR vibrational echo technique should be generally applicable to the study of fast isomerizations, most of which give rise to conformers with necessarily distinct vibrational spectra. For the method to be useful, the isomerization time constant must fall into a time window determined by experimental considerations. The lower time limit of the method is determined by the laser pulse duration, which can be <50 fs and therefore is not a serious restriction. The upper time limit is determined by the vibrational lifetime of the mode used to probe the isomerization. In practice, the isomerization time constant should be less than three times the vibrational lifetime. Some modes, such as the CO stretches of metal carbonyl compounds, can have lifetimes of hundreds of picoseconds. Within the experimental limitations, the method used here should be useful for addressing a variety of important problems.

References and Notes

1. J. March, *Advanced Organic Chemistry* (Wiley, New York, 3rd ed., 1985).
2. W. J. Orville-Thomas, *Internal Rotation in Molecules* (Wiley, New York, 1974).
3. D. Chandler, *J. Chem. Phys.* **68**, 2959 (1978).
4. T. A. Weber, *J. Chem. Phys.* **69**, 2347 (1978).
5. D. Brown, J. H. R. Clarke, *J. Chem. Phys.* **92**, 3062 (1990).
6. R. O. Rosenberg, B. J. Berne, D. Chandler, *Chem. Phys. Lett.* **75**, 162 (1980).
7. R. Edberg, D. J. Evans, G. P. Morris, *J. Chem. Phys.* **87**, 5700 (1987).
8. J. Ramirez, M. Laso, *J. Chem. Phys.* **115**, 7285 (2001).
9. A. Streitwieser, R. W. Taft, *Progress in Physical Organic Chemistry* (Wiley, New York, 1968), vol. 6.
10. L. M. Jackman, F. A. Cotton, *Dynamic Nuclear Magnetic Resonance Spectroscopy* (Academic Press, New York, 1975).
11. I. N. Levine, *Physical Chemistry* (McGraw-Hill, New York, 1978).
12. B. Cohen, S. Weiss, *J. Phys. Chem.* **87**, 3606 (1983).
13. J. J. Turner *et al.*, *J. Am. Chem. Soc.* **113**, 8347 (1991).
14. R. A. MacPhail, H. L. Strauss, *J. Chem. Phys.* **82**, 1156 (1985).
15. H. L. Strauss, *J. Am. Chem. Soc.* **114**, 905 (1992).
16. N. E. Levinger *et al.*, *J. Chem. Phys.* **118**, 1312 (2003).
17. J. Zheng *et al.*, *Science* **309**, 1338 (2005).
18. Y. S. Kim, R. M. Hochstrasser, *Proc. Natl. Acad. Sci. U.S.A.* **102**, 11185 (2005).
19. J. Zheng, K. Kwak, X. Chen, J. B. Asbury, M. D. Fayer, *J. Am. Chem. Soc.* **128**, 2977 (2006).
20. J. B. Asbury, T. Steinel, M. D. Fayer, *J. Lumin.* **107**, 271 (2004).

21. R. G. Parr, W. Yang, *Density Functional Theory of Atoms and Molecules* (Oxford Univ. Press, New York, 1989).
22. M. Khalil, N. Demirdoven, A. Tokmakoff, *J. Chem. Phys.* **121**, 362 (2004).
23. H.-S. Tan, I. R. Piletic, M. D. Fayer, *J. Opt. Soc. Am. B* **22**, 2009 (2005).
24. V. Pophristic, L. Goodman, *Nature* **411**, 565 (2001).
25. F. M. Bickelhaupt, E. J. Baerends, *Angew. Chem. Int. Ed.* **42**, 4183 (2003).
26. We thank X. Chen and J. I. Brauman for insightful discussions. This work was supported by grants from the Air Force Office of Scientific Research (F49620-01-1-0018) and NSF's Division of Materials Research (DMR-0332692).

Supporting Online Material
www.sciencemag.org/cgi/content/full/313/5795/1951/DC1
Figs. S1 to S3
References

6 July 2006; accepted 21 August 2006
10.1126/science.1132178

THE JOURNAL OF CHEMICAL PHYSICS 125, 244508 (2006)

Phenol-benzene complexation dynamics: Quantum chemistry calculation, molecular dynamics simulations, and two dimensional IR spectroscopy

Kijeong Kwac,[a] Chewook Lee, Yousung Jung,[b] and Jaebeom Han
Department of Chemistry, Korea University, Seoul 136-701, Korea and Center for Multidimensional Spectroscopy, Korea University, Seoul 136-701, Korea

Kyungwon Kwak, Junrong Zheng, and M. D. Fayer
Department of Chemistry, Stanford University, Stanford, California 94305

Minhaeng Cho
Department of Chemistry, Korea University, Seoul 136-701, Korea and Center for Multidimensional Spectroscopy, Korea University, Seoul 136-701, Korea

(Received 6 September 2006; accepted 8 November 2006; published online 28 December 2006)

Molecular dynamics (MD) simulations and quantum mechanical electronic structure calculations are used to investigate the nature and dynamics of the phenol-benzene complex in the mixed solvent, benzene/CCl_4. Under thermal equilibrium conditions, the complexes are continuously dissociating and forming. The MD simulations are used to calculate the experimental observables related to the phenol hydroxyl stretching mode, i.e., the two dimensional infrared vibrational echo spectrum as a function of time, which directly displays the formation and dissociation of the complex through the growth of off-diagonal peaks, and the linear absorption spectrum, which displays two hydroxyl stretch peaks, one for the complex and one for the free phenol. The results of the simulations are compared to previously reported experimental data and are found to be in quite reasonable agreement. The electronic structure calculations show that the complex is T shaped. The classical potential used for the phenol-benzene interaction in the MD simulations is in good accord with the highest level of the electronic structure calculations. A variety of other features is extracted from the simulations including the relationship between the structure and the projection of the electric field on the hydroxyl group. The fluctuating electric field is used to determine the hydroxyl stretch frequency-frequency correlation function (FFCF). The simulations are also used to examine the number distribution of benzene and CCl_4 molecules in the first solvent shell around the phenol. It is found that the distribution is not that of the solvent mole fraction of benzene. There are substantial probabilities of finding a phenol in either a pure benzene environment or a pure CCl_4 environment. A conjecture is made that relates the FFCF to the local number of benzene molecules in phenol's first solvent shell. © 2006 American Institute of Physics. [DOI: 10.1063/1.2403132]

I. INTRODUCTION

The nature of organic solutes in liquid solutions is a fundamentally interesting problem that is also of practical importance in chemistry, biology, and materials science. In the simplest view, the solute can be taken to be in a homogenous dielectric continuum.[1] However, a more realistic approach is to consider the radial distribution function of the solvent about a solute.[2] The radial distribution function brings in solvent shells and can account for the influence of solvent structure on the relative positions of solutes[2,3] and diffusion through the structured solvent by including a potential of mean force in a description of transport.[2,3] However, organic solutes and solvents have anisotropic intermolecular interactions. Such interactions may not be negligible compared to thermal energy, k_BT, at room temperature. Therefore, transitory organic solute-solvent complexes can

form and exist for times that depend on the strength of the solute-solvent intermolecular interactions. Such complexes have the potential to influence chemical reaction kinetics by blocking reaction sites on a solute. In effect, the solute-solvent dissociation reaction may have to take place before another chemical reaction with the solute can occur.

Recently, the first direct measurements of organic solute-solvent complex formation and dissociation under thermal equilibrium conditions were made using ultrafast infrared vibrational echo chemical exchange experiments.[4–6] Ultrafast IR methods have been used extensively to study the dynamics of extended hydrogen bonding systems such as water,[7–22] alcohols,[23–32] and nanoscopic water.[33–41] The "complexes" in water and alcohols are extended structures with a very wide variety of geometries and strengths of association. The organic solute-solvent complexes that will be discussed here involve a molecular pair with more or less a single structure and bond strength. Similar to the complexes discussed below is a hydrogen bonded pair of molecules in a solvent. Such a system has been studied using both two color pump-probe spectroscopy[42] and two-dimensional (2D) vibrational chemi-

[a] Present address: Department of Chemistry, University of California at San Diego, La Jolla, CA 92093.
[b] Present address: Department of Chemistry, California Institute of Technology, Pasadena, CA 91125.

244508-2   Kwac *et al.*

J. Chem. Phys. **125**, 244508 (2006)

cal exchange spectroscopy,[43] the experimental method used to obtain the solute-solvent complex data discussed here.

A number of different organic solute-solvent complexes have been studied using 2D vibrational echo chemical exchange experiments to determine the dynamics of complex formation and dissociation.[4,5] The system that is the simplest chemically is the phenol-benzene complex. In the solution, phenol is the low concentration solute in a mixed solvent of benzene and CCl_4. CCl_4 is added to the benzene solvent to shift the equilibrium toward more uncomplexed (free) phenol. In pure benzene, the fraction of phenol that is complexed is >90%, which makes the system difficult to study experimentally. In the mixed solvent employed in the experiments, there is approximately a 50-50 mixture of phenol complex and free phenol.[4] The two species have distinct Fourier transform infrared (FTIR) spectra of the hydroxyl stretch. In the experiments, the hydroxyl H is replaced with D, and the OD hydroxyl stretch frequency of free phenol in the mixed solvent is at 2665 cm^{-1} and the frequency of the complex is at 2631 cm^{-1}. Figure 1(a) shows the FTIR spectrum. Although the spectra of the two species overlap, the two peaks are readily observable.[4] FTIR experiments were used to determine the equilibrium constant and the complex formation enthalpy and entropy.[4] However, the linear absorption experiments cannot provide information on the time dependence of the complex formation and dissociation.

In a 2D IR vibrational echo experiment, three ultrashort IR pulses are tuned to the frequency of the vibrational modes of interest and crossed in the sample. Because the pulses are very short, the hydroxyl stretch of both the complex and free phenol are simultaneously excited. The times between pulses 1 and 2 and pulses 2 and 3 are called τ and T_w, respectively. At a time $\leq \tau$ after the third pulse, a fourth IR pulse is emitted in a unique direction. This is the vibrational echo, the signal in the experiments. The vibrational echo is the infrared vibrational equivalent of the magnetic resonance spin echo[44] and the electronic excitation photon echo.[45] The vibrational echo is combined with another pulse, the local oscillator, and heterodyne detected. Therefore, both amplitude and phase information are obtained. The combined vibrational echo-local oscillator is passed through a monochromator and frequency resolved. The spectrum is an experimental Fourier transform, which provides one of the two Fourier transforms that gives rise to the 2D vibrational echo spectrum. When τ is scanned, an interferogram is produced between the vibrational echo and the local oscillator. One such interferogram is generated at each frequency at which there is vibrational echo emission. The numerical Fourier transforms of these interferograms provide the second Fourier transform for the 2D spectrum. Details of the experimental method used for these experiments have been given previously.[21]

In a dynamic system, the first laser pulse "labels" the initial structures of the species in the sample. The second pulse ends the first time period τ and starts clocking the "reaction time," during which the "labeled" species experience population dynamics. The third pulse ends the population dynamics period of length T_w, and begins a third period of length $\leq \tau$, which ends with the emission of the vibrational echo pulse. The echo signal reads out the information

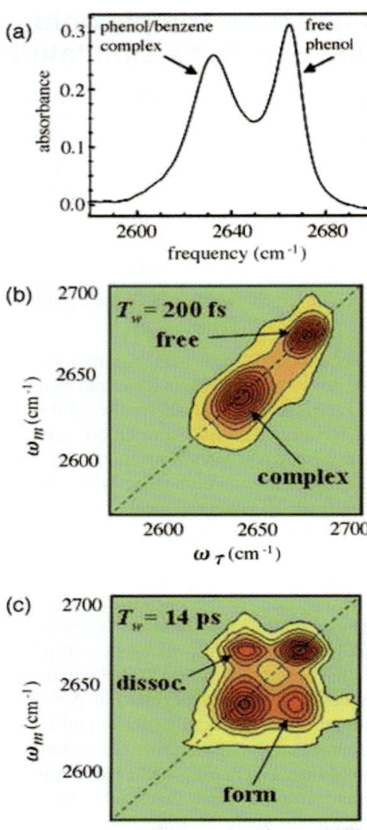

FIG. 1. (Color) (a) FTIR spectra of the OD hydroxyl stretch of phenol for the phenol-benzene complex and free phenol in the benzene/CCl_4 mixed solvent. (b) 2D vibrational echo spectrum at a time (200 fs) short compared to complex formation and dissociation showing two peaks on the diagonal. (c) 2D vibrational echo spectrum at a time (14 ps) long compared to complex formation and dissociation showing two peaks on the diagonal and two addition off-diagonal peaks. The off-diagonal peaks grow in as complex formation and dissociation proceed.

about the final structures of all labeled species. In the chemical exchange problem under consideration here, the two species, complexed and free phenol, are in equilibrium. They are interconverting one to the other without changing the overall number of either species. In an experiment, τ is scanned for fixed T_w. The recorded signals are converted into a 2D vibrational echo spectrum. Then T_w is increased and another spectrum is obtained. Chemical exchange between the complex and free species causes new off-diagonal peaks to grow in as T_w is increased. Figure 1(b) displays a 2D vibrational echo spectrum taken at $T_w=200$ fs, which is a time short compare to the chemical exchange time. The spectrum shows two peaks on the diagonal (dashed line), one is the spectrum of the free phenol, and the other is the complex. Figure 1(c)

244508-3   Phenol-benzene complexation dynamics

J. Chem. Phys. **125**, 244508 (2006)

displays the data at $T_w = 14$ ps, which is a time that is relatively long compared to the chemical exchange time. Now, in addition to the two diagonal peaks, two off-diagonal peaks have grown in, one caused by dissociation of complexes and the other by the formation of complexes from free phenol. When combined with other parameters of the system that are independently measured, the growth of the off-diagonal peaks as T_w is increased from short to long time permits the complex dissociation and formation kinetics to be directly determined.[4,6] Because the complex and free phenol are in equilibrium, the number of complexes per unit time dissociating is equal to the number of complexes forming. Therefore, the process can be characterized by the single parameter, the complex dissociation time, τ_d, which is the inverse of the complex dissociation rate. It was found from the 2D vibrational echo experiments that the phenol-benzene dissociation time, $\tau_d = 8$ ps.[4]

Although the 2D experiments measured the dissociation time for the phenol-benzene complex, the experiments do not give a microscopic picture of the nature of the process. Ultrafast IR experiments on water[7-22] have been greatly augmented by applying molecular dynamics (MD) simulations and other theoretical calculations to understand the implications of the experimental results.[13,14,17-20,46-58] For water, the MD simulations address the dynamic structure of the extended hydrogen bond network and relate the calculations to the experimental observables. Recent applications of MD simulations to the theoretical calculations of one dimensional (1D) and 2D spectra of N-methylacetamide in water showed that the simulation method can provide detailed information on the hydrogen bond making and breaking dynamics of water and methanol molecules in the first solvation shell of N-methylacetamide.[59-62] Here, we will take a similar path for understanding the structure and dynamics of organic solute-solvent complexes. There are a variety of issues to be clarified, which can only be done by theoretical studies. These issues include the existence of stable phenol-benzene complexes, the conformation of the complexes, the dispersive interaction strength, the intermolecular potential energy surface, the set of classical force field (FF) parameters closely mimicking the quantum potential surface, local solvation structures and dynamics, and so on.

In this paper, we will present detailed theoretical descriptions of phenol-benzene complex formation and energetics and the importance of dispersive interaction, using HF, DFT, and MP2 calculation methods. Comparing these different calculation results, a reliable potential energy surface is obtained that is necessary to properly develop the classical FF parameters for MD simulations. Then, using the optimized FF parameters, MD simulations of phenolOD in benzene/CCl$_4$, phenolOD in benzene, and phenolOD in CCl$_4$ solutions are carried out. In order to quantitatively simulate the 1D and 2D vibrational spectra, the electric field (Stark effect) model is employed. In this model, the time dependent OD frequency is linearly proportional to the electric field projected along the direction of the hydroxyl bond at the center of the OD group. The resulting calculated IR absorption and 2D vibrational echo spectra are directly compared

with experimental results.[4,6] Also, the local inhomogeneous environment around the OD chromophore and domain formations in phenolOD in benzene/CCl$_4$ solvent are discussed in detail.

II. QUANTUM CHEMISTRY CALCULATIONS

In this section, we present the results of a variety of quantum chemistry calculations for the complexation of phenol with benzene. Accurate binding energies and harmonic vibrational frequencies of the complex, particularly of the OD stretch mode of phenolOD under different molecular environments, are the main focus of this section.

Long-range electron correlation effects, such as dispersion interactions, are important in describing weakly interacting van der Waals (vdW) systems such as the one under consideration here, but their theoretical treatment is by no means trivial. For example, such effects are absent in Hartree-Fock (HF) or current Kohn-Sham density functional theory (DFT) implementation,[63] which is the most popular electronic structure method used today. One of the extensively studied examples is the benzene dimer, which is predicted to be unstable by HF and almost all standard DFT functionals.[64,65] In spite of the fact that HF and DFT can describe electrostatics fairly well, it has been shown that some vdW complexes are calculated by HF and DFT not to be bound.[63,66,67] However, stable bound complexes are theoretically reproduced only at theoretical levels that include correlation. The differing results obtained with HF and DFT versus theories with correlation are often used to indicate the importance of dispersion interactions in such vdW systems.[64,65]

This limitation of HF and DFT in taking into account the long-range correlation effects has been, in large part, remedied by Møllet-Plesset theory (MP2).[68] MP2 is the simplest wave-function-based method that can correctly describe the long-range correlation effects, although it generally overshoots the binding energies by overestimating such effects. Coupled-cluster with single and double and perturbative triple [CCSD(T)] excitations,[69] on the other hand, *if computationally tractable* for a given system, is probably the most accurate *ab initio* method that is currently available to treat nonbonded vdW systems.

Another computationally useful and relatively less time consuming alternative is the recently proposed scaled opposite spin (SOS) MP2 scheme, in which the opposite spin (OS) correlation energy is scaled up by an empirical factor, c_{OS}, while entirely neglecting the same spin counterpart.[70] Associated computational savings is the reduction of computational scaling by one power, from the fifth to the fourth order, while yielding statistically improved quantitative results over conventional MP2. More recently, SOS-MP2 with $c_{OS} = 1.55$, denoted as SOS-MP2 (1.55), was proposed to specifically study vdW complexes with an aim to reproducing CCSD(T) binding energies with substantially less computational effort.[76] Hence, in this study, in addition to well-known HF and DFT methods we also used MP2, SOS-MP2 (1.55), and CCSD(T) to assess the differences in the results produced by these methods and to obtain accurate binding

244508-4   Kwac *et al.*

J. Chem. Phys. **125**, 244508 (2006)

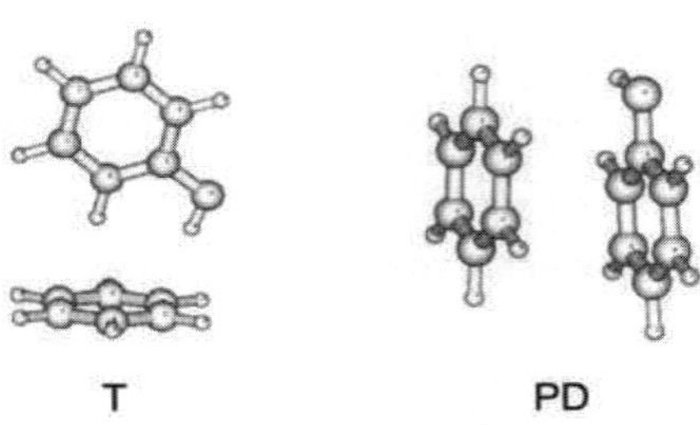

FIG. 2. RI-MP2 optimized T-shaped (T) and parallel-displaced (PD) structures of the phenol-benzene complex. Interplanar distance between benzene and phenol rings in the PD complex is 3.14 Å.

TABLE I. Counterpoise corrected interaction energies (kcal/mol) for the T-shaped (T) and parallel-displaced (PD) configurations of phenol-benzene complex (see also Figs. 1 and 2).

Method	Basis	T	PD
HF	$6\text{-}311\text{+}\text{+}G^{**}$	−1.90	a
B3LYP	$6\text{-}311\text{+}\text{+}G^{**}$	−2.00	a
RI-MP2[b]	CBS[c]	−6.57	−6.28
CCSD(T)[d]	CBS	−5.46	−3.38
SOS-MP2 (1.55)[b,e]	CBS[c]	−5.05	−3.48

[a]PD configuration is unstable at the HF and B3LYP level.
[b]Alhrichs's corresponding auxiliary basis sets, that were designed be used in conjunction with aug-cc-pVXZ regular basis, were used.
[c]Extrapolated to the complete basis set (CBS) limit using Dunning's aug-cc-pV(DT)Z two-point extrapolation scheme for correlation energies.
[d]$\Delta E(\text{CCSD(T)}/\text{CBS}) = \Delta E(\text{RI-MP2}/\text{CBS}) + [\Delta E(\text{CCSD(T)}/6\text{-}31G^*) - \Delta E(\text{RI-MP2}/6\text{-}31G^*)]$.
[e]SOS-MP2/aug-cc-pV(DT)Z with one parameter $c_{OS}=1.55$, which was proposed for a target accuracy of CCSD(T)/CBS (Ref. 76).

energies of the phenol-benzene complex. Efficient resolution of the identity implementation of MP2, namely, RI-MP2,[71] was used for MP2.

All interaction (or complexation) energies reported in this paper were counterpoise corrected for basis set superposition error (BSSE),[72] which usually makes the vdW complex binds too strongly. BSSE is the borrowing of basis functions from the second monomer to improve the quality of basis functions of the first monomer relative to the same monomer in isolation and can occur for any chemical interaction between the fragments that employ finite basis sets. We used two sets of basis functions, $6\text{-}311\text{+}\text{+}G(d,p)$ and aug-cc-pVXZ ($X=D,T,Q$), that include diffuse functions that are important for a good long-range description of weakly interacting nonbonded systems. In particular, the augmented correlation consistent basis sets of Dunning (aug-cc-pVXZ, where $X=D,T,Q$) were chosen because they can be systematically extrapolated to the complete basis set (CBS) limit using the following two-point extrapolation prescription for correlation energies.[73] In this study, we used DT extrapolation.

$$E_{\text{CBS}}[XY] = E_Y^{\text{SCF}} + \frac{X^3 E_X^{\text{Corr}} - Y^3 E_Y^{\text{Corr}}}{X^3 - Y^3}, \quad Y > X. \quad (1)$$

HF and B3LYP density functional theory calculations were performed using GAUSSIAN03,[74] while all the other correlation calculations, RI-MP2, CCSD(T), and SOS-MP2, were carried out using the Q-CHEM3.0 *ab initio* program package.[75]

Two stable configurations were found for the phenol-benzene system, namely, T-shaped (T) and parallel-displaced (PD) structures, which are depicted in Fig. 2. These structures are reminiscent of the benzene dimer, and for that reason we will emphasize some similarities and differences between them when appropriate. The principal results are summarized in Table I. The T-shaped phenol-benzene complex is stable even at the HF and B3LYP levels by −1.90 and −2.0 kcal/mol, respectively, unlike that of benzene dimer.[76] The stability manifested in these calculations suggests relatively strong electrostatic attraction between benzene and phenol, which has a permanent dipole moment. The phenol-benzene dipole-quadrupole interaction is stronger than the interaction between two quadrupole moments of benzene in T-shaped benzene dimer. The strength of the phenol-benzene

electrostatic interactions outweighs the exchange repulsion, resulting in the net binding even without considering dispersion effects. By contrast, the PD configuration, where the dispersion interactions are expected to be the more important source of attraction due to its cofacial geometry but which also causes more repulsion, is not bound at the HF and B3LYP levels, because these methods lack of long-range correlation effects.

Upon incorporating the long-range correlations via MP2, CCSD(T), and SOS-MP2, the PD complex is found to be bound, and also the binding energy of T-shaped configuration becomes even larger at these levels than those obtained by using the HF or DFT methods (Table I). MP2 yields the highest values for the binding energies of the complexes, which are expected to be overestimated, while CCSD(T) yields smaller values. Specifically, at the CCSD(T)/CBS limit, the T-shaped complex is found to be the lowest energy configuration with the association energy of −5.46 kcal/mol, while the PD configuration is predicted to be another stable form with an interaction energy of −3.38 kcal/mol. It is also quite encouraging that the one-parameter SOS-MP2 (1.55) proposed earlier reproduces the CCSD(T)/CBS binding energies very well, with the stability of T configuration slightly underestimated as predicted previously.[76]

The potential energy curves for the T-shaped configuration as a function of phenol-benzene ring center-to-ring center distances are shown in Fig. 3, which graphically illustrates how each method performs relative to CCSD(T)/CBS. Again, the MP2 well depth is too deep, but the SOS-MP2 (1.55) curve overlaps remarkably well with that of CCSD(T)/CBS.

Next, we calculated harmonic vibration frequencies for the optimized T and PD configurations. In particular, the change of hydroxyl stretch mode of phenol for different molecular environments (i.e., free versus complex forms) can serve as a good spectroscopic signature that identifies the structure of the complex when combined with proper calculations. We therefore computed the OD stretch frequencies of phenolOD for the free and benzene-bound forms (T and PD) of phenol, and compared them with the experimental frequencies. The estimated shifts in OD stretch frequency when

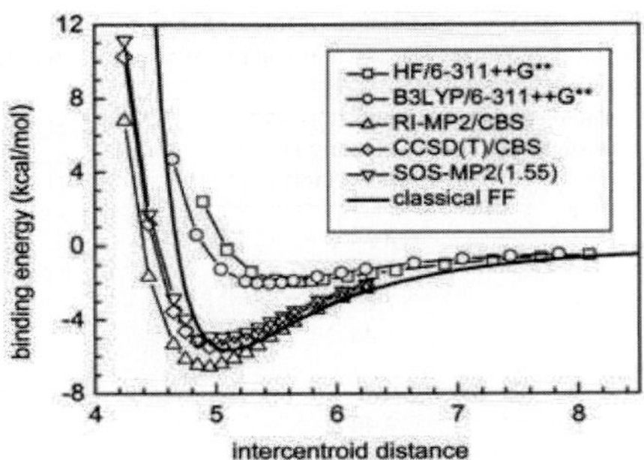

FIG. 3. Potential energy curves for the T-shaped configuration of phenol-benzene complex as a function of a ring-center to ring-center distance between them.

going from the free to complex forms are summarized in Table II. The OD bond strength becomes weaker in accord with the shift of 34 cm^{-1} to lower frequency upon complexation with benzene [see Fig. 1(a)]. Similar trends of weakened OD bond strengths in the complex forms are observed in HF, B3LYP, and MP2 calculations. The frequency shift for the T configuration (49 cm^{-1}) obtained with MP2 (which was the only level of theory among HF, B3LYP, and MP2 employed in this study that also predicted the PD complex to have a minimum) agrees better with experiments (34 cm^{-1}) than that for PD (11 cm^{-1}). Therefore, this fact together with the energetic consideration that the T-shaped complex is the lowest energy minimum by a significant amount suggests that the experimentally observed single complex species is most likely the T-shaped configuration.

The results of the high level electronic structure calculations also demonstrate the importance of the solvent on the energetics of the complex. The structure calculations are for a pair of molecules in the absence of the solvent. The complex enthalpy of formation in the mixed benzene/CCl$_4$ solvent was determined experimentally to be −1.67 kcal/mol.[4] In contrast, the best electronic structure methods yield approximately −5 kcal/mol (see Table I). In the electronic structure calculations, there is no competition between the phenol-benzene interaction and the interaction of the molecules that make up the complex with solvent molecules. For

example, the solvation free energies of benzene and phenol in CCl$_4$ were recently estimated to be −4.2 and −6.6 kcal/mol, respectively.[77] Assuming that both solutes disturb the solvent structure to a similar and only minor extent so that the solvation free energies are dominated by the enthalpy contributions, both solutes have relatively strong interactions with CCl$_4$.

III. THE FORCE FIELD PARAMETERS AND MD SIMULATION METHOD

A. Force field parameters for benzene and phenol

The force field parameters of benzene and phenol molecules are determined using the Antechamber module of AMBER8 molecular dynamics package.[78] We have optimized the structures of benzene and phenol using the GAUSSIAN98 program with the HF/6−31G* basis set. The optimized structures are used as input to the Antechamber program to obtain the partial charges and geometrical parameters for MD simulations. The partial charge parameter for benzene is determined as $q_D = -q_C = 0.1299e$. We denote the carbon atom attached to OD in the phenol molecule as C and denote the other carbon atoms as C1, C2, C3 in the order of distance from C, $q_O = -0.5569e$, $q_D = 0.3781e$, $q_C = 0.4279e$, $q_{C1} = -0.3236e$, $q_{H1} = 0.1777e$, $q_{C2} = -0.0931e$, $q_{H2} = 0.1439e$, $q_{C3} = -0.1996e$, and $q_{H3} = 0.1408e$, where D is bonded to O and H1 is bonded to C1, etc. Only the charge parameters of four out of the six carbons are necessary due to the symmetry of the phenol molecule. We have implemented an equilibration run of the system consisting of a single phenol molecule and 384 benzene molecules under constant temperature and pressure conditions at 298 K and 1 bar for 800 ps. It takes ~200 ps to reach a plateau value of the density. The average value of the density for the last 500 ps trajectory is 0.873 g/cm^3, which is close to the experimental value of 0.879 g/cm^3 for pure benzene.

For the FF parameters of CCl$_4$, we used the OPLS-AA model.[79] The adopted parameters are $r(C-Cl) = 1.769$ Å, $\sigma_C = 3.80$ Å, $\sigma_{Cl} = 3.47$ Å, $\varepsilon_C = 0.050$ kcal/mol, $\varepsilon_{Cl} = 0.266$ kcal/mol, and $q_C = -4q_{Cl} = 0.248e$. We implemented an equilibration run of the system consisting of a single phenol molecule and 645 CCl$_4$ molecules under constant temperature and pressure conditions at 298 K and 1 bar for 800 ps. It

TABLE II. Characteristic OD stretch vibration frequencies (cm^{-1}) of phenol, with and without complexation with benzene. The T-shaped (T) and parallel-displaced (PD) complex configurations were considered. The PD configuration is found unstable at the HF and B3LYP level (see also Table I)

Method	Free	Complex		Difference	
		T	PD	Δ^a (T)	Δ^a (PD)
HF/6-311++G^{**b}	2762	2743	⋯	19	⋯
B3LYP/6-311++G^{**c}	2686	2641	⋯	45	⋯
MP2/6-311++G^{**d}	2680	2631	2669	49	11
Expt.	2665	2631		34	

[a] $\Delta(T, PD) = \nu(\text{free}) - \nu$ (complex: T, PD).
[b] Frequency scaling factor = 0.9051 was used (Ref. 88).
[c] Frequency scaling factor = 0.9614 was used (Ref. 88).
[d] Frequency scaling factor = 0.9500 was used (Ref. 88).

244508-6   Kwac *et al.*   J. Chem. Phys. **125**, 244508 (2006)

TABLE III. Distances between the centers of the rings and the binding energies for the T-shaped configuration of benzene-phenol complex.

Calculation level	Basis set (No. of basis functions)	Shape	Intercentroid distance (Å)	Binding energy (kcal/mol)
HF	6-311G(d) (270)	T	5.607	2.22
	6-311++G(d,p) (370)	Twisted T	5.694	1.90
	cc-pVDZ (242)	T	5.665	2.03
	aug-cc-pVDZ	T	5.645	1.40
B3LYP	6-311G(d) (270)	Tilted T	5.393	2.32
	6-311G(d) (270)	T	5.356	2.46
	D95 (154)	Twisted T	5.438	2.65
	D95++** (332)	T	5.343	2.23
	6-311++G($3df,2pd$) (687)	T	5.413	1.92
	cc-pVDZ (242)	T	5.370	2.09
	aug-cc-pVTZ (874)	T	5.645	1.23
B3PW91	6-311G(d) (270)	Tilted and Twisted T	5.381	1.94
RI-MP2	aug-cc-pVDZ (407)	T	5.045	5.64
	aug-cc-pVTZ (874)	T	4.945	6.25
	CBS	T	4.945	6.56
Classical FF		T	5.099	5.65

takes about 50 ps to reach a plateau value of the density for this system. The average density of the last 500 ps is 1.581 g/cm^3, which is close to the experimental value of 1.594 g/cm^3 for pure CCl$_4$.

B. MD simulation method

We have done the MD simulations of three systems: phenol in benzene, phenol in CCl$_4$, and phenol in the mixed solvent benzene/CCl$_4$. The numbers of molecules used in the phenol in benzene and phenol in CCl$_4$ systems are mentioned in the previous section. To simulate the molecular dynamics of a phenol molecule in the mixture of benzene and CCl$_4$, we use a single phenol molecule and 192 benzene molecules and 480 CCl$_4$ molecules. The mole fraction of benzene is 0.286, which is very close to the experimental conditions.

The Sander module of the AMBER 8 program package[78] was used for the simulations. Each system is placed in a cubic box with a periodic boundary condition. Long range electrostatic interactions are treated by the particle mesh ewald[80] (PME) method and the criterion to switch from direct sum to the calculation by PME is 9 Å. The initial system is minimized by 500 steps of steepest-descent minimization and 500 steps of conjugate-gradient method minimization with the solute molecule fixed. Then, the total system is minimized by 1500 steps of the steepest-descent method and 1000 steps of the conjugate gradient method. The system is equilibrated under the constant temperature and pressure conditions of 298 K and 1 bar for 800 ps and then under constant temperature conditions for 200 ps. After that, a 4 ns production run is implemented under the constant temperature condition of 298 K. All the constant temperature and pressure conditions are implemented using the weak coupling algorithm of Berendsen *et al.*[81] The time step for the equilibration and production runs is 1 fs.

C. Comparisons of the classical force field with the quantum chemistry calculations

As a test for the classical FF parameters, we calculated the interaction energy of the T-shaped phenol-benzene complex as a function of the distance between the ring centers of the two molecules. The result is plotted in Fig. 3 along with the quantum chemistry calculation results (see the solid curve). Also we show the location of the potential minimum and the binding energy as the minimum value of the potential energy curve in Table III.

One of the notable features is that the classical FF (in Fig. 3) produces the potential energy curve which is close to the quantum chemistry methods that correctly describe long-range correlation effects [RI-MP2, CCSD(T), and SOS-MP2 in Fig. 3]. While the classical potential has the steeper repulsive part at short distance, the overall shape and the magnitude of binding energy are well matched to the results of the SOS-MP2 and CCSD(T) method. In the configuration corresponding to the potential minimum under the classical FF, the distance between the ring center of benzene and the D-atom of phenol OD group is 2.39 Å.

We calculated the pair distribution between the ring center of benzene and the D atom of phenol OD group from snapshot structures found in the MD simulation. The result for the phenol in benzene/CCl$_4$ system is shown in Fig. 4. The first maximum is located at 2.55 Å. This value is larger than the value of 2.39 Å for the minimum energy configuration for the phenol-benzene without solvent. As in the comparison of the calculated complex energy and the experimentally determined entropy[4] (see discussion at the very end of Sec. II), the increase in separation demonstrates the influence of the solvent on the complex structure. To make this argument quantitative, using the theorem, $g(r)=\exp(-w(r)/k_BT)$, where $w(r)$ is the potential of mean force (PMF), we calcu-

J. Chem. Phys. **125**, 244508 (2006)

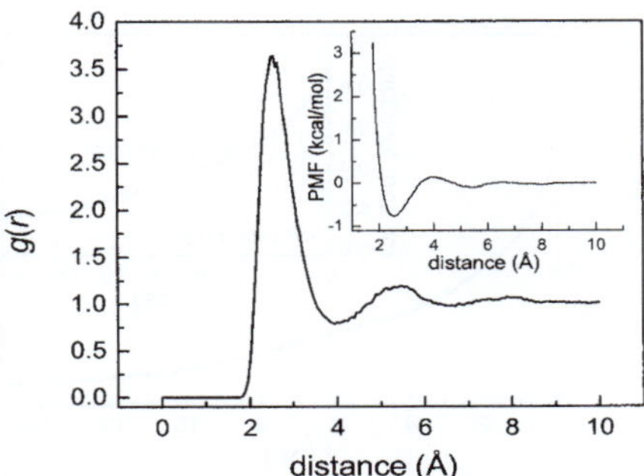

FIG. 4. The pair distribution function for the distance between the center of mass of benzene and the D atom of phenol OD group. The potential of mean force obtained using $g(r) = \exp(-w(r)/k_B T)$ is plotted in the inset.

lated $w(r)$ (see the inset of Fig. 4). The estimated binding energy from PMF is about -0.78 kcal/mol. This is about half of the experimentally measured complex enthalpy of formation, -1.67 kcal/mol.[4]

From the MD trajectories, we analyzed the configuration of the phenol-benzene complex in detail. We picked the benzene molecule which is nearest to the phenol molecule at each snapshot configuration in the MD trajectory for the phenolOD in benzene/CCl$_4$ system. We denote the angle between the two vectors that are normal to the benzene and phenol ring planes as θ, and the distance between the ring center of benzene and the D atom of phenol OD group as R. We plot the population of phenol-benzene geometry with respect to θ and R in Fig. 5. It should be noted that the distribution of θ is broad, indicating the potential energy surface along this angle for a fixed intermolecular distance R is shallow. However, it is clear that the preferred geometry is T shaped as was argued in Sec. II based on electronic structure calculations. The configurations with the $R > 4$ Å correspond to the situation where phenol is solvated by the CCl$_4$ molecules, that is, there is no complex.

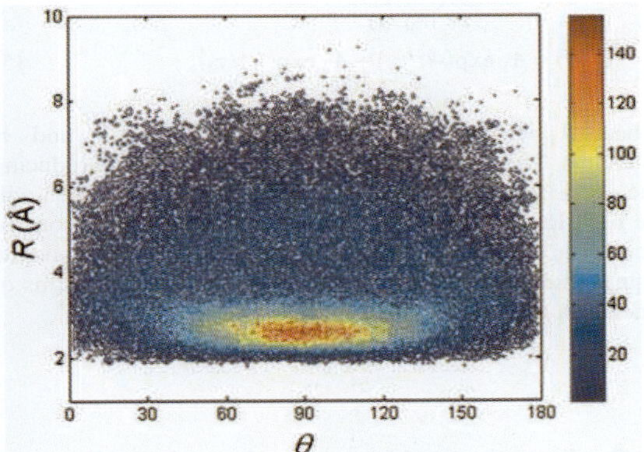

FIG. 5. (Color) Population for the configuration of the benzene molecule nearest to the phenol molecule at each MD snapshot with respect to R and θ, where R denotes the distance between the ring center of benzene and D atom of the OD group of phenol and θ denotes the angle between the normal vectors at the two ring centers of benzene and phenol.

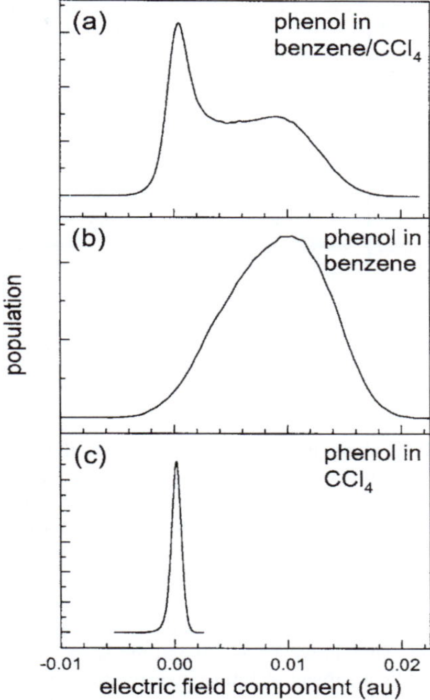

FIG. 6. Distribution of the projection, E, of the electric field in a.u. along the OD bond evaluated at the site of the D atom of the phenol molecule.

IV. VIBRATIONAL SPECTRA: COMPARISON BETWEEN MD SIMULATIONS AND EXPERIMENT

A. OD stretching mode frequency from MD trajectories

To numerically simulate the 1D and 2D IR spectra, it is necessary to obtain the fluctuating transition frequency trajectory from the MD simulations. Here, we will use the vibrational Stark effect theory, where the OD stretching mode frequency is assumed to be linearly proportional to the electric field, E, as[82]

$$\omega_{OD}(t) = \omega_{OD}^0 + \kappa E(t), \qquad (2)$$

where E is the component of the electric field along the OD bond evaluated at the position of the D atom, i.e.,

$$E(t) = \hat{r}_{OD}(t) \sum_{m,i} \frac{q_{mi}}{r_{mi,D}^2(t)} \hat{r}_{mi,D}(t). \qquad (3)$$

The unit vector along the OD bond is denoted as $\hat{r}_{OD}$. q_{mi} and $r_{mi,D}$ ($\hat{r}_{mi,D}$) are the partial charge of the ith atom of the mth solvent molecule and the magnitude (unit vector) of the distance vector pointing from the ith atom of the mth solvent molecule to the D atom of the phenol.

From the three separated MD simulations, phenol in the mixed solvent, in pure benzene, and in pure CCl$_4$, the electric field component distributions were calculated. The results are plotted in Fig. 6 (E is in a.u.). As can be seen in Fig. 6(c), when phenol is dissolved in pure CCl$_4$, the projected

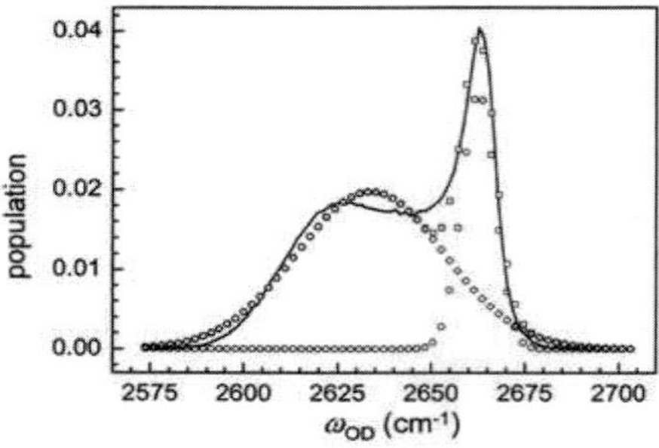

FIG. 7. Population distribution of OD stretch mode frequency. Two fitted Gaussian functions are also plotted (open circles and squares). Total fitting results are plotted as closed squares.

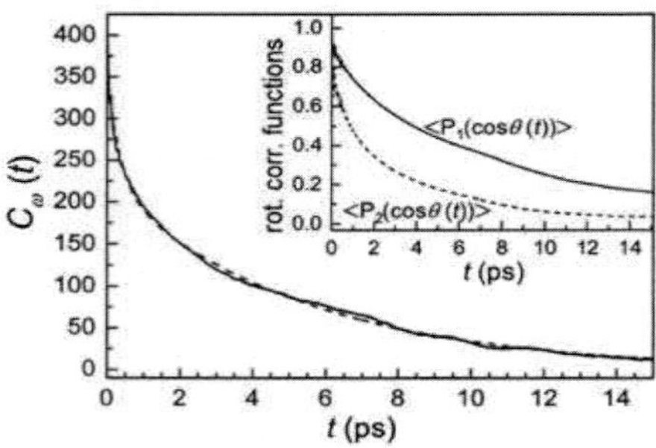

FIG. 8. OD stretch mode frequency-frequency correlation function. The dashed curve is a biexponential fit. The two rotational correlation functions $C_1(t)$ and $C_2(t)$ obtained from MD trajectories are plotted in the inset.

electric field along the phenol OD group at the position of the D is centered virtually at zero, indicating that the solvatochromic OD frequency shift induced by the phenol-CCl$_4$ interaction is exceedingly small. However, it should be mentioned that the OD stretch frequency of phenol in CCl$_4$ would be different from the gas phase phenol OD stretch frequency, indicating that the OD frequency shift can be induced by solute-solvent interaction other than the electric field effect. This, however, is beyond the scope of this work and should be a subject of future investigation. Nevertheless, the electrostatic intermolecular interaction between phenol and benzene can greatly affect the OD frequency and induces a strong redshift. The influence of benzene on the electric field distribution is clearly seen in Figs. 6(a) and 6(b). In Fig. 6(b), phenol in pure benzene, there is a broad electric field distribution shifted to high field. The IR absorption spectrum of phenol in pure benzene shows that phenol exists almost completely as the phenol benzene complex, with very little free phenol.[4,6] In the mixed solvent [Fig. 6(a)], there are two peaks that clearly correspond to the peaks in Figs. 6(b) and 6(c).

The two constants, ω_{OD}^0 and κ, in Eq. (2) are obtained by noting that the high- and low frequency bands in the experimentally measured IR absorption spectrum[4] [see Fig. 1(a)] correspond to the free phenol and the phenol-benzene complex, respectively. The free phenol peak in the mixed solvent is very similar to the IR absorption spectrum of phenol in pure CCl$_4$,[4] while the complexed peak is very similar to the spectrum of phenol in pure benzene. The constant ω_{OD}^0 =2665 cm^{-1} was assigned to the high frequency band, and $\kappa=-4229$ was determined from the frequency difference between the two bands. Using these constants and Eq. (2), the time-dependent frequency $\omega_{OD}(t)$ can be calculated from the MD trajectories. Figure 7 shows the population of frequencies, which is just the probability of having a frequency per unit time obtained from $\omega_{OD}(t)$. The frequency-dependent population cannot be compared directly to the spectrum in Fig. 1(a) not only because the transition dipoles are different for free phenol and the complex but also because the line broadening process is not taken into account.[4] However, it has qualitatively similar features. The peak positions are close to the experimental values and the redshifted peak is

broader than the peak to the blue. Using two Gaussian functions to fit the population distribution in Fig. 7 and considering that the high and low frequency components correspond to the free and complex forms, respectively, we found that the equilibrium constant [complex]$_{eq}$/[free]$_{eq}$ to be about 2.7—note that the experimental value is $\sim$1.[4]

From the time dependence of the OD stretch frequency, one can readily calculate the FFCF, defined as

$$C_\omega(t) = \langle(\omega_{OD}(t) - \langle\omega_{OD}\rangle)(\omega_{OD}(0) - \langle\omega_{OD}\rangle)\rangle, \qquad (4)$$

where the average frequency is found to be 2641.7 cm^{-1}. Although FFCF would not be used to numerically calculate the linear and nonlinear vibrational response functions for the 1D and 2D IR spectroscopies, it will be directly compared with the fluctuations of the inhomogeneously distributed solvent environments around the phenol molecule in the following section. In Fig. 8 the numerically calculated $C_\omega(t)$ (solid curve) is plotted. The dashed curve is a biexponential fit to $C_\omega(t)$ obtained from the simulations, i.e.,

$$C_\omega(t) = A_1 \exp(- t/\tau_1) + A_2 \exp(- t/\tau_2), \qquad (5)$$

where A_1=138 cm^{-2}, τ_1=0.34 ps, A_2=220 cm^{-2}, and τ_2 =5.35 ps. The biexponential does a good job of reproducing the curve and gives a convenient analytical form which will be used later. As will be discussed below, the slow component with τ_2=5.35 ps is directly associated with the dynamic equilibrium process between the free and complex forms of phenol in the benzene/CCl$_4$ solution.

B. Configuration and frequency-dependent OD stretch transition dipole moment

The OD transition dipole moment was shown to be strongly dependent on the local environment. It was determined experimentally that the transition dipole of the

244508-9   Phenol-benzene complexation dynamics   J. Chem. Phys. **125**, 244508 (2006)

phenol-benzene complex is about 1.5 times larger than that of the free phenol.[4] Here, the free phenol approximately corresponds to the case when the phenol is surrounded by CCl_4 molecules. In a real solution, the phenol molecule can have varying local solvation configurations that are neither a perfect complex form nor a perfect free form. Therefore, it is necessary to develop a theoretical method that can be used to quantitatively determine the OD transition dipole moment for a given instantaneous configuration sampled from the MD trajectories. An alternative is to find a relationship between the OD stretch frequency and the transition dipole moment. It should be noted that the OD stretch frequency reflects the surrounding solvent configuration, as can be inferred from eqs. (2) and (3). It is a reasonable assumption that the transition dipole moment of the OD stretch when the phenol is in solution is a function of the electric field along the OD bond, i.e.,

$$\mu_{OD}(E) \cong \mu_f + \left(\frac{\partial \mu_{OD}}{\partial E} \right)_0 E, \tag{6}$$

where μ_f is the transition dipole moment of the free phenol ($\mu_f = 0.96$ D Å^{-1} amu$^{-1/2}$ at MP2/cc-pVDZ). Inserting Eq. (2) into Eq. (6), we find

$$\mu_{OD}(\omega_{OD}) \cong \mu_f + \frac{1}{\kappa} \left(\frac{\partial \mu_{OD}}{\partial E} \right)_0 (\omega_{OD} - \omega_{OD}^0). \tag{7}$$

In order to determine the linear expansion coefficient, $(\partial \mu_{OD}/\partial E)_0$, we chose the phenol-benzene complex with the geometry optimized using the MP2/6-31G* method. Employing the FF partial charges of the benzene molecule, which were used to run the MD simulations, we calculated the E and μ_{OD} values for the configuration that was determined with the QM calculation. We then find $(\partial \mu_{OD}/\partial E)_0$ to be 52.7 D Å^{-1} amu$^{-1/2}$(a.u. E)$^{-1}$. Here it is noted that the experimentally measured transition dipole ratio μ_c/μ_f is $\sim$1.5. Within the assumption that the transition dipole moment of the free phenol is $\mu_f = 0.96$ D Å^{-1} amu$^{-1/2}$, we find that the experimentally estimated $(\partial \mu_{OD}/\partial E)_0$ value is about 60 D Å^{-1} amu$^{-1/2}$(a.u. E)$^{-1}$. In the following numerical simulations of IR absorption and 2D IR spectra, we will use the theoretically calculated value for $(\partial \mu_{OD}/\partial E)_0$.

C. The OD stretch absorption spectrum

The absorption line shape function is given by the Fourier transform of the quantum mechanical dipole correlation function as[3]

$$I(\omega) \sim \int_{-\infty}^{\infty} dt e^{i\omega t} \langle \mu(t) \cdot \mu(0) \rangle, \tag{8}$$

where μ denotes the quantum mechanical dipole operator. $I(\omega)$ can be rewritten in terms of the linear response function $J(t)$,[83]

$$I(\omega) \sim |\mu_{OD}(\omega)|^2 \int_{-\infty}^{\infty} dt e^{i\omega t} \bar{J}(t), \tag{9}$$

where $\bar{J}(t) = J(t) C_1(t) \exp(-i < \omega_{10} > t)$, and

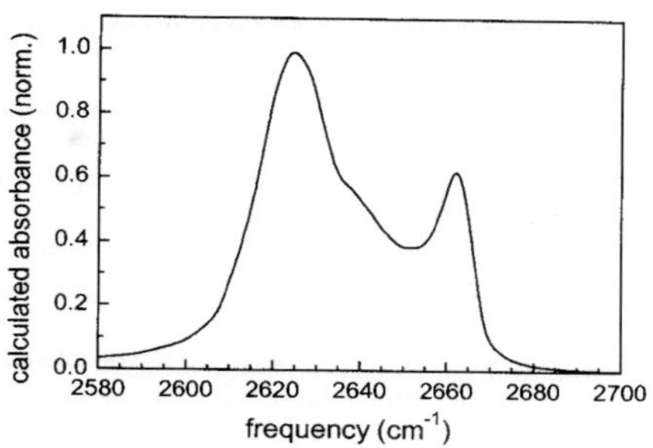

FIG. 9. Calculated OD stretch absorption spectrum.

$$J(t) \equiv \left\langle \exp_+ \left[-i \int_0^t d\tau \delta \hat{\omega}_{10}(\tau) \right] \right\rangle. \tag{10}$$

Here, $\delta \hat{\omega}_{10}(\tau)$ is the fluctuating angular frequency operator in the Heisenberg representation. $\langle \cdots \rangle$ is, in this case, the quantum mechanical trace over the bath eigenstates. $C_1(t)$ is the first-order rotational correlation function that describes the rotational relaxation of the phenol molecule in solution and is defined as $C_1(t) = \langle P_1(\cos \theta(t)) \rangle$, where P_1 is the first-order Legendre polynomial and $\theta(t)$ is the angle between the dipole vector at time zero and that at time t. As is the standard practice, vibration-rotation coupling effects are ignored.[83] As discussed above, the configuration-dependent transition dipole moment is taken into account through the use of Eq. (7).

Because the OD stretch frequency distribution is not Gaussian, one cannot use the second-order truncated cumulant expansion technique[83] to calculate the linear response function, $J(t)$. Therefore, we instead use a classical ensemble averaging method.[59] The linear response function in Eq. (10) is approximated as

$$J_c(t) = \left\langle \exp \left[-i \int_0^t d\tau \delta \omega_{10}(\mathbf{q}, \mathbf{p}, \tau) \right] \right\rangle_c, \tag{11}$$

where the fluctuating part of the frequency is replaced with a classical function, i.e., $\delta \omega_{10} = \delta \omega_{OD} = \omega_{OD} - \langle \omega_{OD} \rangle_c$, in the phase space of the bath degrees of freedom.

The lifetime broadening effect, which is very small, is taken into account by using the normal approach,

$$J_c(t) \rightarrow J_c(t) \exp(-t/2T_1), \tag{12}$$

where the lifetime of the first excited state is taken to be 11 ps, which is approximately the average of the lifetime of the complex (10 ps) and the lifetime of free phenol (12.5 ps).[4]

Figure 9 displays the OD stretch absorption spectrum obtained from the simulations. The calculated spectrum

J. Chem. Phys. **125**, 244508 (2006)

should be compared to the experimental spectrum displayed in Fig. 1(a). The peak positions are close to correct. The calculated peak positions are compared to the experimental values of 2631 and 2665 cm^{-1}. The peak for the complex is wider than for the free phenol, as is true of the experimental spectrum. However, the calculation yields a band for the complex that is somewhat too large relative to the size of the free phenol band. In addition, the calculated spectrum displays a shoulder between the two bands that is not evident in the experiment. However, the shoulder may be obscured in the experiment by the much large size of the free phenol peak. Given the complexity of the system involving two species, the complexed and free phenol, and the large range of local solvent environment, which are discussed in detail below, the agreement between the experimental and calculated spectra is reasonably good.

D. 2D vibrational echoes: Theory

The nonlinear response functions that are directly associated with 2D IR spectroscopy have been presented and discussed in detail.[84] As mentioned above, the frequency distribution of the OD stretch deviates strongly from a Gaussian function so that it is not possible to use the same second-order cumulant approximate expressions. Therefore, to calculate the corresponding nonlinear response functions denoted as $\Phi_j(t_3,t_2,t_1)$,[59,84] we will employ the ensemble averaging procedure used to calculate the linear response function in Sec. IV C. Furthermore, we will assume that $\delta\omega_{21}(t) = \delta\omega_{10}(t)$, which is the harmonic approximation.[60] Then, we have

$$\Phi_1(t_3,t_2,t_1) = -2\exp\{-i\langle\omega_{21}\rangle t_3 + i\langle\omega_{10}\rangle t_1\}$$
$$\times \Psi_A(t_3,t_2,t_1)\Gamma_{TA}(t_3,t_2,t_1)Y(t_3,t_2,t_1),$$

$$\Phi_2(t_3,t_2,t_1) = -2\exp\{-i\langle\omega_{21}\rangle t_3 - i\langle\omega_{10}\rangle t_1\}$$
$$\times \Psi_B(t_3,t_2,t_1)\Gamma_{TA}(t_3,t_2,t_1)Y(t_3,t_2,t_1),$$

$$\Phi_3(t_3,t_2,t_1) = \exp\{-i\langle\omega_{10}\rangle t_3 + i\langle\omega_{10}\rangle t_1\}$$
$$\times \Psi_A(t_3,t_2,t_1)\Gamma_{SE}(t_3,t_2,t_1)Y(t_3,t_2,t_1),$$

$$\Phi_4(t_3,t_2,t_1) = \exp\{-i\langle\omega_{10}\rangle t_3 - i\langle\omega_{10}\rangle t_1\}$$
$$\times \Psi_B(t_3,t_2,t_1)\Gamma_{SE}(t_3,t_2,t_1)Y(t_3,t_2,t_1), \qquad (13)$$

$$\Phi_5(t_3,t_2,t_1) = \exp\{-i\langle\omega_{10}\rangle t_3 + i\langle\omega_{10}\rangle t_1\}$$
$$\times \Psi_A(t_3,t_2,t_1)\Gamma_{GB}(t_3,t_2,t_1)Y(t_3,t_2,t_1),$$

$$\Phi_6(t_3,t_2,t_1) = \exp\{-i\langle\omega_{10}\rangle t_3 - i\langle\omega_{10}\rangle t_1\}$$
$$\times \Psi_B(t_3,t_2,t_1)\Gamma_{GB}(t_3,t_2,t_1)Y(t_3,t_2,t_1),$$

where the dephasing-induced line broadening factors, $\Psi_A(t_3,t_2,t_1)$ and $\Psi_B(t_3,t_2,t_1)$, are defined as

$$\Psi_A(t_3,t_2,t_1) \equiv \left\langle \exp\left\{i\int_0^{t_1} d\tau\,\delta\omega_{10}(\tau)\right\} \right.$$
$$\left. \times \exp\left\{-i\int_{t_1+t_2}^{t_1+t_2+t_3} d\tau\,\delta\omega_{10}(\tau)\right\} \right\rangle,$$

$$\Psi_B(t_3,t_2,t_1) \equiv \left\langle \exp\left\{-i\int_0^{t_1} d\tau\,\delta\omega_{10}(\tau)\right\} \right.$$
$$\left. \times \exp\left\{-i\int_{t_1+t_2}^{t_1+t_2+t_3} d\tau\,\delta\omega_{10}(\tau)\right\} \right\rangle. \qquad (14)$$

The first two contributions, $\Phi_1(t_3,t_2,t_1)$ and $\Phi_2(t_3,t_2,t_1)$, describe the induced transient absorption (TA) between the $v=1$ state and the $v=2$ state, $\Phi_3(t_3,t_2,t_1)$ and $\Phi_4(t_3,t_2,t_1)$ are associated with the stimulated emission (SE) process where the excited state ($v=1$) population evolution is involved, and finally $\Phi_5(t_3,t_2,t_1)$ and $\Phi_6(t_3,t_2,t_1)$ are associated with the ground-state bleaching (GB) contribution where a hole created on the ground state ($v=0$) evolves in time during the population period, t_2. Here, the transition dipole product term was not included in Eq. (13), and the configuration-dependent transition dipole moment will be taken into consideration later in Eq. (19) when the 2D IR spectrum is calculated. The factor of 2 in $\Phi_1(t_3,t_2,t_1)$ and $\Phi_2(t_3,t_2,t_1)$ is included because these contributions involve vibrational transition from the first excited state to the second excited state (two interactions with the radiation field) and within the harmonic approximation the transition dipole for this transition is $\sqrt{2}$ bigger than the $v=0$ to $v=1$ transition dipole.

Denoting the inverse lifetimes of the first and second excited states as γ_1 and γ_2, respectively, we find that the lifetime-broadening factors in Eqs. (13) are given as

$$\Gamma_{TA}(t_3,t_2,t_1) = \exp\left\{-\frac{(\gamma_1+\gamma_2)t_3}{2} - \gamma_1 t_2 - \frac{\gamma_1 t_1}{2}\right\},$$

$$\Gamma_{SE}(t_3,t_2,t_1) = \exp\left\{-\frac{\gamma_1 t_3}{2} - \gamma_1 t_2 - \frac{\gamma_1 t_1}{2}\right\}, \qquad (15)$$

$$\Gamma_{GB}(t_3,t_2,t_1) = \exp\left\{-\frac{\gamma_1 t_3}{2} - \gamma_1 t_2 - \frac{\gamma_1 t_1}{2}\right\}.$$

The lifetimes of the first excited state of the complex and the free phenol were determined experimentally.[4] As discussed above, a value of 11 ps is used. Within the harmonic approximation, the lifetime of the second excited state is a factor of 2 shorter. This approximation is adequate because the lifetime is much longer than the coherence period, t_3.

Finally, the rotational relaxation of phenol molecule in solution contributes to the total nonlinear response function and it is taken into consideration by the auxiliary function $Y(t_3,t_2,t_1)$, defined as[85,86]

$$Y(t_3,t_2,t_1) = \tfrac{1}{9}C_1(t_3)C_2(t_2)C_1(t_1), \qquad (16)$$

where $C_1(t_i)$ was previously defined and

J. Chem. Phys. **125**, 244508 (2006)

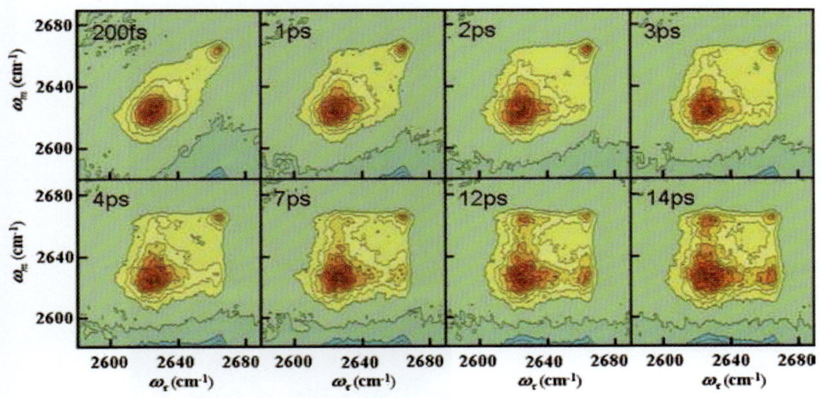

FIG. 10. (Color) 2D vibrational echo spectra calculated from the MD simulations. As T_w increases the off-diagonal chemical exchange peaks grow in. Compared to the experimental results shown in Figs. 1(b) and 1(c).

$$C_2(t_2) = \left(1 + \tfrac{4}{5}\langle P_2(\cos\theta(t_2))\rangle\right). \tag{17}$$

Here, $P_2(x)$ is the second-order Legendre polynomial. In the present numerical simulation of 2D IR spectra, we shall use $C_1(t_i)$ and $C_2(t_2)$ in the inset of Fig. 8, which were obtained from MD trajectories.

To quantitatively determine the 2D IR vibrational echo spectra including chemical exchange between the complex and free phenol, the two dephasing-induced line broadening factors, $\Psi_A(t_3,t_2,t_1)$ and $\Psi_B(t_3,t_2,t_1)$, defined in Eq. (14) are calculated from the MD trajectories. Once these three-dimensional functions, $\Psi_A(t_3,t_2,t_1)$ and $\Psi_B(t_3,t_2,t_1)$, lifetime broadening factors, and rotational relaxation terms are determined, the 2D spectra are obtained using[59]

$$\tilde{\Phi}_j(\omega_1,\omega_3;\tau) = \int_0^\infty dt_3 \int_0^\infty dt_1 \exp(i\omega_3 t_3 - i\omega_1 t_1)$$

$$\times \Phi_j(t_3,t_2=\tau,t_1) \quad \text{(for } j=1,3, \text{ and } 5),$$
$$\tag{18}$$

$$\tilde{\Phi}_k(\omega_1,\omega_3;\tau) = \int_0^\infty dt_3 \int_0^\infty dt_1 \exp(i\omega_3 t_3 + i\omega_1 t_1)$$

$$\times \Phi_k(t_3,t_2=\tau,t_1) \quad \text{(for } k=2,4, \text{ and } 6).$$

Here the 2D vibrational echo spectrum, $S_{2D}(\omega_1,\omega_3;\tau)$, is defined as

$$S_{2D}(\omega_1,\omega_3;\tau) = |\mu_{OD}(\omega_1)|^2 |\mu_{OD}(\omega_3+\Delta)|^2$$

$$\times \text{Re}\left[\sum_{i=1}^{2} \tilde{\Phi}_i(\omega_1,\omega_3;\tau)\right]$$

$$+ |\mu_{OD}(\omega_1)|^2 |\mu_{OD}(\omega_3)|^2$$

$$\times \text{Re}\left[\sum_{i=3}^{6} \tilde{\Phi}_i(\omega_1,\omega_3;\tau)\right], \tag{19}$$

where Δ is the overtone anharmonicity frequency of 91 cm^{-1}. The configuration-dependent transition dipole is considered in this expression. Note that the transition dipole depends on local solvation environment through E (electric field). And then, using the relation between OD stretch mode frequency and E, we can find the frequency-dependent

transition dipole moment. Here, the prefactor, $|\mu_{OD}(\omega_1)|^2 |\mu_{OD}(\omega_3)|^2$, approximately describes the configuration-dependent transition dipole moment in the context of 2D IR nonlinear response function.

E. Simulated 2D vibrational echo spectra and comparison to experiment

Figure 10 displays the 2D vibrational echo spectra calculated from the simulations as described above. The plots display the $v=0-1$ regions of the spectrum. At the shortest T_w (time between pulses 2 and 3), there are only peaks on the diagonal. As T_w increases off-diagonal peaks grow in. The increase in these peaks reflects the chemical exchange in which complexes are formed and dissociate.[59] The spectra at the shortest and longest T_w's can be compared to the experimental data[4] shown in Figs. 1(b) and 1(c). The plots in Fig. 10 capture the main features of the experimental data. However, the peak on the diagonal associated with the complex ($\omega_m = 2630$ cm^{-1}) is too large compared to the diagonal free peak ($\omega_m = 2663$ cm^{-1}) when compared to the experimental results. This is not a failure of the methodology for calculating the nonlinear signal, but rather it is in accord with the equilibrium constant of 2.7 rather than the experimental value of 1 and the linear spectrum which shows that the size of the band for the phenol-benzene complex relative to that for free phenol is too large compared to the experimental spectrum [see Figs. 9 and 1(a)].

It is difficult to compare the calculated and experimental chemical exchange dynamics by comparing the 2D plots directly. This is particularly true because the peak heights are influenced by spectral diffusion, which causes the widths of the peaks to increase and their amplitudes to decrease. However, spectral diffusion does not change the peak volumes. Only population dynamics change the peak volumes. It has been demonstrated that the peak volumes can be used to extract the chemical exchange dynamics.[6] The vibrational relaxation to the ground state and orientational relaxation cause all of the peaks to decrease, while chemical exchange causes the diagonal peaks to decrease but the off-diagonal peaks to grow in. The calculations include all three contributions to the peak volumes. Figures 11(a) and 11(b) show the

244508-12   Kwac *et al.*

J. Chem. Phys. **125**, 244508 (2006)

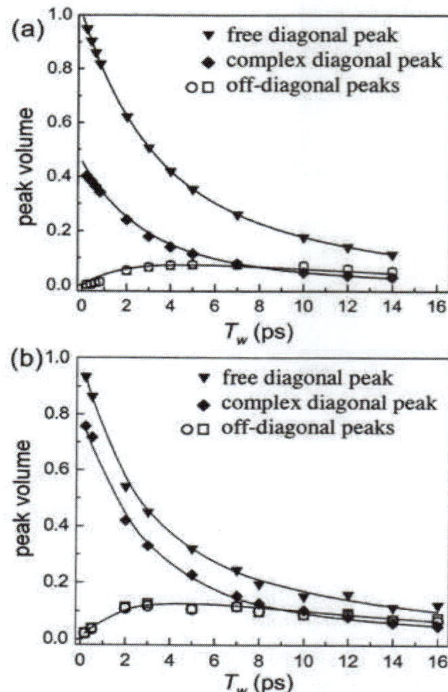

FIG. 11. Calculated (a) and experimental (b) diagonal and off-diagonal peak volumes. The peaks on the diagonal arise from the free phenol (higher frequency) and the phenol-benzene complex (lower frequency). The off-diagonal peaks are formed by formation and dissociation of the complex.

calculated and experimental T_w dependent peak volumes, respectively. The lines through the calculated and experimental data used a single adjustable parameter in the fits, the complex dissociation time, τ_d, which is the inverse of the complex dissociation rate. As discussed in the introduction, τ_d =8 ps from the experimental fits. In the experiments the orientational relaxation rates, the lifetimes, the ratio of the transition dipole for the two peaks, and the equilibrium constant were all measured separately and used as input parameters. In the calculations, the lifetime was taken from the experiments but everything else including the configuration-dependent transition dipole, the orientational relaxation, and the equilibrium between the complex and free phenol are contained in the simulation. These factors determine not only the time dependence of the peaks but the relative amplitudes of the peaks as a function of time. While not perfect, the simulations do a respectable job reproducing the chemical exchange and the other dynamics of the system.

From the simulations and using the theory in Ref. 61 and the fitting procedure used previously to analyze the experimental data,[4–6] we found that the dissociation time constant τ_d is 20 ps, which is two and a half times larger than the experimental value of 8 ps. This value of τ_d is somewhat too slow, but given the complexity of the problem, the simulation describes the system characteristics almost quantitatively. The value of the dissociation time is consistent with

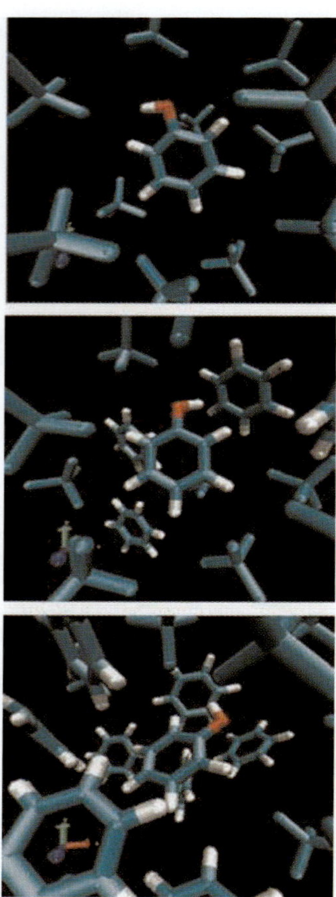

FIG. 12. (Color) Representative configurations extracted from the simulations. Top panel: free phenol surrounded by CCl_4 molecules ($X_b \sim 0$). Middle panel: phenol-benzene complex surrounded by a mix of benzene and CCl_4 molecule ($X_b \sim 0.5$). Bottom panel: phenol-benzene complex surrounded mainly by benzene molecules ($X_b \sim 1$).

the value for the equilibrium $[\text{complex}]_{\text{eq}}/[\text{free}]_{\text{eq}}=2.7$ found in this study. The equilibrium constant is too large compared to experiment and the dissociation time is too slow. Both indicate that the classical force field used in the simulations overestimates the strength of the phenol-benzene complex bond.

V. LOCAL SOLVATION ENVIRONMENT AND RELAXATION

From the MD simulation trajectories, three representative snapshot configurations are shown in Fig. 12. The top panel shows the situation where the free phenol molecule is predominantly surrounded by CCl_4 molecules ($X_b \sim 0$). The middle panel shows a phenol-benzene complex surrounded by a mix of benzene and CCl_4 molecules ($X_b \sim 0.5$), and the bottom panel shows a complex with the surrounding molecules mainly benzenes ($X_b \sim 1$). This suggests that the benzene//CCl_4 mixed solution is not homogeneous at the level of the solute molecule, in this case, phenol and that microscopic solvent domains that are rich in benzene or CCl_4 can exist in this mixed solution. Then, the free-complex dynamical equilibrium process in part involves phenol changing between locally inhomogeneous solvent environments. The chemical exchanges between complex and free phenol are directly probed in the 2D IR spectroscopic measurements. However, another issue that is interesting to study is the local solvation dynamics and inhomogeneity of the local environments around the solute in the mixed solvent. Such dynamics gives rise to spectral diffusion.

Recently an initial experimental analysis of the spectral diffusion of the complex and the free phenol was performed on the benzene phenol system.[6] However, the extraction of the FFCFs of the two species is complicated by the chemical exchange process.[6] Additional experiments are underway on a similar system in which the chemical exchange is much slower than the spectral diffusion, which will greatly simplify the analysis of the spectral diffusion.[87] Here we will address the issue of extracting information on the fluctuation of solvent molecules within the first solvation shell around the solute from vibrational echo spectroscopy. The question arises as to which observable or correlation function can be used to retrieve information on the number of solvent molecules in the vicinity of the solute. In the present section, we will provide a line of theoretical reasoning and plausible answers to these interesting questions.

A. Statistical aspects

To establish the connection between the distribution of microscopically inhomogeneous environments and the spectroscopically measurable OD stretch frequency, we have analyzed the MD trajectories and examined the solvent molecular distribution around the phenol OD chromophore. From the phenol in benzene and phenol in CCl_4 solutions, we calculated the radial distribution functions. We found that the average radius of the first solvation shell is about 5 Å from the center of mass of the phenol OD bond. Now, for the phenol in benzene/CCl_4 solution, we separately counted the number of benzene and CCl_4 molecules within the sphere around the OD bond with a radius of 5 Å. The two numbers are denoted as N_b and N_c. If the center of mass of solvent molecule is inside this solvation shell, that molecule is counted. We found that the average values, $\langle N_b \rangle$ and $\langle N_c \rangle$, are 1.166 and 1.319, respectively. The local number fraction of benzene molecule is then defined as

$$X_b(t) = \frac{N_b(t)}{N_b(t) + N_c(t)}. \tag{20}$$

This value fluctuates in time, and its magnitude is a measure of local inhomogeneity of the solvent molecules in the first solvation shell. $X_b(t)$ differs from the macroscopic mole fraction that is constant in time. Dynamical relaxation of the variables, $X_b(t)$, $N_b(t)$, and $N_c(t)$, will be discussed later in this section.

Because the numbers of benzene and CCl_4 molecules in the first solvation shell are finite and typically less than 5 for

TABLE IV. Local number fraction of benzene molecules X_b and its probability (%).

X_b	Probability (%)
0	15.7
0.2	0.2
0.25	3.9
0.333	17.5
0.4	0.3
0.5	28.7
0.6	0.1
0.667	11.8
0.75	1.2
0.8	0.03
1	20.5

benzene and 6 for CCl_4, X_b values are discrete (see Table IV). The probability distribution of X_b can be obtained from the MD trajectories, and it is plotted in Fig. 13(a). Surprisingly, the distribution is not uniform nor symmetric around the value of macroscopic mole fraction of benzene, ~0.29. If we considered a sphere with much larger radius and count the number of included benzene and CCl_4 molecules within the sphere, the distribution obtained would been broad and close to normal distribution with maximum at ~0.29. In Table IV, the probabilities in percent for varying X_b are summarized. Among these, the most probable X_b value is 0.5, which is larger than the macroscopic benzene mole fraction of 0.286. Furthermore, the probabilities of finding X_b value to be 0.333, 0.667, and 1 are 17.5%, 11.8%, and 20.5%, respectively. This observation suggests (1) that the local sol-

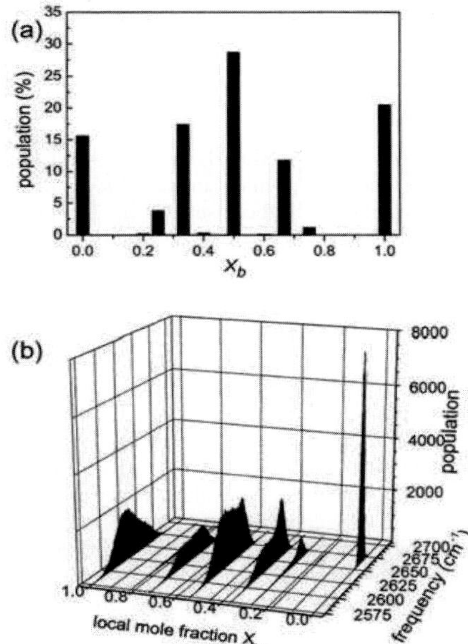

FIG. 13. (a) Population distribution of X_b. (b) Population distribution with respect to X_b and OD stretch mode frequency.

J. Chem. Phys. **125**, 244508 (2006)

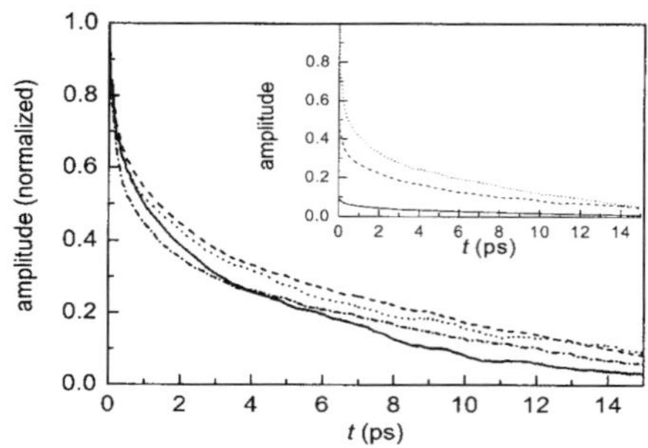

FIG. 14. Normalized correlation functions, $C_{\delta X_b}(t)/C_{\delta X_b}(0)$ (dashed line), $C_{\delta N_b}(t)/C_{\delta N_b}(0)$ (dotted line), and $C_{\delta N_c}(t)/C_{\delta N_c}(0)$ (dash-dotted line), and the normalized OD stretch mode frequency-frequency correlation function, $C_\omega(t)/C_\omega(0)$ (solid line). In the inset, $C_{\delta X_b}(t)$ (solid line), $C_{\delta N_b}(t)$ (dashed line), and $C_{\delta N_c}(t)$ (dotted line).

vation environment in the first solvation shell around the phenol is fairly different from the bulk, (2) that each solute phenol has a discretely different inhomogeneous solvation structure at a given time, and (3) that the phenol is preferentially solvated by benzene. In addition, two limiting cases of $X_b=0$ and $X_b=1$ significantly populate, indicating that the two different solvent molecules can approximately form microscopic domains at least in the vicinity of the phenol, where by microscopic domain we mean a local region where one solvent species is predominantly rich in number.

To find the correlation between X_b and OD stretch mode frequency, we obtained the distributions of the OD frequencies for each X_b. These are plotted in Fig. 13(b). If $X_b=0$, the OD frequency distribution is quite narrow, and its center is around 2670 cm^{-1}. As X_b increases, the distribution becomes broad and its maximum position gradually shifts to lower frequency, as expected.

B. Dynamical aspects

We next consider relaxation dynamics of variables such as $X_b(t)$, $N_b(t)$, and $N_c(t)$, which are reflections of the local solvation environments. In the inset of Fig. 14, the correlation functions $C_{\delta X_b}(t)$ (red), $C_{\delta N_b}(t)$ (blue), and $C_{\delta N_c}(t)$ (green) are plotted, where

$$C_{\delta X_b}(t) = \langle (X_b(t) - \langle X_b \rangle)(X_b(0) - \langle X_b \rangle) \rangle,$$

$$C_{\delta N_b}(t) = \langle (N_b(t) - \langle N_b \rangle)(N_b(0) - \langle N_b \rangle) \rangle, \qquad (21)$$

$$C_{\delta N_c}(t) = \langle (N_c(t) - \langle N_c \rangle)(N_c(0) - \langle N_c \rangle) \rangle.$$

All three correlation functions have a fast and slow decay component, though their initial values differ from one another. To more readily compare the decays of the correlation functions, the normalized correlation functions are plotted in the main part of Fig. 14. As can be seen in the figure, the decays of the three correlation functions are very similar. Furthermore, the normalized FFCF, $C_\omega(t)$ [see Eq. (4)], is also plotted in Fig. 14 as the black curve and found to be quite close to the normalized correlation functions of the local concentrations in the first solvation shell. This observa-

tion is quite important because one might be able to use the experimentally measurable $C_\omega(t)/C_\omega(0)$ function to infer the local solvent dynamics in the first solvation shell, for example, $C_{\delta X_b}(t)/C_{\delta X_b}(0)$.

On the basis of the empirical observations made from Fig. 14,

$$C_\omega(t)/C_\omega(0) \approx C_{\delta X_b}(t)/C_{\delta X_b}(0) \approx C_{\delta N_b}(t)/C_{\delta N_b}(0), \qquad (22)$$

we propose the following ansatz. There is a simple relationship between the projected electric field E and the number of benzene molecules in the first solvation shell, i.e.,

$$E = \gamma N_b. \qquad (23)$$

Then, we have, from Eqs. (2) and (23),

$$C_\omega(t) = \kappa^2 \gamma^2 C_{\delta N_b}(t). \qquad (24)$$

Here, the proportionality constant γ is estimated to be 0.0065 a.u. E/benzene, where the electric field component is in a.u. The relationship in Eq. (24) suggests that by measuring the FFCF one can directly extract information on the solvent molecule concentration dynamics in the immediate vicinity of the solute phenol.

VI. CONCLUDING REMARKS

In this paper MD simulations were used to examine the dynamics of phenol in the mixed benzene/CCl$_4$ solvent. As has been well documented experimentally, phenol forms a complex with benzene, and at room temperature under thermal equilibrium conditions, the complexes are continually forming and dissociating.[4,6] Ultrafast 2D vibration echo experiments have been used to directly measure the chemical exchange between phenol in the complex and free forms. Although the exchange kinetics can be accurately determined from experiments, the experiments do not provide a microscopic picture of the formation and dissociation process.

The combination of the experimental results and the MD simulations amplify both approaches to understand solute-solvent complexes and the nature of solute interactions in complex solvent environments. The experimental results provide benchmarks for the simulations. The calculations of two types of observables from the MD simulations demonstrate that the simulations are of sufficient accuracy to produce usable insights into the details of the system. The simulations were able to do a reasonable job of reproducing both the linear IR absorption spectrum of the phenol hydroxyl stretch and the time-dependent 2D vibrational echo spectra. Of particular importance is that the simulations produced reasonable agreement with the determination of the experimentally measured complex dissociation time.

Perhaps the most interesting feature of the simulation results is the description of the number distribution of solvent molecules in the first solvation shell of the phenol. The number of benzenes on average was not the mole fraction of benzene in the solvent. Furthermore, as shown in Fig. 13, the number distribution of benzenes and CCl$_4$ in the first solvent shell is highly inhomogeneous. In fact, there are significant probabilities of finding a phenol surrounded either completely by benzene or completely by CCl$_4$. Furthermore, it was proposed that the time dependence of the inhomogeneous nature of the solvent environment about the solute can be probed experimentally through the 2D vibrational experiments using a relationship between the fraction of benzenes in the first solvation shell and the frequency-frequency correlation function.

ACKNOWLEDGMENTS

Three of the authors (K.K., J.Z., and M.D.F.) would like to thank the United States Air Force Office of Scientific Research (F49620-01-1-0018) for supporting their contribution to this research. Another author (M.C.) thanks for the financial support from CRIP of MOST, Korea.

[1] C. J. Cramer and D. G. Truhlar, Chem. Rev. (Washington, D.C.) **99**, 2161 (1999).

[2] J. Hansen and I. McDonald, *Theory of Simple Liquids* (Academic, London, 1976).

[3] D. McQuarrie, *Statistical Mechanics* (Harper & Row, New York, 1976).

[4] J. Zheng, K. Kwak, J. Asbury, X. Chen, I. R. Piletic, and M. D. Fayer, Science **309**, 1338 (2005).

[5] J. Zheng, K. Kwak, X. Chen, J. B. Asbury, and M. D. Fayer, J. Am. Chem. Soc. **128**, 2977 (2006).

[6] K. Kwak, J. Zheng, H. Cang, and M. D. Fayer, J. Phys. Chem. B **110**, 19998 (2006).

[7] H. J. Bakker, H. K. Neinhuys, G. Gallot, N. Lascoux, G. M. Gale, J. C. Leicknam, and S. Bratos, J. Chem. Phys. **116**, 2592 (2002).

[8] M. F. Kropman, H.-K. Nienhuys, S. Woutersen, and H. J. Bakker, J. Phys. Chem. A **105**, 4622 (2001).

[9] M. F. Kropman and H. J. Bakker, J. Chem. Phys. **115**, 8942 (2001).

[10] S. Woutersen and H. J. Bakker, Comments Mod. Phys. **2**, D99 (2000).

[11] H. J. Bakker, S. Woutersen, and H. K. Nienhuys, Chem. Phys. **258**, 233 (2000).

[12] C. J. Fecko, J. D. Eaves, J. J. Loparo, A. Tokmakoff, and P. L. Geissler, Science **301**, 1698 (2003).

[13] C. J. Fecko, J. J. Loparo, S. T. Roberts, and A. Tokmakoff, J. Chem. Phys. **122**, 054506 (2005).

[14] J. B. Asbury, T. Steinel, and M. D. Fayer, J. Lumin. **107**, 271 (2004).

[15] S. Yeremenko, M. S. Pshenichnikov, and D. A. Wiersma, Chem. Phys. Lett. **369**, 107 (2003).

[16] S. Yeremenko, M. S. Pshenichnikov, and D. A. Wiersma, Phys. Rev. A **73**, 021804 (2006).

[17] J. D. Eaves, A. Tokmakoff, and P. L. Geissler, J. Phys. Chem. A **109**, 9424 (2005).

[18] J. D. Eaves, J. J. Loparo, C. J. Fecko, A. Tokmakoff, and P. L. Geissler, Proc. Natl. Acad. Sci. U.S.A. **102**, 13019 (2005).

[19] T. Steinel, J. B. Asbury, S. A. Corcelli, C. P. Lawrence, J. L. Skinner, and M. D. Fayer, Chem. Phys. Lett. **386**, 295 (2004).

[20] S. Corcelli, C. P. Lawrence, J. B. Asbury, T. Steinel, M. D. Fayer, and J. L. Skinner, J. Chem. Phys. **121**, 8897 (2004).

[21] J. B. Asbury, T. Steinel, C. Stromberg, S. A. Corcelli, C. P. Lawrence, J. L. Skinner, and M. D. Fayer, J. Phys. Chem. A **108**, 1107 (2004).

[22] J. B. Asbury, T. Steinel, K. Kwak, S. Corcelli, C. P. Lawrence, J. L. Skinner, and M. D. Fayer, J. Chem. Phys. **121**, 12431 (2004).

[23] M. Bonn, H. J. Bakker, A. W. Kleyn, and R. A. van Santen, J. Phys. Chem. **100**, 15301 (1996).

[24] S. Woutersen, U. Emmerichs, and H. J. Bakker, J. Chem. Phys. **107**, 1483 (1997).

[25] M. A. F. H. van den Broek, H. K. Nienhuys, and H. J. Bakker, J. Chem. Phys. **114**, 3182 (2001).

[26] K. Gaffney, I. Piletic, and M. D. Fayer, J. Phys. Chem. A **106**, 9428 (2002).

[27] K. J. Gaffney, P. H. Davis, I. R. Piletic, N. E. Levinger, and M. D. Fayer, J. Phys. Chem. A **106**, 12012 (2002).

[28] J. B. Asbury, T. Steinel, C. Stromberg, K. J. Gaffney, I. R. Piletic, A. Goun, and M. D. Fayer, Chem. Phys. Lett. **374**, 362 (2003).

[29] J. B. Asbury, T. Steinel, C. Stromberg, K. J. Gaffney, I. R. Piletic, A. Goun, and M. D. Fayer, Phys. Rev. Lett. **91**, 237402 (2003).

[30] J. B. Asbury, T. Steinel, C. Stromberg, K. J. Gaffney, I. R. Piletic, and M. D. Fayer, J. Chem. Phys. **119**, 12981 (2003).

[31] K. J. Gaffney, I. R. Piletic, and M. D. Fayer, J. Chem. Phys. **118**, 2270 (2003).

[32] I. R. Piletic, K. J. Gaffney, and M. D. Fayer, J. Chem. Phys. **119**, 423 (2003).

[33] H.-S. Tan, I. R. Piletic, and M. D. Fayer, J. Chem. Phys. **122**, 174501 (2005).

[34] H.-S. Tan, I. R. Piletic, R. E. Riter, N. E. Levinger, and M. D. Fayer, Phys. Rev. Lett. **94**, 057405 (2004).

[35] I. Piletic, H.-S. Tan, and M. D. Fayer, J. Phys. Chem. B **109**, 21273 (2005).

[36] I. Piletic, D. E. Moilanen, D. B. Spry, and M. D. Fayer, J. Phys. Chem. A **110**, 4985 (2006).

[37] D. Cringus, J. Lindner, M. T. W. Milder, M. S. Pshenichnikov, P. Vohringer, and D. A. Wiersma, Chem. Phys. Lett. **408**, 162 (2005).

[38] A. M. Dokter, S. Woutersen, and H. J. Bakker, Phys. Rev. Lett. **94**, 178301 (2005).

[39] Q. Zhong, D. A. Steinhurst, E. E. Carpenter, and J. C. Owrutsky, Langmuir **18**, 7401 (2002).

[40] Q. Zhong, A. P. Baronavski, and J. C. Owrutsky, J. Chem. Phys. **118**, 7074 (2003).

[41] Q. Zhong, A. P. Baronavski, and J. C. Owrutsky, J. Chem. Phys. **119**, 9171 (2003).

[42] S. Woutersen, Y. Mu, G. Stock, and P. Hamm, Chem. Phys. **266**, 137 (2001).

[43] Y. S. Kim and R. M. Hochstrasser, Proc. Natl. Acad. Sci. U.S.A. **102**, 11185 (2005).

[44] E. L. Hahn, Phys. Rev. **80**, 580 (1950).

[45] I. D. Abella, N. A. Kurnit, and S. R. Hartmann, Phys. Rev. **14**, 391 (1966).

[46] C. P. Lawrence and J. L. Skinner, J. Chem. Phys. **117**, 8847 (2002).

[47] A. Piryatinski and J. L. Skinner, J. Phys. Chem. B **106**, 8055 (2002).

[48] C. P. Lawrence and J. L. Skinner, J. Chem. Phys. **118**, 264 (2003).

[49] C. P. Lawrence and J. L. Skinner, Chem. Phys. Lett. **369**, 472 (2003).

[50] A. Piryatinski, C. P. Lawrence, and J. L. Skinner, J. Chem. Phys. **118**, 9664 (2003).

[51] A. Piryatinski, C. P. Lawrence, and J. L. Skinner, J. Chem. Phys. **118**, 9672 (2003).

[52] S. Corcelli, C. P. Lawrence, and J. L. Skinner, J. Chem. Phys. **120**, 8107 (2004).

[53] J. R. Schmidt, S. A. Corcelli, and J. L. Skinner, J. Chem. Phys. **123**, 044513 (2005).

[54] R. Rey and J. T. Hynes, J. Chem. Phys. **104**, 2356 (1996).

[55] R. Rey, K. B. Møller, and J. T. Hynes, J. Phys. Chem. A **106**, 11993 (2002).

[56] K. B. Møller, R. Rey, and J. T. Hynes, J. Phys. Chem. A **108**, 1275 (2004).

[57] R. Rey, K. B. Moller, and J. T. Hynes, Chem. Rev. (Washington, D.C.) **104**, 1915 (2004).

[58] D. Laage and J. T. Hynes, Science **311**, 832 (2006).

[59] K. Kwac, H. Lee, and M. Cho, J. Chem. Phys. **120**, 1477 (2004).

[60] K. Kwac and M. Cho, J. Chem. Phys. **119**, 2256 (2003).

[61] K. Kwac and M. Cho, J. Chem. Phys. **119**, 2247 (2003).

[62] K. Kwac and M. Cho, J. Raman Spectrosc. **36**, 326 (2005).

[63] S. Kristyan and P. Pulay, Chem. Phys. Lett. **229**, 175 (1994).

[64] M. O. Sinnokrot, E. F. Valeev, and C. D. Sherrill, J. Am. Chem. Soc. **124**, 10887 (2002).

[65] S. Tsuzuki, K. Honda, T. Uchimaru, M. Mikami, and K. Tanabe, J. Am. Chem. Soc. **124**, 104 (2002).

[66] J. Cerny and P. Hobza, Phys. Chem. Chem. Phys. **7**, 1624 (2005).

[67] S. Tsuzuki and H. P. Luthi, J. Chem. Phys. **114**, 3949 (2001).

[68] C. Møller and M. S. Plesset, Phys. Rev. **46**, 618 (1934).

[69] K. Raghavachari, G. W. Trucks, J. A. Pople, and M. Head-Gordon, Chem. Phys. Lett. **157**, 479 (1989).

[70] Y. Jung, R. C. Lochan, T. Dutoi, and M. Head-Gordon, J. Chem. Phys. **121**, 9793 (2004).

[71] M. Feyereisen, G. Fitzgerald, and A. Komornicki, Chem. Phys. Lett. **208**, 359 (1993).

[72] S. F. Boys and F. Bernardi, Mol. Phys. **19**, 553 (1970).

[73] T. Helgaker, W. Klopper, H. Koch, and J. Noga, J. Chem. Phys. **106**, 9639 (1997).

[74] M. J. Frisch, G. W. Trucks, H. B. Schlegel et al., GAUSSIAN03, Gaussian, Inc., Pittsburgh, PA, 2003.

[75] Y. Shao, L. F. Molnar, Y. Jung et al., Phys. Chem. Chem. Phys. **8**, 3172 (2006).

244508-16   Kwac *et al.*

J. Chem. Phys. **125**, 244508 (2006)

[76] Y. Jung and M. Head-Gordon, Phys. Chem. Chem. Phys. **8**, 2831 (2006).
[77] P. F. B. Goncalves and H. Stassen, J. Chem. Phys. **123**, 214109 (2005).
[78] D. A. Case, D. A. Pearlman, J. W. Caldwell *et al.*, AMBER 8, University of California, San Francisco, 2004).
[79] E. M. Duffy, D. L. Severance, and W. L. Jorgensen, J. Am. Chem. Soc. **114**, 7535 (1992).
[80] T. Darden, D. York, and L. Pedersen, J. Chem. Phys. **98**, 10089 (1993).
[81] H. J. C. Berendsen, J. P. M. Postma, W. F. v. Gunsteren, A. DiNola, and J. R. Haak, J. Chem. Phys. **81**, 3684 (1984).
[82] J. R. Schmidt, S. A. Corcelli, and J. L. Skinner, J. Chem. Phys. **121**, 8887 (2004).
[83] S. Mukamel, *Principles of Nonlinear Optical Spectroscopy* (Oxford University Press, Oxford, 1995).
[84] M. Cho, J. Chem. Phys. **115**, 4424 (2001).
[85] M. F. DeCamp, L. DeFlores, J. M. McCracken, A. Tokmakoff, K. Kwac, and M. Cho, J. Phys. Chem. B **109**, 11016 (2005).
[86] O. Golonzka and A. Tokmakoff, J. Phys. Chem. B **115**, 297 (2001).
[87] S. Park, K. Kwak, and M. D. Fayer (unpublished).
[88] A. P. Scott and L. Radom, J. Phys. Chem. **100**, 16502 (1996).

Hydrogen Bond Lifetimes and Energetics for Solute/Solvent Complexes Studied with 2D-IR Vibrational Echo Spectroscopy

Junrong Zheng and Michael D. Fayer*

Contribution from the Department of Chemistry, Stanford University, Stanford, California 94305

Received November 3, 2006; E-mail: fayer@stanford.edu

Abstract: Weak π hydrogen-bonded solute/solvent complexes are studied with ultrafast two-dimensional infrared (2D-IR) vibrational echo chemical exchange spectroscopy, temperature-dependent IR absorption spectroscopy, and density functional theory calculations. Eight solute/solvent complexes composed of a number of phenol derivatives and various benzene derivatives are investigated. The complexes are formed between the phenol derivative (solute) in a mixed solvent of the benzene derivative and CCl_4. The time dependence of the 2D-IR vibrational echo spectra of the phenol hydroxyl stretch is used to directly determine the dissociation and formation rates of the hydrogen-bonded complexes. The dissociation rates of the weak hydrogen bonds are found to be strongly correlated with their formation enthalpies. The correlation can be described with an equation similar to the Arrhenius equation. The results are discussed in terms of transition state theory.

I. Introduction

Hydrogen bonds play important roles in chemistry and biology.[1-5] Most hydrogen bonds have dissociation enthalpies in the range of $1-10$ kcal/mol. This range is much smaller than a typical covalent bond enthalpy. For instance, a carbon–carbon single bond has a dissociation enthalpy of $\sim$80 kcal/mol.[1] In spite of being relatively weak, hydrogen bonds are strong enough to have a profound influence on the nature of liquids and material, and on the reactivity of hydrogen-bonded molecules. An important example is water, which is a liquid at room temperature rather than a gas because of its hydrogen-bond network. Another important aspect of hydrogen bonds is that they are weak enough that they can continually dissociate and reform at room temperature. Some of the most significant biological processes, such as DNA replication and protein folding, are made possible because of the reversible nature of hydrogen-bond formation.[4]

The reversibility of hydrogen bonding brings up two interesting and fundamental questions. (1) How long does a hydrogen bond stay "bonded" at room temperature? (2) Is there any correlation between the hydrogen bond lifetime and its strength? In general, it might be expected that stronger hydrogen bonds would dissociate more slowly. However, direct experimental data addressing these issues for relatively weak hydrogen bonds that can readily dissociate at room temperature are rare because there has been a lack of appropriate measurement techniques.

For some strong hydrogen bonds that have dissociation enthalpies greater than 5 kcal/mol, the hydrogen bond lifetimes are sufficiently long that they can be measured with temperature-jump or ultrasonic methods.[2] Measurements of the lifetimes of relatively weak hydrogen bonds (<5 kcal/mol) at room temperature under thermal equilibrium conditions have recently become possible because of the advent of ultrafast IR techniques.[6-11] To the best of our knowledge, no systematic study on the relationship between the lifetimes and strength of hydrogen bonds has been reported.

In this paper, we present a study of the lifetimes of hydrogen bonds between a series of phenol derivatives and benzene derivatives using the recently developed method of ultrafast 2D-IR vibrational echo chemical exchange spectroscopy.[9-15] The experiments described here follow previous work that examined the hydrogen-bond formation and dissociation rates of the phenol/benzene complex in liquid solution of low concentration phenol in the mixed benzene/CCl_4 solvent.[9,10] A phenol molecule forms a weak hydrogen bond, sometimes called "π hydrogen bond," with a benzene molecule. By adding different substituents to either phenol or benzene, the strength of the hydrogen bond can be modified. Using the 2D vibrational echo chemical exchange technique, the dissociation times of a series

(1) Vinogradov, S. N.; Linnell, R. H. *Hydrogen Bonding*; Van Nostrand Reinhold Company: New York, 1971.
(2) Joesten, M. D.; Schaad, L. J. *Hydrogen Bonding*; Marcel Dekker, Inc.: New York, 1974.
(3) Jeffrey, G. A. *An Introduction to Hydrogen Bonding*; Oxford University Press, Inc.: New York, 1997.
(4) Desiraju, G. R.; Steiner, T. *The Weak Hydrogen Bond*; Oxford: New York, 1999.
(5) Pimentel, G. C.; McClellan, A. L. *Annu. Rev. Phys. Chem.* **1971**, *22*, 347–385.
(6) Arrivo, S. M.; Heilweil, E. J. *J. Phys. Chem.* **1996**, *100*, 11975.
(7) Arrivo, S. M.; Kleiman, V. D.; Dougherty, T. P.; Heilweil, E. J. *Opt. Lett.* **1997**, *22*, 1488–1490.
(8) Woutersen, S.; Mu, Y.; Stock, G.; Hamm, P. *Chem. Phys.* **2001**, *266*, 137–147.
(9) Zheng, J.; Kwak, K.; Asbury, J. B.; Chen, X.; Piletic, I.; Fayer, M. D. *Science* **2005**, *309*, 1338–1343.
(10) Zheng, J.; Kwak, K.; Chen, X.; Asbury, J. B.; Fayer, M. D. *J. Am. Chem. Soc.* **2006**, *128*, 2977–2987.
(11) Kim, Y. S.; Hochstrasser, R. M. *Proc. Natl. Acad. Sci. U.S.A.* **2005**, *102*, 11185–11190.
(12) Zheng, J.; Kwac, K.; Xie, J.; Fayer, M. D. *Science* **2006**, *313*, 1951–1955.
(13) Zheng, J.; Kwak, K.; Fayer, M. D. *Acc. Chem. Res.* **2006**, *40*, 75–83.
(14) Kwak, K.; Zheng, J.; Cang, H.; Fayer, M. D. *J. Phys. Chem. B* **2006**, *110*, 19998–20013.
(15) Sanda, F.; Mukamel, S. *J. Chem. Phys.* **2006**, *125*, 014507.

10.1021/ja067760f CCC: $37.00 © xxxx American Chemical Society

of hydrogen-bonded complexes are measured. In addition, temperature-dependent FT-IR absorption experiments are used to determine the formation enthalpies of the hydrogen bonds, and DFT calculations provide reasonable estimates of their structures. The results show a systematic correlation between the hydrogen-bond strengths (formation enthalpy) and the hydrogen-bond dissociation times.

II. Experimental Procedures

Details of the experimental method used for these 2D-IR vibrational echo spectroscopy experiments have been described previously[10,16] and are also included in the Supporting Information for the convenience of the readers. In addition, a diagram of the experimental setup and a schematic representation of how the 2D vibrational echo method is used to measure chemical exchange are provided in the Supporting Information. Very briefly, in a 2D-IR vibrational echo chemical exchange experiment, three ultrashort IR pulses tuned to the frequency range of the vibrational modes of interest are crossed in the sample. Because the pulses are very short, they have broad bandwidths (determined by the uncertainty principle) making it possible to simultaneously excite a number of vibrational bands. The first laser pulse places the vibrational oscillators in coherent superposition states of the ground state (0) and the first excited state (1) and "labels" the initial structures of the species in the sample by setting their initial frequencies along the ω_τ axis. The second pulse ends the first time period τ by transforming the coherences into populations (either in the 0 or 1 states) and starts clocking the reaction time period T_w during which the labeled species undergo chemical exchange, that is hydrogen-bond formation and dissociation. The chemical exchange will be manifested in the 2D spectrum if it changes the vibrational frequencies of the vibrational modes under study. The third pulse ends the population period of length T_w and begins a third period of length $\leq \tau$, which ends with the emission of the vibrational echo pulse with frequencies along the ω_m axis. The vibrational echo pulse is the signal in the experiment. The vibrational echo signal reads out information about the final structures of all labeled species by their ω_m frequencies. During the period T_w between pulses 2 and 3, chemical exchange occurs. The exchange causes new off-diagonal peaks to grow in as T_w is increased.[9–12] The growth of off-diagonal peaks in the 2D spectra with increasing T_w is used to extract the dissociation times (lifetimes) of the hydrogen bonds. (Chemical exchange can also occur during the τ periods, but it does not contribute to the growth of the off-diagonal peaks.) In addition to chemical exchange, other population dynamics such as vibrational relaxation to the ground state and orientational relaxation of the entire molecule will influence the 2D spectrum. These processes are included in the data analysis.

The vibrational population relaxation time constants and rotational relaxation time constants for the samples were measured with the polarization selective IR pump−probe experiments.[10,17] The rotational relaxation time constants were measured in the pure solvents. The value measured in CCl$_4$ was used to determine the rotational time of the free (uncomplexed) species, and the value measured in the pure aromatic solvent was used to determine the rotational time of the complex. These values were corrected with the Stokes−Einstein−Debye (SED) equation for the solvent viscosity differences between the pure aromatic solvent or CCl$_4$ and the mixed solvents used in the chemical exchange experiments. Viscosity measurements were made with Cannon Ubbelohde viscometers at 24 °C, the same temperature used for the vibrational echo and pump−probe measurements.

The hydrogen-bond formation enthalpies of the complexes were determined by measuring the temperature dependence of the equilibrium constant by observing the change in the absorption spectrum of the hydroxyl stretching mode. The ratio of the areas of the peaks for the complex species and the free species, corrected for the differences in extinction coefficients, was used to determine the equilibrium constant at each temperature. The temperature range was 25 to 65 °C. A more detailed description of the measurements and experimental data are given in the Supporting Information.

The hydroxyl groups of all phenol derivatives were deuterated to move the hydroxyl stretching mode frequency to ∼2600 cm^{-1}. The chemicals used to form the hydrogen-bonded complexes are phenol-OD (PH), p-methoxyphenol-OD (pMPH), p-bromophenol-OD (pBrPH), o-methoxyphenol-OD (oMPH), benzene (BZ), toluene (TL), bromobenzene (BrBZ), p-xylene (pX), and mesitylene (MS). The solutions used in the vibrational echo measurements are 0.6 wt % PH in the BZ/CCl$_4$ mix solvent (1:5 wt), 0.6 wt % PH in TL/CCl$_4$ (1:6 wt), 0.6 wt % PH in pX/CCl$_4$ (1:7 wt), 0.6 wt % PH in MS/CCl$_4$ (1:7 wt), 0.6 wt % pBrPH in BZ/CCl$_4$ (1:5 wt), 0.5 wt % pMPH in BZ/CCl$_4$ (1:5 wt), and 3 wt % oMPH in TL. The ratio of the aromatic component of the solvent to the CCl$_4$ component was chosen to make free and complexed species approximately equal in amplitude in the FTIR spectrum of each mixture.

The function of the substituents on the phenol and benzene rings is to adjust the hydrogen-bond strength (enthalpy of formation). Benzene and its derivatives are the hydrogen-bond acceptors. In general, adding an electron-withdrawing group such as bromine to benzene weakens the hydrogen bond by reducing the electron density of the benzene ring. Adding electron-donating groups such as methyl groups to benzene makes the hydrogen bond stronger. Adding an electron-withdrawing group to phenol also makes the hydrogen bond stronger because the hydroxyl, which is the hydrogen bond donor, becomes more positive. The effect of a methoxy group on the phenol ring is position dependent. When it is ortho to the hydroxyl, the hydroxyl group forms an intramolecular hydrogen bond with the methoxy group. The intramolecular hydrogen bond results in an intermolecular π hydrogen bond between the hydroxyl group and aromatic solvent molecule that is very weak.[10] When the methoxy group is in the para position, it has very little effect on the strength of the intermolecular hydrogen bond.

The structures of the hydrogen-bonded complexes were determined with density functional theory (DFT) calculations.[18] The DFT calculations were carried out as implemented in the Gaussian 98 program suite. The level and basis set used were Becke's three-parameter hybrid functional combined with the Lee−Yang−Parr correction functional, abbreviated as B3LYP, and 6-31+G(d,p). All results reported here do not include the surrounding solvent and therefore are for the isolated molecular complexes.

III. Results and Discussions

The DFT calculations show that all of the phenol derivatives form π hydrogen bonds with the aromatic molecules in the mixed solvents and that all of the complexes have a similar T-shaped structure. The structure is shown in Figure 1 for the phenol/toluene complex. Structures for other complexes are provided in the Supporting Information. For the phenol/benzene complex, recent high-level electronic structure calculations on the isolated complex and full molecular dynamic simulations of PH in the BZ/CCl$_4$ solvent confirm the T-shape structure.[19] In addition, the simulations show that the complexes are formed one to one between PH and BZ. The hydroxyl group does not point at the center of the benzene ring but rather at a ring edge for all the systems studied. The structures of the complexes in solution can be different from those predicted by the electronic structure calculations for isolated molecules. However, the consistency between the electronic structure calculations and

(16) Asbury, J. B.; Steinel, T.; Fayer, M. D. *J. Lumin.* **2004**, *107*, 271−286.
(17) Tan, H. S.; Piletic, I. R.; Fayer, M. D. *J. Chem. Phys.* **2005**, *122*, 174501.
(18) Parr, R. G.; Yang, W. *Density Functional Theory of Atoms and Molecules*; Oxford University Press: New York, 1989.
(19) Kwac, K.; Lee, C.; Jung, Y.; Han, J.; Kwak, K.; Zheng, J.; Fayer, M. D.; Cho, M. *J. Chem. Phys.* **2006**, *125*, 244508.

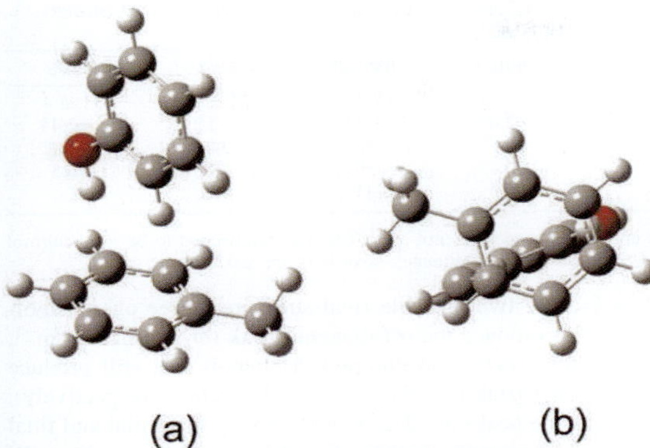

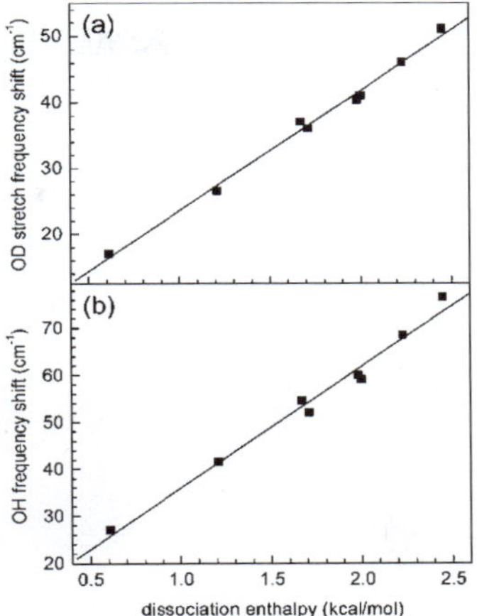

Figure 3. Correlation between the frequency shift (relative to the free species) of (a) OD stretch and (b) OH stretch of complexes of phenol and its derivatives to benzene derivatives and the complex (hydrogen bond) dissociation enthalpies (negative values of formation enthalpies). The two plots show a linear relationship between the parameters.

hydrogen bonds produce greater shifts.[5,20-22] For the complexes studied here, the hydroxyl stretch frequency shifts (relative to each corresponding free species) are also strongly correlated to the strength of the hydrogen bonds. Figure 3 shows that the frequency shift is linearly proportional to the hydrogen-bond dissociation enthalpy (negative of the formation enthalpy) for both the OD stretch with

$$\Delta v \ (\text{cm}^{-1}) = 18.2 \times \Delta H \ (\text{kcal/mol}) + 5.2 \qquad (1)$$

and the OH stretch with

$$\Delta v \ (\text{cm}^{-1}) = 25.9 \times \Delta H \ (\text{kcal/mol}) + 10.0 \qquad (2)$$

These results are similar to some other hydrogen-bonded systems.[2,21,22] The frequency shifts and enthalpy values are listed in Table 1.

The linear IR absorption measurements provide equilibrium thermodynamic data for the π hydrogen-bonded complexes. However, such measurements do not provide information regarding the complex dynamics, that is the dissociation and formation rates. Nonlinear 2D-IR experiments on the hydroxyl stretch permit direct determinations of these rates. The phenol in toluene/CCl_4 system will be used to demonstrate the manner in which the hydrogen bond lifetimes (complex dissociation rates) can be extracted from the vibrational echo 2D-IR chemical exchange measurements. The hydrogen bond lifetimes for all of the other systems are obtained in the same manner.

Figure 4 displays 2D-IR spectra for a very short reaction period, $T_w = 200$ fs (left panel) and a longer period, $T_w = 12$ ps (right panel). The data have been normalized to the largest peak for each T_w. Each contour represents a 10% change in amplitude. (The data has 2% amplitude resolution. The 10% contours are shown here for clarity.) In Figure 4, only the 0−1

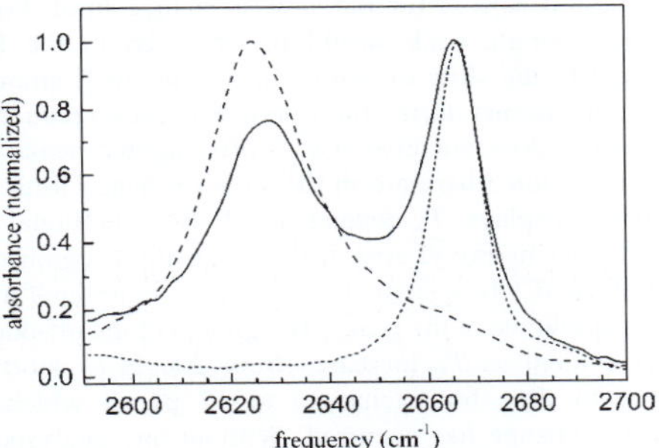

Figure 1. Two views of the hydrogen-bonded structure of the phenol/toluene complex calculated with DFT at the B3LYP/6-31+G(d,p) level for the isolated molecules. The complex's binding energy is found to be −2.3 kcal/mol (with zero-point energy correction) without solvent interactions.

Figure 2. FT-IR absorption spectra of the OD stretch of phenol-OD (hydroxyl H replaced with D) in CCl_4 (free phenol, dotted curve), phenol in toluene (hydrogen-bonded phenol/toluene complex, dashed curve), and phenol in the mixed toluene/CCl_4 solvent, which displays absorptions for both free and complexed phenol (solid curve).

the simulations for the PH/BZ system suggests that the electronic structure calculations for the other complexes provide a reasonable description of their structures.

Experimental evidence for the formation of π hydrogen bonds is the shift of the OD stretch frequency of the phenol derivatives to lower frequencies in aromatic solvents compared to the corresponding frequencies of the phenols in CCl_4. Figure 2 displays FT-IR spectra for PH dissolved in pure CCl_4 (short dash), pure TL (long dash), and the mixed TL/CCl_4 solvent (solid curve).

When phenol is dissolved in CCl_4, there is no hydrogen bond formed. The OD stretch frequency is at 2666 cm^{-1}. When it is dissolved in toluene, a hydrogen-bonded complex is formed between a solute and a solvent molecule. The OD stretch frequency red-shifts to 2625.6 cm^{-1}. When phenol is dissolved in the TL/CCl_4 mixture, some of the phenol molecules form complexes with toluene and some of them are free. The free and complexed phenol molecules are in dynamic equilibrium. As we can see from the spectrum, in the mixed solvent both free phenol and the phenol/toluene complex are present.

It is well-known that hydrogen bonding shifts the hydroxyl stretch frequency to a lower frequency and that stronger

(20) Pimentel, G. C.; McClellan, A. L. *The Hydrogen Bond*; W. H. Freeman and Co.: San Francisco, 1960.
(21) Murphy, A. S. N.; Rao, C. N. R. *Appl. Spectrosc. Rev.* **1968**, 2, 69−191.
(22) Lopes, M. C. S.; Thompson, H. W. *Spectrochim. Acta, Part A* **1968**, 24, 1367−1383.

Table 1. Hydrogen Bond Lifetimes, $\tau_d = 1/k_d$, Bond Formation Enthalpies[a] (ΔH^0), Bond Formation Entropies (ΔS^0),[b] OH and OD Stretch Frequency Shifts of the Hydrogen-Bonded Phenol Derivatives (Relative to the Free Species)

	oMPH/TL	PH/BrBZ	PH/BZ	pMPH/BZ	PH/TL	BrPH/BZ	PH/pX	PH/MS
τ_d (ps)	3 ± 1	6 ± 3	10 ± 3	10 ± 3	15 ± 3	17 ± 2	24 ± 3	31 ± 4
ΔH^0 (kcal/mol)	-0.6	-1.2	-1.7	-1.7	-2.0	-2.0	-2.2	-2.45
ΔS^0 (J/mol K)	-30	-36	-34	-38	-8	-38	-34	-41
OH $\Delta\nu$ (cm^{-1})	27	41.5	54.5	52	60	59	68.5	76.5
OD $\Delta\nu$ (cm^{-1})	17	26.5	37	36	40.4	41	46	51

[a] The ΔH^0 values are relative to phenol in CCl$_4$. The hydrogen-bond formation enthalpy between phenol and CCl$_4$ was determined to be ~0 kcal/mol using anisole as the solvation model compound.[35] [b] The enthalpy error bars are ± 0.2 kcal/mol. The frequency error bars are ± 0.5 cm^{-1}.

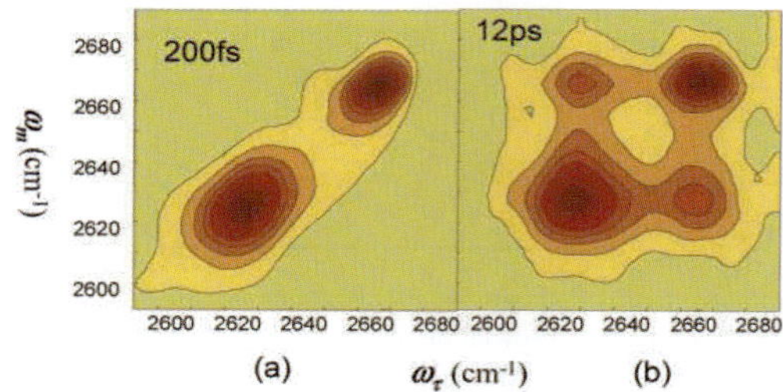

Figure 4. 2D-IR vibrational echo spectra of the OD stretch of phenol in the mixed toluene/CCl$_4$ solvent (only 0−1 transition data are shown). The data have been normalized to the largest peak for each T_w. Each contour represents 10% change in amplitude. (a) Data for $T_w = 200$ fs. (b) Data for $T_w = 12$ ps. At the longer time, additional peaks have grown in because of chemical exchange, that is, the formation and dissociation of the hydrogen-bonded complex of phenol and toluene.

(OD stretch vibrational ground state to the first excited state) transition portions of the spectra are shown. Spectra including both 0−1 and 1−2 transitions for all samples are provided in the Supporting Information. For $T_w = 200$ fs, a short time compared to the exchange time (inverse of the hydrogen-bond dissociation and formation rates), the 2D-IR spectrum consists of two peaks on the upper-right to lower-left diagonal. These peaks arise from the 0−1 transitions of the free and hydrogen-bonded OD stretch (see Figure 2 solid curve). The two peaks are generated in the following manner. As described in the Experimental Procedures section, the first laser pulse simultaneously labels the initial structures of both free and hydrogen-bonded phenol molecules with the initial frequencies ω_τ (the OD stretch frequencies of both species). No exchange has occurred during the 200 fs T_w period. None of the "labeled" species has changed its structure, so the final frequency, ω_m, detected by the vibrational echo is the same as the initial excitation frequency, ω_τ. If the initial frequency (ω_τ) is the same as the final frequency (ω_m), the peak is on the diagonal. Therefore, for very short T_w, there are two diagonal peaks in the 2D-IR spectrum, one peak for the complex and one peak for the free phenol. The complex gives rise to the lower-frequency peak along the ω_m axis because it is the lower-frequency peak in the linear IR spectrum (Figure 2). Within experimental error, both peaks in the 2D spectrum have the same frequencies as those in the FT-IR spectrum in Figure 2.

For $T_w = 12$ ps (Figure 4, right panel), which is long compared to the exchange time for this system, two off-diagonal peaks have grown in. The growth of the additional peaks is caused by the exchange during T_w period. Consider phenols that are complexed with toluenes at the time of the second pulse. During the T_w period, some of these complexed phenols ($\omega_\tau = 2625.6$ cm^{-1}) dissociate, while some remain complexed. The third laser pulse ends the T_w period. The initially complexed

phenols have two possible final structures: free phenol molecules that produce the off-diagonal peak ($\omega_\tau = 2625.6$ cm^{-1}, $\omega_m = 2666$ cm^{-1}) and complexed phenols that still produce the diagonal peak (2625.6 cm^{-1}, 2625.6 cm^{-1}, respectively). Note that the peaks are determined only by the initial and final structures. Any odd number of exchanges, 1, 3, ..., etc., will produce an off-diagonal peak, and any even number of exchanges, 0, 2, ..., etc., will give rise to a diagonal peak. The same reasoning applies to initially free phenols which give rise to the free phenol diagonal peak and the other off-diagonal peak. The discussion here is for the slow exchange limit. For fast exchange, a single peak would be observed in the FT-IR spectrum.[23] In the slow exchange limit, the small amount of exchange that occurs during the τ periods causes a reduction in the signal but does not give rise to off-diagonal peaks.[14] All the systems studied here are in the slow exchange limit.

Figure 5 displays T_w-dependent 2D-IR vibrational echo spectra for the hydroxyl stretch 0−1 transition region of the phenol/toluene/CCl$_4$ system. Each contour represents a 10% change in amplitude of the peaks. The growth of the off-diagonal peaks is evident as T_w increases from 200 fs (a short time compared to time for exchange) to 12 ps, at which time extensive exchange has occurred. Without any analysis, it is clear that the exchange time is slower than ~1 ps and faster than ~100 ps. However, to obtain quantitative rates requires a full analysis of the dynamical processes in the system.

During the T_w period, in addition to chemical exchange other dynamic processes influence the 2D spectrum. These are spectral diffusion, orientational relaxation, and vibrational relaxation.[9,14] None of these produce off-diagonal peaks. Spectral diffusion[24,25] is the result of time-dependent interactions of the vibrational transition with the solvent.[14,26,27] These interactions cause the transition frequency to fluctuate. At short time, the diagonal peaks' line shapes measured in the 2D vibrational echo spectrum are elongated along the diagonal because molecules are in different solvent environments (inhomogeneous broadening). As T_w increases, the solvent configurations around each molecule evolve so that the frequency of each molecule samples an increasingly large fraction of the entire absorption spectrum (spectral diffusion). At sufficiently long T_w, all possible solvent configurations are sampled, and the dynamic line width is equal to the absorption line width. In a 2D spectrum complete spectral diffusion is manifested by a change in the 2D line shape from elongated along the diagonal to symmetrical about the diagonal.

(23) Levinger, N. E.; Davis, P. H.; Behera, P.; Myers, D. J.; Stromberg, C.; Fayer, M. D. *J. Chem. Phys.* **2003**, *118*, 1312−11326.
(24) Walsh, C. A.; Berg, M.; Narasimhan, L. R.; Fayer, M. D. *Chem. Phys. Lett.* **1986**, *130*, 6−11.
(25) Bai, Y. S.; Fayer, M. D. *Phys. Rev. B* **1989**, *39*, 11066.
(26) Asbury, J. B.; Steinel, T.; Stromberg, C.; Corcelli, S. A.; Lawrence, C. P.; Skinner, J. L.; Fayer, M. D. *J. Phys. Chem. A* **2004**, *108*, 1107−1119.
(27) Asbury, J. B.; Steinel, T.; Kwak, K.; Corcelli, S.; Lawrence, C. P.; Skinner, J. L.; Fayer, M. D. *J. Chem. Phys.* **2004**, *121*, 12431−12446.

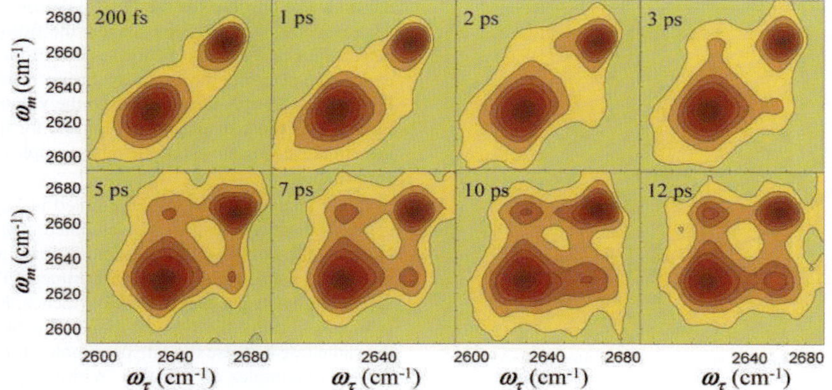

Figure 5. Time dependence of the 2D-IR vibrational echo spectrum in the 0−1 transition region. The data have been normalized to the largest peak for each T_w. Each contour represents 10% change in amplitude. As T_w increases, the off-diagonal peaks grow in because of chemical exchange (formation and dissociation of the toluene/phenol complex). Inspection of the data shows that the time scale for the chemical exchange (growth of the off-diagonal peaks) is a few picoseconds.

This change can be seen most clearly by comparing the central contours of the free peaks at 200 fs and 2 ps in Figure 5. Spectral diffusion changes the shapes of the peaks but preserves their volumes.[9,14] From the data, we can see that spectral diffusion is complete by ∼2 ps. Orientational relaxation and vibrational relaxation causes all peaks to decay in volume in contrast to chemical exchange, which causes the off-diagonal peaks to grow and the diagonal peaks to shrink. Orientational relaxation during the T_w period reduces the peak volumes but does not cause them to decay to zero. Vibrational relaxation does cause the peaks to eventually decay to zero.

To extract the exchange kinetics from the 2D-IR spectra, a kinetic model including all dynamic processes was used.[9,10,14] The method is illustrated schematically for the complex (CC) dissociating to become (CF) free phenol (or a phenol derivative).

$$\underset{\text{decay}}{\overset{\tau_c,\, T_c}{\longleftarrow}} CC \underset{k_a}{\overset{k_d}{\rightleftharpoons}} CF \underset{\text{decay}}{\overset{T_f,\, \tau_f}{\longrightarrow}}$$

CC represents the complexes (hydrogen-bonded species) at the end of the T_w period (the time of the third pulse) that were also complexes immediately after the second pulse (peak (2625.6 cm^{-1}, 2625.6 cm^{-1}) in the right panel of Figure 4). CF represents the free phenol molecules at the end of the T_w period *that were complexes immediately after the second pulse* (peak (2625.6 cm^{-1}, 2666 cm^{-1}) in the right panel of Figure 4). The rate constant for dissociation of complexes is k_d, and the rate constant for the association of a free phenol with benzene or a benzene derivative to form a complex is k_a. During the T_w period, the vibrational echo signals produced by these two populations decrease because of vibrational relaxation (time constant T_c and T_f for complex and free, respectively) and rotational relaxation (time constant τ_c and τ_f for complex and free, respectively). Identical considerations apply to the populations FF and FC, which are the initial diagonal free population that is also free at T_w and the initial free population is complexed at T_w.

The kinetic model gives rise to a set of differential equations for the four populations, CC, CF, FF, and FC. These differential equations have been used previously,[9,10,14] and the complete

solutions have been published.[10,14] In the equations, only k_a and k_d are unknown variables; all of the other parameters are experimentally measured independently. Vibrational and rotational relaxation time constants were measured with pump−probe experiments as described in Experimental Procedures (section II). The time-dependent populations were provided by the peak volumes or intensities of the 2D-IR spectra corrected for transition dipole moment differences.[9,10,14] The transition dipole moment differences of the free and complexed species were measured using FT-IR. The equilibrium constant for the two species was also given by FT-IR measurements. The dissociation rate (number per unit time) of the complex equals the formation rate because the system is in equilibrium (see below).[9] The complex dissociation time constant (also the lifetime of the hydrogen bond), τ_d (given in picoseconds below), is independent of concentration. Therefore, τ_d is reported and discussed. $\tau_d = 1/k_d$, where $k_d = k_a/A$ is the dissociation rate constant. A is the concentration ratio between the complexed and free phenol (or phenol derivative). The dissociation rate is k_d[complex]. Because the concentration ratio is known, *there is a single unknown parameter, τ_d.*

Spectral diffusion changes the shape of the diagonal peaks, reducing their maximum amplitudes without changing their volumes. For this reason in previous publications[9,10,14] peak volumes were determined as a function of T_w to reflect the peak populations. Here we use amplitudes as a function of T_w instead of volumes when fitting the data with the kinetic model for all samples except the oMPH in TL/CCl$_4$ system because the spectral diffusion is fast compared to the exchange rates. Therefore, negligible error is introduced by using the amplitudes, which are more easily determined than the volumes. For oMPH in TL/CCl$_4$ peak volumes were used.

The data were fit in the following manner. For a 2D-IR spectrum taken at $T_w = 200$ fs, the population ratio of the two species is taken to be the same as the equilibrium constant obtained from the FTIR measurements because 200 fs is a very short time compared to the time scale of chemical exchange. All of the free diagonal peaks are divided by the intensity of the free peak at 200 fs to yield the time-dependent populations

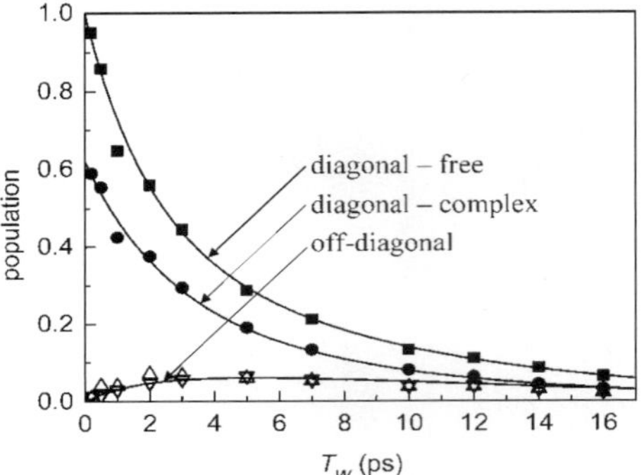

Figure 6. T_w-dependent data (symbols) showing the time dependence of the two diagonal and two off-diagonal peaks, in the 0−1 region of the 2D vibrational echo spectra as in Figure 5. The off-diagonal peaks grow in together because the sample is in thermal equilibrium. The solid curves are from a fit to the data with one adjustable parameter, τ_d, using the kinetic model. Other parameters in the model are determined from independent experiments. For the phenol/toluene system $\tau_d = 15 \pm 3$ ps.

for free phenol molecules (FF). All of the diagonal peaks for the complexes are divided by the intensity of the complex peak at 200 fs and then multiplied with the equilibrium constant to give the time-dependent populations of complexed phenol molecules (CC). This procedure accounts for the fact that the transition dipole moments of the free and complex species are not the same. The diagonal peak amplitudes depend on the fourth power of their respective transition dipoles. The off-diagonal peak amplitudes depend on the product of the squares of the transition dipoles of the free and complex species. To take this into account, the cross-peaks are divided by the square root of the intensity ratio of the complexed peak intensity over the free peak intensity divided by the equilibrium constant. The influence on the peak amplitudes caused by peak overlap is corrected for by fitting the 2D-peaks and subtracting the contributions to a given peak from the other peaks. This is a relatively small correction. The results yield the diagonal populations, FF and CC and the off-diagonal exchanged populations, CF and FC, as a function of T_w.

The data (see Figure 5 as an example) for the 0−1 transition region consist of four time-dependent components: the two diagonal peaks (the complex and the free PH in Figure 5) and the two off-diagonal peaks (dissociation and formation of the complex during the T_w period). As T_w increases, the off-diagonal peaks grow in because of chemical exchange. All four peaks can be reproduced with the single adjustable parameter, τ_d by inputting the known constants and fitting the data with the kinetic equations. For the phenol/toluene system, the input parameters used are $T_c = 10$ ps, $T_f = 9$ ps, $\tau_c = 3.7$ ps, $\tau_f = 2.4$ ps, and the ratio of the complexed and free phenol concentrations [complex]/[free] = 0.62. Because the vibrational and rotational parameters were measured in pure solvents rather than mixtures, we allowed each parameter to vary ∼10−20% in fitting the data and used the results to determine the error bars for τ_d.

Figure 6 shows the peak intensity data for the 0−1 transition region of the spectrum as a function of T_w. The data are normalized to the largest peak at $T_w = 0$. The simultaneous fit (solid curves) to the time dependence of all of the peaks using *a single adjustable parameter*, τ_d, is very good. From the fits,

the dissociation time for the PH/TL complex (hydrogen bond lifetime) is $\tau_d = 15 \pm 3$ ps. To test the fitting procedures, the peak volumes were also fit for several of the samples. Within experimental error, the intensity fitting results for PH/BrBZ, PH/BZ, and PH/pX are identical to the volume fitting results.[9] Fits such as those displayed in Figure 6 for the phenol/toluene system are shown for all of the other samples in the Supporting Information.

In the analysis of the data presented in Figure 6 and for all of the other systems, it was assumed that the equilibrium between the complex and the free species and the exchange rates were not disturbed by the experiment. This assumption is supported by strong experimental evidence. First, if the system is in equilibrium, the two off-diagonal peaks will grow in at the same rate because the rate of complex formation is equal to the rate of complex dissociation. Within experimental error, the off-diagonal peaks grow in at the same rates as can be seen in Figure 6 for the phenol/toluene system and in the Supporting Information for all of the other systems.

Another detailed test showing that the vibrational echo/chemical exchange experiment does not perturb either the thermal equilibrium or the exchange rates has been presented for the phenol/benzene system.[9] The test involves comparing the 0−1 data, such as those shown in Figure 5, to the equivalent data for the 1−2 transition (not shown). There are two quantum pathways that give rise to the 0−1 2D spectrum. For one pathway, the system is in the ground state ($\upsilon = 0$) during the T_w period. For the other pathway the system is in the vibrationally excited state ($\upsilon = 1$) during the T_w period. Fifty per cent of the signal arises from each of these pathways. For the 1−2 portion of the 2D vibrational echo spectrum (see data in Supporting Information), there is only one pathway, and the first two radiation field−matter interactions the same as those that give rise to the vibrationally excited portion of the 0−1 2D spectrum. After the second interaction (second pulse) the system is in the $\upsilon = 1$ state. The third interaction (third pulse) produces a coherence between the $\upsilon = 1$ and $\upsilon = 2$ levels, and the vibrational echo is then emitted at the frequency of the 1−2 transition, which is shifted to lower frequency by the vibrational anharmonicity (∼100 cm^{-1}). The portion of the 2D spectrum that occurs from vibrational echo emission at the 1−2 frequency only involves a pathway in which the system is in the $\upsilon = 1$ state during the T_w period. Therefore, the 0−1 portion of the spectrum involves 50% of the signal in which the dynamics might be influenced by vibrational excitation, whereas the 1−2 portion of the spectrum involves 100% of the signal that might be affected by vibrational excitation. It was observed that the measured dynamics are the same for the 0−1 and 1−2 portions of the spectrum.[9] Therefore, vibrational excitation does not influence the thermal equilibrium/chemical exchange rates. Vibrational excitation of the hydroxyl stretch is actually a very mild perturbation of the system. The hydroxyl stretch is a quantum oscillator. Vibrational excitation to the $\upsilon = 1$ state increases the bond length by a few hundredths of an angstrom. This is a very small change compared to the thermal structural fluctuations of the complex (distance and angular) displayed in the MD simulations.[19] It should also be noted that the energy deposited by vibrational relaxation does not enter into the problem. Once a molecule undergoes vibrational relaxation, it no longer contributes to the signal.

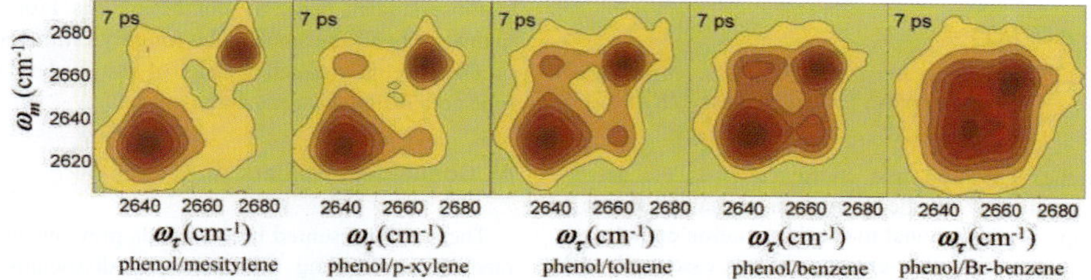

Figure 7. 2D-IR vibrational echo spectra of the OD stretch of phenol in a series of mixed benzene derivative/CCl$_4$ solvents with the same reaction time period, $T_w = 7$ ps. The data have been normalized to the largest peak for each sample. Each contour represents a 10% change in amplitude. From left to right the hydrogen bonds are weaker, and the size of the off-diagonal peaks are larger (more exchange).

The hydrogen bond strength (bond formation enthalpy) of the complex is changed by modifying the chemical structure of the solute molecule, the solvent molecule, or both. It has been qualitatively demonstrated that stronger hydrogen bonds dissociate more slowly.[9] Here a more detailed and quantitative correlation is presented. Before discussing the values of the dissociation times, the nature of the results can be seen qualitatively and directly from the data presented in Figure 7. Figure 7 displays 2D-IR spectra for the same reaction period, $T_w = 7$ ps, for complexes between phenol and five different complex partners in mixed solvents of each partner and CCl$_4$. The data have been normalized to the largest peak for each sample. Phenol forms the strongest hydrogen bond with mesitylene. The hydrogen-bond strength order is phenol/mesitylene > phenol/p-xylene > phenol/toluene > phenol/benzene >phenol/bromobenzene. In Figure 7, moving from left to right, it can be seen that at the single $T_w = 7$ ps, the size of the off-diagonal peaks, which are produced by chemical exchange, increases as the hydrogen-bond strength decreases. For the phenol/mesitylene spectrum, the off-diagonal peaks have just begun to appear. In contrast, in the phenol/bromobenzene spectrum, the off-diagonal peaks have grown in to such an extent that they are now well delineated from the diagonal peaks and the spectrum has the appearance that is approximately square. As discussed above, exchange is the only reason for the growth of the off-diagonal peaks. Faster exchange produces larger off-diagonal peaks (normalized to the largest diagonal peak) for the same reaction period (T_w). From the relative intensity of the cross-peaks (the number of contours), it is straightforward to see that a stronger hydrogen bond dissociates more slowly. Figure 7 is a demonstration that the stronger the hydrogen bond that forms the complex, the slower the dissociation and formation of the complex.

The enthalpies and entropies of complex formations were determined from the temperature dependence of the equilibrium constant using

$$\ln K_{eq} = -\frac{\Delta G^0}{RT} = -\frac{\Delta H^0}{RT} + \frac{\Delta S^0}{R} \qquad (3)$$

The details are discussed in the Supporting Information. ΔG^0, ΔH^0, and ΔS^0 are the standard Gibbs free energy, the enthalpy, and the entropy at room temperature, respectively. The standard state is defined to be a pressure of 1 atm and a temperature of 25 °C. ΔH^0 and ΔS^0 are taken to be temperature independent within the 50 K range used in the studies. Figure 8 shows a very interesting correlation between the lifetimes of the complexes ($1/k_d$, k_d is the hydrogen-bond dissociation rate constant)

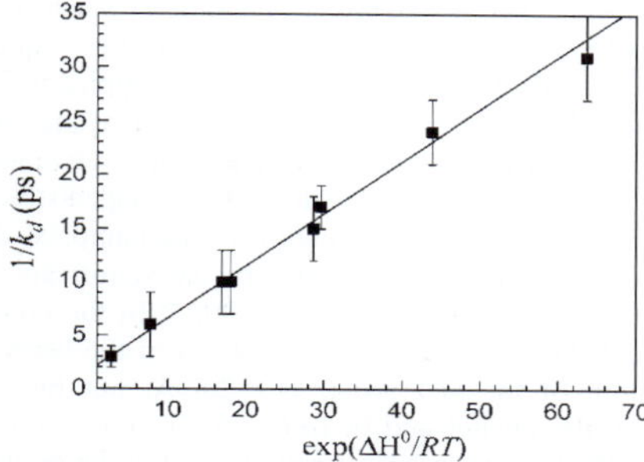

Figure 8. Hydrogen bond lifetimes ($1/k_d$, ps) plotted vs exp($\Delta H^0/RT$)- where ΔH^0 is the hydrogen-bond dissociation enthalpy (negative of the enthalpy of formation), R is the gas constant, and T is the temperature (300 K). The line through the data is given by eq 4.

and the dissociation enthalpies (ΔH^0, negative values of the hydrogen-bond formation enthalpies) for the eight samples studied. The line through the data points is given by

$$1/k_d = B + A^{-1} \exp(\Delta H^0/RT), \qquad (4)$$

where $B = 2.3$ ps and $A^{-1} = 0.5$ ps are constants. The equation can be rearranged to give

$$\frac{1}{1/k_d - B} = A \exp(-\Delta H^0/RT) \qquad (5)$$

Since $B = 2.3$ ps is smaller than or the same as the experimental errors for all of the samples except the fastest one, oMPH/TL, the relationship can be approximated for hydrogen bonds with lifetimes longer than 2.3 ps as

$$k_d = A \exp(-\Delta H^0/RT) \qquad (6)$$

Equation 6 has the form of the Arrhenius equation[28] but with the dissociation enthalpy instead of the activation energy. It is useful to consider the trend seen in Figure 8 heuristically in terms of simple transition state theory.[29] The dissociation rate constant can be written as

(28) Atkins, P. W. *Physical Chemistry*, 5th ed.; W. H. Freeman: New York, 1994.
(29) Chang, R. *Physical Chemistry for the Chemical and Biological Sciences*; University Science Books: Sausalito, 2000.

$$k_{\mathrm{d}} = \frac{k_{\mathrm{B}}T}{h}\exp(-\Delta G^*/RT) =$$

$$\frac{k_{\mathrm{B}}T}{h}\exp(\Delta S^*/R)\exp(-\Delta H^*/RT) \quad (7)$$

where ΔG^* is the activation free energy, ΔS^* is the activation entropy, and ΔH^* is the activation enthalpy. If the activation enthalpy is proportional to the dissociation enthalpy, $\Delta H^* \propto \Delta H^0$, and the activation entropy ΔS^* is essentially a constant, independent of the molecular structure of the complexes, then eq 6 and the behavior displayed in Figure 8 are recovered.

Aside from experimental error, a possible explanation for the nonzero (2.3 ps) intercept in Figure 8 is the time required for the molecules of a hypothetical nonbonded complex to separate diffusively. Consider a "pseudocomplex" that has a bond formation enthalpy of zero but has a phenol derivative molecule and a benzene derivative molecule in the T-shaped structure of a bonded complex. For the phenol/benzene complex, detailed MD simulations show that the structure has significant angular fluctuation, but the average angle is 90°.[19] In the complexed form, the distance between the ring center of the benzene and the D atom of the OD group is $\sim$2.5 Å, but by $\sim$3.5 Å separation, the phenol and benzene are no longer bonded as shown by the lack of angular correlation.[19] For the pseudocomplex with zero bond formation enthalpy, it would still take a finite amount of time for the phenol and benzene to separate from $\sim$2.5 Å to $\sim$3.5 Å. The time to diffuse apart may be responsible for the nonzero intercept in Figure 8. As a very crude check of this idea, using the Stokes−Einstein equation and the viscosity of the benzene/CCl$_4$ solvent, it was determined that the time for a benzene to diffuse 1 Å is $\sim$2 ps, consistent with the intercept. This is a very rough approximation for two reasons that tend to off-set each other. For two molecules in a solvent, the mutual diffusion constant is well approximated by the sum of the diffusion constants.[30] Therefore, the time for a phenol and a benzene to change their separation by 1 Å would be shorter than 2 ps. However, the mutual diffusion constant is only appropriate when the molecules are widely separated. At small separations, the hydrodynamic effect slows their relative diffusion.[31−34] These off-setting effects make the $\sim$2 ps estimate reasonable. The point is that the time scale to simply diffuse apart is not 0.2 ps or 20 ps, but on the order of the intercept in eq 4.

The correlation between the lifetimes of solute−solvent hydrogen bonds and their enthalpies of formation may be useful for estimating the lifetimes of hydrogen bonds that are similar to those studied here. By combining eqs 1 or 2 with eq 6, a hydrogen bond lifetime can be roughly estimated from the OH or OD stretch frequency shift of the system obtained from routine FT-IR measurements.

IV. Concluding Remark

The study presented here extends previous work that examined hydrogen-bond formation and dissociation rates of the phenol/benzene complex.[9,10] Here, eight π hydrogen-bonded complexes of phenol and phenol derivatives with benzene and benzene derivatives were investigated. The series of complexes have a wide range of hydrogen-bond strengths (enthalpies of formation). Both the kinetics and thermodynamics of the complexes were studied. The 2D-IR vibrational echo experiments measure the dissociation and formation rates of the complexes. The chemical exchange dynamics of a complex can be characterized by the dissociation time of the complex (the inverse of the dissociation rate), which is the hydrogen bond lifetime. The measured dissociation times for the eight complexes range from 3 to 31 ps. The bond dissociation enthalpies (negative of the formation enthalpies) range from 0.6 kcal/mol to 2.5 kcal/mol.

It was found that the hydrogen bond lifetimes are correlated with the dissociation enthalpies in a manner akin to the Arrhenius equation. The correlation can be qualitatively understood in terms of transition state theory. In this model, the activation enthalpy scales linearly with the bond dissociation enthalpy, and the activation entropy is essentially independent of the molecular structure of the complex. It was also found that the shift in the hydroxyl stretch frequency of phenol and its derivatives when complexed, compared to that of the free species, is linearly related to the dissociation enthalpy. Thus, it is possible to estimate the hydrogen bond lifetimes of systems that are similar to those studied here from a simple measurement of the hydroxyl stretch frequency in a linear absorption experiment.

Acknowledgment. We thank Sungnam Park, Kyungwon Kwak, and David Ben Spry for their help with some of the pump/probe experiments and Dr. Xin Chen and Professor John I. Brauman for insightful discussions. This work was supported by grants from AFOSR (F49620-01-1-0018), NIH (2 R01 GM-061137-05), and NSF (DMR-0332692).

Supporting Information Available: Vibrational echo 2D-IR experimental setup and details, calculated structures of complexes, thermodynamic data, 2D-IR data and kinetic results of all samples. This material is available free of charge via the Internet at http://pubs.acs.org.

JA067760F

(30) Swallen, S. F.; Fayer, M. D. *J. Chem. Phys.* **1995**, *103*, 8864−8872.
(31) Wolynes, P. G.; Deutch, J. M. *J. Chem. Phys.* **1976**, *65*, 450−454.
(32) Northrup, S. H.; Hynes, J. T. *J. Chem. Phys.* **1979**, *71*, 871−883.
(33) Rice, S. A. *Diffusion-Limited Reactions*; Elsevier: Amsterdam, 1985.
(34) Weidemaier, K.; Tavernier, H. L.; Swallen, S. F.; Fayer, M. D. *J. Phys. Chem. A* **1997**, *101*, 1887−1902.
(35) Fuchs, R.; Peacock, L. A.; Stephenson, W. K. *Can. J. Chem.* **1982**, *60*, 1953−1958.

Direct observation of fast protein conformational switching

Haruto Ishikawa*, Kyungwon Kwak, Jean K. Chung, Seongheun Kim, and Michael D. Fayer[†]

Department of Chemistry, Stanford University, Stanford, CA 94305

Contributed by Michael D. Fayer, April 17, 2008 (sent for review March 21, 2008)

Folded proteins can exist in multiple conformational substates. Each substate reflects a local minimum on the free-energy landscape with a distinct structure. By using ultrafast 2D-IR vibrational echo chemical-exchange spectroscopy, conformational switching between two well defined substates of a myoglobin mutant is observed on the ≈50-ps time scale. The conformational dynamics are directly measured through the growth of cross peaks in the 2D-IR spectra of CO bound to the heme active site. The conformational switching involves motion of the distal histidine/E helix that changes the location of the imidazole side group of the histidine. The exchange between substates changes the frequency of the CO, which is detected by the time dependence of the 2D-IR vibrational echo spectrum. These results demonstrate that interconversion between protein conformational substates can occur on very fast time scales. The implications for larger structural changes that occur on much longer time scales are discussed.

multidimensional IR spectroscopy | myoglobin | protein dynamics | protein structural change | ultrafast IR

A folded protein with a particular structure occupies a minimum on its free-energy landscape (1–3). However, the minimum is frequently a local minimum. Other minima of similar energy can also exist. When a protein occupies any one of these minima, it has a distinct structure. The different structures are substates of the folded protein. Transitions from one minimum to another correspond to dynamical changes in the structure of the protein that take the protein from one substate to another. Under thermal equilibrium conditions, there will be continual conformational switching among substates. In addition to interconversion between substates, proteins undergo continuous structural fluctuations within a particular substate minimum. Such fluctuations within a substate minimum give rise to processes such as small ligand "diffusion" through a protein to an active site (4).

The ability of proteins to undergo conformational switching is central to protein function. When an enzyme binds a substrate, the protein conformation will change (5). On the path of protein folding, a protein will sample many conformations as it progresses toward the native folded structure (6). Proteins can undergo large global conformational changes, which occur on long time scales, milliseconds to seconds. However, these large, slow conformational changes, such as those that occur after substrate binding to an enzyme, involve a vast number of more local elementary conformational steps.

The experimental determination for the time scale of elementary conformational steps is a long-standing problem that has now been addressed successfully by using ultrafast 2D-IR vibrational echo chemical-exchange spectroscopy. The problem of substate switching has been studied extensively for the protein myoglobin (Mb) with the ligand CO bound at the active site (MbCO) (7–9). The Fourier transform IR (FT-IR) spectrum of the heme-ligated CO stretching mode of Mb has three absorption bands, denoted A_0 (1,965 cm^{-1}), A_1 (1,945 cm^{-1}), and A_3 (1,932 cm^{-1}) (10). MbCO interconverts among these three conformational substates under thermal equilibrium. The distal histidine, His-64, has a prominent role in determining the

conformational substates of Mb (Fig. 1*A*). Changes in the configuration of the distal histidine cause its imidazole side group to move relative to the CO (10, 11). When the conformation of the distal histidine results in the imidazole being rotated out of the heme pocket, the interaction of the distal histidine with the ligated CO becomes weak. This out-of-the-pocket configuration gives rise to the A_0 band (9, 12). The distal histidine conformation that places the imidazole in the heme pocket gives rise to two substates, A_1 and A_3, which involve strong interactions of the imidazole side group with the CO ligand. The lower frequency of A_3 compared with A_1 reflects a closer proximity of the imidazole to the CO in A_3 (11, 13, 14). Each A substate exhibits a distinct ligand binding rate (2, 7). Therefore, the three peaks in the FT-IR spectrum of MbCO reflect functionally distinct conformational substates.

By using flash photolysis measurements, the kinetic rate constant for the $A_0 - A_1/A_3$ conformational switching, in which the distal histidine swings its imidazole side group out of and into the heme pocket, is estimated to be in the range of ≈1–10 μs (8, 13). These indirect measurements based on ligand binding rate constants have also been applied to the A_1 and A_3 interconversion by using low-temperature flash photolysis studies. The results were extrapolated to ambient temperature and yield <1 ns for switching between A_1 and A_3 (13). Molecular dynamics (MD) simulations have placed the A_1–A_3 switching time on the order of 100–200 ps (14).

It has been difficult to directly measure fast elementary substate interconversion times, including the A_1-A_3 switching. Here, we report a direct measurement of the A_1–A_3 substate switching time under thermal equilibrium conditions by using 2D-IR vibrational echo chemical-exchange spectroscopy. This method has recently proven useful for studying fast dynamical processes in liquids (15–17). The 2D-IR vibrational echo chemical-exchange experiment is akin to a 2D NMR chemical-exchange experiment, except that it can operate on a picosecond time scale, and it directly probes the structural degrees of freedom through the time evolution of the 2D vibrational spectrum. The A_1 and A_3 peaks in the FT-IR spectrum appear as bands on the diagonal of the 2D-IR vibrational echo spectrum. As time increases, cross peaks grow in because of the interconversion between the A_1 and A_3 substates (chemical exchange). Analysis of the time-dependent growth of the chemical-exchange peaks yields the A_1 going to A_3 interconversion time of 47 ps.

Results and Discussion

The FT-IR spectrum of wild-type MbCO shows three discrete CO stretching bands, the A_0, A_1, and A_3 bands (7, 10, 18). The A_1 substate of wild-type MbCO is predominately populated and

Author contributions: H.I. and M.D.F. designed research; H.I., K.K., J.K.C., and S.K. performed research; H.I., K.K., J.K.C., and M.D.F. analyzed data; and H.I. and M.D.F. wrote the paper.

The authors declare no conflict of interest.

*Present address: Department of Chemistry, Graduate School of Science, Osaka University, Osaka 560-0043, Japan.

†To whom correspondence should be addressed. E-mail: fayer@stanford.edu.

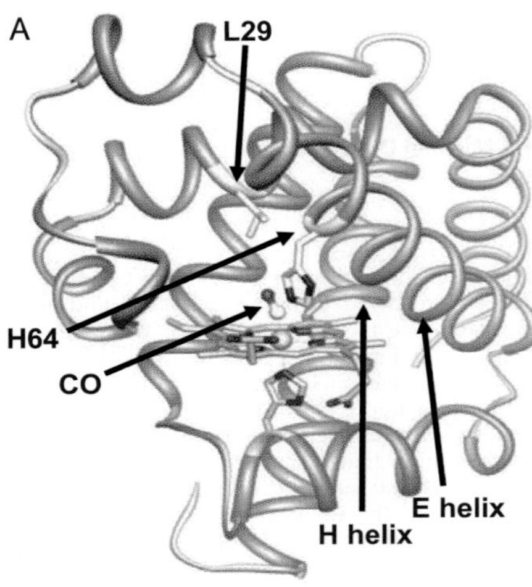

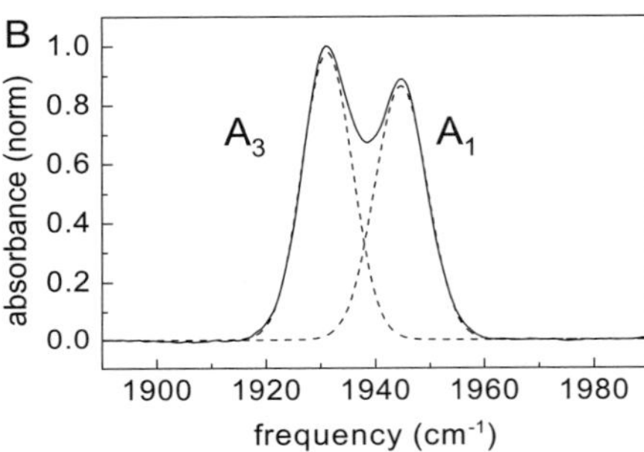

Fig. 1. Structure of CO-bound myoglobin. (*A*) Crystal structure of MbCO taken from the Protein Data Bank (ID code 1MWC). The distal histidine, His-64; the CO; the location of Leu-29; and the E helix are indicated. (*B*) FT-IR spectrum of the CO stretch of heme-ligated CO for the myoglobin mutant L29I (solid curve). The spectrum was fitted by two Gaussians (dashed curves), which represent the absorption bands of the A_1 and A_3 substates.

dominates the absorption spectrum. Because the A_3 absorption peak of wild-type MbCO is relatively small compared with the A_1 peak, we performed the experiments on an Mb mutant, L29I, in which a leucine is replaced with an isoleucine (10). This small change causes the A_1 and A_3 bands in the FT-IR spectrum to be approximately the same size (Fig. 1*B*). Like wild-type MbCO, the amount of A_0 conformation is small in the L29I mutant ($\approx$5%). Changes at position 29 cause a slight modification of the configuration of the distal histidine because the amino acid residue at position 29 interacts with the imidazole side chain of the distal histidine (Fig. 1*A*) (10). In the background-subtracted FT-IR spectrum shown in Fig. 1*B*, the two L29I CO absorption bands are at 1,932 cm^{-1} and 1,945 cm^{-1}. These bands correspond to the A_3 and A_1 substates, respectively (10). The peak positions for the CO absorption bands of L29I are essentially identical to those of wild-type Mb within the reported significant figures (10).

Experimental details for performing 2D-IR vibrational echoes and chemical-exchange experiments have been presented previously (15, 17, 19, 20). Qualitatively, the experimental method works in the following manner. Three ultrashort IR pulses tuned to the frequency of the A_1 and A_3 vibrational modes are crossed in the sample. The IR excitation pulses, 110 fs and 150 cm^{-1}

FWHM, are produced with a Ti:sapphire regenerative amplifier-pumped optical parametric oscillator. Because the pulses are very short, they have a broad bandwidth that makes it possible to simultaneously excite the vibrational modes of both the A_1 and A_3 substates. The first laser pulse "labels" the initial structures of the species by defining their initial frequency, ω_τ. The second pulse ends the first time period τ and starts clocking the reaction time period T_w during which the labeled species undergo substate switching (chemical exchange) and vibrational relaxation to the ground state. The third pulse ends the population dynamics period of length T_w and begins a third period of length $\leq\tau$, which ends with the emission of the vibrational echo pulse of frequency ω_m. The vibrational echo signal reads out information about the final structures of all labeled species by their frequency ω_m.

There are two types of time periods in the experiment. The times between pulses 1 and 2 and between pulse 3 and the vibrational echo pulse are called coherence periods. During these periods, the vibrations are in coherent superpositions of two vibrational states. Fast vibrational oscillator frequency fluctuations induced by fast structural fluctuations of the protein cause dynamic dephasing, which is one contribution to the line shapes in the conventional 1D absorption spectrum. During the period T_w between pulses 2 and 3, called the population period, a vibration is in a vibrational eigenstate. Slower structural fluctuations of the protein cause the CO frequency to evolve in time. This time evolution of the frequency, termed spectral diffusion, contributes to the 2D line shapes. Other processes during the population period, particularly chemical exchange, also produce changes in the 2D spectrum. Chemical exchange occurs when two species in equilibrium interconvert without changing the overall number of either species. Interconversion back and forth between the A_1 and A_3 substates is a type of chemical exchange. In other contexts, it has been demonstrated that chemical exchange causes new cross peaks to grow in as T_w is increased (15–17, 21–23). The growth of the cross peaks with increasing time is directly related to the chemical-exchange rate (15–17, 21–23). Here, the growth of the cross peaks in the 2D vibrational echo spectrum of L29I with increasing T_w is used to extract the A_1–A_3 substate interconversion rate.

Fig. 2*A* shows 2D-IR spectra of CO bound to L29I at several T_w values. The bands in the upper half of the spectrum are positive-going and correspond to the 0–1 vibrational transitions. The bands in the lower half of the spectrum are negative-going, which arise from vibrational echo emission at the frequency of the 1–2 vibrational transition. The negative-going bands are shifted along the ω_m axis by the anharmonicity of CO stretching mode (24, 25). The data have been normalized to the largest peak at each T_w. Consider the spectrum for $T_w = 0.5$ ps. The two bands on the diagonal correspond to the two peaks in the absorption spectrum shown in Fig. 1*B*. These two bands are centered at $(\omega_\tau, \omega_m) = (1,932$ cm^{-1}, 1,932 cm^{-1}) and (1,945 cm^{-1}, 1,945 cm^{-1}) corresponding to the A_3 and A_1 substates, respectively. The off-diagonal band centered at $(\omega_\tau, \omega_m) = (1,932$ cm^{-1}, 1,908 cm^{-1}) and (1,945 cm^{-1}, 1,922 cm^{-1}) are from vibrational echo emission at the 1–2 vibrational transitions of the A_3 and A_1 substates, respectively. At $T_w = 0.5$ ps, there are no cross peaks in the 0–1 and 1–2 regions because 0.5 ps is short relative to the conformational switching time.

In contrast, by $T_w = 48$ ps (Fig. 2*A*) enough time has elapsed for conformational switching to occur to a significant extent. The conformational switching is manifested by the growth of cross peaks (17), which are most apparent in the upper left portion of the 0–1 bands and the lower right portion of the 1–2 bands. These cross peaks correspond to the A_3 switching to A_1 and A_1 switching to A_3, respectively. Because the system is in equilibrium, the rate of A_1 converting into A_3 is equal to the rate of A_3 converting into A_1. If the anharmonicity is sufficiently large, so that the positive-going 0–1 bands do not overlap with the

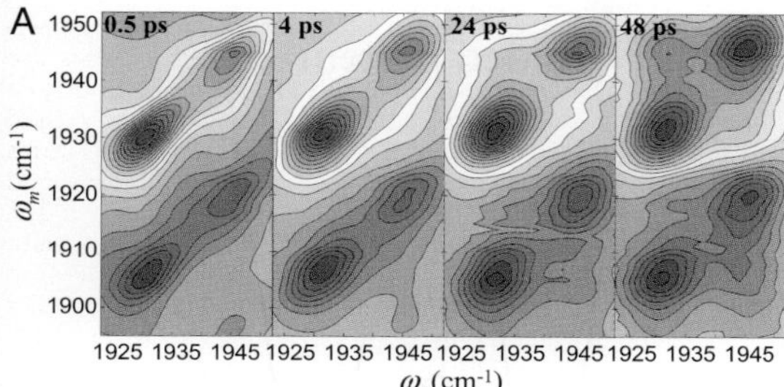

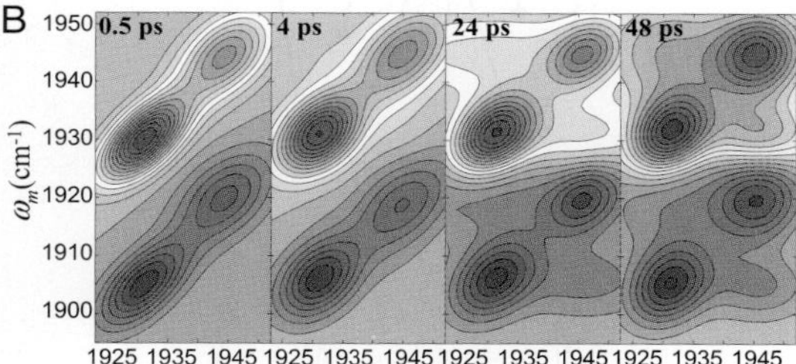

Fig. 2. 2D-IR spectra of CO bound to L29I. (*A*) 2D-IR spectra of L29I-CO at various times, T_w. Each contour corresponds to a 10% signal change. The bands in the upper half of the spectrum (positive-going) correspond to the 0–1 vibrational transition. The bands in the lower half of the spectrum (negative-going) arise from vibrational echo emission at the 1–2 transition frequency. (*B*) Calculated 2D-IR spectra of L29I-CO at various times, T_w, using the known input parameters and the substate switching time constant obtained from fitting the data.

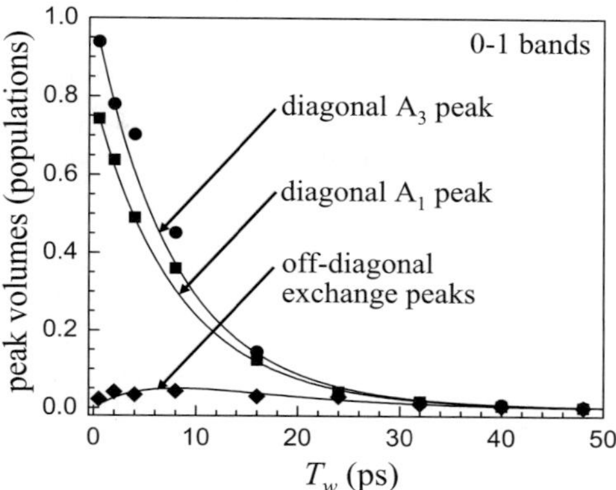

Fig. 3. Peak volume data obtained from the fitting of the 2D-IR spectra of L29I-CO for the 0–1 transition region. The diagonal A_1 and A_3 bands as well as the off-diagonal peaks that grow in because of substate switching are fit with a single adjustable, the substate switching time, $\tau_{13} = 1/k_{13}$. The solid curves are the results of the fits, which yield $\tau_{13} = 47$ ps.

negative-going 1–2 bands, then the cross peaks appear symmetrically about the 0–1 diagonal peaks and the corresponding 1–2 peaks (17). Here, the overlap of the positive- and negative-going bands reduces the amplitudes of the other two cross peaks, that is, the 0–1 A_1 to A_3 cross peak at $(\omega_\tau, \omega_m) = (1{,}945$ cm^{-1}, 1,932 cm^{-1}) and the 1–2 A_3 to A_1 cross peak at $(\omega_\tau, \omega_m) = (1{,}932$ cm^{-1}, 1,920 cm^{-1}).

In addition to the growth of the cross peaks in Fig. 2*A*, the shapes of the bands change with increasing T_w. At short T_w, the 2D line shapes show significant inhomogeneous broadening. Inhomogeneous broadening is evidenced by elongation along the diagonal. As time proceeds, protein structural fluctuations within a given substate structure cause the transition frequency to vary (spectral diffusion) (26–31). The landscape minimum associated with each substate is relatively shallow and has many local minima separated by small barriers (see Fig. 4). Transitions among these minima produce structural fluctuations on various time scales. At sufficiently long time, all structures associated with a substate are sampled and spectral diffusion is complete. All frequencies associated with the absorption band have been accessed. The 2D line shape goes from elongated to symmetrical. Fig. 2*A* shows that, by 48 ps, spectral diffusion is essentially complete. Here, we focus our analysis exclusively on the conformational switching; the protein structural dynamics that give rise to spectral diffusion of the A substates of MbCO have been reported previously (14, 29, 31).

To quantitatively extract the time constant for the conformational switching from the 2D-IR spectra, the integrated peak volumes are fit to obtain the population of each species. The kinetic model described previously is used (15, 17, 20). It is possible to calculate the time evolution of the 2D spectrum including spectral diffusion and chemical exchange by using response function theory (20, 22, 23). However, in a detailed theoretical and experimental study, it was demonstrated that the exchange rate can be extracted by using the simpler method that is used here (20). Because spectral diffusion does not change the peak volumes, only their shapes, at each T_w the peak volumes were determined by fitting all of the peaks to 2D Gaussian functions (17, 20). The resulting fits provide the volume of each peak at each T_w. The conformational exchange causes the original peaks to decrease in volume and the cross peaks to increase in volume. In addition, the vibrational lifetime of the CO stretch, T_1, causes all of the peaks to decrease in volume. In previous applications of the 2D-IR vibrational echo chemical-exchange method to small molecules in liquids, it was also necessary to account for the orientational relaxation of the species (15, 17, 20). Here, the protein is so large that the slow orientational relaxation can be neglected.

Eq. **1** shows the kinetic scheme.

$$\xleftarrow{T_{1A_1}} \; A_1 \underset{k_{31}}{\overset{k_{13}}{\rightleftarrows}} A_3 \; \xrightarrow{T_{1A_3}}$$

By using IR pump–probe experiments, the vibrational lifetimes, T_{1A_1} and T_{1A_3}, were measured. The lifetimes are 25 and 19 ps, respectively. Because the system is in thermal equilibrium, the rate of A_1 going to A_3, $k_{13}[A_1]$, has to equal the rate of A_3 going to A_1, $k_{31}[A_3]$. The ratio of substates determined from FT-IR and IR pump–probe experiment is $[A_1]/[A_3] = 0.9$. The substate switching time constant from A_1 to A_3 is $\tau_{13} = 1/k_{13}$. Therefore, in fitting the data, there is only one adjustable parameter, the substate switching time, τ_{13}. There are eight peaks, four in the 0–1 region and four in the 1–2 region. The details of the fitting procedure have been presented previously (17, 20). The time-dependent volumes of all eight peaks are fit by using the single adjustable parameter, τ_{13}.

Fig. 2*B* shows calculated 2D spectra by using the known input parameters and the results of fitting τ_{13}. Both the measured and the generated spectra are normalized by making the largest peak equal to unity at each time. The calculated 2D spectra are in good agreement with the time-dependent 2D spectra for 0–1 and 1–2 transition regions. The experimental diagonal and off-diagonal peak volumes for the 0–1 region of the spectra are plotted in Fig. 3. The solid lines through the data points are obtained from the fitting procedure with τ_{13} as the single adjustable parameter. The data from the 1–2 transition region can be reproduced with the identical parameters used for the 0–1 region. The fact that both the 0–1 and 1–2 regions can be fit with the same value of τ_{13} demonstrates that the thermal equilibrium of the system is not perturbed by vibrational excitation of the CO stretch (17). The results of the fitting yield $\tau_{13} = 47 \pm 8$ ps. Within experimental error, $\tau_{13} = \tau_{31}$.

Although Mb is one of the most intensively studied proteins and a great deal of attention has been focused on the nature and dynamics of the substates (1–3, 7–9), previously it has not been possible to measure the conformational switching between the A_1 and A_3 substates. The growth of the cross peaks in the 2D-IR vibrational echo spectra provides a direct observation of the conformational exchange. The interconversion time of 47 ps is faster than, but consistent with, the rough estimates that were made previously (13, 14). Because the 3D structure of L29I mutant Mb has not been reported, the impact of the amino acid substitution on the protein structure and dynamics is not clear. However, the peak frequencies of the CO absorption bands of L29I are virtually identical to the peak frequencies of wild-type Mb. The frequencies of the CO peaks in the IR spectrum are very sensitive to the protein structure. The fact that L29I-CO and MbCO have the same peak frequencies demonstrates that the perturbation of the structure of the heme pocket is small.

By using a combination of vibrational echo experiments and MD simulations, aspects of the structural change associated with the conformational switching between the A_1 and A_3 substates were established (14). The structural difference between the substates that is most directly involved in the difference in frequency of the A_1 and A_3 absorption peaks involves the imidazole side group of the distal histidine (N_ε protonated) rotating about the C_β—C_γ (methylene carbon-imidazole ring carbon) bond. Movement of the distal histidine, which brings the N_ε—H into closer proximity to the CO, gives rise to the A_3 substate (11, 14). The A_1 substate has the distal histidine in a configuration that places the N_ε—H further from the CO (14). In the absence of interactions between the imidazole and any other moiety, it would be possible for the imidazole side group of a histidine amino acid to rotate about the C_β—C_γ bond without causing translation or reorientation of the remainder of the molecule. In the crowded environment of the heme pocket, this is not likely to be the case. Switching between the A_1 and A_3 substates changes the interaction of the His-64 imidazole side group with the heme-CO and other moieties (see Fig. 1A). The change in the interactions should cause the His-64 to slightly reorient and translate. The structures of the heme pocket for the A_1 and A_3 substates taken form the MD simulation are shown in figure 5 of ref. 14. In these figures, which are single snap shots of the A_1 and A_3 structures, it is evident that with the rotation of the imidazole ring by $\approx 40°$ around the C_β—C_γ bond there is also a change in the alignment of the C_α—C_β bond relative to the heme. These considerations suggest that there is a change in the geometry of the E helix (see Fig. 1A) in going between the A_1 and A_3 substates.

That significant structural change occurs in the A_1–A_3 interconversion is supported by x-ray experiments. The high-resolution crystal structure of MbCO that contains two conformations has enabled modeling of the structure of A_1 and A_3 substates (11, 32). Although the distal histidine has a critical role in determining the substates of Mb, structural comparison between the A_1 and A_3 substates shows that the A_3 substate contains an additional cavity, Xe3, and another transient cavity found in simulations (32). Xe is used as a probe to identify the locations of cavities in proteins (33, 34). In Mb crystals, four Xe atoms (Xe1, Xe2, Xe3, and Xe4) occupy cavities, which may be involved in gas ligand migration (32, 33). The Xe3 site, which is

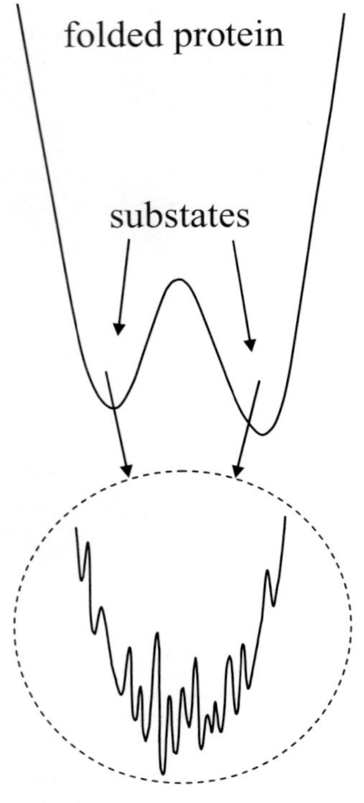

Fig. 4. Schematic illustration of a portion of the energy landscape of a protein, showing two substates and the finer energy landscape that exist in each of the minima. Transitions among the minima in one substate are responsible for structural fluctuations about the substate minimum. Transitions from one substate to another represent distinct structural configurational changes of the protein.

near the surface and far from the iron atom, involves Trp-7 and is located between helices E and H (33). The existence of Xe3 in the A_3 substate but not in the A_1 substate demonstrates that the difference in the substates is significantly more than the rotation of the imidazole side group of the distal histidine.

The fast time constant for interconversion, 47 ps, is in line with the structural difference between the A_1 and A_3 substates. Incoherent quasielastic neutron scattering experiments on native bovine α-lactalbumin observed collective motions on the tens of picosecond to 100-ps time scale (35). The correlation length for such fluctuations was reported to be 18 Å (35). The switching time observed here falls into this time range, which is consistent with a reconfiguration of the E helix in the interconversion between the A_1 and A_3 substates.

Concluding Remarks

NMR techniques probe protein motions in the microsecond, millisecond, and longer time scales (36). Conformational switching processes studied by NMR involve large structural changes that occur, for example, in enzyme catalytic processes (37). Such structural changes require the reconfiguration of many amino acids, helices, and various protein structures. To place in a broader context the time-dependent measurements of substate switching presented here, a single elementary structural change, it is useful to discuss the kinetics of large-scale structural changes in terms of the ideas of linear response. Linear response theory describes how thermal fluctuations and elementary steps bring a system into a new configuration.

The concept of linear response comes from the fluctuation dissipation theorem (38). It states that a system in thermodynamic equilibrium has a response to a small perturbation with a time dependence determined by the equilibrium fluctuations of the system. The time-dependent fluorescence Stokes shift experiment is a well studied example of relaxation to a new equilibrium structure after a perturbation (39). In the experiment, a solute molecule with a small or zero permanent dipole moment in the ground electronic state is excited to the first excited state that has a large dipole moment. The local solvent structure will evolve to produce a net alignment of the solvent dipoles relative to the newly created large solute dipole. The orientational alignment of the solvent dipoles occurs in the same manner as thermal equilibrium orientational relaxation. (Translational relaxation will also occur.) Before the generation of the solute dipole, the solvent molecules are executing angular random walks (orientational diffusion). The sudden presence of the solute dipole slightly biases the random walks toward alignment. The alignment of a single solvent dipole can be viewed as an elementary step in the restructuring of the solvent. The solvation of the solute dipole requires the combined response of many solvent molecules.

Fig. 4 displays a schematic of a portion of the energy landscape of a protein having two substates. Each substate minimum is actually a broad rough landscape with many local minima separated by barriers of varying heights. Transitions between these minimum are responsible for protein structural fluctuations about a specific structure associated with a particular substate minimum (1–3). These fluctuations can vary in time scales from subpicosecond to tens and hundreds of picoseconds to much longer (1–3). The fluctuations range from the fastest motions involving just a few atoms to low-frequency acoustic-type modes of the entire protein. The distinction between structural fluctuations that come from making transitions among the minima on the landscape of a particular substate and transitions between substates is in some sense operational. A substate has a structural aspect that is qualitatively distinct. Fluctuations make interconversions among substates possible by sampling the energy landscape in a particular substate and bringing the protein to the transition state between substates.

When a protein experiences a perturbation such as substrate binding or a temperature jump that induces folding or unfolding, in accord with linear response, it will respond to the perturbation by fluctuation-driven elementary steps (substate changes) that can occur in the absence of the perturbation. However, the sampling of the substates will be biased by the perturbation, and the protein will relax to a new structure. The protein is always executing a multidimensional walk among substates. The walk will be skewed by the perturbation. The skewing can result from the shifting of substate minima and barriers. The relaxation to the new structure can be slow because it requires many elementary steps. A slow response to a perturbation results from structural fluctuation sampling that produces elementary steps, which in turn combine to produce major restructuring.

Here, by using ultrafast 2D-IR vibrational echo chemical-exchange spectroscopy, we have measured the time dependence of a single elementary step, the A_1–A_3 substate switching. The time constant for the substate switching is 47 ps. The concept of conformational substates in proteins was introduced in 1974 (3). In 1987, MD simulations of Mb for 300 ps confirmed the existence of multiple conformational states (40). The rapid advance of computer technology makes simulations of protein dynamics for much longer time scales readily doable. The A_1–A_3 substate switching time reported here will be an important target for future MD simulations because the barrier crossing will need to be simulated accurately. Simulations will determine whether the dynamics of real systems undergoing an elementary structural change can be reproduced, and they will provide details of the structural changes that accompany the A_1–A_3 interconversion.

Materials and Methods

Expression and purification of the mutant sperm whale Mb L29I were performed as described in ref. 41. The CO forms of mutant Mb was prepared according to published protocols (14). For both the linear FT-IR and vibrational echo measurements, ≈ 20 μl of the sample solution was placed in a sample cell with CaF$_2$ windows and a 50-μm Teflon spacer. The 2D-IR vibrational echo experiments were described briefly above, and full details have been published previously (17, 19, 29, 42).

ACKNOWLEDGMENTS. We thank Prof. John S. Olson (Rice University, Houston, TX) for providing the L29I Mb mutant protein. This work was supported by National Institutes of Health Grant 2 R01 GM-061137-05 and Air Force Office of Scientific Research Grant F49620-01-1-0018. H.I. was supported by a fellowship from the Human Frontier Science Program.

1. Frauenfelder H, Parak F, Young RD (1988) Conformational substates in proteins. *Ann Rev Biophys Biophys Chem* 17:451–479.
2. Frauenfelder H, Sligar SG, Wolynes PG (1991) The energy landscapes and motions of proteins. *Science* 254:1598–1603.
3. Austin RH, et al. (1974) Activation-energy spectrum of a biomolecule: Photodissociation of carbonmonoxy myoglobin at low-temperatures. *Phys Rev Lett* 32:403–405.
4. Case DA, Karplus M (1979) Dynamics of ligand-binding to heme-proteins. *J Mol Biol* 132:343–368.
5. Schnell JR, Dyson HJ, Wright PE (2004) Structure, dynamics, and catalytic function of dihydrofolate reductase. *Annu Rev Biophys Biomol Struct* 33:119–140.
6. Oliveberg M, Wolynes PG (2005) The experimental survey of protein-folding energy landscapes. *Q Rev Biophys* 38:245–288.
7. Ansari A, et al. (1987) Rebinding and relaxation in the myoglobin pocket. *Biophys Chem* 26:337–355.
8. Tian WD, Sage JT, Champion PM (1993) Investigation of ligand association and dissociation rates in the "open" and "closed" states of myoglobin. *J Mol Bio* 233:155–166.
9. Muller JD, McMahon BH, Chen EYT, Sligar SG, Nienhaus GU (1999) Connection between the taxonomic substates of protonation of histidines 64 and 97 in carbonmonoxy myoglobin. *Biophys J* 77:1036–1051.
10. Li TS, Quillin ML, Phillips GN, Jr., Olson JS (1994) Structural determinants of the stretching frequency of CO bound to myoglobin. *Biochemistry* 33:1433–1446.
11. Vojtechovsky J, Chu K, Berendzen J, Sweet RM, Schlichting I (1999) Crystal structures of myoglobin-ligand complexes at near atomic resolution. *Biophys J* 77:2153–2174.
12. Yang F, Phillips GN, Jr. (1996) Crystal structures of CO-, deoxy- and met-myoglobins at various pH values. *J Mol Biol* 256:762–774.
13. Johnson JB, et al. (1996) Ligand binding to heme proteins. 6. Interconversion of taxonomic substates in carbonmonoxymyoglobin. *Biophys J* 71:1563–1573.
14. Merchant KA, et al. (2003) Myoglobin-CO substate structures and dynamics: Multidimensional vibrational echoes and molecular dynamics simulations. *J Am Chem Soc* 125:13804–13818.
15. Zheng J, Kwak K, Xie J, Fayer MD (2006) Ultrafast carbon-carbon single bond rotational isomerization in room-temperature solution. *Science* 313:1951–1955.
16. Kim YS, Hochstrasser RM (2005) Chemical exchange 2D IR of hydrogen-bond making and breaking. *Proc Natl Acad Sci USA* 102:11185–11190.
17. Zheng J, et al. (2005) Ultrafast dynamics of solute-solvent complexation observed at thermal equilibrium in real time. *Science* 309:1338–1343.
18. Morikis D, Champion PM, Springer BA, Sligar SG (1989) Resonance Raman investigations of site-directed mutants of myoglobin: Effects of distal histidine replacement. *Biochemistry* 28:4791–4800.
19. Park S, Kwak K, Fayer MD (2007) Ultrafast 2D-IR vibrational echo spectroscopy: A probe of molecular dynamics. *Laser Phys Lett* 4:704–718.
20. Kwak K, Zheng J, Cang H, Fayer MD (2006) Ultrafast two-dimensional infrared vibrational echo chemical exchange experiments and theory. *J Phys Chem B* 110:19998–20013.
21. Zheng J, Kwak K, Chen X, Asbury JB, Fayer MD (2006) Formation and dissociation of intra-intermolecular hydrogen-bonded solute-solvent complexes: Chemical exchange two-dimensional infrared vibrational echo spectroscopy. *J Am Chem Soc* 128:2977–2987.
22. Sanda F, Mukamel S (2006) Stochastic simulation of chemical exchange in two dimensional infrared spectroscopy. *J Chem Phys* 125:014507.

23. Sanda F, Mukamel S (2007) Anomalous lineshapes and aging effects in two-dimensional correlation spectroscopy. *J Chem Phys* 127:154107.
24. Rector KD, *et al.* (1997) Vibrational anharmonicity and multilevel vibrational dephasing from vibrational echo beats. *J Chem Phys* 106:10027.
25. Golonzka O, Khalil M, Demirdoven N, Tokmakoff A (2001) Vibrational anharmonicities revealed by coherent two-dimensional infrared spectroscopy. *Phys Rev Lett* 86:2154–2157.
26. Mukamel S (2000) Multidimensional femtosecond correlation spectroscopies of electronic and vibrational excitations. *Ann Rev Phys Chem* 51:691–729.
27. Mukamel S (1995) *Principles of Nonlinear Optical Spectroscopy* (Oxford Univ Press, New York).
28. Finkelstein IJ, Ishikawa H, Kim S, Massari AM, Fayer MD (2007) Substrate binding and protein conformational dynamics measured via 2D-IR vibrational echo spectroscopy. *Proc Natl Acad Sci USA* 104:2637–2642.
29. Finkelstein IJ, *et al.* (2007) Probing dynamics of complex molecular systems with ultrafast 2D IR vibrational echo spectroscopy. *Phys Chem Chem Phys* 9:1533–1549.
30. Ishikawa H, Kim S, Kwak K, Wakasugi K, Fayer MD (2007) Disulfide bond influence on protein structural dynamics probed with 2D-IR vibrational echo spectroscopy. *Proc Natl Acad Sci USA* 104:19309–19314.
31. Ishikawa H, *et al.* (2007) Neuroglobin dynamics observed with ultrafast 2D-IR vibrational echo spectroscopy. *Proc Natl Acad Sci USA* 104:16116–16121.
32. Teeter MM (2004) Myoglobin cavities provide interior ligand pathway. *Protein Sci* 13:313–318.
33. Tilton RF, Jr., Kuntz ID, Jr., Petsko GA (1984) Cavities in proteins: Structure of a metmyoglobin xenon complex solved to 1.9 Å. *Biochemistry* 23:2849–2857.
34. Doukov TI, Blasiak LC, Seravalli J, Ragsdale SW, Drennan CL (2008) Xenon in and at the end of the tunnel of bifunctional carbon monoxide dehydrogenase/acetyl-coa synthase. *Biochemistry* 47:3474–3483.
35. Bu Z, *et al.* (2000) A view of dynamics changes in the molten globule native folding step by quasielastic neutron scattering. *J Mol Bio* 301:525–536.
36. Palmer AG, III (1997) Probing molecular motion by NMR. *Curr Opin Struct Biol* 7:732–737.
37. Boehr DD, Dyson HJ, Wright PE (2006) An NMR perspective on enzyme dynamics. *Chem Rev* 106:3055–3079.
38. Kubo R, Toda M, Hashitusme N (1985) *Statistical Physics II: Nonequilibrium Statistical Mechanics* (Springer, New York).
39. Fleming GR, Mihnaeng C (1996) Chromophore-solvent dynamics. *Annu Rev Phys Chem* 47:109–134.
40. Elber R, Karplus M (1987) Multiple conformational states of proteins: A molecular dynamics analysis of myoglobin. *Science* 235:318–321.
41. Springer BA, Sligar SG (1987) High-level expression of sperm whale myoglobin in Escherichia coli. *Proc Natl Acad Sci* USA 84:8961-8965.
42. Zheng J, Kwak K, Fayer MD (2007) Ultrafast 2D IR vibrational echo spectroscopy. *Acc Chem Res* 40:75–83.

J. Phys. Chem. B **2008**, *112*, 10221–10227

Solute–Solvent Complex Kinetics and Thermodynamics Probed by 2D-IR Vibrational Echo Chemical Exchange Spectroscopy

Junrong Zheng[†,‡] and M. D. Fayer*[,†]

Department of Chemistry and Pulse Institute, Stanford University, Stanford, California 94305

Received: May 8, 2008; Revised Manuscript Received: June 12, 2008

The formation and dissociation kinetics of a series of triethylsilanol/solvent weakly hydrogen bonding complexes with enthalpies of formation ranging from −1.4 to −3.3 kcal/mol are measured with ultrafast two-dimensional infrared (2D IR) chemical exchange spectroscopy in liquid solutions at room temperature. The correlation between the complex enthalpies of formation and dissociation rate constants can be expressed with an equation similar to the Arrhenius equation. The experimental results are in accord with previous observations on eight phenol/solvent complexes with enthalpies of formation from −0.6 to −2.5 kcal/mol. It was found that the inverse of the solute–solvent complex dissociation rate constant is linearly related to $\exp(-\Delta H_0/RT)$ where ΔH_0 is the complex enthalpy of formation. It is shown here, that the triethylsilanol–solvent complexes obey the same relationship with the identical proportionality constant, that is, all 13 points, five silanol complexes and eight phenol complexes, fall on the same line. In addition, features of 2D IR chemical exchange spectra at long reaction times (spectral diffusion complete) are explicated using the triethylsilanol systems. It is shown that the off-diagonal chemical exchange peaks have shapes that are a combination (outer product) of the absorption line shapes of the species that give rise to the diagonal peaks.

I. Introduction

Solute–solvent intermolecular interactions can lead to well defined solute–solvent complexes.[1] For systems in which solute–solvent complexes exist, the solution cannot be described in terms of a simple radial distribution function[2] with solvent molecules moving diffusively into and out of the first solvation shell of a solute. The typical organic solute–solvent interaction energy is weak, less than 5 kcal/mol.[3] This value suggests that the dissociation time scale of solute–solvent complexes in room temperature solutions will be faster than 1 ns.[4] However, complexes that exist for even short times, ten to a hundred picoseconds, fundamentally change the behavior of a solvent surrounding a solute. It may be necessary for a solute–solvent complex to dissociate before a bimolecular reaction can occur. Therefore, understanding the dynamics of solute–solvent interactions is important for understanding the molecular level reaction mechanisms in solutions.

Because of the short time scales, measuring the kinetics of reactions with activation energy smaller than 5 kcal/mol at room temperature under thermal equilibrium conditions has proven to be difficult.[4] Recently nonlinear IR techniques especially 2D IR methods[5–13] have overcome the difficulties.[14–21] The techniques have an intrinsic fast temporal resolution that can be <100 fs. In addition, perturbations of the dynamics using 2D-IR methods have been shown to be negligible,[17] and there are well defined internal tests to determine if the true equilibrium dynamics are in fact being observed.[17] The basic principle of the methods is to monitor the time evolution of the of the 2D-IR spectrum of a vibration with a frequency that is changed by an equilibrium chemical process. If the vibration has different frequencies for two species that are interconverting, the time

evolution of the 2D spectrum can provide a direct measurement of the rate of interconversion.

Recently we applied the 2D IR chemical exchange method[15–23] to the study of formation and dissociation kinetics of eight phenol–solvent complexes with enthalpies of formation from −0.6 to −2.5 kcal/mol under thermal equilibrium conditions at room temperature.[20] We found that there is a clear correlation between the enthalpies of formation of the complexes and the dissociation time constants (inverse of the dissociation rate constants), and the correlation can be described by an equation similar to the Arrhenius equation.[20,21] It was determined that the solute–solvent complex dissociation time constant is linearly related to $\exp(-\Delta H^0/RT)$ where ΔH^0 is the complex enthalpy of formation, R is the gas constant, and T is the absolute temperature. This observation is interesting and important. If it is general for most weak organic solute–solvent complexes, then the dissociation kinetics of other solute–solvent complexes might be estimated based on their enthalpies of formation. Enthalpies of formation can be obtained from standard tools such as FT-IR through van't Hoff plots.[24] To test the generality of the observed enthalpy/kinetics correlation of the phenol/solvent systems, we studied another series of solute–solvent systems with enthalpies of formation from −1.4 to −3.3 kcal/mol. The vibrational probe is the OD hydroxyl stretch of triethylsilanol with the OH replaced by OD. The OD stretch has a much longer lifetime (~160 ps in CCl₄) than that of the OD stretch of phenol-OD molecules (~15 ps in CCl₄). The substantially longer vibrational lifetime allows us to extend the measurements to stronger complexes (enthalpies of formation more negative than −2.5 kcal/mol). In addition, the longer probe lifetime provides a much longer time window in which to measure the kinetics. In previous studies, the dynamic properties of the phenol-benzene complexes were mostly derived from experiments at relatively short reaction periods (limited by the lifetimes of the probes). Therefore, spectroscopic features of 2D IR chemical exchange spectra at long time have not been

---

* Corresponding author. E-mail:fayer@stanford.edu.
† Department of Chemistry, Stanford University.
‡ Pulse Institute, Stanford University.

10.1021/jp804087v CCC: $40.75 © 2008 American Chemical Society
Published on Web 07/30/2008

10222 *J. Phys. Chem. B, Vol. 112, No. 33, 2008*

Zheng and Fayer

experimentally explored. The new systems offer an opportunity to examine the spectral features at long time and compare them to theoretical predictions.[23] It is found experimentally and theoretically that the 2D chemical exchange off-diagonal line shapes are determined by a combination of the two linear absorption line shapes. Each spectrum is taken as a vector, and the peak shapes are determined by the outer product of the two spectral vectors.

II. Experimental Procedures

The systems are room-temperature solutions of triethylsilanol with the hydroxyl OH replaced by OD (TES) in benzene/CCl_4 (TES:benzene:CCl_4 = 4.6:80:100, molar ratio), toluene/CCl_4 (TES:toluene:CCl_4 = 4:69:100, molar ratio), p-xylene/CCl_4 (TES:p-xylene:CCl_4 = 4.2:38:100, molar ratio), mesitylene/CCl_4 (TES:mesitylene:CCl_4 = 4:34:100, molar ratio) and acetonitrile/CCl_4 (TES:acetonitrile:CCl_4 = 3:8:100, molar ratio) mixed solvents. The low concentrations of TES make sure that vibrational excitation transfer between the TES species is negligible. In the solutions, a good portion of the TES molecules form complexes with the aromatic molecules, while other TES remain free (uncomplexed).

The free and complexed TES species are in thermal equilibrium with the complexes continually dissociating to make the free TES and the free TES continually associating to form complexes. Because the system is in equilibrium, there is no change in the overall number of the free and complexed species. Therefore, the exchange rate cannot be determined from the IR absorption spectrum of the OD stretch. The exchange rates are measured with the 2D IR exchange method,[17] the rotational and vibrational dynamics are measured with pump–probe experiments,[18,25] and the strengths of the triethylsilanol-solvent molecule complexes are measured with the temperature dependent FT-IR absorption spectrum using van't Hoff plots.[21,24] The OD stretch of TES has a very long vibrational lifetime that makes it possible to measure the exchange at long reaction periods. In CCl_4 TES exists only as the free species and has a lifetime of 160 ps. The lifetimes of the complexes are shorter, but still many tens of picoseconds.

Details of the experimental method used for these 2D IR chemical exchange experiments have been described previously.[18,22,26,27] Very briefly, in a 2D IR vibrational echo chemical exchange experiment, three ultrashort IR pulses tuned to the frequency range of the vibrational modes of interest are crossed in the sample. Because the pulses are very short, they have a broad frequency bandwidth that makes it possible to simultaneously excite a number of vibrational modes. The first laser pulse "labels" the initial structures of the species by establishing their initial frequencies, ω_τ. The second pulse ends the first time period τ and starts clocking the reaction time period T_w during which the labeled species undergo chemical exchange, that is, complexes formation and dissociation. The chemical exchange will be manifested in the 2D spectrum by growth of off-diagonal peaks if the exchange changes the vibrational frequency of the vibrational mode under study. In addition to chemical exchange, vibrational relaxation to the ground-state and orientational relaxation will influence the 2D spectrum. The third pulse ends the population period of length T_w and begins a third period of length $\leq \tau$, which ends with the emission of the vibrational echo pulse of frequency ω_m, which is the signal in the experiment. The vibrational echo signal reads out information about the final structures of all labeled species by their frequencies, ω_m. During the period T_w between pulses 2 and 3, chemical exchange occurs when two species in equilib-

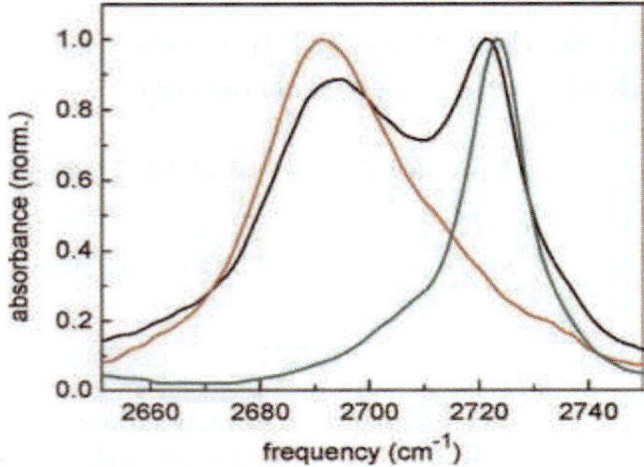

Figure 1. FT-IR absorption spectra of the OD stretch of triethylsilanol (TES, hydroxyl H replaced with D) in CCl_4 (free TES, green curve), TES in benzene (TES–benzene complex, red curve), and TES in the mixed benzene/CCl_4 solvent, which displays absorptions for both free and complexed TES (black curve).

rium interconvert without changing the overall number of either species. The exchange causes new off-diagonal peaks to grow in as T_w is increased. The growth of the off-diagonal peaks in the 2D IR spectra with increasing T_w is used to extract the dissociation and formation time constants of the complexes. The dissociation time constant, τ_d, is the inverse of the dissociation rate constants, k_d. The rates of dissociation and formation are equal because the system is in equilibrium.[17]

The vibrational population relaxation time constants and orientational relaxation time constants were measured with the polarization selective IR pump–probe experiments.[18,28] The orientational relaxation time constants, obtained in pure solvents, were corrected for the solvent viscosity differences between the pure aromatic solvent or CCl_4 and the mixed solvents used in the chemical exchange experiments using the Debye–Stokes–Einstein equation. Viscosity measurements were made with Cannon Ubbelohde Viscometers at 24 °C, the same temperature used for the vibrational echo and pump–probe measurements.

The enthalpies of formation of the solute–solvent complexes were determined by measuring the temperature dependence of the equilibrium constants with FT-IR. The temperature range was 25–65 °C.[20,21]

III. Results and Discussion

A. Chemical Exchange Dynamics. When TES molecules are dissolved in a benzene/CCl_4 mixture, some of the TES form complexes with benzene, while others remain free (not complexed). Experimental evidence for the formation of triethylsilanol/benzene complexes is the shift of the OD stretch frequency of TES to a lower frequency in benzene compared to that of TES in CCl_4. Figure 1 shows FT-IR spectra of triethylsilanol-OD in pure CCl_4, in pure benzene, and a mixed benzene/CCl_4 solvent. When TES is dissolved in CCl_4, no hydrogen bond is formed between the solute and solvent molecules. The OD stretch frequency is at 2723 cm^{-1}. The peak is relatively narrow, and the full width at half-maximum (fwhm) is 14 cm^{-1}. When TES is dissolved in benzene, a π hydrogen bonded complex is formed between a solute and a solvent molecule. The OD stretch frequency red-shifts to 2691 cm^{-1} and the fwhm becomes 25 cm^{-1}. In the mixed benzene/CCl_4 solvent (black curve in Figure 1), both the complexed and free species are present. The relative concentration ratio of the two species can be adjusted by changing the benzene/CCl_4 ratio.

J. Phys. Chem. B, Vol. 112, No. 33, 2008 **10223**

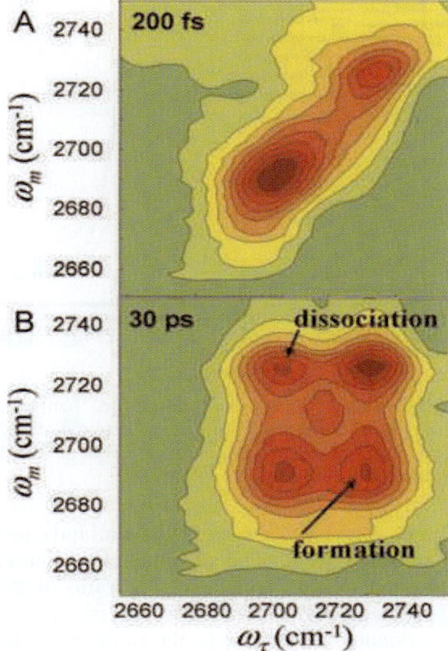

Figure 2. 2D-IR vibrational echo spectra of the OD stretch of TES in the mixed benzene/CCl$_4$ solvent. The data have been normalized to the largest peak for each T_w. Each contour represents 10% change in amplitude. (A) Data for $T_w = 200$ fs. There are only peaks on the diagonal because chemical exchange has yet to occur. (B) Data for $T_w = 30$ ps. At the longer time, additional peaks have grown in because of chemical exchange, that is, the formation and dissociation of the complex of TES and benzene.

At room temperature, the two species of TES are under dynamic equilibrium in the mixed solution. The complexed and free species are constantly exchanging. The TES/benzene complex enthalpy of formation (ΔH^0) in the solution was determined to be −1.4 kcal/mol. On the basis of our previous work on phenol/benzene complexes and other phenol/substituted benzene complexes in the same types of mixed solvents, this ΔH^0 value suggests that the exchange reactions will be very fast.

Figure 2 displays two 2D IR spectra (only the 0−1 transition region is shown) of TES in a benzene/CCl$_4$ mixture for a very short ($T_w = 200$ fs) and a long ($T_w = 30$ ps) reaction period. For the very short reaction time no chemical exchange has occurred. The species (complexed or free TES molecules) in the sample are unchanged. Therefore, in the 2D IR spectrum, the ω_τ frequency (initial frequency) and the ω_m frequency (final frequency) for each peak are unchanged. Only two peaks are present in the spectrum, and they are on the diagonal. These peaks correspond to the absorption bands shown in Figure 1 for the mixed solvent. For the long period, considerable exchange has occurred. During the 30 ps T_w period, some of the complexes that existed at the beginning of the period have dissociated to become free. Some of the free TES have associated with benzenes to become complexes. Two off-diagonal peaks have grown in. The off-diagonal peaks originate from the exchange reactions. The peak labeled as "dissociation" has its ω_τ (initial frequency) at the lower frequency, showing that the initial structure was complexed TES. Its ω_m (final frequency) is the higher frequency, which shows the final structure is free TES. This off-diagonal peak arises from those free TES molecules formed from initially complexed molecules.

The "formation" peak (at 30 ps) arises from complexed TES molecules that were initially free species. The two diagonal peaks are the result of molecules that have either not exchanged or have exchanged an even number of times during the 30 ps reaction period so that they are the same species at the end of the reaction period as they were initially.

The T_w dependent growth of the off-diagonal peaks provides direct information on the time dependence of the formation and dissociation of the TES-benzene complex. As shown in Figure 2A, at very sort time there are no off-diagonal peaks. As T_w increases, the off-diagonal peaks increase in amplitude. By 30 ps (Figure 2B) they are substantial. Illustrations of the growth of the off-diagonal peaks as chemical exchange proceeds have been given previously.[17–19,21] The exchange kinetics can be extracted from analyzing the time dependent peak volumes.[17,21,22] If spectral diffusion is fast compared to the chemical exchange rate, the peak intensities can be used to obtain the chemical exchange dynamics.[17,21]

The 2D IR signal is affected by four dynamic processes: chemical exchange, vibrational relaxation of the OD stretch excitation, orientational relaxation of the species, and the spectral diffusion.[29,30] Chemical exchange causes the two off-diagonal peaks to grow in and the two diagonal peaks shrink. Vibrational relaxation causes all four peaks to decrease in amplitude. Orientational relaxation randomizes the anisotropy of the excited molecules induced by the polarized laser pulses, which reduces all four peaks. Spectral diffusion is the result of time dependent interactions of the vibrational transition with the solvent. Interactions of the vibrational oscillator (OD stretch) cause its transition frequency to evolve. At sufficiently long time, all oscillators will have sampled all frequencies that give rise to the absorption line shape. In the 2D-IR vibrational echo spectrum, at short time the diagonal peaks are elongated along the diagonal, which reflects inhomogeneous broadening of the system. At sufficiently long T_w, all possible solvent configurations have been sampled, and the dynamic line width is equal to the absorption line width. In a 2D spectrum complete spectral diffusion is manifested by a change in the diagonal 2D line shapes from elongated along the diagonal to symmetrical about the diagonal. From the data, it was determined that spectral diffusion is complete by ~4 ps. Spectral diffusion changes the shapes of the peaks but preserves their volumes. Therefore, if the volume of the peaks is used in determining the chemical exchange kinetics, spectral diffusion drops out of the problem.[17,22] In addition, the magnitude of signal of each peak in determined by the transition dipole moments of the species.[31] The diagonal peaks have amplitudes proportional to μ_i^4, where i labels the species, free and complex. The off-diagonal peaks depend on the transition dipoles of both species as $\mu_i^2\mu_j^2$.

By taking into account all of these factors, a kinetic model was constructed to analyze the exchange kinetics.[17,18,21,22] The model is illustrated schematically as

$$\underset{decay}{\overset{T_1^c,\tau_r^c}{\longleftarrow}} C \underset{k_a}{\overset{k_d}{\rightleftharpoons}} F \underset{decay}{\overset{T_1^f,\tau_r^f}{\longrightarrow}} \qquad (1)$$

where the complexed (C) and free (F) species can undergo chemical exchange with dissociation and association rate constants, k_d and k_a, respectively. The free and complex species decay due to vibrational relaxation with time constants T_1^f and T_1^c, respectively, and the species undergo orientational relaxation with time constants τ_r^f and τ_r^c, respectively. The lifetime and orientational relaxation time constants are obtained from pump−probe experiments, the reaction time (T_w) dependent species' concentrations are provided by the 2D IR measurements

10224   *J. Phys. Chem. B, Vol. 112, No. 33, 2008*   Zheng and Fayer

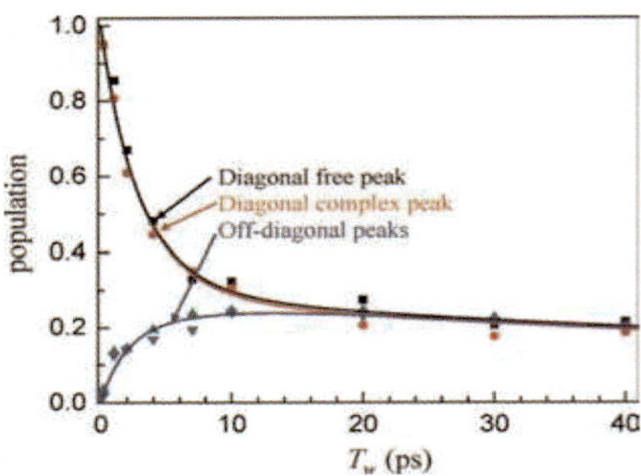

Figure 3. T_w dependent data (symbols) showing the time dependence of the diagonal and off-diagonal chemical exchange peaks, in the 2D IR vibrational echo spectra. The solid lines through the data are the result of the single adjustable parameter fit that yields the TES/benzene dissociation time constant, $\tau_d = 9$ ps.

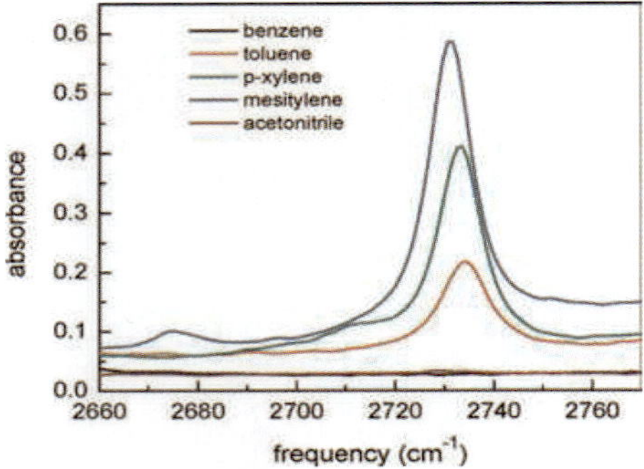

Figure 4. FT-IR spectra of pure acetonitrile, benzene, toluene, p-xylene, mesitylene in the region where the OD stretch of TES absorbs. A peak is apparent in toluene, p-xylene, and mesitylene at ~2730 cm^{-1} but not in benzene or acetonitrile.

after appropriate scaling with the transition dipole moments, and the ratio of the dissociation and association rate constants is determined using the equilibrium constant K obtained from FT-IR measurements.[17,21] Therefore, there is only one adjustable parameter, $\tau_d = 1/k_d$, necessary to fit the experimental data.

The kinetics of the formation and dissociation of TES/benzene complexes were analyzed with the model outlined above and described in more detail previously.[17,21,22] The input parameters used to implement the model shown in equation 1 are $T_1^c = 95$ ps, $T_1^f = 160$ ps, $\tau_r^c = 2.8$ ps, $\tau_r^f = 3.2$ ps, and the ratio of the complexed and free triethylsilanol concentrations [complex]/[free] $= 1$. The ratio of the square of the transition dipole moments is $\mu_c^2/\mu_f^2 = 1.71$.

Figure 3 displays the T_w dependent populations associated with the four peaks shown at 30 ps in Figure 2B. The diagonal peaks decay. The off-diagonal peaks grow in and then decay. Because the system is in equilibrium, the rates of formation and dissociation of the complexes are equal, and the off-diagonal peaks grow in with the same time dependence.[17] The solid lines through the data are calculated with the single adjustable parameter, the dissociation time constant, τ_d. The agreement between the fit and the data is quite good. The results give the dissociation time constant for the TES/benzene complex, $\tau_d = 9$ ps.

The analysis of the TES/benzene data has been described in some detail. We now wish to compare the complex dissociation time obtained for the TES-benzene complex with the other four TES complexes studied here. We will then combine the new results for the five TES complexes with previous results obtained on eight complexes involving the solute phenol rather than TES to investigate the trend in the dissociation times as a function of the enthalpies of formation of the 13 complexes.

The analysis used for the TES/benzene system and the TES/acetonitrile system is identical to that employed previously for phenol/benzene and seven other systems.[20,21] However, in contrast to the other systems studied here and previously, the pump−probe population relaxation decays of the OD stretch of the TES molecules in the pure solvents toluene, p-xylene and mesitylene are biexponentials rather than single exponentials. The precise reason for the biexponential population decay is unknown. However, the spectra of the solvents give an indication of the mechanism. Figure 4 shows the spectra of all five solvents. Benzene and acetonitrile show no absorption in the region of the OD stretch absorption (see Figure 1). In

contrast, toluene, p-xylene, and mesitylene all have absorptions in the OD stretch region. The absorbance increases as the number of methyl groups increases. The ratio of the optical densities of this peak of the three molecules is ~1:2:3 (toluene:p-xylene:mesitylene), very close to the ratio of the number of the methyl groups. Therefore, it is reasonable to assume that the absorbance is a result of a combination band or overtone involving the methyl group. The proximity of the solvent mode frequency to the frequency of the OD stretch of both the complex and free species can provide a pathway for vibrational population decay that involves relaxation directly into these solvent degrees of freedom involving the methyl groups.[32,33] The biexponential decay could arise because only a fraction of the TES are in a configuration relative to the aromatic solvent molecule that permits the solvent pathway to be active.

The biexponential decay can be readily incorporated into the chemical exchange kinetic analysis. The 2D IR kinetic analysis can be separated into two independent single exponential parts. The ratio of the two parts is determined by the prefactors (coefficients) of the biexponential. For each single exponential part the model used above and previously is applied.[17,21] Pump−probe measurements on TES in toluene, p-xylene, and mesitylene yield the biexponential complex lifetimes and amplitudes (see Table 1 for values), but TES in CCl$_4$ is a single exponential.

In the analysis performed above and previously,[21] the lifetime for the complex is the lifetime measured in the pure complexing solvent, and the lifetime for the free species is the lifetime measured in CCl$_4$. Figure 5A shows the results of fitting the T_w dependent 2D IR spectra for the TES/mesitylene-CCl$_4$ system using the model in which the complex has the measured biexponential population relaxation but the free species decays as a single exponential. The analysis yields the only adjustable parameter, $\tau_d = 28$ ps. As seen in Figure 5A, the model does a good job in fitting the TES/mesitylene complex diagonal peak and the two off-diagonal peaks, but it does not reproduce the free diagonal peak very well. The experimental data decay much faster than the best fit. A reasonable explanation for this is that the vibrational population relaxation of the free species in the mixed solvent containing aromatics with methyl groups is biexponential.

Figure 5B shows that same data fit again, but this time taking the population decay of both the free TES and the complex to be biexponential. In this fit there are three adjustable parameters, τ_d and the amplitude and decay time of the fast component of

TABLE 1: Experimentally Determined Parameters[a]

sample	τ_d (ps)	ΔH^0 (kcal/mol)	T_1 (ps)	A	τ_r^f (ps)	τ_r^c (ps)	τ_r^{cp} (ps)
TES			160	1	3.2	3.2	3.2
TES/BZ	9 ± 2	−1.4	95	1	3.3	2.8	3.3
TES/TL	11 ± 2	−1.6	95 (1)	0.92 (0.08)	3.2	3.8	4.2
TES/pX	18 ± 3	−2.1	101 (1.2)	0.87 (0.13)	3.3	4.1	5.1
TES/MS	26 ± 5	−2.4	61 (2.8)	0.80 (0.20)	3.3	6.9	8.7
TES/AN	140 ± 20	−3.3	22	1	3.0	8.5	7

[a] Samples: TES, TES in CCl_4; TES/BZ, TES in benzene/CCl_4; TES/TL, TES in toluene/CCl_4; TES/pX, TES in *p*-xylene/CCl_4; TES/MS, TES in mesitylene/CCl_4; TES/AN, TES in acetonitrile/CCl_4. τ_d: complex dissociation time constant. ΔH^0: complex enthalpy of formation. T_1: population decay time of free TES (160 ps) and TES complexes. For bi-exponential decays, the fast component is given in parentheses. A: normalized amplitude of population decay. For bi-exponential decays, the amplitude of the fast component is given in parentheses. τ_r^f: orientational relaxation time constant of the free species in the mixed solvent. τ_r^c: orientational relaxation time constant of the complex in the mixed solvent. τ_r^{cp}: orientational relaxation time constant of the complex in the pure complexing solvent.

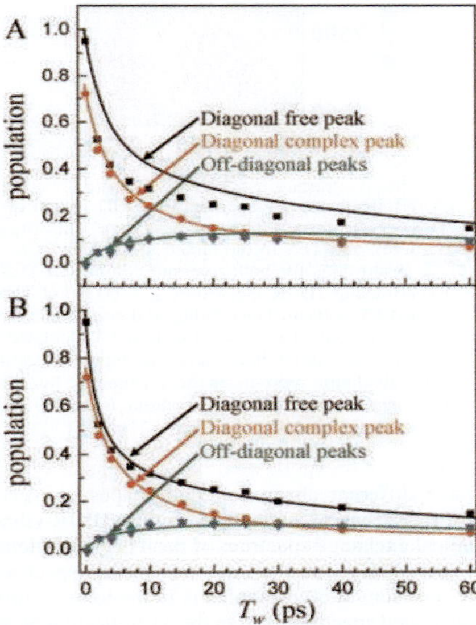

Figure 5. T_w dependent data (symbols) showing the time dependence of the diagonal and off-diagonal 2D IR vibrational echo chemical exchange peaks for the TES/mesitylene−CCl_4 system. The solid curves are from fits to the data. (A) Solid curves are the fitting results for the model in which the OD stretch population of the complex decays as a biexponential as measured on TES in pure mesitylene, but the OD stretch population decay of the free species is a single exponential. The data are fit with one adjustable parameter, τ_d. The fit yields $\tau_d =$ 28 ps. The fit misses the free diagonal peak data but is quite good for the other three peaks. (B) Solid curves are the fitting results for the model in which both the complexed and free TES have biexponential population decays. The fit is substantially improved and yields $\tau_d =$ 23 ps.

the population decay of the free TES in the mysitylene-CCl_4 solvent. As can be seen in Figure 5B, the fit is much better, but there are more adjustable parameters. The results yield $\tau_d = 23$ ps. This value is quite similar to the value obtained from the fitting shown in Figure 5A, indicating that the value obtained for τ_d is robust. The τ_d value is mainly determined by the growth of the off-diagonal peaks. For the three solvents that display

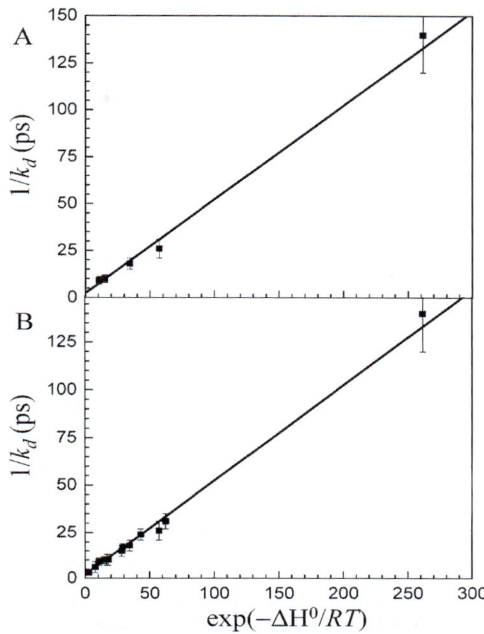

Figure 6. Dissociation times of the complexes, τ_d ($1/k_d$), plotted vs $\exp(-\Delta H^0/RT)$ where ΔH^0 is the enthalpy of formation of the solute−solvent complex. The line through the data is given by equation 2. (A) Data from the five triethylsilanol systems measured here. (B) Data from the five triethylsilanol systems and from the eight phenol systems measured previously.[17,21] The same relation (eq 2) holds for all 13 data points.

the biexponential population decays, we fit the data both ways, that is, the free TES population relaxation is single or biexponential. The τ_d values are taken as the average of the two values. So for TES/mesitylene we obtain $\tau_d = 26 \pm 5$ ps. The τ_d values for the five systems studied are given in Table 1.

B. Correlation between Kinetics and Thermodynamics. Figure 6A shows the correlation between the dissociation time constant τ_d ($1/k_d$) and $\exp(-\Delta H^0/RT)$, where ΔH^0 is the enthalpy of the formation of the complex. The data fall on a line within experimental error. This is the same behavior that was found previously for the eight complexes of phenol-OD. Figure 6B shows the five data points for the TES complexes and the eight data points for the phenol complexes. The line through the data in both panels is

$$1/k_d = B + A^{-1} \exp(-\Delta H^0/RT) \qquad (2)$$

where $B = 2.3$ ps and $A^{-1} = 0.5$ ps are constants. All of the data fall on the same line. The results demonstrate again that there is a clear correlation between the dissociation time constant and the enthalpy of formation of the complexes. The deviation from an intercept of 0, that is the fact that $B = 2.3$ ps, could arise for a number of reasons. Previously we suggested that even in the limit that ΔH^0 goes to zero, a finite time will be required for the pseudocomplex (no binding) to separate.[21] A simple calculation showed that the intercept is the right time scale for diffusive separation.[21] Another possibility could be a systematic error in the determination of the ΔH^0 values, which were obtained with the assumption that the enthalpies and entropies of formation are temperature independent over the temperature range used to determine them.

The important feature of Figure 6B is that all of the data fall on the same line. Transition state theory[24] would suggest that k_d should depend on the activation free energy, ΔG^*, not on the enthalpy of formation, ΔH^0. In terms of simple transition state theory, the dissociation rate constant can be written as

$$k_d = \frac{k_B T}{h} e^{-\Delta G^*/RT} = \frac{k_B T}{h} e^{\Delta S^*/R} e^{-\Delta H^*/RT} \tag{3}$$

where ΔG^* is the activation free energy, ΔS^* is the activation entropy, and ΔH^* is the activation enthalpy. If the activation enthalpy is proportional to the dissociation enthalpy, $\Delta H^* \propto \Delta H^0$, and the activation entropy ΔS^* is essentially a constant, independent of the molecular structure of the complexes and differences in the solvents, then equation 2 and the behavior displayed in Figure 6 are obtained.

The data in Figure 6B has eight complexes of phenol and phenol derivatives with various substituted benzenes[21] and five complexes of triethylsilanol with substituted benzenes and acetonitrile. All of the complexes involve a hydrogen bond with a hydroxyl. However, the complexes vary significantly in their molecular structure. It is not clear how general the correlation between the complex dissociation time and the enthalpy of formation is for other complexes that do not involve a hydrogen bond to a hydroxyl. Nonetheless, equation 2 may serve as a guide for estimating the dissociation times of complexes.

C. 2D IR Chemical Exchange Line Shapes. The long vibrational lifetimes of the OD stretch of both the free (~160 ps) and complexed TES molecules (see Table 1) allow 2D IR vibrational echo chemical exchange data to be acquired at long reaction times (T_w). The data for long reaction periods reveal an important feature of the 2D IR chemical exchange spectra that has not been previously explicated. If T_w is sufficiently long the 2D line shapes are combinations, the outer products, of the linear absorption line shapes of the two exchanging species.

At short T_w, spectral diffusion causes the diagonal 2D line shapes to change with time. At sufficiently long time, spectral diffusion is complete, and the line shapes obtain their long time limit. It has been argued that the off-diagonal peaks generated by chemical exchange are always effectively in the long T_w limit because the frequencies within the inhomogeneous lines of the exchanging species are uncorrelated, and, therefore, the exchange process itself cause complete spectral diffusion.[22] Even if there is frequency correlation in the exchange, at long T_w when spectral diffusion is complete, the diagonal and off-diagonal peaks will no longer change shape with increasing T_w. In the long T_w limit, theory predicts that the line shapes of the off-diagonal peaks can be expressed as,[23]

$$S(\omega_\tau, T_w, \omega_m) \propto W_A(\omega_\tau) \otimes W_B(\omega_m) \tag{4}$$

where $S(\omega_\tau, T_w, \omega_m)$ is the off-diagonal exchange peak signal and $\otimes$ is the outer product between the two vectors, $W_A(\omega_\tau)$ and $W_B(\omega_m)$. $W_A(\omega_\tau)$ is the vector composed of the linear absorption spectrum, that is the intensity at each frequency, of species A, which is initially excited by pulse 1, and $W_B(\omega_m)$ is the vector composed of the linear absorption spectrum of species B, which is the species that emits the vibrational echo following pulse 3. For the diagonal peaks, the signal is proportional to $W_A(\omega_\tau) \otimes W_A(\omega_m)$ and $W_B(\omega_\tau) \otimes W_B(\omega_m)$.

According to theory at long T_w when spectral diffusion is complete, the line shapes of the diagonal peaks are symmetric about the two axes. The line shapes of the off-diagonal chemical exchange peaks will not be symmetric if the two exchanging

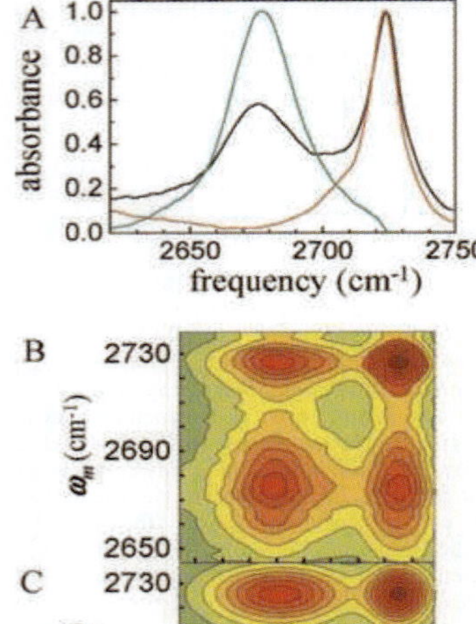

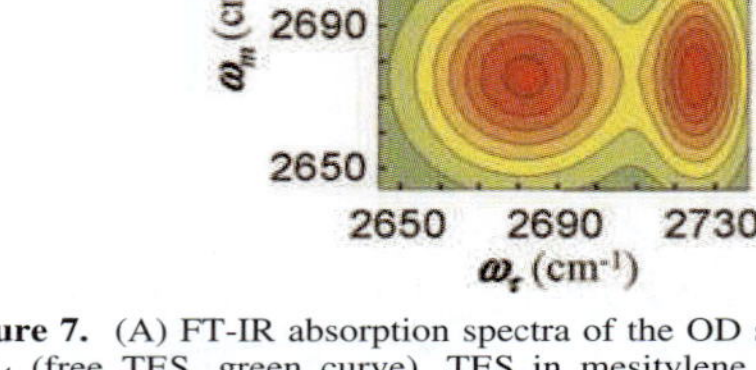

Figure 7. (A) FT-IR absorption spectra of the OD stretch of TES in CCl$_4$ (free TES, green curve), TES in mesitylene (TES/mesitylene complex, red curve), and TES in the mixed mesitylene/CCl$_4$ solvent, which displays absorptions for both free and complexed TES (black curve). (B) Experimental 2D IR spectrum (T_w = 60 ps) of TES in the mixed mesitylene/CCl$_4$ solvent. The off-diagonal peaks have the highly asymmetric shapes predicted by equation 4. (C) Calculated 2D IR spectrum using the absorption line shapes as input in accord with equation 4. The calculation reproduces the asymmetric line shapes of the off-diagonal peaks very well. Each contour is a 10% change in amplitude.

species have different absorption line shapes. A very clear example of this behavior is displayed by the 2D IR vibrational echo chemical exchange spectrum of the TES/mesitylene-CCl$_4$ system. Figure 7 displays the linear absorption spectrum (A), the 2D IR spectrum at 60 ps, which is in the long T_w limit (B), and the calculated spectrum (C). In the TES/mesitylene system, in the absorption spectrum (A) the complex peak is 34 cm^{-1} fwhm, the free peak is 14 cm^{-1} fwhm, and the two peaks are separated by 46 cm^{-1}. The substantial difference in the two linewidths and the large separation of the peaks makes it possible to cleanly observe the behavior in the 2D spectrum.

The 2D IR spectrum shown in Figure 7B displays the behavior embodied in equation 4. The off-diagonal dissociation peak (upper left-hand peak) is broad along the ω_τ axis and narrow along the ω_m axis. The off-diagonal dissociation peak comes from complexes at the time of the first pulse (ω_τ axis), which have a broad spectrum, and free species at the time of echo emission (ω_m axis), which have a narrow spectrum. The formation peak (lower right-hand peak) is narrow along the ω_τ axis and broad along the ω_m axis. The off-diagonal formation peak comes from free species at the time of the first pulse (ω_τ axis), which have a narrow spectrum, and complexes at the time of echo emission (ω_m axis), which have a broad spectrum. The predictions of equation 4 were tested quantitatively as shown in Figure 7C. The linear spectrum of the TES/mesitylene-CCl$_4$

Solute−Solvent Complex Kinetics

system (black curve in Figure 7A) was fit with two Gaussians. These became the two vectors, $W_A(\omega_\tau)$ and $W_B(\omega_m)$ in equation 4. As can be seen in Figure 7C, the calculation does an excellent job of reproducing the experimental 2D spectrum shown in Figure 7B.

IV. Concluding Remarks

The formation and dissociation kinetics of five triethylsilanol solute−solvent complexes at room temperature were investigated with 2D IR vibrational echo chemical exchange spectroscopy. The time dependences of the formation and dissociation of the complexes were measured by observing the growth of the off-diagonal chemical exchange peaks in the 2D IR spectrum. The dissociation time constants range from 9 to 140 ps, and were shown to be directly related to the enthalpy of formation of the complexes. The experiments are in accord with the previous observation from eight phenol/solvent complexes.[17,21] The correlation between the complex enthalpies of formation and dissociation rate constants for all 13 complexes, the five studied here and eight studied previously, can be expressed with an equation similar to the Arrhenius equation.[21]

In addition to the dynamics, theoretical predictions of the shapes of the peaks in the 2D IR chemical exchange spectrum were tested. It was demonstrated that in the long time limit, in which spectral diffusion is complete and the peaks no longer change shape, the shapes of the off-diagonal chemical exchange peaks can be highly asymmetric. The off-diagonal peaks have shapes determined by the outer product of the absorption spectra of the species that give rise to the chemical exchange spectrum.

Organic solute−solvent complexes that have dissociation time constants in the range of 10−100 ps may be quite general. In future publications, experiments will be presented that examine the solute−solvent complex dynamics of the solute chloroform with acetone and dimethyl sulfoxide and a system in which a solute can migrate to different positions on a solvent molecule.

Acknowledgment. This work was support by the Air Force Office of Scientific Research (F49620-01-1-0018) and by the National Science Foundation (DMR 0652232). J.Z. thanks Pulse Institute funded by the Stanford University Dean of Research for a post-doctoral fellowship and the Stanford Graduate fellowship program for a fellowship.

References and Notes

(1) Vinogradov, S. N.; Linnell, R. H. *Hydrogen Bonding*; Van Nostrand Reinhold Company: New York, 1971.

(2) Throop, G. J.; Bearman, R. J. *J. Chem. Phys.* **1965**, *42*, 2408.

(3) Fuchs, R.; Peacock, L. A.; Stephenson, W. K. *Can. J. Chem.* **1982**, *60*, 1953.

(4) Frei, H.; Pimentel, G. C. *Annu. Rev. Phys. Chem.* **1985**, *36*, 491.

(5) Asplund, M. C.; Zanni, M. T.; Hochstrasser, R. M. *Proc. Natl. Acad. Sci. USA* **2000**, *97*, 8219.

(6) Khalil, M.; Demirdoven, N.; Tokmakoff, A. *Phys. Rev. Lett.* **2003**, *90*, 047401. (4)

(7) Khalil, M.; Demirdoven, N.; Tokmakoff, A. *J. Phys. Chem. A* **2003**, *107*, 5258.

(8) Asbury, J. B.; Steinel, T.; Stromberg, C.; Gaffney, K. J.; Piletic, I. R.; Goun, A.; Fayer, M. D. *Phys. Rev. Lett.* **2003**, *91*, 237402.

(9) Steinel, T.; Asbury, J. B.; Corcelli, S. A.; Lawrence, C. P.; Skinner, J. L.; Fayer, M. D. *Chem. Phys. Lett.* **2004**, *386*, 295.

(10) Mukherjee, P.; Kass, I.; Arkin, I.; Zanni, M. T. *Proc. Natl. Acad. Sci. U.S.A.* **2006**, *103*, 3528.

(11) Maekawa, H. T. C.; Moretto, A.; Broxterman, Q. B.; Ge, N. H. *J. Phys. Chem. B* **2006**, *110*, 5834.

(12) Naraharisetty, S. R. G.; Kasyanenko, V. M.; Rubtsov, I. V. *J. Chem. Phys.* **2008**, *128*, 104502.

(13) Nee, M. J.; McCanne, R.; Kubarych, K. J.; Joffre, M. *Opt. Lett.* **2007**, *32*, 713.

(14) Arrivo, S. M.; Heilweil, E. J. *J. Phys. Chem.* **1996**, *100*, 11975.

(15) Kim, Y. S.; Hochstrasser, R. M. *Proc. Natl. Acad. Sci. U.S.A.* **2005**, *102*, 11185.

(16) Woutersen, S.; Mu, Y.; Stock, G.; Hamm, P. *Chem. Phys.* **2001**, *266*, 137.

(17) Zheng, J.; Kwak, K.; Asbury, J. B.; Chen, X.; Piletic, I.; Fayer, M. D. *Science* **2005**, *309*, 1338.

(18) Zheng, J.; Kwak, K.; Chen, X.; Asbury, J. B.; Fayer, M. D. *J. Am. Chem. Soc.* **2006**, *128*, 2977.

(19) Zheng, J.; Kwac, K.; Xie, J.; Fayer, M. D. *Science* **2006**, *313*, 1951.

(20) Zheng, J.; Kwak, K.; Fayer, M. D. *Acc. Chem. Res.* **2007**, *40*, 75.

(21) Zheng, J.; Fayer, M. D. *J. Am. Chem. Soc.* **2007**, *129*, 4328.

(22) Kwak, K.; Zheng, J.; Cang, H.; Fayer, M. D. *J. Phys. Chem. B* **2006**, *110*, 19998.

(23) Sanda, F.; Mukamel, S. *J. Chem. Phys.* **2006**, *125*, 014507.

(24) Chang, R. *Physical Chemistry for the Chemical and Biological Sciences*; University Science Books: Sausalito, CA, 2000.

(25) Tan, H.-S.; Piletic, I. R.; Fayer, M. D. *J.O.S.A. B* **2005**, *22*, 2009.

(26) Asbury, J. B.; Steinel, T.; Fayer, M. D. *J. Lumin.* **2004**, *107*, 271.

(27) Park, S.; Kwak, K.; Fayer, M. D. *Laser Phys. Lett.* **2007**, *4*, 704.

(28) Tan, H.-S.; Piletic, I. R.; Fayer, M. D. *J. Chem. Phys.* **2005**, *122*, 174501. (9)

(29) Bai, Y. S.; Fayer, M. D. *Phys. Rev. B* **1989**, *39*, 11066.

(30) Walsh, C. A.; Berg, M.; Narasimhan, L. R.; Fayer, M. D. *Chem. Phys. Lett.* **1986**, *130*, 6.

(31) Mukamel, S. *Principles of Nonlinear Optical Spectroscopy*; Oxford University Press: New York, 1995.

(32) Oxtoby, D. W. *Adv. Chem. Phys.* **1981**, *47*, 487.

(33) Kenkre, V. M.; Tokmakoff, A.; Fayer, M. D. *J. Chem. Phys.* **1994**, *101*, 10618.

JP804087V

Section 2. The First and Early Ultrafast Vibrational Echo Experiments on Liquids, Glasses, and Proteins

The First and Early Ultrafast Vibrational Echo Experiments on Liquids, Glasses, and Proteins

The papers in this section are the first experiments using ultrafast vibrational echo techniques to study dynamics in condensed matter systems. These first experiments are one dimensional two pulse vibrational echoes. They are the equivalent to the first pulsed NMR experiment, the two pulse spin echo, and the first coherent electronic excited state spectroscopy, the two pulse photon echo. The motivation for performing these first vibrational echo experiments was the desire to extract dynamical information that could not be obtained in any other way from complex molecular condensed matter systems. Pulsed NMR is a very powerful technique, but it operates on relatively slow time scales. The characteristic of NMR that makes it so powerful is NMR spectra are very detailed. The many peaks can be assigned to specific molecular groups. With isotopic labeling, the specificity is further improved. Ultrafast electronic excited state photon echoes can sample the shortest time scales. However, for molecules at room temperature in liquids or systems such as proteins, the linear absorption spectra are very broad. In an ultrafast electronic excited state photon echo experiment, the broad absorption band that is excited, usually many hundreds of wave numbers wide, consists of hundreds of vibronic transitions. Low frequency ground state modes are populated and are excited by the broad spectrum of the ultrashort pulse, which can also excite all of these ground state modes into vibronic bands of the excited state. The vast number of transitions excited in an ultrafast electronic photon echo experiment greatly limits the information content, although some very interesting work has been done with this technique.

Ultrafast vibrational echo experiments have useful characteristics of both NMR and photon echoes. Like NMR, a vibrational spectrum can show well defined peaks that are assigned to specific groups of a molecule, e.g., a hydroxyl stretching mode. Isotopic labeling can also be used to enhance selectivity of exactly which part of a molecule is being excited by the vibrational echo pulse sequence. Like ultrafast photon echoes, the IR pulses used in vibrational echo experiments can be made short enough to observe even the fastest dynamics. The characteristics inherent in the ultrafast vibrational echo technique were the motivation for the first experiments.

The difficulty in 1993, the year of the first experiment, was the lack of a source of tunable short IR pulses. This problem was overcome at Stanford by using the Stanford Free Electron Laser (FEL), which was modified to perform the experiments. The first experiments, although simple by today's standards, were incredibly difficult. First, we did not know that we could do the experiments. Second, the FEL is about 200 m long and takes ten days to turn on. It requires a crew of about ten people to run it. Major advances had to be made in control and manipulation of the FEL beam to perform the experiments. Nonetheless, they proved very successful. These first experiments, some of which are presented in the following papers, taught us a great deal about liquids, glasses, and proteins. In addition, these first experiments set the stage for the ultrafast two dimensional vibrational echo experiments that are performed today by many laboratories around the world. Now we have table top sources of ultrafast IR pulses. We do not need to use an FEL, and the experiments we perform have phase as well as intensity information, permitting true 2D spectroscopy. While the methodology has come a long way, the technique is still advancing rapidly.

VOLUME 70, NUMBER 18   PHYSICAL REVIEW LETTERS   3 MAY 1993

Picosecond Infrared Vibrational Photon Echoes in a Liquid and Glass Using a Free Electron Laser

David Zimdars, A. Tokmakoff, S. Chen, S. R. Greenfield, and M. D. Fayer

Department of Chemistry, Stanford University, Stanford, California 94305

T. I. Smith and H. A. Schwettman

Department of Physics, Stanford University, Stanford, California 94305
(Received 22 December 1992)

The first infrared vibrational photon echo experiments conducted in a liquid and a glass are reported. The experiments were performed on the CO stretching mode of tungsten hexacarbonyl at 5.1 μm (1960 cm^{-1}) in 2-methyltetrahydrofuran over the temperature range 300 to 16 K using picosecond pulses from the free electron laser at Stanford University. In addition, the first vibrational population relaxation measurements spanning a temperature range that takes a system from a liquid to a supercooled liquid to a glass are reported.

PACS numbers: 42.50.Md, 33.70.Fd, 41.60.Cr, 64.70.Pf

Molecular vibrations are involved in a vast number of physical, chemical, and biological processes. Coupling between molecular vibrations and external mechanical degrees of freedom (heat bath) is responsible for the flow of energy into and out of molecules and for thermally activated processes. In spite of the importance of the coupling of molecular vibrations to a heat bath, relatively little is known about the dynamic aspects of molecular vibrations in condensed matter systems. What is known is mainly restricted to vibrational lifetime measurements in crystalline solids at low temperature [1].

Here we report the first infrared (ir) vibrational photon echo experiments conducted in a liquid and glass as well as temperature-dependent pump-probe experiments. The experiments examined the optical dephasing of the vibrational transition and the vibrational population relaxation of the CO asymmetric stretching mode of tungsten hexacarbonyl [W(CO)$_6$] in 2-methyltetrahydrofuran (2-MTHF). The experiments were performed at a wavelength of 5.1 μm (1960 cm^{-1}) at temperatures from 16 to 300 K using the superconducting-linear-accelerator-pumped free electron laser (FEL) at Stanford University.

In general, coupling of molecular vibrations to an external bath is substantially weaker than the coupling of electronic states. This is demonstrated by the fact that vibrational transition energies have a much smaller percentage of gas-to-"crystal" shifts than electronic transitions. The weaker coupling makes it possible to perform photon echoes on vibrational transitions of a molecule dissolved in a liquid using relatively long pulses (3 ps) that have narrow enough bandwidths so that only a well-defined pair of states is coherently coupled by the radiation field. Thus, the observed dephasing of the vibrational transition, with the contribution from population relaxation removed, arises from coupling to the heat bath. By passing through the glass-liquid transition, it is possible to observe how the change of state influences the vibrational dynamics.

The FEL used for these experiments [2,3] is a tunable source of picosecond ir pulses. The FEL emits a 2 ms macropulse at 10 Hz. Each macropulse consists of $\sim$0.5 μJ micropulses at a repetition rate of 11.8 MHz. The micropulses were measured to be Gaussian 2.7 ps transform limited pulses. The micropulse repetition rate of 11.8 MHz was reduced to 50 kHz by an acousto-optic modulator single pulse selector. Thus, the effective experimental repetition rate was 1 kHz.

Both photon echo and pump-probe experiments require two input pulses, one of which is variably delayed. The single ir pulse was beam split. The reflected pulse (the probe pulse or second pulse in the echo sequence) was passed down a computer-controlled stepper-motor delay line. The transmitted pulse (the pump pulse or first pulse in the echo sequence) was chopped at half the single pulse rate by another AOM. The spot size was 200 μm. For the pump-probe experiment, the pump pulse energy was 200 nJ and the probe energy was 20 nJ. The maximum change in transmitted intensity was less than 4%. For the photon echo experiment, the first pulse was 60 nJ and the second pulse was 150 nJ.

The signals were detected using an amplified liquid-nitrogen–cooled indium antimonide detector. In both experiments, the pulses were crossed at a small angle. For the pump-probe experiment, the detector was placed directly in the transmitted probe beam. For the photon echo, the detector was placed at the location determined by the wave vector matching condition $\mathbf{k}_4 = 2\mathbf{k}_2 - \mathbf{k}_1$, where $\mathbf{k}_1$, $\mathbf{k}_2$, and $\mathbf{k}_4$ are the wave vectors of the second pulse, the first pulse, and the photon echo pulse, respectively. The echo signal vanished when either beam was blocked. To check for power artifacts, the intensity of the excitation pulses was reduced by a factor of 2 and no change in the decay constants was observed for either the pump-probe or photon echo experiment. The echo intensity was reduced by a factor of 8 by the factor of 2 reduction in intensity, demonstrating that the echo experiments are in the low flip angle limit. The repetition rate of the pulses was also reduced by a factor of 2. No change in

VOLUME 70, NUMBER 18   PHYSICAL REVIEW LETTERS   3 MAY 1993

signal for either type of experiment was observed.

Experiments were performed using 4.0×10^{-3} M $W(CO)_6$ in 2-MTHF (Aldrich). This concentration gave an optical density of 0.8 at the absorption maximum for the sample thickness of 100 μm. A thin sample was used to minimize the solvent background absorbance. Experiments were also performed at room temperature with more dilute concentrations of 1.0×10^{-3} and 4.0×10^{-4} M and no change in decay rate was observed. Experiments conducted on 2-MTHF with no $W(CO)_6$ gave no signal in either type of experiment. All experiments were performed by rapidly cooling the sample in a variable temperature cryostat to the lowest temperature and then increasing the temperature. Therefore, the glass transition temperature was approached from below. The 2-MTHF was never observed to crystallize.

Figure 1 shows photon echo data taken at 16 K, and the inset shows a semilogarithmic plot of the data. The signal-to-noise ratio is very good, comparable to data taken on electronic transitions using conventional lasers. Data taken at other temperatures are equally good. The echo decay is a single exponential following a very fast feature around $\tau = 0$. The photon echo signal decays as $I(\tau) = I_0 \exp(-4\tau/T_2)$ where T_2 is the homogeneous dephasing time [4]. In Fig. 1, the data decay in 15 ps yielding a T_2 of 60 ps. The homogeneous linewidth $(1/\pi T_2)$ is 5.2 GHz. This is in contrast to the inhomogeneously broadened absorption line which is an 18 cm^{-1} Gaussian at 16 K, narrowing to 15 cm^{-1} at 300 K. The pure dephasing time T_2^* can be found by removing the contribution to T_2 from the population decay of the excited vibrational state using the relation

$$1/\pi T_2 = 1/\pi T_2^* + 1/2\pi T_1 , \tag{1}$$

where T_1 is the excited vibrational state lifetime mea-

sured with the pump-probe experiments. At 16 K, T_1 is 44 ps. Using Eq. (1), this gives a pure dephasing time T_2^* contribution to the homogeneous linewidth $(1/\pi T_2^*)$ of 1.6 GHz.

Figure 2 shows temperature-dependent echo data plotted as linewidths. The open squares are the homogeneous linewidths $1/\pi T_2$, and the filled squares are the pure dephasing linewidths $1/\pi T_2^*$ calculated using Eq. (1) (see Fig. 3 below for T_1 values). The vertical line at 88 K marks the glass-to-liquid transition temperature T_g. Although the time resolution was insufficient to determine the echo decay time above 140 K, it was possible to make photon echo decay measurements in the liquid state 50 K above T_g. The observation of the photon echo signal in the liquid demonstrates that vibrational lines in liquids can be inhomogeneously broadened.

As shown in Fig. 2, the linewidths increase gradually with temperature below T_g. Although there are limited data above T_g, the temperature dependence becomes much steeper. With the available data, it is not clear whether the behavior is discontinuous at T_g. Homogeneous dephasing of *electronic* transitions of chromophores in organic glasses at low temperatures (1 to 20 K) is described by the sum of a power law and an exponentially activated process. The expression used to fit electronic dephasing data is [5,6]

$$1/\pi T_2^* = aT^\alpha + b \exp(-\Delta E/kt)/[1 - \exp(-\Delta E/kt)] . \tag{2}$$

For *electronic* transitions, the power law portion of Eq. (2) arises from the dynamics of the tunneling two-level systems (TLS) in the glass [7–10]. Typical values of α are 1.2 to 1.5 [5,11–13]. The exponentially activated process is due to coupling to low frequency phonons [5,6,14]. Typical values of ΔE for *electronic* transitions fall in the range of 10 to 40 cm^{-1} [5,11,14].

The solid line through the vibrational pure dephasing data was determined by fitting the data by Eq. (2). The

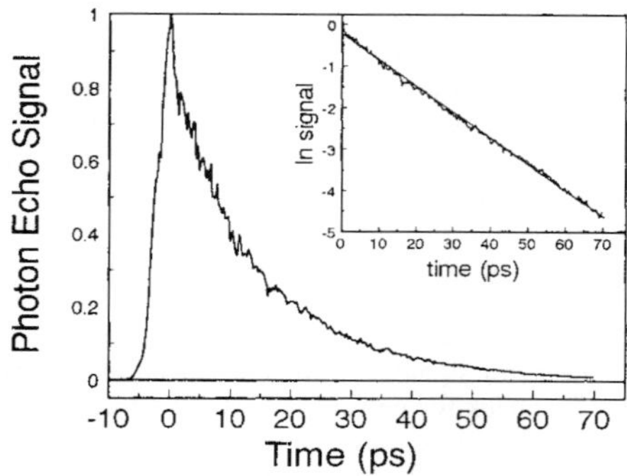

FIG. 1. Vibrational photon echo decay data taken of $W(CO)_6$ in 2-MTHF at 16 K. The inset is a semilogarithmic plot showing the data decay exponentially. The signal decays with a 15 ps time constant, yielding a homogeneous dephasing time $T_2 = 60$ ps and a homogeneous linewidth $(1/\pi T_2)$ of 5.2 GHz.

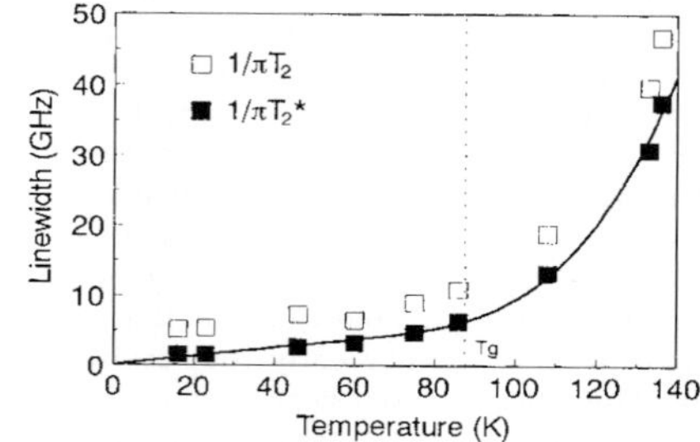

FIG. 2. Temperature-dependent photon echo data displayed as linewidths. The open squares are the total homogeneous linewidths. The filled squares are the pure dephasing linewidths with lifetime contribution removed. The line through the data is a fit by Eq. (2). The vertical line indicates the glass transition temperature of 88 K.

fit shown in Fig. 2 has $\alpha = 0.9$ and $\Delta E = 540$ cm^{-1}. However, given the limited data above T_g, it is also possible to obtain a fit with a somewhat smaller ΔE and $\alpha \approx 1$. The data display power law behavior almost to T_g. The power-law behavior of electronic dephasing at a temperature of a few degrees K is caused by TLS dynamics. As for low temperature heat capacity data, α is approximately 1. The dephasing of the vibrational transition data displays power law behavior to much higher temperature with $\alpha \approx 1$. It is possible that the power law dephasing of the vibrational transition at these elevated temperatures is also caused by TLS dynamics. Here the power law is not masked by the activated processes at elevated temperatures since ΔE is so large.

If Eq. (2) is the appropriate form to describe the data, then the resulting activation energy of 540 cm^{-1} is far too large to correspond to a phonon or pseudo local mode. The magnitude of ΔE is in the range of molecular vibrations. Both $W(CO)_6$ and 2-MTHF have a variety of low frequency modes. By coincidence, both have a mode at ~ 580 cm^{-1} [15,16]. Either of these modes could have energies within the error of the activation energy obtained from the fit. Both $W(CO)_6$ and 2-MTHF have a number of lower frequency modes as well as the 580 cm^{-1} modes. These modes will be thermally populated at lower temperatures. However, to cause dephasing, a mode of either $W(CO)_6$ or 2-MTHF must be quadratically coupled to the CO stretching mode being probed by the photon echo experiments; i.e., excitation of another mode of the system must cause a frequency shift of the CO stretching mode. Thus, the other modes may have such weak coupling to the CO stretching mode that they do not cause pure dephasing.

Figure 3 shows the temperature dependence of the pump-probe data from 16 to 300 K. Below 140 K, the decays are rigorously single exponential and were used as the T_1 values for the pure dephasing linewidth calcula-

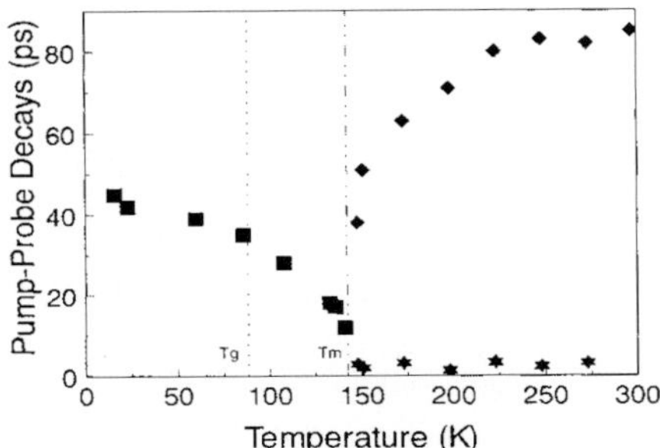

FIG. 3. Temperature-dependent pump-probe vibrational lifetime data. Below ~ 140 K, the decays are single exponentials and are shown as squares. Above ~ 140 K all decays are biexponential and both decay times are shown. The stars show the fast component, and the diamonds show the slower component. The two decay components have equal amplitudes.

tions displayed in Fig. 2. However, above 140 K the data are biexponential decays with the two components having approximately equal amplitudes. Above 140 K, both components of the biexponentials are plotted in Fig. 3. The decay of the pump-probe signal shows a clear discontinuity at ~ 140 K.

If vibrational relaxation directly repopulates the ground state, the decay will be a single exponential. However, if there is an intermediate level in the relaxation pathway, the decay can be biexponential if population of the intermediate causes a shift in the CO stretching frequency. One component is the rate of leaving the excited state into a longer living intermediate state, which turns off the stimulated emission, and the other component is the ground state population recovery. These two components should have equal amplitudes, which is consistent with the data. (If all intermediate states have very short lifetimes, the decay will also appear to be single exponential.)

Because of the limits imposed by the pulse duration, the fast component of the biexponential decay cannot be determined precisely. What is clear is that the slow component speeds up dramatically as the temperature is lowered, and the fast component seems to slow down. The remarkable fact is that the discontinuity in the data takes place at what would be the melting point of crystalline 2-MTHF, ~ 140 K [17]. However, 2-MTHF does not crystallize, but rather supercools and then forms a glass at 88 K [18]. The properties of liquids should be continuous in going from the liquid to the supercooled liquid. Clearly, the vibrational decay is not continuous.

If the fast decay component slows down and the slow component becomes extremely fast below ~ 140 K, the data will take on the appearance of a single exponential, as observed. Generally, one might expect a vibrational decay to become slower as the temperature is decreased because of reduced phonon occupation numbers. Vibrational relaxation requires coupling between internal $W(CO)_6$ modes and the bath modes. It is necessary to convert the vibrational energy into heat. Since the vibrational energy is large, it is unlikely that the vibrational relaxation involves the direct, simultaneous excitation of many bath phonons. A more likely process involves the excitation of at least one vibrational mode of the 2-MTHF and one or more phonons. This requires coupling between the $W(CO)_6$ vibrations and the 2-MTHF vibrations. As the temperature is decreased from room temperature, the viscosity of the liquid increases, and the 2-MTHF rotational diffusion rate slows dramatically. If the matrix element that couples the CO stretching mode to a 2-MTHF vibration is strongly dependent on orientation, then rapid rotation of the 2-MTHF at high temperatures could result in an orientationally averaged matrix element that is small. When the viscosity becomes large, favorable orientations will exist for long enough to enhance the coupling and therefore increase the rate of vibrational relaxation. This will be particularly important if the intermolecular interactions between $W(CO)_6$ and 2-MTHF favor local liquid structures having relative orientations that yield large intermolecular vibrational coupling matrix elments.

In conclusion, vibrational photon echoes, as presented here, are a new method for examining molecular vibrational dynamics in the liquid and glassy states of matter. These initial experiments will be extended to broader ranges of temperature and time and to other solvents and solutes. Experiments will also be conducted on the vibrations of pure liquids and glasses. In addition to photon echoes, stimulated echoes will be employed to examine spectral diffusion, and comparisons of vibrational photon echoes with ir hole burning experiments will aid in elucidating the underlying dynamics that give rise to hole widths. The use of condensed matter vibrational optical coherence experiments will enhance our understanding of molecular vibrations in the same manner as magnetic resonance coherence experiments and optical coherence experiments have broadened our knowledge of spin systems and electronic excitations.

This work was funded by the Medical Free Electron Laser program through the Office of Naval Research (Grant No. N00014-91-C-0170). David Zimdars, A. Tokmakoff, S. Chen, S. R. Greenfield, and M. D. Fayer would like to acknowledge additional support from the National Science Foundation, Division of Material Research (Grant No. DMR90-22675) and the Office of Naval Research, Physics Division (Grant No. N00014-89-J1119).

[1] D. D. Dlott, *Laser Spectroscopy of Solids II*, edited by W. Yen, Topics in Applied Physics Vol. 65 (Springer, Berlin, 1989), p. 167.

[2] R. L. Swent, H. A. Schwettman, and T. I. Smith, in *Short-Wavelength Radiation Sources*, edited by P. Sprangle, SPIE Proceedings Vol. 1552 (SPIE, Bellingham, WA, 1991), p. 24.

[3] D. D. Dlott and M. D. Fayer, IEEE J. Quantum Electron. **27**, 2697 (1991).

[4] I. Abella, N A. Kurnit, and S. R. Hartmann, Phys. Rev. **141**, 391 (1966).

[5] L. R. Narasimhan, K. A. Littau, D. W. Pack, Y. S. Bai, A. Elschner, and M. D. Fayer, Chem. Rev. **90**, 439 (1990).

[6] B. Jackson and R. Silbey, Chem. Phys. Lett. **99**, 331 (1983).

[7] M. Berg, C. A. Walsh, L. R. Narasimhan, K. A. Littau, and M. D. Fayer, J. Chem. Phys. **88**, 1564 (1987).

[8] R. Jankowiak, L. Shu, M. J. Kenney, and G. J. Small, J. Lumin. **36**, 293 (1987).

[9] D. L. Huber, M. M. Broer, and B. Golding, Phys. Rev. Lett. **52**, 2281 (1984).

[10] D. W. Pack, L. R. Narasimhan, and M. D. Fayer, J. Chem. Phys. **92**, 4125 (1990).

[11] S. R. Greenfield, Y. S. Bai, and M. D. Fayer, Chem. Phys. Lett. **170**, 133 (1990).

[12] H. P. H. Thijssen, R. van den Berg, and S. Völker, Chem. Phys. Lett. **103**, 23 (1983).

[13] W. S. Brocklesby, B. Golding, and J. R. Simpson, J. Lumin. **45**, 54 (1990).

[14] W. H. Hesselink and D. A. Wiersma, J. Chem. Phys. **73**, 648 (1980).

[15] P. S. Braterman, *Metal Carbonyl Spectra* (Academic, New York, 1975), p. 182.

[16] *Handbook of Data on Organic Compounds*, edited by R. C. Weast and J. C. Grasselli (CRC Press, Boca Raton, FL, 1985), 2nd ed., p. 2770.

[17] E. V. Whitehead, R. A. Dern, and F. A. Fidler, J. Am. Chem. Soc. **73**, 3632 (1951).

[18] A. C. Ling and J. E. Willard, J. Phys. Chem. **72**, 1918 (1968).

Vibrational photon echoes in a liquid and glass: Room temperature to 10 K

A. Tokmakoff, D. Zimdars, B. Sauter, R. S. Francis, A. S. Kwok, and M. D. Fayer
Department of Chemistry, Stanford University, Stanford, California 94305

(Received 8 March 1994; accepted 5 May 1994)

Picosecond infrared vibrational photon echo experiments were performed on the asymmetric CO stretching mode (1983 cm^{-1}) of tungsten hexacarbonyl in 2-methylpentane from room temperature to 10 K using a free electron laser. This is the first report of a room temperature infrared vibrational photon echo in a liquid.

Molecular vibrations are involved in a vast number of physical, chemical, and biological processes in condensed phases. The coupling between individual molecular vibrations and external degrees of freedom (heat bath) is responsible for fluctuations in the structure and energy levels of a molecule, the flow of energy into and out of molecules, and thermally activated processes. Vibrational dynamics are intimately related to the reactivity of chemical systems. In spite of the importance of the coupling of molecular vibrations to a heat bath, relatively little is known about the temperature-dependent dynamics of molecular vibrations in liquids and glasses.

A fundamental quantity that describes the dynamics of a molecular vibration in liquids or other condensed matter systems is the homogeneous dephasing time, T_2. In principle, T_2 can be obtained from analysis of vibrational lineshapes. However, in liquids and glasses a vibrational absorption line can have a contribution from inhomogeneous broadening. The inhomogeneous line in a glass arises from the differing perturbations of the vibrational energy levels caused by the variety of local solvent structures that surround the molecule of interest. Inhomogeneous broadening masks the dynamic information contained in the homogeneous line. In low temperature glasses, inhomogeneous broadening overwhelms any contribution from the homogeneous line to the observed absorption spectrum. In a liquid, inhomogeneous broadening also comes about from a distribution of environments that result in a spread of vibrational transition energies. Unlike a glass, local structures in a liquid evolve rapidly and assume all possible configurations, so that a spectral line in a liquid will be dynamic on some time scale. However, if the time scale for structural evolution that randomizes the transition energy across the entire absorption line is long compared to other dephasing processes, the homogeneous line is masked. Infrared vibrational photon echo experiments can be used to remove the inhomogeneous contribution to an absorption line and reveal the true homogeneous line shape. As will be shown below, even in room temperature liquids a significant contribution to a vibrational spectral line may come from inhomogeneous broadening.

A description of dephasing dynamics in terms of homogeneous and inhomogeneous broadening implies a separation of time scales between fast fluctuations of the bath, slower dephasing dynamics, and essentially static structural contributions. This is the situation for electronic transitions in low temperature glasses and crystals, where the structure evolves slowly and coupling to the bath is through thermally populated, low frequency modes of the glass.[1] However, in room temperature liquids, the dephasing of *electronic transitions* occurs on a time scale similar to that of the fastest dynamics of the bath.[2,3] Coupling of the electronic states to ballistic motions of the bath results in non-Markovian dynamics.[4,5] The dephasing can arise from dynamics of a liquid that are the same as those responsible for solvation of the electronic excited state.

In general, coupling of vibrations to a bath is substantially weaker than the coupling of electronic states. This is demonstrated by the fact that vibrational transition energies have much smaller percentage gas-to-"crystal" shifts than electronic transitions. The weaker coupling gives rise to longer dephasing times and little or no solvation dynamics. The weaker coupling makes it possible to study dephasing in liquids with relatively long pulses that have narrow enough bandwidths to coherently couple only a well-defined pair of states to the radiation field.

In this communication we report the results of ps infrared vibrational photon echo experiments on a sample that is taken from a room temperature liquid to a glass at 10 K. Measurements were made on the T_{1u} CO stretching mode of $W(CO)_6$ in 2-methylpentane (2-MP), a glass forming liquid. These experiments are the first to follow the evolution of the homogeneous vibrational linewidth from a room temperature liquid to a low temperature glass. Further, it is demonstrated that vibrational transitions in room temperature liquids may be inhomogeneously broadened. This work extends recent observations of the homogeneous vibrational linewidth in a low temperature glass and supercooled liquid.[6] The only other direct measurement of the dephasing times of vibrational transitions in room temperature liquids have been made with Raman echo experiments.[7–9]

The photon echo experiment is a line narrowing experiment that measures T_2. In a vibrational photon echo experiment, two ps IR pulses tuned to a molecular vibrational transition are crossed in the sample. The first pulse induces a coherence that decays due to a combination of inhomogeneous and homogeneous dephasing. A second pulse, delayed by time τ, begins a rephasing of the *inhomogeneous* contribution to the spectral line and results in the emission of an echo pulse at time 2τ. The echo pulse is emitted in a unique direction determined by wave vector matching conditions. The intensity of the echo pulse is measured as a function of τ. The decay of the echo intensity measures the vibrational *homogeneous* dephasing time. The decay of the photon echo is the Fourier transform of the homogeneous spectrum. For example, if the echo decay is an exponential, i.e.,

$$I(t) = I_0 \exp(-4\tau/T_2). \tag{1}$$

The Fourier transform is a Lorentzian homogeneous line having width, $1/\pi T_2$. The echo signal decays four times as fast as the dephasing time, due to the echo rephasing at 2τ, and intensity decaying twice as fast as the polarization.

The photon echo, through the time reversal of the rephasing process, eliminates inhomogeneous contributions to the vibrational transition and is thus the true measure of the homogeneous dephasing time.[10] Although fs coherent anti-Stokes Raman scattering and Raman-induced birefringence measurements can measure dephasing times of low frequency vibrational modes, these experiments do not eliminate inhomogeneity, and thus observe the free induction decay for the transition (Fourier transform of the total inhomogeneously broadening line.[10,11])

The photon echo experiments were performed with $\sim$1.5 ps ($\sim$10 cm^{-1} bandwidth) IR pulses at 5.04 μm (1983 cm^{-1}) generated with the Stanford superconducting-accelerator-pumped free electron laser (FEL). The FEL emits a 2 ms macropulse at a 10 Hz repetition rate. Each macropulse consists of $\sim$0.5 μJ micropulses at a repetition rate of 11.8 MHz. The micropulse repetition rate was reduced to 50 kHz by a germanium AOM single pulse selector, yielding an experimental repetition rate of 1 kHz. The two pulses for the echo pulse sequence were obtained with a ZnSe beam splitter. The data was taken with pulse energies of $\sim$8 nJ for the first pulse and $\sim$50 nJ for the second pulse. Power dependent effects were observed in the low temperature glass for higher energies. The more intense pulse was chopped at 25 kHz by a second AOM. The signal and a reference were measured with two HgCdTe detectors sampled by two gated integrators. The reference detector was used for shot intensity windowing.

Data was taken on 1×10^{-3} M solutions of W(CO)$_6$ in 2-MP (99.9%), corresponding to a mole fraction of $\sim$10^{-4}. The sample was sealed with a 400 μm Teflon gasket between two CaF$_2$ flats, and the temperature was controlled using a closed-cycled He refrigerator. The temperatures were measured to $\pm$0.2 K. The FEL wavelength was tuned at each temperature to follow the small temperature-dependent changes in the position of the absorption maxima of the solution.

Figure 1(a) displays photon echo data taken in the low temperature glass at 10 K. The inset shows a log plot of the data. The decay is exponential (except for a very fast initial transient that vanishes at sufficiently low power), indicating that the homogeneous line is a Lorentzian. At this temperature, the absorption linewidth is 10.5 cm^{-1} (310 GHz). In contrast T_2=240 ps, yielding a homogeneous linewidth of 1.3 GHz. Thus the absorption line in massively inhomogeneously broadened. Figure 1(b) shows temperature dependent photon echo data for W(CO)$_6$ in 2-MP. The photon echo decays are exponential for all the temperatures, although convolutions are necessary to analyze the data above $\sim$160 K. Below 160 K, where the echo decays yield a homoge-

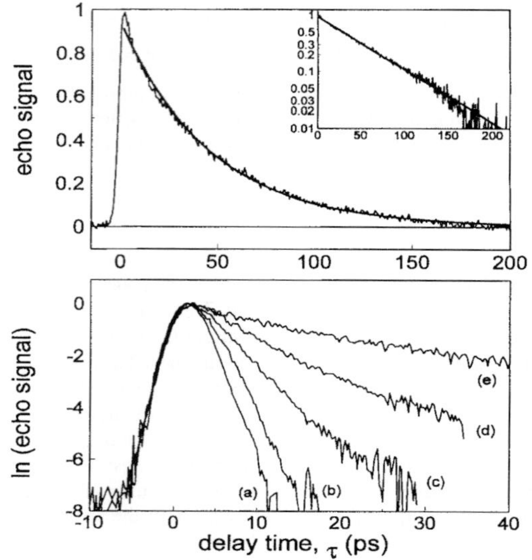

FIG. 1. (Top) Photon echo decay data for the CO asymmetric stretching mode (T_{1u}) of W(CO)$_6$ in 2-methylpentane glass at 10 K. The homogeneous linewidth determined by the echo decay is 1.3 GHz (0.04 cm^{-1}) in contrast to the absorption line which is inhomogeneously broadened to 310 GHz (10.5 cm^{-1}). (Bottom) Semilog plots of photon echo decays as a function of temperature. The decays correspond to (a) 300 K, (b) 125 K, (c) 110 K, (d) 95 K, and (e) 50 K. The glass transition temperature is 88 K. The decays in the liquid demonstrate that the vibrational line is inhomogeneously broadened in liquid solution.

neous linewidth that is much narrower than the width of the absorption spectrum, the results demonstrate conclusively that the vibrational lines of this system are inhomogeneously broadened.

Above 160 K, the dephasing times T_2 approach the pulse width. Under these conditions, the time scale for the rephasing of the echo pulse is shortened by the polarization decay due to homogeneous dephasing. This causes the echo to appear at times between τ and 2τ, and the echo signal at very short delay times decays at a slower rate than the $4/T_2$ given by Eq. (1). For this case, the correct dephasing time must be obtained by convolution of the pulses, with the homogeneous dephasing and the inhomogeneous rephasing, as given by the material response function.[8,10] As the homogeneous dephasing time T_2 decreases, the rephasing of the echo is shifted to shorter times. In the limit that the absorption line is homogeneously broadened, a free induction decay (FID) will be observed along the echo phase matching direction. If the homogeneous line is a Lorentzian, then an exponential decay with a decay constant of $2/T_2$ will be observed.

Figure 2 shows a semi-log plot of the room temperature echo decay of W(CO)$_6$ in 2-MP. A fit of a single exponential with convolution is shown through the data, demonstrating that the echo decay is exponential over at least 5 factors of e. Also shown is the convolution with an instantaneous mate-

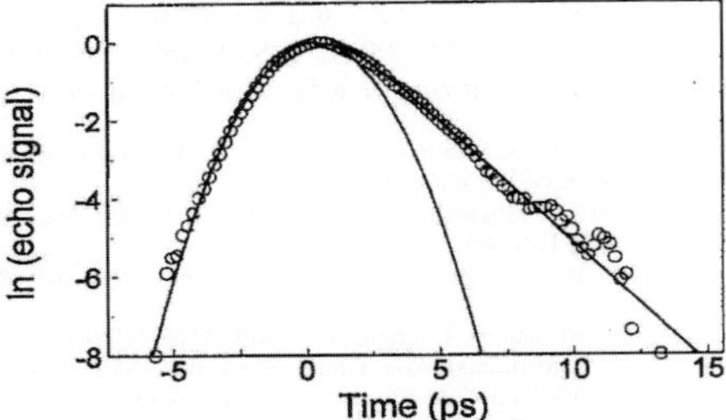

FIG. 2. Semilog plot of the room temperature echo decay, with a single exponential fit, and the instantaneous dephasing response function. The data decays in an exponential manner over >5 factors of e with a decay time of 1.6 ps, corresponding to a homogeneous linewidth of 49 GHz (1.6 cm^{-1}). The absorption linewidth at room temperature is 110 GHz (3.7 cm^{-1}).

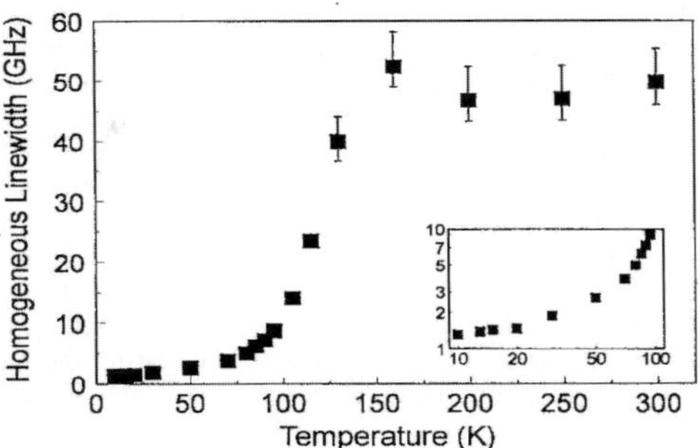

FIG. 3. Temperature-dependent vibrational homogeneous linewidths, $1/\pi T_2$ obtained from the photon echo data using Eq. (1). The inset shows a log–log plot of the data below 100 K. The glass transition temperature is 88 K.

rial response function. If the echo formation in the liquid is in the $4/T_2$ limit, and the absorption line is truly inhomogeneously broadened, then the echo decay of 1.6 ps corresponds to a homogeneous linewidth of 49 GHz. The $2/T_2$ limit, given by the FID, yields a linewidth of 98 GHz, which is similar to the absorption linewidth at room temperature of 110 GHz. From the data shown in Fig. 2 alone, it is not possible to distinguish between a true echo decay and the FID. At this point we assume that the observed data corresponds to a true echo decay described by Eq. (1). The possibility that the room temperature data are a FID is discussed below.

The temperature-dependent homogeneous linewidth for W(CO)$_6$ in 2-MP is shown in Fig. 3. The inset is a log–log plot of the low temperature portion (<100 K). The linewidths, $1/\pi T_2$, were calculated from photon echo data using Eq. (1). The linewidth increases gradually with temperature in the glass. The homogeneous linewidth is <10 GHz throughout the glassy region, compared to the inhomogeneous linewidth of ~300 GHz. The rate of increase of the homogeneous width becomes much steeper after the glass transition, T_g=88 K. However, above 160 K, the temperature dependence of the linewidth appears to level off. Preliminary theoretical calculations indicate that the observed temperature dependence in the liquid could arise through a motional narrowing mechanism. At room temperature, the total linewidth is 3.7 cm^{-1} while the homogeneous linewidth is 1.6 cm^{-1}.

There are several features of the system that suggest that the data points between 160 and 300 K correspond to true echo decays, not FIDs. The absorption line is not Lorentzian. Thus the FID would not be observed as exponential. The data point at 160 K is certainly a true echo decay. The absorption linewidth is 5.5 cm^{-1}, which would give rise to a much faster FID than the observed decay. The decays between 160 and 300 K are very similar in character; a change from an echo decay to a FID would have to occur without a noticeable change in the nature of the data. In addition, using a conventional laser system with 10 ps time resolution, evidence of spectral diffusion has been observed in pump–probe experiments in the liquid up to 180 K. Above this

temperature, the time resolution was inadequate to make further observations. The spectral diffusion, which is much slower than the echo decays, indicates inhomogeneity that exists on a time scale long compared to the homogeneous dephasing in the high temperature liquid.

While the information presented above is suggestive that the room temperature line is inhomogeneously broadened, it is not conclusive. If the room temperature line is in fact inhomogeneous, then spectral diffusion should occur to dynamically broaden the line. Spectral diffusion can be observed by comparing the results of the photon echo with stimulated echo decays.[1] Stimulated echoes will be employed to confirm the inhomogeneous broadening of the room temperature line.

Previously photon echo experiments were performed on the same mode of W(CO)$_6$ in 2-methyltetrahydrofuran (2-MTHF) between 16 and 140 K.[6] In this temperature range, the temperature dependence is qualitatively similar to that reported here for the 2-MP solvent. Below the 2-MTHF glass transition (T_g=86 K) the homogeneous linewidth increases gradually with increasing temperature, becoming much steeper above T_g. However, the homogeneous widths are consistently wider in 2-MTHF. For example, at 16 K in 2-MTHF, $1/\pi T_2$=5.2 GHz, while at the same temperature in 2-MP, $1/\pi T_2$=1.6 GHz. The wider homogeneous lines in 2-MTHF did not permit measurement of the echo decay above 140 K (~50 K above T_g) because of inadequate time resolution. The line is almost certainly still inhomogeneously broadened above this temperature since W(CO)$_6$ in 2-MTHF liquid has an absorption linewidth of 15 cm^{-1}, and the homogeneous width measured at 140 K is ~3 cm^{-1}. The narrower homogeneous widths observed in 2-MP at all temperatures made it possible to perform the experiments up to room temperature.

The photon echo data presented above demonstrates that vibrational lines of solute molecules can be inhomogeneously broadened in liquid solutions as well as glassy solutions. The data suggest the existence of inhomogeneous broadening at room temperature. Raman echo experiments on room temperature neat acetonitrile[7] and neat benzonitrile[9]

Letters to the Editor

have determined that the vibrational bands studied are homogeneously broadened. However, Raman echo experiments on $CH_3I/CDCl_3$ in very concentrated solutions (50/50 mixtures) displayed inhomogeneous broadening at room temperature.[8]

The temperature dependence of the homogeneous linewidth reflects the wide variety of dynamic processes occurring in the liquid. The functional form of the increase in the homogeneous linewidth with temperature is due to the changes in the nature of the dynamical processes of the solvent that are coupled to the internal modes of $W(CO)_6$. Continuing experimental work is aimed at elucidating the nature of the dynamic phenomena responsible for the vibrational dephasing.

The authors acknowledge Professor Alan Schwettman and Professor Todd Smith and their groups at the Stanford FEL whose continuous efforts made these experiments possible. This work was supported by the Medical Free Electron Laser Program (N00014-91-C-0170), the National Science Foundation (DMR90-22675) , and the Office of Naval Research (N00014-92-J-1227-P02). B.S. thanks the Alexander von Humboldt Foundation for a Feodor Lynen Fellowship.

[1] L. R. Narasimhan, K. A. Littau, D. W. Pack, Y. S. Bai, A. Elschner, and M. D. Fayer, Chem. Rev. **90**, 439 (1990).
[2] P. C. Becker, H. L. Fragnito, J. Y. Bigot, C. H. Brito-Cruz, and C. V. Shank, Phys. Rev. Lett. **63**, 505 (1989).
[3] E. T. J. Nibbering, D. A. Wiersma, and K. Duppen, Phys. Rev. Lett. **66**, 2464 (1991).
[4] Y. J. Yan and S. Mukamel, J. Chem. Phys. **89**, 5160 (1988).
[5] L. E. Fried and S. Mukamel, Adv. Chem. Phys. **84**, 435 (1993).
[6] D. Zimdars, A. Tokmakoff, S. Chen, S. R. Greenfield, and M. D. Fayer, Phys. Rev. Lett. **70**, 2718 (1993).
[7] D. Vanden Bout, L. J. Muller, and M. Berg, Phys. Rev. Lett. **67**, 3700 (1991).
[8] L. J. Muller, D. Vanden Bout, and M. Berg, J. Chem. Phys. **99**, 810 (1993).
[9] R. Inaba, K. Tominaga, M. Tasumi, K. A. Nelson, and K. Yoshihara, Chem. Phys. Lett. **211**, 183 (1993).
[10] R. F. Loring and S. Mukamel, J. Chem. Phys. **83**, 2116 (1985).
[11] Y. Tanimura and S. Mukamel, J. Chem. Phys. **99**, 9496 (1993).

Homogeneous vibrational dynamics and inhomogeneous broadening in glass-forming liquids: Infrared photon echo experiments from room temperature to 10 K

A. Tokmakoff[a)] and M. D. Fayer
Department of Chemistry, Stanford University, Stanford, California 94305

(Received 23 December 1994; accepted 9 May 1995)

A study of the temperature dependence of the homogeneous linewidth and inhomogeneous broadening of a high-frequency vibrational transition of a polyatomic molecule in three molecular glass-forming liquids is presented. Picosecond infrared photon echo and pump–probe experiments were used to examine the dynamics that give rise to the vibrational line shape. The homogeneous vibrational linewidth of the asymmetric CO stretch of tungsten hexacarbonyl (~ 1980 cm^{-1}) was measured in 2-methylpentane, 2-methyltetrahydrofuran, and dibutylphthalate from 300 K, through the supercooled liquids and glass transitions, to 10 K. The temperature dependences of the homogeneous linewidths in the three glasses are all well described by a T^2 power law. The absorption linewidths for all glasses are seen to be massively inhomogeneously broadened at low temperature. In the room temperature liquids, while the vibrational line in 2-methylpentane is homogeneously broadened, the line in dibutylphthalate is still extensively inhomogeneously broadened. The contributions of vibrational pure dephasing, orientational diffusion, and population lifetime to the homogeneous line shape are examined in detail in the 2-methylpentane solvent. The complete temperature dependence of each of the contributions is determined. For this system, the vibrational line varies from inhomogeneously broadened in the glass and low temperature liquid to homogeneously broadened in the room temperature liquid. The homogeneous linewidth is dominated by the vibrational lifetime at low temperatures and by pure dephasing in the liquid. The orientational relaxation contribution to the line is significant at some temperatures but never dominant. Restricted orientational relaxation at temperatures below ~ 120 K causes the homogeneous line shape to deviate from Lorentzian, while at higher temperatures the line shape is Lorentzian. © *1995 American Institute of Physics.*

I. INTRODUCTION

Vibrational line shapes in condensed phases contain the details of the interactions of a normal mode with its environment. These interactions include the important microscopic dynamics, intermolecular couplings, and time scales of solvent evolution that modulate the energy of a transition, in addition to essentially static structural perturbations. An infrared absorption spectrum or Raman spectrum gives frequency-domain information on the ensemble-averaged interactions that couple to the states involved in the transition.[1–3] Line shape analysis of vibrational transitions has long been recognized as a powerful tool for extracting information on molecular dynamics in condensed phases.[4,5] The difficulty with determining the microscopic dynamics from a spectrum arises because linear spectroscopic techniques have no method for separating the various contributions to the vibrational line shape. The IR absorption or Raman line shape represents a convolution of the various dynamic and static contributions to the observed line shape. In some cases, polarized Raman spectra can be used to separate orientational and vibrational dynamics from the line shape, yet as with all linear spectroscopies, contributions from inhomogeneous broadening cannot be eliminated.[6]

In order to completely understand a vibrational line shape, a series of experiments are required to characterize each of its static and dynamic components. These experiments can be effectively accomplished in the time domain, where well-defined techniques exist for measuring the various quantities. Nonlinear vibrational spectroscopy can be used to eliminate static inhomogeneous broadening from IR and Raman line shapes.[6] Techniques such as the infrared photon echo[7–9] and Raman echo[10–13] can determine the homogeneous vibrational line shape which contains the important microscopic dynamics when this line shape is masked by inhomogeneous broadening. Yet nonlinear time-domain techniques alone cannot separate the various dynamic contributions to the homogeneous vibrational line shape. Traditional time-domain experiments, such as transient absorption pump–probe experiments, are needed to observe population dynamics such as the vibrational lifetime and orientational relaxation. To further characterize the long time-scale dynamics of spectral diffusion within the line, additional experiments, such as stimulated photon echoes[14] or transient hole burning,[15] are required.

In this paper we present a detailed study of the temperature dependence of the infrared line shape of a high-frequency vibrational transition of a solute in three molecular glass-forming liquids between 10 and 300 K using picosecond infrared photon echo and pump–probe experiments. These experiments are the first to follow the evolution with temperature of the dynamics that comprise the vibrational

---

a)Present address: Physik Department E11, Technische Universität München, D-85748 Garching, Germany.

line shape in amorphous condensed phases and to quantify the degree of inhomogeneous broadening for these systems. The dynamics cover temperature regions from the low temperature glass, through the glass transition and supercooled liquid, to the room temperature liquid. The results are of significance for the understanding of various vibrational dynamic phenomena in condensed phases, and have relevance to the study of fast dynamics in the passage through the glass transition. These results also form a basis for comparison with a number of theories of dephasing in liquids,[4,16–18] and in glasses.[19,20]

Temperature-dependent studies of condensed matter vibrational dynamics are of importance for understanding the coupling of vibrational modes with the external degrees of freedom of the solvent. The population and density of states of low frequency modes of a solvent and the coupling of these modes to internal molecular vibrations dictate the vibrational dynamics. Despite their importance in a wide variety of fields of chemistry, biology, and physics, relatively little is known about the temperature-dependent dynamics of molecular vibrations in polyatomic liquids and glasses. Although a great deal of work has been done using line shape analysis of vibrational transitions, there have only been a few studies that unambiguously eliminated inhomogeneous broadening and examined the temperature dependence of the homogeneous dynamics. Raman echo measurements of dephasing of C–H stretching modes in ethanol liquid and glass[13] have demonstrated that the line is homogeneously broadened at all temperatures between 10 and 300 K. It has been proposed that in this system fast intramolecular vibrational energy redistribution determines the linewidth. Measurements of line narrowed vibrational line shapes in glasses and crystals have been performed with persistent infrared hole burning.[21–24] While hole burning and Raman echoes are vibrational line narrowing techniques, in general they cannot distinguish among the contributions to the homogeneous vibrational line shape from lifetime, rotation, and pure dephasing. In addition, hole burning cannot discriminate between homogeneous dephasing and additional line broadening contributions from longer time scale spectral diffusion.[14] Many of these problems are being overcome with time-resolved resonant infrared experiments. Temperature-dependent picosecond infrared studies of discrete vibrations in liquids and glasses have observed homogeneous vibrational dephasing with photon echoes[7–9] and vibrational population lifetimes with pump–probe experiments[7,15,25,26] and picosecond transient hole burning.[27]

Below we present the temperature-dependent vibrational dynamics of the triply degenerate T_{1u} CO stretching mode of tungsten hexacarbonyl [$W(CO)_6$] in the molecular glass-forming liquids 2-methyltetrahydrofuran (2-MTHF), 2-methylpentane (2-MP), and dibutylphthalate (DBP). Two aspects of the vibrational line shape in these systems are discussed in detail. Initially, we compare the behavior of the homogeneous linewidths in the glasses and the transition to the room temperature liquids for the three glasses, using ps IR photon echo experiments. The temperature dependence of the vibrational dephasing and the degree of inhomogeneity as the liquids approach room temperature are described. The

temperature dependence of the homogeneous vibrational linewidth in the three glassy solvents is T^2 within experimental error, but the behavior is distinct in each of the liquids. While in 2-MP the vibrational line is homogeneously broadened at room temperature, the line in DBP is massively inhomogeneously broadened in the room temperature liquid.

Following the comparison of the temperature-dependent dephasing in the three solvents, one of the systems, $W(CO)_6$ in 2-MP, is analyzed in greater detail. The contributions to the vibrational line shape from different dynamic processes are delineated by combining the results of photon echo measurements of the homogeneous line shape[8] with pump–probe measurements of the lifetime and reorientational dynamics.[15] This combination of measurements allows the decomposition of the total homogeneous vibrational line shape into the individual components of pure-dephasing (T_2^*), population relaxation (T_1), and orientational relaxation. The results demonstrate that each of these can contribute significantly, but to varying degrees at different temperatures.

$W(CO)_6$ was chosen as a well-understood vibrational chromophore to use as a dilute probe of intermolecular solvent interactions while discriminating against solute–solute effects. It is soluble in a wide variety of solvents, and it is stable under normal experimental conditions. The CO stretch of $W(CO)_6$ has a particularly strong transition dipole moment in the infrared, allowing very dilute solutions to be probed and strong photon echo signals to be observed. The absorption linewidth for this transition is very narrow, permitting dynamics to be characterized with picosecond pulses. Also, several aspects of the dynamics of this probe molecule have previously been characterized in a number of liquids and glasses.[7,8,9,15,25,28,29] The degenerate nature of the transition studied makes the analysis of the orientational relaxation more involved, yet ultimately contributes more information on the temperature-dependent dynamics.

The paper is divided as follows. In Sec. II, the dipole correlation function formalism that relates the microscopic dynamics of the vibrational dipole to the infrared line shape and the infrared photon echo experiment is discussed. This material is presented because the influence of orientational relaxation and restricted orientational relaxation on the photon echo observable has not been discussed previously. The results are given in the text while the full treatments are reserved for the appendices. The experimental procedures and results are described in Secs. III and IV. The discussion is divided into two sections. Section V A is a comparison of the temperature dependences of vibrational dephasing in the three glasses. Section V B is a detailed discussion of the liquid 2-MP, in which the temperature dependence of all dynamic contributions to the vibrational line shape are delineated. A summary and concluding remarks are given in Sec. VI.

II. THEORETICAL BACKGROUND

A. Correlation function description of the vibrational line shape

For a dilute solution of a vibrational chromophore, the homogeneous vibrational linewidth, Γ, generally has contri-

2812   A. Tokmakoff and M. D. Fayer: Infrared photon echo in glass-forming liquids

butions from the rate of population relaxation (lifetime), $1/T_1$, the rate of pure dephasing, $1/T_2^*$, and the rate of orientational relaxation, Γ_{or}. T_1 processes are caused by the anharmonic coupling of the vibrational mode to the bath, which includes other vibrational modes of the solute and the solvent and the low frequency continuum of intermolecular solvent modes or phonons.[17,25,30,31] Population relaxation is the only dynamic process that can contribute to the vibrational linewidth in the limit that $T \rightarrow 0$ K.[31,32] Pure dephasing describes the adiabatic modulation of the vibrational energy levels of a transition caused by thermal fluctuations of its environment.[18,33] Measurement of this quantity provides detailed insight into the fast dynamics of the system. Although generally equated with physical rotation of the dipole, orientational relaxation is defined as any process that causes the loss of angular correlation of an ensemble of dipoles. For the T_{1u} mode of W(CO)$_6$ studied here, orientational relaxation occurs through the time evolution of the coefficients of the three states in the superposition state created by the initial excitation of the triply degenerate T_{1u} mode. Both pure dephasing and orientational relaxation will vanish in the limit that $T \rightarrow 0$ K, since they are thermally induced processes.

The infrared absorption line shape is related to these microscopic dynamics through the Fourier transform of the two-time transition dipole correlation function[1,2,3,33]

$$I(\omega) = (2\pi)^{-1} \int_{-\infty}^{+\infty} dt\, e^{i\omega t} \langle \boldsymbol{\mu}(t) \cdot \boldsymbol{\mu}^*(0) \rangle. \tag{1}$$

The infrared absorption spectrum gives dynamic information on $\langle P_1(\mu(t)\mu^*(0)) \rangle$, where P_l is the Legendre polynomial of order l. A similar equation is valid for Raman or light scattering experiments where the time correlation function of the molecular polarizability is considered; these experiments give information in the form of P_2.[34]

If the vibrational and rotational motions are decoupled, then the dipole correlation function can be factored

$$\langle \boldsymbol{\mu}(t) \cdot \boldsymbol{\mu}^*(0) \rangle = \langle \mu(t)\mu^*(0) \rangle \langle \hat{\mu}(t) \cdot \hat{\mu}(0) \rangle. \tag{2}$$

The first term in Eq. (2) describes only the time dependence of the nuclear motions and is often written as[35]

$$\langle \mu(t)\mu^*(0) \rangle = |\mu|^2 \left\langle \exp\left[-i \int_0^t dt'\, \Delta(t') \right] \right\rangle. \tag{3}$$

Here Δ describes the variations in the transition energies for the ensemble of vibration transition dipoles and includes any inhomogeneous broadening.[14] The second term in Eq. (2) is the ensemble averaged correlation function for unit vectors along the dipole direction and describes the orientational motion of the dipole.

In the Markovian limit, in which these quantities are described by independent exponentially relaxing dipole correlation functions, the total dipole correlation function described by Eq. (2) decays exponentially at a rate of $1/T_2$, the dephasing time. Equation (1) gives a Lorentzian line shape, and contributions to the full line width at half-maximum (FWHM) are additive

$$\Gamma = 1/\pi T_2 = 1/\pi T_2^* + 1/2\pi T_1 + \Gamma_{or}. \tag{4}$$

The orientational contribution to the infrared line shape due to isotropic rotational diffusion of the dipole is determined by $\langle P_1(\hat{\mu}(t) \cdot \hat{\mu}(0)) \rangle$ and, as shown in Appendix A, is given by

$$\Gamma_{or} = 2D_{or}/\pi, \tag{5}$$

where D_{or} is the orientational diffusion constant. This form arises from the general relation of the time-dependent decay of $\langle P_l(\hat{\mu}(t) \cdot \hat{\mu}(0)) \rangle$ to the expansion coefficients of the Green's function solution to the orientational diffusion equation[36,37]

$$c_l^m(t) = e^{-l(l+1)D_{or}t}. \tag{6}$$

Equation (4) allows the contribution of pure dephasing to the vibrational line shape to be determined from a knowledge of the homogeneous linewidth, the vibrational lifetime, and the orientational diffusion constant.

B. The infrared photon echo experiment

The infrared photon echo[7–9] is a time-domain nonlinear technique that allows inhomogeneity to be removed from the vibrational line shape. It is the infrared vibrational equivalent of photon echoes conducted on electronic states with visible pulses and spin echoes in magnetic resonance. Two short IR pulses, tuned to the molecular vibration of interest, are crossed in the sample. The first pulse creates a coherent state that begins to dephase due to both its interactions with the dynamic environment and the static inhomogeneity. A second pulse, delayed by time τ, initiates rephasing of the inhomogeneous contributions to the vibrational transition; however, homogeneous dephasing is irreversible. This rephasing results in a macroscopic polarization that is observed as an echo pulse in a unique direction at time 2τ. The integrated intensity of the echo pulse is measured as a function of τ, and its decay time with τ is proportional to the homogeneous dephasing time of the vibrational transition. The rephasing of the inhomogeneous broadening eliminates its contribution to the signal.

The infrared absorption line shape, including any inhomogeneity, is governed by a two-time correlation function. The two-pulse photon echo, and other line narrowing techniques such as the stimulated photon echo and hole burning, eliminate the inhomogeneous contribution to the line shape and are described by four-time correlation functions[6] of the form

$$C = \langle \boldsymbol{\mu}^*(t_3 + t_2 + t_1) \cdot \boldsymbol{\mu}(t_2 + t_1) \cdot \boldsymbol{\mu}(t_1) \cdot \boldsymbol{\mu}^*(0) \rangle, \tag{7}$$

where t_1, t_2, and t_3 refer to three consecutive time intervals. We have written this form[38] of the correlation function for direct comparison to Eq. (1). Just as the quantum mechanical correlation function in Eq. (1) can also be written in terms of the anticommutator $\langle \{\boldsymbol{\mu}(t), \boldsymbol{\mu}(0)\}/2 \rangle$, the correlation function in Eq. (7) is normally written as a trace over three nested commutators of a dipole operator with the density matrix using the Heisenberg representation.[39] This form yields four correlation functions (and their complex conjugate) in the form of Eq. (7) that contribute to the observed signal, two of which describe photon echo experiments.[40,41] The two-pulse photon echo is a third-order nonlinear experiment that uses three input fields E_i resonant with a vibrational transition to generate an output echo signal described by $C(t_3,t_2,t_1)$. The first field interaction is during the first pulse, and the second and third interactions are with the second pulse. For a pulse separation τ, the experiment is described by correlation functions of the form $\langle \boldsymbol{\mu}^*(2\tau)\cdot\boldsymbol{\mu}(\tau)\cdot\boldsymbol{\mu}(\tau)\cdot\boldsymbol{\mu}^*(0)\rangle$.

As with the two-time correlation function, the assumption that vibrational and rotational degrees of freedom are independent allows the four-time correlation function C to be separated as $C=C_V C_{\mathrm{OR}}$. The vibrational contribution to the four-time correlation function C_V is given by[6,39,42]

$$C_V(t_3,t_2,t_1)=|\mu|^4\left\langle \exp\left[i\int_{t_1+t_2}^{t_1+t_2+t_3}dt'\Delta(t')\right.\right.$$

$$\left.\left.-i\int_0^{t_1}dt'\Delta(t')\right]\right\rangle. \tag{8}$$

Unlike Eq. (3), Eq. (8) allows for time reversal processes to remove inhomogeneity. The contribution of orientational relaxation to the correlation function C_{OR} can be written as

$$C_{\mathrm{OR}}(t_3,t_2,t_1)=\langle \hat{\mu}(t_3+t_2+t_1)\cdot\hat{\mu}(t_2+t_1)$$

$$\cdot\hat{\mu}(t_1)\cdot\hat{\mu}(0)\rangle, \tag{9}$$

which is the ensemble averaged four-time correlation function for unit vectors along the transition dipole direction. C_{OR} can be treated as a classical probability average when the orientational relaxation is diffusive.

The description of the third-order nonlinear polarization that governs infrared photon echo experiments in terms of the dynamics of lifetime, pure dephasing, and orientational diffusion is discussed in Appendix A. For the case that the relaxation processes that contribute to the line shape are separable, the photon echo signal with delta function pulses decays exponentially as

$$I(\tau)/I(0)=\exp[-4\tau(1/T_2^*+1/2T_1+2D_{\mathrm{OR}})] \tag{10a}$$

$$=\exp[-4\tau/T_2]. \tag{10b}$$

The signal decays at a rate four times faster than the decay of the homogeneous dipole correlation function. The decay of the photon echo in this limit can be written as proportional to $|\langle\boldsymbol{\mu}(\tau)\cdot\boldsymbol{\mu}^*(0)\rangle|^4$, and thus gives the homogeneous vibrational line shape through Eq. (1). It is important to note that due to the separability of the rotational and vibrational degrees of freedom and the Markovian form of the correlation function decay, the four-time correlation function can be written as the product of two-time correlation functions.

III. EXPERIMENTAL PROCEDURES

Vibrational photon echoes were performed with infrared pulses at approximately 5 μm generated by the Stanford superconducting-accelerator-pumped free electron laser (FEL). The FEL generates Gaussian pulses that are transform limited with pulse length (bandwidth) variable from about 0.7 to 2.0 ps. The wavelength is tunable in 0.01 μm increments; active frequency stabilization allows wavelength

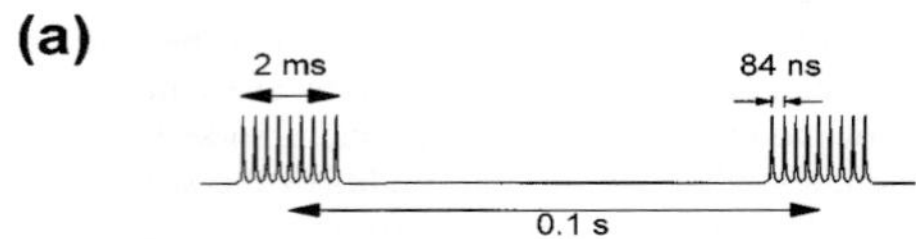

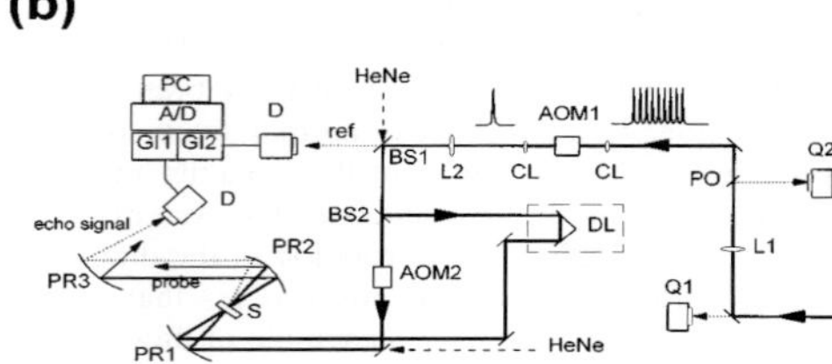

FIG. 1. (a) FEL pulse train structure. The macropulse repetition rate is 10 Hz, with a 2 ms envelope. The micropulse repetition rate is 12 MHz within the macropulse. (b) FEL experimental apparatus. Pulse selection is accomplished with AOM1, while AOM2 is used for chopping of pulse 2. AOM, acousto-optic modulator; A/D, analog-to-digital converter; BS, ZnSe beam splitter; CL, cylindrical lens; D, detector; DL, optical delay line; GI, gated integrator; L, lens; PC, personal computer; PO, pick-off; PR, off-axis parabolic reflector; Q, position sensitive detector; S, sample.

drifts to be limited to $<0.01\%$, or <0.2 cm^{-1}. The pulse length and spectrum are monitored continuously with an autocorrelator and grating monochromator.

Acquisition of data requires manipulation of the unusual FEL pulse train shown in Fig. 1(a). The FEL emits a 2 ms macropulse at a 10 Hz repetition rate. Each macropulse consists of the ps micropulses at a repetition rate of 11.8 MHz (84 ns). The micropulse energy at the input to experimental optics is ~0.5 μJ, and the corresponding energy in the full FEL beam is 120 mW. In vibrational experiments, virtually all power absorbed by the sample is deposited as heat. To avoid sample heating problems, micropulses are selected out of the macropulse at a reduced frequency.

The experimental apparatus is shown in Fig. 1(b). The infrared beam enters the experimental area roughly collimated with a 16 mm diam. Beam pointing stability is monitored with two position sensitive detectors. A 1:6$\times$telescope (L1 & L2) reduces the beam size. At the focus of the telescope is a Ge acousto-optic modulator (AOM) for pulse selection, within a 1:1 cylindrical telescope using CaF$_2$ lenses. Micropulses are selected out of each macropulse at a repetition rate of 50 kHz by the AOM single pulse selector. The cylindrical telescope makes the AOM rise time less than the interpulse separation. This pulse selection yields an effective experimental repetition rate of 1 kHz, and an average power <0.5 mW. A ZnSe beam splitter allows 1% of the IR beam to be directed into a HgCdTe reference detector. The reference detector was used for shot intensity windowing; all data from pulses with intensities outside of a 10% window were discarded.

The two pulses for photon echo or pump–probe experiments were obtained with a 10% ZnSe beam splitter. The 10% beam (first pulse in echo sequence and probe pulse) is sent through a computer-controlled stepper motor delay line.

The remaining portion (second echo pulse or pump pulse) is chopped at 25 kHz by a second Ge AOM. A HeNe beam is made collinear with each IR beam for alignment purposes. The two pulses were focused in the sample to 220 μm diameter using an off-axis parabolic reflector, for achromatic focusing of the IR and HeNe. The beams and echo signal were recollimated with a second parabolic reflector, and focused into a HgCdTe signal detector with a third parabolic reflector. By selecting the desired beam with an iris between the second and third parabolic reflectors, either the photon echo or pump–probe signal could be observed. The photon echo signal and an intensity reference signal were sampled by two gated integrators and digitized for collection by computer.

Careful studies of power dependence and repetition rate dependence of the data were performed. It was determined that there were no heating or other unwanted effects when data were taken with pulse energies of $\sim$15 nJ for the first pulse, $\sim$80 nJ for the second pulse, and the effective repetition rate of 1 kHz (50 kHz during each macropulse).

Vibrational photon echo and pump–probe data were taken on the triply degenerate T_{1u} asymmetric CO stretching mode of $W(CO)_6$. Solutions of $W(CO)_6$ in the glass-forming liquids were made to give a peak optical density of 0.8 with thin path length. Concentrations varied from 0.6 mM with a 200 μm path length for 2-MP, to 4 mM with a 400 μm path length for DBP. (The difference arises because of the much broader inhomogeneous line found in DBP, which is discussed below.) These solutions correspond to mole fractions of $\leq 10^{-4}$, and are dilute enough to eliminate Förster resonant energy transfer.[28] 2-MP ($>$99.9%), DBP ($>$99%), and $W(CO)_6$ (99%) were purchased and used without further purification. 2-MTHF (99%) was distilled in order to remove peroxide inhibitors added by the vendor. The temperature of the sample was controlled to $\pm$0.2 K using a closed-cycled He refrigerator. After equilibration, the temperature gradient across the sample never exceeded $\pm$0.2 K.

The temperature dependences of the vibrational absorption spectra were characterized by infrared absorption spectra. Fourier transform infrared spectra of the 2-MTHF and 2-MP samples were taken with 0.25 cm^{-1} resolution using the same closed cycled helium refrigerator. Spectra on the DBP solution were taken with 2 cm^{-1} resolution using a variable temperature liquid nitrogen cryostat.

IV. RESULTS

The results of temperature-dependent photon echo experiments on the T_{1u} mode of $W(CO)_6$ in the three glass-forming liquids are shown in Fig. 2. Examples of the actual echo decay data have been shown elsewhere.[7–9] The homogeneous linewidth, $\Gamma = 1/\pi T_2$, is shown, derived from fits assuming that the photon echo signal decays exponentially as Eq. (10b). Data on 2-MTHF were taken from 10 up to 120 K, at which point the decay rate exceeded the instrument response. This data, although very similar, supersedes an earlier report on this liquid.[7] Echo data in the other liquids were observable up to room temperature. A preliminary description of the data for 2-MP has been given previously.[8] The data in DBP, taken with 0.7 ps pulses, has been discussed

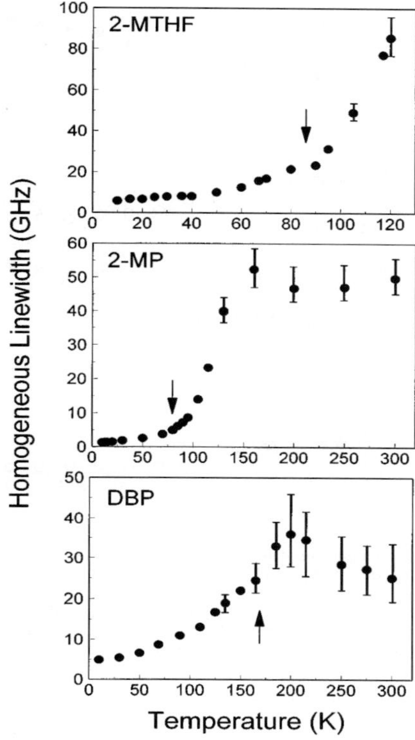

FIG. 2. Temperature dependence of the homogeneous linewidths of the T_{1u} CO stretching mode of $W(CO)_6$ in 2-MTHF, 2-MP, and DBP, determined from infrared photon echo experiments using Eq. (10b), i.e., the data decays exponentially at four times the homogeneous dephasing rate. Arrows mark the glass transition temperature. Note the different temperature and linewidth scales.

recently up to 150 K.[9] At temperatures above 150 K, rapid damping of additional decay processes from higher vibrational levels gave rise to the large error bars in the Fig. 2.

The temperature dependence of the homogeneous linewidth increases monotonically in the glass. Above the glass transition temperature, the linewidth in 2-MTHF (T_g=86 K) and 2-MP (T_g=80 K) rises rapidly in a manner that appears thermally activated. In 2-MP, the rapid increase in dephasing rate slows above 130 K and becomes temperature independent. In DBP (T_g=169 K), the linewidth appears to decrease with temperature above 200 K although the error bars are large enough that it is possible that the temperature dependence is essentially flat.

Pump–probe measurements of the population dynamics in 2-MP and 2-MTHF taken with parallel pump and probe polarizations are shown in Fig. 3. A detailed analysis of the population dynamics and their temperature dependences for the 2-MP data are given in Ref. 15. The interpretation is identical for 2-MTHF. The data at all temperatures are biex-

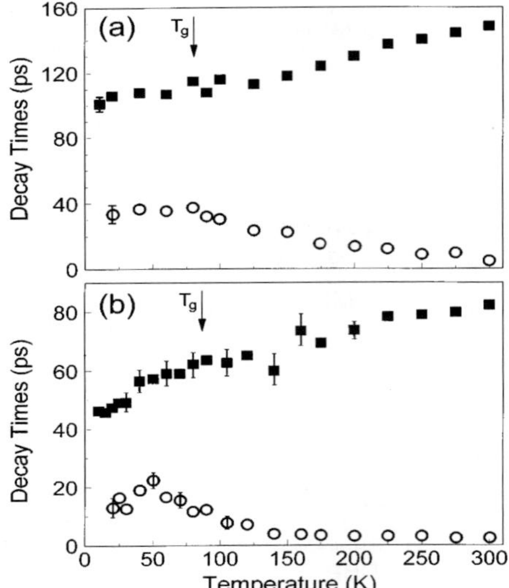

FIG. 3. Temperature-dependent population relaxation dynamics of the T_{1u} of W(CO)$_6$ in (a) 2-MP and (b) 2-MTHF. Each set of data shows the two decay components of the biexponential data from parallel-polarized pump–probe experiments. A detailed analysis of the dynamics and their temperature dependences for 2-MP are given in Ref. 15. The interpretation is identical for 2-MTHF. The solid squares represent the population lifetime T_1. The open circles are a fast component that is due to orientational relaxation of the chromophore.

ponential. The decay times of the two decay components are given in Fig. 3. The long decay component is due to vibrational population relaxation (lifetime), and decays with an exponential decay time of T_1. Both liquids have a counterintuitive temperature dependence, with the rate of relaxation decreasing as the temperature increases. This phenomenon has recently been observed in crystallizing and glass-forming liquids.[15,25] The fast component in the pump–probe data is due to orientational relaxation of the chromophore, and decays exponentially at a rate of $(6D_{or}+1/T_1)$. With magic angle probing to eliminate orientational relaxation effects, the fast component vanished from the decay, leaving only the long component. The amplitudes of the decay components are consistent with complete orientational relaxation above $\sim$120 K and restricted relaxation at lower temperatures. The pump–probe data on 2-MTHF supersede the preliminary data reported on this system.[7]

The absorption linewidths for the T_{1u} mode of W(CO)$_6$ in the three glass-forming liquids are shown in Fig. 4. The absorption linewidth in DBP is 26 cm^{-1} and is temperature independent. In 2-MTHF, the linewidth varies somewhat from 16 cm^{-1} at room temperature to 19 cm^{-1} at 10 K. The spectrum in 2-MP shows the most change with temperature. In the glass, the absorption line is $\sim$10 cm^{-1} wide. Above the glass transition, the line narrows rapidly to a minimum linewidth of 3.2 cm^{-1} at 250 K and broadens slightly at higher temperatures. The absorption linewidths for all solu-

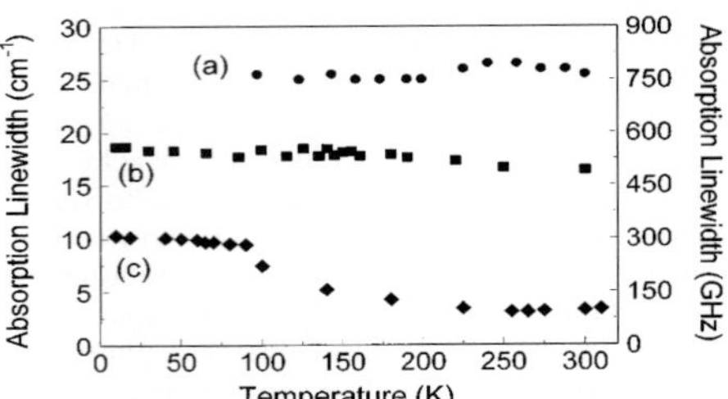

FIG. 4. Temperature dependence of the infrared absorption linewidth of the T_{1u}CO stretching mode of W(CO)$_6$ in (a) DBP, (b) 2-MTHF, and (c) 2-MP.

tions are wider than the homogeneous linewidths measured with the echo experiments at all temperatures with the exception of 2-MP above 200 K. As will be discussed below, proper interpretation of the echo decays above 200 K demonstrates that the absorption line in 2-MP is homogeneously broadened.

The temperature dependences of the homogeneous vibrational linewidths in the three glasses are compared using a reduced variable plot in Fig. 5. Although the absolute linewidths for each system may vary, this is a reflection of the strength of coupling of the transition dipole to the bath, and can be removed by normalization to the linewidth at the glass transition temperature. Likewise, using a reduced temperature T/T_g allows thermodynamic variables that contribute to determining the glass transition to be normalized. Such normalization allows comparison of the functional form of the temperature dependences, independent of differences in T_g and coupling strengths. Figure 5 shows that the temperature dependences of the homogeneous linewidths are identical in the three glasses and are well described by a power law of the form

$$\Gamma(T) = \Gamma_0 + AT^\alpha. \tag{11}$$

The offset at 0 K, Γ_0, represents the linewidth due to the low temperature vibrational lifetime. A fit to Eq. (11) for all tem-

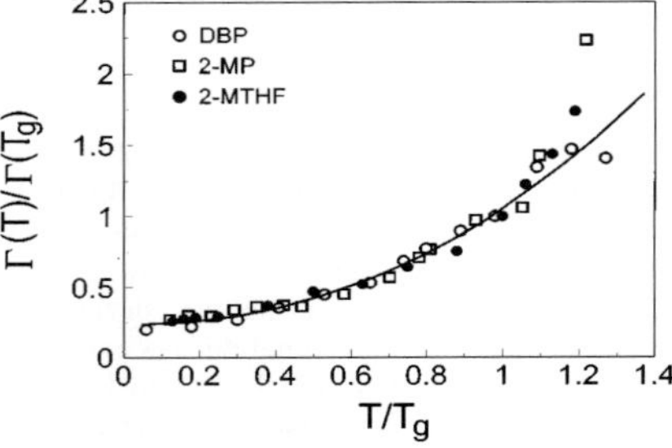

FIG. 5. Comparison of the homogeneous infrared linewidth in three organic glasses. The homogeneous linewidth normalized to the linewidth at the glass transition temperature to remove the coupling strength is plotted against the reduced temperature. The temperature dependence follows a power law with exponent $\alpha = 2.1 \pm 0.2$.

    A. Tokmakoff and M. D. Fayer: Infrared photon echo in glass-forming liquids

TABLE I. Fit parameters for all glasses to Eq. (11).

Liquid	T_g (K)	Power law fit to T_g $1/\pi T_2$			Power law fit to T_g $1/\pi T_2 - 1/2\pi T_1$	
		log(A) (GHz)	α	$T_2(0\ K)$ (ps)	log(A) (GHz)	α
2-MP	79.5	-3.2 ± 0.4	2.0 ± 0.3	124 ± 2	-2.5 ± 0.3	2.2 ± 0.2
2-MTHF	86	-3.2 ± 0.4	2.3 ± 0.3	26 ± 2	-2.9 ± 0.3	1.9 ± 0.2
DBP	169	-3.3 ± 0.3	2.0 ± 0.1	33 ± 2	$\cdots$	$\cdots$
All liquids (Fig. 5)		0.82 ± 0.03	2.1 ± 0.1	$(0.24\pm0.02)\times\Gamma(T_g)$	$\cdots$	$\cdots$

peratures below the glass transition is shown in Fig. 5, and yields an exponent of $\alpha=2.1\pm0.2$ and $\Gamma_0/\Gamma(T_g)=0.24\pm0.02$. At temperatures above $\sim1.2T_g$, the linewidths of the three liquids diverge from one another. All of the liquids were individually fit to Eq. (11). The results are summarized in Table I, and demonstrate that the temperature dependence is well described by a power law of the form T^2.

Because of its high glass transition temperature, DBP allows the largest temperature range over which to observe the power law. The data are presented in two ways in Fig. 6. The solid circles are the data and the line through them is a fit to Eq. (11). To show more clearly the power law temperature dependence, the open circles are the data with the low temperature linewidth, Γ_0 [corresponding to a vibrational lifetime, $T_1(0\ K)=33$ ps] subtracted out. The line through the data is T^2. It can be seen that the power law describes the data essentially perfectly over a change of linewidth of ~500 from 10 to 200 K. This temperature dependence is similar to that observed for infrared vibrational transitions in inorganic glasses with infrared hole burning.[22]

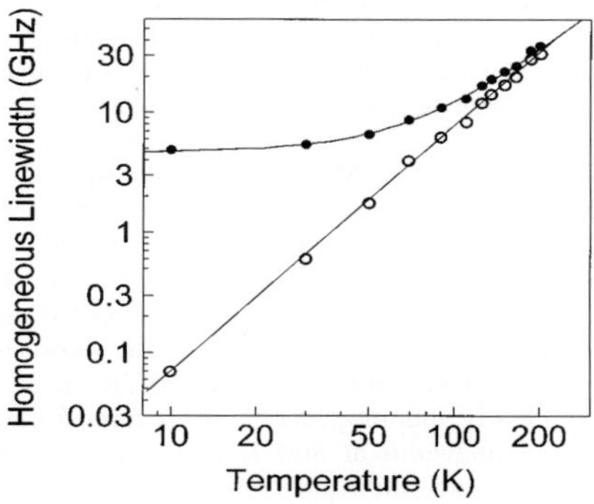

FIG. 6. Homogeneous linewidth of the T_{1u} mode of W(CO)$_6$ in DBP between 10 and 200 K. The homogeneous linewidth is shown in solid circles, and the corresponding data with the low temperature lifetime of $T_1(0\ K)$ =33 ps removed are shown with open circles. The open circle data fit a power law of $\alpha=2.0\pm0.1$.

V. DISCUSSION

A. Comparison of vibrational dephasing in 2-MP, 2-MTHF, and DBP

1. Temperature dependence of dephasing in the glasses

The dynamics of optical dephasing in glasses[19] are generally interpreted within the model of two-level systems, initially proposed to describe their anomalous low temperature heat capacities.[43,44] In this model, pure dephasing dynamics are described in terms of phonon-induced tunneling between two structural potential wells. Above the Debye frequency of the glass, a power law of T^2 is theoretically predicted for dephasing due to two-phonon (Raman) scattering processes between potential wells.[19] Huber has pointed out that in glasses, this temperature dependence is expected above an effective Debye temperature, which can often be two to ten times lower than the true Debye temperature.[45] The $\alpha\approx2$ temperature dependence is almost universally seen for the high temperature dephasing of electronic transitions in a variety of glasses.[19] The temperature dependencies of vibrational linewidths using infrared hole burning have been observed to have similar temperature dependencies between 10 and 60 K. These include $\alpha=2.3$ for N–D stretches in mixed ammonium salts,[23] $\alpha=2.2$ and 2.0 for S–H and CO$_2$ defects in chalcogenide glasses,[22] and $\alpha=2.0$ for CN$^-$ defects in CsCl crystals.[24] The same temperature dependence, $\alpha=2.08$, is seen for ReO$_4^-$ impurity vibrations in alkali halide crystals.[20] Hole burning measures a linewidth with contributions from both homogeneous broadening and spectral diffusion.

Figure 5 demonstrates that the underlying temperature dependence of the vibrational homogeneous dephasing measured with the photon echoes is, in fact, identical in the three glasses studied here, and is described by a power law T^2. The results presented here are the first to examine the temperature dependence of vibrational dephasing in organic glasses. The observation of the T^2 dependence suggests that these glasses are in the high temperature limit above 10 K and that the temperature dependence of vibrational pure dephasing may be universal in all high temperature glasses. This would be consistent with dephasing caused by Raman (two quantum) phonon scattering.

Fits to Eq. (11) assume that the vibrational lifetime is

constant with temperature. However, T_1 can vary by $\sim 50\%$ over the temperature range. This variation of T_1 with temperature is relatively small compared to the power law temperature dependence. Equation (11) was fit to the 2-MP and 2-MTHF data after using the lifetime data of Fig. 3 to subtract out the temperature-dependent lifetime contribution to the linewidth according to Eq. (10). The parameters obtained from these fits, given in Table I, are slightly different but remain within the error bars of the original fit.

2. Temperature dependence above the glass transition

The temperature dependences of the rate of dephasing in 2-MP and 2-MTHF (Fig. 2) both show a gradual increase with temperature until shortly after the glass transition, at which point a rapid increase in dephasing is observed. Coincidentally, the glass transition temperatures for 2-MP ($T_g = 80$ K) and 2-MTHF ($T_g = 86$ K) are very similar. Thus, the rapid increase in dephasing can be explained by two possible phenomena: either by the emergence of additional solvent dynamics above the glass transition, or by thermally activated dephasing through coupling to a low frequency intramolecular mode of $W(CO)_6$. If one of the low frequency modes of $W(CO)_6$ is quadratically coupled to the T_{1u} mode, thermal population of this low frequency mode would result in thermally activated dephasing of the T_{1u} mode. For this mechanism, the break in the functional form of the temperature dependence would be independent of T_g. The temperature dependence of the T_{1u} vibrational linewidth in DBP shows no dramatic change near 90 K, demonstrating that such a mechanism is not the cause of the rapid increase in dephasing rates near 90 K in the other glasses. Rather, since the mechanism is dependent on the solvent, the dephasing is dictated by the particular temperature-dependent dynamics of the solvent.

At temperatures above the glass transition, the normalized linewidths in the three liquids diverge from each other, indicating the emergence of distinct relaxation processes in the various liquids. However, the temperature dependences observed in the glasses do not deviate from each other until $T/T_g > 1.15$. In 2-MTHF and 2-MP, the rate of homogeneous dephasing increases rapidly above the glass transition temperature, in a manner that appears thermally activated. The temperature dependences for the two liquids are in fact very similar. Apart from coupling strength they are identical until ~ 130 K. At this temperature, the rate of increase of the dephasing in 2-MP slows as it approaches its high temperature behavior. The temperature dependence observed for DBP also deviates from the universal behavior observed in the glassy state for $T/T_g > 1.15$, i.e., 30 degrees above the glass transition temperature.

The observation of a temperature for crossover between short time- and long time-scale structural relaxation processes in supercooled liquids is predicted by mode-coupling theory to occur slightly above the glass transition temperature.[46] This crossover temperature has been observed in many organic liquids to be $T_c = 1.18 T_g$.[47] T_c is regarded as the temperature below which relaxation processes bifurcate into two branches, α relaxations and β relaxations. α processes are longer time-scale structural evolution of a su-

percooled liquid which appear to freeze out on the relevant experimental time scale at the glass transition. β processes involve high frequency, more or less localized, molecular motions. The α relaxation processes are described by transport processes that can be related to the viscosity, translational and rotational diffusion, and dielectric relaxation. β relaxation processes are interpreted in terms of microscopic molecular fluctuations that exist at temperatures above and below the glass transition.[47] If β processes are coupled to the vibrational transition, they can contribute to the homogeneous vibrational dephasing. Although a change occurs in the temperature dependence of the vibrational dephasing at $\sim T_c$ above the glass transition, this may be a coincidence, not related to the bifurcation point. The onset of α relaxations have been reported near the glass transition using neutron scattering[48] and light scattering[49] experiments. The infrared photon echo is sensitive to fast dephasing dynamics (perhaps β relaxations) while it should be insensitive to slower structural processes, such as α relaxations near T_c. However, there is clearly a change in the dephasing dynamics in the range of temperatures around T_c. It is possible that the merging of the two relaxation branches at T_c modifies the coupling to the intramolecular vibration of the high frequency part of the fluctuation spectrum. The additional coupling might contribute substantially to the dephasing.

In contrast to the solvents 2-MP and 2-MTHF, there is no rapid increase in the homogeneous linewidth above the glass transition temperature in DBP. This observation shows that the temperature dependence is sensitive to the individual dynamics of the solvent. The differences in the dephasing data in the three solvents above T_g suggest that the rates of dephasing cannot be related to the solvent viscosity, even though the changes in the rates of dephasing are clearly due to the onset of additional relaxation processes above T_g. The additional processes may be new or involve dynamics that were too slow to cause dephasing below T_g. A fragility plot[50] of the viscosities of the three liquids[51–53] shows they superimpose. The temperature dependence of the viscosities, and thus the ability to assume structural configurations over longer time scales, is virtually identical for the three liquids. Thus differences in the temperature dependence of the dephasing in the three solvents at temperatures greater than $T/T_g = 1.15$ cannot be ascribed to differences in the solvent fragilities.

In 2-MTHF and in 2-MP below 150 K, the echo decays yield a homogeneous linewidth that is much narrower than the width of the absorption spectrum. These results demonstrate that the vibrational lines of these system are inhomogeneously broadened in the glass and supercooled liquid. As shown below, in 2-MP the $W(CO)_6$ T_{1u} line becomes homogeneously broadened at room temperature. However, in DBP the line is clearly inhomogeneous at all temperatures. The homogeneous linewidth at 300 K is ~ 1 cm^{-1}, while the absorption spectrum linewidth is 26 cm^{-1}. This is the first conclusive evidence for intrinsic inhomogeneous broadening of a vibrational line in a room temperature liquid. Previously, measurements of room temperature homogeneous vibrational dephasing of the C–H stretch of neat acetonitrile,[10] the C–H stretch of 1,1-d_2-ethanol,[13] and C$\equiv$N stretch of neat

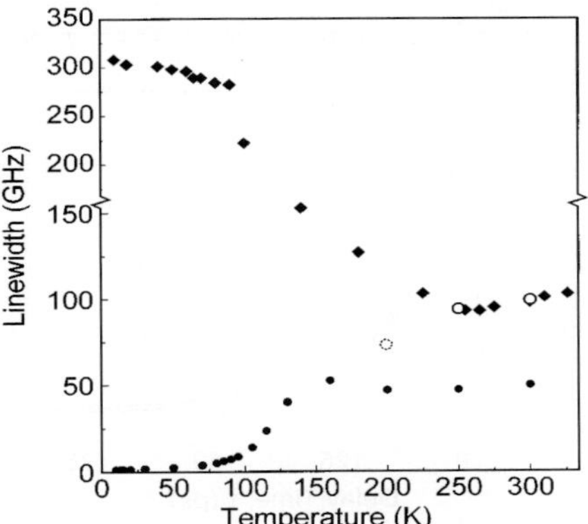

FIG. 7. Temperature dependence of the homogeneous and absorption line-widths for the T_{1u} mode of $W(CO)_6$ in 2-MP. The homogeneous linewidths (solid circles) are taken from Fig. 2 and the absorption linewidths (diamonds) are taken from Fig. 3(c). The "echo" data at 250 and 300 K are actually free induction decays that match the observed absorption line (open circles), showing the spectra are homogeneously broadened. At 200 K (dotted circle), the homogeneous and inhomogeneous contributions to the absorption line are of equal magnitude. This point is only an interpolation between the 160 K and 250 K points. Below 200 K, the homogeneous linewidth is much narrower than the inhomogeneous width.

benzonitrile[12] using Raman echo experiments have determined that these vibrational bands are homogeneously broadened. Inhomogeneous broadening at room temperature has been observed with Raman echo experiments on the C–H stretch in $CH_3I/CDCl_3$ mixtures.[11] The source of inhomogeneity in this case is due to concentration fluctuations of the two types of molecules within the first solvation shell.

B. Total analysis of contributions to the homogeneous line shape in 2-MP

1. Transition to a homogeneously broadened line

The measured homogeneous vibrational linewidth obtained from infrared photon echo measurements (Fig. 2) assumed that the echo decays as Eq. (10b). This equation is valid in the inhomogeneous limit, when the width of the inhomogeneous distribution of homogeneous lines Δ far exceeds the homogeneous linewidth Γ. The inhomogeneous limit implies that a separation of time scales exists between the fast fluctuations of homogeneous dephasing and long time-scale inhomogeneous structural evolution.

Above 150 K, the homogeneous linewidth begins to approach the measured absorption linewidth, as illustrated in Fig. 7. Under these conditions, the time scale for the rephasing of the echo pulse is shortened by the polarization decay due to homogeneous dephasing. This causes the echo to rephase at times between τ and 2τ. Thus, the echo signal decays at a slower rate than the $4/T_2$ given by Eq. (10b).[54,55] In the limit that the absorption line is homogeneously broadened, a free induction decay (FID) will be observed along the echo phase matching direction. If the homogeneous line is a Lorentzian, then an exponential decay with a decay constant of $2/T_2$ will be observed.[55]

The data above 150 K reflect this transition to a homogeneously broadened line. If near room temperature the measured echo data are actually FIDs from a homogeneously broadened line, the decay constants $2/T_2$ yield a linewidth given by the open circles in Fig. 7. The FID linewidths match the measured absorption linewidths exactly, demonstrating that the line is homogeneously broadened for temperatures $\geqslant 250$ K. Notice that the absorption linewidth, which narrows for temperatures up to 200 K actually broadens slightly for higher temperatures, in a manner that precisely follows the echo data. Without experimentally time resolving the rephasing of the echo pulse, it is not possible to determine from an echo experiment alone if the observable is a true echo decay, a FID, or an intermediate case. However, a comparison to the absorption spectrum provides a test. Since treating the point at 160 K as a FID still results in a linewidth that is narrower than the absorption line, this point corresponds to a true photon echo. The point at 200 K represents the transition between a homogeneous and inhomogeneous absorption line shape, and is merely added as an interpolation.

2. Orientational relaxation and restricted orientational relaxation

Pump–probe studies of the population dynamics of the T_{1u} mode of $W(CO)_6$ in 2-MP have observed dynamics that contribute to the infrared absorption line shape.[15] In addition to pure dephasing, population relaxation and orientational relaxation of the initially excited dipole contribute to the homogeneous linewidth. The contributions to the homogeneous vibrational linewidth measured with the photon echo are given by Eqs. (4) and (5). Using previous pump–probe measurements of the orientational diffusion constant D_{or} and the population relaxation time T_1,[15] the contribution due to pure dephasing can be determined from the homogeneous linewidth.

Although the viscosity of 2-MP changes over 14 orders of magnitude between 300 and 80 K, orientational relaxation slows by less than one order of magnitude over this range.[15] The orientational relaxation is distinctly nonhydrodynamic. This is because the orientational relaxation is not due to physical rotation of the molecule, but rather an evolution of the nature of the initially excited superposition of the triply degenerate T_{1u} modes.[15] The initial excitation creates a superposition state of the three basis states taken to be asymmetric CO stretches along the three molecular axes. The amplitudes, or coefficients, of a given basis state in the initial superposition is given by the projection of the E-field onto the direction of the basis state. Fluctuations of the environment cause time evolution of these coefficients, resulting in orientational randomization of the initially excited ensemble of dipoles. This reorientation is well described as isotropic Brownian orientational diffusion of the dipole.[15] Pump–probe experiments allow the determination of the orientational diffusion constant in Eq. (5).

Although this describes the high temperature ($T > 125$ K) behavior of the 2-MP system, at lower temperatures, full orientational relaxation does not occur. Rather, the orientation is restricted and can be interpreted as diffusion within a cone of semiangle θ_0.[56-58] Although the interpretation of pump–probe data within this model is well understood,[15] no such theory exists for the analysis of photon echo experiments. The influence of restricted orientational relaxation on the echo data is not dramatic at most temperatures. However, near the glass transition, echo decays are distinctly nonexponential, leading to non-Lorentzian homogeneous line shapes. In Appendix B, we discuss the details of a treatment of restricted orientational relaxation for the photon echo experiment. As demonstrated in the following section, this nonexponentiality does not invalidate the discussion of the previous sections in terms of exponential fits since the deviations from Lorentzian line shapes are small.

As with the treatment of complete orientational relaxation, we consider the case where the orientational and vibrational degrees of freedom are decoupled, so that the dipole correlation function can be written as Eq. (2). In the limit of small cone angles ($\theta_0 < 60°$), the two-time orientational correlation function decays as

$$C_{OR}(t_i) = (S_1)^2 + (1 - (S_1)^2) c_{\nu_1^1}^1(t_i), \tag{12}$$

where $S_1 = (1 + \cos\theta_0)/2$. The exponential term $c_{\nu_n^m}^m$ is defined as in Eq. (6), but for the noninteger factors ν_n^m. The factors ν_n^m are functions of θ_0, and are discussed in Appendix B. In the limit of complete orientational relaxation, $S_1 = 0$ and $\nu_n^m \to l$, and the expression for complete orientational relaxation is obtained. Equation (12) shows the primary feature of restricted orientational relaxation: the correlation function decays to a nonzero value related to the degree of restriction, i.e., the cone angle. For the two-time restricted orientational correlation function, the long time value ($t = \infty$) is given by $\langle\mu\rangle^2 = S_1^2$ which is the constant offset in Eq. (12). For the four-time correlation function, the long time offset is $\langle\mu\rangle^4 = S_1^4$.

If, as with the complete orientational relaxation, the four-time correlation function can be described as the product of two two-time correlation functions, then the echo signal (see Appendix B) is given by

$$I(\tau)/I(0) = |C_V C_{OR}|^2 = \exp(-4\tau(1/T_2^* + 1/2T_1))$$

$$\times [(S_1)^2 + (1 - (S_1)^2) c_{\nu_1^1}^1(t_i)]^4. \tag{13}$$

In the limit that the cone angle grows to 180°, $S_1 = 0$ and $\nu_1^1 \to l = 1$, and the signal for complete orientational relaxation in Eq. (10a) is recovered. Likewise for the case when no orientational relaxation occurs at all, $\theta_0 = 0°$, and the exponential decay of the signal is due only to the vibrational terms. For intermediate cases, it is observed that the effect of restricted orientational relaxation on the echo signal decay is relaxation to an offset of (S_1),[8] the result predicted for the square of the four-time correlation function. This high order dependence demonstrates the sensitivity of the echo technique to restricted orientational relaxation. The offset $(S_1)^8$ allows the cone angle to be determined, and the cone angle determines ν_n^m for the fit of the orientational diffusion constant.

From Eq. (13) it is clear that when the dynamics of restricted orientational relaxation are faster than those for

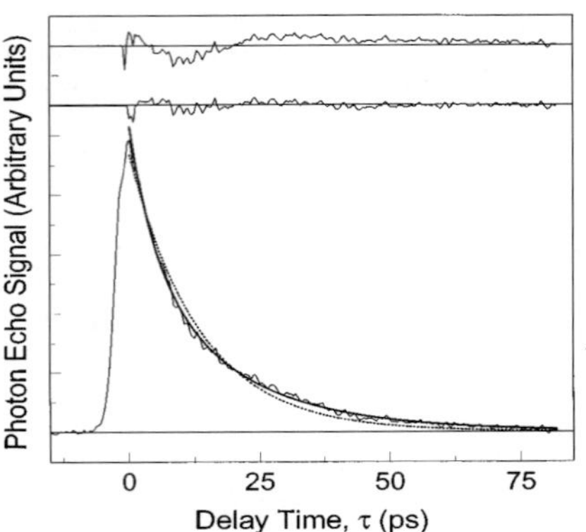

FIG. 8. Photon echo data, fits, and residuals for the T_{1u} mode of W(CO)$_6$ in 2-MP at 85 K. To demonstrate that the data are not exponential, two fits are shown: a single exponential fit (dashed line) and a fit to the model including restricted orientational relaxation (solid line) given by Eq. (13). The residuals for the exponential (top) and Eq. (13) (bottom) fits are given to allow comparison. The nonexponential echo decay means that the homogeneous line shape is not Lorentzian.

population relaxation and pure dephasing, the echo decay allows the orientational contribution to be separated from the other dynamics influencing the echo decay. Once the orientational dynamics have decayed, the residual decay is due only to the vibrational terms. A fit to the tail of the decay determines $(1/T_2^* + 1/2T_1)$ independently from the orientational dynamics. This is particularly useful in the small cone angle limit. As the cone angle shrinks, the orientational decay to $(S_1)^2$ becomes faster. This rapid orientational decay, followed by the exponential relaxation due to the vibrational term, is observed as a clear division of time scales in the decay.

An example of a fit of Eq. (13) to echo data at 85 K is shown in Fig. 8 (solid line). Also shown is an exponential fit (dashed line) and the residuals of the fits. The data are clearly nonexponential, with an initial contribution due both to the vibrational and orientational dynamics, followed by a tail that will decay at a rate determined by the vibrational correlation function. The quality of the fit obtained with Eq. (13) is good for all temperatures. At low temperatures, it fits the fast but small amplitude restricted orientational decay component, which is observed as a small deviation from exponentiality. At intermediate temperatures, given the good signal-to-noise ratio, the deviations from exponential decays are pronounced, as in Fig. 8. At higher temperatures, when the cone angle approaches complete orientational relaxation, obtaining accurate values of the orientational parameters solely from the echo decay becomes difficult, since the rate of pure dephasing far exceeds the orientational contribution.

The functional form of Eq. (13) is correct, i.e., the restricted orientational relaxation will cause a decay to a nonzero value. It can be used to extract the time dependence of the orientational and vibrational contributions to homogeneous dephasing. However, there are clear indications that it is not quantitative. A quantitative relationship would permit the cone angle and the orientational diffusion constant to be obtained from the decay constant and magnitude of the decay caused by restricted orientational relaxation. Analyzing the echo data with Eq. (13) does not yield the cone angles and diffusion constants obtained for this system from the analysis of the pump–probe experiments.[15] In the glass, the cone angles obtained from the echo data are consistently smaller by ~50% than the cone angles obtained from pump–probe data. The diffusion constants obtained from the echo data are equal within error bars to those derived from pump–probe data at low temperatures, but are anomalously high between 70 and 100 K.

Since the theoretical analysis used to analyze the pump–probe experiments is well defined,[15] the lack of agreement suggests uncertainty in the more complex problem of the theoretical description of the influence of restricted orientational relaxation on the echo signal. The separability of the four-time correlation function is based on assuming a Markovian process. In general, higher-order correlation functions can be expressed in their two-time equivalents if the probability function for a given time interval is independent of the history of previous processes.[59] This is not the case for restricted orientational relaxation. The true form needs to be determined using the full four-time correlation function. For complete orientational relaxation, which occurs at the higher temperatures, the decay is exponential, and the four-time correlation function can be factored. The derivation of the echo decay for complete orientational relaxation is given in Appendix A and for restricted orientational relaxation is given in Appendix B.

3. Contributions to the homogeneous vibrational line shape

By combining the photon echo and pump–probe data, all of the dynamics that contribute to the homogeneous line shape are known. At higher temperatures, pump–probe experiments demonstrate that there is full orientational relaxation and can be used to measure the orientational diffusion constant.[15] Full orientational relaxation results in an exponential decay of the correlation function, and thus a Lorentzian contribution to the line shape. Therefore, all contributions to the echo decay are exponential, and the homogeneous line shape is Lorentzian with width given by Eqs. (4) and (5). The homogeneous infrared line shape in the absence of inhomogeneous broadening is given by Eq. (1).

As described in Appendix B, the two-time correlation function for restricted orientational relaxation has been given by Wang and Pecora.[57] The small cone angle result is Eq. (12). The photon echo data at long times give $1/T_2^*$ + $1/2T_1$, and pump–probe experiments give T_1, D_{or}, and θ_0.[15] Since these are the dynamic parameters that contribute to the line shape, the homogeneous line shape can be determined from Eq. (1), where

$$\langle \mu(t)\mu^*(0)\rangle = \exp(-t(1/T_2^* + 1/2T_1))$$

$$\times [(S_1)^2 + (1-(S_1)^2)c_{\nu_1}^1(t)] \qquad (14)$$

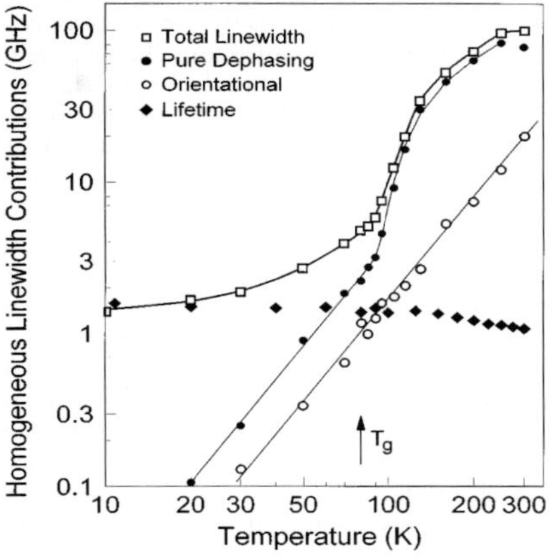

FIG. 9. Log–log plot of the dynamic contributions to the homogeneous vibrational linewidth of the T_{1u} mode of W(CO)$_6$ in 2-MP. The total linewidth—squares; lifetime contribution—diamonds; pure dephasing contribution—solid circles; orientational contribution—open circles. The lifetime dominates the homogeneous linewidth at the lowest temperatures and is only mildly temperature dependent. At high temperatures, pure dephasing dominates the homogeneous linewidth. Orientational relaxation never dominates but makes significant contributions to the linewidth at intermediate temperatures. The total homogeneous linewidths were determined directly from the dynamic variables T_1, T_2^*, D_{or}, and θ_0, and are identical to the data in Fig. 7 within experimental error. Lifetime contributions are taken from Fig. 3. The orientational contribution, shown with a $T^{2.2}$ power law line, is continuous over all temperatures. The line through the pure dephasing data is a fit to Eq. (16). Error bars on the total linewidth, lifetime, and pure dephasing are approximately the size of the symbols at most temperatures. The orientational error bars vary from the size of the symbols at high temperature to ±50% at 30 K.

in the small cone angle limit. For data below 125 K, Eq. (14) applies,[15] and the homogeneous line shape is given by the sum of two Lorentzians.

As seen from Eq. (14), one of the Lorentzian contributions to the line shape is due purely to the vibrational relaxation terms ($1/T_2^* + 1/2T_1$), and the other is a convolution of the orientational and vibrational terms. Combining the photon echo and pump–probe results allows the individual dynamic contributions to the total linewidth to be quantified. At temperatures >125 K, the line is Lorentzian with contributions to the linewidth (FWHM) described by Eq. (4). Using the T_1 and D_{or} measured with the pump–probe experiment, the contribution of pure dephasing to the line can be obtained. At lower temperatures, the vibrational line is non-Lorentzian, but all dynamics contribute to the FWHM linewidth based on their amplitudes. Using the linewidth from line shapes obtained from Eq. (14), T_1 from the pump–probe experiments, and T_2^* obtained from the echo data, the orientational contribution can be determined by difference.

In Fig. 9, the various contributions to the homogeneous linewidths are given. Although the manner of analysis of the dynamics changes at 125 K, the temperature dependence of each of the contributions is continuous. The total linewidth, obtained directly from the dynamic variables T_1, T_2^*, D_{or}, and θ_0, match the numbers from an exponential fit to the echo data in Fig. 2 almost exactly. There is only a slight deviation ($<10\%$) for the temperatures between 80 and 95 K. The single exponential fit, although inadequate for determining the line shape in regions of restricted orientational relaxation provides a very good number for the FWHM linewidth. This conclusion provides the basis for the discussion of the temperature dependence of the homogeneous linewidths of the other liquids given in Secs. IV and V A.

Figure 9 displays the decomposition of the temperature dependence of the homogeneous vibrational linewidth into its three dynamic components. At low temperatures, the lifetime is the dominant contribution. At high temperatures, pure dephasing is the dominant contribution. Only at intermediate temperatures does orientational relaxation make a substantial contribution. The linewidth contribution from lifetime broadening remains significant until $\sim$50 K. The mild temperature dependence of T_1 shows that it is adequate to subtract a constant, Γ_0, for the lifetime contribution in Eq. (11). Above 50 K, the contributions to the linewidth from pure dephasing and orientational relaxation dominate. By 100 K, the pure dephasing is a substantially larger contribution than either the lifetime or the orientational relaxation, and the overwhelmingly dominant component of the homogeneous linewidth above $\sim$150 K.

At low temperature, where the contributions from pure dephasing and orientational relaxation are negligible, the contribution to the linewidth from the lifetime (from pump–probe data) and the linewidth determined from the decay of the echo are equal within error. This is the expected low temperature limit for the homogeneous vibrational linewidth where processes caused by thermal fluctuations disappear, and only lifetime broadening is possible.[31,32]

From low temperature to slightly above T_g, both pure dephasing and orientational relaxation have power law temperature dependences. The earlier discussion of the combined dynamics in 2-MP and the other two solvents showed a T^2 temperature dependence when the lifetime contribution was taken out. Here, the contributions from each mechanism are shown to follow the same T^2 power law behavior in the glass. Thus the total temperature dependence, excluding T_1, is T^2, as shown in Figs. 5 and 6. In addition, the power law observed for the orientational relaxation in the glass is observed to continue into the liquid. In Fig. 9, a power law fit of $\alpha=2.2\pm0.2$ is shown. The orientational dynamics are independent of the glass transition and distinctly nonhydrodynamic. The orientational relaxation does not have the temperature dependence of the viscosity. This is consistent with the results of pump–probe experiments.[15]

In Fig. 9 it is seen that there is a rapid increase in pure dephasing above the glass transition temperature. This implies the onset of an additional mechanism operating on the appropriate time scale to cause pure dephasing and linked to the glass-to-liquid transition. The onset of dynamic processes near the glass transition is often described with a Vogel–Tammann–Fulcher equation[50,60,61]

$$\tau = \tau_0 \exp(B/(T-T_0)). \qquad (15)$$

This equation describes a process characterized by a time, τ,

with a temperature-dependent activation barrier that diverges at a temperature, T_0, below the nominal glass transition temperature $[\eta(T_g)=10^{13}$ poise$]$. T_0 can be linked thermodynamically to an "ideal" glass transition temperature that would be measured with an ergodic observable.[50] This equation describes the temperature dependence of the viscosity of 2-MP well, and gives $T_0=59$ K.

If the Vogel–Tammann–Fulcher (VTF) equation applies to pure dephasing near and above the glass transition, then the full temperature dependence is given by the sum of the low temperature power law and a VTF term:

$$\Gamma^*(T)=A_1 T^\alpha + A_2 \exp(-B/(T-T_0^*)). \qquad (16)$$

A fit to Eq. (16) for all temperatures below 300 K is shown as the line through the pure dephasing data in Fig. 9. The fit describes the entire temperature dependence exceedingly well, but yields a reference temperature of $T_0^* = 80$ K. This reference temperature matches the laboratory glass transition temperature T_g exactly not the ideal glass transition temperature T_0. We can thus infer that the onset of the dynamics that cause the rapid increase in homogeneous dephasing in 2-MP is closely linked with the onset of structural processes near the laboratory glass transition temperature. This may be a manifestation of the short time scale of the measurement and a reflection of the nonergodicity of the system.

The temperature dependence given by Eq. (16) does not describe the decrease in the pure dephasing linewidth observed for the 300 K point. This may be explained by motional narrowing of the line,[35] which is consistent with the observed transition to a homogeneously broadened absorption line at 250 K.

4. Relationship to theories of pure dephasing in liquids

The data presented here can be compared with the predications of a number of theories for the temperature dependence of vibrational dephasing in liquids.[4,16,17,33] A number of the theories for vibrational pure dephasing in liquids are based on vibrational–translational coupling and collision-induced frequency perturbations and yield results that are related to viscosity. Lynden-Bell relates vibrational dephasing to translational diffusion within the liquid potential and obtains a temperature dependence $1/T_2^* \propto \rho\eta/T$.[62] The isolated binary collision (IBC) model of Fischer and Lauberau[63] relates the linewidth to dephasing collision probability yielding $1/T_2^* \propto \eta T/\rho$. The hydrodynamic model of Oxtoby[64] obtains results similar to the IBC model, $1/T_2^* \propto \eta T$. In these expressions, the temperature dependence of the viscosity, η, is the dominant factor.

None of these theories can describe the data in this study. Although the viscosity in liquid 2-MP increases by approximately 14 orders of magnitude between 80 and 300 K, the rate of pure dephasing changes by less than 2 orders of magnitude. The problem may be that the viscosity does not form an adequate approximation for the collision frequency,[4] or that an explicit mechanism for dephasing in a particular system may be necessary to explain the temperature dependence.

Schweizer and Chandler have calculated the vibrational dephasing due to fast interactions with the repulsive wall of the potential based on an Enskog collision time.[18] This theory predicts a temperature dependence of $1/T_2^*$ $\propto \rho T^{3/2} g(\sigma)$, where $g(\sigma)$, the contact value of the radial distribution function for a spherical solvent with diameter σ, may be mildly temperature dependent. This temperature dependence comes much closer to describing the pure dephasing of $W(CO)_6$ in 2-MP between 130 and 275 K, but does not reflect the rapid increase in the viscous, supercooled liquid between 80 and 130 K.

A description of vibrational dephasing above the glass transition that is valid for viscous fluids is clearly necessary. Such a theory might benefit from some of the approaches to amorphous media that are used for the theories of high temperature dephasing in glasses. The data presented here have demonstrated the existence of continuous T^2 behavior through the glass transition with the rapid onset of an additional or modified dephasing process above the glass transition at a temperature.

VI. CONCLUDING REMARKS

The infrared photon echo and pump–probe experiments presented above have allowed a complete characterization of the temperature dependence of the infrared vibrational homogeneous line shape of the T_{1u} CO stretching mode of $W(CO)_6$ in three molecular glass forming liquids. This work has quantified the degree of inhomogeneous broadening, the dynamic contributions to the homogeneous vibrational line shape, and the temperature dependences of these quantities. Results are followed from the nonequilibrium low temperature glass, through the glass transition and supercooled liquid, to the equilibrium room temperature liquid.

To completely characterize the infrared line shape, a combination of experimental methods is necessary to elucidate each of the dynamic and static contributions to the line. Picosecond infrared photon echo experiments were used to measure the homogeneous line shape and remove inhomogeneity. Infrared pump–probe experiments were used to measure the vibrational lifetime and orientational relaxation. The total vibrational line shape was determined by absorption spectroscopy. Together, these experiments yield a complete characterization of the dynamics that make up the homogeneous line, and the extent of inhomogeneity of the infrared absorption line.

Several key finding of this study are worth stating. At low temperatures, the homogeneous linewidth is determined by the vibrational lifetime. From 10 K to slightly above the glass transition, a T^2 temperature dependence for the vibrational pure dephasing is observed in all three solvents, indicating dephasing through a two-phonon Raman scattering process. In the glass and supercooled liquid, orientational relaxation is incomplete, leading to nonexponential echo decays and non-Lorentzian homogeneous line shapes. However, the orientational relaxation contribution to the total homogeneous line shape is never dominant, and the deviations from Lorentzian are not great. In the liquid, above $\sim$125 K, orientational relaxation is complete and, therefore, the homo-

geneous line shape is Lorentzian. The role of orientational relaxation and the effect of restricted orientational relaxation on photon echo observables has been examined in some detail theoretically in the Appendices. While the dynamics are universal in the three glasses studied, they differ in the liquids for temperatures above $\sim 1.2T_g$. The homogeneous dephasing in the liquids are independent of their hydrodynamic characteristics, having a much milder temperature dependence than the viscosity. At room temperature, the vibrational line is homogeneous in the solvent 2-methylpentane but substantially inhomogeneous in dibutylphthalate.

The homogeneous line shape in the 2-methylpentane solvent was decomposed into its three components, pure dephasing, orientational relaxation, and vibrational lifetime, over the entire temperature range of 10–300 K. The vibrational lifetime has a very mild temperature dependence and dominates the linewidth at low temperature. The orientational relaxation contribution has a T^2 temperature dependence over the entire range with no discontinuity near the glass transition temperature. Its contribution, while significant in some temperature ranges, never dominates the linewidth. The temperature dependence of the pure dephasing contribution to the linewidth is described by a T^2 power law in the glass but becomes much steeper above the glass transition. Its temperature dependence can be fit well to the sum of the power law and a Vogel–Tammann–Fulcher equation. However, the Vogel–Tammann–Fulcher divergence temperature is T_g rather than the "ideal" glass transition temperature 20 K below T_g.

While the study presented here gives detailed results from three solvents, it examines only one mode of one molecule. Clearly the time-domain methods that have been employed here to obtain a complete determination of the temperature-dependent vibrational dynamics need to be applied to other systems. Vibrational photon echoes combined with pump–probe experiments will allow accurate decomposition of vibrational line shapes that are convolutions of multiple dynamic and static contributions acting on widely variable time scales.

ACKNOWLEDGMENTS

The authors gratefully acknowledge David Zimdars, Dr. Randy Urdahl, Dr. Bernd Sauter, Rick Francis, and Dr. Alfred Kwok who contributed to the acquisition of the data discussed in this paper. The authors thank Professor Alan Schwettman and Professor Todd Smith of the Department of Physics at Stanford University and their groups for the opportunity to use the Stanford Free Electron Laser. We also thank Dr. Camilla Ferrante for many helpful discussions. This work was supported by the National Science Foundation (DMR93-22504), the Office of Naval Research (N00014-92-J-1227), the Medical Free Electron Laser Program (N00014-91-C-0170), and the Air Force Office of Scientific Research (F49620-94-1-0141).

APPENDIX A: CONTRIBUTION OF ORIENTATIONAL RELAXATION TO THE THIRD-ORDER POLARIZABILITY

Here we describe the effect of orientational relaxation on the stimulated photon echo response function for an inhomogeneously broadened line comprised of motionally narrowed homogeneous lines, i.e., the Bloch limit. The photon echo is the limiting case of the stimulated echo in which the second and third pulses are simultaneous. Generally, the third-order material response[39,65] is written for isotropic media, or unpolarized light. A proper treatment of orientational anisotropy in time-domain third-order (four-wave mixing) experiments requires a tensorial nonlinear response function formalism. This formalism has been presented by Cho et al.,[66] but only derived for pump–probe experiments.

The density matrix description of resonant third-order nonlinear experiments in terms of four-time correlation functions has been given by others,[6,39–42] and is not discussed here. The experimental four-time correlation functions are given by the third-order response function $\mathbf{R}$, and are related to the signal through the third-order polarization $\mathbf{P}$[39,40,41,66]

$$\mathbf{P}_\eta(\mathbf{k},t) = (-i^3) \int_0^\infty dt_3 \int_0^\infty dt_2 \int_0^\infty dt_1 \mathbf{R}_{\eta\alpha\beta\gamma}(t_1,t_2,t_3)$$
$$\times \exp(i(\omega_1 + \omega_2 + \omega_3)t_3 + i(\omega_1 + \omega_2)t_2$$
$$+ i\omega_1 t_1) E_{3\alpha}(t-t_3) E_{2\beta}(t-t_3-t_2)$$
$$\times E_{1\gamma}(t-t_3-t_2-t_1). \tag{A1}$$

Here $\mathbf{R}_{\eta\alpha\beta\gamma}(t_1,t_2,t_3)$ is the third-order tensorial nonlinear response function. This equation describes the signal generated at time t with wave vector $\mathbf{k}$ and polarization component η, due to three input pulses E_1, E_2, and E_3 with polarization components γ, β, and α. In Eq. (A1), t_1 and t_2 refer to the time intervals between interactions with the fields E_1, E_2, and E_3; t_3 is the time interval between E_3 and the observation time t.[39]

To greatly simplify this problem, it is assumed that the orientational motion is separable from the conventional isotropic response, i.e., the vibrational and orientational motions are not coupled in the system. Further, the orientational behavior of the excited coherent state is assumed to be similar to that of the ground state. These assumptions (discussed below) allow us to write

$$\mathbf{R}_{\eta\alpha\beta\gamma}(t_1,t_2,t_3) = R_{\mathrm{OR}}(t_1,t_2,t_3) R_V(t_1,t_2,t_3), \tag{A2}$$

where R_{OR} is the orientational contribution to the response function, and R_V is the conventional isotropic vibrational response function. R_{OR} contains only the time-dependent orientational information and the dependence on the input field polarizations, α, β, and γ, while the remaining dipole-coupling terms are contained in R_V.

To describe the contribution of orientational diffusion to the third-order response function in the Bloch limit, we consider the case of three delta function pulses incident along the x laboratory axis, with parallel linear z axis polarizations. The generalization to any polarization is straightforward. We consider an ensemble of spherical rotors with nondegenerate dipole transitions with orientations Ω. We write Ω to

refer to the spherical coordinates (θ,ϕ), with $\int d\Omega = \int_0^{2\pi} d\phi \int_0^\pi \sin\theta d\theta$. The classical orientational diffusion of such a dipole is well known.[36,37] Orientational diffusion is described by $G(\Omega,t|\Omega_0)$, the probability evolution Green's function that satisfies the orientational diffusion equation, which gives the probability that if μ has direction Ω_0 at time $t=0$, then it will have direction Ω at time t. $G(\Omega,t|\Omega_0)$ is a well-known function that is given by an expansion in spherical harmonics[67]

$$G(\Omega_i,t_i|\Omega_0) = \sum_{l=0}^\infty \sum_{m=-l}^l c_l^m(t_i)[Y_l^m(\Omega_0)]^* Y_l^m(\Omega_i), \tag{A3}$$

where the expansion coefficients c_l^m are given by Eq. (6). Evaluation of probability functions for linear spectroscopies are greatly simplified by the orthogonality of spherical harmonics. Such correlation functions can generally be expressed in the form of $\langle P_l(\hat{\mu}(t) \cdot \hat{\mu}(0)) \rangle$, where P_l is the lth order Legendre polynomial ($m=0$), and can be expressed in terms of the expansion coefficients [Eq. (A6)]. In particular, it can be shown that[67]

$$\langle P_1(\hat{\mu}(t) \cdot \hat{\mu}(0)) \rangle = c_1(t), \tag{A4a}$$

$$\langle P_2(\hat{\mu}(t) \cdot \hat{\mu}(0)) \rangle = c_2(t). \tag{A4b}$$

The first equality is of importance in dielectric relaxation measurements and infrared line shapes, while the second is used for Raman scattering, light scattering, pump–probe experiments, and fluorescence depolarization.[3,34]

For the stimulated echo, the orientational contribution to the third-order polarizability is given by a four-time joint probability function[66]

$$R_{\mathrm{OR}}(t_3,t_2,t_1) = \int d\Omega_3 \int d\Omega_2 \int d\Omega_1 \int d\Omega_0 (\hat{\mu}_3 \cdot \hat{\epsilon}_\eta)$$
$$\times G(\Omega_3,t_3|\Omega_2)(\hat{\mu}_2 \cdot \hat{\epsilon}_\alpha) G(\Omega_2,t_2|\Omega_1)$$
$$\times (\hat{\mu}_1 \cdot \hat{\epsilon}_\beta) G(\Omega_1,t_1|\Omega_0)(\hat{\mu}_0 \cdot \hat{\epsilon}_\gamma) P(\Omega_0), \tag{A5}$$

where $P(\Omega_0)$ is the initial dipole distribution and $(\hat{\mu}_i \cdot \hat{\epsilon}_j)$ represents the interaction of the field with the distribution of dipoles at time t_i. For an initially isotropic sample $P(\Omega_0) = 1/4\pi$. Here we consider all incident pulses polarized parallel along the z axis, so that

$$(\hat{\mu}_i \cdot \hat{\epsilon}_j) = \cos\theta_i = P_1(\cos\theta_i). \tag{A6}$$

Substitution of Eqs. (A3) and (A6) into Eq. (A5), with parallel detection polarization, yields

$$R_{\mathrm{OR}}(t_1,t_2,t_3) = \tfrac{1}{9} c_1(t_1)[1 + \tfrac{4}{5}c_2(t_2)]c_1(t_3). \tag{A7}$$

This result shows that the polarization of the two pulse photon echo experiment decays with time exponentially as $\exp[-2D_{\mathrm{or}}t]$. Equation (A7) also reproduces all polarization dependence of all parallel polarized four-wave mixing experiments. For pump–probe and transient grating experiments, $t_1 = t_3 = 0$, and the well-known expression for the parallel-probed orientational signal is obtained. The orientational component of the signal decays exponentially as $\exp[-6D_{\mathrm{or}}t]$ to a constant offset of 0.56. Notice that the four-time correlation function given by Eq. (A7) can be writ-

 A. Tokmakoff and M. D. Fayer: Infrared photon echo in glass-forming liquids

ten, using Eq. (A4), as the product of two-time correlation functions. The two-pulse photon echo signal ($t_2=0$) gives the orientational dynamics in terms of $\langle P_1(\hat{\mu}(t)\cdot\hat{\mu}(0))\rangle$, just as represented in the infrared line shape.

The vibrational contribution to the response function R_V has been evaluated for a number of models,[6,42,68] and is not discussed here. For the dephasing and relaxation of the vibrational modes studied here, the fast modulation (Bloch) limit applies, and the vibrational response function has been given as[11]

$$R_V(t_1,t_2,t_3)=\exp[-(t_1+t_3)/T_2^*]$$
$$\times\exp[-(t_1+2t_2+t_3)/2T_1]$$
$$\times\exp[-\Delta^2(t_3-t_1)^2/2]. \qquad (A8)$$

Here any dipole operator terms that reflect the amplitude of the signal have been suppressed. The excitation fields span the inhomogeneous line, which is described by a Gaussian distribution of homogeneous packets, $G(\omega)$, with a standard deviation in time of Δ. Notice that in the Markovian limit, where $t_1=t_3$, the four-time vibrational response function can be factored into the product of three two-time correlation functions. Thus, the entire four-time correlation function in the Bloch limit can be expressed as two-time correlation functions.[38]

For delta-function pulses, $E_i(t-t')=\delta(t-t')$, the polarization is given by the response function. For the two-pulse photon echo, $t_1=t_3=\tau$, $t_2=0$, and $\Delta=0$, and the signal is calculated from

$$I(2k_2-k_1,\tau)\propto|P(2k_2-k_1,\tau)|^2. \qquad (A9)$$

For R_{OR} and R_V given by Eqs. (A7) and (A8), respectively, the echo signal is given by Eq. (10). For an echo signal decay due only to rotational diffusion, the echo signal decays exponentially at rate of eight times the orientational diffusion constant. This is in contrast to the anisotropy decay in pump–probe or fluorescence depolarization experiments, which decay as $6D_{or}$. Notice that for this Markovian limit, in which the correlation functions decay exponentially, the signal is proportional to $|\langle\boldsymbol{\mu}(\tau)\cdot\boldsymbol{\mu}^*(0)\rangle|^4$.

The derivation of the expressions used here for the effect of orientational relaxation is based on the assumption that it is uncorrelated with the vibrational dynamics of the dipole. This assumption is often invoked to simplify the description of rotational motion in complex systems.[34,38,66] This assumption may be considered if a separation of time scales exists for the dynamics of vibrational dephasing and orientational relaxation. The rapid fluctuations of the bath allow its interaction with the vibrational coherence to be considered as a Markovian process in the weak coupling limit. Further, the influence of these fast bath fluctuations on the motion of the dipole allow the orientational motion to be treated as Brownian diffusion.

These generalizations are different for the current case, when we consider that the orientational motion of the initially excited dipole is not due to the physical rotation of the molecule. In fact, the orientational relaxation occurs through the evolution of the initially excited superposition state of the T_{1u} mode.[15] This process is independent of the solvent vis-

cosity, and occurs due to thermal fluctuations in the environment that cause the coefficients of the basis states to evolve. However, the observation that reorientational motion is linked to spectral diffusion in the line[15] allows us to decouple these dynamics from the pure dephasing. Spectral diffusion, observed for temperatures below 200 K, is the evolution of the homogeneous transition frequencies in an inhomogeneous line on a time scale longer then that of the homogeneous pure dephasing.

A further assumption in the description of the role of rotational diffusion on the echo experiment is that the rotational behavior of the evolving excited and ground vibrational states are similar. This allows the rotational characteristics to be treated as uniform, and as classical diffusion. This assumption would most likely not be valid for electronic spectroscopy, when the differing electronic configurations and change of size or shape of the electronic cloud could dramatically effect the rotational characteristics. For the case of ground electronic state vibrational spectroscopy on stretching modes, this is not unreasonable. Vibrational excitation leads to an increased root-mean-squared displacement of the oscillating dipole, but this should not alter the rotational characteristics greatly.

In this treatment of orientational relaxation of the vibrational line shape, all expressions have been written for a well-defined dipole transition direction. It has been assumed that the results derived above can be applied to the degenerate T_{1u} mode. The basis for this assumption is the influence of orientational relaxation on the pump–probe experiment. For parallel excitation, the pump–probe data show a zero-time orientational amplitude of 0.44, which yields a polarization anisotropy of $r(0)=0.4$, as predicted for the isotropic reorientation of a nondegenerate system. The applicability of the results to a degenerate system needs to be confirmed by a full theoretical calculation, which is very complex.

APPENDIX B: RESTRICTED ORIENTATIONAL RELAXATION

This section describes the influence of restricted orientational relaxation on the orientational correlation function C_{OR}, and its effect on the two pulse photon echo experiment. Instead of diffusion on the surface of a sphere, restricted orientational diffusion is often modeled as Brownian diffusion of the dipole within a cone of semiangle θ_0 centered about the z axis ($0°\leq\theta_0\leq180°$). In this discussion, we follow the treatment of two-time correlation functions for restricted orientation relaxation within a cone by Wang and Pecora.[57] As shown previously, the description of the infrared line shape requires a knowledge of $C_{OR}=P_1\langle\hat{\mu}(t)\cdot\hat{\mu}(0)\rangle$. The treatment of restricted orientational relaxation for $P_2\langle\hat{\mu}(t)\cdot\hat{\mu}(0)\rangle$ has been given in a number of studies.[56,58,69,70]

In general, the two-time dipole correlation function for the orientational relaxation within a cone will decay to a nonzero value, $(S_l)^2$, which gives a measure for the extent of orientational relaxation.[69] The factor S_l, the generalized order parameter, satisfies the inequality $0\leq S_l\leq1$, where $S_l=0$ describes unrestricted reorientation, while $S_l=1$ is no orientational motion. The value of $(S_l)^2$ can be determined from the relationship[69]

$$\lim_{t\to\infty}\langle A(0)B(t)\rangle=\langle A\rangle\langle B\rangle. \tag{B1}$$

For the two-time dipole correlation function $P_1\langle\hat{\mu}(t_i)\cdot\hat{\mu}(0)\rangle$, the decay should be to

$$\left[(2\pi(1-x_0))^{-1}\int d\Omega P_1(x_0)\right]^2=((1+x_0)/2)^2=(S_1)^2, \tag{B2}$$

where $x_0\equiv\cos\theta_0$. The factor $[2\pi(1-x_0)]^{-1}$ normalizes the expression to ensure a uniform distribution within the cone at $t=\infty$. The functional form of the decay of the correlation function to S_l^2 cannot be written in closed form, but can be approximated by a number of methods.[57,58,71] In the limit of small cone angles, the decay is well approximated by an exponential-with-offset of the form

$$C_{\mathrm{OR}}(t)=S_1^2+(1-S_1^2)\exp(-\kappa D_{\mathrm{or}}t), \tag{B3}$$

where κ is an proportionality constant that relates the effective decay time to the orientational diffusion constant. In the limit of complete orientational relaxation, $S_1=0$ and $\kappa=l(l+1)$, and Eq. (6) is recovered.

The probability evolution function $G(\Omega,t|\Omega_0)$ for the diffusion within a cone is given by an expansion in spherical harmonics of noninteger degree ν_n^m

$$G(\Omega_i,t_i\Omega_0)=(2\pi(1-x_0))^{-1}\sum_{n=1}^{\infty}\sum_{m=-\infty}^{\infty}c_{\nu_n^m}^m(t_i)$$

$$\times[Y_{\nu_n^m}^m(\Omega_0)]^*Y_{\nu_n^m}^m(\Omega_i), \tag{B4}$$

where the expansion coefficients are given by

$$c_{\nu_n^m}^m(t_i)=\exp(-\nu_n^m(\nu_n^m+1)D_{\mathrm{or}}t_i). \tag{B5}$$

The parameter ν_n^m is a function of the cone angle, yet cannot be obtained in closed form as such. The calculation of ν_n^m from the diffusion equation and its relation to the cone angle has been outlined elsewhere.[57]

Wang and Pecora have calculated the two-time dipole correlation function $\langle\hat{\mu}(t)\cdot\hat{\mu}(0)\rangle$ for first and second degree spherical harmonics. These are given by an infinite sum of exponentials of the form[57]

$$\langle\hat{\mu}(t)\cdot\hat{\mu}(0)\rangle\propto\sum_{n=1}^{\infty}D_n^m c_{\nu_n^m}^m(t), \tag{B6}$$

where the cone angle determines the coefficients D_n^m and the factors ν_n^m. Dipole correlation functions can be approximated by truncating the sum in Eq. (B6). For first degree spherical harmonics, the dipole correlation function is well approximated for any cone angle by[57]

$$P_1\langle\hat{\mu}(t_i)\cdot\hat{\mu}(0)\rangle=D_1^0+D_2^0\exp(-\nu_2^0(\nu_2^0+1)D_{\mathrm{or}}t_i)$$

$$+D_1^1\exp(-\nu_1^1(\nu_1^1+1)D_{\mathrm{or}}t_i). \tag{B7}$$

For small cone angles ($\theta_0<60°$), $D_2^0\approx0$ and the two-time dipole correlation function is observed to decay exponentially to a constant value D_1^0. In this limit, comparison to Eq.

(B3) shows that $(S_1)^2=D_1^0$ and the exponential decay is given by $c_{\nu_1^1}^1(t_i)$. In this limit, the parameter ν_1^1 is well approximated by $\nu_1^1(\nu_1^1+1)=24/7\,\theta_0^2$. In the limit of complete orientational relaxation ($\theta_0=180°$), the parameters ν_1^1 and ν_2^0 converge toward $l=1$ and $D_1^0=0$, so that Eq. (6) is recovered. For purposes of fitting in this work, calculated values of ν_n^m as a function of θ_0 were fit to empirical functions.[72]

The four-time correlation function for restricted orientational relaxation will likewise decay to a constant offset given by

$$\langle P_1(x_0)\rangle^4=((1+x_0)/2)^4=(S_1)^4. \tag{B8}$$

The four-time correlation function for two-pulse photon echo experiments has been shown in Appendix A to be written in the Markovian limit as a product of two-time correlation functions. Thus, the four-time orientational correlation function for the photon echo experiment would be written as the product of two-time correlation functions

$$C_{\mathrm{OR}}(t_1,t_3)=P_1\langle\hat{\mu}(t_1)\cdot\hat{\mu}(0)\rangle P_1\langle\hat{\mu}(t_1+t_3)\cdot\hat{\mu}(t_1)\rangle. \tag{B9}$$

Within this description, the two-pulse echo signal in the small cone angle limit is described by Eq. (13). Notice Eq. (13) shows the correct limits by decaying to S_1^4. In the limit of complete orientational relaxation, $S_1=0$ and $\nu_1^1=l=1$, and the result given by Eq. (10) is obtained. For completely restricted orientational relaxation, $S_1=1$, and orientational diffusion does not contribute to the signal. In each of these limits, the Lorentzian line shape is obtained. The limitations of this descriptions are pointed out in Sec. V B 2.

[1] R. G. Gordon, J. Chem. Phys. **43**, 1307 (1965).
[2] R. G. Gordon, Adv. Magn. Reson. **3**, 1 (1968).
[3] B. J. Berne, in *Physical Chemistry: An Advanced Treatise*, edited by D. Henderson (Academic, New York, 1971), Vol. VIII B.
[4] J. Yarwood, Annu. Rep. Prog. Chem., Sec. C **76**, 99 (1979).
[5] W. G. Rothschild, *Dynamics of Molecular Liquids* (Wiley, New York, 1984).
[6] R. F. Loring and S. Mukamel, J. Chem. Phys. **83**, 2116 (1985).
[7] D. Zimdars, A. Tokmakoff, S. Chen, S. R. Greenfield, and M. D. Fayer, Phys. Rev. Lett. **70**, 2718 (1993).
[8] A. Tokmakoff, D. Zimdars, B. Sauter, R. S. Francis, A. S. Kwok, and M. D. Fayer, J. Chem. Phys. **101**, 1741 (1994).
[9] A. Tokmakoff, A. S. Kwok, R. S. Urdahl, R. S. Francis, and M. D. Fayer, Chem. Phys. Lett. **234**, 289 (1995).
[10] D. Vanden Bout, L. J. Muller, and M. Berg, Phys. Rev. Lett. **67**, 3700 (1991).
[11] L. J. Muller, D. Vanden Bout, and M. Berg, J. Chem. Phys. **99**, 810 (1993).
[12] R. Inaba, K. Tominaga, M. Tasumi, K. A. Nelson, and K. Yoshihara, Chem. Phys. Lett. **211**, 183 (1993).
[13] D. Vanden Bout, J. E. Freitas, and M. Berg, Chem. Phys. Lett. **229**, 87 (1994).
[14] L. R. Narasimhan, K. A. Littau, D. W. Pack, Y. S. Bai, A. Elschner, and M. D. Fayer, Chem. Rev. **90**, 439 (1990).
[15] A. Tokmakoff, R. S. Urdahl, D. Zimdars, A. S. Kwok, R. S. Francis, and M. D. Fayer, J. Chem. Phys. **102**, 3919 (1995).
[16] D. W. Oxtoby, Adv. Chem. Phys. **40**, 1 (1979).
[17] J. Chesnoy and G. M. Gale, Adv. Chem. Phys. **70**, 297 (1988).
[18] K. S. Schweizer and D. Chandler, J. Chem. Phys. **76**, 2296 (1982).
[19] R. M. Macfarlane and R. M. Shelby, J. Lumin. **36**, 179 (1987).
[20] W. E. Moerner, A. R. Chraplyvy, and A. J. Sievers, Phys. Rev. B **29**, 6694 (1984).
[21] M. Dubs and H. H. Guenthard, J. Mol. Struct. **60**, 311 (1980).
[22] S. P. Love and A. J. Seivers, J. Lumin. **45**, 58 (1990).
[23] H.-G. Cho and H. L. Strauss, J. Chem. Phys. **98**, 2774 (1993).

[24] M. Schrempel, W. Gellermann, and F. Luty, Phys. Rev. B **45**, 9590 (1992).

[25] A. Tokmakoff, B. Sauter, and M. D. Fayer, J. Chem. Phys. **100**, 9035 (1994).

[26] U. Happek, J. R. Engholm, and A. J. Seivers, Chem. Phys. Lett. **221**, 279 (1994).

[27] H. Graener, T. Loesch, and A. Laubereau, J. Chem. Phys. **93**, 5365 (1990).

[28] E. J. Heilweil, R. R. Cavanagh, and J. C. Stephenson, Chem. Phys. Lett. **134**, 181 (1987).

[29] E. J. Heilweil, M. P. Casassa, R. R. Cavanaugh, and J. C. Stephenson, Annu. Rev. Phys. Chem. **40**, 143 (1989).

[30] S. Califano, V. Schettino, and N. Neto, *Lattice Dynamics of Molecular Crystals* (Springer, Berlin, 1981).

[31] V. M. Kenkre, A. Tokmakoff, and M. D. Fayer, J. Chem. Phys. **101**, 10618 (1994).

[32] S. Velsko and R. M. Hochstrasser, J. Phys. Chem. **89**, 2240 (1985).

[33] D. W. Oxtoby, Annu. Rev. Phys. Chem. **32**, 77 (1981).

[34] B. J. Berne and R. Pecora, *Dynamic Light Scattering* (Krieger, Malabar, FL, 1990).

[35] R. Kubo, in *Fluctuation, Relaxation, and Resonance in Magnetic Systems*, edited by D. Ter Haar (Oliver and Boyd, London, 1962).

[36] P. Debye, *Polar Molecules* (Dover, New York, 1945).

[37] A. Carrington and A. D. McLachlan, *Introduction to Magnetic Resonance* (Harper and Row, New York, 1967).

[38] S. Bratos and J.-C. Leicknam, J. Chem. Phys. **101**, 4536 (1994).

[39] S. Mukamel and R. F. Loring, J. Opt. Soc. Am. B **3**, 595 (1986).

[40] Y. J. Yan and S. Mukamel, J. Chem. Phys. **89**, 5160 (1988).

[41] Y. J. Yan and S. Mukamel, J. Chem. Phys. **94**, 179 (1991).

[42] R. F. Loring and S. Mukamel, Chem. Phys. Lett. **114**, 426 (1985).

[43] P. W. Anderson, B. I. Halperin, and C. M. Varma, Philos. Mag. **25**, 1 (1972).

[44] W. A. Phillips, J. Low Temp. Phys. **7**, 351 (1972).

[45] D. L. Huber, J. Non-Cryst. Solids **51**, 241 (1982).

[46] J.-L. Barrat and M. L. Klein, Annu. Rev. Phys. Chem. **42**, 23 (1991).

[47] E. Roessler, Phys. Rev. Lett. **65**, 1595 (1990).

[48] U. Buchenau, C. Schönfeld, D. Richter, T. Kanaya, K. Kaji, and R. Wehrmann, Phys. Rev. Lett. **73**, 2344 (1994).

[49] N. J. Tao, G. Li, and H. Z. Cummins, Phys. Rev. Lett. **66**, 1334 (1991).

[50] C. A. Angell, J. Phys. Chem. Solids **49**, 863 (1988).

[51] 2-MP viscosity data was compiled from the following sources: A. C. Ling and J. E. Willard, J. Phys. Chem. **72**, 1918 (1968); H. Greenspan and E. Fischer, J. Phys. Chem. **69**, 2466 (1965); C. Alba, L. E. Busse, D. J. List, and C. A. Angell, J. Phys. Chem. **92**, 617 (1990).

[52] 2-MTHF viscosity data were compiled from the following sources: A. C. Ling and J. E. Willard, J. Phys. Chem. **72**, 1918 (1968); D. Nicholls, C. Sutphen, and M. Szwarc, J. Phys. Chem. **72**, 1021 (1968).

[53] DBP viscosity data were compiled from the following sources: A. C. Ling and J. E. Willard, J. Phys. Chem. **72**, 1918 (1968); A. J. Barlow, J. Lamb, and A. J. Matheson, Proc. R. Soc. London Ser. A **292**, 322 (1966).

[54] M. Cho and G. R. Fleming, J. Chem. Phys. **98**, 2848 (1993).

[55] T. Joo and A. C. Albrecht, Chem. Phys. **176**, 233 (1993).

[56] K. Kinosita, Jr., S. Kawato, and A. Ikegami, Biophys. J. **20**, 289 (1977).

[57] C. C. Wang and R. Pecora, J. Chem. Phys. **72**, 5333 (1980).

[58] G. Lipari and A. Szabo, Biophys. J. **30**, 489 (1980).

[59] D. A. McQuarrie, *Statistical Mechanics* (Harper and Row, New York, 1976).

[60] C. A. Angell, J. Phys. Chem. **86**, 3845 (1982).

[61] G. H. Fredrickson, Annu. Rev. Phys. Chem. **39**, 149 (1988).

[62] R. M. Lynden-Bell, Mol. Phys. **33**, 907 (1977).

[63] S. F. Fischer and A. Laubereau, Chem. Phys. Lett. **35**, 6 (1975).

[64] D. W. Oxtoby, J. Chem. Phys. **70**, 2605 (1979).

[65] S. Mukamel, Annu. Rev. Phys. Chem. **41**, 647 (1990).

[66] M. Cho, G. R. Fleming, and S. Mukamel, J. Chem. Phys. **98**, 5314 (1993).

[67] T. Tao, Biopolymers **8**, 609 (1969).

[68] Y. J. Yan and S. Mukamel, Phys. Rev. A **41**, 6485 (1990).

[69] G. Lipari and A. Szabo, J. Am. Chem. Soc. **104**, 4546 (1982).

[70] A. Szabo, J. Chem. Phys. **81**, 150 (1984).

[71] G. Lipari and A. Szabo, J. Chem. Phys. **75**, 2971 (1981).

[72] Using the results of Wang and Pecora, it can be shown that a relationship given by Lipari and Szabo (Ref. 58) describes the parameter ν_1^1 in the limit of small cone angles: $\nu_1^1(\nu_1^1+1)=24/7\,\theta_0^2$. An empirical relationship that gives ν_1^1 within 1% for $50°<\theta_0<150°$ is $\nu_1^1 = 10^{1.676}\theta_0^{-0.8334}$, where θ_0 is in degrees. For $\theta_0<170°$, $\nu_0^2 = 10^{2.496}\theta_0^{-1.122}$.

13310   *J. Phys. Chem.* **1995**, *99*, 13310−13320

FEATURE ARTICLE

Infrared Vibrational Photon Echo Experiments in Liquids and Glasses

A. Tokmakoff,[†] D. Zimdars, R. S. Urdahl, R. S. Francis, A. S. Kwok, and M. D. Fayer*

Department of Chemistry, Stanford University, Stanford, California 94305

Received: May 12, 1995[⊗]

The vibrational dynamics of polyatomic solutes in polyatomic liquid and glassy solvents are examined using picosecond infrared photon echo experiments and pump−probe experiments from room temperature to 10 K. The photon echo experiments measure T_2, the homogeneous dephasing time (homogeneous line shape), while the pump−probe experiments measure the vibrational lifetime, T_1, and the orientational relaxation dynamics. By combining these measurements, a complete analysis of vibrational dynamics is obtained in the liquid, in the supercooled liquid, through the glass transition, and in the glass. Experiments were conducted on the asymmetric CO stretching mode of tungsten hexacarbonyl ($\sim$1980 cm^{-1}) in 2-methylpentane (2-MP), 2-methyltetrahydrofuran, dibutyl phthalate (DBP), carbon tetrachloride, and chloroform. The experiments were conducted using the picosecond IR pulses from a superconducting-accelerator-pumped free electron laser. The absorption line widths for all glasses are massively inhomogeneously broadened at low temperature. In the room temperature liquids, while the vibrational line in 2-MP is homogeneously broadened, the line in DBP is still extensively inhomogeneously broadened. The temperature dependences of the homogeneous line widths in the three glasses are a T^2 power law. The contributions of vibrational pure dephasing, orientational relaxation, and population lifetime to the homogeneous line shape are examined in detail in the 2-MP solvent. The complete temperature dependence of each of the contributions is determined. In addition, the temperature dependence of T_1 is observed to be "inverted" in most of the solvents; i.e., the lifetime becomes longer as the temperature is increased. Analysis shows that this is caused by temperature dependence of the anharmonic coupling matrix elements.

I. Introduction

Vibrational spectra of even large molecules reveal detailed structure, a large number of lines that can be assigned to the various types of motions that occur in a molecule. The position of the spectroscopic peak associated with a particular mode yields the energy of the vibration. However, in general, even a well-resolved vibrational line does not provide information on dynamics. In principle, dynamical information is contained in a spectroscopic line shape. Vibrational line shapes in condensed phases contain all of the details of the interactions of a normal mode with its environment. These interactions include the important microscopic dynamics, intermolecular couplings, and involve the time scales of solvent evolution that modulate the energy of a transition. However, the line shape can also include essentially static structural perturbations associated with the distribution of local solvent configurations (inhomogeneous broadening). An infrared absorption spectrum or Raman spectrum gives frequency-domain information on the ensemble-averaged interactions that couple to the states involved in the transition.[1−3] Line shape analysis of vibrational transitions has long been recognized as a powerful tool for extracting information on molecular dynamics in condensed phases.[4,5] The difficulty with determining the microscopic dynamics from a spectrum arises because linear spectroscopic techniques have no method for separating the various contributions to the

vibrational line shape. The IR absorption or Raman line shape represents a convolution of the various dynamic and static contributions to the observed line shape. In some cases, polarized Raman spectra can be used to separate orientational and vibrational dynamics from the line shape, yet as with all linear spectroscopies, contributions from inhomogeneous broadening cannot be eliminated.[6]

To completely understand a vibrational line shape, a series of experiments are required to characterize each of its static and dynamic components. These experiments can be effectively accomplished in the time domain, where well-defined techniques exist for measuring the various quantities. Nonlinear vibrational spectroscopy can be used to eliminate static inhomogeneous broadening from IR and Raman line shapes.[6] Techniques such as the infrared photon echo[7−9] and Raman echo[10−13] can determine the homogeneous vibrational line shape which contains the important microscopic dynamics, when this line shape is masked by inhomogeneous broadening.

Below we present the temperature-dependent vibrational dynamics of the triply degenerate T_{1u} CO stretching mode ($\sim$1980 cm^{-1}) of tungsten hexacarbonyl (W(CO)$_6$) in the molecular glass-forming liquids 2-methyltetrahydrofuran (2-MTHF), 2-methylpentane (2-MP), and dibutyl phthalate (DBP). Two aspects of the vibrational line shape in these systems are discussed in detail. Initially, we compare the behavior of the homogeneous line widths in the glasses and the transition to the room temperature liquids for the three solvents, using picosecond IR photon echo experiments. The temperature dependence of the vibrational dephasing and the degree of

* To whom correspondence should be addressed.
† Present address: Physik Department, Technische Universität München, D85748 Garching, Germany.
⊗ Abstract published in *Advance ACS Abstracts*, August 15, 1995.

Feature Article

inhomogeneity as the liquids approach room temperature are discussed. The temperature dependence of the homogeneous vibrational line width in each of the three glassy solvents is T^2, but the behavior is distinct in each of the liquids. While in 2-MP the vibrational line is homogeneously broadened at room temperature, the line in DBP is massively inhomogeneously broadened in the room temperature liquid.

Following the comparison of the temperature-dependent dephasing in the three solvents, one of the systems, W(CO)$_6$ in 2-MP is analyzed in greater detail. The contributions to the vibrational line shape from different dynamic processes are delineated by combining the results of photon echo measurements of the homogeneous line shape[8] with pump—probe measurements of the lifetime and reorientational dynamics.[14] This combination of measurements allows the decomposition of the total homogeneous vibrational line shape into the individual components of pure dephasing (T_2^*), population relaxation (T_1), and orientational relaxation. The results demonstrate that each of these can contribute significantly, but to varying degrees at different temperatures.

In measuring T_1, it was found that in a number of solvents the lifetimes become longer as the temperature is increased.[14-17] This counterintuitive temperature dependence is examined more closely in the solvents carbon tetrachloride (CCl$_4$) and chloroform (CHCl$_3$).[17] Analysis of the influence of the temperature dependence of the occupation numbers of the solute and solvent modes[16,18] and of the solvent density of states[16] shows that these cannot account for the "inverted" temperature dependence. This suggests that the temperature dependence of the liquid density, resulting in changes in the anharmonic potential, is responsible for the form of the temperature dependence.[16]

II. Infrared Vibrational Photon Echo Experiments

A. The Nature of the Experiment. The picosecond infrared vibrational photon echo experiment is a time domain nonlinear experiment that can extract the homogeneous vibrational line shape even when the inhomogeneous line width is thousands of times wider than the homogeneous width. The echo technique was originally developed as the spin echo in magnetic resonance in 1950.[19] In 1964, the method was extended to the optical regime as the photon echo.[20] Since then, photon echoes have been used extensively to study electronic excited state dynamics in many condensed matter systems. For experiments on vibrations, a source of picosecond IR pulses is tuned to the vibrational transition of interest. The echo experiment involves a two-pulse excitation sequence. The first pulse puts the solutes' vibrations into superposition states, which are mixtures of the $v = 0$ and $v = 1$ vibrational levels. Each superposition has a microscopic electric dipole associated with it. This dipole oscillates at the vibrational transition frequency. Immediately after the first pulse, all of the microscopic dipoles in the sample oscillate in phase. Because there is an inhomogeneous distribution of vibrational transition frequencies, the individual dipoles oscillate with some distribution of frequencies. Thus, the initial phase relationship is very rapidly lost. This is referred to as the free induction decay. After a time, τ, a second pulse traveling along a path making an angle θ with that of the first pulse passes through the sample. This second pulse changes the phase factors of each vibrational superposition state in a manner that initiates a rephasing process. At time τ after the second pulse, the sample emits a third coherent pulse of light along a unique path which makes an angle 2θ with the direction of the first pulse (see Figure 1a). The third pulse is the photon echo. It is generated when the ensemble of microscopic dipoles are rephased at time 2τ. The phased array of microscopic

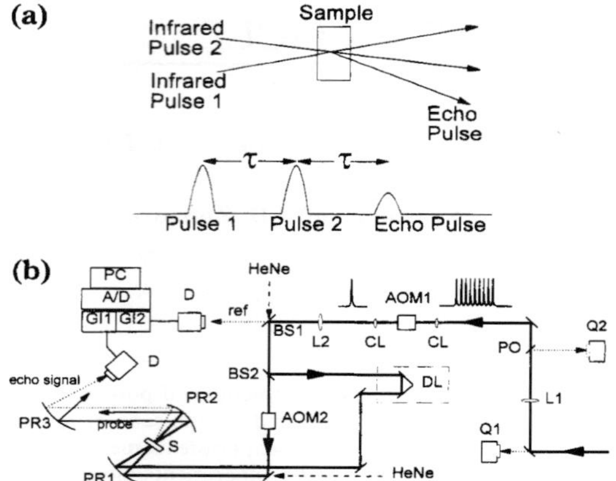

Figure 1. (a) Schematic diagram of the infrared vibrational photon echo experiment. Two picosecond IR pulses tuned to the vibrational transition frequency enter the sample crossed at a small angle, θ. Because of wave vector matching, the echo pulse emerges from the sample in a unique direction, 2θ. Pulse 1 and pulse 2 are separated by time τ. The echo is formed a time 2τ after pulse 1. (b) Experimental apparatus. Pulse selection is accomplished with AOM1, while AOM2 is used for chopping of pulse 2. Abbreviations: AOM, acoustooptic modulator; A/D, analog-to-digital converter; BS, ZnSe beam splitter; CL, cylinderical lens; D, detector; DL, optical delay line; GI, gated integrator; L, lens; PC, personal computer; PO, pick-off; PR, off-axis parabolic reflector; Q, position-sensitive detector; S, sample.

dipoles behaves as a macroscopic oscillating dipole, which generates an IR pulse of light. A free induction decay (inhomogeneous frequency distribution) again destroys the phase relationships, so only a short pulse of light is generated.

The rephasing at 2τ has removed the effects of the inhomogeneous broadening. However, fluctuations due to coupling of the vibrational mode to the heat bath (solvent) cause the oscillation frequencies to fluctuate. Thus, at 2τ there is not perfect rephasing. As τ is increased, the fluctuations produce an increasingly large accumulated phase error, and the size of the echo is reduced. A measurement of the echo intensity vs τ, the delay time between the pulses, is called an echo decay curve. Thus, the echo decay is related to the fluctuations in the vibrational frequencies, not the inhomogeneous spread in frequencies. The Fourier transform of the echo decay is the homogeneous line shape.[21] For example, if the echo decay is an exponential, the line shape is a Lorentzian with a width $1/\pi T_2$ determined by the exponential decay constant. The vibrational photon echo makes the vibrational homogeneous line shape an experimental observable. In fact, the echo measures directly the decay of the system's off-diagonal density matrix elements and is the most fundamental observable.

B. Correlation Function Description of the Vibrational Line Shape. For a dilute solution of a vibrational chromophore, the homogeneous vibrational line width, Γ, generally has contributions from the rate of population relaxation (lifetime), $1/T_1$, the rate of pure dephasing, $1/T_2^*$, and the rate of orientational relaxation, Γ_{or}. T_1 processes are caused by the anharmonic coupling of the vibrational mode to the bath. The bath includes other vibrational modes of the solute and the solvent and the low-frequency continuum of intermolecular solvent modes.[17,18,22,23] Population relaxation is the only dynamic process that can contribute to the vibrational line width in the limit that $T \rightarrow 0$ K.[18,24] Pure dephasing describes the adiabatic modulation of the vibrational energy levels of a

13312   *J. Phys. Chem., Vol. 99, No. 36, 1995*   Tokmakoff et al.

transition caused by thermal fluctuations of its environment.[25,26] Measurement of this quantity provides detailed insight into the fast dynamics of the system. Although generally equated with physical rotation of the dipole, orientational relaxation is defined as any process that causes the loss of angular correlation of an ensemble of dipoles. For the T_{1u} mode of $W(CO)_6$ studied here, orientational relaxation occurs through the time evolution of the coefficients of the three states in the superposition created by the initial excitation of the triply degenerate T_{1u} mode.[14,15] Both pure dephasing and orientational relaxation will vanish in the limit that $T \rightarrow 0$ K, since they are thermally induced processes.

The infrared absorption line shape is related to these microscopic dynamics through the Fourier transform of the two-time transition dipole correlation function[1-3,26]

$$I(\omega) = (2\pi)^{-1} \int_{-\infty}^{+\infty} dt \, e^{i\omega t} \langle \mu(t) \cdot \mu^*(0) \rangle \qquad (1)$$

The infrared absorption spectrum gives dynamic information on $\langle P_1(\mu(t) \cdot \mu^*(0)) \rangle$, where P_1 is the first Legendre polynomial. If the vibrational and rotational motions are decoupled, then the dipole correlation function can be factored

$$\langle \mu(t) \cdot \mu^*(0) \rangle = \langle \mu(t) \, \mu^*(0) \rangle \langle \hat{\mu}(t) \cdot \hat{\mu}(0) \rangle \qquad (2)$$

The first term in eq 2 describes the time dependence of the vibrational motions and is often written as[35]

$$\langle \mu(t) \, \mu^*(0) \rangle = |\mu|^2 \langle \exp[-i \int_0^t dt' \, \Delta(t')] \rangle \qquad (3)$$

Here Δ describes the variations in the transition energies for the ensemble of vibration transition dipoles and includes any inhomogeneous broadening.[27] The second term in eq 2 is the ensemble-averaged correlation function for unit vectors along the dipole direction and describes the orientational motion of the dipole.

In the Markovian limit, in which these quantities are described by independent exponentially relaxing dipole correlation functions, the total transition dipole correlation function described by eq 2 decays exponentially at a rate of $1/T_2$, where T_2 is the dephasing time. This gives a Lorentzian line shape, and contributions to the full line width at half-maximum (fwhm) are additive

$$\Gamma = 1/\pi T_2 = 1/\pi T_2^* + 1/2\pi T_1 + \Gamma_{or} \qquad (4)$$

The orientational contribution to the infrared line shape due to isotropic rotational diffusion of the dipole is determined by $\langle P_1(\hat{\mu}(t) \cdot \hat{\mu}(0)) \rangle$ and,

$$\Gamma_{or} = 2D_{or}/\pi \qquad (5)$$

where D_{or} is the orientational diffusion constant.[15,28] Equation 4 allows the contribution of pure dephasing to the vibrational line shape to be determined from a knowledge of the homogeneous line width, the vibrational lifetime, and the orientational diffusion constant.

C. The Infrared Photon Echo Experiment. The infrared photon echo[7-9] is a time-domain nonlinear technique that allows inhomogeneity to be removed from the vibrational line shape. The two-pulse photon echo, and other line narrowing techniques such as the stimulated photon echo and hole burning, eliminate the inhomogeneous contribution to the line shape and are described by four-time correlation functions[6] of the form

$$C = \langle \mu^*(t_3 + t_2 + t_1) \, \mu(t_2 + t_1) \, \mu(t_1) \, \mu^*(0) \rangle \qquad (6)$$

where t_1, t_2, and t_3 refer to three consecutive time intervals. We have written this form[29] of the correlation function for direct comparison to eq 1. The two-pulse photon echo is a third-order nonlinear experiment that uses three input fields E_i resonant with a vibrational transition to generate an output echo signal described by $C(t_3, t_2, t_1)$. The first field interaction is during the first pulse, and the second and third interactions are with the second pulse. For a delta function pulses with separation τ, the experiment is described by correlation functions of the form $\langle \mu^*(2\tau) \cdot \mu(\tau) \cdot \mu(\tau) \cdot \mu^*(0) \rangle$.

As with the two-time correlation function, the assumption that vibrational and rotational degrees of freedom are independent allows the four-time correlation function C to be separated as $C = C_v C_{or}$. The vibrational contribution to the four-time correlation function C_v is given by[6,30,31]

$$C_v(2\tau, \tau, \tau) = |\mu|^4 \langle \exp[i \int_\tau^{2\tau} dt' \, \Delta(t') - i \int_0^\tau dt' \, \Delta(t')] \rangle \qquad (7)$$

Unlike eq 3, eq 7 allows for time reversal processes to remove inhomogeneity. If the Δ's in eq 7 are strictly static energies (inhomogeneous broadening only), then the two integrals are identical and cancel. Thus, inhomogeneous broadening does not contribute to the echo decay. Random, homogeneous energy fluctuations will not be identical in the two time intervals $0 \rightarrow \tau$ and $\tau \rightarrow 2\tau$. Therefore, the two integrals will not cancel identically, and homogeneous dephasing will contribute to the echo decay.

The contribution of orientational relaxation to the correlation function C_{or} can be written as

$$C_{or}(2\tau, \tau, \tau) = \langle \hat{\mu}(2\tau) \cdot \hat{\mu}(\tau) \cdot \hat{\mu}(\tau) \cdot \hat{\mu}(0) \rangle \qquad (8)$$

which is the ensemble-averaged four-time correlation function for unit vectors along the transition dipole direction. C_{or} can be treated as a classical probability average when the orientational relaxation is diffusive.

The description of the third-order nonlinear polarization that governs infrared photon echo experiments in terms of the dynamics of lifetime, pure dephasing, and orientational diffusion has been presented.[15] For the case that the relaxation processes that contribute to the line shape are separable, the photon echo signal with delta function pulses decays exponentially as

$$I(\tau)/I(0) = \exp[-4\tau(1/T_2^* + 1/2T_1 + 2D_{or})] \qquad (9a)$$

$$= \exp[-4\tau/T_2] \qquad (9b)$$

The signal decays at a rate 4 times faster than the decay of the homogeneous dipole correlation function. The decay of the photon echo in this limit can be written as proportional to $|\langle \mu(\tau) \cdot \mu^*(0) \rangle|^4$ and thus gives the homogeneous vibrational line shape through eq 1. It is important to note that due to the separability of the rotational and vibrational degrees of freedom and the Markovian form of the correlation function decay, the four-time correlation function can be written as the product of two-time correlation functions.

D. Experimental Procedures. Vibrational photon echo experiments require tunable IR pulses with durations of ~ 1 ps and energies of ~ 1 μJ. These can be produced with systems based on conventional picosecond lasers using an optical parametric amplifier (frequency difference mixing) to generate the IR pulses. However, the experiments described below were performed using a different approach, i.e., with infrared pulses at ~ 5 μm generated by the Stanford superconducting-accelerator-pumped free electron laser (FEL). The FEL generates

Gaussian pulses that are transform limited with pulse duration that is adjustable between 0.7 and 2 ps. Active frequency stabilization allows wavelength drifts to be limited to <0.01% or <0.2 cm^{-1}. The pulse length and spectrum are monitored continuously with an autocorrelator and grating monochromator.

The FEL emits a 2 ms macropulse at a 10 Hz repetition rate. Each macropulse consists of the picosecond micropulses at a repetition rate of 11.8 MHz (84 ns). The micropulse energy at the input to experimental optics is ~0.5 μJ, and the corresponding energy in the full FEL beam is 120 mW. In vibrational experiments, virtually all power absorbed by the sample is deposited as heat. To avoid sample heating problems, micropulses are selected out of the macropulse at a reduced frequency.

The experimental apparatus is shown in Figure 1b. The infrared beam enters the experimental area roughly collimated with a 16 mm diameter. A 1:6× telescope (L1, L2) reduces the beam size. At the focus of the telescope is a Ge acoustooptic modulator (AOM) for pulse selection, within a 1:1 cylindrical telescope using CaF$_2$ lenses. Micropulses are selected out of each macropulse at a repetition rate of 50 kHz by the AOM single-pulse selector. The cylindrical telescope makes the AOM rise time less than the interpulse separation. This pulse selection yields an effective experimental repetition rate of 1 kHz and an average power of <0.5 mW. A ZnSe beam splitter allows 1% of the IR beam to be directed into a HgCdTe reference detector. The reference detector was used for shot intensity windowing; all data from pulses with intensities outside of a 10% window were discarded.

The two pulses for photon echo or pump−probe (transient absorption to measure the vibrational lifetime) experiments were obtained with a 10% ZnSe beam splitter. The 10% beam (first pulse in echo sequence and probe pulse) is sent through a computer-controlled stepper motor delay line. The remaining portion (second echo pulse or pump pulse) is chopped at 25 kHz by a second Ge AOM. A HeNe beam is made collinear with each IR beam for alignment purposes. The two pulses were focused in the sample to 220 μm diameter using an off-axis parabolic reflector for achromatic focusing of the IR and HeNe. The beams and echo signal were recollimated with a second parabolic reflector and focused into a HgCdTe signal detector with a third parabolic reflector. By selecting the desired beam with an iris between the second and third parabolic reflectors, either the photon echo or pump−probe signal could be observed. The photon echo signal and a intensity reference signal were sampled by two gated integrators and digitized for collection by computer.

Careful studies of power dependence and repetition rate dependence of the data were performed. It was determined that there were no heating or other unwanted effects when data were taken with pulse energies of ~15 nJ for the first pulse and ~80 nJ for the second pulse and the effective repetition rate of 1 kHz (50 kHz during each macropulse).

Vibrational photon echo and pump−probe data were taken on the triply degenerate T$_{1u}$ asymmetric CO stretching mode of W(CO)$_6$. Solutions of W(CO)$_6$ in the glass-forming liquids were made to give a peak optical density of 0.8 with thin path length. These solutions correspond to mole fractions of ≤10^{-4}. The temperatures of the samples were controlled to ±0.2 K using a closed-cycled He refrigerator.

Pump−probe experiments conducted in the solvents CCl$_4$ and CHCl$_3$ were performed with the 1 kHz output of an optical parametric amplifier using doubled YAG and a dye laser as inputs. This system has been described in detail.[32]

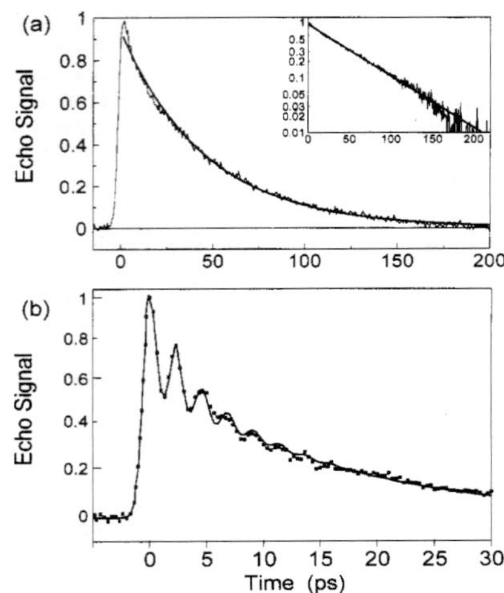

Figure 2. (a) Photon echo decay data for the CO asymmetric stretching mode (T$_{1u}$) of W(CO)$_6$ in 2-methylpentane glass at 10 K. The homogeneous line width determined by the echo decay is 1.3 GHz (0.04 cm^{-1}) in contrast to the absorption line which is inhomogeneously broadened to 310 GHz (10.5 cm^{-1}). (b) Vibrational photon echo decay and fit for asymmetric CO stretching mode of W(CO)$_6$ in dibutyl phthalate in the glass at 10 K. The fit is to eq 10 and represents the homogeneous dephasing of the three-level coherence with beating at the anharmonic vibrational frequency splitting.

III. Experimental Results

A. Vibrational Photon Echo Decays. Figure 2a displays photon echo data for W(CO)$_6$ in 2-MP taken in the low-temperature glass at 10 K.[8,15] The inset shows a log plot of the data. The decay is exponential, indicating that the homogeneous line is a Lorentzian. At this temperature, the absorption line width is 10.5 cm^{-1} (310 GHz). In contrast, $T_2 = 240$ ps, yielding a homogeneous line width of 1.3 GHz. Thus, the absorption line is massively inhomogeneously broadened.

Figure 2b displays data taken using 0.7 ps pulses in the solvent DBP.[9] When the pulse duration is made shorter, the associated bandwidth, Ω, of the transform limited pulse is larger. If the bandwidth of the excitation pulses exceeds the vibrational anharmonicity, then population can be excited to higher vibrational levels. For the case where $\Delta \approx \Omega$, short pulse excitation will create a three-level coherence involving the $v = 0$, 1, and 2 vibrational levels. The expected photon echo signal can be described for an unequally spaced three-level system using a semiclassical perturbative treatment of the third-order nonlinear polarizability in the Bloch limit.[30,33] The three-level system is spaced by the frequencies ω_{10} and ω_{21}, where $\omega_{10} = \omega_{21} + \Delta$ and $\Delta \ll \omega_{10}$ and ω_{21}. The transition frequencies ω_{10} and ω_{21} lie within the bandwidth of the pulses. For a finite pulse bandwidth, where the E-field amplitude differs at ω_{10} and ω_{21}, the decay is given by[9]

$$I(\tau) =$$
$$\exp(-2\gamma_{10}\tau)[E_{10}{}^2 \exp(-2\gamma_{10}\tau) + E_{21}{}^2 \exp(-2\gamma_{21}\tau) -$$
$$2E_{10}E_{21} \exp(-(\gamma_{10} + \gamma_{21})\tau) \cos(\Delta\tau + \phi)] \quad (10)$$

Here, E_{10} and E_{21} are excitation E-field amplitudes for the $v = 0 \rightarrow 1$ and $v = 1 \rightarrow 2$ transitions, respectively. Equation 10 shows that the echo signal envelope decays in proportion to the dephasing rates of the $v = 0 \rightarrow 1$ and $v = 1 \rightarrow 2$ transitions,

13314 *J. Phys. Chem., Vol. 99, No. 36, 1995*

Tokmakoff et al.

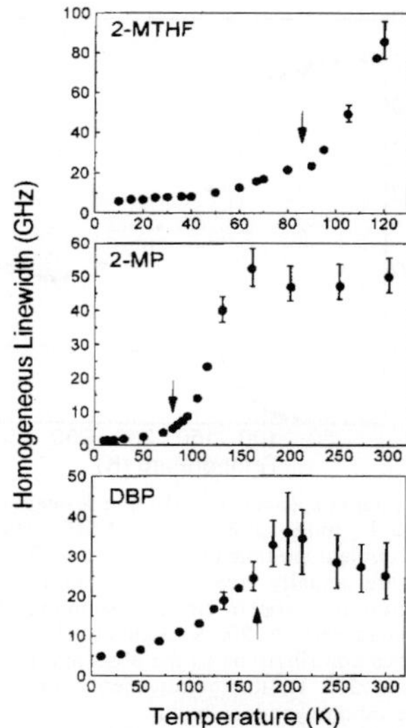

Figure 3. Temperature dependence of the homogeneous line widths of the T_{1u} CO stretching mode of $W(CO)_6$ in 2-MTHF, 2-MP, and DBP, determined from infrared photon echo experiments using eq 9b. Arrows mark the glass transition temperatures. Note the different temperature and line width scales.

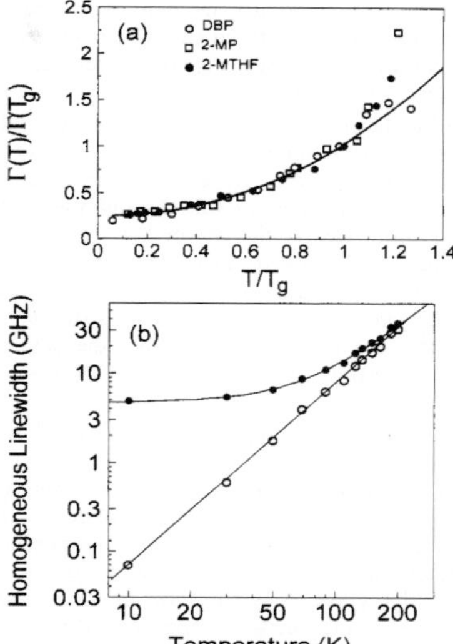

Figure 4. (a) Comparison of the homogeneous infrared line width in three organic glasses. The homogeneous line width normalized to the line width at the glass transition temperature to remove the coupling strength is plotted against the reduced temperature. The temperature dependence follows a power law with exponent $\alpha = 2.1 \pm 0.2$. (b) Homogeneous line width of the T_{1u} mode of $W(CO)_6$ in DBP between 10 and 200 K. The homogeneous line widths are shown in solid circles, and the corresponding data with the low-temperature lifetime of $T_1(0$ K$) = 33$ ps removed are shown with open circles. The data fit a power law of $\alpha = 2.0 \pm 0.1$.

with exponentially damped beats observed at the frequency splitting, Δ. The dephasing rates for the two transitions are γ_{10} and γ_{21}. For the narrow bandwidth case ($E_{21} \approx 0$), eq 9b is recovered.

As can be seen from Figure 2b, the decay is consistent with the expected decay of a three-level vibrational coherence. The decay is modulated at a 2.3 ps frequency, which is constant within error over all temperatures. Based on the average of several data sets, the vibrational anharmonic splitting is $\Delta = 14.7 \pm 0.3$ cm^{-1}.[9] This splitting is in accord with the value of 15 ± 1 cm^{-1} subsequently obtained by Heilweil and co-workers from observation of the $v = 1 \rightarrow 2$ and $v = 2 \rightarrow 3$ transitions of the asymmetric CO stretching mode of $W(CO)_6$ in hexane using transient infrared absorption.[34] The agreement between the anharmonicity obtained from the beat frequency and that obtained by transient absorption confirms the interpretation of the beats as arising from the multilevel coherence of the anharmonic oscillator. The echo decay data also provide the homogeneous dephasing times for the two transitions involved in the multilevel coherence.[9]

B. Temperature Dependence of Vibrational Dephasing.
1. Below the Glass Transition Temperature. The results of temperature-dependent photon echo experiments on the T_{1u} mode of $W(CO)_6$ in the three glass-forming liquids[15] are shown in Figure 3. The homogeneous line width, $\Gamma = 1/\pi T_2$, is shown, derived from fits with the photon echo signal decaying exponentially as eq 9b. Data on 2-MTHF were taken from 10 to 120 K, at which point the decay rate exceeded the instrument response. With the same time resolution, echo data in the other liquids were observable up to room temperature. The data on DBP (discussed above) were taken with 0.7 ps pulses. At temperatures above 150 K, where the decays are fast, the presence of the beats on the decays introduced some uncertainty into the determination of the vibrational line width for the $v = 0 \rightarrow 1$ level.

The temperature dependence of the homogeneous line width increases monotonically in the glass. Above the glass transition temperature, the line widths in 2-MTHF ($T_g = 86$ K) and 2-MP ($T_g = 80$ K) rise rapidly in a manner that appears thermally activated. In 2-MP, the rapid increase in dephasing rate slows above 130 K and becomes temperature independent. In DBP ($T_g = 169$ K), the line width appears to decrease with temperature above 200 K although the error bars are large enough that it is possible that the temperature dependence is essentially flat.

The temperature dependences of the homogeneous vibrational line widths in the three glasses are compared using a reduced variable plot[15] in Figure 4a. Although the absolute line widths for each system may vary, this is a reflection of the strength of coupling of the transition dipole to the bath and can be removed by normalization to the line width at the glass transition temperature. Likewise, using a reduced temperature, T/T_g, allows thermodynamic variables that contribute to determining the glass transition to be normalized. Such normalization allows comparison of the functional form of the temperature dependences, independent of differences in T_g and coupling strengths. Figure 4a shows that the temperature dependences of the homogeneous line widths are identical in the three glasses and are well described by a power law of the form

$$\Gamma(T) = \Gamma_0 + AT^\alpha \qquad (11)$$

The offset at 0 K, Γ_0, represents the line width due to the low-temperature vibrational lifetime. A fit to eq 11 for all temperatures below the glass transition is shown in Figure 4a and yields an exponent of $\alpha = 2.1 \pm 0.2$ and $\Gamma_0/\Gamma(T_g) = 0.24 \pm 0.02$. At temperatures above $\sim 1.2 T_g$, the line widths of the

J. Phys. Chem., Vol. 99, No. 36, 1995 **13315**

three liquids diverge from one another. The data from each sample were individually fit to eq 11. The results demonstrate that the temperature dependence is well described by a power law of the form T^2.

Because of its high glass transition temperature, DBP allows the largest temperature range over which to observe the power law. The data are presented in two ways in Figure 4b. The solid circles are the data and the line through them is a fit to eq 11. To show more clearly the power law temperature dependence, the open circles are the data with the low-temperature line width, Γ_0 (corresponding to a vibrational lifetime, $T_1(0\ K)$ = 33 ps) subtracted out. The line through the data is T^2. It can be seen that the power law describes the data essentially perfectly over a change of line width of ~500 from 10 to 200 K.

Electronic excited state dephasing in glasses[35] is generally interpreted within the model of two-level systems, initially proposed to describe the anomalous low-temperature heat capacities of glasses.[36,37] In this model, pure dephasing dynamics are described in terms of phonon-induced tunneling between structural potential wells. At temperatures above the Debye frequency of the glass, a power law of T^2 is theoretically predicted for dephasing due to two-phonon (Raman) scattering processes between potential wells.[35] Huber has pointed out that in glasses this temperature dependence is expected above an effective Debye temperature, which can often be 2−10 times lower than the true Debye temperature.[38] The $\alpha \approx 2$ temperature dependence is almost universally seen for the high-temperature dephasing of electronic transitions in a variety of glasses.[35]

Figure 4 demonstrates that the underlying temperature dependence of the vibrational homogeneous dephasing measured with the IR photon echoes is, in fact, identical in the three glasses studied here and is described by a power law T^2. The results presented here were the first to examine the temperature dependence of vibrational dephasing in organic glasses.[7−9,15] The observation of the T^2 dependence suggests that these glasses are in the high-temperature limit above 10 K and that temperature dependence of vibrational pure dephasing may be universal in all high-temperature glasses. This would be consistent with dephasing caused by Raman (two quantum) phonon scattering.

2. Above the Glass Transition Temperature. The temperature dependences of the rate of dephasing in 2-MP and 2-MTHF (Figure 3) both show a gradual increase with temperature until shortly after the glass transition, at which point a rapid increase in dephasing is observed. Coincidentally, the glass transition temperatures for 2-MP (T_g = 80 K) and 2-MTHF (T_g = 86 K) are very similar. The temperature dependence in DBP shows no dramatic change near 90 K, demonstrating that the mechanism is dependent on the solvent; the dephasing is dictated by the particular temperature-dependent dynamics of the solvent. At temperatures above the glass transition, the normalized line widths in the three liquids diverge from each other, indicating the emergence of distinct relaxation processes in the various liquids.[15]

In 2-MTHF and in 2-MP below 150 K, the echo decays yield a homogeneous line width that is much narrower than the width of the absorption spectrum. These results demonstrate that the vibrational lines of these systems are inhomogeneously broadened in the glass and supercooled liquid. As shown below, in 2-MP the W(CO)$_6$ T$_{1u}$ line becomes homogeneously broadened at room temperature. However, in DBP the line is clearly inhomogeneous at all temperatures. The homogeneous line width at 300 K is ~1 cm^{-1}, while the absorption spectrum line

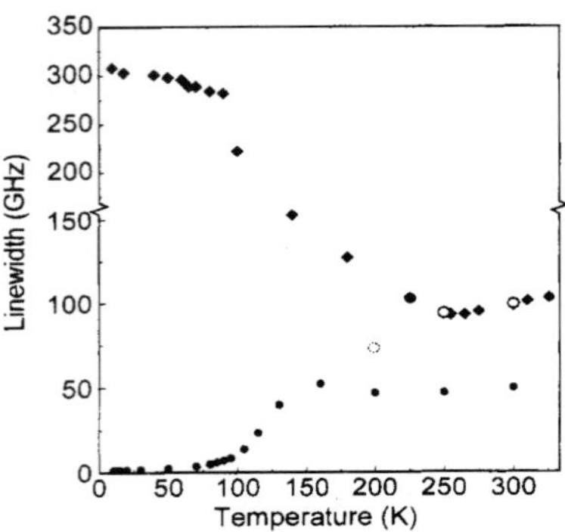

Figure 5. Temperature dependence of the homogeneous and absorption line widths for the T$_{1u}$ mode of W(CO)$_6$ in 2-MP. The homogeneous line widths (solid circles) are taken from Figure 4. The "echo" data at 250 and 300 K are actually free induction decays that match the observed absorption line (open circles), showing the spectra are homogeneously broadened. At 200 K (dotted circle), the homogeneous and inhomogeneous contributions to the absorption line are of equal magnitude. Below 200 K, the homogeneous line width is much narrower than the inhomogeneous width.

width is 26 cm^{-1}. This is the first conclusive evidence for intrinsic inhomogeneous broadening of a vibrational line in a room temperature liquid. Thus, even in room temperature liquids, it is not safe to assume that dynamics can be obtained by taking a vibrational spectrum and analyzing it assuming it is homogeneously broadened.

C. Contributions to the Homogeneous Line Shape in 2-MP. *1. Transition to a Homogeneously Broadened Line.* The measured homogeneous vibrational line width obtained from infrared photon echo measurements (Figure 3) assumed that the echo decays as eq 9b. This equation is valid in the inhomogeneous limit, when the width of the inhomogeneous distribution of homogeneous lines far exceeds the homogeneous line width. This inhomogeneous limit implies a separation of time scales exists between the fast fluctuations of homogeneous dephasing and long time-scale inhomogeneous structural evolution. Such is the case below 150 K, where the echo decays yield a homogeneous line width that is much narrower than the width of the absorption spectrum. At these temperatures, the data demonstrate with certainty that the T$_{1u}$ vibrational line in 2-MP is substantially inhomogeneously broadened.

Above 150 K, the homogeneous line width begins to approach the measured absorption line width,[14,15] as illustrated in Figure 5. Under these conditions, the time scale for the rephasing of the echo pulse is shortened by the polarization decay due to homogeneous dephasing. This causes the echo to rephase at times between τ and 2τ. Thus, the echo signal decays at a slower rate than the $4/T_2$ given by eq 9b.[39,40] As the homogeneous dephasing time T_2 decreases, the rephasing of the echo is shifted to shorter times. In the limit that the absorption line is homogeneously broadened, a free induction decay (fid) will be observed along the echo phase matching direction. If the homogeneous line is a Lorentzian, then an exponential decay with a decay constant of $2/T_2$ will be observed.[40]

The data above 150 K reflect this transition to a homogeneously broadened line. If near room temperature the measured echo data are actually fid's from a homogeneously broadened line, the decay constants $2/T_2$ yield a line width given by the open circles in Figure 5. The FID line widths match the measured absorption line widths exactly, demonstrating that the line is homogeneously broadened for temperatures ≥250 K.

13316 *J. Phys. Chem., Vol. 99, No. 36, 1995*

Tokmakoff et al.

Notice that the absorption line width, which narrows for temperatures up to 200 K, actually broadens slightly for higher temperatures, in a manner that precisely follows the echo data. Without experimentally time-resolving the rephasing of the echo pulse, it is not possible to determine from an echo experiment alone whether the observable is a true echo decay, a fid, or an intermediate case. However, a comparison to the absorption spectrum provides a test. Since treating the point at 160 K as an fid still results in a line width that is narrower than the absorption line, this point corresponds to a true photon echo. The point at 200 K represents the transition between a homogeneous and an inhomogeneous absorption line shape.

In liquids, the distinction between homogeneous and inhomogeneous broadening is a matter of definitions and time scales. A photon echo experiment measures the homogeneous line shape, which arises from fast fluctuations of the medium that cause rapid fluctuations of the vibrational energy levels. In a liquid, the homogeneous line width can be much narrower than the inhomogeneous line width even at room temperature, as shown above for $W(CO)_6$ in DBP. The slower dynamic processes, which are essentially static on the time scale of the homogeneous dephasing, appear as part of the quasi-static inhomogeneous background. This quasi-static distribution of vibrational energies is rephased in a photon echo experiment and does not contribute to the homogeneous line width.

2. Components of the Homogeneous Line. Pump–probe studies of the population dynamics of the T_{1u} mode of $W(CO)_6$ in 2-MP have observed dynamics that contribute to the infrared absorption line shape.[14,15] In addition to pure dephasing (energy fluctuations), population relaxation and orientational relaxation contribute to the homogeneous line width. Pump–probe measurements of the population dynamics in 2-MP and other solvents were performed.[14,15,17] A detailed analysis of the population dynamics and their temperature dependences for the 2-MP data is given in ref 14. The data taken with parallel pump and probe polarizations are biexponential at all temperatures. The long decay component is due to vibrational population relaxation (lifetime) and decays with an exponential decay time of T_1. The data display a counterintuitive temperature dependence, with the rate of relaxation decreasing as the temperature increases. This phenomenon has also been observed in crystallizing and other glass-forming liquids[15,17] and is discussed in some detail below.[16] The fast component in the pump–probe data is due to orientational relaxation of the chromophore and decays exponentially at a rate of $(6D_{or} + 1/T_1)$. With magic angle probing to eliminate orientational relaxation effects, the fast component vanished from the decay, leaving only the long component.

Equation 9 gives a Lorentzian line shape, and contributions to the full line width at half-maximum are additive, given in eqs 4 and 5. Using the pump–probe measurements of the orientational diffusion constant D_{or} and the population relaxation time T_1,[14,15] the contribution due to pure dephasing can be determined from the homogeneous line width. (There are a number of subtle issues involved in considering the influence of orientational relaxation on the photon echo decay. A full discussion of these in relation to the material presented below is given in ref 15.)

Figure 6 displays the decomposition of the temperature dependence of the homogeneous vibrational line width into its three dynamic components.[15] At low temperatures, the lifetime is the dominate contribution. At high temperatures, pure dephasing is the dominate contribution. Only at intermediate temperatures does orientational relaxation make a substantial contribution. The line width contribution from lifetime broad-

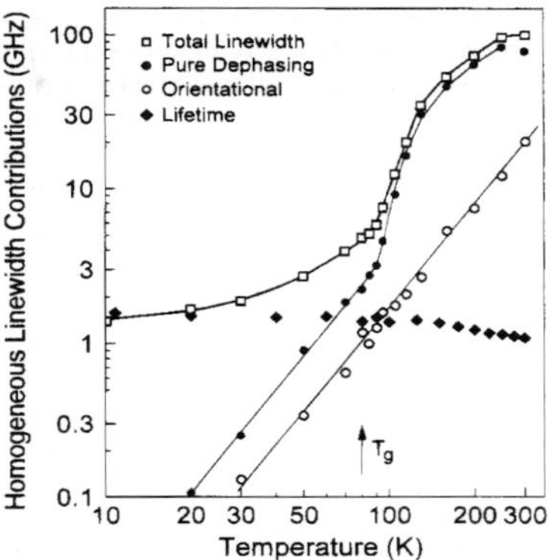

Figure 6. A log−log plot of the dynamic contributions to the homogeneous vibrational line width of the T_{1u} mode of $W(CO)_6$ in 2-MP: total line width, squares; lifetime contribution, diamonds; pure dephasing contribution, solid circles; orientational contribution, open circles. The vibrational lifetime dominates the homogeneous line width at the lowest temperatures and is only mildly temperature dependent. At high temperatures, pure dephasing dominates the homogeneous line width. Orientational relaxation never dominates but makes significant contributions to the line width at intermediate temperatures. The orientational contribution, shown with a T^2 power law line, is continuous over all temperatures. The temperature dependence of the pure dephasing is the same as the orientational relaxation in the glass. The line through the pure dephasing data is a fit to eq 13 with $\alpha = 2.2$ and $T_0^* = 80$ K. Error bars on the total line width, lifetime, and pure dephasing are approximately the size of the symbols at most temperatures. The orientational error bars vary from the size of the symbols at high temperature to $\pm50\%$ at 30 K.

ening remains significant until $\sim$50 K. Above 50 K, the contributions to the line width from pure dephasing and orientational relaxation dominate. Orientational relaxation does not dominate in any temperature range but makes its largest percentage contribution around 100 K. By 100 K, the pure dephasing is a substantially larger contribution than either the lifetime or the orientational relaxation. Above $\sim$150 K, pure dephasing is the overwhelmingly dominate component of the homogeneous line width.

At low temperature, where the contributions from pure dephasing and orientational relaxation are negligible, the contribution to the line width from the lifetime (from pump–probe data) and the line width determined from the decay of the echo are equal within error. This is the expected low-temperature limit for the homogeneous vibrational line width where processes caused by thermal fluctuations disappear, and only lifetime broadening is possible.

From low temperature to slightly above T_g, both pure dephasing and orientational relaxation have power law temperature dependences. The earlier discussion of the combined dynamics in 2-MP and the other two solvents showed a T^2 temperature dependence when the lifetime contribution was taken out. Here, the contributions from both pure dephasing and orientational relaxation are shown to follow the same T^2 power law behavior in the glass. Thus, the total temperature dependence, excluding T_1, is T^2, as shown in Figure 4. In addition, the power law observed for the orientational relaxation in the glass is observed to continue into the liquid. In Figure 6, a power law fit of $\alpha = 2.2 \pm 0.2$ is shown. The orientational dynamics are independent of the glass transition and distinctly nonhydrodynamic.[15] This is consistent with the results of pump–probe experiments.[14]

J. Phys. Chem., Vol. 99, No. 36, 1995 **13317**

In Figure 6 it is seen that there is a rapid increase in pure dephasing beginning slightly above the glass transition temperature. This implies that an additional mechanism or a change in the nature of the mechanism for pure dephasing turns on, and it is linked to the glass-to-liquid transition. The onset of dynamic processes near the glass transition is often described with a Vogel–Tammann–Fulcher (VTF) equation[41–43]

$$\tau = \tau_0 \exp(B/(T - T_0)) \tag{12}$$

This equation describes a process characterized by a time, τ, with a temperature-dependent activation energy that diverges at a temperature, T_0, below the nominal glass transition temperature. T_0 can be linked thermodynamically to an "ideal" glass transition temperature that would be measured with an ergodic observable.[41] This equation describes the temperature dependence of the viscosity of 2-MP well and gives $T_0 = 59$ K.[15]

If the VTF equation applies to pure dephasing near and above the glass transition, then the full temperature dependence would be the sum of the low-temperature power law plus a VTF term. To test this idea, the temperature dependence of pure dephasing was fit to

$$\Gamma^*(T) = A_1 T^\alpha + A_2 \exp(-B/(T - T_0^*)) \tag{13}$$

for all temperatures below 300 K, and is shown as the line through the pure dephasing data in Figure 6. The fit describes the entire temperature dependence exceedingly well but yields a reference temperature of $T_0^* = 80$ K. This reference temperature matches the laboratory glass transition temperature T_g exactly but is not related to the ideal glass transition temperature T_0. We can thus infer that the onset of the dynamics that cause the rapid increase in homogeneous dephasing in 2-MP is closely linked with the onset of structural processes near the laboratory glass transition temperature. This may be a manifestation of the short time scale of the measurement and a reflection of the nonergodicity of the system.

D. Vibrational Lifetimes Measured with IR Pump–Probe Experiments. The population dynamics of the vibrations of a polyatomic solute molecule in a polyatomic solvent can involve the internal vibrational modes of the solute, the vibrational modes of the solvent, and the low-frequency continuum of solvent modes.[17,18] An initially excited high-frequency vibrational mode of a solute molecule can relax by transferring vibrational energy to a combination of lower frequency internal vibrations and solvent vibrations. Vibrational energy can also be transferred to modes of higher frequency, but this process is generally less efficient than comparable downward pathways.[18] In general, a combination of lower frequency vibrations will not match the initial vibrational frequency. Therefore, one or more quanta of the continuum will also be excited (or annihilated) to make up for the mismatch in the vibrational frequencies and conserve energy. The low-frequency solvent continuum can be described in terms of instantaneous normal modes (INM).[44–48] Not all of the modes are bound, so that INM's have both real and imaginary frequencies. The imaginary frequency modes are related to the structural evolution of the liquid.[49]

Vibrational relaxation involves a cubic or higher order anharmonic process. The "order" of the process refers to the number of quanta involved in the relaxation. In the simplest cubic anharmonic process, the initial excited vibration is annihilated, a lower frequency internal mode or solvent mode is excited, and a phonon (INM) is excited to conserve energy. For a high-frequency mode to relax by a cubic process, there

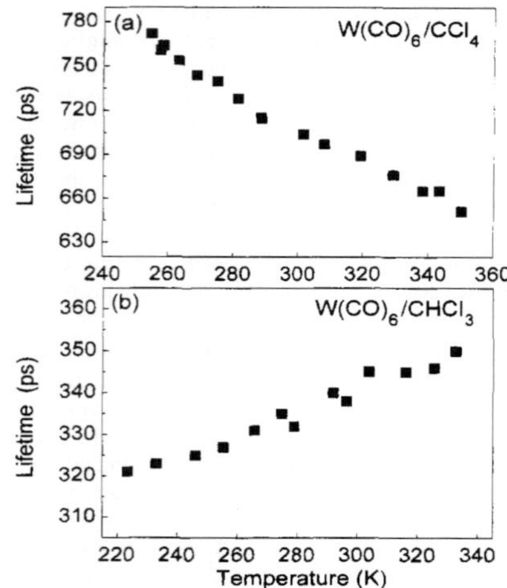

Figure 7. (a) Temperature dependence of the vibrational relaxation time, T_1, for the T_{1u} CO stretching mode of $W(CO)_6$ in CCl_4. The data were taken from the melting point to the boiling point. (b) Temperature dependence of T_1 in $CHCl_3$. The data were taken from the melting point to the boiling point. The lifetime actually becomes slower as the temperature is increased (inverted temperature dependence).

must be another high-frequency mode close enough in energy for the energy mismatch to fall within the phonon bandwidth. In a quartic or higher order process, the initial vibration is annihilated, two or more lower frequency vibrations are created, and one or more phonons are created to conserve energy. Unless there is a coincidence or Fermi resonance in which energy can be conserved by the creation and annihilation of discrete vibrational modes alone, at least one mode of the low-frequency continuum of states will be involved in vibrational population dynamics.

The vibrational relaxation times (T_1) as a function of temperature for T_{1u} mode of $W(CO)_6$ in CCl_4[17] are shown in Figure 7a. Although the change in the decay with temperature is not large, given the excellent signal-to-noise ratio of the data,[17] the differences are readily discernible. From the melting point to the boiling point, the lifetimes decrease monotonically by 19%. The values range from 775 ps at the melting point (250 K) to 650 ps at the boiling point (350 K).

The results of the pump–probe lifetime measurements in $CHCl_3$[17] are shown in Figure 7b. The decay times are substantially different from those in CCl_4 although the solvents differ only by substitution of a hydrogen for a chlorine. Even more significant is that the basic nature of the temperature dependence is different. Vibrational lifetimes for the T_{1u} mode of $W(CO)_6$/ $CHCl_3$ actually become *longer as the temperature is increased,* changing by 9% from 322 ps at the melting point (210 K) to 350 ps at the boiling point (334 K). This is analogous to the temperature dependence of T_1 seen in 2-MP (see Figure 6).

$W(CO)_6$ has a variety of internal modes that are lower in frequency than the initially excited T_{1u} CO stretching frequency.[16,17] Both CCl_4 and $CHCl_3$ have a number of lower energy vibrational modes than the initially excited mode at $\sim$1980 cm^{-1}.[16,17] $CHCl_3$ also has a CH stretching mode at higher energy, $\sim$3020 cm^{-1}, which is too high in energy to participate in the vibrational dynamics. The simplest relaxation pathway would involve the deposition of the initial vibrational energy into a single lower frequency vibration and one phonon.

13318 *J. Phys. Chem., Vol. 99, No. 36, 1995*

Tokmakoff et al.

However, as shown by INM calculations, neither CCl_4 nor $CHCl_3$ has INM bandwidths that extend past 180 cm^{-1}.[16] Therefore, in $CHCl_3$, it is necessary for the initial vibration to relax into at least two vibrations and a phonon and in CCl_4 to relax into at least three vibrations and a phonon. This difference most likely is responsible for the generally longer relaxation times observed in CCl_4. It is possible that it is necessary to excite more than one phonon to conserve energy. For simplicity, the following discussion assumes that only one phonon is involved, although this will not influence the conclusions that are reached.

The rate of vibrational relaxation, K, of the initially excited mode is generally described by Fermi's Golden Rule[18,26,50]

$$K = (2\pi/\hbar)\sum_{r,r'} \varrho_{r,r'}|\langle\sigma',r'|V|\sigma,r\rangle|^2 \qquad (14)$$

In eq 14, σ and σ' denote the initial and final state of the initially excited vibration (the T_{1u} mode in this case), while r and r' refer to the receiving (or reservoir) modes. The ket $|\sigma,r\rangle$ is the initial state, described by thermal occupation numbers of the various modes of the system, in addition to unit occupation of the state initially excited by the IR pump. The bra $\langle\sigma',r'|$ is the final state with the initially populated state having occupation number 0 after relaxation and other states having increased occupation numbers. ϱ is the density of states of the reservoir modes for the relaxation step and is often written as a delta function to denote the energy conservation requirement. The summation in eq 14 denotes the fact that the true relaxation rate is a sum over contributions from all possible pathways;[18] however, in the following discussion we will describe relaxation through a single anharmonic path involving one phonon.

In eq 14, V is the potential that describes the system–reservoir interaction.[18] The potential energy surface for the system and reservoir is expanded about the potential minima of the various INM coordinates. $V^{(i)}$ is the *i*th matrix element that describes the interactions which couple *i* modes. The anharmonic terms, $i \geq 3$, govern relaxation processes involving the coupling of multiple vibrational modes. The magnitudes of the matrix element expansion coefficients decrease with order, leading to decreased relaxation rates with higher order processes.

Given the vibrational energies of the solutes and the solvents, the relaxation of the initially excited mode results in the excitation of at least two other vibrations and a phonon. For such a process, the anharmonic coupling matrix element is at least fourth order, or quartic. For the following discussion, we use one of the quartic relaxation paths that contributes to the relaxation of the T_{1u} CO stretch in $CHCl_3$.[16] The CO stretch relaxes by transferring energy to the W–C–O bending motion of the $W(CO)_6$, the H–C–Cl bending motion of the $CHCl_3$, and a ~160 cm^{-1} solvent phonon.

The quartic anharmonic matrix element $\langle V^{(4)}\rangle$ contains the magnitude of the quartic anharmonic coupling term, $|V^{(4)}|$, and combinations of raising and lowering operators that describe the anharmonic relaxation step.[17,18] If the operator a annihilates the initially excited vibration and the operator b describes the change in the reservoir modes A, B, and the phonon, then the interaction is described by

$$a(b_A + b_A^+)(b_B + b_B^+)(b_{ph} + b_{ph}^+) \qquad (15)$$

This interaction leads to seven possible quartic relaxation pathways,[18] some of which are unphysical. The relaxation pathway being considered, a simple cascade process in which the energy relaxes only to lower energy modes, is described by one term arising from eq 15, $ab_A^+b_B^+b_{ph}^+$. Once substituted into

eq 14, a raising operator brings out a factor of $\sqrt{n+1}$ and a lowering operator brings out a factor of $\sqrt{n}$, where n is the occupation number of the particular mode involved in the fourth-order process. This allows eq 14 to be written as

$$K = (2\pi/\hbar)\varrho_{ph}|\langle V^{(4)}\rangle|^2 (n_A + 1)(n_B + 1)(n_{ph} + 1) \qquad (16)$$

where n is the thermally averaged occupation number,

$$n_i = (\exp(\hbar\omega_i/kT) - 1)^{-1} \qquad (17)$$

ω_i is the frequency of the vibrational or phonon mode. We take $\varrho_{ph} = \langle\varrho(\omega_i)\rangle$; i.e., the "phonon" density of states is given by the ensemble-averaged density of INM.

Clearly, if the reservoir modes are high frequency ($\hbar\omega \gg kT$ and, therefore, $n \ll 1$), such as the discrete vibrational modes of the solute and solvent, eq 16 is

$$K = (2\pi/\hbar)\varrho_{ph}|\langle V^{(4)}\rangle|^2(n_{ph} + 1) \qquad (18)$$

Although the discussion of the derivation of this relaxation rate expression has been qualitative, the same results can be shown rigorously.[18] In general, the expression for the relaxation rate along a given one-phonon, *i*th-order anharmonic pathway is given by the product of the phonon density of states, the magnitude of the anharmonic coupling matrix element squared, and occupation number factors for the receiving modes. If a reservoir mode is created in the relaxation step, it contributes a factor of $(n + 1)$, whereas if it is annihilated, it contributes a factor of n. This is a simple, yet rigorous, method for describing even complex relaxation pathways.

Considering only the occupation number in eq 18, K should become larger, and the observed decay times should become shorter as the temperature is increased. If more than one thermally occupied phonon were involved in the relaxation pathway or if a vibrational occupation number changes significantly, the temperature dependence would be even steeper. If the phonon occupation number is the only factor responsible for the temperature dependence, then for $\hbar\omega \ll kT$, K would increase linearly with temperature. For $\hbar\omega \gg kT$, K goes as $\exp(-\hbar\omega/kT)$ if a phonon is annihilated as part of the relaxation process, and K goes as $1 + \exp(-\hbar\omega/kT)$ if a phonon is created. In either limit and for intermediate situations, the temperature dependence of the occupation number(s) will always yield a decrease in the vibrational lifetime with increasing temperature. Only near $T \approx 0$ K (only phonon emission processes are possible) where $1 > n_{ph}$ will the temperature dependence vanish, and the vibrational lifetime will become temperature independent. Above $T \approx 0$ K, an inverted temperature dependence cannot be explained by considering occupation numbers, regardless of the pathways or number of modes involved.[18]

The experimental data for $W(CO)_6$ in $CHCl_3$ displayed in Figure 7b, with its inverted temperature dependence, show that the temperature dependence is influenced by temperature-dependent factors in addition to the occupation numbers. A competition among these factors will yield the observed temperature dependence.[16,17] The factors that cause the $W(CO)_6$/$CHCl_3$ vibrational lifetimes to become longer as the temperature is increased are operative in $W(CO)_6$ in CCl_4 and presumably other systems, even though they do display vibrational lifetimes that decrease with increasing temperature.[16]

In examining eq 16 or 18, there are two other factors besides the occupation numbers that can contribute to the temperature dependence of the decay constant. They are the density of states, ϱ, and the magnitude of the anharmonic coupling matrix element, $V^{(i)}$. First, consider the density of states.[16,17] The

Feature Article

densities of states of the internal vibrational modes of $W(CO)_6$ do not change with temperature. However, the densities of states of the solvent modes will change with temperature.

To gain insights into the temperature dependences of the vibrational lifetimes in the two solvents, the temperature-dependent low-frequency INM spectra, $\langle \varrho(\omega) \rangle$, of CCl_4 and $CHCl_3$ were calculated.[16] The calculations employed a detailed potential that included intermolecular and intramolecular components. While the low-frequency intermolecular modes are of interest here, the potential is able to do a reasonable job of reproducing the vibrational spectrum of the liquids as well. The use of the full potential proved important. Calculations with an accurate Lennard-Jones (LJ) potential do not generate the high-frequency "rotational" part of the INM spectrum even though the LJ potential yields the correct melting point of the crystal. The real and imaginary components of the INM spectrum were calculated at three temperatures: near the melting point, at room temperature, and near the boiling point for both solvents. As the temperature is increased, there is a small decrease in the density of states across most of the real part of the spectra with a corresponding increase in the imaginary part. The changes in the two solvents were very similar, and detailed analysis[16] showed that the changes were far too small to account for the observed inverted temperature dependence.

These results suggest that a significant temperature dependence exists for the magnitude $V^{(i)}$. The temperature dependence of this term must be opposite to that of the phonon occupation numbers and must be sufficiently large to offset the change in occupation numbers to yield the inverted temperature dependence. Although $V^{(i)}$ is not explicitly temperature dependent, it can vary with density. As the density decreases with increasing temperature, intermolecular separations are increased on average. The important point is that the region of the intermolecular potential that is sampled changes, and therefore the anharmonic coupling matrix elements can change. If this causes the matrix elements to become smaller with increasing temperature, then this decrease will work to offset the increase in occupation numbers. If this effect is sufficiently large, the observed inverted temperature dependence of $W(CO)_6$/$CHCl_3$ would be observed. This is the most likely explanation for the inverted temperature dependence.

Three factors were discussed that will influence the vibrational lifetime of the metal carbonyls studied and, presumably, other systems as well. They are (1) the temperature-dependent occupation number of the phonon(s) or other very low-frequency modes excited in the relaxation of the initially pumped vibration, (2) the temperature dependence of the liquid's density of INM states, and (3) the temperature dependence of the magnitude of the anharmonic coupling matrix elements responsible for vibrational relaxation. The results briefly discussed here demonstrate that understanding the temperature dependence of vibrational lifetimes requires consideration of the interplay of the temperature dependences of the three factors. Observation of a normal (noninverted) temperature dependence does not indicate that all three factors are not involved. Fitting a normal temperature dependence as a simple activated process can yield a highly flawed indication of the frequency of the low-frequency mode that is involved.

IV. Concluding Remarks

In principle, analysis of a vibrational line shape can provide a great deal of information on dynamics and intermolecular interactions in condensed matter systems. However, to completely characterize an infrared line shape, a combination of experimental methods is necessary to elucidate each of the dynamic and static contributions to the line. Picosecond infrared photon echo experiments were used to measure the homogeneous line shape and remove inhomogeneity. Infrared pump–probe experiments were used to measure the vibrational lifetime and orientational relaxation. The total vibrational line shape was determined by absorption spectroscopy. Together, these experiments yield a complete characterization of the dynamics that make up the homogeneous line and the extent of inhomogeneity of the infrared absorption line.

The development of picosecond vibrational photon echo experiments to examine condensed matter systems represents a significant extension of the field of vibrational spectroscopy. Recently, we have also applied other IR nonlinear experimental methods to study vibrations. We have used stimulated photon echoes and transient grating experiments. All of these methods are types of four-wave mixing experiments. The application of four-wave mixing experiments to the study of electronic states and nonresonant phenomena in the visible part of the spectrum has had explosive growth in the 1980s and 1990s. Vibrational spectroscopy has always gained importance because of its selectivity, i.e., its ability to look at specific, well-defined mechanical degrees of freedom of molecules. The advent of IR vibrational four-wave mixing experiments, such as the photon echo experiments described here, will provide important tools to greatly increase our understanding of the dynamics and intermolecular interactions in molecular systems of interest in chemistry, biology, and materials science.

Acknowledgment. The authors thank Professor Alan Schwettman and Professor Todd Smith of the Department of Physics, Stanford University, and their groups for the opportunity to use the Stanford Free Electron Laser. This work was supported by the National Science Foundation (DMR93-22504), the Office of Naval Research (N00014-92-J-1227), the Medical Free Electron Laser Program (N00014-91-C-0170), and the Air Force Office of Scientific Research (F49620-94-1-0141).

References and Notes

(1) Gordon, R. G. *J. Chem. Phys.* **1965**, *43*, 1307.
(2) Gordon, R. G. *Adv. Magn. Reson.* **1968**, *3*, 1.
(3) Berne, B. J. In *Physical Chemistry: An Advanced Treatise*; Henderson, D., Ed.; Academic Press: New York, 1971; Vol. VIIIB.
(4) Yarwood, J. *Annu. Rep. Prog. Chem., Sect. C* **1979**, *76*, 99.
(5) Rothschild, W. G. *Dynamics of Molecular Liquids*; John Wiley and Sons: New York, 1984.
(6) Loring, R. F.; Mukamel, S. *J. Chem. Phys.* **1985**, *83*, 2116.
(7) Zimdars, D.; Tokmakoff, A.; Chen, S.; Greenfield, S. R.; Fayer, M. D. *Phys. Rev. Lett.* **1993**, *70*, 2718.
(8) Tokmakoff, A.; Zimdars, D.; Sauter, B.; Francis, R. S.; Kwok, A. S.; Fayer, M. D. *J. Chem. Phys.* **1994**, *101*, 1741.
(9) Tokmakoff, A.; Kwok, A. S.; Urdahl, R. S.; Francis, R. S.; Fayer, M. D. *Chem. Phys. Lett.* **1995**, *234*, 289.
(10) Vanden Bout, D.; Muller, L. J.; Berg, M. *Phys. Rev. Lett.* **1991**, *67*, 3700.
(11) Muller, L. J.; Vanden Bout, D.; Berg, M. *J. Chem. Phys.* **1993**, *99*, 810.
(12) Inaba, R.; Tominaga, K.; Tasumi, M.; Nelson, K. A.; Yoshihara, K. *Chem. Phys. Lett.* **1993**, *211*, 183.
(13) Vanden Bout, D.; Freitas, J. E.; Berg, M. *Chem. Phys. Lett.* **1994**, *229*, 87.
(14) Tokmakoff, A.; Urdahl, R. S.; Zimdars, D.; Kwok, A. S.; Francis, R. S.; Fayer, M. D. *J. Chem. Phys.* **1995**, *102*, 3919.
(15) Tokmakoff, A.; Fayer, M. D. *J. Chem. Phys.*, in press.
(16) Moore, P.; Tokmakoff, A.; Keyes, T.; Fayer, M. D. *J. Chem. Phys.*, in press.
(17) Tokmakoff, A.; Sauter, B.; Fayer, M. D. *J. Chem. Phys.* **1994**, *100*, 9035.
(18) Kenkre, V. M.; Tokmakoff, A.; Fayer, M. D. *J. Chem. Phys.* **1994**, *101*, 10618.
(19) Hahn, E. L. *Phys. Rev.* **1950**, *80*, 580.
(20) Kurnit, N. A.; Abella, I. D.; Hartmann, S. R. *Phys. Rev. Lett.* **1964**, *13*, 567. Abella, I. D.; Kurnit, N. A.; Hartmann, S. R. *Phys. Rev.* **1966**, *141*, 391.

13320 *J. Phys. Chem., Vol. 99, No. 36, 1995*   Tokmakoff et al.

(21) Farrar, T. C.; Becker, D. E. *Pulse and Fourier Transform NMR*; Academic Press: New York, 1971. Skinner, J. L.; Andersen, H. C.; Fayer, M. D. *J. Chem. Phys.* **1981**, *75*, 3195.

(22) Chesnoy, J.; Gale, G. M. *Adv. Chem. Phys.* **1988**, *70*, 297.

(23) Califano, S.; Schettino, V.; Neto, N. *Lattice Dynamics of Molecular Crystals*; Springer-Verlag: Berlin, 1981.

(24) Velsko, S.; Hochstrasser, R. M. *J. Phys. Chem.* **1985**, *89*, 2240.

(25) Schweizer, K. S.; Chandler, D. *J. Chem. Phys.* **1982**, *76*, 2296.

(26) Oxtoby, D. W. *Annu. Rev. Phys. Chem.* **1981**, *32*, 77.

(27) Narasimhan, L. R.; Littau, K. A.; Pack, D. W.; Bai, Y. S.; Elschner, A.; Fayer, M. D. *Chem. Rev.* **1990**, *90*, 439.

(28) Berne, B. J.; Pecora, R. *Dyanmic Light Scattering*; R. E. Krieger Publishing: Malabar, FL, 1990.

(29) Bratos, S.; Leicknam, J. C. *J. Chem. Phys.* **1994**, *101*, 4536.

(30) Mukamel, S.; Loring, R. F. *J. Opt. Soc. B* **1986**, *3*, 595.

(31) Loring, R. F.; Mukamel, S. *Chem. Phys. Lett.* **1985**, *114*, 426.

(32) Tokmakoff, A.; Marshall, C. D.; Fayer, M. D. *J.O.S.A. B* **1993**, *10*, 1785.

(33) Yan, Y. J.; Mukamel, S. *J. Chem. Phys.* **1991**, *94*, 179.

(34) Arrivo, S. M.; Dougherty, T. P.; Grubbs, W. t.; Heilweil, E. J. *Chem. Phys. Lett.* **1995**, *235*, 247.

(35) Macfarlane, R. M.; Shelby, R. M. *J. Lumin.* **1987**, *36*, 179.

(36) Anderson, P. W.; Halperin, B. I.; Varma, C. M. *Philos. Mag.* **1972**, *25*, 1.

(37) Phillips, W. A. *J. Low Temp. Phys.* **1972**, *7*, 351.

(38) Huber, D. L. *J. Non-Cryst. Solids* **1982**, *51*, 241.

(39) Cho, M.; Fleming, G. R. *J. Chem. Phys.* **1993**, *98*, 2848.

(40) Joo, T.; Albrecht, A. C. *Chem. Phys.* **1993**, *176*, 233.

(41) Angell, C. A. *J. Phys. Chem. Solids* **1988**, *49*, 863.

(42) Angell, C. A. *J. Phys. Chem.* **1982**, *86*, 3845.

(43) Fredrickson, G. H. *Annu. Rev. Phys. Chem.* **1988**, *39*, 149.

(44) Seeley, G.; Keyes, T. *J. Chem. Phys.* **1989**, *91*, 5581.

(45) Xu, B. C.; Stratt, R. M. *J. Chem. Phys.* **1990**, *92*, 1923.

(46) Wu, T. M.; Loring, R. M. *J. Chem. Phys.* **1992**, *97*, 8568.

(47) Moore, P.; Keyes, T. *J. Chem. Phys.* **1993**, *100*, 6709.

(48) Cho, M.; Fleming, G. R.; Saito, S.; Ohmine, I.; Stratt, R. M. *J. Chem. Phys.* **1994**, *100*, 6672.

(49) Keyes, T. *J. Chem. Phys.* **1994**, *101*, 5081.

(50) Oxtoby, D. W. *Adv. Chem. Phys.* **1981**, *47*, 487.

JP951323Y

VOLUME 77, NUMBER 8   PHYSICAL REVIEW LETTERS   19 AUGUST 1996

Vibrational Echo Studies of Protein Dynamics

C. W. Rella,[1] Alfred Kwok,[2] Kirk Rector,[2] Jeffrey R. Hill,[3] H. A. Schwettman,[1] Dana D. Dlott,[3,*] and M. D. Fayer [2,*]

[1]*Stanford Free Electron Laser Center, Hansen Experimental Physics Laboratory, Stanford University,
Stanford, California 94305-4085*
[2]*Department of Chemistry, Stanford University, Stanford, California 94305*
[3]*School of Chemical Sciences, University of Illinois at Urbana-Champaign, Urbana, Illinois 61801*
(Received 10 April 1996)

The first picosecond infrared vibrational echo experiments on a protein, myoglobin-CO, are described. The experiments were performed at temperatures ranging from 60 to 300 K with a midinfrared free electron laser tuned to 1945 cm^{-1}. Below ~185 K, the pure dephasing, T_2^*, displays a power law temperature dependence, $T^{1.3}$. This behavior is reminiscent of that associated with the properties of low temperature glasses (<5 K) but is observed here at much higher temperatures. Above the solvent glass transition temperature, T_2^* is exponentially activated.   [S0031-9007(96)00882-4]

PACS numbers: 87.15.He

We present the first vibrational echo experiments performed on a protein. These experiments provide a new method for examining protein dynamics. The vibrational echo [1] is the vibrational analog of the spin echo of NMR [2] and the photon echo of visible and UV spectroscopy [3]. The advent of spin echo in 1950 ushered in a new dimension in magnetic resonance spectroscopy [2]. Like the spin echo, the vibrational echo is expanding the scope of infrared vibrational spectroscopy. The vibrational echo experiment gains its importance because it permits the homogeneous vibrational line shape, which is the result of dynamics, to be extracted from an inhomogeneously broadened vibrational spectrum.

The experiments were performed on the vibrational stretching mode of carbon monoxide (CO) bound to the active site of myoglobin (Mb). By combining vibrational echo measurements with vibrational pump-probe lifetime measurements, the pure dephasing time T_2^* is determined [1]. Previous optical coherence experiments performed on proteins examined the dephasing of electronic transitions at a few degrees K [4]. In contrast, the vibrational echo experiments make it possible to use optical coherence methods to study protein dynamics at physiologically relevant temperatures.

Myoglobin is used in the storage and transport of dioxygen (O_2) in muscle tissue [5]. It consists of a prosthetic group called protoheme [Fe(II)protoporphyrin IX] embedded in a protein. The protein modifies the chemical reactivity of the Fe binding site, allowing Mb to function properly in a biological setting [5].

When bound at the Mb active site, the CO frequency is strongly redshifted from its gas phase value and separated into four distinct bands, labeled A_0-A_3 in order of decreasing stretch frequency [6]. In principle, information about vibrational dynamics can be obtained from the Fourier transform of the linear absorption spectra. Finite lifetime and vibrational dephasing due to time dependent perturbations of the transition frequency lead to homogeneous broadening. In a disordered condensed matter sys-

tem such as Mb-CO, however, even a well resolved vibrational spectrum does not provide information on dynamics. The spectral line is inhomogeneously broadened by the distribution of protein conformations that result in a distribution of protein-CO interactions. The IR absorption spectrum is a convolution of the various dynamic and static contributions to the observed line shape. Inhomogeneous broadening cannot be eliminated with linear spectroscopy [7].

The infrared vibrational echo experiment is a time domain nonlinear method which can extract the homogeneous vibrational line shape from inhomogeneously broadened lines [1,8]. In 1964, the magnetic resonance spin echo method was extended to the visible optical regime as the photon echo [3]. Recently, vibrational echoes have been used to examine vibrational dynamics in liquids and glasses [1,8].

In the vibrational echo experiment, the sample is irradiated by an intense pulse which coherently drives the ensemble of CO oscillators. These oscillators lose phase coherence due to homogeneous and inhomogeneous broadening. A second intense pulse, incident at time τ after the first, causes a partial rephasing. The rephasing generates a macroscopic polarization at time 2τ, and, thus, the emission of a third pulse of light, the echo. The echo signal, which is emitted in a unique direction, is measured as a function of delay time τ.

An exponential vibrational echo decay corresponds to a Lorentzian line shape with a width, Γ, given by

$$\Gamma = \frac{1}{\pi T_2} = \frac{1}{\pi T_2^*} + \frac{1}{2\pi T_1}, \tag{1}$$

where T_2 is the homogeneous dephasing time determined from the echo decay constant, and T_1 is the vibrational lifetime determined from pump-probe experiments. Measurements of T_2 and T_1 permit the determination of T_2^*, the pure dephasing time. The vibrational echo decay signal $S(t)$ is given by

$$S(t) = S_0 e^{-4\tau/T_2}. \tag{2}$$

VOLUME 77, NUMBER 8   PHYSICAL REVIEW LETTERS   19 AUGUST 1996

The experiments were conducted using ps infrared pulses from the Stanford Free Electron Laser (FEL). The FEL pulse train consists of a macropulse of about 3 ms length repeating at 10 Hz, within which is contained a series of micropulses repeating at 11.8 MHz. Each micropulse has an energy of ~ 1 μJ and is nearly a transform-limited Gaussian 1.7 ps in duration. The laser was tuned to the A_1 band at 1945 cm^{-1}, and was actively stabilized to within $2 \times 10^{-2}\%$ of the center frequency. Both the autocorrelation and the spectrum were monitored continuously during the experiments.

The vibrational echo and pump-probe experiments were performed using essentially the same experimental apparatus illustrated in Fig. 1. Micropulses were selected from the macropulse at a 50 kHz repetition rate using a fast germanium acousto-optic modulator (AOM). A fraction of the beam (50% for the echo, 10% for the pump probe) was picked off using a beam splitter and sent directly to the sample. The remainder of the beam passed through a second AOM, which chopped the beam to enable the detection electronics to suppress noise due to scattered light and intensity fluctuations. This beam was sent down a computer controlled optical delay line and then to the sample. The two infrared beams were focused to a diameter of 100 μm on the myoglobin sample. Echo pulse and pump pulse energies of 150–300 nJ were typical, with probe pulse energy ~ 20 nJ.

The sample was a 15 mM solution of wild type horse heart myoglobin in buffered 95% glycerol/5% H$_2$O saturated with CO. The sample had a path length of 125 μm and was cooled with a helium flow cryostat. The Mb-CO absorption is a peak with an absorbance of 0.3 on a very broad background of absorbance $\cong 1$. Despite

this, it is possible to obtain high quality vibrational echo data because the nonlinear nature of the method rejects contributions from the background, which arises from a large number of weak transitions.

Vibrational echo measurements were taken from 60 to 300 K. Several decay curves were taken at each temperature. Figure 2 displays a typical echo decay curve taken at 80 K; the inset is a semilogarithmic plot of this same data. The signal-to-noise ratio is quite good in spite of the large background absorption of the sample. The solid line through the data in the inset is the result of an exponential fit. The echo decays at all temperatures are fit well by a single exponential except for the 300 K point for which the fit included convolution with the Gaussian instrument response. The homogeneous line shape is Lorentzian at all temperatures.

The vibrational dephasing arises from fluctuations in the protein structure, not from the surrounding solvent. Linear vibrational spectroscopy [6,9] shows that the spectrum is extremely sensitive to structural changes of the protein, e.g., the position of the distal histidine, and is insensitive to changing the solvent. Because of the strong coupling between the protein structure and the CO frequency, protein fluctuations produce significant variations in the CO frequency and are the dominant source of vibrational dephasing. A variety of other factors, which will be discussed in a subsequent publication [10], give additional support to the concept that protein fluctuations are responsible for pure dephasing.

The mechanism that couples the protein to the CO, described in detail elsewhere [10], involves fluctuations in the back donation of heme π electron density into the CO π^* antibonding molecular orbital. Static changes in back donation are responsible for shifts in the CO vibrational frequency, and these are caused by changes in protein structure [11]. Thus, the pure dephasing can be caused by dynamic changes in protein structure inducing

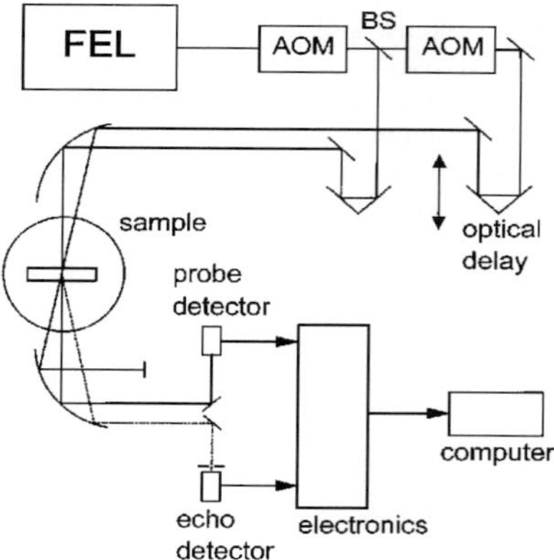

FIG. 1. Schematic of the experimental apparatus used for both the vibrational echo and pump-probe experiments. The only significant differences between the two setups are in the timing of the acousto-optic modulators (AOMs) and in the choice of beam splitter (BS): 50% reflectance for the echo and 10% reflectance for the pump probe.

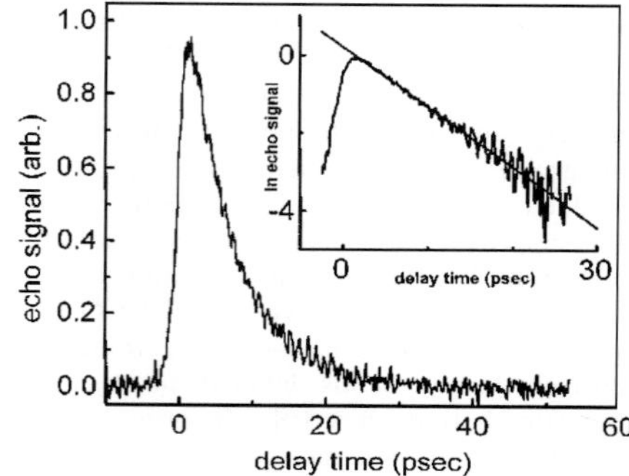

FIG. 2. Vibrational echo data on Mb-CO in glycerol/water at 80 K. The inset displays the same data on a semilogarithmic plot along with the fit to the data, which shows that the echo decay is exponential over a wide dynamic range.

VOLUME 77, NUMBER 8   PHYSICAL REVIEW LETTERS   19 AUGUST 1996

fluctuations in back donation, and, therefore, variations in the vibrational frequency.

At 80 K, the echo data yield $T_2 = 26.5 \pm 0.6$ ps, corresponding to a homogeneous linewidth of 0.40 cm^{-1}. The width of the absorption spectrum at 80 K is 12 cm^{-1}. Therefore, the line is massively inhomogeneously broadened, with the two widths differing by a factor of 30. At room temperature, the homogeneous linewidth is 2.7 ± 0.5 cm^{-1}, which is still approximately 5 times narrower than the 13 cm^{-1} width of the room temperature absorption spectrum. The observation of inhomogeneous broadening at room temperature allows us to conclude that on the echo time scale the protein exists in many different conformational substates, each characterized by different transition frequency of the CO stretch.

Pump-probe lifetime measurements were made over the same range of temperatures. The temperature dependent T_1 and T_2 data, as well as the pure dephasing times, T_2^*, obtained using Eq. (1), are displayed in Fig. 3. T_1 and T_2 could be determined from the data within $\pm 3\%$ error. The error in the derived quantity T_2^* is approximately $\pm 5\%$.

Figure 4 displays the pure dephasing rate, $1/T_2^*$, on a logarithimic plot. The temperature dependence of the pure dephasing rate is less steep at low temperatures. There is a break in the temperature dependence at ~185 K, which is within the range of temperatures associated with the glass transition temperature of the glycerol/water solvent. Below this transition, the data fall on a straight line, indicating power law behavior of the form aT^α, where $\alpha = 1.3 \pm 0.05$. This fit is indicated by the dashed line in Fig. 4.

The $T^{1.3}$ dependence is reminiscent of the temperature dependence that has been observed for the pure dephasing rate of electronic transitions of molecules in low temperature glasses [12]. Several experiments, including ligand recombination studies [13] and pressure relaxation experiments [14], suggest that protein behavior is similar to that of glasses in many ways. Furthermore, protein simulations demonstrate the existence of many conformations that involve only small structural changes [15]. It is, therefore, reasonable to analyze the protein dephasing data using the techniques developed for the study of glasses.

Although there have been several theoretical treatments of glasses, the most successful by far has been the tunneling two-level system (TLS) model [16]. The TLS model postulates that some atoms or molecules (or groups of atoms or molecules) can reside in either of two minima of the local potential surface. Each side of the double well potential represents a distinct local configuration of the glass. The bulk glass material contains an ensemble of these two-level systems characterized by a broad distribution of energy differences and tunneling parameters. Transitions occur via tunneling through the potential barriers, or, at sufficiently high temperatures, by activation over these barriers. Thus, the complex potential surface is modeled as a collection of double well potentials.

The low temperature pure dephasing linewidths can be calculated using the TLS uncorrelated sudden jump model [12,17]. It is found that the temperature dependence of the homogeneous dephasing rate is a power law, T^α, where α is determined by the probability distribution, $P(E)$, for the TLS energy difference, E. If $P(E)$ equals a constant, i.e., there is an equal probability of all TLS splittings, and the distribution of tunneling parameters is also flat, then $\alpha = 1$. In general, for $P(E) \propto E^\mu$, $\alpha = 1 + \mu$.

The $T^{1.3}$ temperature dependence observed in Mb-CO can be understood in terms of a tunneling protein two-level system (PTLS) model. A protein is a nonequilibrium system with many possible conformations. Conformational changes can be viewed as occurring via motion on a multidimensional potential surface. The PTLS model represents this complex potential surface as a

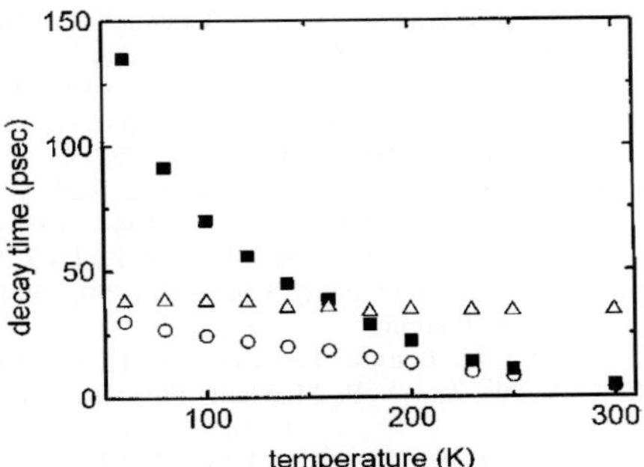

FIG. 3. Plot of twice the measured excited state lifetime $2T_1(\triangle)$, the total dephasing time $T_2(\bigcirc)$, and the pure dephasing $T_2^*(\blacksquare)$, derived from the first two quantities using Eq. (1). At high temperatures, the total dephasing, T_2, is dominated by pure dephasing; at low temperatures, the total dephasing arises mainly from T_1.

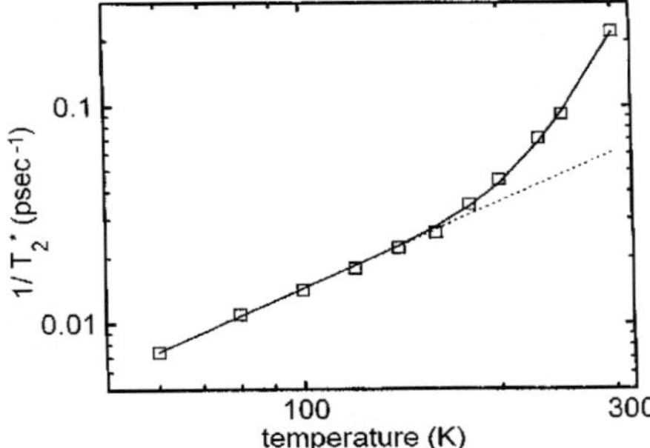

FIG. 4. Plot of the natural logarithm of $1/T_2^*$ vs the natural logarithm of temperature. Below the solvent's glass transition temperature (~185 K), the data follow a power law, $T^{1.3}$ dependence, indicated by the dashed line in the figure. Above the glass transition, the data are exponentially activated with $\Delta E \approx 1000$ cm^{-1}.

Volume 77, Number 8   PHYSICAL REVIEW LETTERS   19 August 1996

collection of double well potentials, each side representing a different protein conformation. Each protein molecule contains many PTLS, which are associated with the possible conformational changes that can occur. The wells have a variety of energy differences, barrier heights, and tunneling parameters.

In the context of the uncorrelated sudden jump model, the observed $T^{1.3}$ temperature dependence implies that $P(E) = E^{0.3}$. This is a very flat distribution of energies. The PTLS model of the Mb protein indicates that the protein has an energy landscape on which the distribution of energy differences between conformations is broad and almost flat. The energies involved are much greater than those invoked to explain dynamics in low temperature glasses. Experiments on proteins at low temperature (2 K) indicate that proteins also have very low energy potential barriers [4]. The results presented here suggest that there are also higher barriers in Mb that play an important role at temperatures on the order of 100 K. Tunneling through barriers in the 100 K range is in contrast to glasses in which dynamics at such elevated temperatures are dominated by phonons or activation over low energy barriers [12,16].

The PTLS model is capable of explaining the observed vibrational dephasing temperature dependence below the solvent glass transition temperature. It builds on the extensive history of considering proteins in terms of concepts that have been applied to glasses [14]. A theoretical investigation is currently in progress to determine if a less restrictive model of protein fluctuations can account for the observed power law temperature dependence.

Above the solvent glass transition, the Mb-CO dephasing dynamics exhibit a clear deviation from the low temperature power law behavior. This change can likely be attributed to the softening of the boundary condition placed on protein motions by the solvent. Barriers to conformational change that were insurmountable with the rigid glassy boundary condition will become lower when the solvent becomes a liquid permitting conformational changes via activation over barriers. We can model this process by fitting the dephasing rate with an activation term, $be^{-\Delta E/kT}$, where ΔE is the barrier height. Thus,

$$\frac{1}{T_2^*} = aT^\alpha + be^{-\Delta E/kT}. \qquad (3)$$

The solid line in Fig. 4 is a result of the fit by this equation, with $\Delta E = 1250 \pm 200$ cm^{-1}. It is possible that the tunneling term is less significant above the transition temperature. This would reduce the ΔE. The net result is that the high temperature data are fit by an activation energy on the order of 1000 cm^{-1}.

We have presented the first vibrational echo experiments performed on a protein. Vibrational echo experiments are a new method for examining protein dynamics and have already yielded some intriguing insights into how protein dynamics are transmitted to the active site of myoglobin. Future experiments will examine mutant Mb and other heme-CO proteins to investigate how specific structural features influence protein dynamics at the active site.

This research was supported by the Medical Free Electron Laser Program, through the Office of Naval Research, Contract No. N00014-94-1-1024 (C.W.R., M.D.F., A.K., H.A.S.). Additional support was provided by the Office of Naval Research, Biology Division, Contract No. N00014-95-1-0259, and the National Science Foundation, Division of Materials Research Grant No. DMR94-04806 (D.D.D., J.R.H.), and the National Science Foundation, Division of Materials Research, Grant No. DMR93-22504 (K.R., M.D.F.).

*To whom correspondence should be addressed.

[1] D. Zimdars, A. Tokmakoff, S. Chen, S. R. Greenfield, M. D. Fayer, T. I. Smith, and H. A. Schwettman, Phys. Rev. Lett. **70**, 2718 (1993); A. Tokmakoff, D. Zimdars, R. S. Urdahl, R. S. Francis, A. S. Kwok, and M. D. Fayer, J. Phys. Chem. **99**, 13 310 (1995).

[2] E. L. Hahn, Phys. Rev. **80**, 580 (1950).

[3] N. A. Kurnit, I. D. Abella, and S. R. Hartmann, Phys. Rev. Lett. **13**, 567 (1964).

[4] D. T. Leeson and D. A. Wiersma, Phys. Rev. Lett. **74**, 2138 (1995).

[5] E. B. M. Antonini, *Hemoglobin and Myoglobin in Their Reactions with Ligands* (North Holland, Amsterdam, 1971).

[6] A. Ansari *et al.*, Biophys. Chem. **26**, 337 (1987).

[7] R. F. Loring and S. Mukamel, J. Chem. Phys. **83**, 2116 (1985).

[8] A. Tokmakoff and M. D. Fayer, J. Chem. Phys. **102**, 2810 (1995).

[9] Dmitri Ivanov, J. Timothy Sage, Markus Keim, Jay R. Powell, Sanford A. Asher, and Paul M. Champion, J. Am. Chem. Soc. **116**, 4139 (1994).

[10] C. W. Rella, Kirk Rector, Alfred Kwok, Jeffrey R. Hill, H. A. Schwettman, Dana D. Dlott, and M. D. Fayer, J. Phys. Chem. (to be published).

[11] X. Y. Li and T. G. Spiro, J. Am. Chem. Soc. **110**, 6024 (1988). K. D. Park, K. Guo, F. Adebodun, M. L. Chiu, S. G. Sligar, and E. Oldfield, Biochemistry **30**, 2333 (1991); E. Oldfield, K. Guo, J. D. Augspurger, and C. E. Dykstra, J. Am. Chem. Soc. **113**, 7537 (1991).

[12] M. Berg, C. A. Walsh, L. R. Narasimhan, Karl A. Littau, and M. D. Fayer, J. Chem. Phys. **88**, 1564 (1988); L. R. Narasimhan, K. A. Littau, Dee William Pack, Y. S. Bai, A. Elschner, and M. D. Fayer, Chem. Rev. **90**, 439 (1990), and references therein.

[13] R. H. Austin, K. Beeson, L. Eisenstein, H. Frauenfelder, I. C. Gunsalus, and V. P. Marshall, Phys. Rev. Lett. **32**, 403 (1974).

[14] I. E. T. Iben *et al.*, Phys. Rev. Lett. **62**, 1916 (1989); H. Frauenfelder *et al.*, J. Phys. Chem. **94**, 1024 (1990).

[15] R. Elber and M. Karplus, Science **235**, 318 (1987).

[16] P. W. Anderson, B. I. Halperin, and C. M. Varma, Philos. Mag. **25**, 1 (1972). W. A. Phillips, J. Low Temp. Phys. **7**, 351 (1972).

[17] Y. S. Bai and M. D. Fayer, Phys. Rev. B **39**, 11 066 (1989).

15620

J. Phys. Chem. **1996,** *100,* 15620–15629

Vibrational Echo Studies of Myoglobin—CO

C. W. Rella,[†] **K. D. Rector,**[‡] **Alfred Kwok,**[‡] **Jeffrey R. Hill,**[§] **H. A. Schwettman,**[†]
Dana D. Dlott,*,[§] **and M. D. Fayer***,[‡]

*Stanford Free Electron Laser Center, Hansen Experimental Physics Laboratory, Stanford University,
Stanford, California 94305-4085, Department of Chemistry, Stanford University, Stanford, California 94305,
and School of Chemical Sciences, University of Illinois at Urbana—Champaign, Urbana, Illinois 61801*

Received: April 17, 1996; In Final Form: June 27, 1996[⊗]

The first picosecond infrared vibrational echo experiments on a protein, myoglobin—CO, are described. These vibrational dephasing experiments examine the influence of protein dynamics on the CO ligand bound to the active site of the protein at physiologically relevant temperatures. The experiments were performed with a mid-IR free electron laser tuned to the CO stretch mode at 1945 cm^{-1}. The vibrational echo results are combined with infrared pump—probe measurements of the CO vibrational lifetime to yield the homogenous pure dephasing, the Fourier transform of the homogeneous line width with the lifetime contribution removed. The measurements were made from 60 to 300 K. The results show that the CO vibrational spectrum is inhomogeneously broadened, even at room temperature. Above the glycerol/water solvent's glass transition temperature, ∼185 K, the temperature dependence can be fit as an activated process with $\Delta E \approx 1000$ cm^{-1}. Below 185 K, the pure dephasing displays a power law temperature dependence, $T^{1.3}$. This temperature dependence is reminiscent of that associated with the properties of low-temperature glasses (<5 K) but is observed at much higher temperatures. A two-level system model of protein dynamics is considered. The nature of the temperature dependence and the mechanism of the coupling of the protein fluctuations to the CO vibrational transition energy are discussed.

I. Introduction

In this paper, we present the first vibrational echo experiments performed on a protein. The vibrational echo is a time domain experiment that measures the Fourier transform of the homogeneous vibrational line shape.[1] By combining vibrational echo measurements with vibrational pump—probe lifetime measurements, the dynamical contributions to a vibrational transition can be elucidated.[1] Previous optical coherence experiments performed on proteins examined the dephasing of electronic transitions.[2] Because of the very rapid dephasing (broad homogeneous line widths) of electronic transitions, these experiments can only be performed at very low temperatures, i.e., a few degrees K. The vibrational echo experiments make it possible to use optical coherence methods to study protein dynamics at physiologically relevant temperatures.

From a vibrational absorption spectrum, it is not possible to determine if a line shape is inhomogeneously or homogeneously broadened. The vibrational echo makes the homogeneous line a direct experimental observable and, with lifetime measurements, permits the pure dephasing (energy level fluctuations) and the lifetime contributions to be separated. The experiments presented here were performed on the vibrational stretching mode of carbon monoxide (CO) bound to the active site of myoglobin (Mb).

Mb, a protein found in muscle tissue, is used in the storage and transport of dioxygen (O$_2$).[3] It consists of a prosthetic group called protoheme [iron(II) protoporphyrin IX] embedded in a protein shell. The protein modifies the chemical reactivity of the Fe binding site, allowing Mb to function properly in a biological setting.[3]

A substantial literature exists that explores the structure and binding kinetics of CO bound at the active site of Mb using a variety of techniques, including X-ray crystallography,[4] ^{13}C NMR,[5] time-dependent visible optical spectroscopy,[6] Raman spectroscopy,[7] and linear mid-IR spectroscopy.[8] Mid-IR spectroscopy of the CO stretching frequency has a demonstrated utility, supplying insights into how the structure of the protein modifies the nature of the binding site. When bound to Mb, the CO gas phase frequency is substantially red-shifted and separated into several distinct bands, which are labeled A$_0$—A$_3$ in order of decreasing carbonyl frequency. Although the intensity and widths of these bands are sensitive to temperature, mild pressure changes, and pH, their peak frequencies remain largely unchanged.

In general, methods such as IR spectroscopy and X-ray crystallography yield a wealth of information about the equilibrium structure of Mb—CO but can provide only indirect dynamical information. Room temperature proteins are dynamic. Molecular dynamics simulations of Mb suggest a flexible structure in constant motion rather than a rigid, scaffold-like structure.[9] Such motions can be on a relatively small scale involving few of the constituent Mb atoms, such as the torsion of an amino acid residue, or they can be large-scale motions involving entire regions of the protein backbone. Simulations over a period of 300 ps indicate that Mb samples thousands of local energy minima of approximately equal energy, separated by barriers of varying height.[9] These minima correspond to different conformational states of the protein. It has been proposed that this characteristic of proteins is analogous to the energy landscape in glasses.[10]

In this paper, a detailed temperature-dependent vibrational echo study of the CO stretching mode of Mb—CO is presented. The experiments were conducted using a free electron laser as the source of tunable picosecond IR pulses. In addition to the echo measurements, the temperature dependence of the vibra-

---

* Authors to whom correspondence should be addressed.
† Stanford Free Electron Laser Center.
‡ Department of Chemistry, Stanford University.
§ School of Chemical Sciences, University of Illinois at Urbana—Champaign.
⊗ Abstract published in *Advance ACS Abstracts,* August 15, 1996.

S0022-3654(96)01129-X CCC: $12.00

tional lifetime was measured using IR pump−probe experiments. At the lowest temperature studied (60 K), the homogeneous line is dominated by the lifetime contribution (T_1), although the pure dephasing ($T_2{}^*$) still makes a measurable contribution. At all temperatures, the vibrational echo decays are exponential. Thus, the homogeneous line is Lorentzian in shape. By room temperature, the homogeneous line width arises almost completely from pure dephasing. The homogenous line width (2.7 cm^{-1}) is substantially less than the width of the absorption spectrum (13.1 cm^{-1}), demonstrating that the line is inhomogeneously broadened, even at room temperature.

Above the glycerol/water solvent's glass transition ($\sim$185 K), the liquid solvent provides a soft boundary condition for the protein motions. The protein can undergo large amplitude as well as small amplitude conformational changes. These changes in protein conformation couple to the CO bound to the active site and cause the CO vibrational energy levels to fluctuate, giving rise to pure dephasing. The data are well fit by an activated process with an activation energy on the order of 1000 cm^{-1}. This activation energy is associated with potential barrier heights for protein conformational changes.

Below 185 K, the temperature dependence of the pure dephasing line width ($1/\pi T_2{}^*$) is a power law, $T^{1.3}$. This temperature dependence is the same as that observed for electronic excited state dephasing[11] and heat capacities[12] in very low temperature ($<$5 K) glasses, where a tunneling two-level system (TLS) model is used to explain the data.[13] However, this is the first time that a power law with such a small exponent has been observed for a pure dephasing process at such elevated temperatures. At higher temperatures ($>$10 K) in glasses, dephasing has either a T^2 or exponentially activated dependence on temperature, and heat capacities have a T^3 dependence. The observation of a $T^{1.3}$ dependence for vibrational dephasing of Mb−CO may suggest a tunneling TLS process for the protein. Protein dynamics dominated by tunneling can be shifted to higher temperature because of high energy barriers for conformational changes induced by the rigid boundary condition imposed by the glassy solvent. This possibility is discussed in terms of a protein two-level system (PTLS) model.

The nature of the coupling of the protein fluctuations to the CO vibrational transition energy is also addressed. A model is suggested in which protein fluctuations modulate the heme electron density, causing fluctuations in the back-bonding (back donation of electron density from the heme π electron system to the CO π^* antibonding molecular orbital). Fluctuations in back-bonding give rise to energy level fluctuations and thus $T_2{}^*$. Therefore, the vibrational echo experiments provide an observable for the communication of protein dynamics to a ligand bound to the active site of the protein.

II. The Vibrational Echo Method and Experimental Procedures

In disordered, condensed matter systems, in general, even a well-resolved vibrational line does not provide information on dynamics. Vibrational line shapes in condensed phases contain the details of the dynamic interactions of a normal mode with its environment.[14] However, the line shape can also include essentially static, structural perturbations associated with the distribution of local configurations of the environment, i.e., inhomogeneous broadening. The IR absorption or Raman line shape represents a convolution of the various dynamic and static contributions, and contributions from inhomogeneous broadening cannot be eliminated.[15]

The picosecond IR vibrational echo experiment is a time domain, nonlinear method that can extract the homogeneous

vibrational line shape from inhomogeneously broadened lines.[1,16] The echo technique was originally developed as the spin echo in magnetic resonance in 1950.[17] In 1964, the method was extended to the visible optical regime as the photon echo.[18] Since then, photon echoes have been used extensively to study electronic excited state dynamics in many condensed matter systems. Recently, vibrational echoes have been used to examined vibrational dynamics in liquids and glasses.[1,16] For experiments on vibrations, a source of picosecond IR pulses is tuned to the vibrational transition of interest. The vibrational echo experiment involves a two-pulse excitation sequence. The first pulse places each molecule's vibration into a superposition state, which is a coherent mixture of the $v = 0$ and $v = 1$ vibrational levels. Immediately after the first pulse, the vibrational dipoles oscillate in phase. Because there is an inhomogeneous distribution of vibrational transition frequencies, the individual dipoles oscillate with some distribution of frequencies. Thus, the initial phase relationship is very rapidly lost. This is the vibrational free induction decay. After a time, τ, a second pulse, traveling along a path making an angle θ with that of the first pulse, passes through the sample. This second pulse changes the phase factors of each vibrational superposition state in a manner that initiates a rephasing process. At time τ after the second pulse, the sample emits a third coherent pulse of light. The emitted pulse propagates along a path that makes an angle 2θ with the path of the first pulse. The third pulse is the vibrational echo. It is generated when the ensemble of microscopic dipoles is rephased at time 2τ.

The rephasing at 2τ has removed the effects of the inhomogeneous broadening. However, fluctuations due to coupling of the vibrational mode to the environment cause the oscillation frequencies to fluctuate. Thus, at 2τ there is not perfect rephasing. As τ is increased, the fluctuations produce increasingly large accumulated phase errors among the microscopic dipoles, and the intensity of the echo is reduced. A measurement of the echo intensity vs τ, the delay time between the pulses, is called an echo decay curve. The Fourier transform of the echo decay is directly related to the homogeneous line shape.[19] An exponential vibrational echo decay corresponds to a Lorentzian line shape with a width, Γ, given by

$$\Gamma = \frac{1}{\pi T_2} = \frac{1}{\pi T_2{}^*} + \frac{1}{2\pi T_1} \tag{1}$$

T_2 is the homogeneous dephasing time determined from the echo decay constant. T_1 is the vibrational lifetime determined from pump−probe experiments. Measurements of T_2 and T_1 permit the determination of $T_2{}^*$, the pure dephasing contribution to the line width. In liquids, there is an additional contribution from orientational relaxation.[16] However, in the experiments presented below, orientational relaxation does not occur on the time scale of the experiments because of the large size of the protein and the high viscosity of the solvent. This was confirmed with polarization selective pump−probe experiments. T_2 is determined from the vibrational echo decay signal, $S(\tau)$, given by

$$S(\tau) = S_\mathrm{o}e^{-4\tau/T_2} \tag{2}$$

The vibrational echo and pump−probe experiments were performed at the Stanford Free Electron Laser (FEL) Center. The electron beam from a superconducting linear accelerator is directed into a magnetic wiggler at the center of an optical resonator. The electron beam interacts with the intracavity radiation, emitting photons coherently at an energy corresponding to the Doppler-shifted spatial frequency of the wiggler. The

15622 *J. Phys. Chem., Vol. 100, No. 38, 1996*

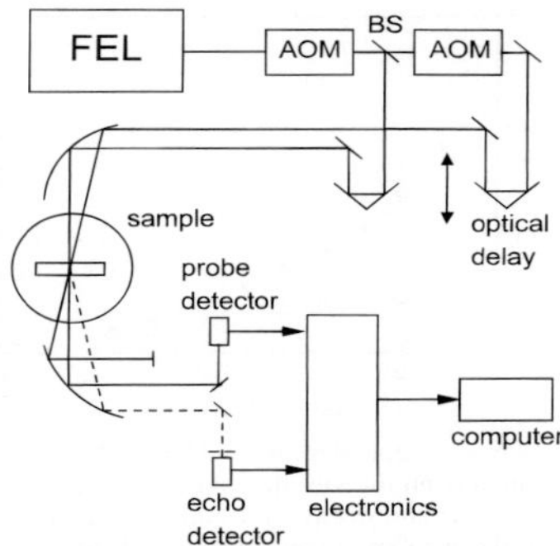

Figure 1. Schematic of the apparatus used to perform the echo and pump−probe experiments. AOM denotes Ge acousto-optic modulators used as pulse selectors. BS denotes beam splitter. The path of the echo signal emitted by the sample is indicated by the dashed line. The differences between the two setups involve the choice of beam splitter (50% R for echo, 10% R for pump−probe), in the optical pulse sequence selected by the AOMs, and in the detectors.

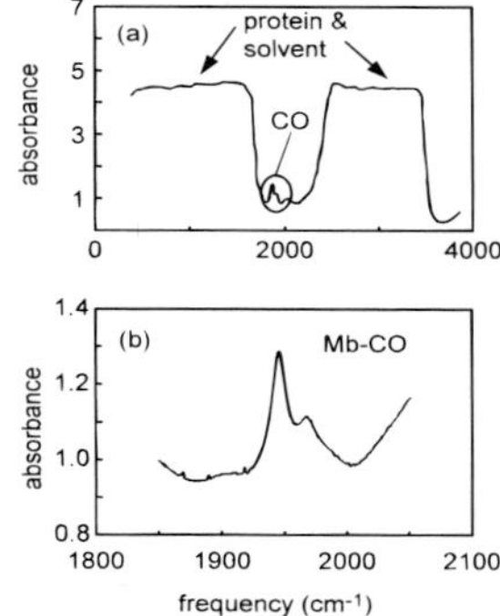

Figure 2. (a) Mid-IR spectrum of Mb−CO in glycerol/water. The circled region indicates the location of the CO vibrational fundamental transition on top of a broad background from protein and solvent. (b) Blow up of the CO stretching transition. The experiments are conducted on the large peak (A_1). The small peak to higher frequency is the A_0 band associated with another Mb−CO conformer.

FEL pulse train consists of a macropulse of about 3 ms length repeating at 10−20 Hz, within which is contained a series of micropulses repeating at 11.8 MHz. Each micropulse has an energy of ∼1 μJ. These micropulses are transform-limited Gaussians. In the experiments, the pulse duration was ∼1.7 ps. The FEL frequency, tuned to 1945 cm^{-1} for the experiments, is actively stabilized to within 2×10^{-2}% of the center frequency. Both the autocorrelation and the spectrum are monitored continuously. Although the experiments were performed with an FEL, a source based on conventional lasers having similar characteristics may also be used.

A schematic describing the experimental apparatus used for the photon echo and pump−probe experiments is shown in Figure 1. In order to reduce the heating of the sample while retaining high peak power, single micropulses are selected from the macropulse at a reduced repetition rate of 60 kHz, using a germanium acousto-optic modulator (AOM). Each micropulse is split into two roughly equal parts to become the two input pulses in the vibrational echo sequence. One beam is sent through a motorized optical delay line after passing through a second AOM that chops it at 30 kHz. The chopping is used to do background subtraction. The two beams are subsequently focused to 100 μm diameter spots in the sample using off-axis paraboloidal mirrors. The energies of the two pulses at the sample are ∼250 and ∼150 nJ. For noncollinear input beams, phase-matching conditions cause the echo signal to emerge from the sample in a unique direction. A liquid nitrogen-cooled InSb detector is used to detect the echo signal. The entire apparatus is contained in a nitrogen purge box to remove the effects of atmospheric water absorption, thus increasing the total pulse energy at the sample as well as preventing any temporal pulse distortion effects.

The apparatus used for the pump−probe vibrational lifetime measurements is nearly identical to that used for the vibrational echo. Unlike the zero-background photon echo experiment, the pump−probe experiment requires accurate measurements of small changes in the probe amplitude. For this reason, the first AOM is used to pick two adjacent micropulses, instead of just one. These two pulses are highly correlated in amplitude, temporal width, and frequency due to the relatively high $Q(30)$ of the FEL cavity. Ten percent of this beam is split off to become the probe beam. The remainder of the beam travels through the delay line and the second AOM, where the second

of the two pulses is selected to become the pump pulse. These two beams are crossed in the sample. A fast, liquid nitrogen-cooled MCT detector is used to detect the individual micropulses of the probe beam. The amplitude of the first of the two probe pulses, which arrives prior to the pump pulse, is used as a reference for the second pulse. The signals from the two pulses are subtracted, and the results are recorded as a function of delay time.

The samples used for the experiments were wild type horse heart myoglobin−CO dissolved in a buffered pH = 7 glycerol/water (95%/5% w/w). The protein concentration of these samples is typical of that used in infrared experiments,[8] but the glycerol concentration is greater than the 75%/25% solutions typically used. The greater glycerol concentration improved the optical quality of the solvent glass at low temperatures, which is desirable because light from the excitation pulses scattered by the solvent glass hinders detection of the echo pulse. The samples were prepared by slowly adding lyophilized myoglobin (Sigma) to 1 mL of pure glycerol with continuous stirring until the concentration of the solution was 15 mM. The frothy solution was then allowed to settle for a few hours. The solution was saturated with CO by bubbling gas through it. Then 50 μL of a solution of sodium dithionite in pH = 7 phosphate buffer, also saturated with CO, was added to reduce the protein to allow it to bind CO. The sample was placed inside a cell with CaF$_2$ windows, a path length of 125 μm, and a Mb−CO concentration of 15 mM. The sample cell was placed on the cold finger of a Janis helium flow cryostat. Temperature stability was better than ±0.5 K.

Photon echo and pump−probe measurements were made approximately every 20 K from 60 to 300 K. Careful studies of the power dependence of pump−probe lifetime measurements and vibrational echo dephasing measurements were made; in both experiments, the measured decay times were constant for all input intensities. The FEL micropulse with duration 1.7 ps has a corresponding spectral width of 8.6 cm^{-1}. Polarization selective pump−probe experiments demonstrated no dependence of the decay on the probe polarization. Therefore, orientation relaxation is unimportant on the time scale of the measurements.

III. Results and Discussion

Figure 2 displays an IR spectrum of Mb−CO in the vicinity of the CO vibrational absorption, taken at ambient temperature,

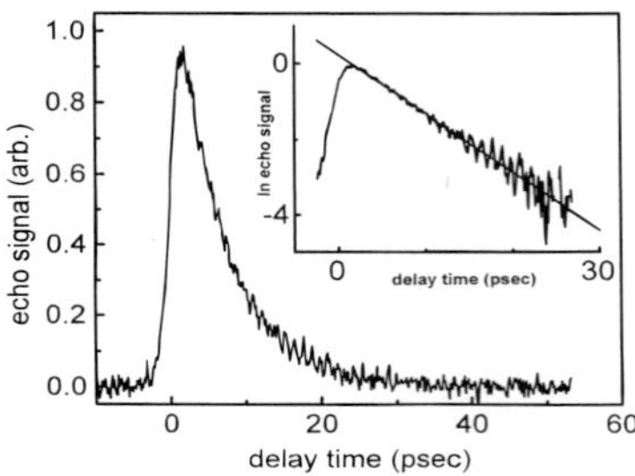

Figure 3. Vibrational echo data on Mb−CO in glycerol/water at 80 K. Inset: semilog plot (base e) shows the echo decay is exponential over a wide dynamic range. An exponential echo decay means that the homogeneous line shape is Lorentzian. At 80 K, the vibrational echo data gives a value for the dephasing time constant of $T^2 = 26.2$ ps. Therefore the homogeneous line width is 12.1 GHz (0.4 cm^{-1}). This is a factor of ~30 narrower than the inhomogeneously broadened absorption spectrum line width.

with 2 cm^{-1} resolution. The upper panel shows a broad range of wavelengths. The enormous number of vibrational states associated with the protein and the solvent produce a significant background with an absorbance of ~1. The Mb−CO absorption appears as a smaller feature against this background. The lower panel shows an expanded view of the Mb−CO spectrum. The CO absorption is readily seen above the broad background. It has been shown this absorption spectrum can be decomposed into four subbands.[8b] The large peak in the lower panel (~1945 cm^{-1}) corresponds to the subband conventionally labeled A$_1$. The smaller peak (~1970 cm^{-1}) corresponds to the subband conventionally labeled A$_0$. The intensity at ambient temperature of the A$_0$ subband, relative to the A$_1$ subband, tends to increase as the viscosity of the solvent increases,[8b] and in this 95% glycerol solvent, the A$_0$ transition in this spectrum is more prominent than in the more usual 75% glycerol solutions. In previous vibrational echo experiments, a system was investigated with an essentially background free spectrum.[1] However, as shown below, it is possible to obtain high-quality vibrational echo data even from a system with a congested spectrum like Mb−CO. This is because the nonlinear nature of the method eliminates the background, which arises from a large number of weak transitions.

Figure 3 shows an echo decay taken at 80 K. The main figure is a linear plot. It can be seen that the signal-to-noise ratio is very good in spite of the large amplitude of the background in the spectrum. The inset displays a semilog plot of the data. The data fall on a straight line over 4 factors of e. Therefore, the data decay exponentially, demonstrating that the homogeneous line shape is Lorentzian. The echo decays observed at all temperatures are exponential. The T_2 obtained from the 80 K data is 26.2 ps, yielding a homogeneous line width of 0.4 cm^{-1}. The width of the absorption spectrum is ~12 cm^{-1}. Therefore, the line is massively inhomogeneously broadened; the two widths differ by a factor of ~30.

The temperature dependence of the vibrational echo decays as well as the pump−probe lifetime measurements were obtained from 60 to 300 K. Each decay curve was fit to an exponential using a Levenberg−Marquardt fitting routine.[20] For the high-temperature echo decays, the decay times are approaching the pulse width. Therefore, the data were fit to the convolution of a Gaussian and an exponential. Care was taken in each case to verify the stability of the fit, and for most temperatures, the decay constant could be determined within ±3% for the echo data and ±5% for the pump−probe data.

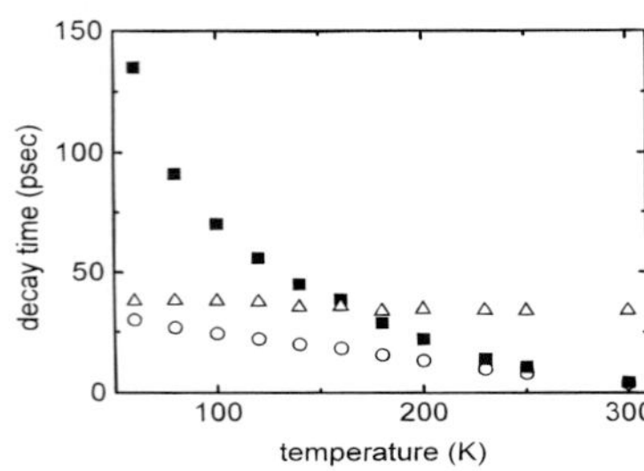

Figure 4. Temperature-dependent data. The circles are the measured values of T_2 obtained from the vibrational echo decays using eq 2. The triangles are $2T_1$. T_1 is the decay constant measured in the pump−probe vibrational lifetime experiments. From eq 1, the relevant quantity is $2T_1$. The squares are T_2*, the pure dephasing time, obtained from T_2 and $2T_1$ using eq 1.

Figure 4 shows temperature-dependent data. There are three sets of points in the figure. The circles are the measured values of T_2 obtained from the vibrational echo decays using eq 2. The triangles are $2T_1$. T_1 is the decay constant measured in the pump−probe vibrational lifetime experiment. From eq 1, it is seen that the relevant quantity is twice the lifetime, which is $2T_1$. The squares are T_2*, the pure dephasing time, obtained from T_2 and $2T_1$ using eq 1. T_1 has a very mild-temperature dependence. T_2 has a steeper temperature dependence, changing by about an order of magnitude between 60 and 300 K. The pure dephasing, T_2*, which arises from vibrational energy level fluctuations, has a very steep temperature dependence. At the lowest temperatures, the lifetime is the major contributor to the homogeneous dephasing (inverse of the line width). By room temperature, pure dephasing completely dominates the homogeneous dephasing. At all temperatures, the line is inhomogeneously broadened. At room temperature, the homogeneous line width is 2.7 cm^{-1} while the absorption spectrum width is ~13 cm^{-1}, i.e., the absorption spectrum is ~5 times wider than the homogeneous line width. (The absorption line width was obtained from the data in Figure 2 by taking the half-width at half-height on the red side of the line and doubling it.) The observation of inhomogeneous broadening at room temperature is important. It allows us to conclude that, on the echo time scale, the protein exists in many different distinguishable conformational substates that induce different transition frequencies of the carbonyl stretch.

The photon echo experiment is nonlinear. The signal depends on the cube of the intensity and the square of the concentration. Because of the strong dependence on intensity and concentration, when the FEL is tuned to the peak of the A$_1$ line, there is no contribution from the large A$_0$ shoulder or the small A$_2$ and A$_3$ peaks.[8b] If the FEL were tuned to the peak of the A$_0$ line, the signal would be down by a factor of ~5. Because the experiments are conducted in the low-power limit, only the portion of the pulse spectrum that is resonant with a vibration can generate signal. The half-width of the pulse spectrum is 4.3 cm^{-1} and the A$_0$ line is shifted from the A$_1$ line by 25 cm^{-1}. Therefore, the intensity is so far down that no signal is generated from A$_1$. The FEL was tuned ±4 cm^{-1} around the peak of A$_1$. This change will still result in a negligible contribution from the other peaks. The vibrational echo decays and the pump−probe decays were unchanged by the shifts in wavelength, demonstrating that the dynamics are independent of the position in the inhomogeneous line over this range of wavelengths.

Figure 5 displays the pure dephasing rate, $1/T_2$*, on a log plot. It is clear from this plot that the temperature dependence

15624 *J. Phys. Chem., Vol. 100, No. 38, 1996*

Rella et al.

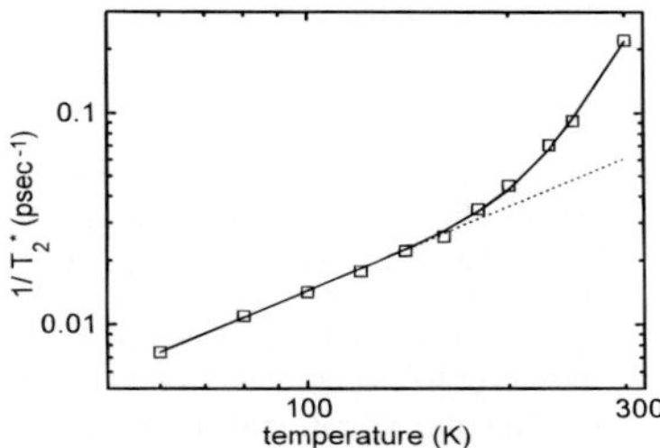

Figure 5. Plot of the natural log of $1/T_2^*$ vs the natural log of temperature. The solid line in the figure is a fit to the sum of a power law and an exponentially activated process, eq 3. The arrow indicates the position of the glass transition at ~185 K. Below this temperature, the data follow a power law, $T^{1.3}$ dependence, which appears linear on the log−log plot. Above ~185 K, an exponentially activated process describes the data with $\Delta E \approx 1000$ cm^{-1}.

of the pure dephasing rate is much milder at low temperatures, with a break in the temperature dependence at ~185 K, which is approximately the glass transition temperature of the glycerol/water solvent. The low-temperature data fall precisely on a straight line, indicating that the temperature dependence is described by a power law, T^α. The exponent of the power law can be obtained directly from the slope of the linear portion of the data, yielding $\alpha = 1.3 \pm 0.1$. A variety of functional forms were tried in the data analysis. The line through the data is a fit to the functional form of a power law plus an exponentially activated process,

$$\frac{1}{T_2^*} = aT^\alpha + be^{-\Delta E/kT} \tag{3}$$

Of the many functional forms that were tried, eq 3 is the only one that provides good agreement with the data. There are two ways to obtain a fit, and they yield distinctly different activation energies but the same power law exponent. One is to assume that both terms in eq 3 are active at all temperatures, while the other assumes a change in mechanism at the solvent's glass transition. If eq 3 is used as written, as the sum of the two functions, $\Delta E = 1250$ cm^{-1}. The activated process makes little contribution at the lower temperatures. However, the power law contributes 25% at room temperature. In the second approach, at low temperature, the dephasing is given only by the power law. This does not change the value of α. At the glass transition, the dephasing becomes exponentially activated. Therefore, there is no contribution from the power law at high temperature. Using this approach, the activated process fits the data with $\Delta E = 800$ cm^{-1}. It needs to be emphasized that, while eq 3 fits the data extremely well, it cannot be proven to be unique.

The data raise several important issues: 1. Why is there a change in the form of the temperature dependence near the solvent glass transition temperature? 2. What are possible explanations for the power law temperature dependence and the exponentially activated temperature dependence? 3. What is the mechanism that couples the protein fluctuations to the CO stretching frequency, causing the pure dephasing?

In the following discussion, some initial ideas and relevant information will be presented. It is not possible to provide definitive answers to the three questions that are proposed, but it is possible to indicate directions for constructive thought. After the discussion of the three questions, some future lines of research that should be fruitful in narrowing the possibilities will be described.

1. Change in the Form of the Temperature Dependence. The helical chains that make up Mb can undergo both large-amplitude and small-amplitude motions under biological conditions.[21] However, at low temperatures, below the solvent glass transition temperature, the glassy solvent provides a rigid boundary condition for the protein. This rigid boundary condition will suppress conformational changes that would change the shape of the protein surface. Molecular dynamics simulations have demonstrated a protein's ability to sample thousands of minor conformational changes on fast time scales.[9] These are classical, room temperature calculations. As discussed below, the low-temperature dynamics may involve tunneling. The simulations, nonetheless, demonstrate the existence of many conformations that involve only small structural changes. Once the temperature is raised above the glass transition temperature, the liquid solvent provides a much softer boundary condition. The softer boundary will facilitate more significant conformational changes that need not preserve the surface shape of the protein. It will lower barriers that are insurmountable in the rigid, glassy solvent. Recent experiments using other solvents, which will be presented subsequently,[22] show that the change from a power law to an exponentially activated process follows the solvent glass transition temperature.

Evidence exists to support the supposition that Mb undergoes a transition, called the "slaved glass transition", near the glass transition temperature of the solvent.[10,23] Calorimetric measurements of Mb in glycerol/water, ethylene-glycol/water, and PVA indicate that the temperature at which this transition occurs, T_{sg}, is generally a few degrees warmer than the glass transition of the solvent itself. Studies of CO to Mb rebinding kinetics, infrared line shape, and small molecule diffusion all imply a sharp change in dynamics occurring around T_{sg}. The dynamics are characterized below T_{sg} by slow, markedly nonexponential time dependences, similar to effects seen in glasses, and, above T_{sg}, by much faster kinetics. These experiments imply a change in the nature of the dynamics at T_{sg}. It should be noted that experiments, such as rebinding kinetics, operate on a time scale of tens of microseconds to thousands of seconds, while the echo measurements are examining dynamics on the 10 ps time scale. The pure dephasing data does not show a sharp change near the slaved glass transition, but, rather, a change in the functional form of the temperature dependence.

2. The Temperature Dependence of the Pure Dephasing. In the lower temperature range (from 60 to ~185 K), the pure dephasing line width has a temperature dependence of $T^{1.3}$. This is reminiscent of the temperature dependence that has been observed for the pure dephasing of electronic transitions of molecules in low-temperature glasses using photon echo experiments.[11] In low-temperature (from 0.1 to 10 K) glasses such as ethanol, glycerol, and PMMA, the pure dephasing line widths of dye molecules display temperature dependences of T^α, where the most common value for α is 1.3, but α ranges from 1.2 to 1.5 for various glasses.[11] The similarity is so striking that it is worth discussing the nature of the model that gives rise to this temperature dependence. Comparing proteins to glasses has a significant history. The two basic hallmarks of a glassy system, the long-range disorder and the multiple energy minima, are found in protein systems as well. There exists experimental evidence that is used to support the comparisons. Measurements of the nonexponential time dependence of the rebinding kinetics of O_2 and CO at low temperature were interpreted, early on, in terms of glassy behavior in proteins.[24] Later, measurements found that the low-temperature (1 K) heat capacities of proteins mimic the near linear behavior of glasses.[25] More recent measurements of the pressure and temperature dependence of

J. Phys. Chem., Vol. 100, No. 38, 1996 **15625**

the mid-IR absorption bands of Mb−CO have shown nonexponential time dependence and non-Arrhenius temperature dependence of the pressure relaxation as a function of temperature.[26] These observations have all been interpreted as evidence for a similarity between proteins and glasses.

Glasses are systems which, unlike crystals, possess no long-range translational or rotational order and are not in a state of thermodynamic equilibrium.[27] Glass properties are determined by a combination of the dynamics of quasi-phonons (the disordered equivalent of phonons in a crystal) and by the additional dynamics introduced by the evolution of local glass conformations. Extensive experimental studies have clarified the differences between glasses and crystals.[12] The heat capacity in a crystal obeys the Debye T^3 dependence at low temperature; in a glass, there is an additional term, approximately linear in temperature below a few degrees K. In a crystal, the thermal conductivity varies as T^3, while in a glass, the temperature dependence at low temperature is T^2. Several theoretical models have been developed to explain the various observations. However, the two-level system (TLS) model has enjoyed the most success.[13]

The possible local conformations of a glass are associated with an extremely complex, multidimensional potential surface. The TLS model simplifies the description of dynamics on this potential surface. The TLS model postulates that some atoms or molecules (or groups of atoms or molecules) can reside in either of only two minima of the local potential surface. Each side of the double-well potential represents a distinct local structural configuration of the glass. The bulk glass material contains an ensemble of these two-level systems, having a broad distribution of energy differences and tunneling parameters. At low temperatures, transitions occur via tunneling through the potential barriers or, at sufficiently high temperatures, by activation over these barriers. Thus, the complex potential surface is modeled as a collection of double-well potentials.

Using this simple physical picture, the TLS model simulates the rugged, multidimensional energy potential surfaces that exist in glasses. The agreement of the predictions of the theory with experimental data is striking; the model reproduces the temperature dependence of the heat capacity and thermal conductivity.

Optical dephasing of chromophores in low-temperature glasses is caused by the coupling of the optical centers to the TLS dynamics. When the TLS undergoes transitions via tunneling between the two wells, the associated changes in local structures cause the transition energy of a chromophore to fluctuate. Because there is a wide range of tunneling parameters, λ, and energy differences, E, between the sides of the two wells, there is a broad range of rates of transitions between the wells. This broad distribution of tunneling rates produces a broad range of rates of chromophore energy fluctuations. The temperature dependence of the electronic excited state pure dephasing line width measured in photon echo experiments on molecules in glasses at low temperatures can be calculated using the TLS uncorrelated sudden jump model.[28] More complex models give similar results.[29] It is found that the temperature dependence is

$$\frac{1}{T_2^*} = aT^\alpha \tag{4}$$

where α is determined by $P(E)$, the probability distribution for the TLS energy difference E. If $P(E)$ equals a constant, i.e., there is equal probability of all TLS splittings and the distribution of tunneling parameters is also flat, then $\alpha = 1$. In general, for

$$P(E) \propto E^\mu \tag{5a}$$

$$\alpha = 1 + \mu \tag{5b}$$

Therefore, the typical temperature dependence observed for electronic dephasing in the temperature range below 5 K indicates an almost flat distribution of TLS energy splittings and an essentially flat distribution of tunneling parameters. These distributions are consistent with the observed heat capacities, which are described by the same model and parameters.

Prior to discussing the vibrational dephasing of Mb−CO caused by protein fluctuations using the PTLS model, which is analogous to the glassy TLS model, it is important to present the strong evidence that Mb−CO dephasing is not caused by the solvent.

For the solvent to cause dephasing, it must couple to the transition frequency of the CO. When molecules go from the gas phase to a condensed matter environment, there is a shift of the transition frequency. This is true of both electronic transitions and vibrational transitions. This shift is referred to as the solvent shift. Molecular interactions with the condensed matter environment are responsible for line broadening as well as the solvent shift. These two are closely related. The line broadening can be static, giving rise to an inhomogeneous line, or dynamic, giving rise to a homogeneous line. In either case, variations in the solvent shift cause line broadening.

In Mb−CO, the nature of the solvent itself has little effect on the CO vibrational transition frequency. The Mb−CO transition frequency is virtually identical whether the solvent is water (1944 cm^{-1}),[8b] 75% glycerol/water (1945 cm^{-1}),[8b] poly-(vinyl alcohol) (1946 cm^{-1}),[8b] trehalose (1945 cm^{-1}),[30a] or a Mb crystal (1946 cm^{-1}).[30b] These values are all within the error of the determination of the peak center. The solvent shift is unaffected by the medium surrounding the protein even when the change is from a liquid solvent to a protein crystal. In contrast, the frequency difference between the Mb−CO A$_0$ and A$_1$ lines is 25 cm^{-1}.[8b] This difference is caused by a change in the distal histidine. Changes in the protein structure have a major influence on the transition frequency while changes in the solvent have a negligible influence. This leads to the reasonable conclusion that fluctuations of the protein structure will cause homogeneous dephasing while fluctuations of the solvent structure will not. Of course, the solvent provides a heat bath and a boundary condition that are intimately involved in the protein fluctuations and the dephasing, but the argument made above strongly supports the idea that the dephasing does not arise from direct coupling of the solvent dynamics to the CO transition frequency.

The TLS properties of glasses are only manifested at very low temperatures. In the 60−185 K temperature range under discussion, TLS dynamics are overwhelmed by the quasi-phonon dynamics of a glass.[11,31] The barrier heights in glasses are too low for the dynamics to still involve tunneling processes at the elevated temperatures of these experiments,[31] and it is tunneling that gives rise to the power law temperature dependence with $\alpha \approx 1$.

The experiments that are most directly comparable to those presented in this paper are vibrational echo measurements conducted on the asymmetric CO stretching mode (1980 cm^{-1}) of W(CO)$_6$ in three liquid/glass solvents, 2-methylpentane, 2-tetrahydrofuran, and dibutyl phthalate.[16] While W(CO)$_6$ is not the same as CO bound to Fe−heme, the experiments provide vibrational dephasing information for a CO stretching mode of a metal carbonyl in glassy solvents in the same temperature

15626 *J. Phys. Chem., Vol. 100, No. 38, 1996*   Rella et al.

range as the Mb−CO experiments. Although the three glasses are quite different from each other, $W(CO)_6$ vibrational dephasing has an identical T^2 dependence in all three. The T^2 temperature dependence is consistent with a dephasing mechanism that occurs in glasses at high temperatures. It arises in the high-temperature limit from activation over potential barriers. It is a two-phonon scattering process in which one phonon is annihilated to take the system above a barrier and a second phonon is created when the system relaxes into the other side. In inorganic glasses, optical dephasing of electronic states at elevated temperatures is also observed to have a T^2 temperature dependence.[32] This is the hallmark of activation over barriers at high temperatures, rather than tunneling.

$W(CO)_6$ vibrational dephasing was also studied above the glass transition temperature.[16] There is a very sharp increase in the slope of the temperature dependence at the glass transition temperature. However, unlike the results presented here, the temperature dependence for $W(CO)_6$ vibrational dephasing in 2-methylpentane liquid is not exponentially activated, but rather obeys a Vogel−Tammann−Fulcher (VTF) equation.[33] The VTF equation has the form

$$f(T) = \beta e^{-D/(T-T^*)} \qquad (6)$$

where $f(T)$ is a temperature-dependent quantity, in this case the pure dephasing time, and T^* is a critical temperature. The functional form is much steeper than an activated process since it diverges at the glass transition temperature. It was found that a VTF form cannot fit the data in Figure 4, regardless of the choice of parameters. It is far too steep.

Thus, vibrational dephasing of a metal carbonyl in glass forming liquids has a temperature dependence that is much steeper both in the glass and in the liquid than the Mb−CO dephasing over the same temperature range. The observed differences help confirm that the vibrational pure dephasing of Mb−CO is caused by protein fluctuations rather than direct coupling to the solvent dynamics.

The $T^{1.3}$ pure dephasing temperature dependence of Mb−CO in the lower temperature portion of Figure 5 is consistent with a tunneling TLS model of the protein. To apply a protein-tunneling TLS model requires describing the ensemble of protein molecules in the following manner. A protein in the ensemble is a nonequilibrium system with many possible conformations. Conformational changes can be viewed as occurring via motion on a multidimensional potential surface. Such a surface has frequently been referred to as the energy landscape. Three types of dynamics can occur. Motions that do not move the system from one point to another on the surface are vibrations. These can be high-frequency vibrations of small groups such as a C−H stretch or low-frequency motions that involve larger sections of the protein backbone. These motions are fluctuations about a local minimum on the surface. They do not change the protein conformation. Other motions move the system from one point on the energy landscape to another. These are conformational changes and can occur via tunneling through a potential barrier from one minimum to another, or they can occur by activation over a barrier. In either case, the system will pass from one local minimum to another.

If the system is in one local minimum of the multidimensional potential surface, at relatively low temperature the vast majority of the barriers to movement to other minima may be so high that both tunneling through or activation over the barriers is impossible. However, from this point on the landscape there may be one other conformation that can be reached by tunneling. The PTLS model represents this complex potential surface as a collection of double well potentials. There is a broad distribu-

tion of differences in energy E between the two sides of the double wells. Each side of a double well represents a conformation on the energy landscape. There is also a distribution of barrier heights V and tunneling parameters λ. (For a barrier modeled as an inverted parabola, the probability of tunneling goes as $e^{-\lambda}$, with $\lambda = d(2mV/h^2)^{1/2}$; d is the distance through the barrier, and m is the mass.) The tunneling PTLS model describes the fluctuations in protein conformations as tunneling between the two sides of the PTLS. There are many PTLSs, and each transition of a particular PTLS represents a specific conformational change of the protein.

Although activation does not occur over barriers, the thermal energy of the protein still plays a role because the two sides of the PTLS, in general, have different energies. To move from the low-energy side to the high-energy side requires the absorption of a phonon from the protein. To move from the high-energy side to the low-energy side requires emission of a phonon. In the protein, the continuum of very low frequency vibrations[34] serves the role of the phonon continuum in a glass.

Since there are many PTLSs in each protein, there are many conformational fluctuations occurring with a distribution of rates, $P(R)$, determined by the distributions of E and λ, i.e., $P(E)$ and $P(\lambda)$. As the temperature is raised, the occupation numbers of the proteins' low-frequency vibrational modes increase. The increase causes faster transitions between the sides of the active PTLS and also allows PTLSs that were inactive because E was too large to become active. Fluctuations in the protein structure, which correspond to transitions between sides of the PTLS, are coupled to the vibrational frequency of the CO. These fluctuations in structure cause fluctuations in the vibrational frequency, giving rise to pure dephasing.

In a protein, there will be a large number of PTLSs each undergoing jumps between two states. If these jumps are uncorrelated, i.e., a transition in one PTLS does not influence the probability of a transition in another PTLS, then theoretical work developed to describe dynamics in low-temperature glasses can be applied to the vibrational dephasing problem in Mb−CO.[28] In the context of the uncorrelated sudden jump model, from eqs 4 and 5, the observed $T^{1.3}$ temperature dependence implies that $P(E) \propto E^{0.3}$.

This is a very flat distribution of energies. The PTLS model of the Mb protein indicates that the protein has an energy landscape on which conformational changes occur via tunneling, and the distribution of energy differences between conformations is very broad and almost flat. The tunneling rate must be sufficiently fast to cause vibrational pure dephasing on a 10 ps time scale. The essentially flat energy landscape does not have to extend over all energies to be consistent with the data. PTLSs with energies splittings greater than ~ 2 kT contribute little to the dephasing. Thus, there can be a deviation from $E^{0.3}$ or even a cut off at high energy which will not influence the observed temperature dependence. Furthermore, there can be a distribution of very low energy barriers that will not come into play. Electronic dephasing experiments conducted on heme proteins at very low temperature (2 K) have been interpreted in terms of very low barriers to structural changes.[2] In the current experiments, such a set of barriers will be so low in energy that they do not impose constraints on the dynamical motions in the 100 K range. The set of low barriers can be viewed as slight roughness of the bottoms of the potential wells that are involved in the dynamics at higher temperatures. The low-temperature and high-temperature results suggest there is a set of very low barriers involved in the very low temperature dynamics and a set of relatively high barriers that is involved in the higher temperature dynamics and a gap in between. If the distribution of barrier heights were continuous, then, at the elevated temperatures of these experiments, a significant amount of the conformational dynamics would involve activation over the intermediate barriers. Such a situation is inconsistent with the PTLS model that gives rise to the observed $T^{1.3}$ temperature dependence.

The PTLS model of the Mb protein is consistent with molecular dynamics simulations which show many conformations with transitions occurring on a picosecond time scale.[9] However, the simulations are classical and at high temperature. They do not address the question of tunneling. Nonetheless, they do indicate a complex energy landscape with many local minima, a prerequisite for the PTLS model.

Within the context of the uncorrelated sudden jump model of vibrational dephasing, two other restrictions are placed on the system to be consistent with the experimental data.[28] As shown in Figure 3, the vibrational echo data decay exponentially. To obtain the observed temperature dependence and an exponential decay requires that the coupling of the PTLS to the CO vibrational transition frequency falls off as $1/r^3$, where r is the distance of a PTLS to the CO or probably the heme (see below). Also the distribution of PTLS fluctuation rates, $P(R) \propto 1/R$, where R is the rate of fluctuation (jumps).[28] Other theoretical descriptions of TLS dynamics and optical dephasing in low-temperature glasses place similar restrictions on the coupling and rate distribution.[29] Dipolar coupling and a $\sim 1/R$ fluctuation rate distribution are believed to occur in glasses at low temperatures. A theoretical investigation is currently in progress to determine if a less restrictive, non-PTLS model of protein fluctuations can account for the observed power law temperature dependence.[35] This model still involves an energy landscape with a broad distribution of barriers and transitions among a limited set of protein conformations, but it does not require a specific coupling mechanism or a $1/R$ distribution of fluctuation rates.

Above the solvent glass transition temperature, the vibrational pure dephasing line width appears to be exponentially activated. As discussed above, the change in the temperature dependence may be attributed to the softening of the boundary condition placed on protein motions by the solvent. The rigid glass boundary may permit only conformational changes that preserve the surface shape of the protein. When the solvent is a liquid, protein conformational changes that involve changes in the surface shape become possible. Barriers to conformational change that were insurmountable with the rigid glassy boundary condition become lower when the solvent is a liquid. The result is a distinct change in the energy landscape. Conformational changes that were impossible below the solvent glass transition can occur via activation over barriers. The activated conformational changes can replace the $T^{1.3}$ process as the dominant mechanism for protein fluctuations. The activated process does not arise from excitation of a particular vibrational mode of the protein since the break in the functional form of the temperature dependence tracks the glass transition temperature when the solvent is changed.[22]

As discussed above in connection with the fit of the data to eq 3, it is not possible to determine if there is a change from the $T^{1.3}$ mechanism to activated barrier crossing or if the power law portion persists to high temperatures. If the activated pathways involve conformational changes that cannot occur by tunneling, then the number of tunneling events per unit time will decrease and possibly become unimportant. Therefore, there is a range of possible activation energies that fall between $\Delta E = 1250$ cm^{-1} (power law continues to high temperature) and $\Delta E = 800$ cm^{-1} (power law unimportant at high temper-

atures). The net result is that $\Delta E \approx 1000$ cm^{-1}, but this number is uncertain to several hundred cm^{-1}. Since the activated temperature dependence is observed over a relatively narrow range of temperature, the single activation energy could actually represent a distribution of ΔE values. It is important to emphasize that only those structural changes that influence the CO vibrational energy are manifest in the vibrational echo pure dephasing temperature dependence.

3. Mechanism of Coupling Protein Fluctuations to the CO Vibrational Frequency. Vibrational pure dephasing is caused by fluctuations in the vibrational transition energy.[16] For protein conformational changes to cause pure dephasing there must be a mechanism that couples the protein dynamics to the CO vibrational energy.

In the literature of the vibrational spectroscopy of CO bound to metalloporphyrins, model hemes, myoglobin, and mutant myoglobins, it is well established that static shifts in the vibrational frequency of the CO stretch are caused by changes in back-donation of electron density from the extended metal−macrocycle π system to the CO π^* antibonding molecular orbital.[36] The back-donation is referred to as back-bonding. The π system of the macrocycle is a superposition of the metal d_π and ring nitrogen and carbon p_π orbitals. Electron density is transferred to the empty CO π^* orbital. Electron density in the π^* orbital reduces the strength of the CO bond and, therefore, lowers the vibrational frequency. In model compounds, when Fe is replace by Ru, and then by Os, the vibrational frequency shifts progressively to lower energy.[37] As the metal atom becomes larger, the d orbitals are larger and more diffuse, resulting in more back-bonding. Changes in substituents on the porphyrin are also correlated with changes in back bonding and frequency shifts. In Mb and mutant Mb's, changes in particular amino acids influence back-bonding via changes in local electric fields that influence the macrocycle electron density and therefore the back-bonding.[5]

Recently, back-bonding has been shown to be intimately involved in CO vibrational lifetime dynamics in model hemes and in Mb and mutant Mb's.[37,38] The vibrational lifetimes of CO bound to 37 model compounds and 7 Mb's were found to be related to the vibrational frequency of the CO. As the frequency shifts to the red, the rate of vibrational relaxation becomes larger. The relationship between frequency and relaxation rate is essentially linear. A 3% change in vibrational frequency results in an approximately 3-fold change in the vibrational relaxation rate. As the back-bonding increases, the coupling of the CO to the heme increases, and the vibrational relaxation rate increases.

Static changes in back-bonding cause changes in the CO vibrational frequency and cause changes in the vibrational lifetime. Therefore, we propose that the basic mechanism for the Mb−CO vibrational pure dephasing involves fluctuations in the back-bonding. Fluctuations in back-bonding will cause fluctuations in the CO vibrational frequency just as static changes in back-bonding cause static shifts in the CO energy. We suggest two possible paths by which fluctuations in protein structure influence the back-bonding and, therefore, cause the vibrational pure dephasing. One we term the electric field mechanism and the other, the mechanical mechanism.

The electric field mechanism depends on the fact that the heme−CO subsystem is surrounded by the protein, which contains polar groups, e.g., the distal histidine. Fluctuations in the protein conformation will produce time varying electric fields at the heme. These time dependent fields will in turn cause modulations of the macrocycle electron density distribution. The time-dependent variations in electron density will cause fluctua-

15628 *J. Phys. Chem., Vol. 100, No. 38, 1996*

Rella et al.

tions in the back-bonding and, therefore, CO vibrational transition frequency fluctuations, i.e., pure dephasing.

The mechanical mechanism involves the fact that the heme— CO subsystem is coupled to the protein by only one covalent bond. The proximal histidine, which is the ligand bound to the metal on the opposite side of the heme from the CO, makes the only covalent bond to the protein. From CO photodissociation experiments, it is known that movement of the Fe out of the plane of the macrocycle is mechanically coupled to the protein through the proximal histidine.[7] In the mechanical pure dephasing mechanism, as the protein undergoes conformational fluctuations, the proximal histidine, like a piston, pushes and pulls the Fe in and out of the plane of the heme ring system. The Fe motion produces modifications of the π system's electron density distribution. The electron density fluctuations will, in turn, cause the back-bonding to fluctuate and therefore induce pure dephasing.

Both of the proposed mechanisms involve protein conformational motions that modulate the heme back-bonding. Both mechanisms provide the coupling between the protein motions and the CO vibrational transition energy necessary to cause pure dephasing. At this time, it is not possible to determine which of the mechanisms is dominant. However, it should be possible to address the nature of the mechanism in the future by performing temperature-dependent vibrational echo experiments on a mutant Mb in which the proximal histidine is removed and replaced by a ligand not covalently bonded to the protein and on a mutant Mb in which the distal histidine is replaced with valine. Changes in the pure dephasing of the mutants will provide insights into the coupling mechanism.

IV. Concluding Remarks

In this paper, we have described the first application of vibrational echoes to the study of protein dynamics. The vibrational echo experiment makes it possible to remove the inhomogeneous broadening from a vibrational spectrum and obtain the dynamical information contained in the homogeneous spectrum. Using the Stanford FEL as a source, we have performed vibrational echo and lifetime relaxation measurements on CO bound to the active site of myoglobin. Combining the results of the two types of experiments, we have obtained the pure dephasing time, T_2^*, at a series of temperatures from 60 K to room temperature. T_2^* is a measure of the vibrational energy fluctuations induced by conformational fluctuations of the protein. It is found that the CO vibrational line is inhomogeneously broadened at all temperatures including room temperature. Thus, even at room temperature, the ensemble of protein molecules exists in a distribution of conformational substates that interconvert slowly compared to the 10 ps time scale of the echo experiments.

At low temperatures, the pure dephasing contribution to the homogeneous line width increases as a power law, $T^{1.3}$. Above the solvent glass transition temperature, the temperature dependence becomes exponentially activated. The change in the functional form of the temperature dependence is attributed to the softening of the boundary condition imposed on protein chain motions when the solvent (glycerol/water) becomes liquid. This view will be tested in subsequent experiments by using other solvents with different glass transitions temperatures. A protein two-level system (PTLS) model of protein dynamics, equivalent to the TLS model of glasses, was discussed. The characteristics of protein dynamics required to give rise to the observed temperature dependence are described in the context of the PTLS dynamics. This model places considerable restrictions on the nature of the protein dynamics if all

observations are to be explained. A theoretical investigation is in progress to determine if a less restrictive model, which does not involve PTLS dynamics, can reproduce the observed results.

A general mechanism was proposed to explain how conformational fluctuations of the protein couple to the vibrational transition energy to cause pure dephasing. It is known that static changes in heme back-bonding to the CO antibonding π^* molecular orbital result in changes in the transition energy of the CO vibration. It was proposed that protein conformational fluctuations cause fluctuations in the back-bonding and, therefore, cause pure dephasing. Two possible mechanisms through which the protein conformational fluctuations can cause modulation of the back-bonding were put forward. One is how protein induced fluctuating electric fields influence the heme, and the other is how protein induced fluctuations of the proximal histidine position causes structural perturbations of the heme. Future experiments will address these mechanisms by performing vibrational echoes on mutant Mb's.

These first vibrational echo experiments on proteins have provided a new method for examining protein dynamics at biologically relevant temperatures and have already yielded some intriguing insights into how protein dynamics are transmitted to the active site of myoglobin. The vibrational echo is the ps IR vibrational analog of the spin echo of NMR. The advent of the spin echo in 1950 ushered in a new dimension in magnetic resonance spectroscopy. The spin echo is the simplest coherent pulse sequence in magnetic resonance, yet it is the precursor to all of the important pulse sequences that are now used routinely in magnetic resonance. Without coherent pulse sequences, NMR would not be the powerful probe of structure and dynamics that it is today. As the spin echo did for NMR, the vibrational echo is expanding scope of IR vibrational spectroscopy.

Acknowledgment. This research was supported by the Medical Free Electron Laser Program, through the Office of Naval Research, Contract N00014-94-1-1024 (C.W.R., M.D.F., A.K., H.A.S.). Additional support was provided by the Office of Naval Research, Biology Division, Contract N00014-95-1-0259, the National Science Foundation, Division of Materials Research Grant DMR94-04806 (D.D.D., J.R.H.), and the National Science Foundation, Division of Materials Research, Grant DMR93-22504 (K.R., M.D.F.). We thank Professor Steven G. Boxer, Stanford University, Professor Robin Hochstrasser, University of Pennsylvania, and Professor Kenneth Suslick, University of Illinois at Urbana, for informative conversations pertaining to this research.

References and Notes

(1) Tokmakoff, A.; Zimdars, D.; Urdahl, R. S.; Francis, R. S.; Kwok, A. S.; Fayer, M. D. *J. Phys. Chem.* **1995**, *99*, 13310. Tokmakoff, A.; Fayer, M. D. *Acc. Chem. Res.* **1995**, *20*, 437.

(2) Leeson, D. T.; Wiersma, D. A. *Phys. Rev. Lett.* **1995**, *74*, 2138.

(3) Antonini, E. B. *Hemoglobin and Myoglobin in their Reactions with Ligands*; North Holland: Amsterdam, 1971.

(4) Kuriyan, J. W. S.; Karplus, M.; Petsko, G. A. *J. Mol. Biol.* **1986**, *192*, 133. Quillin, M. L.; Arduini, R. M.; Olson, J. S.; Phillips, G. N., Jr. *J. Mol. Biol.* **1993**, *234*, 140.

(5) Oldfield, E., Guo, K.; Augspurger, J. D.; Dykstra, C. E. *J. Am. Chem. Soc.*, **1991**, *113*, 7537.

(6) DeBrunner, P. G.; Frauenfelder, H. *Annu. Rev. Phys. Chem.* **1982**, *33*, 283. Frauenfelder, H.; Parak, F.; Young, R. D. *Annu. Rev. Biophys. Biophys. Chem.* **1988**, *17*, 451.

(7) Petrich, J. W.; Martin, J. L. In *Time-Resolved Spectroscopy*; Clark, R. J. H., Hester, R. E., Eds.; Wiley: New York, 1989; p 335. Friedman, J. M.; Rousseau, D. L.; Ondrias, M. R. *Annu. Rev. Phys. Chem.* **1982**, *33*, 471

(8) (a) Alben, J. O.; Caughy, W. S. *Biochemistry* **1968**, *7*, 175. Caughey, W. S.; Shimada, H.; Choc, M. C.; Tucker, M. P. *Proc. Natl. Acad. Sci. U.S.A.* **1981**, *78*, 2903. Alben, J. O.; Beece, D.; Bowne, S. F.; Doster, W.;

Vibrational Echo Studies of Myoglobin—CO

Eisenstein, L.; Frauenfelder, H.; Good, D.; McDonald, J. D.; Marden, M. C.; Moh, P. P.; Reinisch, L.; Reynolds, A. H.; Shyamsunder, E.; Yue, K. T. *Proc. Natl. Acad. Sci. U.S.A.* **1982**, *79*, 3744. (b) Ansari, A.; Berendzen, J.; Braunstein, D.; Cowen, B. R.; Frauenfelder, H.; Hong, M. K.; Iben, I. E. T.; Johnson, J. B.; Ormos, P.; Sauke, T. Scholl, R.; Schulte, A.; Steinbach, P. J.; Vittitow, J.; Young, R. D. *Biophys. Chem.* **1987**, *26*, 337.

(9) Elber, R.; Karplus, M. *Science* **1987**, *235*, 318.

(10) Iben, I. E. T.; Braunstein, D.; Doster, W.; Frauenfelder, H.; Hong, M. K.; Johnson, J. B.; Luck, S.; Ormos, P.; Schulte, A.; Steinbach, P. J.; Xie, A.; Young, R. D. *Phys. Rev. Lett.* **1989**, *62*, 1916.

(11) Narasimhan, L. R.; Littau, K. A.; Pack, D. W.; Bai, Y. S.; Elschner, A.; Fayer, M. D. *Chem. Rev.* **1990**, *90*, 439 and references contained therein.

(12) Phillips, W. A., Ed.; *Amorphous Solids. Low Temperature Properties, Topics in Current Physics*; Springer, Berlin, 1981; Vol. 24. Stevels, J. M. The Structural and Physical Properties of Glass. In *Encyclopedia of Physics*, Flügge, S., Ed.; (Thermodynamics of Liquids and Solids). Springer-Verlag: Berlin, 1962; Vol. 13.

(13) Anderson, P. W.; Halperin, B. I.; Varma, C. M. *Philos. Mag.* **1972**, *25*, 1. Phillips, W. A. *J. Low Temp. Phys.* **1972**, *7*, 351.

(14) Gordon, R. G. *J. Chem. Phys.* **1965**, *43*, 1307. Gordon, R. G. *Adv. Magn. Reson.* **1968**, *3*, 1. Berne, B. J. In *Physical Chemistry: An Advanced Treatise*; Henderson, D., Ed.; Academic Press: New York, 1971; Vol. VIIIB.

(15) Loring, R. F.; Mukamel, S. *J. Chem. Phys.* **1985**, *83*, 2116.

(16) Tokmakoff, A.; Fayer, M. D. *J. Chem. Phys.* **1995**, *102*, 2810.

(17) Hahn, E. L. *Phys. Rev.* **1950**, *80*, 580.

(18) Kurnit, N. A.; Abella, I. D.; Hartmann, S. R. *Phys. Rev. Lett.* **1964**, *13*, 567. Abella, I. D.; Kurnit, N. A.; Hartmann, S. R. *Phys. Rev.* **1966**, *141*, 391.

(19) Farrar, T. C.; Becker, D. E. *Pulse and Fourier Transform NMR* , Academic Press: New York, 1971. Skinner, J. L.; Andersen, H. C.; Fayer, M. D. *J. Chem. Phys.* **1981**, *75*, 3195.

(20) Marquardt, D. W. *J. Soc. Ind. Appl. Math.* **1963**, *2*, 431.

(21) Frauenfelder, H.; Petsko, G. A.; Tsernoglou, D. *Nature* **1979**, *280*, 558. Brooks, C. L.; Karplus, M.; Pettitt, B. M. *Advances Physical Chemistry*; Wiley: New York, 1988; Vol. LXXI. Li, H.; Elber, R.; Straub, J. E. *J. Biol. Chem.* **1993**, *268*, 17908. Elber, R; Karplus, M. *J. Am. Chem. Soc.* **1990**, *112*, 9161. Straub, J. E.; Karplus, M. *Chem. Phys.* **1991**, *158*, 221.

(22) Rella, C. W.; Kwok, A.; Rector, K.; Hill, J. R.; Schwettman, H. A.; Dlott, D. D.; Fayer, M. D. *J. Chem. Phys.* **1996**, in preparation.

(23) Frauenfelder, H.; Alberding, N. A.; Ansari, A.; Braunstein, D.; Cowen, B. R.; Hong, M. K.; Iben, I. E. T.; Johnson, J. B.; Luck, S.; Marden, M. C.; Mourant, J. R.; Ormos, P.; Reinisch, L.; Scholl, R.; Schulte, A.; Shyamsunder, E.; Sorensen, L. B.; Steinbach, P. J.; Xie, A.; Young, R. D.; Yue, K. T. *J. Phys. Chem.* **1990**, *94*, 1024.

(24) Austin, R. H.; Beeson, K.; Eisenstein, L.; Frauenfelder, H.; Gunsalus, I. C.; Marshall, V. P. *Phys. Rev. Lett.* **1974**, *32*, 403.

(25) Phillips, W. A. *Rep. Prog. Phys.*, **1987**, *50*, 1657.

(26) Iben, I. E. T.; Baunstein, D.; Doster, W.; Frauenfelder, H.; Hong, M. K.; Johnson, J. B.; Luck, S.; Ormos, P.; Schulte, A.; Steinbach, P. J.; Xie, A. H.; Young, R. D. *Phys. Rev. Lett.* **1989**, *62*, 1916.

(27) Zeller, R. C.; Pohl, R. O. *Phys. Rev. B* **1971**, *4*, 2029.

(28) Berg, M.; Walsh, C. A.; Narasimhan, L. R.; Littau, Karl A.; Fayer, M. D. *J. Chem. Phys.* **1988**, *88*, 1564. Bai, Y. S.; Fayer, M. D. *Phys. Rev. B* **1989**, *39*, 11066.

(29) Silbey, R., private communication, to be published.

(30) (a) Hill, J. R.; Dlott, D. D.; Fayer, M. D., unpublished results. (b) Ivanov, D.; Sage, J. T.; Keim, M.; Powell, J. R.; Asher, S. A.; Champion, P. M. *J. Am. Chem. Soc.* **1994**, *116*, 4139.

(31) Freidrich, J.; Harrer, D. In *Optical Spectroscopy of Glasses*; Zschokke, I., Ed.; R. Reidel Publishing Company: Dordrecht, The Netherlands, 1986.

(32) Macfarlane, R. M.; Shelby, R. M. *J. Lumin.* **1987**, *36*, 179.

(33) Angell, C. A. *J. Phys. Chem. Solids* **1988**, *49*, 863. Angell, C. A. *J. Phys. Chem.* **1982**, *86*, 3845. Fredrickson, G. H. *Annu. Rev. Phys. Chem.* **1988**, *39*, 149.

(34) Go, N.; Noguti, T.; Nishikawa, T. *Proc. Natl. Acad. Sci. U.S.A.* **1983**, *80*, 3696. Roitberg, A.; Gerber, R. B.; Elber, R.; Ratner, M. A. *Science* **1995**, *268*, 1319.

(35) Rella, C. W.; Fayer, M. D., to be published.

(36) Cotton, F. A.; Wilkinson, G. *Advanced Inorganic Chemistry*; Wiley-Interscience: New York, 1988; p 57. Boldt, N. J.; Goodwill, K. E.; Bocian, D. F. *Inorg. Chem.* **1988**, *27*, 1188. Spiro, T. G. In *Iron Porphyrins*; Lever, A. B. P., Gray, H. B., Eds.; Addison-Wesley: Reading, MA, 1983; Vol. II. Li, X. Y. Spiro, T. G. *J. Am. Chem. Soc.* **1988**, *110*, 6024.

(37) Hill, Jeffrey R.; Dlott, D. D.; Fayer, M. D.; Peterson, K. A.; Rella, C. W.; Rosenblatt, M. M.; Suslick, K. S.; Ziegler, C. J. *Chem. Phys. Lett.* **1995**, *244*, 218.

(38) Hill, J. R.; Tokmakoff, A.; Peterson, K. A.; Sauter, B.; Zimdars, D. A.; Dlott, D. D.; Fayer, M. D. *J. Phys. Chem.* **1994**, *98*, 11213. Hill, J. R.; Dlott, D. D.; Rella, C. W.; Peterson, K. A.; Decatur, S. M.; Boxer, S. G.; Fayer, M. D. *J. Phys. Chem.* **1996**, accepted. Hill, J. R.; Dlott, D. D.; Rella, C. W.; Smith, T. I.; Schwettman, H. A.; Peterson, K. A.; Kwok, A.; Rector, K.; Fayer, M. D. *Biospec.* **1996**, submitted for publication. Hill, J. R.; Dlott, D. D.; Fayer, M. D.; Rella, C. W.; Rosenblatt, M. M.; Suslick, K. S.; Ziegler, C. J. *J. Phys. Chem.* **1996**, accepted for publication.

JP961129R

Vibrational anharmonicity and multilevel vibrational dephasing from vibrational echo beats

K. D. Rector, A. S. Kwok,[a] C. Ferrante,[b] and A. Tokmakoff[c]
Department of Chemistry, Stanford University, Stanford, California 94305

C. W. Rella
Hansen Experimental Physics Laboratory, Stanford University, Stanford, California 94305

M. D. Fayer[d]
Department of Chemistry, Stanford University, Stanford, California 94305

(Received 22 January 1997; accepted 21 March 1997)

Vibrational echo experiments were performed on the IR active CO stretching modes (~ 2000 cm^{-1}) of rhodium dicarbonylacetylacetonate [Rh(CO)$_2$acac] and tungsten hexacarbonyl [W(CO)$_6$] in dibutylphthalate and a mutant of myoglobin-CO (H64V-CO) in glycerol–water using ps IR pulses from a free electron laser. The echo decays display pronounced beats and are nonexponential. The beats and nonexponential decays arise because the bandwidths of the laser pulses exceed the vibrational anharmonicities, leading to the excitation and dephasing of a multilevel coherence. From the beat frequencies, the anharmonicities are determined to be 14.7, 13.5, and 25.4 cm^{-1}, for W(CO)$_6$, Rh(CO)$_2$acac, and H64V-CO, respectively. From the components of the nonexponential decays, the vibrational dephasing at very low temperature of both the $v=0-1$ and $v=1-2$ transitions are determined. At the lowest temperatures, $T_2 \approx 2T_1$, so the $v=2$ lifetimes are obtained for the three molecules. These are found to be significantly shorter than the $v=1$ lifetimes. Although the $v=1$ lifetimes are similar for the three molecules, there is a wide variation in the $v=2$ lifetimes. © *1997 American Institute of Physics.* [S0021-9606(97)02324-6]

I. INTRODUCTION

The measurement of vibrational spectra using Fourier transform infrared (FTIR) spectroscopy provides a wealth of information about the equilibrium structures of molecules and the effects of solvents on normal modes. However, important types of measurements, i.e., vibrational homogeneous linewidth and the vibrational anharmonicity, are either difficult or impossible to perform with ordinary linear techniques. Measurements of vibrational anharmonicities provide information about the shape of the vibrational potential surface. Vibrational anharmonicities are not readily obtained using a conventional FTIR because the $v=0\rightarrow2$ transition is forbidden, and, therefore, the absorption at the frequency corresponding to the $v=0\rightarrow2$ transition is weak. In addition, the spectral region around a $v=0\rightarrow2$ transition is frequently congested with combination bands, making it difficult to identify the correct peak. A direct measurement of the $v=1\rightarrow2$ transition is not usually feasible for the higher frequency modes (1000–4000 cm^{-1}) because the thermal population at ambient temperatures of the $v=1$ state is down five to twenty factors of e compared to the $v=0$ state.

Another technique for measuring vibrational anharmonicity involves the use of time resolved, two color vibrational pump–probe spectroscopy.[1,2] In this method, the $v=0\rightarrow1$ transition is pumped with an intense ps IR pulse. On a time scale short compared to the vibrational lifetime, a second IR probe, generally having a wide bandwidth, is passed through the sample and spectrally resolved. The $v=1\rightarrow2$ transition appears as an absorption of the probe that is not present in the absence of the pump. This method has been successfully applied,[1,2] but it requires two ps IR pulses with different characteristics, i.e., wavelength and bandwidth.

An alternative approach is to perform a vibrational echo[3–6] using pulses having sufficiently wide bandwidth to establish a multilevel coherence among the $v=0$, 1, and 2 states.[7] The multilevel coherence results in a vibrational echo decay with beats. The beats occur at the frequency of the difference between the $v=0\rightarrow1$ and $v=1\rightarrow2$ transition energies, i.e., at the frequency of the vibrational anharmonic splitting. This is not a conventional quantum beat. In a quantum beat, a state is coupled by the radiation field directly to two other states that fall within the bandwidth of the pulse. In the vibrational echo anharmonic beat (VEB) experiment, the radiation field couples $v=0\rightarrow1$ and then $v=1\rightarrow2$. There is no direct coupling between $v=0\rightarrow2$.

In addition to measuring the vibrational anharmonicity through the echo decay beat frequency, the echo decay provides the homogeneous dephasing time (homogeneous linewidth) of both the $v=0-1$ and $v=1-2$ transitions. The homogeneous linewidth provides information on the dynamical interactions of a vibration with its environment. Vibrational echo experiments on liquids,[4,5] glasses,[4,5] and proteins[6,8] have shown that vibrational spectra may be

[a]Permanent address: Department of Physics, Franklin and Marshall College, Lancaster, PA 17604.
[b]Present address: Department of Chemistry, University of Munich, 80290 Munich Germany.
[c]Present address: Department of Chemistry, University of Chicago, Chicago, IL 60637.
[d]Author to whom correspondence should be sent.

inhomogeneously broadened, even at room temperature. Therefore, vibrational echo experiments are necessary to extract the homogeneous linewidth from the observed spectroscopic line. The VEB experiment makes it possible to compare the homogeneous linewidths of the $v=0-1$ and $v=1-2$ transitions. At the polarization level for Lorentzian homogeneous lines, the echo decay is a bi-exponential squared with beats. One exponential corresponds to the decay of the $v=1-2$ coherence and the other exponential corresponds to the decay of the $v=0-1$ coherence. Therefore, the VEB experiment allows both the homogeneous linewidths and the vibrational anharmonicity to be experimental observables.

In this paper, VEB experiments on CO stretching modes of three molecules are presented: rhodiumdicarbonylacetylacetonate [Rh(CO)$_2$acac] and tungsten hexacarbonyl [W(CO)$_6$] in dibutylphthalate (DBP), and CO bound to the active site of a myoglobin protein in a 95:5 mixture of glycerol:water. The myoglobin protein is a mutant of wild type myoglobin in which the distal histidine (position 64) is replaced with a valine (H64V-CO).[6] The protein pocket around the CO in H64V is markedly different than either of the inorganic compounds. In the protein, the CO is bound to an octahedral Fe which has four bonds to a large aromatic macrocycle. The last bond of the Fe is to the proximal histidine (position 93).

The vibrational echo experiments on Rh(CO)$_2$acac are the first for this molecule. Vibrational echo studies of W(CO)$_6$ in several liquid and glassy solvents have been reported,[3,5] including the first observation of vibrational echo anharmonic beats on W(CO)$_6$ in DBP.[7] Further, extensive vibrational echo studies and lifetime measurements of the $v=0-1$ transition have been presented previously for Mb–CO and H64V-CO.[6,8]

In the following, Rh(CO)$_2$acac data is presented as a function of the excitation frequency. As the frequency is decreased, the beats become more pronounced. These results are compared to an approximate theoretical calculation, and the trends are found to be in accord. The $v=0-1$ and $v=1-2$ dephasing times and the anharmonicity are also obtained. For H64V-CO, the anharmonicity is much greater than in the two organometallic compounds. By tuning to lower frequencies than the $v=0\rightarrow1$ transition, beats are observed although the pulse bandwidth is insufficient to produce beats when centered on the $v=0\rightarrow1$ line. Again, the $v=0-1$ and $v=1-2$ dephasing times and the anharmonicity are obtained. These results are compared to prior measurements on W(CO)$_6$ in DBP.

II. THE NATURE OF THE EXPERIMENT AND PROCEDURES

In a vibrational echo experiment, two IR pulses, tuned to the frequency of the molecular vibration of interest, are crossed in the sample. The first pulse creates an ensemble of coherent superposition states that begin to dephase because of inhomogeneous broadening. A second pulse, delayed by time τ, is incident on the sample at an angle θ with respect to the first pulse and initiates rephasing of the inhomogeneous contributions to the vibrational spectral line. This rephasing results in a macroscopic polarization that is observed as an echo pulse at time 2τ. The echo emerges from the sample at an angle 2θ with respect to the first beam due to wave vector matching conditions. The integrated intensity of the echo pulse is measured as a function of τ.

For the case in which the laser bandwidth is narrow with respect to the vibrational anharmonicity, such that the vibrational coherence involves only the $v=0-1$ transition, the experiment is modeled well by a two level system. For a Lorentzian homogeneous lineshape with linewidth $\Gamma=1/\pi T_2$, the echo decays as

$$I(\tau)=I_0\exp(-4\gamma\tau), \tag{1}$$

where $\gamma=1/T_2$ for the $v=0-1$ transition. The homogeneous dephasing time has contributions from the pure dephasing time, T_2^*, and the vibrational lifetime, T_1,

$$\frac{1}{\pi T_2}=\frac{1}{\pi T_2^*}+\frac{1}{2\pi T_1}. \tag{2}$$

There can also be a contribution from orientational relaxation.[5] However, the experiments reported here were conducted in low temperature glassy solvents, so the contribution from orientational relaxation is negligible. The lifetime can be measured with a pump–probe experiment, and when combined with a vibrational echo measurement of T_2, the pure dephasing contribution to the homogeneous linewidth can be determined. For the systems studied here, the lifetime components of the homogeneous lines are only slightly temperature dependent. The pure dephasing components are more strongly temperature dependent.

When the laser bandwidth is similar to the anharmonicity, short pulse excitation will create a three level coherence involving the $v=0$, 1, and 2 vibrational levels. Previous theoretical work on a three level echo described an equally spaced system.[9] The derivation of the VEB signals for three level systems has a large contribution from experiments and theory studying coherent oscillations in semiconductor structures.[10,11] The echo signal can be described for an unequally spaced three-level system using a semiclassical diagrammatic perturbation theory treatment of the third-order nonlinear polarizability.[12,13] (See Appendix for details.) The three-level system is spaced by the frequencies ω_{01} and ω_{12}, where $\omega_{01}=\omega_{12}+\Delta$, and $\Delta\ll\omega_{01}$ and ω_{12}. Δ is the anharmonic vibrational energy splitting. The transition frequencies ω_{01} and ω_{12} lie within the bandwidth of the pulses. For such a system, three independent resonant pathways (diagrams) exist that result in rephasing and the generation of the echo pulse. (See Fig. 6 in the Appendix.) In addition to the two that describe a rephasing in a two level system,[13] a third diagram accounts for the possibility of rephasing the $v=1-2$ coherence. As discussed in the appendix, for a finite pulse bandwidth, where the E-field amplitude differs at ω_{01} and ω_{12}, the decay is given by[7]

$$I(\tau)=I(0)\exp(-2\gamma_{01}\tau)\{(E_{01}\cdot\mu_{01})^2\exp(-2\gamma_{01}\tau)$$

$$+(E_{12}\cdot\mu_{12})^2\exp(-2\gamma_{12}\tau)-2(E_{01}\cdot\mu_{01})$$

$$\times(E_{12}\cdot\mu_{12})\exp[-(\gamma_{01}+\gamma_{12})\tau]\cos(\Delta\tau+\phi)\}.$$

$$(3)$$

Here, E_{01} and E_{12} are amplitudes of the electric fields at the respective transitions and μ_{01} and μ_{12} are the respective dipole transition moments which are constant. γ_{01} and γ_{12} are the corresponding homogeneous dephasing decay constants, and Δ is the vibrational anharmonic splitting frequency (the beat frequency in the signal). $I(0)$ contains all the factors that determine the strength of the signal but are not involved in either the time dependent decays or the wavelength dependence of the beats. The dephasing rates for the two transitions, γ_{01} and γ_{12}, are phenomenological; no model has been assumed for the coupling of these modes to the bath. For the narrow bandwidth case ($E_{12}=0$), Eq. (1) is recovered. The phase factor in Eq. (3) does not arise from the derivation, i.e., $\phi=0$. As is standard in such derivations for which analytical results can be obtained, the pulse is taken to be a delta function in duration. Therefore, it has infinite bandwidth, and the E-field amplitude is independent of frequency. The derivation was modified by specifying different E-fields at the two transition frequencies. However, it has been shown that inclusion of the finite bandwidth of the pulses leads to a phase factor that need not be zero.[14] Therefore, the phase factor was included, and it was found to be an aid in fitting the data discussed below.

To compare Eq. (3) to data requires a convolution to account for the finite pulse duration. To extract an echo decay that is on the same time scale as the pulse duration requires full consideration of the three time ordered interactions of the radiation fields with the vibrations. In these experiments, the echo decays are long compared to the pulse durations, so this procedure is unnecessary. However, the beat frequency is comparable to the pulse duration. Therefore, in the data, the beats appear with much less depth of modulation than they would have in the absence of a finite instrument response. To account for this, Eq. (3) is convolved with the echo time dependence that would be observed for a sample with delta function response. This is defined as the instrument response function. For finite instrument responses, the data are fit by

$$S(\tau)=\mathfrak{T}^{-1}\{\mathfrak{T}[p(\sqrt{\tfrac{3}{2}}\tau)]x\mathfrak{T}[I(\tau)]\},\qquad(4)$$

where $\mathfrak{T}$ and $\mathfrak{T}^{-1}$ are Fourier transform and inverse Fourier transform, respectively. $I(t)$ is from Eq. (3), and $p(\tau)$ is the laser pulse envelope at the intensity level. In these experiments, the pulse envelope is an essentially transform limited Gaussian. The factor of $\sqrt{3/2}$ arises from the three interactions of the two Gaussian pulses used in the vibrational echo experiment. This procedure is the convolution of the sample response function with $\sqrt{3/2}$ times the Gaussian pulse envelope. There are two interactions with the second pulse, which gives rise to the pulse envelope squared. This is convolved

with the first pulse. The net result is the square of the Gaussian pulse is convolved with the Gaussian pulse, yielding the $\sqrt{3/2}$ factor.

The vibrational echo experiments were performed using the Stanford Free Electron Laser (FEL). The FEL pulse train consists of a macropulse having a duration of ~3 ms and repeating at 10 Hz. Within a macropulse is a series of micropulses repeating at 11.8 MHz. Each transform limited Gaussian micropulse has an energy of ~0.5 μJ. The pulse duration was ~1 ps (see below). The FEL frequency is actively stabilized to within 0.02% of the center frequency. Both the autocorrelation and the spectrum were monitored continuously during experiments.

The experimental apparatus is a slightly modified version of one reported previously.[6,8] Very briefly, the IR beam is split into two equal parts to become the two input pulses in the vibrational echo sequence. To reduce sample heating while retaining high peak power, single micropulses are selected from the macropulse at a reduced repetition rate using germanium acousto-optic modulators (AOM). One pulse is selected at ~60 kHz, while the other is chopped to ~30 kHz by a second AOM and sent through a motorized optical delay line. The chopping is performed to do background subtraction. The two beams are subsequently focused to ~50 μm diameter spots in the sample using an off-axis paraboloidal mirror. The echo signal is subsequently focused onto an IR detector. The echo signal is acquired with a gated integrator and digitized by computer as the computer scans a stepper motor delay line.

The $W(CO)_6$ and H64V samples were prepared as reported previously.[6,7] The $Rh(CO)_2$acac sample was prepared as follows: $Rh(CO_2)$acac (Aldrich, 99%) was added to DBP (Aldrich, 99.9%) to a molarity of $\sim1\times10^{-3}$. The sample was placed in a custom built copper housing between CaF_2 windows and an optical path of 400 μm. The samples were cooled using a He flow cryostat and the temperature was controlled to within 0.2 K.

III. RESULTS AND DISCUSSION

Figure 1(a) displays data taken on $Rh(CO)_2$acac in DBP at 3.4 K using a frequency of 2004.0 cm^{-1}. The center of the $v=0\rightarrow1$ transition is at 2010.1 cm^{-1}. So this data is taken with the frequency tuned somewhat to the red of the line center. The pulse bandwidth is 8 cm^{-1} FWHM, which corresponds to a 1.3 ps pulse duration. A fit to the data is also shown. The fit uses Eq. (3) convolved with a 1.6 ps FWHM instrument response. Figure 1(b) shows the same data but with a fit to Eq. (3) that holds the cosine term constant at one, i.e., the decay kinetics are the same but there are no beats in the fitting function. The inset in Fig. 1(b) are the residuals, which contain only the beats. Δ, the anharmonicity (the difference between the $v=0\rightarrow1$ and the $v=1\rightarrow2$ transition frequencies) can be obtained from the fit in Fig. 1(a) or it can be read off directly from the inset in Fig. 1(b). The results yield $\Delta=13.5\pm0.2$ cm^{-1}.

In the fit to Eq. (3), the fast component corresponds to the dephasing time of the $v=1-2$ vibrational coherence

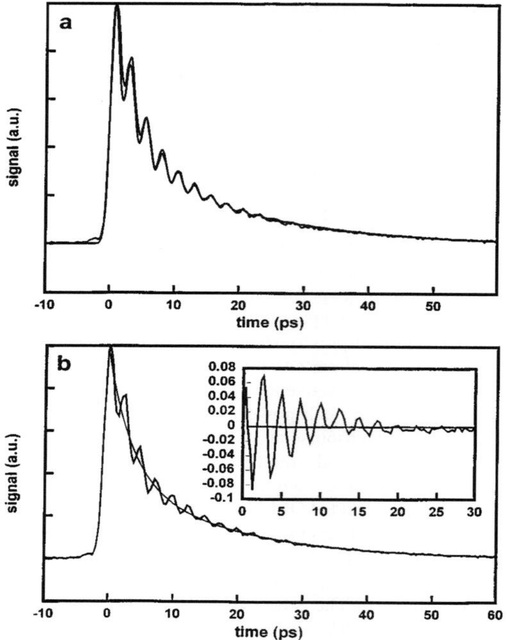

FIG. 1. (a) $Rh(CO)_2$acac in DBP at 3.4 K and laser frequency of 2004.0 cm^{-1} and fit using Eqs. (3) and (4). The center of the $v=0\rightarrow1$ transition is at 2010.1 cm^{-1}. The pulse bandwidth is 8 cm^{-1} FWHM, which corresponds to a 1.3 ps pulse duration. The fit to the data uses Eq. (3) convolved with a 1.6 ps FWHM instrument response and yields T_2s of 15 and 92 ps for the $v=1\rightarrow2$ and $v=0\rightarrow1$ levels, respectively. (b) The same data but with a fit with Eq. (3) that holds the cos term constant at 1. The inset in (b) are the residuals. This is a method of displaying only the beats. Δ, the anharmonicity (the difference between the $v=0\rightarrow1$ and the $v=1\rightarrow2$ transition frequencies) can be obtained from the fit. The results yield $\Delta=13.5\pm0.2$ cm^{-1}.

$(T_2(12))$ and the slow component corresponds to the dephasing time of the $v=0-1$ vibrational coherence $(T_2(01))$. As can be seen from both Figs. 1(a) and 1(b), the fits to the data are quite good. The decay constants yield homogeneous dephasing times of $T_2(12)=15$ ps and $T_2(01)=92$ ps, respectively. In fitting the data, the phase factor ϕ in Eq. (3) was varied in addition to the other parameters. While the data can be fit with $\phi=0$, the fitting routine consistently returned values of $\phi\approx\pi/2$.

Equation (3) predicts that the magnitudes of the components of decay and the amplitude of the beats are related to the strengths of the E-fields at the two transition frequencies. For a finite bandwidth pulse, as the excitation frequency is moved from around the peak of the $v=0\rightarrow1$ transition to lower energies, the beats will become more pronounced, and the component of the decay corresponding to the relaxation of the $v=1-2$ coherence will become larger. Figure 2(a) displays $Rh(CO)_2$acac vibrational echo data taken at 3.4 K at a variety of frequencies. The center of the $v=0\rightarrow1$ transi-

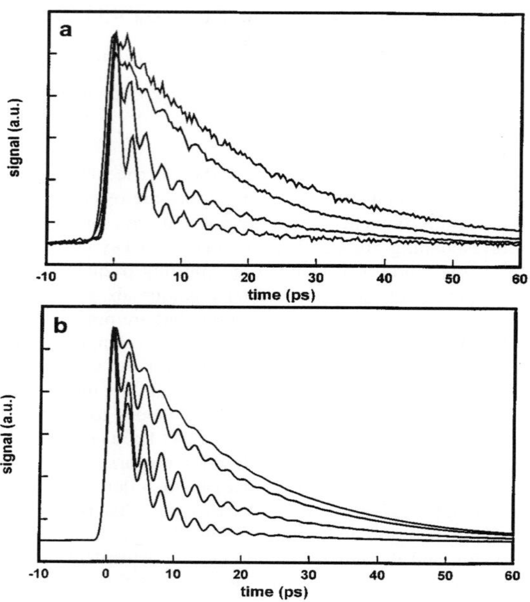

FIG. 2. (a) $Rh(CO)_2$acac vibrational echo data taken at 3.4 K at laser frequencies, from top to bottom, 2020.2, 2012.1, 2004.0, and 1996.0 cm^{-1}. The center of the $v=0\rightarrow1$ transition is 2010.1 cm^{-1}. The dephasing times of the $v=1\rightarrow2$ and $v=0\rightarrow1$ transitions are 15 and 92 ps, respectively. (b) Calculated echo decays obtained using Eq. (3) for several ratios of the E-field amplitudes. These curves use lifetimes, beats frequencies, and phases determined from the $Rh(CO)_2$acac data, and have been convolved with an 1.6 ps FWHM instrument response. Lines, from top to bottom in (b), have ratios of the E fields at the $v=0\rightarrow1$ and $v=1\rightarrow2$ transitions of 99.5/0.5, 95/5, 67/33, and 20/80, respectively. The calculation presented in (b) are qualitatively very similar to the data presented in (a).

tion is 2010.1 cm^{-1}. The first data set is at 2020.2 cm^{-1}. In this data set, there are no apparent beats. These data can be fit with a single exponential, Eq. (1). At this laser frequency, there is little or no overlap of the pulse bandwidth with the $v=1\rightarrow2$ transition. The single exponential decay yields γ_{01} only. Tuning to lower energy, 2012.1 cm^{-1}, there are only very low amplitude beats, which are almost lost in the noise. However, the data cannot be fit well to a single exponential decay. To obtain a good fit, Eqs. (3) and (4) are needed. This frequency is still to the blue of the line center of the $v=0\rightarrow1$ transition, so the overlap of the pulse bandwidth with the $v=1\rightarrow2$ transition is small. The next data set, at 2004.0 cm^{-1}, definitely display beats. In this case, there is significant overlap of the pulse bandwidth with the $v=1\rightarrow2$ transition, and the beat amplitude and fast decay component magnitude increase markedly. Finally, the 1996.0 cm^{-1} data set shows significant beats as there is now extensive overlap of the excitation bandwidth with the $v=1\rightarrow2$ transition. All data sets indicate that the dephasing times of the $v=1-2$ and $v=0-1$ transitions are 15 and 92 ps, respectively. There is a clear phase shift in the data of

Fig. 2(a) that should be noticed. The phase shift may be explained by a full theoretical analysis of this problem using finite bandwidth pulses.[14]

Figure 2(b) shows calculated echo decays obtained using Eq. (3) for several ratios of the E-field amplitudes. These curves use dephasing times, beats frequencies, and phases determined from the $Rh(CO)_2acac$ data, and have been convolved with an 1.6 ps FWHM instrument response. Lines A, B, C, and D in Fig. 2(b) have ratios of the $v=0\rightarrow1$ and $v=1\rightarrow2$ transition E-fields of 99.5/0.5, 95/5, 67/33, and 20/80, respectively. In the calculations, $\mu_{12}=\sqrt{2}\mu_{01}$, as is the case for a harmonic oscillator. As the quotient of the two E fields decreases, the amplitudes of the beats increase, and there is an increased short time decay of the signal because the contribution of the γ_{12} portion of the signal is increased. The magnitude of the beats in the simulation is also a function of the instrument response compared to the beat frequency. For all of the data presented in this paper, the instrument responses are in the 1 ps range. Since the beat frequencies are 2–5 ps, depending on the sample, the instrument response significantly decreases the observed magnitude of the beats. The calculation presented in Fig. 2(b) are qualitatively very similar to the data presented in Fig. 2(a). This demonstrates the basic validity of the description of the multilevel coherence and its frequency dependence. However, as discussed above and in detail in the Appendix, the analytical expression given in Eq. (3) was derived for a delta function duration pulse, but with the standard type of derivation modified to include different E fields at the two transition frequencies. A quantitative theoretical description of this problem cannot be obtained analytically because it will include a finite pulse duration with a finite bandwidth. This produces a complicated numerical problem that is under investigation.[14] As shown above, Eq. (3) provides a good description of the results. Because it is analytical, it is very useful in data fitting. It gives all parameters correctly except the depth of the beats.

Previously, an excitation frequency dependent study of vibrational beats on $W(CO)_6$ in DBT was reported.[7] These data display a frequency dependence which is substantially different from the data and calculations presented in Fig. 2. This difference probably arises because of the triple degeneracy of the T_{1u} CO transition of the $W(CO)_6$. The triple degeneracy is undoubtedly broken in the glassy environment. This will lead to many more pathways for the interaction of the radiation field with the system than are contained in the derivation of Eq. (3). The $Rh(CO)_2acac$ CO stretching mode on which the experiments were performed is nondegenerate, and Eq. (3) provides a good description of the nondegenerate problem.

VEB measurements can be made on a large variety of systems. Figure 3 shows data taken on H64V in glycerol–water at 3.4 K with the laser tuned to 1957 cm^{-1}. The center of the $v=0\rightarrow1$ transition for this line is 1969 cm^{-1}. The excitation bandwidth for these data was ~16 cm^{-1} FWHM. When the echo decay is measured on line center, the data are a single exponential decay without beats. IR pump–probe experiments measured the $v=0\rightarrow1$ lifetime at this tempera-

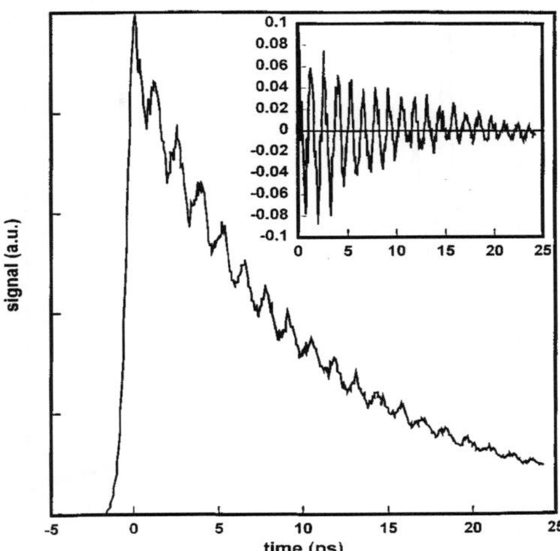

FIG. 3. Vibrational echo data on H64V in glycerol–water at 3.4 K with the laser tuned to 1957 cm^{-1}. The center of the $v=0\rightarrow1$ transition for this line is 1969 cm^{-1}. The excitation bandwidth was ~16 cm^{-1} FWHM. The dephasing times of the $v=1\rightarrow2$ and $v=0\rightarrow1$ transitions are 20 and 65 ps, respectively.

ture as 35 ps.[15] Comparison of the line center echo decay and the pump–probe data demonstrate that at these low temperatures, the homogeneous dephasing time is approximately twice the lifetime $(2T_1)$, i.e., there is no significant pure dephasing. From the fit to the data, the decay constants, γ_{12} and γ_{01}, yield homogeneous dephasing times of $T_2(12)=20$ ps and $T_2(01)=65$ ps, respectively. The fit also gives the beat frequency. The inset in Fig. 3 displays only the beats, obtained in the same manner as described for Fig. 1(b). The data are modulated with a 1.3 ps beat which corresponds to an anharmonicity of 25.4 ± 0.2 cm^{-1}. This can be compared to the ~26 cm^{-1} anharmonicity reported for Mb-CO, which was measured using two color pump–probe experiments.[2] While H64V-CO has a shift in frequency and a 20% decrease in the rate of pure dephasing,[6] within experimental error, replacing the distal histidine with a valine does not change the CO vibrational anharmonicity.

These data on H64V-CO show that it is possible to use the VEB method to obtain vibrational anharmonicities which are substantially larger than previously reported for $W(CO)_6$. Metal carbonyls with two or more equivalent carbonyls have very small anharmonicities. By using a pulse duration of 670 fs FWHM and tuning to lower frequency than the $v=0\rightarrow1$ line center, it was possible to measure the moderate anharmonicity of H64V-CO. Currently, it is not possible to produce significantly shorter pulses than 600 fs with the Stanford FEL. However, conventional laser/OPA systems can produce IR pulses which are substantially shorter. Therefore, using available technology, it should not be difficult to measure large anharmonicities of 100 cm^{-1} or more.

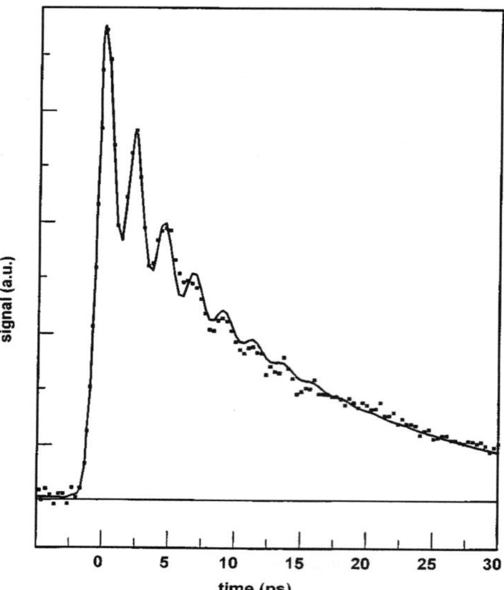

FIG. 4. Vibrational echo decay and fit W(CO)$_6$ in dibutylphthalate at 10 K and 1976.3 cm^{-1}. The beats are separated by 2.3 ps, corresponding to a vibrational anharmonic splitting of $\Delta = 14.7$ cm$^{-1} \pm 0.3$ cm^{-1}.

Figure 4 shows a vibrational echo decay for asymmetric CO stretching mode of W(CO)$_6$ in DBP at 10 K with the laser tuned to the center of the $v = 0 \rightarrow 1$ transition 1976.3 cm^{-1}, as well as a fit using Eq. (3).[7] The beats are separated by 2.3 ps, corresponding to a vibrational anharmonic splitting of $\Delta = 14.7$ cm$^{-1} + \pm 0.3$ cm^{-1}. This splitting is in accord with the value of 15 cm$^{-1} \pm 1$ cm^{-1} subsequently obtained by Heilweil and co-workers from observation of the $v = 1 \rightarrow 2$ and $v = 2 \rightarrow 3$ transitions of W(CO)$_6$ in hexane using two color pump–probe experiments.[1]

The experiments on W(CO)$_6$ were performed as a function of temperature.[7] From the fits to Eq. (3), the homogeneous dephasing times for the $v = 0 \rightarrow 1$ and $v = 1 \rightarrow 2$ transitions were obtained for temperatures between 10 and 150 K. These are displayed in Fig. 5. Detailed analysis of the temperature dependence of the $v = 0 - 1$ dephasing demonstrates that by 10 K the homogeneous linewidth is dominated by the vibrational lifetime, i.e., $T_2 \approx 2T_1$.[5] From the fit to Eq. (3) of the data at 10 K and using results from Ref. 7 for the $v = 0 \rightarrow 1$ transition, the decay constants, γ_{12} and γ_{01}, yield homogeneous dephasing times of $T_2(12) = 5.0$ ps and $T_2(01) = 66$ ps, respectively. Above 10 K, the $v = 0 - 1$ homogeneous linewidth increases as a power law, $T^{1.3}$.[5] However, the temperature dependence of the $v = 1 - 2$ dephasing has a very different character. It is temperature independent up to ~ 90 K. This demonstrates that the $v = 1 - 2$ dephasing is dominated by T_1 with $T_2 \approx 2T_1$ since a pure dephasing contribution to T_2 always displays a temperature dependence. Thus, at 10 K, both $T_2(01)$ and $T_2(12)$ are measures of the vibrational lifetimes of the two states.

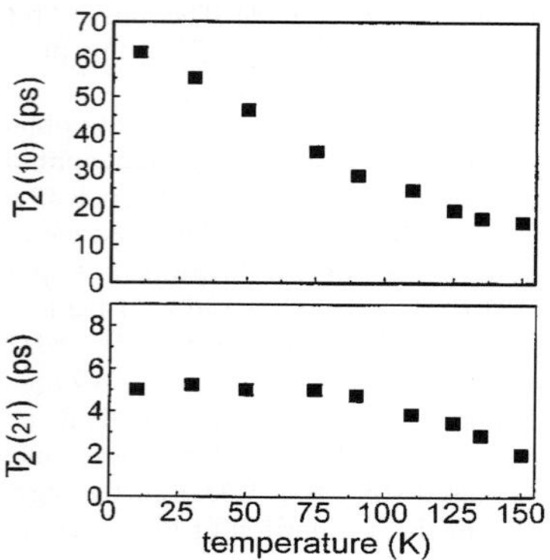

FIG. 5. Decay times $T_2(01) = 1/\gamma_{01}$ and $T_2(12) = 1/\gamma_{12}$ of W(CO)$_6$ in DBT for temperatures between 10 and 150 K obtained from fits to Eq. (3). The lack of temperature dependence of $T_2(12)$ at the lower temperatures shows that T_2 is lifetime limited.

The data for W(CO)$_6$ in DBP at 10 K demonstrates that $T_2 \approx 2T_1$ for both transitions. It is reasonable to assume that $T_2 \approx 2T_1$ for both transitions of Rh(CO)$_2$acac in DBP and for H64V-CO as well. These samples were studied at 3.4 K. It is known that $T_2 \approx 2T_1$ for the $v = 0 \rightarrow 1$ transitions in both samples at this temperature.[6,15] At these very low temperatures, the samples are in the low temperature limit. The thermal fluctuations of the heat bath are insufficient to cause significant pure dephasing. Even electronic transitions of large molecules in low temperature glasses have very slow pure dephasing times of ~ 1 ns.[16] These are only measurable because the electronic excited state lifetimes are long compared to the vibrational lifetimes discussed here. The coupling of vibrational transitions to the medium is much weaker than the coupling of electronic transitions as evidenced by much smaller gas to solvent shifts in transition frequencies and much slower pure dephasing at elevated temperatures. In addition, there is no fundamental reason why the pure dephasing of the $v = 1 - 2$ transition should be different than the $v = 0 - 1$ pure dephasing. For these reasons, we will take the T_2s of both transitions of all three samples to be lifetime limited at the lowest temperature.

The lifetimes, $T_1(1)$ and $T_1(2)$ of $v = 1$ and $v = 2$, respectively, and dephasing times $T_2(01)$ and $T_2(12)$ of the $v = 0 - 1$ and $v = 1 - 2$ transitions, respectively, are summarized in Table I. The approximate errors for $T_1(1)$ and $T_2(01)$ is 3%. The approximate error for $T_2(12)$ is 5%. Also reported in Table I are the ratios of lifetimes, $R_{T_1} = T_1(1)/T_1(2)$ and dephasing times, $R_{T_2} = T_2(01)/T_2(12)$.

The value of $T_1(1)$ is highly dependent on the choice of the solute–solvent systems studied. The lifetime of W(CO)$_6$ is solvent dependent. At room temperature, its lifetime in CCl$_4$ is ~ 700 ps, in CHCl$_3$ is ~ 340 ps, in 2-methylpentane is 150 ps. In the solvents CHCl$_3$ and 2-methylpentane, the lifetimes display an inverted temperature dependence,[17–19] i.e., the lifetimes become faster as the

TABLE I. Experimental results. Summary of the lifetimes, $T_1(1)$ and $T_1(2)$ of $v=1$ and $v=2$, respectively, and dephasing times $T_2(01)$ and $T_2(12)$ of the $v=0-1$ and $v=1-2$ transitions, respectively. The approximate errors of $T_1(1)$ and $T_2(01)$ is 3%. The approximate error for $T_2(12)$ is 5%. The approximate error for $T_1(2)$ is 10%. Also reported are the ratios of lifetimes, $R_{T_1}=T_1(1)/T_1(2)$ and dephasing times, $R_{T_2}=T_2(01)/T_2(12)$.

Sample	Anharmonicity	$T_2(01)$	$T_2(12)$	R_{T_2}	$T_1(1)$	$T_1(2)$	R_{T_1}
$W(CO)_6$	$14.7\ cm^{-1}$	66 ps	5 ps	13	33 ps	3 ps	11
$Rh(CO)_2acac$	$13.5\ cm^{-1}$	92 ps	15 ps	6	49 ps	9 ps	5
H64V-CO	$25.4\ cm^{-1}$	65 ps	20 ps	3.3	35 ps	14 ps	2.5

temperature is decreased. In 2-methylpentane, the lifetime at 10 K is 100 ps. The change in lifetime with solvent shows that the solvent is intimately involved in the relaxation pathway. The initial CO vibration relaxes into a combination of solute and solvent modes plus a mode of the solvent continuum (solvent phonon) as necessary to conserve energy.[17–19] As the complexity of the solvent increases, more modes are available to provide a low-order pathway for the relaxation. In CCl_4, a fifth-order process is required that involves the annihilation of the initial excitation and the creation of three solute–solvent vibrational excitations and a phonon.[17–19] In $CHCl_3$, the addition of the CH bending mode at $\sim 1250\ cm^{-1}$ reduces the order of the process to fourth-order, and the lifetime is reduced.[17–19] 2-methylpentane has a wider variety of modes, and DBP an even wider variety. Presumably, these considerations are very similar for $Rh(CO)_2acac$.

However, H64V-CO, other Mb-COs, and a large variety of model heme-COs have been shown to relax in a different manner.[20–22] The relaxation is essentially solvent independent.[21,23,24] Relaxation occurs through deposition of the CO vibrational energy into the enormous density of states provided by the heme. The coupling has been shown to be via π-bond interactions. Back bonding from the metal–heme π electron system into the CO π^* antibonding molecular orbital couples the CO vibration directly to the vast number of heme vibrational states. An increase in the back bonding decreases the vibrational lifetime. Mb-CO has a lifetime of 16 ps. Replacing the distal histidine with a valine to form H64V-CO reduces the back bonding and slows the lifetime to 35 ps. Therefore, by selecting the particular form of Mb, it is possible to have a heme-CO system with a $v=1$ lifetime that is similar to the simpler metal carbonyls. This facilitates the comparison of the $v=1$ lifetimes and the $v=2$ lifetimes.

A striking feature of Table I is the trend in the ratios (R_{T_1}) of $T_1(1)$ to $T_1(2)$ for the three samples. In all cases, relaxation from $v=2$ is substantially faster than relaxation from $v=1$. In the most basic analysis, it would be expected that ratio $R_{T_1}=T_1(1)/T_1(2)$ would be 2.[25] This will occur if the density of bath states and the strength of coupling to the bath for the relaxation process are independent of the initial vibrational quantum number, v, and if no additional relaxation pathways become available for the higher v initial state. The action of a lowering operator contained in the vibrational relaxation coupling matrix element for a $v \rightarrow v-1$ relaxation brings out a factor of $\sqrt{v}$. The relaxation

rate scales as the square of the matrix element. Therefore, when the initial state is 2, the $\sqrt{v}$ factor squared yields 2, and when the initial state is 1, the $\sqrt{v}$ factor squared yields 1, giving an R of 2. So from this consideration alone, the $v=1$ lifetime should be twice as long as the $v=2$ lifetime.

At low temperature, where there is no pure dephasing, T_2 is determined by $2T_1$. However, when looking at dephasing, if R_{T_1} is 2, the ratio, $R_{T_2}=T_2(01)/T_2(12)$ should be 3. The dephasing rate is the average of the rates of population decay out of the two levels involved. For dephasing of the $v=0-1$ transition,

$$\frac{1}{T_2(01)}=\frac{\gamma(1)+\gamma(0)}{2}=\frac{1}{2T_1(1)}, \tag{5}$$

where $\gamma(0)=1/T_1(0)$ and $\gamma(1)=1/T_1(1)$. Since the rate out of the ground state is 0, the common result for a $v=0-1$ vibrational echo of $T_2(01)=2T_1(1)$ is obtained. Rearranging Eq. (5) yields

$$\gamma(1)=\frac{2}{T_2(01)}. \tag{6}$$

The upper state dephasing is similar in that it is the average of the rates out of the two levels as given in Eq. (7)

$$\frac{1}{T_2(12)}=\frac{\gamma(1)+\gamma(2)}{2}. \tag{7}$$

The $v=2$ lifetimes displayed in Table I were calculated from measured values of the $v=1$ lifetimes ($\gamma(1)s$) and the $v=1-2$ dephasing rates, using Eq. (7). The approximate error for these calculated lifetimes is 10%. As stated above, the minimum ratio of the upper and lower dephasing rates is 3. This follows from the simplification of Eq. (7) using the Landau–Teller model for which $\gamma(2)=2\gamma(1)$,

$$\frac{1}{T_2(12)}=\frac{\gamma(1)+\gamma(2)}{2}=\frac{3\gamma(1)}{2}. \tag{8}$$

Using the results of Eq. (6) in Eq. (8) yields

$$\frac{1}{T_2(12)}=\frac{3}{T_2(01)}. \tag{9}$$

Clearly, these factors of 2 and 3 alone do not account for the R_{T_1} and R_{T_2} listed in Table I, although they come very close for H64V-CO. The vibrational relaxation in the two metal carbonyls depends strongly on interactions with the solvent. The $v=1 \rightarrow 2$ transition is shifted to lower energy by ~ 14 cm^{-1} for both molecules. The relaxation pathway almost certainly includes a phonon mode of the glassy solvent continuum. For a non-hydrogen bonding organic like DBP, the continuum extends to ~ 200 cm^{-1}.[26] Therefore, if the phonon mode is to the high energy side of the peak of the density of states, a 14 cm^{-1} shift to lower energy could result in an increase in the density of states, and therefore an increase in the vibrational relaxation rate for $v=2$. Also, an increase in the strength of coupling results from the participation of a different phonon. However, such a change is likely to only account for a portion of the R_{T_1} for W(CO)$_6$ and Rh(CO)$_2$acac. H64V-CO, which does not utilize the solvent in its relaxation would not have this effect.

A factor which may be the major contributor to the large R_{T_1} values of the simple metal carbonyls is the opening of a new relaxation pathway for relaxation from the $v=2$ state. The new pathway is direct relaxation from $v=2$ to $v=0$. The $v=2\rightarrow0$ transitions for the systems studied have energies ~ 3950 cm^{-1}. It is possible for the numerous C–H transitions, which have energies in the $3300-2900$ cm^{-1} range, to participate in the vibrational relaxation. If the $v=2\rightarrow0$ relaxation is a low order path, perhaps not involving a phonon because of the large number of C–H modes covering a wide range of frequencies, it could substantially augment the rate of relaxation out of the $v=2$ level. The experiment measures the decay constant, γ_2, which is independent of the pathway for leaving the $v=2$ level. Relaxation from $v=2\rightarrow0$ or from $v=2\rightarrow1$ have the same effect on the decay of the coherence. While both W(CO)$_6$ and Rh(CO)$_2$acac have the $v=2\rightarrow0$ opened, its influence on R_{T_1} will depend on the details of the coupling matrix elements and differences in the internal molecular modes that can participate in the relaxation. H64V-CO already has an enormous density of coupled states provided by the heme, and therefore, the availability of an additional pathway apparently makes less difference.

The ratio R_{T_1} in Table I can have at least three contributions: the factor of 2 which comes from the quantum number dependence through the lowering operator, a change in the phonon density of states, and the additional $v=2\rightarrow0$ pathway. While all three molecules have similar $v=1$ lifetime, the experiments demonstrate that there can be a wide range of rates for decay out of the $v=2$ level. This range depends on the relative contribution of the three factors discussed above.

IV. CONCLUDING REMARKS

In this paper, we have investigated multilevel vibrational coherences observed in vibrational echo experiments in three systems, Rh(CO)$_2$acac in dibutylphthalate, H64V in glycerol–water, and W(CO)$_6$ in dibutylphthalate. The beats in the echo decay provide a direct measure of the vibrational anharmonicity. The observations in the three systems show that the VEB experiment can be of utility for measuring vibrational anharmonicities provided the coherence decay is sufficiently slow to permit observation of the beats. Equation

(3) provides a good description of the observables. It permits extraction of the important physical parameters, i.e., the anharmonicity and the homogeneous linewidths of the $v=0-1$ transition and the $v=1-2$ transition. The frequency dependent Rh(CO)$_2$acac data demonstrates how the amplitudes of the beats and, therefore, the extent of the multilevel coherence changes with the laser excitation frequency. Equation (3) can reproduce the general nature of the frequency dependence, but it does not provide a quantitative description because the derivation of the analytical expression does not include the finite duration of the excitation pulses.

At the very low temperatures of the experiments, the homogeneous dephasing times are determined by the vibrational lifetimes, $T_2\approx 2T_1$. The three molecules studied have similar $v=1$ lifetimes, but the decays of the multilevel coherences observed in the vibrational echo experiments show that the $v=2$ lifetimes of all three molecules are significantly faster and that they vary from one molecule to another. Possible mechanisms for the decrease in the lifetimes and the variations were discussed.

To observe the VEB signal, it is necessary to have the transform limited bandwidth of the excitation pulses comparable to the anharmonic splitting. As demonstrated, the amplitude of the beats can be enhanced by tuning to lower frequency so the that the pulse bandwidth can more effectively overlap both the $v=0\rightarrow1$ and $v=1\rightarrow2$ transitions. However, it is important to emphasize that the shortest possible pulse duration (widest bandwidth) is not always desirable in a vibrational echo experiment. If the aim is to understand the dynamics of the a liquid, glass, or protein system by studying the pure dephasing of the $v=0-1$ transition, the multilevel coherence may interfere by turning a single exponential decay into a multiexponential decay with beats. At high temperatures, where the echo decay time can be comparable to the beat period, it can be difficult to extract the true echo decay. Therefore, it is desirable to use a pulse duration that is appropriate to obtain the information that is being sought.

ACKNOWLEDGMENTS

The authors would like to thank Professor Alan Schwettman and Professor Todd Smith, Physics Department, Stanford University, and their research groups at the Stanford Free Electron Laser Center whose efforts made these experiments possible. We would also like to thank Professor Dana D. Dlott and Professor Stephen G. Sligar, and Dr. Jeffrey R. Hill and Dr. Ellen Y. T. Chien for providing us with the H64V-CO sample used in these experiments. This research was supported by the Office of Naval Research, (N00014-94-1-1024), by the National Science Foundation, Division of Materials Research (DMR93-22504), and by the Office of Naval Research, (N00014-92-J-1227).

APPENDIX

We present a derivation of Eq. (3) in this Appendix. We use a standard Markovian model in the low temperature limit. This treatment is similar to that of Fourkas *et al.* where the signal was derived assuming $\omega_{12}=\omega_{01}$.[9]

There are three possible resonant quantum mechanical pathways that contribute to the vibrational echo signal (after $t=0$) when excitation bandwidth overlaps with three levels. These are shown in Figs. 6(a) and 6(b). Figure 6(a) is the ladder diagram representation of the resonant pathways. The solid arrows denote interactions on the ket of the density matrix, while dotted arrows denote interactions on the bra of the density matrix. The signal is indicated with the dashed arrows. The first two pathways are the standard ones for the vibrational echo involving two levels. The third pathway involves excited-state absorption, and hence, interferes with the first two.[9]

The general polarization is a sum of the individual response functions integrated over the excitation fields

$$P^{(3)} = \int_0^\infty d\omega\, G(\omega) \int_0^\infty dt_3 \int_0^\infty dt_2 \int_0^\infty dt_1\ \bar{E}_1^* \bar{E}_2 \bar{E}_3$$

$$\times \sum_i R_i(t_1, t_2, t_3), \tag{A1}$$

where t_1, t_2, t_3 are the times between each of the pulses in the vibrational echo sequence and R_i are the response functions. The distribution of transition frequencies, ω, is reflected by the inhomogeneous distribution function, $G(\omega)$. For delta function pulses

$$\bar{E}_i = |E_i|\, \delta(t - \tau_i) \exp(-i\omega t). \tag{A2}$$

The response functions are given by

$$R_1 = R_2 = \mu_{01}^4 \exp(i\omega_{01}t_1 - i\omega_{01}t_3)\exp(-\gamma_{01}t_1 - \gamma_{01}t_3), \tag{A3}$$

$$R_3 = -\mu_{01}^2\mu_{12}^2 \exp(i\omega_{01}t_1 - i\omega_{12}t_3)\exp(-\gamma_{01}t_1 - \gamma_{12}t_3), \tag{A4}$$

where $I_{ij}(t) = \exp(-i\omega_{ij}t - \gamma_{ij}t)$ and $\mu_{ij} = |\langle i|\mu|j\rangle|$ and $\omega_{ij} = \omega_i - \omega_j$. Note that the complete derivation depends also on the t_2 time, which is the time between the second and third interactions. In the echo experiment, the second pulse provides both the second and third interactions so the time between them is zero for delta function pulses. Therefore, terms containing t_2 in Eq. (A3) and (A4) (and henceforth) have been dropped.

Since for a purely harmonic system, $\mu_{01} = \sqrt{2}\mu_{12}$, the third path will completely cancel the echo at $t=0$ and will cancel the echo signal at all time if $\omega_{12} = \omega_{01}$ and $\gamma_{01} = \gamma_{12}$.[9] However, in real systems, the oscillator is not harmonic, therefore $\omega_{12} \neq \omega_{01}$. As shown below, the beating in the data will arise from the difference between ω_{12} and ω_{01} and the decay of the beats will be the average of the γ_{01} and γ_{12} times.

To include inhomogeneous broadening, we take the distribution function to be a Gaussian centered at ω_{ij}°:

$$G(\omega_{ij}) = \frac{1}{\sqrt{2\pi}\sigma} \exp\left(\frac{-(\omega_{ij} - \omega_{ij}^\circ)^2}{2\sigma^2}\right) = G(\delta_{ij}), \tag{A5}$$

where δ_{ij} is the deviation from the center of the line. We will further assume that the distribution function for the ω_{01} transition is the same as the ω_{12} transition. Also, the inhomoge-

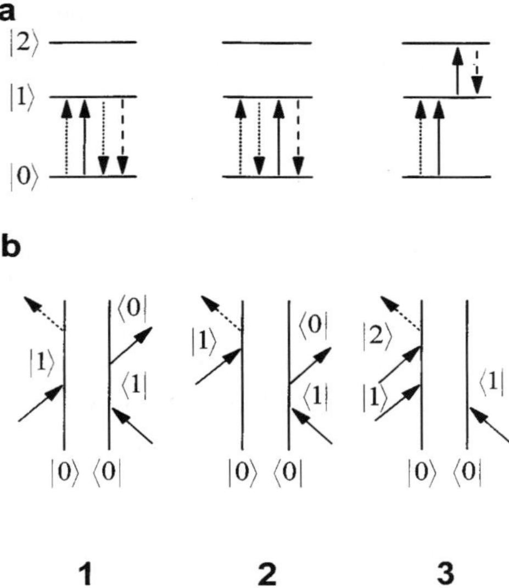

FIG. 6. (a) Latter diagram of the three pulse sequences in a VEB experiment. The first two diagrams are the two-level vibrational echo diagrams. The solid arrows denote interactions on the ket of the density matrix, while dotted arrows denote interactions on the bra of the density matrix. The signal is indicated with the dashed arrows. (b) Another representation of the three pulse sequences of in the VEB experiment. Time is taken to increase moving up the parallel set of lines.

neous broadening is taken to be perfectly correlated in the two transitions. Further, all transitions for a given distribution have the same μ_{ij} and γ_{ij}.

In the delta function pulse limit, the polarization is given by the response function

$$P^{(3)}(t_3, t_1) = \int_0^\infty G(\delta_{ij}) \left[\sum_i R_i(t_i, t_3, \delta)\right] |E_1^*||E_2||E_3| d\delta_{ij}, \tag{A6}$$

substituting $\omega_{ij} = \omega_{ij}^0 + \delta_{ij}$ yields

$$R_1 = R_2 = \mu_{01}^4 \exp[i\omega_{01}(t_1 - t_3)]$$
$$\times \exp[-\gamma_{01}(t_1 + t_3)]\exp[i\delta(t_1 - t_3)], \tag{A7}$$

$$R_3 = -\mu_{01}^2\mu_{12}^2 \exp(i\omega_{01}t_1 - i\omega_{12}t_3 - \gamma_{01}t_1 - \gamma_{12}t_3)$$
$$\times \exp[i\delta(t_1 - t_3)]. \tag{A8}$$

Substituting Eqs. (A7) and (A8) into Eq. (A6) yields the third-order polarization as

$$P^{(3)}(t_3, t_1) = |E_1^*||E_2||E_3| \exp\left(\frac{-(t_3 - t_1)^2 \sigma^2}{2}\right)$$
$$\times \{2\mu_{01}^4 \exp[i\omega_{01}(t_1 - t_3)]$$
$$\times \exp[-\gamma_{01}(t_1 + t_3)] - \mu_{01}^2\mu_{12}^2$$
$$\times \exp(i\omega_{01}t_1 - i\omega_{12}t_3 - \gamma_{01}t_1 - \gamma_{12}t_3)\}. \tag{A9}$$

This shows the echo characteristic of being sharply peaked at $t_1 = t_3$, if the distribution function is very broad.

The observed integrated signal is

$$S(t) = \int_{-\infty}^{+\infty} dt_3 |P(t_1 - t_3)|^2. \tag{A10}$$

Defining the intensity as the square of the absolute value of the E-field and substituting Eq. (A9) into Eq. (A10) yields

$$S(t) = I^3 [4\mu_{01}^8 \exp(-4\gamma_{01}t) + \mu_{01}^4\mu_{12}^4 \exp(-2\gamma_{01}t -2\gamma_{12}t) - 2\mu_{01}^6\mu_{12}^2 \exp(-3\gamma_{01}t - \gamma_{12}t) \times \{\exp[i(\omega_{01}-\omega_{12})t] + \exp[-i(\omega_{01}-\omega_{12})t]\}, \tag{A11}$$

which simplifies to

$$S(t) = I^3 [\exp(-2\gamma_{01}t)][\mu_{01}^8 \exp(-2\gamma_{01}t) + \mu_{01}^4\mu_{12}^4 \exp(-2\gamma_{12}t) + \mu_{01}^6\mu_{12}^2 \times \exp(-\gamma_{01}t - \gamma_{12}t)\cos(\Delta t)]. \tag{A12}$$

This derivation assumed delta function pulses with infinite bandwidth. Therefore, the E fields have no frequency dependence. There is no frequency dependence in Eq. (A12). In particular, the depth of the modulation occurring at frequency, Δ, is independent of the laser frequency. The depth depends only on the values of the μ_{01} and μ_{12}, which are constants. This is in contradiction to the data in Fig. 2(a). The use of a delta function duration pulse makes it possible to obtain an analytical expression. In the experiments, the laser pulses have finite durations and, therefore, finite bandwidths. The finite bandwidth results in different driving E fields, E_{01} and E_{12}, at the two transition frequencies. When the frequency is changed, the values of E_{01} and E_{12} change, giving rise to the frequency dependence displayed in Fig. 2(a). For a two level system, the coherence is produced by the coupling of the radiation field to the transition dipole as $(E \cdot \mu)$. To include the effect of finite bandwidth pulses and yet preserve the analytical expression, we associate E_{ij} with μ_{ij} where appropriate. This results in Eq. (A13),

$$S(t) = (E_{01} \cdot \mu_{01})^4 [\exp(-2\Gamma_{01}t)][4(E_{01} \cdot \mu_{01})^2\mu_{01}^2 \times \exp(-2\Gamma_{01}t) + (E_{12} \cdot \mu_{12})^2\mu_{12}^2 \exp(-2\Gamma_{12}t) - 4(E_{01} \cdot \mu_{01})^2(E_{12} \cdot \mu_{12})^2\mu_{01}\mu_{12} \exp(-\Gamma_{01}t - \Gamma_{12}t)\cos(\Delta t)]. \tag{A13}$$

As can be seen from Figs. 6(a) and 6(b), all three pathways have the first two interactions involving only the $v = 0-1$ transition. These interactions give rise to the first term. The extra factors of μ^2 not multiplied by E^2 give rise to the signal.

[1] S. M. Arrivo, T. P. Dougherty, W. T. Grubbs, and E. J. Heilweil, Chem. Phys. Lett. **235**, 247 (1995).
[2] J. C. Owrutsky, M. Li, B. Locke, and R. M. Hochstrasser, J. Phys. Chem. **99**, 4842 (1995).
[3] D. Zimdars, A. Tokmakoff, S. Chen, S. R. Greenfield, and M. D. Fayer, Phys. Rev. Lett. **70**, 2718 (1993).
[4] A. Tokmakoff, D. Zimdars, R. S. Urdahl, R. S. Francis, A. S. Kwok, and M. D. Fayer, J. Phys. Chem. **99**, 13310 (1995).
[5] A. Tokmakoff and M. D. Fayer, J. Chem. Phys. **102**, 2810 (1995).
[6] K. D. Rector, C. W. Rella, A. S. Kwok, J. R. Hill, S. G. Sligar, E. Y. T. Chien, D. D. Dlott, and M. D. Fayer, J. Phys. Chem. B **101**, 1468 (1997).
[7] A. Tokmakoff, A. S. Kwok, R. S. Urdahl, R. S. Francis, and M. D. Fayer, Chem. Phys. Lett. **234**, 289 (1995).
[8] C. W. Rella, K. Rector, A. Kwok, J. R. Hill, H. A. Schwettman, D. D. Dlott, and M. D. Fayer, J. Phys. Chem. **100**, 15620 (1996).
[9] J. T. Fourkas, H. Kawashima, and K. A. Nelson, J. Chem. Phys. **103**, 4393 (1995).
[10] X. Zhu, M. S. Hybertsen, P. B. Littlewood, and M. C. Nuss, Phys. Rev. B **50**, 11915 (1994).
[11] S. T. Cundiff, Phys. Rev. A **49**, 3114 (1994).
[12] S. Mukamel and R. F. Loring, J. Opt. Soc. B **3**, 595 (1986).
[13] Y. J. Yan and S. Mukamel, J. Chem. Phys. **94**, 179 (1991).
[14] J. Fourkas (private communication).
[15] K. D. Rector and M. D. Fayer (unpublished).
[16] L. R. Narasimhan, K. A. Littau, D. W. Pack, Y. S. Bai, A. Elschner, and M. D. Fayer, Chem. Rev. **90**, 439 (1990).
[17] A. Tokmakoff, B. Sauter, and M. D. Fayer, J. Chem. Phys. **100**, 9035 (1994).
[18] V. M. Kenkre, A. Tokmakoff, and M. D. Fayer, J. Chem. Phys. **101**, 10618 (1994).
[19] P. Moore, A. Tokmakoff, T. Keyes, and M. D. Fayer, J. Chem. Phys. **103**, 3325 (1995).
[20] J. R. Hill, D. D. Dlott, M. D. Fayer, C. W. Rella, M. M. Rosenblatt, K. S. Suslick, and C. J. Ziegler, J. Phys. Chem. **100**, 218 (1996).
[21] J. R. Hill, M. M. Rosenblatt, C. J. Ziegler, K. S. Suslick, D. D. Dlott, C. W. Rella, and M. D. Fayer, J. Phys. Chem. **100**, 18023 (1996).
[22] D. D. Dlott, M. D. Fayer, J. R. Hill, C. W. Rella, K. S. Suslick, and C. J. Ziegler, J. Am. Chem. Soc. **118**, 7853 (1996).
[23] A. Ansari, J. Beredzen, D. Braunstein, B. R. Cowen, H. Frauenfelder, M. K. Hong, I. E. T. Iben, J. B. Johnson, P. Ormos, T. Sauke, R. Schroll, A. Schulte, P. J. Steinback, J. Vittitow, and R. D. Young, Biophys. Chem. **26**, 337 (1987).
[24] D. Ivanov, J. T. Sage, M. Keim, J. R. Powell, S. A. Asher, and P. M. Champion, J. Am. Chem. Soc. **116**, 4139 (1994).
[25] L. Landau and E. Teller, Phys. Z. Sov. **10**, 34 (1936).
[26] Y. J. Chang and E. W. Castner, Jr., J. Phys. Chem. **100**, 3330 (1996).

Vibrational dephasing mechanisms in liquids and glasses: Vibrational echo experiments

K. D. Rector and M. D. Fayer
Department of Chemistry, Stanford University, Stanford, California 94305

(Received 16 September 1997; accepted 27 October 1997)

Picosecond vibrational echo studies of the asymmetric stretching mode (2010 cm^{-1}) of (acetylacetonato)dicarbonylrhodium(I) $[\text{Rh(CO)}_2\text{acac}]$ in liquid and glassy dibutyl phthalate (DBP) (3.5 K to 250 K) are reported and compared to previous measurements of a similar mode of tungsten hexacarbonyl $[\text{W(CO)}_6]$. The $\text{Rh(CO)}_2\text{acac}$ pure dephasing shows a T^1 dependence on temperature at very low temperature with a change to an exponentially activated process $(\Delta E \cong 400 \text{ cm}^{-1})$ above ~ 20 K. There is no change in the functional form of the temperature dependence in passing from the glass to the liquid. It is proposed that the T^1 dependence arises from coupling of the vibration to the glass's tunneling two level systems. The activated process arises from coupling of the high-frequency CO stretch to the 405 cm^{-1} Rh–C stretch. Excitation of the Rh–C stretch produces changes in the back donation of electron density from the rhodium d_π orbital to the CO π^* antibonding orbital, shifting the CO stretching transition frequency and causing dephasing. In contrast, W(CO)_6 displays a T^2 dependence below T_g in DBP and two other solvents. Above T_g, there is a distinct change in the functional form of the temperature dependence. In 2-methylpentane, a Vogel–Tammann–Fulcher-type temperature dependence is observed above T_g. It is proposed that the triple degeneracy of the T_{1u} mode of W(CO)_6 is broken in the glassy and liquid solvents. The closely spaced levels that result give rise to unique dephasing mechanisms not available in the nondegenerate $\text{Rh(CO)}_2\text{acac}$ system. © *1998 American Institute of Physics.* [S0021-9606(98)51705-9]

I. INTRODUCTION

A molecule in a condensed matter medium, such as liquid or a glass, is influenced by intermolecular interactions with the surrounding solvent. The shift in vibrational frequencies in going from the gas phase to a condensed phase is an indication of the effect of the solvent on internal mechanical degrees of freedom. The static shift in vibrational absorption energies is a reflection of the average force exerted by solvent on the molecular oscillators. The medium also exerts fluctuating forces on the internal degrees of freedom of a solute molecule, which are responsible for fluctuations in molecular structure. The structural fluctuations are manifested in time-dependent vibrational eigenstates, and, thus, time-dependent vibrational energy eigenvalues. Fluctuating forces are involved in a wide variety of chemical and physical phenomena, including thermal chemical reactions, promotion of a molecule to a transition state, electron transfer, and energy flow into and out of molecular vibrations.

The dynamics of the bath of modes, which cause time evolution of the vibrational energy eigenvalues, give rise to fluctuations in vibrational energy level separations. The bath contains bulk solvent degrees of freedom arising from translational and orientation motions of the solvent molecules as well as the internal vibrational degrees of freedom of the solvent. The bath also contains the solute's vibrational modes other than the oscillator of interest. In a glass, bath frequency fluctuations range from very high frequency to essentially static. For a pair of energy levels, e.g., $v = 0$ and $v = 1$, the very fast fluctuations produce homogeneous pure dephasing, which, for an exponential decay of the off-diagonal density matrix elements (Lorentzian homogeneous line shape), can be characterized by an ensemble average pure dephasing time, T_2. The total homogeneous dephasing time, T_2, also has contributions from the vibrational lifetime, T_1. Evolution of the system on time scales substantially slower than T_2 appear as inhomogeneous broadening. In a glass, the time scale of the slowest system evolution may be so long that there is essentially truly static inhomogeneous broadening. However, there are also slow fluctuations that do not contribute to homogeneous pure dephasing but give rise to spectral diffusion. Spectral diffusion has been observed frequently in electronic excitation dephasing experiments in glasses,[1–3] and has also been observed for vibrational transitions.[4,5]

Identical considerations apply to homogeneous dephasing in liquids. There is a range of high-frequency fluctuations that give rise to homogeneous pure dephasing. Compared to this time scale, there can be inhomogeneous broadening arising from more slowly evolving components of the liquid structure. However, unlike a glass, there are no essentially static local environments that give rise to permanent inhomogeneous broadening. In a liquid, spectral diffusion will cause all possible transition energies to be sampled by an oscillator on some time scale.

In principle, information on dynamical intermolecular interactions of an oscillator with its environment can be obtained from vibrational absorption spectra. The vibrational line shape, line width, and their dependences no temperature and the nature of the solvent depend on the forces experienced by the oscillator. However, a vibrational absorption spectrum reflects the full range of broadening of the vibrational transition energies, both homogeneous and inhomogeneous. In glasses and in liquids, inhomogeneous broadening can exceed the homogeneous line width. Under these circumstances, measurement of the absorption spectrum does not provide information on vibrational dynamics.

The ultrafast infrared (IR) vibrational echo experiment eliminates inhomogeneous broadening and provides a direct measurement of homogeneous dephasing (homogeneous spectrum). The vibrational echo is the equivalent of the magnetic resonance spin echo developed in 1950[6] and the electronic excitation photon echo developed in 1964.[7,8] Using vibrational echoes to measure the homogeneous dephasing time (T_2) and IR pump-probe experiments to measure the vibrational lifetime (T_1) (and orientational relaxation if it occurs), the homogeneous pure dephasing time (T_2^*) can be obtained.[9–11] Pure dephasing describes the adiabatic modulation of the vibrational energy levels of a transition caused by thermal fluctuations of its environment.[12,13] Measurement of this quantity provides detailed insight into the fast dynamics of the system. Thus, working in the time domain using nonlinear vibrational experiments, it is possible to determine the homogeneous spectrum and all dynamical contributions to it.

In this paper, detailed vibrational echo studies of (acetylacetonato)dicarbonylrhodium(I) [$Rh(CO)_2acac$] in dibutyl phthalate (DBP) above and below the solvents glass transition temperature ($T_g = 169$ K) are presented and compared to prior results for $W(CO)_6$ in several solvents including DBP.[14] In both metal carbonyls, the asymmetric CO stretching mode at ~ 2000 cm^{-1} is examined over a wide range of temperatures, temperatures at which the solvent is a low-temperature glass, passes through the glass transition, and is a liquid well above T_g. Of particular interest is the pure dephasing time, T_2^*, which reflects the medium induced transition energy fluctuations of the CO oscillator.

The CO asymmetric stretching modes of $Rh(CO)_2acac$ and $W(CO)_6$ are different in a manner that appears to be important. The $Rh(CO)_2acac$ mode (A_1 of molecular point group C_{2v}) is nondegenerate while the $W(CO)_6$ mode (T_{1u} of molecular point group O_h) is formally triply degenerate in the gas phase. In a condensed phase, the local solvent structure will be anisotropic. In general, there will be different solute/solvent interactions along the molecular x, y, and z axes. These anisotropic interactions will break the triple degeneracy of the T_{1u} mode of $W(CO)_6$, yielding three modes with small energy splittings.

The results presented below show that $Rh(CO)_2acac$ has temperature-dependent pure dephasing with a different functional form than $W(CO)_6$ even when the solvent, DBP, is the same. At low temperature, in glassy DBP, the $Rh(CO)_2acac$ pure dephasing temperature dependence is linear, T^1, and is exponentially activated at higher temperatures.[14] In contrast, $W(CO)_6$ has a T^2 temperature dependence in three different glassy solvents up to their corresponding T_g's.[9] Above T_g in all three solvents, there is a change in the form of the temperature dependence. In 2-methylpentane (2MP), $W(CO)_6$ pure dephasing makes an abrupt transition from T^2 to a Vogel–Tammann–Fulcher (VTF) type temperature

dependence.[9] Mechanisms are proposed to explain the temperature-dependent pure dephasing of $Rh(CO)_2acac$, and differences between it and $W(CO)_6$ including the role of the different degeneracies of the modes of interest.

II. THE VIBRATIONAL ECHO METHOD AND EXPERIMENTAL PROCEDURES

A. The vibrational echo method

The vibrational echo experiment is a time domain third-order nonlinear experiment that can extract the homogeneous vibrational line shape even when the inhomogeneous line width is thousands of times wider than the homogeneous width.[9,10,15]

A source of ps IR pulses is tuned to the vibrational transition of interest to provide the vibrational echo two pulse excitation sequence. The first pulse excites each solute molecule into a superposition state, which is a mixture of the $v = 0$ and $v = 1$ vibrational states. Each superposition has a microscopic electric dipole associated with it. This dipole oscillates at the vibrational transition frequency. Immediately after the first pulse, all of the microscopic dipoles in the sample oscillate in phase. Because there is an inhomogeneous distribution of vibrational transition frequencies, the dipoles oscillate with some distribution of frequencies. Thus, the initial phase relationship is very rapidly lost. This effect is the free induction decay. After a time, τ, a second pulse, traveling along a path making an angle, θ, with that of the first pulse, passes through the sample. This second pulse changes the phase factors of each vibrational superposition state in a manner that initiates a rephasing process. At time 2τ, the ensemble of superposition states is rephased. The phased array of microscopic dipoles behaves as a macroscopic oscillating dipole which generates an additional IR pulse of light called the vibrational echo. The vibrational echo pulse propagates along a path that makes an angle, 2θ, with that of the first pulse. Subsequently, a free induction decay again destroys the phase relationships, so only a temporally short pulse of IR light is generated.

The rephasing at 2τ has removed the effects of the inhomogeneous broadening.[16] However, fast fluctuations due to coupling of the vibrational mode of interest to the bath cause the oscillation frequencies to fluctuate. Thus, at 2τ there is not perfect rephasing. As τ is increased, the fluctuations produce increasingly large accumulated phase errors among the microscopic dipoles, and the signal amplitude of the vibrational echo is reduced. A measurement of the vibrational echo intensity versus τ is an echo decay curve. Thus, the vibrational echo decay is related to the fluctuations in the vibrational frequencies, not the inhomogeneous spread in frequencies. The Fourier transform of the echo decay is the homogeneous line shape.[16,17] The vibrational echo makes the vibrational homogeneous line shape an experimental observable.

In the experiments on $Rh(CO)_2acac$ in DBP, orientational relaxation does not occur on the time scale of the vibrational echo experiments because of the high viscosity of the solvent at all temperatures studied. This fact was verified experimentally using polarization selective pump-probe experiments. Therefore, orientational relaxation does not contribute to $Rh(CO)_2acac$ homogeneous dephasing. The role of orientational relaxation in vibrational echo experiments of $W(CO)_6$ has been previously discussed[5] and reanalyzed below.

The IR absorption line shape is related to these microscopic dynamics through the Fourier transform of the two-time transition dipole correlation function[10,13,18–20] which depends upon variations in the transition energies for the ensemble of vibrational transition dipoles and includes inhomogeneous broadening.[3] In the Markovian limit, and in the absence of inhomogeneous broadening, the transition dipole correlation function decays exponentially at a rate of $1/T_2$, where T_2 is the homogeneous dephasing time. The Fourier transform gives a Lorentzian line shape, and contributions to the full line width at half maximum are additive

$$\Gamma = \frac{1}{\pi T_2} = \frac{1}{\pi T_2^*} + \frac{1}{2\pi T_1}. \tag{1}$$

Equation (1) allows the contribution of pure dephasing to the vibrational homogeneous line shape to be determined from a knowledge of the homogeneous line width and the vibrational lifetime.

The description of the third-order nonlinear polarization that governs vibrational echo experiments in terms of the dynamics of lifetime and pure dephasing has been presented.[9] For a Lorentzian homogeneous line, the vibrational echo signal pulses decays exponentially as

$$I(\tau)/I(0) = \exp[-4\tau(1/T_2^* + 1/2T_1)] \tag{2a}$$

$$= \exp[-4\tau/T_2]. \tag{2b}$$

The signal decays at a rate four times faster than the decay of the homogeneous dipole correlation function. If the pulse duration is comparable to T_2, a detailed calculation of the third-order polarization can be performed to extract T_2.[21]

B. Experimental procedures

The vibrational echo experiments require tunable IR pulses with durations of ~ 1 ps and energies of ~ 1 μJ. The experiments described below were performed using IR pulses near ~ 5 μm generated by the Stanford superconducting-accelerator-pumped free electron laser (FEL).[9] The FEL generates nearly transform-limited pulses with a pulse duration that is adjustable between 0.7 and 2 ps. Active frequency stabilization allows wavelength drifts to be limited to $<0.01\%$, or <0.2 cm^{-1}. The pulse duration, spectrum, and peak power are monitored continuously.

The FEL produces a 2 ms macropulse at a 10 Hz repetition rate. Each macropulse consists of the ps micropulses at a repetition rate of 11.8 MHz (84.7 ns). The micropulse energy at the input to experimental optics is ~ 0.5 μJ.

To avoid sample heating problems, micropulses are selected out of each macropulse by Ge acousto-optic modulator (AOM) single pulse selectors.[9] This pulse selection yields an effective experimental repetition rate of 1 kHz, and an average power <0.5 mW.

The two pulses for the vibrational echo or the pump-probe (transient absorption to measure the vibrational lifetime) experiments were obtained using a 10% beamsplitter. The 10% beam (first pulse in vibrational echo sequence or probe pulse) is a single pulse selected using a Ge AOM and sent through a computer-controlled stepper motor delay line. The remaining portion (second echo pulse or pump pulse) is single pulse selected by a second Ge AOM. The AOMs can select pulses at either 25 or 50 kHz. For the vibrational echo experiment, the lower energy beam is selected at 25 kHz for background subtraction. For pump-probe, background subtraction is performed by selecting two adjacent micropulses. The two pulses were focused into the sample to ~ 50 μm diameter using an off-axis parabolic reflector. The signals are measured with HgMgTe detectors, gated integrators, and digitized for collection by computer.

Careful studies of power dependence and repetition rate dependence of the data were performed. It was determined that there were no heating or other unwanted effects when vibrational echo experiments were performed with pulse energies of ~ 15 nJ for the first pulse and ~ 80 nJ for the second pulse and the effective repetition rate of 1 kHz (50 kHz during each macropulse).

Vibrational echo and pump-probe data were taken on the A_1 asymmetric CO stretching mode of $Rh(CO)_2acac$. Solutions in DBP were made to give a peak optical density of 0.8 in a 400 μm path length cell. These solutions correspond to mole fractions of $\leqslant 10^{-4}$. The temperatures of the samples were controlled to ± 0.2 K using a constant flow He cryostat.

The $Rh(CO)_2acac$ vibrational echo data had two forms due to its small vibrational anharmonicity (13.7 cm^{-1}).[22,23] For the short pulses used in these experiments, it is possible for the pulse bandwidth to exceed the anharmonic splitting. In this case, a three-level coherence is formed of the $v = 0, 1$, and 2 levels. This produces a nonexponential vibrational echo decay with beats at the anharmonic splitting frequency.[22,23] When the laser wavelength is tuned to the center of the inhomogeneous line and the laser bandwidth is narrow, single exponential decays are measured. Further, single exponentials can be obtained with wider bandwidths by tuning the laser to the blue. In the decays with beats, the slowest component of the decay corresponds to the dephasing of the $v = 0-1$ transition only.[22,23] Both types of data were recorded and are reported here. Within experimental error, no differences in dephasing times were observed for the single exponential (narrow bandwidth) decays or the longest component of the nonexponential (wide bandwidth) decays. (The $v = 1-2$ dephasing time can be extracted from the fast component of the nonexponential decay.[22,23])

III. RESULTS

Vibrational echo experiments were conducted on the CO asymmetric stretching mode of $Rh(CO)_2acac$ (2010 cm^{-1}) in DBP from 3.4 K to 250 K. In addition, vibrational pump-

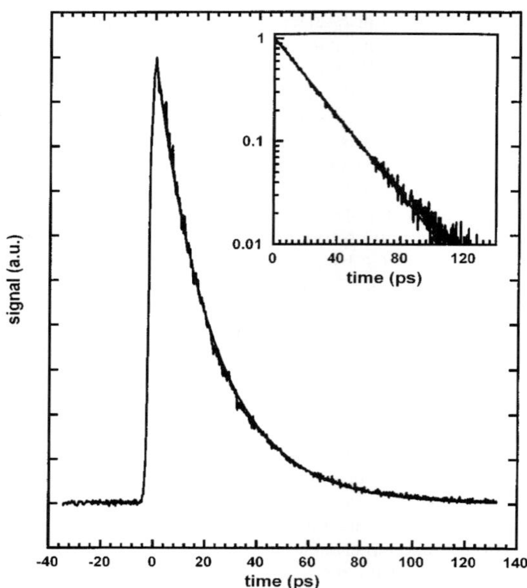

FIG. 1. Vibrational echo decay data of the asymmetric CO stretch mode of Rh(CO)2acac in DBP at 3.4 K. The decay is single exponential. The inset is a semilog plot of the data and fit. The decay constant is 23.8 ps, which yields a homogeneous line width of 0.11 cm^{-1}. The absorption spectrum has a line width of $\sim$15 cm^{-1} at this temperature, demonstrating that the line is massively inhomogeneously broadened.

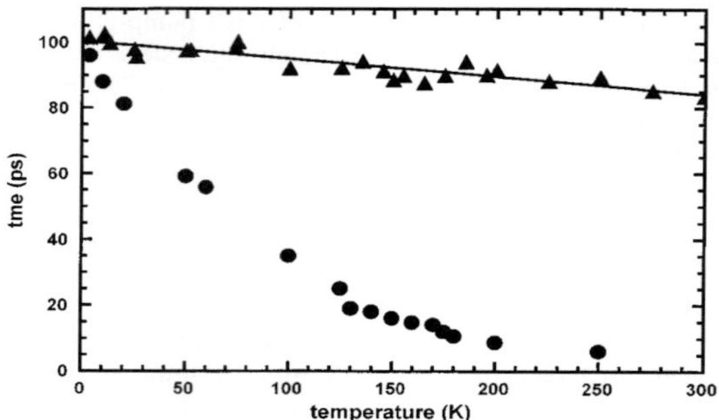

FIG. 2. Vibrational echo (circles) and pump-probe (triangles) data for the asymmetric CO stretch mode of Rh(CO)$_2$acac in DBP. The pump-probe results are plotted as $2T_1$, in accordance with Eq. (1). The solid line through the T_1 data is the best fit to the temperature dependence. Using these results, the temperature-dependent pure dephasing rates can be calculated from Eq. (1).

probe experiments were performed on the same transition from 3.4 K to 300 K. Figure 1 shows vibrational echo data taken at 2020 cm^{-1} which is 10 cm^{-1} to the blue of the center line at 3.4 K and a fit to an exponential decay [Eq. (2)]. The laser bandwidth for this data set is $\sim$14 cm^{-1}, so overlap with $v=1-2$ negligible. Within experimental uncertainty, the decay is a single exponential. Therefore, the homogeneous line shape is a Lorentzian. The decay constant $T_2=95.2$ ps, yielding a homogeneous line width of 0.11 cm^{-1}. For comparison, the absorption spectrum has a line width of $\sim$15 cm^{-1} at this temperature. The absorption line is massively inhomogeneously broadened. The vibrational echo experiments show that the absorption line is inhomogeneously broadened at all temperatures studied, including 250 K [$1/(\pi T_2)=1.5$ cm^{-1}], which is $\sim$80 K above T_g.

Figure 2 displays the results of the temperature-dependent vibrational echo (circles) and pump-probe (triangles, plotted as $2T_1$) experiments. As with studies made on W(CO)$_6$ in a number of solvents, the temperature dependence of $2T_1$ is very mild, and the temperature dependence of T_2 is much steeper. Using Eq. (1) and the $2T_1$ and T_2 values obtained from the experiments, the pure dephasing, T_2^* can be obtained. There is a small amount of scatter in the pump-probe data. The scatter is insignificant at the higher temperatures where pure dephasing totally dominates the homogeneous line width. The solid line through the data is a fit to a straight line, which accurately reflects the temperature dependence of the T_1 data over the full range of temperatures. To reduce scatter in the values of T_2^* obtained by removing the contribution from the lifetime, the T_1 values at each temperature were obtained from the linear fit to all of the T_1 data.

Figure 3 displays the values of the pure dephasing time versus inverse temperature on semilog plot.[14] The solid line through the data is a fit to the form

$$\frac{1}{T_2^*}=a_1 T^\alpha + a_2 e^{-\Delta E/kT} \tag{3}$$

with $\alpha=1$ and $\Delta E=385$ cm^{-1}. Note that there is no break in the temperature dependence at $T_g=169$ K (see inset). Also shown are dotted and dashed lines, in which α is fixed at 0.7 and 1.3, respectively, and the other parameters are allowed to float. Within experimental error, $\alpha=1.0$ gives the best fit to the data, and the value of α has only a very minor effect on the activation energy. By considering a variety of fits such as those displayed in Fig. 3, the best values for α and ΔE are $\alpha=1.0\pm0.1$ and $\Delta E=385\pm50$ cm^{-1}.

The Rh(CO)$_2$acac pure dephasing temperature dependence is different from that previously observed for the pure dephasing of the T_{1u} mode of W(CO)$_6$ in three glass forming solvents. Figure 4 shows $1/(\pi T_2^*)$ for W(CO)$_6$ in DBP from 10 K to just above T_g on a log plot.[9,10] The straight line through the data yields a temperature dependence of T^2 over the entire temperature range. Clearly the temperature dependence of the pure dephasing of the two compounds in the glass is fundamentally different even though the pure dephasing is for a CO asymmetric stretching mode of both molecules at $\sim$2000 cm^{-1} in the same solvent.

Figure 5 shows a reduced variable plot of the homogeneous dephasing of the T_{1u} mode of W(CO)$_6$ in three glassy solvents, DBP, 2MP, and 2-methyltetrahydrofuran (2MTHF).[9] The solid line through the data is T^2. Within

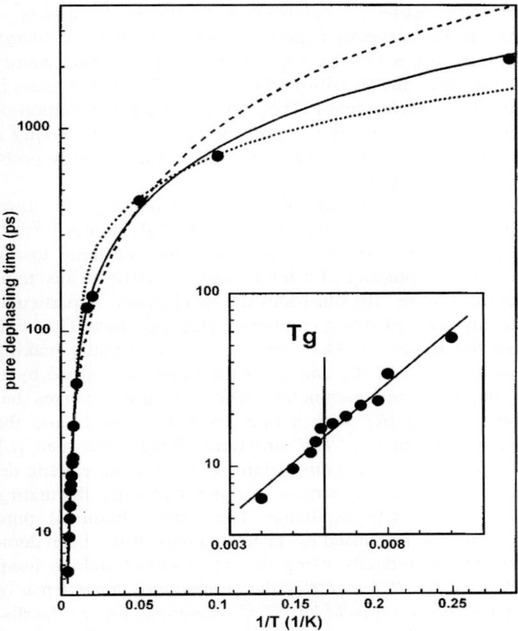

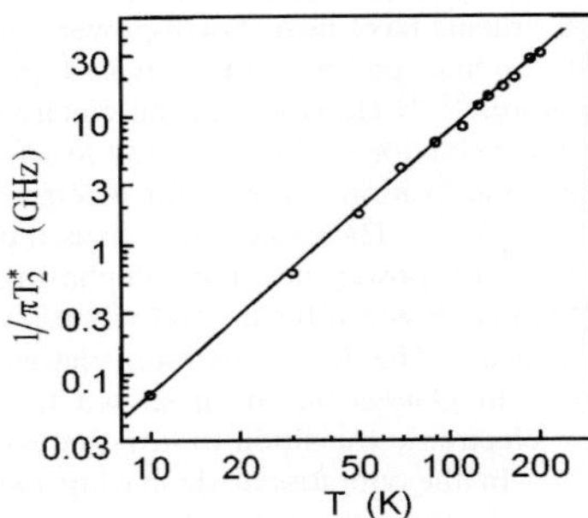

FIG. 3. Pure dephasing time of the asymmetric CO stretch mode of $Rh(CO)_2acac$ in DBP versus inverse temperature on a semi-log plot. The solid line through the data is a fit to Eq. (3), the sum of a power law and an exponentially activated process. Note that there is no break at the experimental T_g, 169 K. The inset is an expanded view at high temperatures showing that the process is activated. The best fit has the power law exponent, $\alpha = 1.0$ and $\Delta E = 385$ cm^{-1}. Dotted and dashed lines are with α in Eq. (3) fixed at 0.7 and 1.3, respectively, and the other parameters of Eq. (3) are allowed to float.

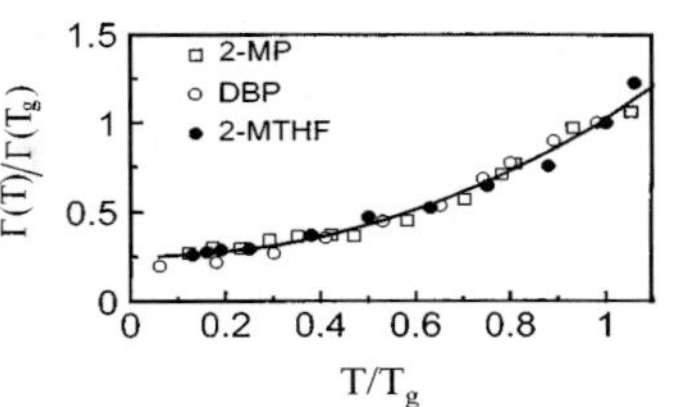

FIG. 5. Reduced variable plot of the homogeneous dephasing of $W(CO)_6$ in DBP, 2MP, and 2MTHF. The abscissa is the homogeneous line width divided by the homogeneous line width at T_g. The ordinate is T scaled by T_g. All data sets fall on the T^2 line.

$W(CO)_6$ in 2MP displays orientational relaxation because of a mechanism that depends on the degeneracy of the T_{1u} mode (see below). The temperature dependences of the orientational relaxation as well as T_2 and T_1 have been measured.[10] Using these, the temperature dependence of the pure dephasing was obtained and is displayed in Fig. 6 on a log plot. Below T_g, the temperature dependence is T^2, as discussed above. Above T_g, there is a dramatic break in the temperature dependence. The solid line through the data is a fit to Eq. (4),

$$\frac{1}{\pi T_2^*} = a_1 T^\alpha + a_2 \exp(-B/(T-T_0)),\qquad(4)$$

where α is 2.0 ± 0.1. The first term is the T^2 temperature dependence observed in the low-temperature glass. The second term has the form of the VTF equation which is often used to describe the onset of dynamic processes near the T_g.[24–26] The VTF equation describes a process with a temperature-dependent activation energy that diverges at a

experimental error, the homogeneous dephasing has a T^2 temperature dependence in all three solvents in spite of the fact that the three solvents are quite different.

A full temperature dependence of the pure dephasing of $W(CO)_6$ is only available in the solvent 2MP.[9,10] The

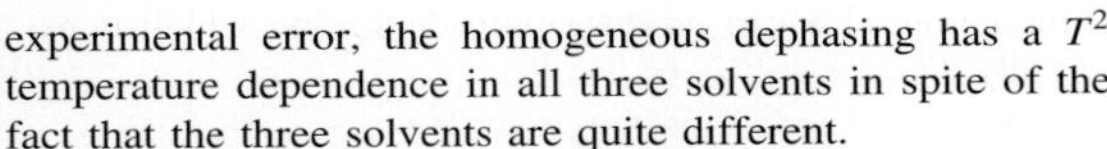

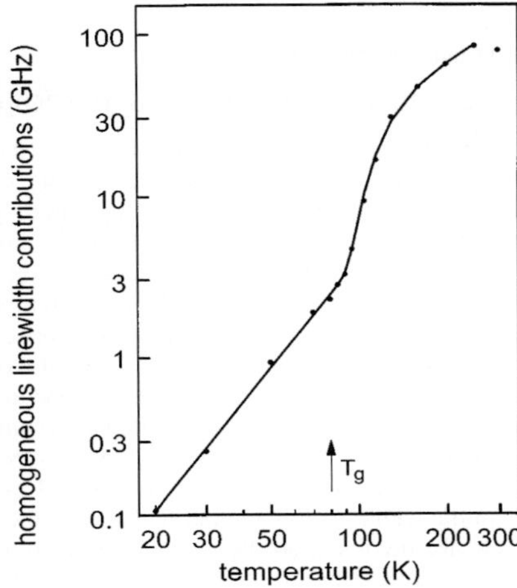

FIG. 4. Pure dephasing rate of $W(CO)_6$ in DBP versus temperature on a log plot. The straight line through the data is a fit which yields a T^2 dependence.

FIG. 6. Temperature dependence of the pure dephasing of the asymmetric CO stretch of $W(CO)_6$ in 2MP. The line through the pure dephasing data is a fit to Eq. (4).

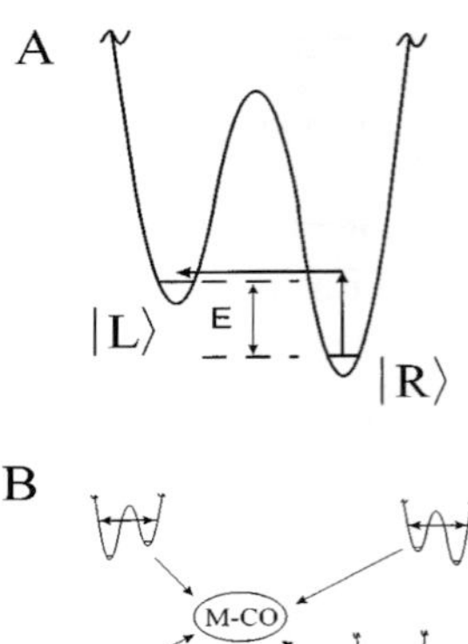

FIG. 7. (a) An illustration of a two-level system (TLS) making a transition from a lower energy local structure in the glass to a higher energy structure. A glass is modeled as having many TLS with a broad distribution of tunneling splittings, E. (b) A schematic of the CO oscillator coupled to a number of TLSs. The TLS transitions produce fluctuating forces at the oscillator causing pure dephasing.

temperature, T_0, below the nominal T_g. T_0 can be linked thermodynamically to an "ideal" T_g.[24] This equation describes the temperature dependence of the viscosity of 2MP well, and gives $T_0 = 59$ K.[9] The fit of Eq. (4) to the pure dephasing data describes the entire temperature dependence exceedingly well up to 250 K, but yields a reference temperature of $T_0 = 80$ K. This reference temperature matches the laboratory T_g, not the ideal T_g. The onset of the dynamics that cause the rapid increase in homogeneous dephasing of $W(CO)_6$ in 2MP is apparently linked with the onset of structural processes near the laboratory T_g.

IV. DEPHASING MECHANISMS

A. Low-temperature pure dephasing of Rh(CO)₂acac

Pure dephasing of the form T^{α} where $\alpha \approx 1$ has been seen repeatedly for homogeneous pure dephasing of electronic transitions of molecules in low-temperature glasses using photon echoes[3,27,28] and hole burning spectroscopy.[1,29-31] The electronic dephasing has been described using the two-level system (TLS) model of low-temperature glass dynamics[3,29,32,33] that was originally developed in the early 70s to explain the anomalous heat capacity of low-temperature glasses, which is approximately linear in T.[34,35] Even at low temperatures, glasses are continuously undergoing structural changes. The complex potential surface on which local structural dynamics occur is modeled as a collection of double wells. At low temperatures, only the lowest energy levels are involved, so these are referred to as TLS. Figure 7(a) is an illustration of a TLS. $|L\rangle$ represents a particular local structure of the glass. $|R\rangle$ represents a different local structure. Transitions can be made between $|L\rangle$ and $|R\rangle$ by phonon-assisted tunneling. At very low temperatures,

where the Debye T^3 contribution to the heat capacity is small, the heat capacity is dominated by the uptake of energy in going from a lower energy structure to a higher energy structure, e.g., the transition $|R\rangle \Rightarrow |L\rangle$ in Fig. 7(a). A glass is modeled as having many TLSs with a broad distribution of tunnel splittings, E. If the probability, $P(E)$, of having a splitting E is constant, $P(E) = C$, (all E's are equally probable), then the heat capacity is T^1.

The description of electronic dephasing in low-temperature glasses is based on the TLS dynamics.[3,29,32,33] We propose that identical considerations can apply to the vibrational dephasing of $Rh(CO)_2acac$ in DBP at low temperature. Figure 7(b) illustrates the mechanism. A particular molecule is coupled to a number of TLS. For those TLS with E not too large ($E \lesssim 2kT$), the TLS are constantly making transition between $|L\rangle$ and $|R\rangle$ with a rate determined by E and the tunneling parameter.[2] The structural changes between $|L\rangle$ and $|R\rangle$ with a rate determined by E and the tunneling parameter.[2] The structural changes between $|L\rangle$ and $|R\rangle$ produce fluctuating strains or fluctuating electric dipoles. The fluctuating strains or dipoles produce fluctuating forces on the CO oscillator. Thus, the vibrational pure dephasing can be caused by TLS dynamics. It has been demonstrated theoretically using the uncorrelated sudden jump model that for $P(E) = CE^{\mu}$, the temperature dependence of the pure dephasing is $T^{1+\mu}$.[3,36-38] Therefore, for the flat distribution, $\mu = 0$, the pure dephasing temperature dependence is T^1. T^1 and somewhat steeper temperature dependences, e.g., $T^{1.3}$, have been observed in electronic dephasing experiments in low-temperature glasses.[1,3,27,36] Recent theoretical work that has examined the problem in more detail suggests that even the apparent super linear temperature dependences may arise from an energy distribution $P(E) = C$.[39] Other theoretical work has investigated the influence of coupled TLS.[40] Regardless of the theoretical approach, the qualitative results are the same. Coupling of a transition to a distribution of tunneling TLS can produce pure dephasing which is essentially T^1.

B. High-temperature pure dephasing of Rh(CO)₂acac

Above ~ 20 K, the T^1 vibrational pure dephasing is dominated by the exponentially activated process. Electronic dephasing experiments have also shown power law temperature dependences that go over to activated processes at higher temperatures.[41-43] However, in the electronic experiments, power law behavior is observed only to a few degrees K because typical activation energies for electronic dephasing are $15 - 30$ cm^{-1}.[42,43] Therefore, the activated process begins to dominate the power law pure dephasing at lower temperatures than is observed for the CO vibrational dephasing of $Rh(CO)_2acac$. The low activation energy for electronic dephasing in glasses has been shown to arise from coupling of the electronic transition to low-frequency modes of the glass.[42,43] In the vibrational dephasing experiments, the $\Delta E \cong 400$ cm^{-1}. Thus, the power law component of the temperature dependence is not masked until higher temperature.

A

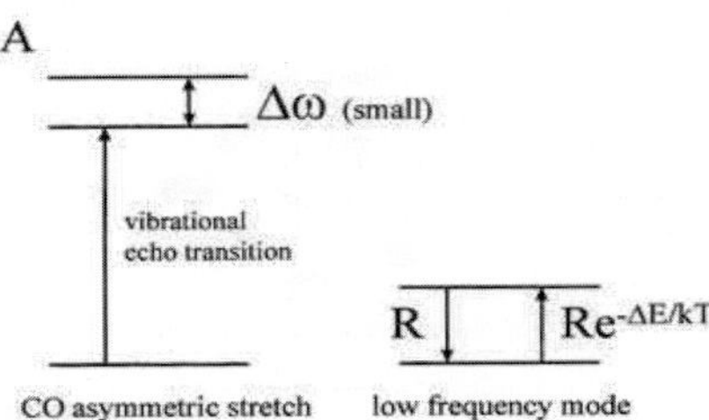

B

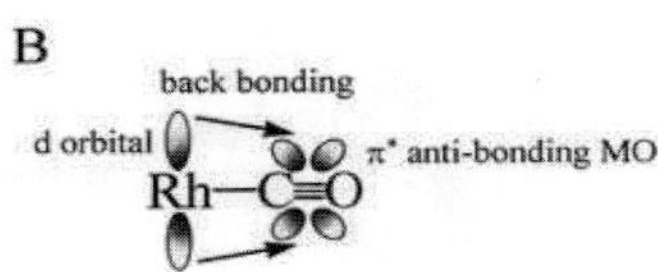

FIG. 8. (a) Proposed dephasing mechanism of the asymmetric CO stretching mode of Rh(CO)$_2$acac at high temperature. Thermal activation of a low-frequency mode causes a small change in the transition frequency, $\Delta\omega$, of the high-frequency mode. During the time when the low-frequency mode is excited, the high-frequency mode develops a phase error. (b) Electron donation from the d_π orbital of the Rh atom to the CO π^* antibonding molecular orbital of CO. Thermal excitation of the Rh–C stretch will lengthen the bond which decreases the magnitude of backbonding, producing a shift to higher energy of the CO transition.

In the Rh(CO)$_2$acac system, the temperature dependence of the pure dephasing changes rapidly above ~ 20 K. By 100 K, the temperature dependence is well described by the activated process alone (see inset in Fig. 3). There is no break in the pure dephasing data as the sample passes through T_g. The activation energy, $\Delta E = \sim 400$ cm^{-1}, is well above the typical cutoff for phonon modes of organic solids.[44] Furthermore, the far-IR absorption spectra of neat DBP shows no strong transitions in the region around 400 cm^{-1}, indicating that there is no specific mode of the solvent that might couple strongly to the CO mode. These facts suggest that the high temperature Arrhenius dephasing process is not caused by a motion associated with the glass/liquid solvent, but rather that the dephasing arises from coupling of the CO mode to an internal mode of Rh(CO)$_2$acac.

The proposed mechanism is illustrated in Fig. 8(a). Thermal excitation of a low-frequency mode causes the CO stretching mode transition frequency to shift a small amount, $\Delta\omega$. The lifetime of the low-frequency mode is $\tau = 1/R$. During the time period in which the low-frequency mode is excited, the initially prepared CO superposition state precesses at a higher frequency. Thus, a phase error develops. For a small $\Delta\omega$ and a short τ, the phase error is on the order of $\tau\Delta\omega < 1$. This is the slow or intermediate exchange limit.[45,46] Repeated excitation and relaxation of the low-frequency mode will produce homogeneous dephasing.[45,46]

Over a temperature range in which the energy of the low-frequency transition, ΔE, is large compared to kT, the rate of excitation of the low-frequency mode increases exponentially with temperature, i.e., the rate of excitation is

$Re^{-\Delta E/kT}$. Over this same temperature range, the downward rate will be temperature independent or have a weak temperature dependence. From the fit, $\Delta E \cong 400$ cm^{-1} and the highest experimental temperature corresponds to ~ 170 cm^{-1}, so this condition is met. This system is in the weak coupling limit, i.e., the change in the CO frequency, $\Delta\omega$, with excitation of the low-frequency mode is small compared to the CO frequency. In addition, $\hbar|\Delta\omega|/kT \ll 1$. For these conditions, the pure dephasing contribution to the linewidth from repeated excitation and relaxation of the low-frequency mode is[45]

$$\frac{1}{\pi T_2^*} = \frac{1}{\tau}\left[\frac{(\Delta\omega\tau)^2}{1+(\Delta\omega\tau)^2}\right]e^{-\Delta E/kT}. \tag{5}$$

Equation (5) shows that this contribution to the homogeneous linewidth is exponentially activated. The right-hand side of Eq. (5) is consistent with Eq. (3), which was used to fit the data. The factor multiplying the exponential is the constant a_2 in Eq. (3). This term dominates the temperature dependence at high temperature.

For the proposed mechanism to account for the observed high temperature pure dephasing, a mode of ~ 400 cm^{-1} must couple nonnegligibly to the asymmetric CO stretch so that $\Delta\omega$ is significant. The Rh–C asymmetric stretching mode has a transition energy of 405 cm^{-1}.[47] The closest other modes of Rh(CO)$_2$acac are outside of the error bars on the activation energy.[47] There is a reasonable explanation why the Rh–C stretch couples significantly to the CO mode, but modes of lower frequency, which would become populated at lower temperature, do not. The explanation is illustrated in Fig. 8(b). The Rh(CO)$_2$acac has significant back donation of electron density from the Rh d_π to the CO p_{π^+} antibonding orbital (back bonding) that weakens the CO bond and red shifts the transition energy about 100 to ~ 2045 cm^{-1}. (The splitting of the symmetric and asymmetric linear combination of the 2 CO stretches further shifts the asymmetric mode to the observed value of 2010 cm^{-1}.) Thus, back bonding plays a significant role in determining the transition frequency. It is well known that in metal-carbonyl compounds, the CO frequency is very sensitive to changes in back bonding.[48] Also, a combination of isotope substitution spectroscopic experiments and calculations show that for metal-carbonyls, there is substantial coupling between the M–C stretch and the C–O stretch.[49] When the Rh–C mode is thermally excited from the $v=0$ state to the $v=1$ state, the average bond length will increase. The increase in the sigma bond length will decrease the Rh d_π–CO p_{π^+} orbital overlap, and, therefore, decrease the magnitude of the back bonding. Thus, excitation of the Rh–C mode causes a blue shift of the CO stretching frequency by decreasing the back bonding.[14]

Although data is not available for Rh(CO)$_2$acac, IR absorption measurements on transition metal hexacarbonyls support this mechanism.[49] For the equivalent mode of M(CO)$_6$ (M=Mo, Cr), the combination absorption band of the M–C asymmetric stretch and the CO asymmetric stretch is ~ 20 cm^{-1} higher in energy than the sum of the two fundamental energies.[49] Thus, $\Delta\omega \approx 20$ cm^{-1}. The change in back bonding upon excitation of the Rh–C mode provides a direct mechanism for coupling excitation of the Rh–C stretch to the CO stretch transition frequency. Other low-frequency modes, such as a methyl rocking mode of the acac ligand, will not have this type of direct coupling, and, therefore, will not cause dephasing even though they may be thermally populated.

If the proposed mechanism is valid, it should be possible to reproduce the constant a_2 in Eq. (3) using Eq. (5) with reasonable values of the other parameters:

$$a_2 = \frac{1}{\pi\tau}\left[\frac{(\Delta\omega\tau)^2}{1+(\Delta\omega\tau)^2}\right]. \tag{6}$$

The value of a_2 is obtained from the data in Fig. 3; $a_2 = 1.2\times10^{12}$ Hz. As discussed above, based on compounds similar to Rh(CO)$_2$acac, $\Delta\omega \approx 20$ cm^{-1}. Using the values for a_2 and $\Delta\omega$ yields a value of $\tau \approx 0.75$ ps is obtained for the vibrational lifetime of the 405 cm^{-1} Rh–C stretch. Direct measurements of the lifetime of this mode or of any low frequency vibrations have not been made. However, 0.75 ps is a plausible number. The Rh–C mode can relax via a cubic anharmonic process involving the annihilation of the original Rh–C excitation and the creation of two lower frequency modes.[47] The Rh(CO)$_2$acac has several internal lower frequency modes.[47] One possible relaxation pathway is to create one internal mode, e.g., 300 cm^{-1}, and create a mode of the solvent continuum, assuring conservation of energy. Another possible pathway is relaxation into two modes of the solvent continuum. For a nonhydrogen bonding solvent such as DBP, the continuum of translational and orientational modes (instantaneous normal modes[50,51]) will extend to several hundred cm^{-1}.[52] For either possibility, the low-order cubic anharmonic processes available for the relaxation and the high density of states provided by the solvent continuum will cause rapid relaxation of the Rh–C vibration. In the future, it may be possible to perform a far IR pump-probe experiment to make a direct measurement of the Rh–C lifetime.

C. Dephasing of W(CO)$_6$

As can be seen from a comparison of Figs. 3, 4, and 6, the temperature-dependent pure dephasing of Rh(CO)$_2$acac is fundamentally different from that of W(CO)$_6$ at all temperatures even though an asymmetric CO stretch at $\sim$2000 cm^{-1} was studied in both molecules. It is proposed that the W(CO)$_6$ pure dephasing is different because of the high symmetry of the T_{1u} asymmetric CO stretching mode, which makes it triply degenerate in the gas phase,[49] while the Rh(CO)$_2$acac mode is not degenerate. The W(CO)$_6$$T_{1u}$ mode consists of all six CO's moving in concert, one pair along the molecular x axis, one pair along y, and one pair along z. In a liquid or glass, the local solvent structure is anisotropic. In general, there will be different solute/solvent interactions (forces exerted on the oscillator) along x, y, and z. These interactions will break the triple degeneracy, yielding three

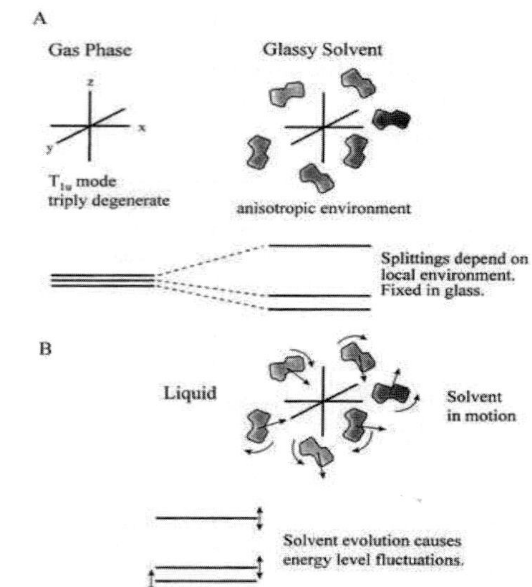

FIG. 9. (a) In the gas phase, the T_{1u} mode of W(CO)$_6$ is triply degenerate. In a condensed phase, such as a liquid or glass, the anisotropic environment breaks the degeneracy, producing three levels with small energy splittings. (b) Solvent molecular motions in the liquid state produce a local, time-dependent anisotropic structure. Fluctuations in local structure couple to the three levels, causing energy level fluctuations and pure dephasing.

modes with small energy splittings. This model is illustrated in Fig. 9(a). There will be a range of such splittings reflecting the range of local solvent structures.

In a glass, the local solvent structure about W(CO)$_6$ is essentially fixed on the time scale of the pure dephasing. The T^2 temperature dependence in the three glassy solvents studied suggests a two phonon process in the high-temperature limit, i.e., $kT>\hbar\omega_p$, where ω_p is a typical phonon frequency involved in the two phonon process. Two phonon elastic scattering that causes fluctuations in the anisotropic local solvent structure surrounding W(CO)$_6$ will induce fluctuations of the level splittings and can cause pure dephasing.

A second mechanism that would result in a T^2 temperature dependence is two phonon scattering from one level to another. This mechanism is referred to as inelastic two phonon scattering. This second possibility can be ruled out. The nondegenerate levels will be approximately the molecular x, y, and z basis modes of the triply degenerate state, although there may be some mixing caused by the anisotropic nature of the solvent perturbation. Scattering among the levels would result in orientational relaxation since it will take the oscillating dipole from, for example, x to y or z. In both the glass and liquid states of the solvent, the contribution of orientational relaxation to the total homogeneous line width has been analyzed, and it is small compared the other contributions.[9] If two phonon inelastic scattering among the three nondegenerate levels were responsible for the pure dephasing, it would occur at the same rate as the orientational relaxation.

In the previous studies, orientational relaxation, which was shown not to depend on physical rotation of the molecule, was ascribed to the triply degenerate nature of the T_{1u} state.[9,10] However, it was implicitly assumed that the state was actually degenerate, and that the radiation field would excite some particular coherent superposition of the x, y, and z basis states.[9] Because the degeneracy is broken by the anisotropic environment, a single state will be excited. The probability of exciting x, y, or z will depend on the projection of the excitation E-field onto the molecular x, y, and z axes. Then, the overall proposal is that orientational relaxation is caused by two-phonon inelastic scattering among the three nondegenerate states, and, in the glass, pure dephasing is caused by two-phonon elastic-scattering-induced fluctuations of the x, y, and z levels. Because the T_{1u} mode is initially degenerate, any anisotropic perturbation will break the degeneracy. The resulting splitting are very sensitive to even small phonon-induced fluctuations of the local solvent environment. This is not a mechanism that is available to $Rh(CO)_2$acac. Apparently, the two-phonon elastic scattering mechanism available to $W(CO)_6$ dominates the mechanisms responsible for pure dephasing of $Rh(CO)_2$acac in the glass and liquid states.

The temperature dependence of the total homogeneous line widths of $W(CO)_6$ changes abruptly above T_g in the three solvents, DBP, 2MTHF, and 2MP.[9,10] In DBP and 2MTHF, there is evidence of motional narrowing.[9,10] The pure vibrational dephasing above T_g in 2MP has the very steep VTF temperature dependence (see Fig. 6). In the liquid, the solvent structure surrounding $W(CO)_6$ is no longer static on the homogeneous time scale [see Fig. 9(b)]. Translation and rotational motions of the solvent will produce fluctuating forces that do not occur below T_g. Thus, there is a change from a two-phonon elastic scattering mechanism with essentially fixed solvent structure below T_g to a mechanism that involves the evolution of the local anisotropic solvent structure above T_g.

The evolution of the local solvent structure will cause the splittings of the three closely spaced levels to evolve in time, inducing pure dephasing [see Fig. 9(b)]. Thus, the triply degenerate nature of the $W(CO)_6$ T_{1u} mode is also intimately involved in the pure dephasing in the liquid, but the nature of the solvent dynamics changes above T_g, causing the abrupt change in the temperature dependence. In the proposed mechanism, the pure dephasing of $W(CO)_6$ in liquid 2MP is caused by the very-high-frequency solvent motions that are ultimately responsible for longer time scale processes such as translational and rotational diffusion and dielectric relaxation. The observed VTF temperature dependence would seem to be consistent with this dephasing mechanism.

V. CONCLUDING REMARKS

Vibrational echo experiments have made it possible to perform a detailed examination of the dynamics of inter- and intramolecular interactions that give rise to the homogeneous line widths and pure dephasing of the asymmetric stretching modes of $Rh(CO)_2$acac and $W(CO)_6$ in liquid and glassy solvents. Even when $Rh(CO)_2$acac and $W(CO)_6$ are in the same solvent, the functional forms of their temperature-dependent pure dephasing are very different. At low temperature (3.5 K to $\sim 20\,K$), $Rh(CO)_2$acac pure dephasing goes as T^1. This is interpreted as the result of coupling of the vibrational mode to the dynamical two-level systems of the glassy DBP solvent. Above ~ 20 K, the pure dephasing becomes exponentially activated with an activation energy of ~ 400 cm^{-1}. There is no change in the functional form of the temperature dependence in passing from the glass to the liquid. These results suggest that the activated process arises from coupling of the high-frequency CO stretch to the internal 405 cm^{-1} Rh–C asymmetric stretching mode. Excitation of the Rh–C stretch produces changes in the back donation of electron density from the metal d_π orbitals to the CO π^* anti-bonding orbital, shifting the CO stretching transition frequency and causing dephasing.

The pure dephasing of $W(CO)_6$ has a T^2 temperature dependence in DBP, 2MP, and 2MTHF glasses. In all three solvents, there is an abrupt change in the functional form of the temperature dependence in going from the glass to the liquid. In liquid 2MP, the pure dephasing has a VTF temperature dependence. The major differences in the temperature-dependent pure dephasing of the asymmetric stretching modes of $Rh(CO)_2$acac and $W(CO)_6$ are attributed to the difference in the degeneracy of the modes. The $Rh(CO)_2$acac mode is nondegenerate while the $W(CO)_6$ mode is triply degenerate in the gas phase. When $W(CO)_6$ is placed in a glass or liquid solvent, the local anisotropic solvent structure beaks the degeneracy, yielding three modes with small energy splittings. The results indicate that these splittings are very sensitive to local fluctuations in the solvent, giving rise to dephasing mechanisms that are not available to $Rh(CO)_2$acac. The abrupt change in the temperature dependence of the $W(CO)_6$ vibrational pure dephasing above T_g occurs because the nature of the local solvent structural fluctuations change in going from a glass to a liquid.

It is interesting and useful to note that the two molecules studied provide probes of different aspects of solvent dynamics. At low temperature, the vibrational pure dephasing of $Rh(CO)_2$acac is sensitive to the glass's structural evolution produced by two level system dynamics, while that of $W(CO)_6$ is not. However, above T_g, the broken degeneracy of the T_{1u} mode of $W(CO)_6$ causes its vibrational pure dephasing to be sensitive to the dynamics of the supercooled liquid, while the nondegenerate mode of $Rh(CO)_2$acac is not.

The observed differences between the vibrational pure dephasing of the asymmetric stretching modes of $Rh(CO)_2$acac and $W(CO)_6$ have been attributed to the difference in the mode degeneracy. Future vibrational echo experiments on other metal carbonyls and vibrations of other classes of molecules will continue to explore the nature of vibrational pure dephasing in liquids, glasses, and proteins.[53,54]

ACKNOWLEDGMENTS

We would like to thank Professor James Skinner, Department of Chemistry, University of Wisconsin at Madison, for very useful conversations pertaining to this work. We would also like to thank Dr. C. W. Rella, Dr. A. S. Kwok, and Dr. C. Ferrante for their assistance in performing some of the experiments, and Professors Alan Schwettman and Todd Smith, Physics Department, Stanford University, and their research groups at the Stanford FEL Center whose efforts made these experiments possible. This research was supported by the National Science Foundation, Division of Materials Research (DMR-9610326), and the Office of Naval Research (N00014-94-1-1024, N00014-92-J-1227-P00006, and N00014-97-1-0899.

[1] H. P. H. Thijssen, A. I. M. Dicker, and S. Völker, Chem. Phys. Lett. **92**, 7 (1982).

[2] D. Harrer, *Photochemical Hole-Burning in Electronic Transitions* (Springer-Verlag, Berlin, 1988).

[3] L. R. Narasimhan, K. A. Littau, D. W. Pack, Y. S. Bai, A. Elschner, and M. D. Fayer, Chem. Rev. **90**, 439 (1990).

[4] S. Fei, G. S. Yu, H. W. Li, and H. L. Strauss, J. Chem. Phys. **104**, 6398 (1996).

[5] A. Tokmakoff, R. S. Urdahl, D. Zimdars, R. S. Francis, A. S. Kwok, and M. D. Fayer, J. Chem. Phys. **102**, 3919 (1994).

[6] E. L. Hahn, Phys. Rev. **80**, 580 (1950).

[7] N. A. Kurnit, I. D. Abella, and S. R. Hartmann, Phys. Rev. Lett. **13**, 567 (1964).

[8] I. D. Abella, N. A. Kurnit, and S. R. Hartmann, Phys. Rev. Lett. **14**, 391 (1966).

[9] A. Tokmakoff and M. D. Fayer, J. Chem. Phys. **102**, 2810 (1995).

[10] A. Tokmakoff, D. Zimdars, R. S. Urdahl, R. S. Francis, A. S. Kwok, and M. D. Fayer, J. Phys. Chem. **99**, 13 310 (1995).

[11] A. Tokmakoff and M. D. Fayer, Acc. Chem. Res. **28**, 437 (1995).

[12] K. S. Schweizer and D. Chandler, J. Chem. Phys. **76**, 2240 (1982).

[13] D. W. Oxtoby, Annu. Rev. Phys. Chem. **32**, 77 (1981).

[14] K. D. Rector, A. S. Kwok, C. Ferrante, R. S. Francis, and M. D. Fayer, Chem. Phys. Lett. **276**, 217 (1997).

[15] D. Zimdars, A. Tokmakoff, S. Chen, S. R. Greenfield, and M. D. Fayer, Phys. Rev. Lett. **70**, 2718 (1993).

[16] T. C. Farrar and D. E. Becker, *Pulse and Fourier Transform NMR* (Academic, Press, New York, 1971).

[17] J. L. Skinner, H. C. Anderson, and M. D. Fayer, J. Chem. Phys. **75**, 3195 (1981).

[18] R. G. Gordon, J. Chem. Phys. **43**, 1307 (1965).

[19] R. G. Gordon, Adv. Magn. Reson. **3**, 1 (1968).

[20] B. J. Berne, *Physical Chemistry: An Advanced Treatise* (Academic, New York, 1971).

[21] D. A. Zimdars, Graduate thesis, Stanford University, 1996.

[22] K. D. Rector, A. S. Kwok, C. Ferrante, A. Tokmakoff, C. W. Rella, and M. D. Fayer, J. Chem. Phys. **106**, 10027 (1997).

[23] A. Tokmakoff, A. S. Kwok, R. S. Urdahl, R. S. Francis, and M. D. Fayer, Chem. Phys. Lett. **234**, 289 (1995).

[24] C. A. Angell, J. Phys. Chem. Solids Suppl. **49**, 863 (1988).

[25] C. A. Angell, J. Phys. Chem. **86**, 3845 (1982).

[26] G. H. Fredrickson, Annu. Rev. Phys. Chem. **39**, 149 (1988).

[27] L. W. Molenkamp and D. A. Wiersma, J. Chem. Phys. **83**, 1 (1985).

[28] Y. G. Vainer, R. I. Personov, S. Zilker, and D. Haarer, in *Fifth International Meeting on Hole Burning and Related Spectroscopies: Science and Applications*, edited by G. J. Small (Gordon and Breach, Brainerd, MN, 1996), Vol. 291, pp. 51.

[29] H. W. H. Lee, A. L. Huston, M. Gehrtz, and W. E. Moerner, Chem. Phys. Lett. **114**, 491 (1985).

[30] J. M. Hayes, R. P. Stout, and G. J. Small, J. Chem. Phys. **74**, 4266 (1981).

[31] R. M. Macfárlane and R. M. Shelby, Opt. Commun. **45**, 46 (1983).

[32] R. M. Macfárlane and R. Shelby, J. Lumin. **36**, 179 (1987).

[33] J. Friedrich, H. Wolfrum, and D. Haarer, J. Chem. Phys. **77**, 2309 (1982).

[34] W. A. Phillips, J. Low Temp. Phys. **7**, 351 (1972).

[35] P. W. Anderson, B. I. Halperin, and C. M. Varma, Philos. Mag. **25**, 1 (1972).

[36] J. M. Hayes, R. Jankowiak, and G. J. Small, in *Persistent Spectral Hole Burning: Science and Applications*, edited by W. E. Moerner (Springer-Verlag, Berlin, 1988), Vol. 44, pp. 153.

[37] D. L. Huber, J. Non-Cryst. Solids **51**, 241 (1982).

[38] D. L. Huber, M. M. Broer, and B. Golding, Phys. Rev. Lett. **52**, 2281 (1984).

[39] E. Geva and J. L. Skinner, J. Chem. Phys. **107**, 7630 (1997).

[40] K. Kassner and R. Silbey, J. Phys. C **1**, 4599 (1989).

[41] P. M. Selzer, D. L. Huber, D. S. Hamilton, W. M. Yen, and M. J. Weber, Phys. Rev. Lett. **36**, 813 (1976).

[42] A. Elschner, L. R. Narasimhan, and M. D. Fayer, Chem. Phys. Lett. **171**, 19 (1990).

[43] S. R. Greenfield, Y. S. Bai, and M. D. Fayer, Chem. Phys. Lett. **170**, 133 (1990).

[44] A. I. Kitaigorodsky, *Molecular Crystals and Molecules* (Academic, New York, 1973).

[45] D. Hsu and J. L. Skinner, J. Chem. Phys. **83**, 2097 (1985).

[46] R. M. Shelby, C. B. Harris, and P. A. Cornelius, J. Chem. Phys. **70**, 34 (1978).

[47] D. M. Adams and W. R. Trumble, J. C. S. Dalton, 690 (1974).

[48] D. M. Adams, *Metal-Ligand and Related Vibrations* (St. Martin's, New York, 1968).

[49] L. H. Jones, R. S. McDowell, and M. Goldblatt, Inorg. Chem. **8**, 2349 (1969).

[50] R. M. Stratt, Acc. Chem. Res. **28**, 201 (1995).

[51] T. Keyes, J. Phys. Chem. A **101**, 2921 (1997).

[52] Y. J. Chang and E. W. Castner, Jr., J. Phys. Chem. A **100**, 3330 (1996).

[53] C. W. Rella, K. D. Rector, A. S. Kwok, J. R. Hill, H. A. Schwettman, D. D. Dlott, and M. D. Fayer, J. Phys. Chem. **100**, 15 620 (1996).

[54] K. D. Rector, C. W. Rella, A. S. Kwok, J. R. Hill, S. G. Sligar, E. Y. P. Chien, D. D. Dlott, and M. D. Fayer, J. Fayer, J. Phys. Chem. B **101**, 1468 (1997).

J. Phys. Chem. A **1999**, *103*, 2381–2387

A Dynamical Transition in the Protein Myoglobin Observed by Infrared Vibrational Echo Experiments

K. D. Rector,[†] J. R. Engholm,[‡,⊥] C. W. Rella,[‡,∥] J. R. Hill,[§,¶] D. D. Dlott,[§] and M. D. Fayer*,[†]

Department of Chemistry, Stanford University, Stanford, California 94305; Stanford Free Electron Laser Center, Stanford University, Stanford, California 94305; and School of Chemical Sciences, University of Illinois at Urbana–Champaign, Urbana, Illinois 61801

Received: September 30, 1998; In Final Form: January 27, 1999

Ultrafast infrared vibrational echo measurements of the temperature-dependent pure dephasing of the A_1 CO stretching mode of myoglobin–CO (Mb-CO) were performed in the solvents trehalose and 50:50 ethylene glycol:water. The results are compared to previously reported data in 95:5 glycerol:water. The temperature dependence (11–300 K) of the pure dephasing in trehalose (a glass at all temperatures studied) is a power law, $T^{1.3}$, below $T \cong 200$ K, while at higher temperature it becomes dramatically steeper. The change in functional form occurs although the solvent does not go through its glass transition. In the other two solvents, the breaks in the temperature dependences occur at lower temperatures, and the temperature dependences are even steeper above the power law region. The results are discussed in terms of a combination of a temperature and viscosity dependence of protein dynamics.

I. Introduction

The dynamics of proteins on a wide variety of time scales are intimately related to protein function. Fast fluctuations (picosecond time scale) of protein structure enable a protein to sample the complex protein conformational energy landscape. Fast movement on the energy landscape gives rise to the slower processes associated with protein function. Molecular dynamics simulations have shown that a protein can sample thousands of conformations on a very short time scale.[1] Understanding such dynamics is key to determining the connection between protein structure, as measured by X-ray,[2,3] NMR,[4,5] or other experimental techniques,[6–10] or theory[1] and protein function.

Recently, the ultrafast infrared vibrational echo technique has been applied to the study of protein dynamics.[11–13] Unlike other ultrafast techniques,[14–18] which involve electronic excitation of chromophores, the vibrational echo experiments have directly examined the CO ligand bound to the active site of myoglobin (Mb-CO) in the ground electronic state. The vibrational echo is sensitive to ground state fluctuations of protein structure that are communicated to the ligand bound at the active site. Studies of the myoglobin mutant H64V suggest that the protein fluctuations communicated to the active site are global in nature. The protein fluctuations are coupled to the CO transition frequency by producing fluctuating electric fields which act on the heme.[12]

Myoglobin has been extensively studied by a wide variety of methods. Myoglobin is a 154 amino acid protein, which has the primary biological function of the reversible binding and transport of O_2 in muscle tissues. Myoglobin's ability to bind O_2, and other biologically relevant ligands, such as CO or NO, is due to a non-peptide prosthetic group, heme, which is located in the protein's "pocket" and covalently bound at the proximal histidine of the globin. The Mb structure has no gaps for ligands to pass through. For ligands to move in and out of the pocket, they must traverse the intervening protein. Traversing the protein is made possible by the protein dynamics, which open paths for ligand diffusion through the protein.[19] Because of the ability of ligands to move through the protein, in some sense, the protein is considered to have liquidlike character.

A glass transition for a bulk material is recognized to be dynamical in nature.[20] Below the glass transition temperature, T_g, the system can no longer interconvert on a reasonable time scale among the full range of structural configurations accessed in the liquid. Previous temperature-dependent experiments on Mb-CO have revealed a type of dynamical transition in the protein at ∼200 K that has been referred to as a "protein glass transition".[21–23] This is not a glass transition in the normal sense, since a transition from a liquid to a glass as temperature is decreased is a phenomenon associated with a bulk material. In contrast to a bulk material, a protein is a single molecule. However, its complexity is so great that it undergoes continual structural evolution among a vast number of configurations. Below the "protein glass transition temperature", T_g^P, sampling of protein configurations is greatly slowed. Thus, like a true liquid/glass, a protein may undergo a type of dynamical transition. (When the term protein glass transition is used in discussing Mb, it is meant in the sense discussed in this paragraph.)

The myoglobin dynamics near ∼200 K, which suggest a protein glass transition, have been the subject of considerable investigation previously. For example, inelastic neutron scattering of hydrated Mb below 180 K observes only vibrational motion and, above 180 K, there is a dynamical transition, which is interpreted as the onset of torsional jumps between states.[24,25]

* To whom correspondence should be addressed. E-mail: fayer@fayerlab.stanford.edu.
[†] Department of Chemistry, Stanford University.
[‡] Stanford Free Electron Laser Center.
[§] School of Chemical Sciences, University of Illinois at Urbana–Champaign.
[⊥] Permanent address: Reveo, Inc., 8 Skyline Dr., Hawthorne, New York 10532.
[∥] Present address: FOM-Institute AMOLF, Kruislaan 407, 1098 SJ Amsterdam, The Netherlands.
[¶] Present address: Department of Chemistry, Coe College, Cedar Rapids, IO 52402.

10.1021/jp983923d CCC: $18.00   © 1999 American Chemical Society
Published on Web 03/02/1999

2382 *J. Phys. Chem. A, Vol. 103, No. 14, 1999* Rector et al.

MD simulations of the torsional transitions of the dihedral angles of Mb in water shows that the anharmonic mean square displacements change at 200 K, which is indicative of a glasslike transition.[26] In addition, IR[27−29] and visible[30] transition frequencies, dielectric relaxation,[27] specific heat of water in Mb crystals,[27] and Mössbauer spectra of $^{57}Fe-Mb$[31] all show breaks near 200 K. Some experiments have suggested that this is a "slaved" glass transition; i.e., the protein undergoes a transition induced by the true glass transition of the solvent.[21−23]

In this paper, vibrational echo measurements on Mb-CO are presented as a function of temperature from temperatures well below the Mb protein glass transition to well above it. We begin with the idea that the many experimental and theoretical studies conducted previously (some of which have been cited above) have established that there is a dynamical transition in Mb in the vicinity of 200 K. The aim here is to use vibrational echo experiments to provide insights into the nature of fast protein dynamics at temperatures below and above Mb-CO's glass transition and how the solvent environment of the protein influences the protein glass transition. In the absence of the wealth of other data on the Mb glass transition, it would be possible to envision a variety of scenarios other than a protein glass transition that could give rise to the data presented below. However, given the many previous studies, it is reasonable to discuss the vibrational echo results in the context of a protein glass transition. There is evidence from experiments on $W(CO)_6$ in the glass-forming liquid, 2-methylpentane (2MP)[32] (discussed briefly below) that the homogeneous dephasing time, T_2, measured in a vibrational echo experiment, can be sensitive to a glass transition, but the general question of the fundamental relationship between vibrational dephasing and a dynamical transition, like a glass transition, is not addressed.

A previous vibrational echo study of Mb-CO in 95:5 mixture of glycerol:water displayed a change in the functional form of the temperature dependence of the pure dephasing at ∼180 K.[11] Because the break in temperature-dependent pure dephasing occurred near the solvent T_g, the change was discussed in terms of a slaved protein glass transition.[11] In this paper, new temperature-dependent vibrational echo experiments on Mb-CO in trehalose and 50:50 ethylene glycol:water are presented. Trehalose is a glass at room temperature, with $T_g = 353$ K. Nonetheless, in trehalose, a change in the functional form of the Mb-CO pure dephasing temperature dependence is observed at ∼200 K. Below 200 K, the pure dephasing rate has a $T^{1.3}$ temperature dependence. Above 200 K, the data can be fit as an exponentially activated process with $\Delta E \cong 650$ cm^{-1}, although other forms, discussed below, can also fit the data for $T > 200$ K. The results demonstrate that a change in the nature of the Mb-CO pure dephasing occurs although the solvent remains a glass; i.e., a change in the nature of the solvent dynamics does not occur. In the solvents 95:5 glycerol:water and 50:50 ethylene glycol:water, changes in the temperature dependence also occur but at progressively lower temperatures. In all three solvents at low temperatures, the pure dephasing rate temperature dependence is the same power law, $T^{1.3}$. The results are discussed in terms of a transition in the nature of the protein dynamics. Both the transition temperature and the temperature dependence above T_g^P are influenced by the solvent viscosity.

II. Experimental Procedures

A. Vibrational Echo Method. The infrared vibrational echo experiments were performed at the Stanford Free Electron Laser (FEL) Center. The FEL produces tunable, picosecond, mid-IR pulses. The experimental setup is discussed in detail elsewhere.[11,33] The IR pulses had an energy of ∼0.5 μJ and were nearly transform-limited Gaussians, 1.2 ps in duration. Both the autocorrelation of the IR pulse and the spectrum were monitored continuously during the experiments. The spot size (ω_0) was ∼100 μm. The energies in the two pulses used in the echo sequence were ∼150 and ∼50 nJ, respectively. Pump−probe experiments were also conducted to measure the vibrational lifetime, T_1.

A vibrational echo is a two-pulse time domain technique which measures the Fourier transform of the homogeneous vibrational spectrum. In these experiments, the Mb-CO absorption spectra (1945 cm^{-1}) are inhomogeneously broadened, even at room temperature. Therefore, the vibrational absorption spectrum alone cannot provide information on the protein dynamics.

In the vibrational echo experiment, two picosecond IR pulses, tuned to the Mb-CO transition frequency, are crossed in a sample at an angle θ with a variable time delay, τ, between them. The vibrational echo pulse, which is formed at 2τ, propagates along a path that makes an angle 2θ with that of the first pulse. As τ is increased, the protein dynamics cause increasingly large accumulated phase errors among the ensemble of CO oscillators, and the signal amplitude of the vibrational echo is reduced. A measurement of the vibrational echo intensity vs τ is an echo decay curve. The Fourier transform of the echo decay is the homogeneous line shape.[34,35] The vibrational echo makes the vibrational homogeneous line shape an experimental observable in spite of inhomogeneous broadening.

For an exponential decay of the off diagonal density matrix elements (Lorentzian homogeneous line shape), the echo decay is given by

$$S(\tau) = S(0)e^{-4\tau/T_2} \tag{1}$$

where T_2 is the ensemble-averaged homogeneous dephasing time. The homogeneous line width is $1/\pi T_2$.[34] Combining the vibrational echo measurement of T_2 with a pump−probe measurement of T_1 provides a determination of the ensemble average homogeneous pure dephasing time, T_2^*, through

$$\frac{1}{T_2} = \frac{1}{T_2^*} + \frac{1}{2T_1} \tag{2}$$

T_2^* is determined by the fluctuations in the transition frequency caused by the protein dynamics.[11]

B. Sample Preparation. The experiments were performed on native horse heart myoglobin (Sigma, used without further purification). The trehalose sample was prepared by making a solution ∼10 wt % of trehalose in 0.1 M pH 7 phosphate buffer and dissolving lyophilized metmyoglobin in it to obtain a final protein concentration of 1−2 mM. The solution was saturated with CO, and a 10 molar excess of sodium dithionite was added. A few drops of the resulting solution were placed on a sapphire window and allowed to dry under CO atmosphere for a few days. The final protein concentration was ∼20 mM. To improve optical clarity, the sample was annealed at 90 °C for 1 h. IR absorption measurements verified that the annealing of the sample did not denature the protein. A second sapphire window was placed on the sample and mild pressure was applied to obtain good thermal contact. The final sample thickness was approximately 100 μm. The sample was inserted into a copper sample holder attached to the coldfinger of an optical cryostat.

The ethylene glycol sample was prepared by adding a 15 mM aqueous solution of lyophilized metMb to 50:50 (v:v) ethylene

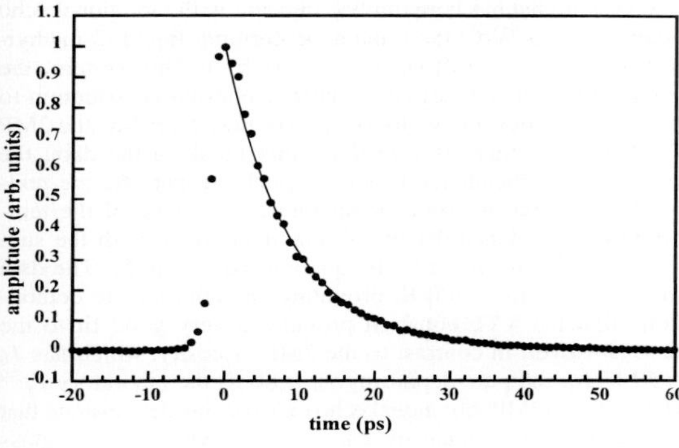

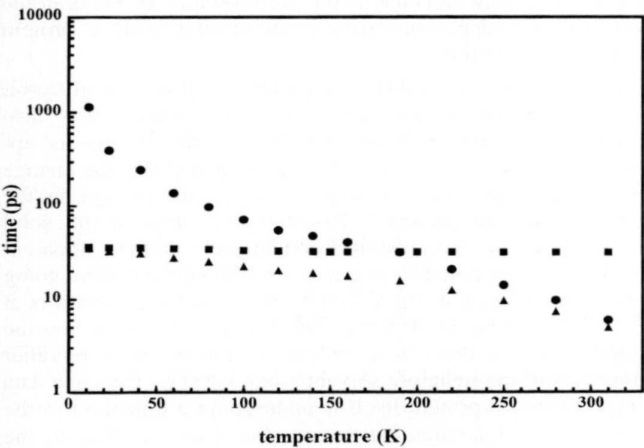

Figure 1. (a, top) An example of a vibrational echo decay of Mb-CO in trehalose glass at 11K. These data are fit well by an exponential curve as are all of the decays at all temperatures. The decay constant is 7.0 ps. This value gives a homogeneous dephasing time of 28 ps and a homogeneous width of ~0.3 cm^{-1} using eqs 1 and 2. (b, bottom) The temperature-dependent data for Mb-CO in trehalose. The triangles are the homogeneous dephasing times, T_2, measured with vibrational echo experiments. The squares are twice the vibrational lifetime, $2T_1$, measured with IR pump–probe experiments. The lifetime is plotted as $2T_1$ in accordance with eq 2. The circles are the pure dephasing times, as determined at each temperature from eq 2.

glycol/0.1M pH 7 phosphate buffer. The resulting solution was then stirred under CO atmosphere for 8 h before being reduced by a 10-fold molar excess of dithionite. A copper sample cell with CaF$_2$ windows was filled with the resulting solution. The path length was 200 μm.

III. Results and Discussion

Figure 1a shows a vibrational echo decay of Mb-CO in trehalose glass at 11 K. The signal-to-noise ratio is excellent in spite of the fact there is a large background absorption by the protein and the trehalose. The data are fit well by a 7.0 ps decay constant exponential curve, which corresponds to a homogeneous dephasing time of 28 ps and a homogeneous width of 0.38 cm^{-1}. At room temperature, the homogeneous width increases to ~2 cm^{-1}. The absorption line width is ~15 cm^{-1} fwhm at all temperatures. Therefore, the Mb-CO absorption line is inhomogeneously broadened at all temperatures studied.

Figure 1b is a semilog plot of the temperature dependence in trehalose of T_2 (triangles), $2T_1$ (squares), and T_2^* (circles). T_2^* was computed from the other two quantities using eq 2. T_1 has

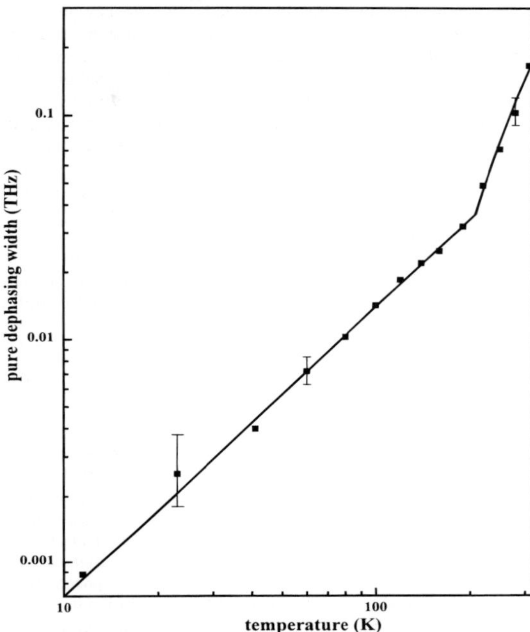

Figure 2. Temperature-dependent pure dephasing rate of Mb-CO in trehalose. The error bars on the data are larger for the $T < 50$ K data because of the potential error caused by subtracting two similar numbers (see Figure 1b). The data are fit with a power law, $T^{1.3}$, below ~200 K and with an exponentially activated process with $\Delta E = 650$ cm^{-1} above 200 K.

only a slight, essentially linear temperature dependence. It can be seen that at low temperatures the homogeneous line width is nearly lifetime limited. By room temperature, the homogeneous line width is dominated by pure dephasing.

The manner in which the global Mb protein fluctuations are communicated to the CO bound at the active site to produce pure dephasing has been discussed in detail previously.[12,13] Figure 2 shows the pure dephasing contribution to the line width, $1/\pi T_2^*$, vs temperature on a log plot for Mb-CO in trehalose. On a log plot, a power law graphs as a straight line. As can be seen in Figure 2, between 11 and ~200 K, the functional form of the data is a power law

$$\frac{1}{\pi T_2^*} = aT^{1.3} \qquad (3)$$

where the prefactor $a = 3.5 \times 10^7 \pm 0.1 \times 10^7$ Hz/(deg K)$^{1.3}$. The error bar on the power law exponent is ±0.1. This power law has been reported previously for Mb-CO in glycerol:water.[11] However, the trehalose data are for a much broader range of temperatures and leave little doubt as to the functional form of the data. The same power law is observed in ethylene glycol: water (see below).

In Figure 2, there is a change in the functional form of the data at ~200 K. The points above ~200 K can be fit with

$$\frac{1}{\pi T_2^*} = 3.3 \times 10^{12}\, e^{-650/k_B T} \text{ Hz} \qquad (4)$$

where k_B is Boltzmann's constant, $k_B T$ has units of cm^{-1}, and the error bars on the prefactor and activation energy are $\pm 0.2 \times 10^{12}$ Hz and ± 25 cm^{-1}, respectively. It is clear that there is a change in the functional form of the temperature dependence at ~ 200 K. However, it is important to emphasize that the form of eq 4 is not unique given the small number of points. It is possible to describe the data with the power law plus another function without an abrupt switching of the functional form. If this is done with an exponentially activated process, the value of the exponent changes, but the power law is identical. A very good fit is obtained if the data are fit to a power law plus a Vogel−Tamman−Fulcher (VTF) type equation.[20,36,37] A VTF equation describes many processes, such as viscosity, in supercooled liquids as they approach the glass transition temperature. The VTF equation for the homogeneous line width is

$$\frac{1}{\pi T_2^*} = b e^{-E/k_B(T-T_0)} \tag{5}$$

For a true glass-forming liquid, T_0 is the "ideal" glass transition temperature. It typically has a value a few tens of degrees below the laboratory T_g.[20,36,37] A fit to the data with eq 3 plus eq 5 yields a T_0 of ~ 180 K and an E corresponding to a temperature of ~ 230 K. These parameters can vary somewhat about the given values because of the wide range of fits that can be achieved when fitting four points with three parameters. However, the power law is always identical, independent of the form used to fit the points above ~ 200 K. If the exponential fit and the VTF fit are extended to higher temperatures, they do not become distinguishable below 500 K. Therefore, experiments at temperatures below the Mb denaturation temperature cannot distinguish these two forms. Regardless of the form that is used to fit the data, it is clear that there is a sudden change in the nature of the temperature dependence of the pure dephasing.

Switching from a power law to an activated process or the power law plus the VTF function is appropriate if there is a transition in the fundamental nature of the dynamics. A power law plus an activated process does not require a change in nature of the dynamics, but only that the more rapidly increasing activated process overtakes the power law. In fact, this overtaking of a power law by an activated process has been demonstrated to occur in the electronic excited state dephasing of chromophores in low-temperature glasses very far below the glass transition temperature of the solvent.[38−42] In the absence of other information, these vibrational echo experiments alone cannot distinguish between a fundamental change in the nature of the protein dynamics or one process simply overwhelming another.

However, given the many other types of studies (discussed in the Introduction) that have observed changes in the Mb dynamics at ~ 200 K and that have been interpreted as evidence of a protein glass transition,[11] it is reasonable to assume that the vibrational echo data do, indeed, display a manifestation of a change in the basic nature of the protein dynamics; i.e., they reflect the protein glass transition. As has been discussed previously, Mb-CO pure vibrational dephasing arises from global fluctuations of the protein structure.[11−13] The CO dephasing is caused by electric field fluctuations produced by overall motions of the protein, rather than very local protein dynamics near the CO. A change in the nature of the protein dynamics, which influences the various observables that have been studied previously, can also produce a change in the nature of the temperature dependence of the vibrational pure dephasing.

A system that has been studied in detail with vibrational echo experiments is W(CO)$_6$ in the glass forming liquid, 2-methyl-pentane (2MP).[32] 2MP has $T_g = 80$ K. In this system, the temperature dependence of the pure dephasing contribution to the homogeneous line width is a power law, T^2, below the 2MP T_g. Above T_g, there is a very distinct break in the data; the temperature dependence becomes much steeper. As the temperature approaches room temperature, the slope of the temperature dependence decreases. The data are fit with the sum of the power law and a VTF equation (eqs 3 and 5). The data are measured up to 300 K, providing enough range to demonstrate that the VTF equation provides a very good fit to the data. However, in contrast to the 2MP viscosity, which has $T_0 = 59$ K, for the pure dephasing $T_0 = 80$ K, the laboratory T_g.[32] The W(CO)$_6$/2MP vibrational echo experiments demonstrate that vibrational pure dephasing can be sensitive to a true glass transition, adding credence to the proposal that the break in the Mb-CO pure dephasing data is associated with a protein dynamical transition.

In a previous vibrational echo study of Mb-CO in glycerol:water, a change in the temperature dependence was also observed at ~ 180 K.[11] In that system, the change is approximately at the solvent T_g. It was suggested that the change in solvent properties was responsible for the change in the dynamics of the protein.[11] However, the data in the solid trehalose sample demonstrates conclusively that the apparent break in the functional form can occur without the solvent going through its glass transition. The trehalose glass transition is at $T_g = 353$ K. The break in the data occurs ~ 150 K below the trehalose T_g, so there is no change in the viscosity or other properties of the trehalose solvent around 200 K. Thus, the data suggest that the protein itself is undergoing a transition in the nature of its dynamics that is not slaved to a change in the solvent dynamics.

Another issue is that of time scales. Vibrational echo measurements of the homogeneous dephasing time are sensitive to fluctuations on a time scale of approximately an order of magnitude faster than T_2 to an order of magnitude slower than T_2. In Mb-CO at ~ 200 K, this range spans $\sim 1-100$ ps. The picosecond time scale is in contrast to the time scale of some processes in supercooled liquids, such as long-range spatial diffusion, which become infinitely slow as T_g is approached from above. It has been observed in a wide variety of low-temperature glasses, far below T_g, that the temperature dependence of the optical dephasing of electronic transitions is independent of the time scale. Photon echo experiments on a 100 ps time scale and optical hole burning experiments on a 100 s time scale produce different pure dephasing widths.[38−42] The hole burning experiments yield broader pure dephasing line widths because of spectral diffusion caused by very slow processes that do not contribute to the photon echo experiments.[40,43,44] However, both types of experiments display the same $T^{\sim 1.3}$ temperature dependence of the pure dephasing contribution to the line width.[38−42] (The exponent varies somewhat depending on the glass.) In the W(CO)$_6$/2MP system, vibrational pure dephasing in the supercooled liquid exhibits VTF behavior, which has its onset at T_g. The dephasing's sensitivity to the glass transition has a shift to higher temperature than viscosity, possibly because dephasing is sensitive to short time scale fluctuations rather than the slow processes associated with viscosity. In the vibrational echo study of W(CO)$_6$/2MP, it was suggested that the VTF type dephasing turns on because of the onset of local solvent molecule orientational and translational motions above T_g.[32] Thus, dephasing can display

Dynamical Transition in the Protein Myoglobin

the same or similar temperature dependences as the long time scale process and can be sensitive to processes which are generally associated with long time scales.

The Mb-CO vibrational absorption line is inhomogeneously broadened both below and above the proposed protein dynamical transition temperature, T_g^P. At first glance, this may seem inconsistent with making a transition from a glassy like state to a liquidlike state. In a glass, the time scale for the system to sample all possible structural configurations is, essentially, infinitely long. In contrast, in a liquid, the time scale is relatively short. Vibrational echo experiments on metal carbonyls in true glass-forming liquids, e.g. $W(CO)_6$ in 2-methylpentane, demonstrate that vibrational lines can be inhomogeneously broadened well above the glass transition temperature, T_g.[32] The vibrational echo measures the homogeneous dephasing time. In a liquid, on some relatively short time scale, all structural configurations will be sampled. However, the time scale may be long compared to the homogeneous dephasing time, and, therefore, the vibrational line is inhomogeneously broadened. On a time scale longer than T_2, spectral diffusion will eventually result in the transition frequency sampling all possible values. The fact that spectroscopic lines can be inhomogeneously broadened above T_g is true for electronic transitions as well. Ultrafast photon echo experiments, even in room temperature liquids, have shown that the electronic transitions of chromophores in glass-forming liquids can be inhomogeneously broadened.[45-48] Thus, passing from a glassy state to a liquid state ensures that a chromophore will sample the full range of transition energies contained in the spectroscopic line, but the time scale for the sampling can be long compared to the homogeneous dephasing time measured in a vibrational echo experiment or a photon echo experiment, resulting in an inhomogeneously broadened absorption line.

Figure 3 shows pure dephasing line widths as a function of temperature for Mb-CO in three solvents. The diamonds are the trehalose data shown in Figure 2; the triangles are data for Mb-CO in the solvent 95:5 glycerol:water;[11,12] and the circles are data for Mb-CO in the solvent 50:50% ethylene glycol:water. The line through the trehalose data is the same as in Figure 2. The lines through the other data are guides to facilitate discussion.

In all three solvents, at temperatures below their respective break points, the data fall on the same $T^{1.3}$ power law line. The fact that the vibrational dephasing comes solely from protein fluctuations, and not from the solvent, has been discussed in detail previously.[11-13] The identical power law temperature dependences, which have the same slopes and the same values of the dephasing in three solvents, is another demonstration that the vibrational pure dephasing is a measure of protein dynamics. The power law temperature dependence is reminiscent of the temperature dependence of processes in low-temperature glasses (< a few K). Power laws have been seen in measurements of the heat capacities of glasses[49,50] and optical[43] and vibrational[51] dephasing of chromophores in glasses. Power law temperature dependences of heat capacities, optical and vibrational dephasing, as well as other observables have been explained in terms of the tunneling two-level system (TLS) model of glasses.[52,53] Generally, the TLS model is only invoked up to a few K. In optical dephasing experiments in organic glasses below a few K, T^{α} with values of α in the range of 1.1–1.5 are frequently observed with $\alpha = 1.3$ the most common value. A power law temperature dependence of the vibrational pure dephasing line width was observed for a solute in an organic glass up to ~20 K.[51] TLS in glasses arise from slight differences in local

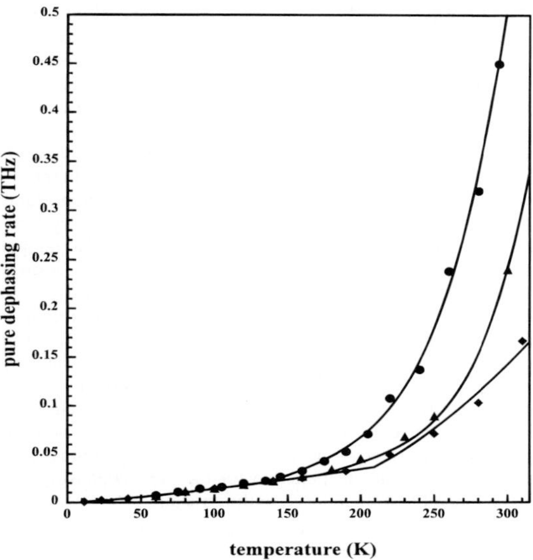

Figure 3. Pure dephasing of Mb-CO in three solvents. The squares are the trehalose data shown in Figure 2. The triangles are data in 95:5 glycerol:water.[11] The circles are data in 50:50 ethylene glycol:water. Below ~150 K, the dynamics for all three solvents are fit with an identical $T^{1.3}$ power law. Above ~200 K, the trehalose data is fit with an exponentially activated process. The glycerol:water and ethylene glycol:water data have both a temperature dependence and a viscosity dependence at the higher temperatures.

structures. The complex structural energy landscape is modeled as a distribution of double wells having a broad distribution of energy differences between the two sides of the double well. The dynamics in glasses at low temperatures are caused by phonon-assisted tunneling among local structures modeled as transitions between the two sides of the double wells.

In the Mb-CO vibrational dephasing, the $T^{1.3}$ temperature dependence is observed to much higher temperatures than in true glasses. One possible explanation of the power law temperature dependence is thermally assisted tunneling among slightly different protein configurations. Small internal protein structural changes might be described in terms of protein two-level systems (PTLS).[11,12] The PTLS are akin to the two-level systems of very low temperature glasses except the protein energy landscape would have to be such that tunneling is the dominant process even at temperatures up to 200 K. If this is the case, the same statistical mechanics machinery used to describe the low temperature (~1 K) optical dephasing of electronic transitions of chromophores in low-temperature glasses[43,54] can be used to describe the PTLS induced vibrational dephasing of Mb-CO at much higher temperatures (~100 K). Alternatively, the power law temperature dependence could arise from activation over barriers rather than tunneling if there is the appropriate energy landscape to provide the necessary broad distribution of activation energies.[55,56] In either case, the $T^{1.3}$ temperature-dependent pure dephasing can occur because of motion on a broad protein energy landscape in a manner analogous to dynamics in true glasses. While an entirely different mechanism cannot be ruled out, the similarity of the vibrational echo data for Mb-CO to dephasing and other measurements on low-temperature glasses is suggestive of a type of glassy-like state of the protein.

2386 *J. Phys. Chem. A, Vol. 103, No. 14, 1999*

Rector et al.

Considering Figure 3, it can be seen that the breaks in the dephasing in the three solvents do not occur at the same temperature and that the temperature dependences are not the same in the three solvents at higher temperatures. At all temperatures studied in trehalose, the viscosity is essentially infinite. The observed temperature-dependent pure dephasing comes from protein fluctuations in a solid medium in which the topology of the protein/surface interface is fixed. In the two liquid solvents, the situation is quite different. As the temperature is increased, the viscosities of the solvents decrease. Very recent vibrational echo studies on Mb-CO conducted at room temperature as a function of solvent viscosity show that there is a strong viscosity dependence to the Mb-CO pure dephasing.[57] At constant temperature, as the viscosity of the solvent is decreased, the Mb-CO vibrational pure dephasing rate increases. Therefore, the temperature dependences observed in the glycerol:water and ethylene glycol:water solvents are actually combinations of a pure temperature dependence and a viscosity dependence. For this reason, the data in these solvents were not fit with the functions given in eq 4 or 5. The trehalose data characterizes the protein dynamics responsible for the pure dephasing with the protein/solvent boundary condition static.

The order of the break in the functional form of the temperature dependences displayed in Figure 3, from lowest temperature to highest temperature, is ethylene glycol:water ($\sim$150 K), glycerol:water ($\sim$180 K), and trehalose ($\sim$200 K). This is also the order of the solvents' glass transition temperatures. From Figure 2, it is clear that the dynamical transition displayed in the vibrational echo data does not depend on the solvent undergoing a glass transition. However, the data in Figure 3 show that if the solvent goes through its glass transition at a temperature below T_g^P, then T_g^P is reduced.

A way to visualize what might be happening is to consider a hypothetical phase-separated polymer blend in which the minority component, A, is contained in nanometer size domains,[58] which are essentially pure A, embedded in the host polymer, B. If $T_g^A < T_g^B$, then the nanodomains can undergo the glass to liquid transition while B is still solid. Above T_g^A, the liquidlike nanodomains will have a fixed nanodomain/host surface boundary. However, if $T_g^A > T_g^B$, then the host will become liquid at a temperature at which the nanodomains are still solidlike. This changes the boundary condition and could very well reduce the nanodomain glass transition temperature because, for a nanometer-sized object, the surface region has a substantial influence on the interior. There is some evidence to support this picture. A very thin film of a polymer (large surface to volume ratio) on a substrate with one surface unconstrained has a lower T_g than the bulk polymer.[59] The extra mobility afford by the free surface reduces T_g.

We suggest that the protein in a glassy solvent might be in some sense like the nanodomains in the polymer blend. In trehalose, $T_g^P < T_g^S$, where T_g^S is the glass transition temperature of the solvent. In this situation, the protein glass transition occurs with a fix protein/solvent interface, giving the highest T_g^P. In glycerol:water, $T_g^P > T_g^S$. The protein glass transition occurs in a very viscous liquid, but the interface is no longer fixed. This provides the protein with extra mobility, reducing T_g^P. In ethylene glycol:water, T_g^S is even lower. The liquid is less viscous as T_g^P is approached, and T_g^P is even lower. This is not a slaved protein glass transition, but, rather, a protein/solvent boundary condition influence on T_g^P.

While the vibrational echo experiments presented here and many other experiments[11] display a change in Mb dynamics at

$\sim$200 K, there have been other experiments that do not see a change near 200 K. The temperature dependence of the X-ray crystal structure has been determined, and the structure at 80 K does not seem to be different than that at room temperature.[60,61] In addition, the IR spectra of low-humidity samples are independent of temperature between 15 and 300 K and have a transition, depending on the degree of hydration, between 180 and 270 K.[27] Differential scanning calorimetry shows no unique glass transition of water associated with the protein but, rather, many transitions between 150 K and denaturation.[62,63]

The results presented here are consistent with the previous measurements, both those that show protein transitions and those that do not. The vibrational echo detects a change in the nature of the dynamics. Such a change does not require a change in the protein structure. The trehalose samples studied here contain significant amounts of water. It would be interesting to see if the degree of hydration changes the nature of the vibrational echo temperature dependence. The vibrational echo experiments detect the protein fluctuations that are communicated to the CO bound at the active site. Apparently, such fluctuations are not significantly influenced by changes in the state of the water bound at the protein surface that are detected by calorimetry.

IV. Concluding Remarks

Vibrational echo experiments can provide new insights into the nature of protein dynamics. In the experiments described above, Mb protein fluctuations that are communicated to the CO bound at the active site were examined. Because the experiments involve only the ground electronic state and involve time scales only out to a few tens of picoseconds, they should be amenable to simulations. The results show that, independent of the solvent, the low-temperature pure dephasing rate is a power law, $T^{1.3}$. This power law temperature dependence is suggestive of glasslike behavior, but at much higher temperatures than typically observed in true glasses. The results show that, even at low temperatures, fast protein structural fluctuations occur.

The observation of the change in the form of the temperature dependence of the pure dephasing rate at $\sim$200 K in the solid solvent trehalose is consistent with the protein undergoing a transition in the nature of its dynamics that does not depend on a change in the solvent properties. If the power law pure vibrational dephasing temperature dependence below $\sim$200 K is due to motion on a glasslike energy landscape, then the data above $\sim$200 K could result from activation above the top of the landscape.[20] Another possibility is that the break in the temperature dependence indicates that the top of the landscape has been reached, and the ΔE in eq 4 arises from a barrier or a narrow ranges of barrier heights that are higher in energy than the top of the landscape. The fact that the temperature at which the break in the temperature dependence occurs depends on the solvent T_g and the temperature dependence above the break in liquid solvents depends on the solvent viscosity is indicative of the important role that the protein/solvent boundary condition plays in protein dynamics and, presumably, the nature of the energy landscape. The relationship of the vibrational dephasing to the solvent viscosity will be discussed in detail in a subsequent publication.

Acknowledgment. We thank Professors Alan Schwettman and Todd Smith and their students and staff at the Stanford Free Electron Laser Center for making this work possible. The Stanford Free Electron Laser Center is supported by the Office of Naval Research (N00014-94-1-1024). M.D.F. acknowledges support from the National Science Foundation, Division of Materials Research DMR-9610326. D.D.D. acknowledges support from Office of Naval Research N00014-95-1-0259, and National Science Foundation, Division of Materials Research DMR-9714843.

References and Notes

(1) Elber, R.; Karplus, M. *Science* **1987**, *235*, 318.

(2) Quillin, M. L.; Arduini, R. M.; Olson, J. S.; Phillips, G. N., Jr. *J. Mol. Biol.* **1993**, *234*, 140.

(3) Barrick, D. *Biochemistry* **1994**, *33*, 6546.

(4) Tsuda, S. *Crystallogr. Soc. Jpn.* **1996**, *38*, 84.

(5) Clore, G. M.; Gronenborn, A. M. *Prog. in Nucl. Magn. Reson. Spectrosc.* **1991**, *23*, 43.

(6) Braunstein, D. P.; Chu, K.; Egeberg, K. D.; Frauenfelder, H.; Mourant, J. R.; Nienhaus, G. U.; Ormos, P.; Sligar, S. G.; Springer, B. A.; Young, R. D. *Biophys. J.* **1993**, *65*, 2447.

(7) Jackson, T. A.; Lim, M.; Anfinrud, P. A. *Chem. Phys.* **1994**, *180*, 131.

(8) Janes, S. M.; Dalickas, G. A.; Eaton, W. A.; Hochstrasser, R. M. *Biophys. J.* **1988**, *54*, 545.

(9) Oldfield, E.; Guo, K.; Augspurger, J. D.; Dykstra, C. E. *J. Am. Chem. Soc.* **1991**, *113*, 7537.

(10) Surewicz, W. K.; Mantsch, H. H. Infrared Absorption Methods for Examining Protein Structure. In *Spectroscopic Methods for Determining Protein Structure in Solution*; Havel, H. A., Ed.; VCH Publishers: New York, 1996; p 135.

(11) Rella, C. W.; Rector, K. D.; Kwok, A. S.; Hill, J. R.; Schwettman, H. A.; Dlott, D. D.; Fayer, M. D. *J. Phys. Chem.* **1996**, *100*, 15620.

(12) Rector, K. D.; Rella, C. W.; Kwok, A. S.; Hill, J. R.; Sligar, S. G.; Chien, E. Y. P.; Dlott, D. D.; Fayer, M. D. *J. Phys. Chem. B* **1997**, *101*, 1468.

(13) Rector, K. D.; Engholm, J. R.; Hill, J. R.; Myers, D. J.; Hu, R.; Boxer, S. G.; Dlott, D. D.; Fayer, M. D. *J. Phys. Chem. B* **1998**, *102*, 331.

(14) Leeson, D. T.; Wiersma, D. A. *Phys. Rev. Lett.* **1995**, *74*, 2138.

(15) Thijssen, H. P. H.; Dicker, A. I. M.; Völker, S. *Chem. Phys. Lett.* **1982**, *92*, 7.

(16) Owrutsky, J. C.; Li, M.; Locke, B.; Hochstrasser, R. M. *J. Phys. Chem.* **1995**, *99*, 4842.

(17) Hill, J. R.; Dlott, D. D.; Rella, C. W.; Peterson, K. A.; Decatur, S. M.; Boxer, S. G.; Fayer, M. D. *J. Phys. Chem.* **1996**, *100*, 12100.

(18) Hill, J. R.; Dlott, D. D.; Rella, C. W.; Smith, T. I.; Schwettman, H. A.; Peterson, K. A.; Kwok, A. S.; Rector, K. D.; Fayer, M. D. *Biospectroscopy* **1996**, *2*, 227.

(19) Case, D. A.; Karplus, M. *J. Mol. Biol.* **1979**, *132*, 343.

(20) Angell, C. A. *J. Phys. Chem. Solids* **1988**, *49*, 863.

(21) Iben, I. E. T.; Basunstein, D.; Doster, W.; Frauenfelder, H.; Hong, M. K.; Johnson, J. B.; Luck, S.; Ormos, P.; Schulte, A.; Steinback, P. J.; Xie, A.; Young, R. D. *Phys. Rev. Lett.* **1989**, *62*, 1916.

(22) Parak, F.; Frauenfelder, H. *Physica A* **1993**, 332.

(23) Frauenfelder, H.; Sligar, S. G.; Wolynes, P. G. *Science* **1991**, *254*, 1598.

(24) Doster, W.; Cusack, S.; Petry, W. *Nature* **1989**, *337*, 754.

(25) Loncharich, R. J.; Brooks, B. R. *J. Mol. Biol.* **1990**, *215*, 439.

(26) Steinbach, P. J.; Brooks, B. R. *PNAS* **1993**, *90*, 9135.

(27) Doster, W.; Bachleitner, A.; Dunau, R.; Hiebl, M.; Luscher, E. *Biophys. J.* **1986**, *50*, 213.

(28) Hong, M. K.; Draunstein, D.; Cowen, B. R.; Frauenfelder, H.; Iben, I. E. T.; Mourant, J. R.; Ormos, P.; Scholl, R.; Schulte, A.; Steinbach, P. J.; Xie, A.; Young, R. D. *Biophys. J.* **1990**, *58*, 429.

(29) Mayer, E. *Biophys. J.* **1994**, *67*, 862.

(30) Cordone, L.; Cupane, A.; Leone, M.; Vitrano, E. *J. Mol. Biol.* **1988**, *199*, 213.

(31) Parak, F.; Knapp, E. W.; Kucheida, D. *J. Mol. Biol.* **1982**, *161*, 177.

(32) Tokmakoff, A.; Fayer, M. D. *J. Chem. Phys.* **1995**, *103*, 2810.

(33) Tokmakoff, A.; Zimdars, D.; Urdahl, R. S.; Francis, R. S.; Kwok, A. S.; Fayer, M. D. *J. Phys. Chem.* **1995**, *99*, 13310.

(34) Farrar, T. C.; Becker, D. E. *Pulse and Fourier Transform NMR*; Academic Press: New York, 1971.

(35) Skinner, J. L.; Anderson, H. C.; Fayer, M. D. *J. Chem. Phys.* **1981**, *75*, 3195.

(36) Angell, C. A. *J. Phys. Chem.* **1982**, *86*, 3845.

(37) Fredrickson, G. H. *Annu. Rev. Phys. Chem.* **1988**, *39*, 149.

(38) Walsh, C. A.; Berg, M.; Narasimhan, L. R.; Fayer, M. D. *J. Chem. Phys.* **1987**, *86*, 77.

(39) Walsh, C. A.; Berg, M.; Narasimhan, L. R.; Fayer, M. D. *Chem. Phys. Lett.* **1986**, *130*, 6.

(40) Berg, M.; Walsh, C. A.; Narasimhan, L. R.; Littau, K. A.; Fayer, M. D. *J. Chem. Phys.* **1988**, *88*, 1564.

(41) Thijssen, H. P. H.; van den Berg, R.; Volker, S. *Chem. Phys. Lett.* **1985**, *120*, 503.

(42) Thijssen, H. P. H.; Volker, S. *Chem. Phys. Lett.* **1985**, *81*, 3915.

(43) Narasimhan, L. R.; Littau, K. A.; Pack, D. W.; Bai, Y. S.; Elschner, A.; Fayer, M. D. *Chem. Rev.* **1990**, *90*, 439.

(44) Bai, Y. S.; Fayer, M. D. *Phys. Rev. B* **1989**, *39*, 11066.

(45) Becker, P. C.; Fragnito, H. L.; Bigot, J. Y.; Cruz, C. H. B.; Fork, R. L.; Shank, C. V. *Phys. Rev. Lett.* **1989**, *63*, 505.

(46) Nibbering, E. T. J.; Wiersma, D. A.; Duppen, K. *Phys. Rev. Lett.* **1991**, *66*, 2464.

(47) de Boeji, W. P.; Pshenichnikov, M. S.; Wiersma, D. A. *Chem. Phys. Lett.* **1995**, *238*, 1.

(48) Vohringer, P.; Arnet, D. C.; Yang, T. S.; Scherer, N. F. *Chem. Phys. Lett.* **1995**, *237*, 387.

(49) Phillips, W. A. *Amorphous Solids. Low-Temperature Properties, Top. Curr. Phys.* Springer: Berlin, 1981.

(50) Stevels, J. M. The Structural and Physical Properties of Glass. In *Thermodynamics of Liquids and Solids*; Flügge, S., Ed.; Springer-Verlag: Berlin, 1962; p 13.

(51) Rector, K. D.; Fayer, M. D. *J. Chem. Phys.* **1998**, *108*, 1794.

(52) Anderson, P. W.; Halperin, B. I.; Varma, C. M. *Philos. Mag.* **1972**, *25*, 1.

(53) Phillips, W. A. *J. Low. Temp. Phys.* **1972**, *7*, 351.

(54) Leeson, D. T.; Wiersma, D. A.; Fritsch, K.; Friedrich, J. *J. Phys. Chem. B* **1997**, *101*, 6331.

(55) Gilroy, K. S.; Phillips, W. A. *Philos. Mag. B* **1981**, *43*, 735.

(56) Kohler, W.; Zollfrank, J.; Friedrih, J. *Phys. Rev. B* **1989**, *39*, 5414.

(57) Rector, K. D.; Sengupta, D.; Fayer, M. D., manuscript in preparation.

(58) Marcus, A. H.; Hussey, D. M.; Diachun, N. A.; Fayer, M. D. *J. Chem. Phys.* **1995**, *103*, 8189.

(59) Forrest, J. A.; Dalnoki-Veress, K.; Dutcher, J. R. *Phys. Rev. E* **1997**, *56*, 5705.

(60) Frauenfelder, H.; et al. *Biochemistry* **1987**, *26*, 254.

(61) Hartmann, H.; Parak, F.; Steigemann, W.; Petsko, G. A.; Ponzi, D. R.; Frauenfelder, H. *Proc. Natl. Acad. Sci. U.S.A.* **1982**, *79*, 4967.

(62) Sartor, G.; Hallbrucker, A.; Mayer, E. *Biophys. J.* **1995**, *69*, 2679.

(63) Sartor, G.; Mayer, E.; Johari, G. P. *Biophys. J.* **1994**, *66*, 249.

Section 3. Ultrafast 2D IR Vibrational Echo Spectral Diffusion Studies of Liquids and Proteins

Ultrafast 2D IR Vibrational Echo Spectral Diffusion Studies of Liquids and Proteins

In the written part of this monograph and in the first section of published papers, the main experimental method is ultrafast vibrational echo chemical exchange spectroscopy. In the systems discussed, two species interconvert. At short time there are two peaks on the diagonal of the 2D IR vibrational echo spectrum. At long time, two off-diagonal peaks grow in. The analysis of the rate of growth of the off-diagonal peaks yields the exchange rate between the two species. In many systems there are not two species but an infinite number. For example, water has a vast number of hydrogen bonding configurations. The frequency of the hydroxyl stretching mode of water depends on the number and the strengths of the hydrogen bonds. Because of the wide range of hydrogen bonding configurations, the hydroxyl stretch spectrum is very broad. The hydrogen bonds are constantly changing. Strong hydrogen bonds are becoming weak and vice versa. Hydrogen bonds are being formed and broken. These changes in water hydrogen bonding configurations cause the frequency of the hydroxyl stretch to evolve in time. The time evolution of the frequency of the hydroxyl stretch is called spectral diffusion, and it is directly related to the time evolution of the hydrogen bond structure of water. The time dependent hydrogen bond structure cannot be obtained from the linear hydroxyl stretch absorption spectrum, but it can be determined by analysis of ultrafast 2D IR vibrational echo spectra.

In general, a vibration of a system like water or a protein is inhomogeneously broadened. Inhomogeneous broadening reflects the fact that there are a wide variety of configurations of the system, and the differences in the local structure associated with the different configurations influence the frequency of the vibration under study. The width of the absorption band reflects the distribution of environments but does not tell how rapidly the environments interconvert. At room temperature, a liquid or a protein will sample all possible configurations. The configuration sampling causes spectral diffusion, which manifests itself in the 2D IR vibrational echo spectrum through the time dependent change in the shape of the spectrum. At short time, the spectrum will be elongated along the diagonal. As time increases the spectrum will evolve to symmetrical about the diagonal. The time dependent change in shape of the spectrum is related to the frequency-frequency correlation function (FFCF), which is the joint probability distribution that the frequency has a certain initial value $\omega(0)$ at $t = 0$ and another value $\omega(t)$ at a later time t. The FFCF is the connection between the 2D IR observables and microscopic models of the systems dynamics. The FFCF can be calculated from molecular dynamics simulations of the system. Comparison of the

FFCF determined experimentally and the FFCF determined from simulation permits the validity of the simulation to be tested.

The time evolution of the 2D IR vibrational echo line shape is non-trivial to determine. However, we have made considerable progress in accurately determining an FFCF from the time dependent 2D IR line shape. Determination of the FFCF has been made more robust by recent theoretical advances that show how to extract the FFCF from data. In the following papers, a variety of systems are investigated by measurements of spectral diffusion with 2D IR vibrational echo experiments. The last two papers included below are the recent theoretical developments that are making the method more accurate and easier to use.

VOLUME 86, NUMBER 17   PHYSICAL REVIEW LETTERS   23 APRIL 2001

Two-Dimensional Time-Frequency Ultrafast Infrared Vibrational Echo Spectroscopy

K. A. Merchant, David E. Thompson, and M. D. Fayer

Department of Chemistry, Stanford University, Stanford, California 94305
(Received 15 September 2000)

2D spectrally resolved ultrafast ($<$200 fs) IR vibrational echo experiments were performed on Rh(CO)$_2$acac [(acetylacetonato)dicarbonylrhodium (*I*)]. The 2D spectra display features that reflect the 0-1 and 1-2 transitions and the combination band transition of the symmetric (*S*) and antisymmetric (*A*) CO stretching modes. Three oscillations in the data arise from the frequency difference between the *S* and *A* modes (quantum beats) and the *S* and *A* anharmonicities. A new explanation is given for these "anharmonic" oscillations. Calculations show that spectral resolution enables the 0-1 and 1-2 dephasing to be measured independently.

DOI: 10.1103/PhysRevLett.86.3899   PACS numbers: 78.47.+p, 78.30.Jw

I. Introduction.—The development of pulsed NMR [1] and the extension to two-dimensional (2D) methods [2], has vastly expanded the usefulness of NMR in many areas of science. The advent of the photon echo, the electronic excited state equivalent of the spin echo [3], and ultrafast 2D photon echo experiments [4,5] have extended NMR-like techniques into the visible and the UV. Recently, photon echoes were extended to the study of molecular vibrations in condensed matter as ultrafast IR vibrational echoes [6]. Vibrational echoes [7–13] make possible the direct study of molecular structural degrees of freedom using the types of sensitive probes that have been available for the study of spins and electronic excited states.

In this Letter, 2D (time-frequency) ultrafast IR vibrational echo spectroscopy (VES) experiments are presented. The ultrashort pulses required to study vibrational dynamics [14] with vibrational echoes can have bandwidths that produce coherent excitation of multiple excited vibrational states resulting in complex oscillatory nonexponential decay curves even if each transition would give rise to a single exponential decay. By spectrally resolving the ultrafast vibrational echo and obtaining a 2D echo spectrum, the time resolution is preserved, the nature of the signal can be more readily elucidated, and additional information can be obtained. Spectrally resolved ultrafast VES is an addition to an increasing collection of new 2D ultrafast IR spectroscopic methods [10,11,15–18].

II. Experimental procedures.—The experiments [12] use mid-IR pulses centered at 2050 cm^{-1}, 180 fs in duration, and 90 cm^{-1} in bandwidth. The pulses were produced using an amplified Ti:sapphire pumped optical parametric amplifier. A 3.2 μJ pulse was split into two pulses having wave vectors $\vec{k}_1$ and $\vec{k}_2$ that were crossed and focused to a spot size of ~200 μm at the sample. The echo pulse emerged in the $2\vec{k}_2-\vec{k}_1$ direction. The second pulse was delayed by a time τ. Vibrational echo spectra (resolution 2 cm^{-1}) were measured by focusing the echo pulse into a 1 m scanning monochromator with a 150 lines/mm grating, and were recorded as a function of the excitation pulse delay.

The sample of (acetylacetonato)dicarbonylrhodium (*I*) [Rh(CO)$_2$acac] (Aldrich) in poly(methyl methacrylate) (PMMA) (Aldrich) was prepared under oxygen-free conditions by solvent (methylene chloride) evaporation in a vacuum desiccator for several days. The resulting high optical quality 300 μm thick film was mounted on a sample holder in a continuous flow cryostat. A silicon diode thermometer was bonded to the sample surface. Data were collected at 110 K [Fig. 1(a)] and 150 K [Fig. 1(b)]. IR spectra show that the peak positions are temperature independent. The symmetric (*S*) and antisymmetric (*A*) CO absorption lines are centered at 2082 cm^{-1} (FWHM = 12.8 cm^{-1}) and 2010 cm^{-1} (FWHM = 15.4 cm^{-1}), respectively.

III. Results and discussion.—There are six energy levels that are sampled in the experiment: the ground state, the *S* fundamental and overtone, the *A* fundamental and overtone, and the *S-A* combination band. Figure 1(a) displays VES data taken with low time resolution (500 fs per step). The inset shows the IR absorption spectrum. The higher energy VES band is from the *S* mode, and the lower frequency component is from the *A* mode. Each band has contributions from the ground state ($v = 0$) to first excited state ($v = 1$) transition and from the $v = 1$ state to the second excited state ($v = 2$, overtone) transition. The 1-2 contributions are on the red side of each band. Both features display oscillations at the frequency of the anharmonic shift, that is, the difference between the 0-1 and 1-2 transition frequencies (10.7 cm^{-1} for *S* and 13.7 cm^{-1} for *A*) [7,14]. The VES spectrum is background-free although there is significant solvent absorption in the linear IR spectrum (see inset). The ability of the VES coherent pulse sequence to eliminate unwanted absorptions [9,15] is akin to methods used in pulsed NMR to eliminate unwanted spectral features [19,20].

Figure 1(b) is a more detailed display of the *S* band. The time resolution is increased to 32 fs per step, and additional features are evident. The inset is a time slice of the data at 2084 cm^{-1}. The high frequency oscillations, which can be seen both in the main plot and in the inset, are at 71.1 cm^{-1}, the nominal splitting between the

VOLUME 86, NUMBER 17 — PHYSICAL REVIEW LETTERS — 23 APRIL 2001

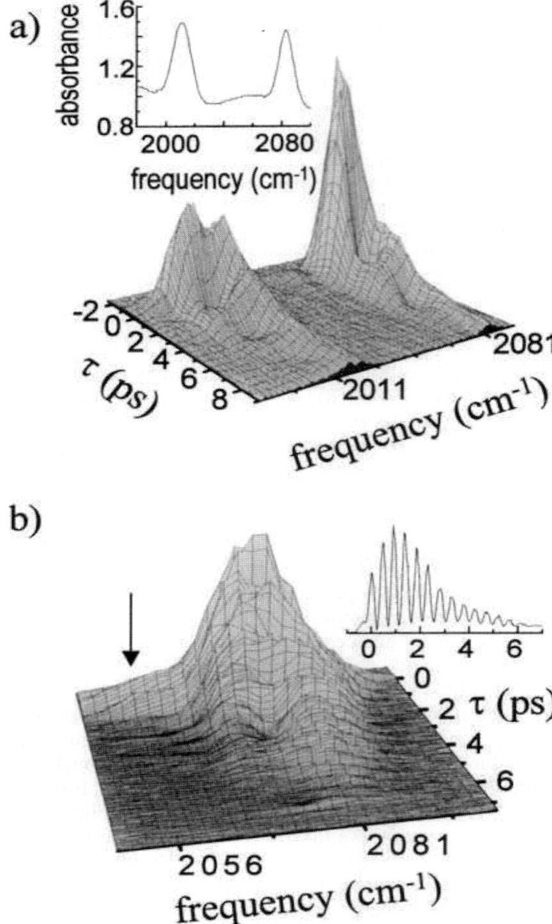

FIG. 1. 2D vibrational echo spectra of the two CD stretching modes in $Rh(CO)_2(acac)$. (a) Low time resolution showing both modes. The data show pronounced oscillations in the middle of each peak. The oscillation frequency is equal to the anharmonicity of the particular transition. The inset shows the IR spectrum of the sample. The broad solvent features at 1975 and 2050 cm^{-1} are absent in the vibrational echo spectrum. (b) High time resolution of the symmetric peak. The high frequency oscillations are at the difference in frequency between the two modes. The inset is a time cut at 2084 cm^{-1}. The smaller peak on the red side of the main band (arrow) is produced by the combination band transition and is shifted 25 cm^{-1}, reflecting the coupling between the two modes.

S and A modes. These oscillations are quantum beats arising because the transition is branching. A lower amplitude feature (marked with an arrow) is visible on the red side of the band. It arises from combination band transitions, that is, transitions in which both the S and A modes

are excited. In linear spectroscopy, the combination band absorption occurs at ~ 4067 cm^{-1}, the sum of the S and A energies shifted by an amount (25 cm^{-1} to the red) that reflects the anharmonic coupling between the two modes. In the vibrational echo experiment, the combination band is accessed through excited state absorption [i.e., initial interaction at 2082 (2010) cm^{-1}, followed by an interaction at 1985 (2057) cm^{-1}], and appears as two peaks that are shifted to the red of each fundamental transition by 25 cm^{-1}. The decay of the spectrum reflects the homogeneous dephasing of the system. The 0-1 transition (blue side of the line) decays somewhat more slowly than the 1-2 transition (red side of the main band), and the combination band peak decays very quickly. A complete analysis of the dephasing linewidth will not be given here. However, we can estimate the decay time of the 0-1 and 1-2 levels to be ~ 2–3 ps, with corresponding pure dephasing linewidths of ~ 0.9–1.3 cm^{-1}, indicating that, at this temperature, the system is massively inhomogeneously broadened. Additionally, in a simple model, the decay of the combination band peak would be closely related to the homogeneous dephasing of the S and A modes. The rapid decay of the combination band peak suggests that fluctuations of the anharmonic coupling of the modes may play an important role in the dephasing.

The lower frequency anharmonic oscillations [Fig. 1(a) and 1(b)] are not normal quantum beats or other types of beats seen in coherence experiments on a variety of multilevel systems [21,22]. The origin of the oscillations is different from the one proposed previously for experiments without frequency resolution [7,23] and applies to experiments with and without spectral resolution. The current interpretation of the anharmonic oscillations is that distinct frequencies, the 0-1 and 1-2 frequencies that arise from distinct Feynman diagrams, interfere with each other [7]. The fact that oscillations are seen in the spectrally resolved VES data (2 cm^{-1} resolution) shows that this description of the process is incomplete.

A normal quantum beat requires a branching transition. The beat arises because there are pairs of diagrams in which the first interaction in one diagram takes the system to a state $|i\rangle$ and the first interaction in the second diagram takes the system to a state $|j\rangle$. (Alternatively, the ground state may be composed of two states.) The anharmonic oscillation also involves two sets of diagrams describing single and double excitations of a single state, namely, two-level ($|0\rangle$ and $|1\rangle$) and three-level ($|0\rangle$, $|1\rangle$, and $|2\rangle$) diagrams. However, in the diagrams that produce the anharmonic oscillations, the first two interactions involve only the $|0\rangle$ and $|1\rangle$ states; there is no branching. In a quantum beat, the first interactions couple the ground and singly excited states of *two different modes,* but the emission occurs from the *singly excited state of a single mode.* In anharmonic oscillations, the first interactions involve ground and single excited states of *a single mode,* but the emission occurs from both *doubly and singly excited*

VOLUME 86, NUMBER 17   PHYSICAL REVIEW LETTERS   23 APRIL 2001

states of the same mode; the emission arises at the same frequency only because of inhomogeneous broadening.

The "two-level" diagrams, which have transitions only between the ground state and one of the two fundamentals, have a coherence between the $|0\rangle$ and $|1\rangle$ levels after the first pulse, that is, a density matrix element of $\rho_{01}(\omega)$ that evolves at frequency ω. The system interacts with the second pulse twice; the first interaction takes $\rho_{01}(\omega)$ into either $\rho_{00}(0)$ or $\rho_{11}(0)$ which have no phase evolution, and the second interaction takes both these elements into the $\rho_{10}(-\omega)$ coherence, which then evolves at frequency $-\omega$ (the negative frequency implies rephasing). The phase differences caused by inhomogeneous broadening that occurred in the time τ between the first and second pulse is rephased by time 2τ. Summing ω over all the frequencies in the inhomogeneous line produces a macroscopic polarization that dephases as a free induction decay (FID) after the first interaction, and that rephases and dephases again as an FID at a time τ after the second pulse, generating the echo pulse. One diagram involving the $|0\rangle$, $|1\rangle$, and $|2\rangle$ levels is responsible for the vibrational echo signal contribution from a 1-2 coherence. The density matrix elements corresponding to this diagram are $\rho_{00}(0) \rightarrow \rho_{01}(\omega) \rightarrow \rho_{11}(0) \rightarrow \rho_{21}(-(\omega - \Delta))$, where Δ is the anharmonic frequency shift of the 1-2 transition relative to the 0-1 transition and $\rightarrow$ represents an interaction with an E field.

The monochromator measures the signal from all oscillators emitting in a narrow bandwidth centered at frequency ω_o. There are two subensembles of molecules that can emit at ω_o. Molecules with $\rho_{10}(-\omega_o)$ (third interaction) and the corresponding $\rho_{01}(\omega_o)$ (first interaction) will emit at ω_o (two-level diagrams). Molecules with $\rho_{21}(-\omega_o)$ (third interaction) and the corresponding $\rho_{01}(\omega_o + \Delta)$ (first interaction) will also emit at ω_o (three-level diagram). The 1-2 transition frequency of this second set of molecules is at $(\omega_o + \Delta) - \Delta$; that is, their 0-1 transition frequency is $\omega_o + \Delta$, and the 1-2 emission frequency is redshifted by the anharmonic shift, Δ, bringing it to ω_o.

The frequency at which the oscillators emit, ω_o, is the frequency at which the oscillators rephase to form the echo pulse. In the diagrams involving only the $|0\rangle$ and $|1\rangle$ levels, the density matrix evolves as $\rho_{00}(0) \rightarrow \rho_{01}(\omega) \rightarrow \rho_{00}(0)$ [or $\rho_{11}(0)] \rightarrow \rho_{10}(-\omega_o)$. The two-level diagrams dephase and rephase at ω_o. In the diagram involving the $|0\rangle$, $|1\rangle$, and $|2\rangle$ levels, the situation is different. For a molecule with 0-1 transition frequency $(\omega_o + \Delta)$, the density matrix evolves as $\rho_{00}(0) \rightarrow \rho_{01}(\omega_o + \Delta) \rightarrow \rho_{11}(0) \rightarrow \rho_{21}\{-[(\omega_o + \Delta) - \Delta]\} = \rho_{21}(-\omega_o)$. Thus, the coherence produced by the three-level diagram dephases at $(\omega_o + \Delta)$ but rephases at ω_o, producing a polarization at ω_o. At τ, the time of the second pulse, there is a phase difference between $\rho_{01}(\omega_o)$ and $\rho_{01}(\omega_o + \Delta)$ equal to $\Delta \times \tau$. After the second pulse, both ensembles of oscillators begin rephasing at ω_o.

Although the vibrational echo polarization produced by both types of diagrams have the same frequency, the two polarizations do not have the same phase. As τ is increased, the phase difference advances, giving rise to the oscillations in the "single" frequency detected signal. The analysis is identical for a three pulse stimulated vibrational echo. The oscillations can occur only when the first pulse can prepare a 0-1 coherence at both frequencies ω_o and $\omega_o + \Delta$; if the inhomogeneous linewidth is much smaller than the anharmonicity, then no oscillations will be detected at third order. The mechanism applies whether or not the echo is spectrally resolved. The oscillations come from the evolution of the phase relationship of two distinct subensembles of molecules that emit at the same frequency. Unlike a quantum beat, a branching transition is not involved in the first interaction, and, unlike a quantum beat, a sufficiently broad inhomogeneous line is required for the oscillation to occur.

Figure 2 presents calculations of the VES for a single anharmonic mode for two cases. Eight diagrams contribute to the signal detected in the phase matched direction; five of these are not rephasing diagrams, and contribute only

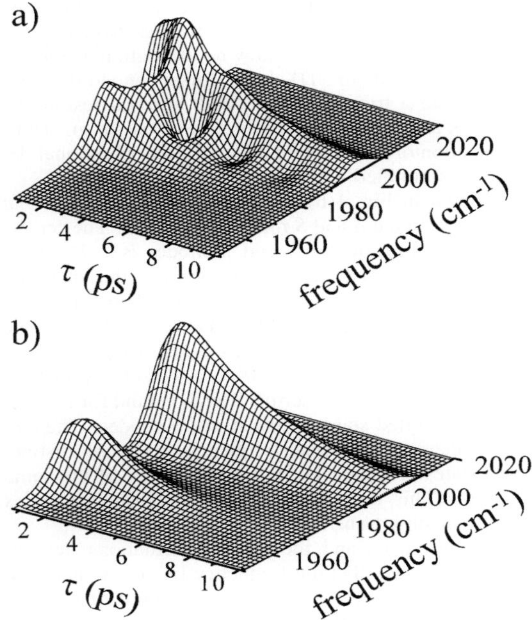

FIG. 2.   Calculations of spectrally resolved VES data of a single mode for two situations.   (a) The inhomogeneous linewidth is equal to the anharmonic shift.   The 0-1 and 1-2 spectroscopic lines overlap, and oscillations occur at the anharmonic shift frequency. (b) The anharmonic shift is twice the inhomogeneous linewidth. The 0-1 and 1-2 transitions are well resolved, and there are no oscillations.

VOLUME 86, NUMBER 17   PHYSICAL REVIEW LETTERS   23 APRIL 2001

to the signal near $\tau = 0$. All eight terms have been included in the calculations. The calculated vibrational echo pulse for each delay time was Fourier transformed, and the one-sided power spectrum was computed. The 0-1 and 1-2 inhomogeneous lines were taken to have the same Gaussian inhomogeneous frequency distribution, and the anharmonicity is constant across the inhomogeneous line. Dephasing for each transition is taken to be exponentially damped. Figure 2(a) is for an inhomogeneous linewidth equal to the anharmonicity, and Fig. 2(b) is for an anharmonicity twice the inhomogeneous linewidth. Figure 2(a) resembles the antisymmetric mode data in Fig. 1(a). When the 0-1 and 1-2 inhomogeneous lines overlap [Fig. 2(a)], there are pronounced oscillations present at the anharmonic shift frequency. However, as can be seen from the figure, the blue side of the band yields the time dependence of the 0-1 transition, and the red side of the band yields the time dependence of the 1-2 transition. In a regular 1D vibrational echo decay (not spectrally resolved), there are oscillations and cross terms. Even if each transition has a distinct exponential decay, the resulting 1D curve is a triexponential with oscillations. If the decays are more complex than exponentials and the form is not known in advance [13], then without using this or other 2D methods reliable extraction of the dynamics of the individual transitions is virtually impossible. In Fig. 2(b), the two peaks are well resolved and show no oscillations. Without spectral resolution, the 1D decay for exponential damping would be a triexponential. With spectral resolution, the decays are isolated and there is no cross term. Thus, determination of the functional form of the vibrational dephasing is simplified with the 2D approach.

As shown in Fig. 1(b) and the inset, a quantum beat occurs between the A and S mode frequencies. The center-to-center splitting of the A and S modes is ~ 71 cm^{-1}. This is the nominal beat frequency for the quantum beats shown in the inset of Fig. 1(b). Three pairs of diagrams give rise to the beats. For example, for one pair the density matrix evolves as $\rho_{00}(0) \rightarrow \rho_{0A}(\omega_A) \rightarrow \rho_{AA}(0) \rightarrow \rho_{A0}(-\omega_A)$ and $\rho_{00}(0) \rightarrow \rho_{0S}(\omega_S) \rightarrow \rho_{00}(0) \rightarrow \rho_{A0}(-\omega_A)$, where the subscripts A and S stand for the first vibrational excited states of the A and S modes. The two paths dephase at different frequencies because distinct excited states are involved, but they rephase at the same frequency. This is a standard quantum beat that arises from a branching transition.

The experiments and theoretical calculations demonstrate that ultrafast 2D time-frequency IR vibrational echo experiments can extract a substantial amount of spectroscopic and dynamic information from complex molecular systems. The analysis presents an alternative interpretation for the anharmonic oscillations that is distinct from conventional quantum beats. For complex molecular systems, in which the 1D decay is a composite of the dynamics of many transitions, spectroscopically resolving the

vibrational echo can greatly simplify data interpretation and make it possible to understand important molecular dynamical processes.

This work was supported by NIH (1RO1-GM61137), NSF (DMR-0088942), and AFOSR (F49620-94-1-0141).

[1] E. L. Hahn, Phys. Rev. **80**, 580 (1950).
[2] G. Bodenhausen, R. Freeman, G. A. Morris, and D. L. Turner, J. Magn. Reson. **31**, 75 (1978).
[3] N. A. Kurnit, I. D. Abella, and S. R. Hartmann, Phys. Rev. Lett. **13**, 567 (1964).
[4] W. P. deBoeij, M. S. Pshenichnikov, and D. A. Wiersma, Chem. Phys. **1998**, 287 (1998).
[5] S. M. Gallagher, A. W. Albrecht, T. D. Hybl, B. L. Landin, B. Rajaram, and D. M. Jonas, J. Opt. Soc. Am. B **15**, 2338 (1998).
[6] D. Zimdars, A. Tokmakoff, S. Chen, S. R. Greenfield, M. D. Fayer, T. I. Smith, and H. A. Schwettman, Phys. Rev. Lett. **70**, 2718 (1993).
[7] K. D. Rector, A. S. Kwok, C. Ferrante, A. Tokmakoff, C. W. Rella, and M. D. Fayer, J. Chem. Phys. **106**, 10027 (1997).
[8] K. D. Rector, C. W. Rella, A. S. Kwok, J. R. Hill, S. G. Sligar, E. Y. P. Chien, D. D. Dlott, and M. D. Fayer, J. Phys. Chem. B **101**, 1468 (1997).
[9] K. D. Rector and M. D. Fayer, Int. Rev. Phys. Chem. **17**, 261 (1998).
[10] P. Hamm, M. Lim, W. F. Degrado, and R. M. Hochstrasser, J. Chem. Phys. **112**, 1907 (2000).
[11] M. C. Asplund, M. T. Zanni, and R. M. Hochstrasser, Proc. Natl. Acad. Sci. U.S.A. **97**, 8219 (2000).
[12] K. D. Rector, D. E. Thompson, K. Merchant, and M. D. Fayer, Chem. Phys. Lett. **316**, 122 (2000).
[13] M. A. Berg, K. D. Rector, and M. D. Fayer, J. Chem. Phys. **113**, 3233 (2000).
[14] J. D. Beckerle, M. P. Casassa, R. R. Cavanagh, E. J. Heilweil, and J. C. Stephenson, Chem. Phys. **160**, 487 (1992).
[15] K. D. Rector, M. D. Fayer, J. R. Engholm, E. Crosson, T. I. Smith, and H. A. Schwettmann, Chem. Phys. Lett. **305**, 51 (1999).
[16] K. D. Rector, D. A. Zimdars, and M. D. Fayer, J. Chem. Phys. **109**, 5455 (1998).
[17] W. M. Zhang, V. Chernyak, and S. Mukamel, J. Chem. Phys. **110**, 5011 (1999).
[18] M. C. Asplund, M. Lim, and R. M. Hochstrasser, Chem. Phys. Lett. **323**, 269 (2000).
[19] X. Yang and L. W. Jelinski, J. Magn. Reson. B **107**, 1 (1995).
[20] S. Mani, J. Pauly, S. Conolly, C. Meyer, and D. Nishimura, Magn. Reson. Med. **37**, 898 (1997).
[21] A. I. Lvovsky and S. R. Hartmann, Laser Phys. **6**, 535 (1996).
[22] M. Koch, J. Feldmann, G. von Plessen, E. O. Gobel, P. Thomas, and K. Kohler, Phys. Rev. Lett. **69**, 3631 (1992).
[23] P. Hamm, M. Lim, M. Asplund, and R. M. Hochstrasser, Chem. Phys. Lett. **301**, 167 (1999).

VOLUME 91, NUMBER 23   PHYSICAL REVIEW LETTERS   week ending
5 DECEMBER 2003

Hydrogen Bond Dynamics Probed with Ultrafast Infrared Heterodyne-Detected Multidimensional Vibrational Stimulated Echoes

John B. Asbury, Tobias Steinel, C. Stromberg, K. J. Gaffney, I. R. Piletic, Alexi Goun, and M. D. Fayer

Department of Chemistry, Stanford University, Stanford, California 94305, USA
(Received 24 February 2003; published 3 December 2003)

Hydrogen bond dynamics are explicated with exceptional detail using multidimensional infrared vibrational echo correlation spectroscopy with full phase information. Probing the hydroxyl stretch of methanol-OD oligomers in CCl_4, the dynamics of the evolving hydrogen bonded network are measured with ultrashort (< 50 fs) pulses. The data along with detailed model calculations demonstrate that vibrational relaxation leads to selective hydrogen bond breaking on the red side of the spectrum (strongest hydrogen bonds) and the production of singly hydrogen bonded photoproducts.

DOI: 10.1103/PhysRevLett.91.237402   PACS numbers: 78.30.Cp, 78.47.+p, 82.30.Rs, 82.50.Bc

Hydrogen bonding liquids, such as alcohols and water, have complex structures and dynamics. Such liquids have generated a great deal of experimental [1–5] and theoretical [6–9] study because of their importance as solvents in chemical and biological systems. Properties of hydrogen bond networks can be studied using infrared spectroscopy of the hydroxyl stretching mode because of the influence of the number and strength of hydrogen bonds on the hydroxyl stretch frequency [10]. A water molecule can form up to four hydrogen bonds, and the various number and types of hydrogen bonds are not spectroscopically resolvable under the broad hydrogen bonded hydroxyl stretch band [6]. In contrast, molecular dynamics simulations demonstrate that methanol predominantly forms two hydrogen bonds, whether in the pure liquid [8] or as oligomers in CCl_4 [9]. The description of the structure and evolution of hydrogen bonded networks in methanol is greatly simplified compared to water by the preponderance of a single conformation in which each methanol is a single hydrogen bond donor and a single acceptor. These are called δ, which absorb at ~ 2490 cm^{-1} (FWHM ~ 150 cm^{-1}). For oligomers that are not rings, each oligomer has two ends. One end is a hydrogen bond donor but not an acceptor called γ, which absorbs at ~ 2600 cm^{-1} (FWHM ~ 80 cm^{-1}). The other end is a hydrogen bond acceptor but not a donor called β, which absorbs at ~ 2690 cm^{-1} (FWHM ~ 20 cm^{-1}) [3]. All three of these species are spectroscopically resolvable. Consequently, methanol oligomers in CCl_4 are an important system for studying the dynamics of hydrogen bonded networks [2].

Here we report the first application of ultrafast heterodyne detected multidimensional stimulated vibrational echo correlation spectroscopy with full phase information to the study of the dynamics of hydrogen bonding liquids. The development of the ultrafast infrared vibrational echo technique [11,12] and the recent extension to multidimensional vibrational echo methods [13] provide a new approach for the study of condensed matter systems. By using the shortest mid-IR pulses produced to date (<50 fs or <4 cycles of light), it is possible to simultaneously examine the entire broad hydroxyl stretching band of methanol-OD (MeOD, deuterated hydroxyl) oligomers in CCl_4 solution despite its >400 cm^{-1} width. Data are obtained with correct phase relationships across the entire spectrum, which permits accurate separation of the absorptive and dispersive contributions to the spectra. As a result, the 2D IR correlation spectra are akin to 2D NMR spectra [14]. The experimental results and model calculations demonstrate the selective breaking of the strongest hydrogen bonds through a nonequilibrium relaxation pathway that precedes vibrational energy equilibration.

For the experiment, the <50 fs transform limited IR pulses are generated using a Ti:sapphire regeneratively amplified laser/optical parametric amplifier system. The IR beam is split into five beams. Three of the beams are the excitation beams for the stimulated vibrational echo. A fourth beam is the local oscillator (LO) used to heterodyne detect the vibrational echo signal. The vibrational echo signal combined with the LO is passed through a monochromator and detected by a 32 element mercury cadmium telluride array. The monochromator is stepped to cover the entire spectrum. The sample, 10% MeOD in CCl_4, was held in a sample cell of CaF_2 flats with a spacing of 50 μm. A fifth beam is the probe in pump-probe experiments.

The multidimensional stimulated vibrational echoes were measured as a function of one frequency variable, ω_m (monochromator/array), and two time variables, τ (time between pulses 1 and 2) and T_w (time of third pulse after the first two pulses). The measured signal is the absolute value squared of the sum of the vibrational echo signal electric field, S, and the local oscillator electric field, L. The spectrum of the cross term, $2LS$, is the ω_m frequency axis. Scanning τ generates an interferogram. Numerical Fourier transformation converts the interferogram into the frequency variable ω_τ, which contains both the absorptive and dispersive components of the signal. However, two sets of quantum pathways

     0031-9007/03/91(23)/237402(4)$20.00

can be measured independently by appropriate time ordering of the pulses [13]. By adding the Fourier transforms of the interferograms from the two pathways, the dispersive component cancels, leaving only the absorptive component.

Lack of perfect knowledge of the timing of the pulses and consideration of chirp on the vibrational echo pulse requires a "phasing" procedure to be used. The projection slice theorem [14] is employed to generate the absorptive 2D correlation spectrum that is compared to the experimental pump-probe spectrum. The procedure requires a comprehensive mathematical formulation and implementation that will be detailed subsequently.

Figure 1 displays some of the 2D correlation spectra, four contour plots at $T_w = 125$, 1.2, 1.8, and 5.0 ps for MeOD. The maximum positive signal has been normalized to unity in each plot. The contours represent equal 10% graduations. The ω_τ axis is the axis of the first radiation field interaction. The ω_m axis is the axis of the emission of the echo pulse. The positive going band on the diagonal in the 125 fs spectrum corresponds to the 0–1 transition of the δ band with some contribution from the 0–1 γ band on the blue end. The center frequencies of the δ and γ 0–1 transitions are indicated by horizontal and vertical lines. The negative going band below the main diagonal band arises from the 1–2 transition of the δ band. The 1–2 band is off diagonal because the emission is shifted to lower frequency by the vibrational anharmonicity (~ 150 cm^{-1}).

As the T_w delay increases, the correlation spectra displayed in Fig. 1 exhibit changes that provide insight into the mechanism of hydrogen bond evolution. By $T_w = 1.2$ ps, the correlation spectrum has changed dramatically due to vibrational relaxation and hydrogen bond breaking. A good fraction of the initially produced excited state population has decayed to the ground state because the vibrational lifetime of the δ OD stretch is ~ 0.5 ps [2]. By $T_w = 1.8$ ps, the δ excited state has completely decayed. Because the lifetime of the γ band (~ 1 ps) is longer than that of the δ band, the diagonal γ 0–1 band is uncovered as the δ band decays. By $T_w = 1.8$ ps, the diagonal γ peak is almost gone.

As will be demonstrated below, when the δ OD excited state relaxes, hydrogen bonds break [1,2], producing photoproduct γ ODs. The resulting photoproduct γ peak, located above the δ band, is visible by $T_w = 1.2$ ps. By $T_w = 5.0$ ps, the only remaining features are the preserved δ and the photoproduct γ bands. Comparison of the center of the preserved δ band at $T_w = 5.0$ ps and the center of the initially excited δ band (indicated by the horizontal line at 2490 cm^{-1}) demonstrates that the preserved δ band is contracted to the red and shifted off diagonal along the ω_m axis. For times longer than ~ 5 ps, the peaks do not change shape significantly, but slowly decay in magnitude on a time scale of tens of ps because of hydrogen bond recombina-

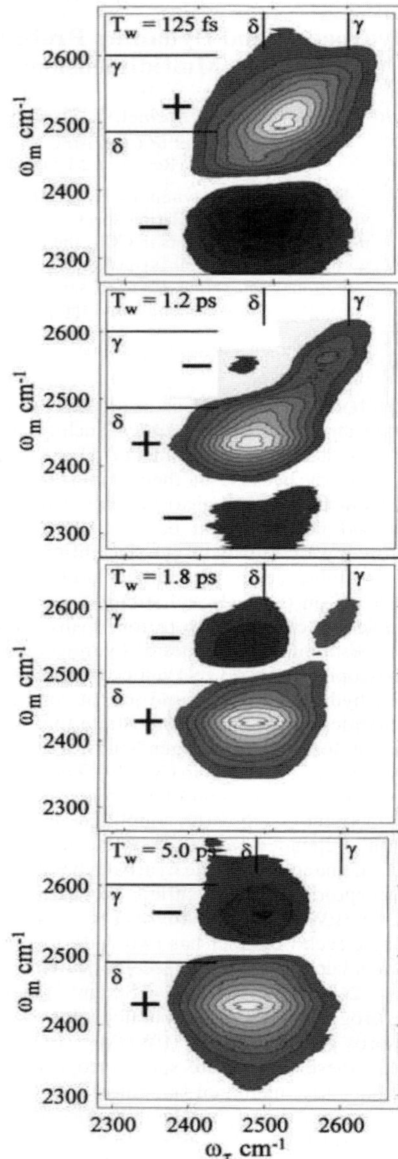

FIG. 1. Correlation spectra for several T_w delays. See text for details.

tion [1,2]. An important aspect of the 2D correlation spectrum is that the photoproduct γ is spectrally separated in the plane from the equilibrium γ band (see Fig. 1, $T_w = 1.2$ and 1.8 ps), in contrast to 1D spectroscopies.

The T_w dependence of the correlation spectra raises a number of questions. First, why does a correlation spectrum exist for times long compared to the vibrational lifetime? Normally, the decay of the excited state into the ground state causes the vibrational echo signal to decay. However, following vibrational relaxation, a few tens of percent of the initially excited δs break a hydrogen bond in $\sim$200 fs, removing them as δ absorbers and creating new photoproduct γ absorbers [2]. Therefore, the decay of the excited state does not completely eliminate the contribution to the signal from the ground state. The ground state signal remains for the hydrogen bond recombination time, which is tens of ps [1,2].

Second, a fundamentally important question that will be addressed here is whether vibrational relaxation simply "heats" the hydrogen bond network following vibrational relaxation or is there another mechanism that gives rise to the time dependence of the bands shown in Fig. 1? To address this question, we have performed detailed calculations to model the correlation spectra. The four models will be described in detail in a future publication. For brevity, we give a very brief overview of the modeling procedure and discuss only the two models that gave the best results. The correlation spectra are fit using the sum of 2D Gaussian band shapes for the δ and γ, 0–1 and 1–2 bands. For the $T_w = 125$ fs correlation spectrum, the widths and positions of the bands are set equal to the equilibrium δ and γ widths and positions as determined from linear spectroscopy of the sample. The $T_w = 125$ fs correlation spectrum is very well described by this model. The calculations use results of the fits to the 125 fs data and factors that are unique to each model to calculate the 5 ps data. The 1–2 bands do not contribute because 5 ps is much longer than the lifetime.

The experiments also give the ratios of the amplitudes of the 0–1 bands at 125 fs to those at 5 ps. From comparison of the fit to the $T_w = 125$ fs correlation spectrum with the fit to the 5 ps correlation spectrum using each model, the ratios of the amplitudes are determined for each model.

Model 1.—Following vibrational relaxation of the initially excited OD stretch, the deposited energy (heat) causes hydrogen bond weakening. Weakening of hydrogen bonds shifts their absorption to the blue (higher energy) [10]. The δ and γ bleach responsible for the long lived vibrational echo signal is not filled in by the vibrational relaxation because the population is shifted to the blue. The center positions and widths of the preserved δ and γ bleach bands are described by their equilibrium spectra. The weakened δ and γ photoproducts have spectra that are shifted to the blue along the ω_m axis. In the calculations, the magnitude of the blueshift is allowed to vary to give the best fit to the 5 ps data ($R^2 = 1.1$, residuals squared). The correlation spectrum calculated from model 1 is displayed in Fig. 2(a). The model does a reasonable job of reproducing aspects of the

data (compare to lowest panel of Fig. 1). However, it does not have the correct shape. The preserved band (positive going) is elongated to the upper right because of the preserved γ portion of the spectrum. Even more significant, to obtain this best fit, the amplitudes of the bands are unphysical. To get the features in the correct positions along the ω_m axis requires a very small shift (5 cm^{-1}) of the weakened band produced by vibrational relaxation. Because of the small shift, the preserved δ bleach band strongly overlaps and nearly cancels the weakened photoproduct band. To obtain the correct ratio of the amplitude of the $T_w = 5.0$ ps data relative to the 125 fs data, model 1 requires the magnitude of the preserved δ bleach band to be $\sim$9 times larger than the magnitude of the initial δ bleach, which would require the concentration of initially excited MeODs *increasing* by a factor of $\sim$9 between $T_w = 125$ fs and 5.0 ps. Obviously this result is not possible. Model 1 cannot describe the data. Model 1 is essentially a thermal equilibration model. Detailed measurements of the pump-probe spectrum show that it takes ≥ 30 ps for the system to reach thermal equilibrium (after which the pump-probe spectrum is the same as the static temperature difference spectrum). The time dependent pump-probe spectrum again demonstrates that the mechanism that gives rise to the 5 ps data is not thermal equilibration.

Model 2.—Vibrational relaxation of initially excited OD stretches selectively breaks hydrogen bonds on the red side of the δ band and produces redshifted γ photoproducts. In the calculations, the spectral region of hydrogen bond breaking is allowed to vary to give the best fit to the 5 ps data ($R^2 = 0.13$). The simulated correlation spectrum calculated from model 2 is displayed in Fig. 2(b). This model provides an excellent description of the $T_w = 5.0$ ps experimental data. The shape of the spectrum is correct, and the R^2 value is a factor of $\sim$8 smaller for model 2 compared to model 1. In addition to having the correct shape and a better R^2, model 2 predicts a physically reasonable magnitude for the amplitudes of

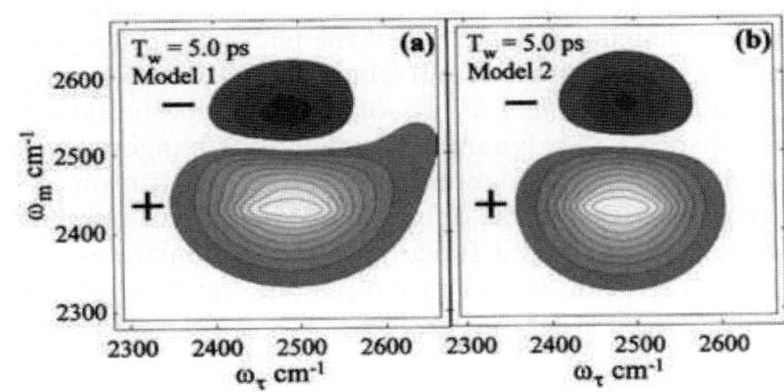

FIG. 2. Calculated correlation spectra from two models of hydrogen bond evolution following vibration relaxation. See text for details.

the bands at 5 ps relative to 125 fs. Model 2 gives the preserved δ bleach to ~60% of the initial δ bleach, that is, 60% of the initially excited δ MeODs break a hydrogen bond, which is within a few tens of percent of the value determined by independent measurements of the quantum yield of hydrogen bond breaking in MeOD oligomers [2].

The results presented above demonstrate that the δ band is contracted to the red and the photoproduct γ band is generated because vibrational relaxation selectively breaks the stronger hydrogen bonds (the red side of the line). The breaking of the strongest hydrogen bonds is at first surprising. However, this nonequilibrium mechanism can occur if vibrational relaxation on the low energy side of the line populated different modes than vibrational relaxation on the high energy side of the line, and the modes populated through the lower energy relaxation pathway are more effective at breaking hydrogen bonds. Significant support for the proposed nonequilibrium mechanism comes from the IR pump/Raman probe experiments of Iwaki and Dlott, who observed that the pathway for vibrational relaxation was different on the red and blue sides of the hydroxyl stretching band in OH methanol [15]. Excitation on the red side led to vibrational relaxation that produced substantially more relative population of hydroxyl bends compared to excitation on the blue side that put much more population in the CH bends and other modes [15]. The hydroxyl bends would be expected to be strongly coupled to the hydrogen bond in contrast to CH stretches. This necessary feature of the proposed mechanism is indeed observed experimentally.

Finally, two types of frequency correlation between the initially excited δ and photoproduct γ bands are observed. These will be discussed in detail subsequently, but, briefly, the first type, which we call coarse correlation, is observed in the redshift along the ω_m axis of the preserved δ and photoproduct γ bands at $T_w = 5.0$ ps (compare to the equilibrium δ and γ band positions marked in Fig. 1). The redshift indicates that breaking the strongest hydrogen bonds associated with δs creates photoproduct γs that retain coarse "memory" of the hydrogen bond strengths. However, detailed analysis of the time dependent 2D shape of the photoproduct γ peak demonstrates that a fine correlation also exists. Briefly, the width along the ω_τ axis of the 2D photoproduct γ peak displays the dynamical linewidth. Changes in the dynamical linewidth report the spectral diffusion dynamics. Analysis of the dynamical linewidth of the photoproduct γ peak as a function of T_w demonstrates that spectral diffusion occurs in the photoproduct γ peak on

the tens of ps time scale. Observation of spectral diffusion dynamics in the photoproduct γ band proves that the broken oligomers retain a "fine" as well as a coarse memory of the structural degrees of freedom of the network prior to hydrogen bond breaking.

Here we have presented the first application of vibrational echo correlation spectroscopy to the study of hydrogen bond network dynamics. The experiments covered the entire very broad spectral region with full phase information using ultrashort midinfrared pulses. For brevity, the analysis focused on the nonequilibrium hydrogen bond breaking dynamics following vibrational relaxation that results in selective breaking of the stronger hydrogen bonds. The data also provide unique information, which will be presented subsequently, on spectral diffusion that describes hydrogen bond evolution prior to and following hydrogen bond breaking.

This work was supported by the AFOSR (F49620-01-1-0018), DOE (DE-FG03-84ER13251), and NSF (DMR-0088942). T. S. thanks the Emmy Noether program of the DFG for partial support.

---

[1] R. Laenen and C. Rausch, J. Chem. Phys. **106**, 8974 (1997).

[2] K. J. Gaffney *et al.*, J. Phys. Chem. A **106**, 12012 (2002).

[3] S. Woutersen, U. Emmerichs, and H. J. Bakker, J. Chem. Phys. **107**, 1483 (1997).

[4] G. M. Gale *et al.*, Phys. Rev. Lett. **82**, 1068 (1999).

[5] S. Yeremenko, M. S. Pshenichnikov, and D. A. Wiersma, Chem. Phys. Lett. **369**, 107 (2003).

[6] C. P. Lawrence and J. L. Skinner, J. Chem. Phys. **118**, 264 (2003).

[7] R. Rey, K. B. Møller, and J. T. Hynes, J. Phys. Chem. A **106**, 11993 (2002).

[8] D. M. zum Buscheenfelde and A. Staib, Chem. Phys. **236**, 253 (1998).

[9] R. Veldhuizen and S. W. de Leeuw, J. Chem. Phys. **105**, 2828 (1996).

[10] A. Novak, in *Structure and Bonding*, edited by J. D. Dunitz (Springer-Verlag, Berlin, 1974), Vol. 18, p. 177.

[11] D. Zimdars *et al.*, Phys. Rev. Lett. **70**, 2718 (1993).

[12] P. Hamm, M. Lim, and R. M. Hochstrasser, Phys. Rev. Lett. **81**, 5326 (1998).

[13] M. Khalil, N. Demirdöven, and A. Tokmakoff, J. Phys. Chem. A **107**, 5258 (2003).

[14] R. R. Ernst, G. Bodenhausen, and A. Wokaun, *Nuclear Magnetic Resonance in One and Two Dimensions* (Oxford University Press, Oxford, 1987).

[15] L. K. Iwaki and D. D. Dlott, J. Phys. Chem. A **104**, 9101 (2000).

Myoglobin-CO Substate Structures and Dynamics: Multidimensional Vibrational Echoes and Molecular Dynamics Simulations

Kusai A. Merchant,[†] W. G. Noid,[‡] Ryo Akiyama,[‡] Ilya J. Finkelstein,[†] Alexei Goun,[†] Brian L. McClain,[†] Roger F. Loring,[‡] and M. D. Fayer*,[†]

Contribution from the Department of Chemistry, Stanford University, Stanford, California 94305, and Department of Chemistry and Chemical Biology, Baker Laboratory, Cornell University, Ithaca, New York 14853

Received April 16, 2003; E-mail: fayer@stanford.edu

Abstract: Spectrally resolved infrared stimulated vibrational echo data were obtained for sperm whale carbonmonoxymyoglobin (MbCO) at 300 K. The measured dephasing dynamics of the CO ligand are in agreement with dephasing dynamics calculated with molecular dynamics (MD) simulations for MbCO with the residue histidine-64 (His64) having its imidazole ϵ nitrogen protonated (N_ϵ−H). The two conformational substate structures B_ϵ and R_ϵ observed in the MD simulations are assigned to the spectroscopic A_1 and A_3 conformational substates of MbCO, respectively, based on the agreement between the experimentally measured and calculated dephasing dynamics for these substates. In the A_1 substate, the N_ϵ−H proton and N_δ of His64 are approximately equidistant from the CO ligand, while in the A_3 substate, the N_ϵ−H of His64 is oriented toward the CO, and the N_δ is on the surface of the protein. The MD simulations show that dynamics of His64 represent the major source of vibrational dephasing of the CO ligand in the A_3 state on both femtosecond and picosecond time scales. Dephasing in the A_1 state is controlled by His64 on femtosecond time scales, and by the rest of the protein and the water solvent on longer time scales.

I. Introduction

Understanding protein dynamics is essential to establishing the relationships between protein structure and protein function.[1-5] Protein dynamics occur on a continuum of time scales, from the subpicosecond to the microsecond and longer. Such dynamics can be investigated with a variety of spectroscopic techniques. However, spectroscopic techniques with sufficient time resolution to probe the shortest time scale dynamics typically cannot provide the necessary structural specificity to assign these dynamics to particular atomic motions. Computational techniques such as molecular dynamics (MD) simulations can provide a view of protein motions at an atomic level of detail over a short range of time scales. The application of molecular simulation to interpret spectroscopic measurements requires the direct modeling of the experimental observable, as performed here.

Carbonmonoxymyoglobin (MbCO) (see Figure 1) has been extensively studied both experimentally[6-19] and computationally.[20-26] Myoglobin (Mb) is a small globular heme protein

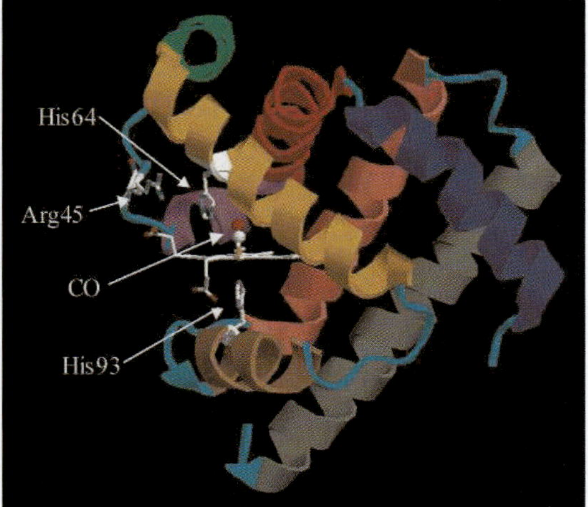

Figure 1. Crystal structure of MbCO taken from the Protein Data Bank (1A6G). The heme group and three residues are indicated explicitly. The heme iron is covalently bound to the protein via His93. His64 is thought to play a major role in the conformational substates of the protein. Secondary structure is color-coded: helix A, blue; helix B, red; helix C, purple; helix D, green; helix E, orange; helix F, brown; helix G, pink; helix H, gray. All loops are colored cyan.

weighing approximately 17 kDa that is found in mammalian muscle tissue. The prosthetic heme group can reversibly bind a number of small molecule ligands such as O_2, CO, and NO. The crystal structure of Mb has been known for 40 years[27] and

† Stanford University.
‡ Cornell University.
(1) Frauenfelder, H.; Sligar, S. G.; Wolynes, P. G. *Science* **1991**, *254*, 1598−1603.
(2) Hong, M. K.; Braunstein, D.; Cowen, B. R.; Frauenfelder, H.; Iben, I. E. T.; Mourant, J. R.; Ormos, P.; Scholl, R.; Schulte, A.; Steinbach, P. J.; Xie, A.; Young, R. D. *Biophys. J.* **1990**, *58*, 429−436.
(3) Andrews, B. K.; Romo, T.; Clarage, J. B.; Pettitt, B. M.; Phillips, G. N., Jr. *Structure* **1998**, *6*, 587−594.
(4) Campbell, B. F.; Chance, M. R.; Friedman, J. M. *Science* **1987**, *238*, 373−376.
(5) Frauenfelder, H.; McMahon, B. H.; Austin, R. H.; Chu, K.; Groves, J. T. *Proc. Natl. Acad. Sci. U.S.A.* **2001**, *98*, 2370−2374.

10.1021/ja035654x CCC: $25.00 © 2003 American Chemical Society

has been continually refined over the last four decades to atomic resolution.[28,29] Because of its small size, relatively long MD simulations are tractable.[21,22] The CO stretching mode of MbCO has a strong transition dipole, making its infrared absorption an easily monitored experimental observable that can be used to track kinetics and dynamics in the protein. Studies on Mb, and in particular on MbCO, serve as tests for many of the ideas on the relationship between structure and function in proteins.

The infrared (IR) spectrum of the CO stretching mode of MbCO has at least three major absorption bands,[6,7,12-14] denoted A_0 (~1965 cm^{-1}), A_1 (~1944 cm^{-1}), and A_3 (~1932 cm^{-1}). It is well known that these substates have different ligand binding rates,[1,6,13] but the nature of the structural differences between these substates is still in question. It has been suggested that different electrostatic environments in the heme pocket arising from different conformations of heme pocket residues are largely responsible for the observed bands.[15,19,25,26,30] Examples of previously hypothesized structures for the A substates are shown in Figure 2. Mutant studies have shown that the distal histidine, His64, plays a prominent role in determining the CO stretching frequency,[15,31] but the tautomerization and orientation of this residue remain unclear. His64 has two titratable nitrogens, N_δ and N_ϵ, either of which can be oriented toward the ligand through rotation of the imidazole ring. This residue is also fairly mobile and has been reported at a wide variety of distances (from 2 to 7 Å) from the CO ligand in crystal structures.[28,29,32] At low pH, His64 is thought to be doubly protonated and has been observed in a low pH crystal structure[32] and in a resonance Raman study[11] to be rotated out of the heme binding pocket

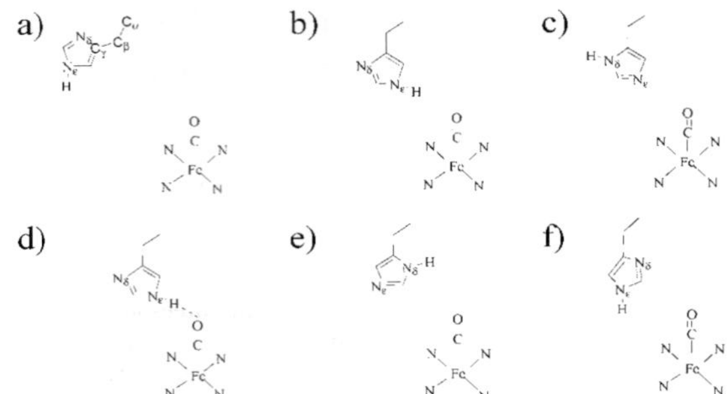

Figure 2. Possible conformers and tautomers of His64 that have been assigned to conformational substates. (a) A_0.[29,32] The distal His64 is swung out of the pocket away from the ligand. The carbons of the imidazole side chain are labeled C_α–C_γ. (b) A_1.[19,20,29] The N_ϵ–H is in the pocket and oriented toward the ligand. (c) A_1.[28,34] The N_δ–H is oriented toward the solvent, and the lone pairs of the N_ϵ interact with the ligand. (d) A_3.[29,45,80] The N_ϵ–H is in the pocket and is very close to and interacts strongly with the ligand. There may be a hydrogen bond between N_ϵ–H and the CO. (e) A_1[21,34] or A_3.[19] The N_δ–H is in the pocket near the ligand. (f) A_1.[45] Both N_ϵ–H and N_δ are in the pocket. There is moderate interaction between His64 and the ligand.

away from the CO ligand in an "open" configuration. This conformer, with little interaction between His64 and the ligand, is thought to correspond to the A_0 substate, because at low pH the A_0 absorption line is the most intense. In addition, mutations of His64 to apolar residues produce a band at approximately the same frequency as the A_0 line.[31] At pH greater than 6, the A_1 and A_3 substates are the most populated at room temperature.[12,14] Kinetic hole burning studies on the MbCO A substates strongly suggest that the A_1 and A_3 substates are the result of a single tautomerization state of His64 because the time scale of interconversion between the A_1 and A_3 substates (~1 ns) is much faster than the time scale of His64 tautomerization (~1 μs).[13] Two recent atomic resolution crystal structures at neutral pH indicate that His64 is rotated into the heme pocket and is much closer to the ligand than in the A_0 substate, but the exact conformation of His64 remains uncertain.[28,29] In addition, these two crystal structures assign different tautomerization states to the singly protonated His64.

Both MD methods[21,33] and ab initio structural calculations[20] have been used to address the issue of the structural origins of the A substates, as well as the tautomerization state of His64.[34] Rovira et al.[20] performed mixed quantum mechanical/molecular mechanical calculations for various possible conformers for the A substates. On the basis of shifts of the CO vibrational frequency, they concluded that His64 exists as the N_ϵ–H tautomer with the proton pointed toward the ligand. However, they were unable to assign specific structures to the A_1 and A_3 substates. Other computational studies have proposed structural assignments for the A substates using the N_δ–H tautomer,[21,33] which also reproduce the trends in the IR frequency shift. These studies illustrate that computations of CO vibrational frequencies alone cannot unambiguously identify the configurations associated with the spectroscopic substates.

In this paper, we present measured and calculated spectrally resolved infrared stimulated vibrational echo data on sperm

(6) Ansari, A.; Berendzen, J.; Braunstein, D.; Cowen, B. R.; Frauenfelder, H.; Hong, M. K.; Iben, I. E. T.; Johnson, J. B.; Ormos, P.; Sauke, T. B.; Scholl, R.; Schulte, A.; Steinbach, P. J.; Vittitow, J.; Young, R. D. *Biophys. Chem.* **1987**, *26*, 337–355.
(7) Caughey, W. S.; Shimada, H.; Choc, M. G.; Tucker, M. P. *Proc. Natl. Acad. Sci. U.S.A.* **1981**, *78*, 2903–2907.
(8) Rector, K. D.; Engholm, J. R.; Rella, C. W.; Hill, J. R.; Dlott, D. D.; Fayer, M. D. *J. Phys. Chem. A* **1999**, *103*, 2381–2387.
(9) Springer, B. A.; Sligar, S. G.; Olson, J. S.; Phillips, G. N., Jr. *Chem. Rev.* **1994**, *94*, 699–714.
(10) Tian, W. D.; Sage, J. T.; Champion, P. M.; Chien, E.; Sligar, S. G. *Biochemistry* **1996**, *35*, 3487–3502.
(11) Zhu, L.; Sage, J. T.; Rigos, A. A.; Morikis, D.; Champion, P. M. *J. Mol. Biol.* **1992**, *224*, 207–215.
(12) Shimada, H.; Caughey, W. S. *J. Biol. Chem.* **1982**, *257*, 1893–1900.
(13) Johnson, J. B.; Lamb, D. C.; Frauenfelder, H.; Müller, J. D.; McMahon, B.; Nienhaus, G. U.; Young, R. D. *Biophys. J.* **1996**, *71*, 1563–1573.
(14) Müller, J. D.; McMahon, B. H.; Chen, E. Y. T.; Sligar, S. G.; Nienhaus, G. U. *Biophys. J.* **1999**, *77*, 1036–1051.
(15) Phillips, G. N., Jr.; Teodoro, M. L.; Li, T.; Smith, B.; Olson, J. S. *J. Phys. Chem. B* **1999**, *103*, 8817–8829.
(16) Fayer, M. D. *Annu. Rev. Phys. Chem.* **2001**, *52*, 315–356.
(17) Morikis, D.; Champion, P. M.; Springer, B. A.; Sligar, S. G. *Biochemistry* **1989**, *28*, 4791–4800.
(18) Nienhaus, G. U.; Mourant, J. R.; Chu, K.; Frauenfelder, H. *Biochemistry* **1994**, *33*, 13413–13430.
(19) Oldfield, E.; Guo, K.; Augspurger, J. D.; Dykstra, C. E. *J. Am. Chem. Soc.* **1991**, *113*, 7537–7541.
(20) Rovira, C.; Schulze, B.; Eichinger, M.; Evanseck, J. D.; Parrinello, M. *Biophys. J.* **2001**, *81*, 435–445.
(21) Schulze, B. G.; Evanseck, J. D. *J. Am. Chem. Soc.* **1999**, *121*, 6444–6454.
(22) Meller, J.; Elber, R. *Biophys. J.* **1998**, *74*, 789–802.
(23) Elber, R.; Karplus, M. *Science* **1987**, *235*, 318–321.
(24) Vitkup, D.; Ringe, D.; Petsko, G. A.; Karplus, M. *Nat. Struct. Biol.* **2000**, *7*, 34–38.
(25) Kushkuley, B.; Stavrov, S. S. *Biophys. J.* **1996**, *70*, 1214–1229.
(26) Kushkuley, B.; Stavrov, S. S. *Biophys. J.* **1997**, *72*, 899–912.
(27) Kendrew, J. C.; Dickerson, R. E.; Strandberg, B. E.; Hart, R. G.; Davies, D. R.; Phillips, D. C.; Shore, V. C. *Nature* **1960**, *185*, 422–427.
(28) Kachalova, G. S.; Popov, A. N.; Bartunik, H. D. *Science* **1999**, *284*, 473–476.
(29) Vojtechovsky, J.; Chu, K.; Berendzen, J.; Sweet, R. M.; Schlichting, I. *Biophys. J.* **1999**, *77*, 2153–2174.
(30) Franzen, S. *J. Am. Chem. Soc.* **2002**, *124*, 13271–13281.
(31) Li, T. S.; Quillin, M. L.; Phillips, G. N., Jr.; Olson, J. S. *Biochemistry* **1994**, *33*, 1433–1446.
(32) Yang, F.; Phillips, G. N., Jr. *J. Mol. Biol.* **1996**, *256*, 762–774.
(33) Schulze, B. G.; Grubmuller, H.; Evanseck, J. D. *J. Am. Chem. Soc.* **2000**, *122*, 8700–8711.
(34) Jewsbury, P.; Kitagawa, T. *Biophys. J.* **1994**, *67*, 2236–2250.

whale MbCO at 300 K. The ultrafast infrared vibrational echo[8,35,36] and the multidimensional vibrational echo[37–46] measure the fast structural dynamics of molecules through their effects on vibrational line shapes. The current work expands the findings from our previous report on horse heart MbCO[40] using a two-pulse spectrally resolved vibrational echo experiment. The three-pulse measurement (stimulated echo) permits the variation of the time delay T_w between the second and third pulses, providing a probe of dynamics on longer time scales than possible with the two-pulse echo. The three-pulse vibrational echo provides a family of dynamical line shapes rather than the single dynamical line shape provided by a two-pulse vibrational echo. Therefore, a more rigorous comparison between the experiments and the calculations is made possible. More recently, we presented the results of a preliminary study[45] on horse heart MbCO using spectrally resolved stimulated echo measurements with a limited set of detection frequencies. Comparing these measurements on horse heart MbCO to MD simulations of sperm whale MbCO with the N_ϵ–H tautomer of His64, we provisionally assigned the structural origins of the A_1 and A_3 substates of MbCO. Here, we present the spectrally resolved stimulated vibrational echo response at a wide range of detection frequencies for the substates of sperm whale MbCO. These measurements are compared to molecular dynamics results for sperm whale MbCO with both possible tautomerization states of His64. Using a time-dependent Stark effect model to derive a frequency-frequency correlation function (FFCF) for the CO ligand from the MD simulations, spectrally resolved stimulated vibrational echo spectra are computed from the FFCF for each tautomer and compared to the data.

The comparisons between the calculated vibrational echo data for the two His64 tautomers and the measured vibrational echo data over a wide range of frequency and T_w indicate that the correct tautomer state of His64 is the N_ϵ–H structure and confirm the assignment of the structural origins of the A_1 and A_3 substates of MbCO made in our previous study.[45] In addition, analysis of the MD results reveals the origins of the dephasing dynamics for the A_1 and A_3 substates in terms of contributions from particular portions of the protein and the water solvent.

II. The Vibrational Echo

Time-dependent interactions between a vibrational mode and its environment give rise to spectral line broadening. The linear absorption spectrum of a vibrational spectroscopic transition contains contributions from dynamics occurring on all time scales that are relevant to the interactions between the transition and its environment. In principle, all dynamical information about the interaction of a vibrational mode with its environment can be deduced from the absorption line shape alone.[47] In practice, however, dynamics on long time scales often dominate the shape of the linear absorption spectrum, masking the precise nature of the interactions between an oscillator and its environment.

The stimulated vibrational echo is a nonlinear spectroscopic technique capable of separating the dynamical processes contributing to an absorption spectrum line width by selectively eliminating the contribution of dynamics that have time scales slower than an experimentally controlled waiting time.[48–53] In a stimulated vibrational echo experiment, three resonant pulses with variable delay τ between pulses 1 and 2 and variable delay T_w between pulses 2 and 3 propagate along three different paths and are crossed in the sample. Interactions in the sample with the three applied pulses lead to the generation of a fourth vibrational echo pulse in a unique phase-matched direction. In a typical experiment, the value of T_w is fixed, and the intensity of the vibrational echo signal is recorded as a function of the scanned time delay τ. The stimulated vibrational echo eliminates spectral line broadening contributions from any dynamical processes that occur on time scales slower than the experimentally controlled time T_w.[48,49,51,52] By measuring stimulated vibrational echoes decays for a series of T_w times, the dynamical spectral broadening that occurs on different time scales is systematically mapped out. The stimulated vibrational echo thus probes the dynamics of the interaction of a vibrational mode and its environment. Thorough descriptions of the vibrational echo can be found in several recent reviews by Fayer and co-workers[16,54] and by Hamm and Hochstrasser.[52]

III. Experimental and Computational Methods

A. Experimental Methods. The current experimental setup has been modified extensively from that reported previously.[16,40] An additional beam line for performing stimulated vibrational echoes has been added, as well as an array detector that greatly enhances the data acquisition rate. In the stimulated vibrational echo experiments reported here, an IR pulse (center frequency = 1940 cm^{-1}, bandwidth = 125 cm^{-1}, pulse duration = 125 fs) produced from a regeneratively amplified Ti: Sapphire pumped OPA (Spectra Physics) is beam split once, and each daughter pulse is split again with three 50%/50% (3.5–6.5 μm) ZnSe beam splitters (II–VI Inc.) with broadband antireflection coating to produce four equal-energy pulses ($\sim$700 nJ), three of which are used in the experiments. The three pulses have wave vectors $\vec{k}_1$, $\vec{k}_2$, and $\vec{k}_3$ with variable delay time τ between the pulses with wave vector $\vec{k}_1$ and $\vec{k}_2$ and with variable delay time T_w between pulses with wave vector $\vec{k}_2$

(35) Zimdars, D.; Tokmakoff, A.; Chen, S.; Greenfield, S. R.; Fayer, M. D.; Smith, T. I.; Schwettman, H. A. *Phys. Rev. Lett.* **1993**, *70*, 2718–2721.
(36) Tokmakoff, A.; Fayer, M. D. *Acc. Chem. Res.* **1995**, *28*, 437–445.
(37) Hamm, P.; Lim, M.; Hochstrasser, R. M. *Phys. Rev. Lett.* **1998**, *81*, 5326–5329.
(38) Merchant, K. A.; Thompson, D. E.; Fayer, M. D. *Phys. Rev. Lett.* **2001**, *86*, 3899–3902.
(39) Thompson, D. E.; Merchant, K. A.; Fayer, M. D. *J. Chem. Phys.* **2001**, *115*, 317–330.
(40) Merchant, K. A.; Thompson, D. E.; Xu, Q.-H.; Williams, R. B.; Loring, R. F.; Fayer, M. D. *Biophys. J.* **2002**, *82*, 3277–3288.
(41) Golonzka, O.; Khalil, M.; Demirdoven, N.; Tokmakoff, A. *Phys. Rev. Lett.* **2001**, *86*, 2154–2157.
(42) Golonzka, O.; Khalil, M.; Demirdoven, N.; Tokmakoff, A. *J. Chem. Phys.* **2001**, *115*, 10814–10828.
(43) Zanni, M. T.; Gnanakaran, S.; Stenger, J.; Hochstrasser, R. M. *J. Phys. Chem. B* **2001**, *105*, 6520–6535.
(44) Zanni, M. T.; Asplund, M. C.; Hochstrasser, R. M. *J. Chem. Phys.* **2001**, *114*, 4579–4590.
(45) Merchant, K. A.; Noid, W. G.; Thompson, D. E.; Akiyama, R.; Loring, R. F.; Fayer, M. D. *J. Phys. Chem. B* **2003**, *107*, 4–7.
(46) Khalil, M.; Demirdoven, N.; Tokmakoff, A. *Phys. Rev. Lett.* **2003**, *90*, 047401(047404).

(47) Mukamel, S. *Principles of Nonlinear Optical Spectroscopy*; Oxford University Press: New York, 1995.
(48) Berg, M.; Walsh, C. A.; Narasimhan, L. R.; Littau, K. A.; Fayer, M. D. *J. Chem. Phys.* **1988**, *88*, 1564–1587.
(49) Joo, T. H.; Jia, Y. W.; Yu, J. Y.; Lang, M. J.; Fleming, G. R. *J. Chem. Phys.* **1996**, *104*, 6089–6108.
(50) de Boeij, W. P.; Pshenichnikov, M. S.; Wiersma, D. A. *Chem. Phys.* **1998**, *233*, 287–309.
(51) Hamm, P.; Lim, M.; Hochstrasser, R. M. *J. Phys. Chem. B* **1998**, *102*, 6123–6138.
(52) Hamm, P.; Hochstrasser, R. M. In *Ultrafast Infrared and Raman Spectroscopy*; Fayer, M. D., Ed.; Marcel Dekker: New York, 2001; Vol. 26, pp 273–347.
(53) Asbury, J. B.; Steinel, T.; Stromberg, C.; Gaffney, K. J.; Piletic, I. R.; Goun, A.; Fayer, M. D. *Chem. Phys. Lett.* **2003**, *374*, 362–371.
(54) Merchant, K. A.; Xu, Q.-H.; Thompson, D. E.; Fayer, M. D. *J. Phys. Chem. A* **2002**, *106*, 8839–8849.

and $\vec{k}_3$. The parent IR laser pulse was compressed by carefully balancing the amount of CaF_2 and Ge the beam passed through prior to being split. The beams were crossed and focused (spot size $\sim$150 μm) in the sample with a 6 in. focal length 90° gold coated off-axis parabolic reflector (Janos Technologies), and the vibrational echo signal was detected in the $\vec{k}_s = \vec{k}_2 + \vec{k}_3 - \vec{k}_1$ phase-matched direction. Another off-axis parabolic reflector was used to collimate the emitted vibrational echo beam, and a HeNe laser beam was coaligned onto the vibrational echo beam to aid in further alignments and routing of the vibrational echo beam.

The vibrational echo signal was dispersed in a SPEX 0.5 m monochromator (210 lines/mm) and detected with a 32-element HgCdTe array detector (Infrared Associates/Infrared Systems Development) with a resolution of $\sim$1.2 cm^{-1} per element. The data were recorded in blocks of 32 wavelengths by scanning τ at a fixed value of T_w, and then stepping the value of T_w. Two blocks of 32 wavelengths were necessary to span the spectral range of the data. One pixel from adjacent frequency blocks was overlapped during the acquisition to ensure the consistency of the data. The frequency spectrum of the array was calibrated by comparing the spectrum of atmospheric water absorptions to data from the HITRAN96 molecular database. The response of each pixel was normalized to account for variations in the pixel signal output. The second pulse in the echo experiment was chopped at 500 Hz to permit subtraction of scattered light and to correct for drifts in the array detector baseline.

Recombinant sperm whale Mb in pH 8.0 Tris-chloride buffer ($\sim$5 mg/mL) was obtained from Sigma. The solution was concentrated and flushed several times with 0.1 M phosphate pH 7 buffer (VWR Scientific) using a Microcon centrifugal filter device and was then concentrated to 15 mM. The sample was filtered with a 0.45 μm cellulose acetate filter (Osmonics Laboratories) prior to the final concentration step. The Mb sample was reduced with excess sodium dithionite under a CO atmosphere. The sample was stirred under a CO atmosphere for an hour after reduction to ensure that all Mb had a CO ligand before being loaded into a 50 μm gastight custom IR sample cell with 3 mm thick CaF_2 windows. Detailed power and concentration studies were performed, and the vibrational echo decay was found to be independent of both the laser intensity and the concentration used in the experiments.

B. Molecular Dynamics Computations. Molecular dynamics simulations at constant energy were performed using the MOIL software package[55] on one molecule of sperm whale MbCO, 2627 rigid water molecules, one SO_4^{2-} present in the crystal structure, and two Na^+ added to ensure electroneutrality,[22] at $T \approx 300$ K. Simulations were run for both the N_ϵ–H and the N_δ–H tautomers of His64. Calculations for N_δ–H began with an equilibrated structure employed by Meller and Elber.[22] Treatment of the N_ϵ–H structure required modification of the original MOIL force field.[56] Dynamics reported here for the N_ϵ–H structure are determined from 39 production trajectories totaling 12.7

ns in duration. Dynamics for the N_δ–H tautomer of His64 are calculated from 30 production trajectories totaling 12.0 ns in duration.

The CO transition frequency is highly sensitive to the electric field in the heme pocket.[15,25,26,30,31] Therefore, the pure dephasing dynamics in both tautomers of His64 were modeled as arising from time-dependent Stark effect perturbations of the transition frequency induced by the fluctuating electric field at the CO caused by protein and solvent motions.[40,45,57] The frequency fluctuations take the form

$$\delta\omega(t) = \lambda[\vec{u}(t) \cdot \vec{E}(t) - \langle\vec{u} \cdot \vec{E}\rangle] \tag{1}$$

where $\vec{u}(t)$ is a unit vector along the electric dipole moment direction of CO, $\vec{E}(t)$ is the instantaneous electric field at the midpoint of the CO bond, and λ is the Stark coupling constant, discussed further below. We calculated $\vec{E}(t)$ at CO using Coulomb's law in a vacuum, the atomic partial charges in the MOIL force field, and the atomic positions generated by the MD. As described below, the spectrally resolved three-pulse vibrational echo signal and the linear absorption spectrum were calculated from the frequency–frequency correlation function (FFCF)

$$C(t) = \langle\delta\omega(t)\delta\omega(0)\rangle \tag{2}$$

The angular brackets in eqs 1 and 2 denote an equilibrium ensemble average.

C. Vibrational Echo Computations. We calculate the vibrational echo signal from third-order perturbation theory in the radiation–matter interaction. The third-order nonlinear polarization that generates the vibrational echo signal is a convolution of the material response function $\sum_i R_i(t_3,t_2,t_1)$ with the applied optical fields $E_j(t)$:

$$P^{(3)}(\tau,T_w,t) \propto \int_0^\infty dt_3 \int_0^\infty dt_2 \int_0^\infty dt_1 \left(\sum_i R_i(t_3,t_2,t_1)\right) \times$$
$$E_3(t - t_3)E_2(T_w + t - t_3 - t_2)E_1^*(\tau + T_w + t - t_3 - t_2 - t_1) \tag{3}$$

The index i in eq 3 labels contributions to the response function within third-order perturbation theory,[47] which are described in Appendix A. The material system is modeled as a quantum-mechanical three-level system, coupled to a classical solvent that is treated within a second-order cumulant approximation. Details of the calculation of the nonlinear polarization $P^{(3)}(\tau,T_w,t)$ are given in Appendix A.

The third-order nonlinear response function for MbCO is calculated by treating each of the conformational substates as a distinct species that retains its identity for the duration of the echo measurement. Each nonlinear response function is convolved with the temporal profiles of the excitation fields[58] as shown in eq 3 to yield the nonlinear polarization for that substate, and the total nonlinear polarization is computed from a concentration weighted sum of polarizations associated with each conformer

$$P_{total}^{(3)}(\tau,T_w,t) = \sum_\alpha c_\alpha P_\alpha^{(3)}(\tau,T_w,t) \tag{4}$$

where c_α is the concentration, $P_\alpha^{(3)}$ is the nonlinear polarization, and α is the index labeling the conformational substates. The transition dipole moment and anharmonicity were assumed to be the same for each substate. Determination of the concentrations c_α is described in Section IVB and in Appendix B. The total nonlinear polarization was Fourier transformed with respect to the time variable t, and the power spectrum was computed as a function of the delay times τ and T_w and was evaluated at the detection frequencies for comparison with vibrational echo data. The linear absorption spectrum of the CO vibration is calculated within the same approximations used to calculate the nonlinear polarization by summing the concentration weighted contri-

(55) Elber, R.; Roitberg, A.; Simmerling, C.; Goldstein, R.; Li, H.; Verkhivker, G.; Keasar, C.; Zhang, J.; Ulitsky, A. *Comput. Phys. Commun.* **1994**, *91*, 159−189.
(56) MOIL (ref 55) defines force fields for two histidine structures: a neutral structure with the histidine protonated at N_δ and a singly charged structure with both N_δ and N_ϵ protonated. Simulating the N_ϵ–H tautomer with MOIL required defining the associated force field parameters. MOIL derives atomic partial charges and Lennard-Jones parameters from the OPLS force field (Jorgensen; Tirado-Rives. *J. Am. Chem. Soc.* **1988**, *110*, 1666−1671), it obtains parameters describing bond lengths, bond angles, and torsional angles from the AMBER force field (Brooks; et al. *J. Comput. Chem.* **1983**, *4*, 187−217), and it obtains parameters describing improper torsional interactions from the CHARMM force field (Weiner; et al. *J. Am. Chem. Soc.* **1984**, *106*, 765−784). We obtained force field parameters for the N_ϵ–H tautomer from the OPLS and AMBER force fields, for consistency with MOIL. An additional approximation was introduced in the case of the improper torsion interaction of C_β, C_γ, C_δ, and the unprotonated N_δ. This interaction was approximated using the parameters for the improper torsion interaction of C_β, C_γ, C_δ, and the protonated N_δ of the doubly protonated histidine residue defined in MOIL. A complete listing of our modifications of MOIL force field files, which would allow our MD simulations to be reproduced, is given in the Supporting Information.

(57) Williams, R. B.; Loring, R. F.; Fayer, M. D. *J. Phys. Chem. B* **2001**, *105*, 4068−4071.
(58) The laser pulses in the calculation are transform-limited Gaussian line shape pulses with a full-width-half-maximum (fwhm) of 125 fs at the intensity level.

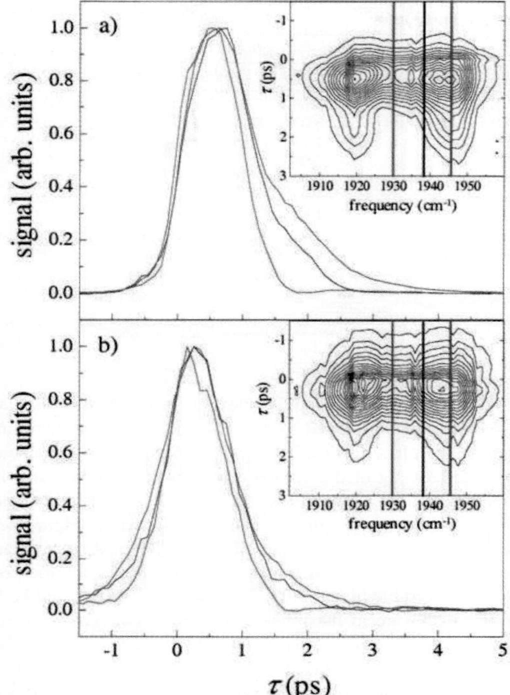

Figure 3. Spectrally resolved stimulated vibrational echo signals for (a) $T_w = 0$ ps and (b) $T_w = 8$ ps. Slices through the vibrational echo decay data are shown in the main plots at frequencies 1946 cm^{-1} (blue), 1938 cm^{-1} (black), and 1930 cm^{-1} (red). The full spectral response for the two T_w values is shown in the inset, along with the frequencies of each slice. The maximum amplitude of each slice has been normalized for comparison. The vibrational echo decays are highly frequency dependent for early T_w, but become independent of frequency for long T_w. The trends shown in the data at these T_w are representative of the trends seen in the full spectrally resolved vibrational echo data at other T_w.

butions of each substate to the total linear polarization. The details of the calculation of the linear spectrum are presented in Appendix A.

IV. Results and Discussion

A. Multidimensional Vibrational Echo Data. Multidimensional stimulated vibrational echo data are presented in Figure 3. Slices through the data at three frequencies and $T_w = 0$ (Figure 3a) and 8 ps (Figure 3b) are shown in the main plots, while the full spectral response at these T_w values is shown in the insets as τ-frequency contour plots. Contour plots of the spectrally resolved stimulated vibrational echo signal on MbCO at other T_w values can be found in the Supporting Information. It is clear from the data in Figure 3a that the measured vibrational echo decay is frequency dependent. The vibrational echo data around 1945 cm^{-1} (blue curves) reflect contributions from the 0–1 transition of the A$_1$ substate, while the vibrational echo data around 1932 cm^{-1} (red curves) reflect contributions from the 0–1 transition of the A$_3$ substate. The black curves show the echo decay for an intermediate frequency value of 1938 cm^{-1}. The vibrational echo decay for the 1–2 transition of the A$_1$ substate appears at ~1920 cm^{-1} (see insets) red-shifted from the 0–1 transition by the vibrational anharmonicity of the

CO stretching frequency, 25.3 cm^{-1}. The vibrational echo decay for the 1–2 transition of the A$_3$ line appears to the red of the A$_1$ 1–2 transition at ~1910 cm^{-1}. Because of the bandwidth and center frequency of the excitation pulses, the A$_3$ 1–2 transition is a small contribution to the data.

There are a number of significant qualitative features of the dependence of the spectrally resolved vibrational echo decays on τ, T_w, and frequency. First, the vibrational echo decays faster around 1932 cm^{-1} (red curves) than around 1945 cm^{-1} (blue curves). Second, comparisons between Figure 3a and Figure 3b show that the difference between the vibrational echo decay dephasing dynamics at different frequencies decreases as T_w is increased. The vibrational echo decays become more uniform as a function of frequency. Next, the vibrational echo signal decays faster as T_w is increased, and the peak of the vibrational echo signal shifts to earlier time as T_w is increased. These effects are the result of the inclusion of more dephasing processes in the T_w period as T_w is increased to encompass slower and slower dynamics. More dynamical broadening occurs as the system is allowed to evolve for longer and longer periods of time. Additional line broadening (frequency domain) results in a faster vibrational echo decay (time domain).

Our calculations of spectrally resolved stimulated vibrational echoes using the formalism of Appendix A reveal that these qualitative features are significantly influenced by the dynamics of the several overlapping vibrational transitions. However, the detailed shapes of the FFCFs used in the calculations control the quantitative agreement between the experimental data and calculated vibrational echo signal (see below). Several approaches have been previously proposed and applied to extract an FFCF directly from echo data on electronic and vibrational transitions.[37,49,59] Implementation of these strategies is problematic for MbCO, which is characterized by several overlapping transitions, none of which is massively inhomogeneously broadened. A calculation of relevant FFCFs from a molecular model, as described in the following section, is required to fully interpret spectrally resolved echo data for such a system.

B. Comparison of Measured and Simulated Dephasing Dynamics from MD. MD calculations of the instantaneous Stark shift $\delta\omega(t)$ in eq 1 for the N$_\epsilon$–H structure show pronounced two-state behavior with a state R$_\epsilon$ characterized by a redder CO vibrational frequency and a state B$_\epsilon$ characterized by a bluer CO vibrational frequency. The Stark shifts of these two states fluctuate about mean values that differ by 9.3λ in units of cm^{-1}, where λ is the value of the Stark coupling constant expressed in units of cm^{-1}/(MV/cm). In 13 trajectories of the 39 production trajectories of the N$_\epsilon$–H tautomer, the system began and ended in the state R$_\epsilon$ without making a transition to the B$_\epsilon$ state, and in 3 trajectories, MbCO began and ended in B$_\epsilon$ without making a transition to R$_\epsilon$. In total, the R$_\epsilon$ state was observed 47 times for 8.96 ns of simulation. The B$_\epsilon$ state was observed 30 times for 3.24 ns of simulation. The time duration of a transition between the two conformers was ~500 fs. From this limited number of jumps, the average substate lifetimes of R$_\epsilon$ and B$_\epsilon$ are on the order of 100–200 ps. The limited time scale of the simulation prevents us from assigning quantitatively correct lifetimes or statistical weights to the R$_\epsilon$ and B$_\epsilon$ states. However, the lifetimes of the states are not required for the

(59) Piryatinski, A.; Skinner, J. L. *J. Phys. Chem. B* **2002**, *106*, 8055–8063.

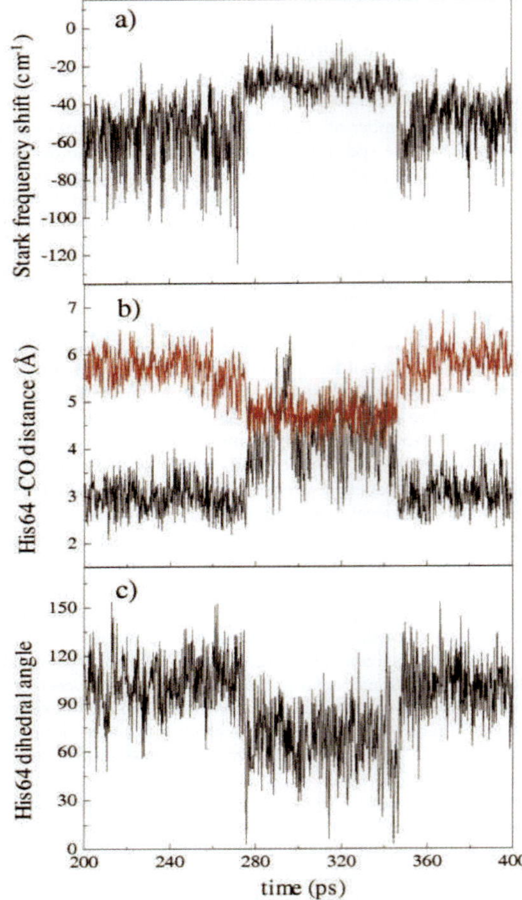

Figure 4. An example of a MD trajectory showing the transition between the R_ϵ state and the B_ϵ state. A transition occurs from R_ϵ to B_ϵ at $\sim$280 ps, and then back to the R_ϵ state at $\sim$345 ps. (a) The average value of the Stark frequency shift changes abruptly during the transition. (b) The distance between N_ϵ−H (black) and N_δ of His64 (red) during the trajectory. In the R_ϵ state, N_ϵ−H is much closer to the ligand than N_δ. In the B_ϵ state, the distances of N_ϵ−H and N_δ of His64 to the CO are about the same. (c) The dihedral angle of the C_β−C_δ bond changes during the transition from R_ϵ to B_ϵ at $\sim$280 ps and then back at $\sim$345 ps.

Table 1. Coordinates for Heme Pocket Atoms for the R_ϵ and B_ϵ Configurations

	distance (Å)		variance (Å)^2	
	R_ϵ state	B_ϵ state	R_ϵ state	B_ϵ state
$R(N_\delta-O)$ (His64)	5.69	4.46	0.205	0.142
$R(HN_\epsilon-O)$ (His64)	3.07	4.63	0.300	0.618
$R(C-O)$	1.13	1.13	0.000	0.000
$R(Fe-C)$	1.90	1.90	0.001	0.001
$R(Fe-N_a)$ (Heme)	1.97	1.97	0.001	0.001
$R(Fe-N_b)$	1.97	1.97	0.001	0.001
$R(Fe-N_c)$	1.97	1.97	0.001	0.001
$R(Fe-N_d)$	1.97	1.97	0.001	0.001
$R(Fe-N_\epsilon)$ (His93)	2.21	2.21	0.004	0.004

	angle (deg)	angle (deg)	R_ϵ state ((deg)^2)	B_ϵ state ((deg)^2)
$(Fe-C-O)$	173.38	173.33	11.97	12.20

4.8 Å, respectively. The angle relevant to this rotation is that between the plane formed by C_α, C_β, and C_γ of the histidine arm, and the plane formed by C_γ, C_δ, and N_δ of the imidazole ring. The labeling of the carbon atoms of the histidine arm is defined in Figure 2a. The time dependence of this dihedral angle for the trajectory of Figure 4a and 4b is shown in Figure 4c. This angle rotates from a mean value of 108° in R_ϵ to a mean value of 68° in B_ϵ. In the B_ϵ state, the N_ϵ−H bond is not oriented directly toward the ligand but rather toward the heme, and both N_ϵ−H and N_δ are in the interior of the protein. The state R_ϵ has a structure qualitatively similar to Figure 2b, while B_ϵ qualitatively resembles Figure 2f. Table 1 summarizes some of the geometric differences between the B_ϵ and R_ϵ states. Representative snapshots of the R_ϵ and B_ϵ configurations are shown in Figure 5a and 5b, respectively.

Frequency−frequency correlation functions were calculated for the two substates from eqs 1 and 2 and are shown in Figure 6a by the red (R_ϵ) and blue (B_ϵ) curves. The FFCF for R_ϵ has a larger initial amplitude than that for B_ϵ, indicating a larger root-mean-squared Stark shift fluctuation for the former state, which may also be seen qualitatively from the frequency fluctuations in Figure 4a. The FFCF for R_ϵ has an initial rapid decay on the time scale of $\sim$100 fs, followed by slower decay on the picosecond time scale. This FFCF shows pronounced ringing during the first few picoseconds. The B_ϵ state FFCF, while of smaller initial amplitude, shows similar rapid and slower decays, but does not display the fine structure of the FFCF for R_ϵ. Neither FFCF is well described as a sum of two exponential decays.

The FFCFs for the R_ϵ and B_ϵ states were used to compute spectrally resolved vibrational echo decays for particular values of T_w. The IR line shape of MbCO shows two dominant features: the redder A_3 peak at $\sim$1934 cm^{-1} and the bluer A_1 peak at 1945 cm^{-1}. The R_ϵ state FFCF was used to model the dynamics of the A_3 substate, and the B_ϵ state FFCF was used to model the dynamics of the A_1 substate. We assumed that the simulation did not show the A_0 substate, because His64 never adopted a swung out configuration seen in low pH crystal structures.[32] Also, the inverse rate constant for interconversion between the A_1 and A_3 substates has been estimated to be $\sim$1 ns, while the inverse rate constant for interchanging between the A_1 and A_0 substates is $\sim$1 μs.[10,13] Transitions between the R_ϵ and B_ϵ states occurred numerous times during the 12.8 ns of

analysis presented below. A representative trajectory of $\delta\omega(t)$ of duration 200 ps showing both states is plotted in Figure 4a.

The change in the CO vibrational frequency shown in Figure 4a is associated with the motion of the distal histidine, His64, between two well-defined configurations. Figure 4b shows the fluctuating distance between the N_ϵ−H proton and the midpoint of the CO bond (black) and the fluctuating distance between N_δ and the CO (red) for the same MD trajectory depicted in Figure 4a. In state R_ϵ, the N_ϵ−H bond points toward the CO ligand with a mean distance from H to CO midpoint of 3.2 Å. The mean distance from N_δ to CO is 5.8 Å. In the B_ϵ state, His64 has rotated about the C_β−C_γ bond relative to the R_ϵ state, such that both the N_ϵ proton and the N_δ are approximately the same distance from the ligand, with mean distances of 4.6 and

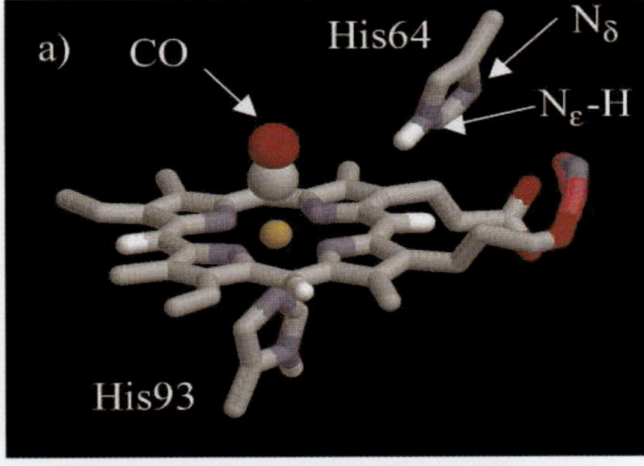

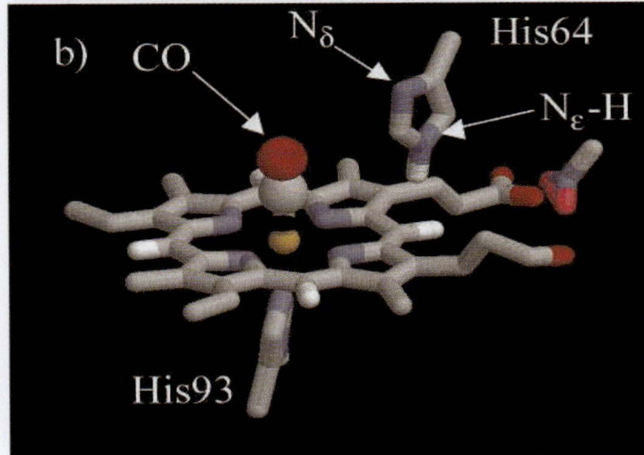

Figure 5. Snapshots of the heme pocket[45] during the MD trajectory shown in Figure 4. (a) The R_ϵ configuration has N_ϵ–H oriented toward the CO, and the N_δ oriented toward the solvent. (b) In the B_ϵ configuration, the imidazole ring of His64 has undergone a rotation around the C_β–C_δ bond so that N_ϵ–H is oriented away from the CO ligand and N_δ is inside the heme pocket.

MD simulations, consistent with the rate constant between A_1 and A_3.

Calculation of the spectrally resolved stimulated vibrational echo decay curves from the formalism of section III requires knowledge of the following quantities: the peak frequencies of the substates, the concentrations of the substates (c_α in eq 4), the vibrational lifetimes of each substate, and the Stark coupling constant λ in eq 1. We include all three A substate peaks to reproduce the features seen in the data. The approximate inclusion of the A_0 state, not observed in the simulation, is described in Appendix B. The line center for the A_1 substate was measured by taking a vibrational echo spectrum at $\tau = 3.0$ ps, $T_w = 0.25$ ps, shown in Figure 7. At this value of τ, vibrational echo contributions from the A_3 substate are negligible, leaving only the 0–1 and 1–2 transition peaks of the A_1 substate. The vibrational echo spectrum in Figure 7 is an example of vibrational echo peak suppression, where differences in dephasing times are used as a means of enhancing certain peaks relative to other peaks.[60,61] Further details used

(60) Rector, K. D.; Zimdars, D. A.; Fayer, M. D. *J. Chem. Phys.* **1998**, *109*, 5455–5465.
(61) Rector, K. D.; Fayer, M. D.; Engholm, J. R.; Crosson, E.; Smith, T. I.; Schwettmann, H. A. *Chem. Phys. Lett.* **1999**, *305*, 51–56.

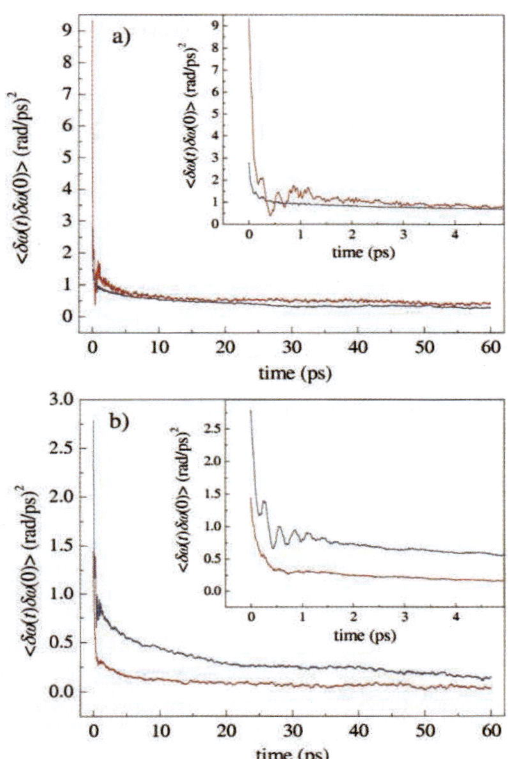

Figure 6. Frequency–frequency correlation functions (FFCFs) for the two tautomers of MbCO derived from MD simulations used to calculate vibrational echo signals. (a) The N_ϵ–H tautomer of His64. The R_ϵ conformer (red) and B_ϵ conformer (blue) both have a fast initial decay over the first few picoseconds, and then continue to decay slowly. The R_ϵ substate has a much larger initial amplitude. The value of the Stark coupling constant λ is $\lambda = 2.1$ cm^{-1}/(MV/cm). (b) The N_δ–H tautomer of His64. The B_δ conformer (blue) has a larger initial amplitude than the R_δ (red) conformer. The value of λ is $\lambda = 1.4$ cm^{-1}/(MV/cm).

in the calculation of the spectrally resolved vibrational echo data for the A substates are presented in Appendix B. However, it is important to note that the substate concentrations, c_α in eq 4, and center line frequencies are determined by fitting the linear absorption spectrum to Voigt profiles. Therefore, the determination of the center frequencies and concentrations does not depend on any of the MD simulations used to analyze the stimulated vibrational echo data. The same line center frequencies and substate concentrations are used in the comparisons of the MD simulations for the two tautomers. All parameter values employed in the linear spectrum and vibrational echo calculations are summarized in Table 2.

Measured vibrational echo decays for four frequencies and three values of T_w are shown in Figure 8 in black. The red curves show vibrational echoes calculated using FFCFs for the N_ϵ–H tautomer. The FFCFs calculated from the MD simulations can be used to fit the experimental vibrational echo data for all measured values of T_w.[62] In the fits, there is only one adjustable parameter, the Stark coupling constant, λ. The best fit value for the Stark coupling constant was found to be $\lambda =$

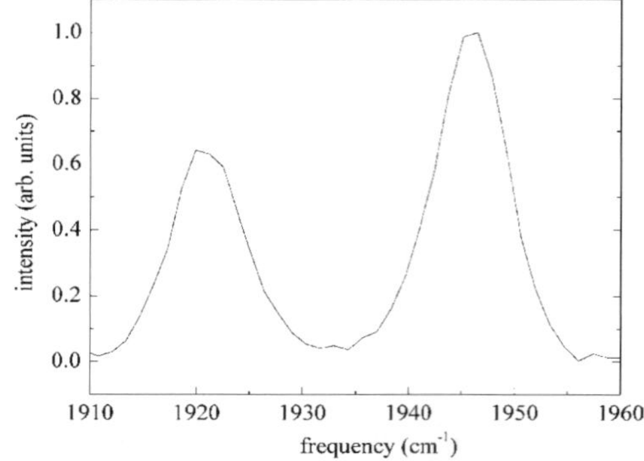

Figure 7. Vibrational echo spectrum at $\tau = 2.1$ ps, $T_w = 0$ ps. The 0–1 and 1–2 transitions of the A_1 substate are clearly seen at 1945 and 1920 cm^{-1}. Contributions to the echo spectrum from the A_3 substate have been eliminated using T_2 selectivity. The splitting between the two peaks gives an anharmonicity of $\Delta = 25.3$ cm^{-1}.

Table 2. Summary of Parameters Used in Vibrational Echo Calculations[a]

MbCO substate	center frequency (cm^{-1})	concentration (relative)	lifetime (ps)	FFCF (N_ϵ–H tautomer)	FFCF (N_δ–H tautomer)
A_0	1965	0.125	16.5	$0.64 \times B_\epsilon$	$0.64 \times B_\delta$
A_1	1945	1.4	16.5	B_ϵ	B_δ
A_3	1934	1	14.7	R_ϵ	R_δ

[a] The transition dipole moments for all A substates are assumed to be the same. An anharmonicity of $\Delta = 25.4$ cm^{-1} is assumed for all transitions. The vibrational lifetimes for the A_1 and A_3 substates have been measured previously,[40] and the lifetime of the A_0 substate is assumed to be the same as the lifetime of the A_1 substate. The laser pulse is assumed to be a transform-limited Gaussian pulse centered at 1940 cm^{-1} with a fwhm = 125 fs. The FFCFs for the N_ϵ–H tautomer were used in the calculations shown in Figures 8 and 9, and the FFCFs for the N_δ–H tautomer were used in the calculations shown in Figures 10 and 11. See Appendix B for further details.

2.1 cm^{-1}/(MV/cm).[63] The value for λ has been determined experimentally by Boxer and co-workers using vibrational Stark spectroscopy on MbCO and a number of other heme-CO systems.[64,65] They found the Stark coupling constant to be $\lambda = 1.8$–2.2 cm^{-1}/(MV/cm).[66] Our value of λ is in excellent agreement with these results. Thus, the only adjustable parameter in the fit of the experimental data using the FFCF determined by the MD simulations is, within experimental error, the same as the measured value. Spectrally resolved stimulated vibrational echo decays are very sensitive to protein structure and dynamics. The agreement between experiment and theory provides strong support for the assignment of the A_1 and A_3 structures.

Figure 9 displays the measured linear absorption spectrum (solid curve) and the calculated linear spectrum (dashed curve) employing the same parameter values used for the calculated

vibrational echo decays in Figure 8, and listed in Table 2. The center frequencies and relative substate concentrations derived from the model independent fits to the linear spectrum are used and are not adjustable parameters. The simulated linear spectrum is slightly wider than the measured line shape, and the amplitude of the A_0 peak is too large. Overall, the agreement with the linear spectrum is reasonably good. However, the MD results overestimate the splitting between the A_1 and A_3 substates by ~8 cm^{-1}, predicting a splitting of 19 cm^{-1} between the substates. While the stimulated vibrational echo is only sensitive to dynamics on short time scales, the linear absorption spectrum is also sensitive to dynamics on all possible time scales. Very long time scale dynamics may not be accurately modeled using only ~13 ns of MD simulation. Nonetheless, the linear spectrum agrees reasonably well and shows, along with the dynamical calculations in Figure 8, that the simulated FFCFs for the N_ϵ–H tautomer are consistent with the measured dynamics of the A_1 and A_3 substates of MbCO.

Trajectories of the fluctuating CO Stark shift for the N_δ–H structure, analogous to the calculation for N_ϵ–H shown in Figure 4a, also show pronounced two-state behavior, with a blue state B_δ (qualitatively similar to Figure 2c) and a red state R_δ (qualitatively similar to Figure 2e) having vibrational frequencies separated by 18.1 λ in units of cm^{-1}.[67] The state B_δ has a structure similar to R_ϵ. In both structures, N_ϵ points in the heme pocket toward the CO ligand. However, whereas in R_ϵ the proton is directed toward the ligand, in B_δ the proton is on the surface of the protein and directed into the solvent. The mean dihedral angle defined for Figure 4b has the same value of 108° for B_δ as for R_ϵ. In B_δ, the distances from N_ϵ and the N_δ–H proton to the midpoint of the CO bond are 5.6 and 9.5 Å, respectively. The red state R_δ is structurally similar to B_ϵ, but with the imidazole proton directed into the heme pocket, and with a mean dihedral angle of 80°, as compared to 68° for B_ϵ. The mean distances from N_ϵ and the N_δ–H proton to the midpoint of the CO bond for the R_δ structure are 7.2 and 6.5 Å, respectively.

The FFCFs associated with these states are shown in Figure 6b. The FFCF for B_δ resembles that for R_ϵ in having a relatively large initial value and in displaying ringing on the picosecond time scale. Similarly, the FFCF for R_δ resembles that for B_ϵ in having a relatively small initial value and lacking fine structure. These FFCFs were used to compute the vibrational echo decays and absorption line shape with the same procedure, line centers, and concentrations used for the N_ϵ–H calculation. Details of the fitting procedure are in Appendix B.

The computed vibrational echo signal for the N_δ–H tautomer is shown in Figure 10 for the same frequencies and values of T_w as shown in Figure 8. The calculated linear spectrum (dashed curve) using the FFCFs for the N_δ–H tautomer is compared to the experimental linear spectrum for sperm whale MbCO (solid

(62) At large values of T_w, an additional small signal around $\tau = 0$ is observed. This signal is due to a nonresonant contribution from the solvent background that results from the breaking of hydrogen bonds in the solvent. Correcting the vibrational echo data to remove this signal contribution gives agreement between the calculated and measured signals at large values of T_w comparable to the agreement shown in Figure 8.

(63) The best fit value of λ for MbCO at 300 K was previously reported as $\lambda = 1.9$ cm^{-1}/(MV/cm) (ref 45). The vibrational echo decay at the concentrations used in those experiments was concentration-dependent and became faster as the concentration was reduced. We attribute the slower vibrational echo decays in ref 45 to the higher viscosity of the more concentrated protein sample.

(64) Park, E. S.; Boxer, S. G. *J. Phys. Chem. B* **2002**, *106*, 5800–5806.

(65) Park, E. S.; Boxer, S. G. *J. Phys. Chem. B* **2002**, *106*, 8910–8910.

(66) The range in the value of λ results from uncertainties in the local electric field correction factor for the heme pocket associated with applying an external electric field to a dielectric medium. See refs 64–65.

(67) Our first reported echo calculations for MbCO (refs 40 and 57) were based on simulations of the N_δ–H structure. In those simulations, we adopted a criterion suggested by Schulze and Evanseck (ref 21) and associated with the A_0 state all structures in which the distance between the midpoint of the CO bond and the proton of N_δ–H exceeds 6.5 Å. Such structures were not investigated in those calculations. We have therefore performed new simulations of MbCO with the N_δ–H tautomer of His64 to examine the FFCFs associated with all spectroscopically distinct substates generated in the simulation. In addition, we have corrected an error in our previous calculations for the N_δ–H tautomer of the electric field at the CO generated by the solvent.

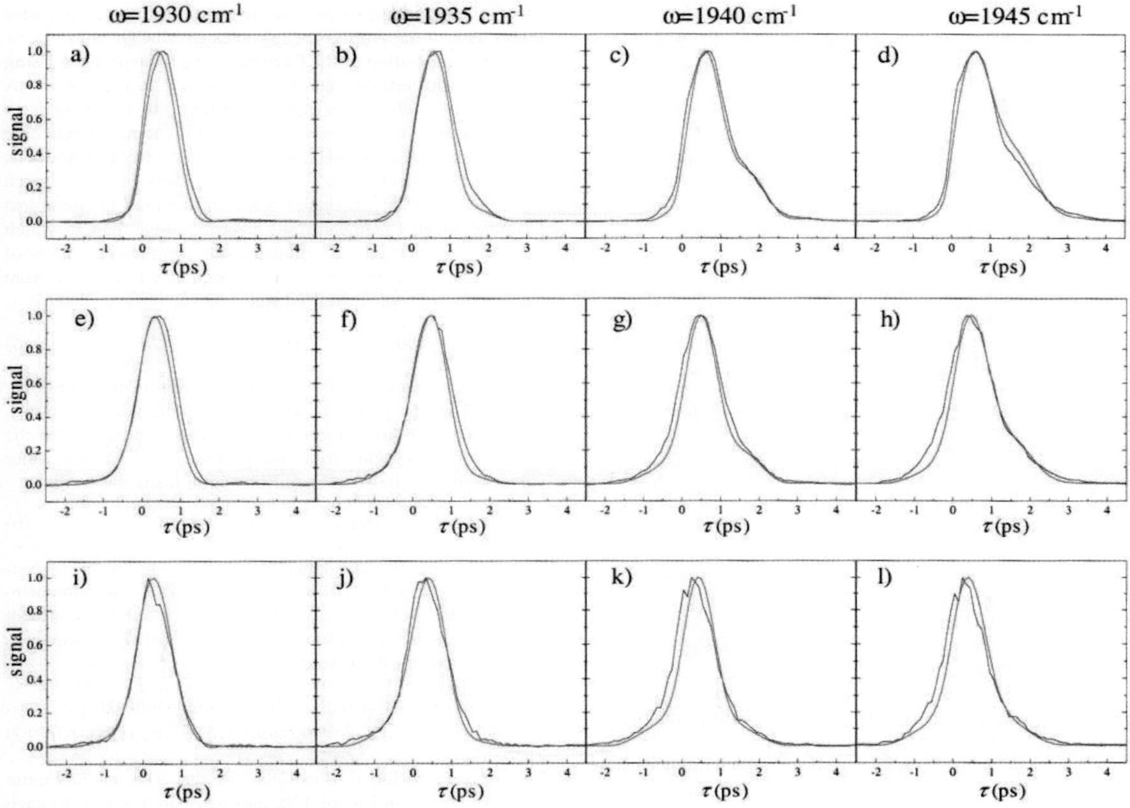

Figure 8. Comparison of the measured vibrational echo data (black) with the calculated vibrational echo data (red) for the N_ϵ–H tautomer of His64. The vibrational echo signals were calculated using the FFCFs shown in Figure 6a. The parameters used in the vibrational echo calculation are listed in Table 2. Spectrally resolved vibrational echo decays are shown for the frequencies 1930, 1935, 1940, and 1945 cm⁻¹, and $T_w = 0$ ps (a–d), 2 ps (e–h), and 8 ps (i–l). Vibrational echo decays for a particular frequency as a function of T_w are in columns. The oscillations in the echo decays in panels c and d are due to anharmonic accidental degeneracy beats[38,77,81] between the 0–1 transition of the A₁ line and the 1–2 transition of the A₀ line.

Table 3. Stark Effect Dephasing Contributions from MD Simulations[a]

	R_ϵ state		B_ϵ state	
	Stark frequency shift (cm⁻¹)	FFCF ($t = 0$) ((rad/ps)²)	Stark frequency shift (cm⁻¹)	FFCF ($t = 0$) ((rad/ps)²)
all	−48.8	9.35	−29.6	2.80
protein	−51.7	8.45	−31.4	2.12
His64	−6.7	7.75	14.0	1.03
solvent	2.8	2.01	1.8	1.93

[a] The Stark effect frequency shift contributions and initial amplitudes of the FFCFs for various parts of the protein and solvent for the N_ϵ–H tautomer simulations. The differences in the average frequencies for B_ϵ and R_ϵ are almost entirely due to the electric field contribution of His64. The Stark effect frequency shift and the initial amplitudes of the FFCF for the solvent are essentially the same. The Stark frequency shifts are additive; the initial amplitudes of the FFCFs are not.

curve) in Figure 11. The parameters used in the calculations shown in Figures 10 and 11 are listed in Table 2. The agreement between the calculated and measured vibrational echo signals and linear spectra shown in Figures 10 and 11 is not very good when using the FFCFs derived from the N_δ–H tautomer of His64. There are pronounced systematic differences in both the

fits to the dynamical line shapes and the calculation of the linear spectrum. In addition, the best fit value for the Stark coupling constant was found to be $\lambda = 1.4$ cm⁻¹/(MV/cm), which lies well outside the experimental range for λ measured by Boxer and co-workers.[64,65] The calculated frequency splitting between the B_δ and R_δ state is 25 cm⁻¹, a significant overestimate of the measured splitting 11 cm⁻¹. A summary of the parameters used in the vibrational echo calculation for the N_δ–H tautomer is given in Table 2.

Comparison of Figures 8 and 10 shows that the differences between the FFCFs for the two tautomers of His64 shown in Figure 6 give rise to very different spectrally resolved stimulated vibrational echo signals. The agreement between the measured vibrational echo data and the vibrational echo signals calculated using the N_ϵ–H tautomer of His64 is quite good, while the agreement with the measured vibrational echo data using the N_δ–H tautomer of His64 is significantly worse. Figure 11 shows that the predicted absorption spectrum for the N_δ–H tautomer also agrees poorly with the measured result. In addition, the value of the Stark coupling constant λ necessary to give reasonable agreement between the calculated vibrational echo data for the N_δ–H tautomer and the measured vibrational echo

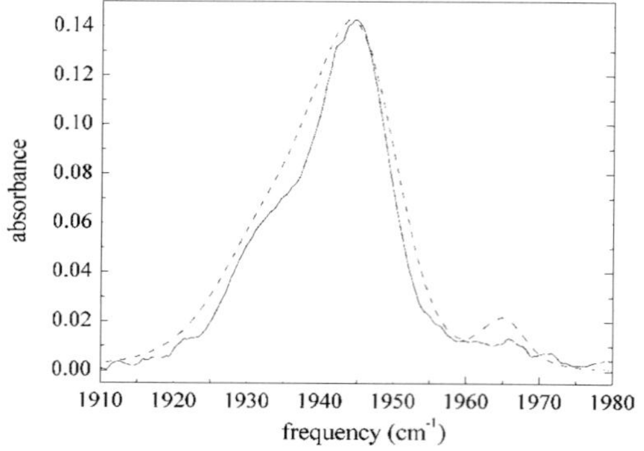

Figure 9. Comparison of the measured IR absorption spectrum of the CO stretch in MbCO (solid line) with that calculated using FFCFs for the N_ϵ–H tautomer of His64 (dashed line). The measured spectrum shows three bands at ~1965, 1945, and 1934 cm^{-1}. The calculated linear spectrum for the N_ϵ–H tautomer is slightly wider than the measured spectrum, but generally agrees reasonably well with the overall shape of the experimental spectrum.

data is unphysically small. The best fit values of λ are 2.1 cm^{-1}/(MV/cm) for the N_ϵ–H tautomer and 1.4 cm^{-1}/(MV/cm) for the N_δ–H tautomer of His64. A value of $\lambda = 1.4$ cm^{-1}/(MV/cm) implies an extremely high (approaching infinitely high) dielectric constant for the heme pocket, inconsistent with what is known about the structure of the hydrophobic heme pocket.[15,31,64,65,68] The value of $\lambda = 2.1$ cm^{-1}/(MV/cm) obtained from fitting the vibrational echo data for the N_ϵ–H tautomer lies within the range established by Park et al. using vibrational Stark effect spectroscopy.[64,65,68] It is possible to improve the quality of the agreement between the measured vibrational echo signals and those calculated using the N_δ–H tautomer by increasing the A_1/A_3 concentration ratio from 1.4 to around 4. However, a ratio of 4 is completely outside any error in the model independent fit to the linear spectrum used to determine the concentrations. In addition, using a ratio of 4 does not change the unphysical best fit value of $\lambda = 1.4$ cm^{-1}/(MV/cm) for the N_δ–H tautomer.

Our finding that the vibrational echo data are best modeled by the dynamics of MbCO with the N_ϵ–H structure is consistent with other measurements and calculations that have indicated that the N_ϵ–H tautomer is associated with the A_1 and A_3 substates. While two recent high-resolution X-ray crystal structures[28,29] disagree on the exact position and orientation of His64, both structures have N_ϵ inside the pocket, close to the ligand. Given that this is the orientation of the His64 imidazole, if the tautomerization state of His64 were N_δ–H, lone pair interactions between N_ϵ and the CO ligand would blue-shift the CO transition frequency.[20] Rovira et al.[20] calculated the expected transition frequency of the CO stretching mode in MbCO for a number of different configurations and tautomerizations and concluded that only the N_ϵ–H tautomer could reproduce the frequency shift trends seen experimentally. In addition, a recent electron nuclear double resonance (ENDOR) experiment[69] on MbNO assigned the tautomerization state of His64 to N_ϵ–H. Our assignment of the tautomerization state of His64 to N_ϵ–H

is consistent with the existing strong case for this tautomer state giving rise to the A_1 and A_3 substates of MbCO.

C. Decomposition of the Contributions to Dephasing Using the MD Simulations. Once the structural origins of the A substates of MbCO have been identified, the next step is to determine the source of the dynamical differences between these states. We can use the MD simulations to identify contributions to vibrational dephasing from different parts of the solvated MbCO system. The fluctuating Stark shift of the CO vibrational frequency in eq 1 is linear in the electric field at the CO, which may be decomposed into contributions from various parts of the system. The contribution to the frequency fluctuation from a collection of atoms denoted j is

$$\delta\omega_j(t) = \lambda[\vec{u}(t) \cdot \vec{E}_j(t) - \langle \vec{u} \cdot \vec{E}_j \rangle] \tag{5}$$

where $\vec{E}_j$ is the electric field at the midpoint of the CO bond, generated by the partial charges on the atoms in collection j. For example, we may decompose the total electric field at the CO into contributions from the protein and from the water solvent, so that the frequency fluctuation takes the form

$$\delta\omega(t) = \delta\omega_p(t) + \delta\omega_s(t) \tag{6}$$

with subscripts p and s labeling protein and solvent, respectively.[70] The FFCF, $C(t)$, is then decomposed into autocorrelation functions of frequency fluctuations induced by the protein and by the solvent and cross-correlations between frequency fluctuations from those sources:

$$\langle \delta\omega(t)\delta\omega(0) \rangle = \langle \delta\omega_p(t)\delta\omega_p(0) \rangle + \langle \delta\omega_s(t)\delta\omega_s(0) \rangle + \langle \delta\omega_p(t)\delta\omega_s(0) \rangle + \langle \delta\omega_s(t)\delta\omega_p(0) \rangle \tag{7}$$

This decomposition is illustrated for the A_3 state in Figure 12a, and for A_1 in Figure 12b, and listed in Table 3. In each panel, $C(t)$ is shown in black, $C_{pp}(t) = \langle \delta\omega_p(t)\delta\omega_p(0) \rangle$ is shown in red, $C_{ss}(t) = \langle \delta\omega_s(t)\delta\omega_s(0) \rangle$ is shown in green, and the sum of the cross-correlation functions, $C_{ps}(t) + C_{sp}(t) = \langle \delta\omega_p(t)\delta\omega_s(0) \rangle + \langle \delta\omega_s(t)\delta\omega_p(0) \rangle$ is shown in blue. The black curve is thus the sum of the red, blue, and green curves. For the A_3 substate in Figure 12a, $C_{pp}(t)$ is very similar to $C(t)$, including the fine structure on the picosecond time scale. This similarity does not arise because the solvent contribution $C_{ss}(t)$ is negligible, but because of a near cancellation between $C_{ss}(t)$ and the sum of solvent–protein cross-correlation functions, $C_{sp}(t) + C_{ps}(t)$. Each of these cross-correlation functions is negative, indicating that electric field fluctuations from protein and solvent are anticorrelated. The solvent contribution $C_{ss}(t)$ lacks fine structure and is very similar for the two states. Figure 12b shows that $C_{pp}(t)$ is also very similar to $C(t)$ for the A_1 substate, but that on the time scale of a few picoseconds, $C(t)$ decays more slowly than $C_{pp}(t)$, because the cancellation between $C_{ss}(t)$ and $C_{sp}(t) + C_{ps}(t)$ is not as complete as for the A_3 state.

We have further separated the protein contribution $C_{pp}(t)$ into contributions from the eight helices and connecting loops that compose Mb (see Figure 1). With the exception of helix E, which contains His64, the autocorrelation functions of induced frequency fluctuations for all the helices and loops are generally

(68) Park, E. S.; Andrews, S. S.; Hu, R. B.; Boxer, S. G. *J. Phys. Chem. B* **1999**, *103*, 9813–9817.

(69) Flores, M.; Wajnberg, E.; Bemski, G. *Biophys. J.* **2000**, *78*, 2107–2115.

(70) In this decomposition, the subscript p includes the heme, but the FFCF for fluctuations in the CO vibrational frequency induced only by the heme is very similar for R_ϵ and B_ϵ and contributes little to the time dependence of the total FFCF in either state.

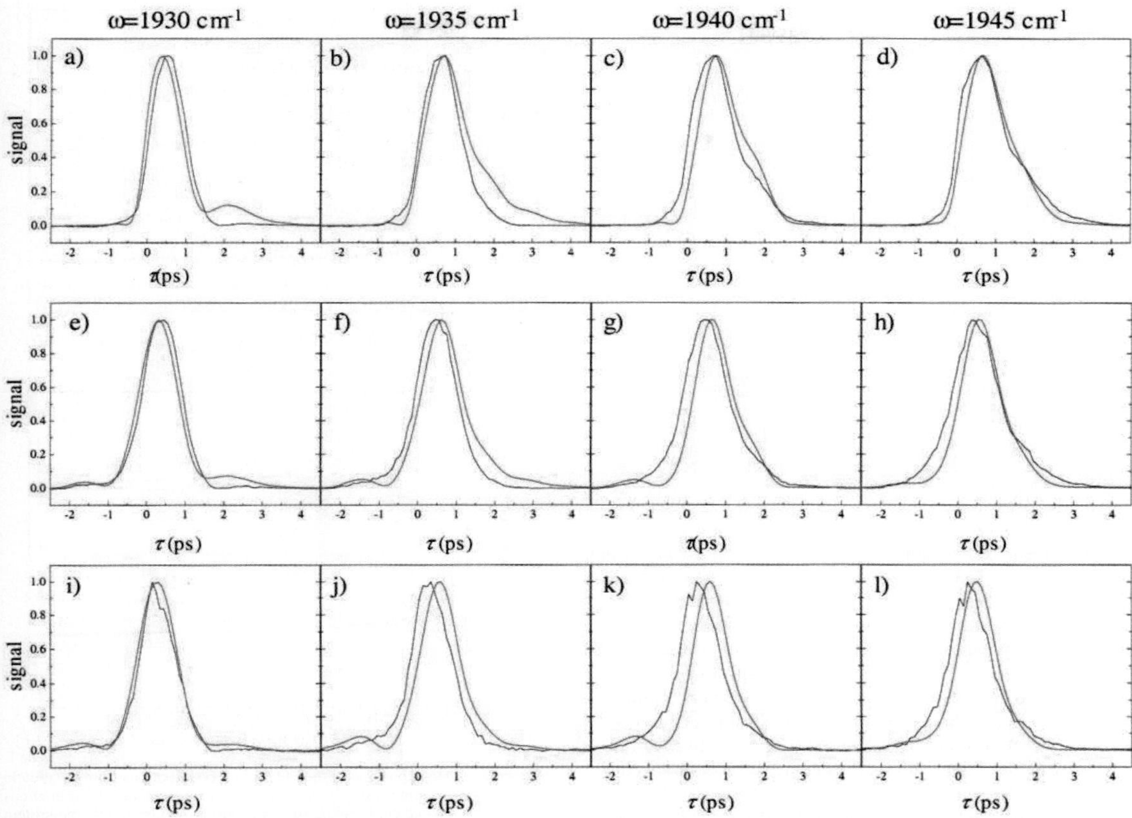

Figure 10. Comparison of the measured vibrational echo data (black) with the calculated vibrational echo data (red) for the $N_\delta-H$ tautomer of His64. Frequencies and values of T_w are the same as those shown in Figure 8. The vibrational echo signals were calculated using the FFCFs shown in Figure 6b. The parameters used in the vibrational echo calculation are listed in Table 2. The agreement is poorer for the $N_\delta-H$ tautomer of His64 than for the $N_\epsilon-H$ tautomer of His64 shown in Figure 8.

small. However, we calculate systematic differences between the A_3 state and the A_1 state in the contributions of the B, C, and D helices to the FFCF. The dephasing contribution for all three of these helices is larger in the A_1 state. The cross-correlations between different helices are negligible, indicating that the helices move independently on these time scales. Dynamics of the residue Arg45 (see Figure 1) have been proposed to play a role in the transition between states A_1 and A_3.[21] We have computed the autocorrelation function of frequency fluctuations of the CO vibration from the electric field of Arg45 and find that these correlation functions are very similar for the two states, implying that the dynamics of this residue do not contribute to dephasing differences between A_3 and A_1.

We have carried out corresponding analysis to examine the contribution to $C(t)$ from interactions between CO and the heme ring, including the Fe atom. While the heme ring contributes substantially to the average electric field at the CO, and hence to the total Stark shift of the CO frequency, heme dynamics[71] do not contribute significantly to the electric field fluctuations

at the CO on the time scales relevant to the vibrational echo. Moreover, contributions to $C(t)$ from heme dynamics are nearly identical for the A_1 and A_3 substates. We have also ascertained that the mean bond length between Fe and the proximal histidine and the fluctuations in that bond length are identical in the A_1 and A_3 substates, as can be seen from Table 1. This finding suggests that, within the Stark effect model employed here, the proximal histidine is not a major contributor to dephasing differences between the two states. This classical mechanical model does not include polarization effects and other quantum mechanical phenomena that may play a role in the laboratory. However, as shown in Section IVB, the time-dependent Stark effect model is sufficient to describe the dephasing dynamics probed by the vibrational echo, and the magnitude of the Stark coupling constant is consistent with independent experimental measurements.[64,65]

The role of His64 in dephasing the CO vibration is investigated with the decomposition

$$\delta\omega(t) = \delta\omega_h(t) + \delta\omega_r(t) \qquad (8)$$

in which the first term denotes the contribution from His64 and

(71) Sage, J. T.; Schomacker, K. T.; Champion, P. M. *J. Phys. Chem.* **1995**, *99*, 3394−3405.

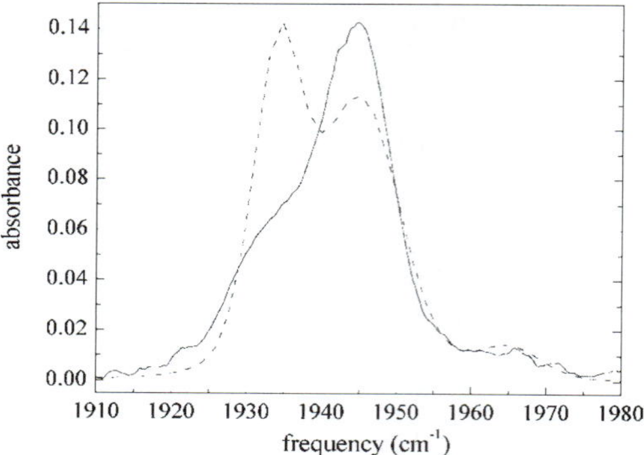

Figure 11. Comparison of the measured IR absorption spectrum (solid line) of the CO stretch in MbCO with that calculated using FFCFs for the N_δ−H tautomer of His64 (dashed line). The linear spectrum for the N_δ−H tautomer is calculated using the same line centers and concentrations used in the calculation of the linear spectrum for the N_ϵ−H tautomer of His64 shown in Figure 9. The agreement between the measured and calculated linear spectrum for the N_δ−H tautomer is quite poor.

the second term represents the contribution from the rest of the system, solvent and protein. This decomposition is shown in Figure 13a for A_3 and in Figure 13b for A_1, and listed in Table 3. In each panel, $C(t)$ is shown in black, the His64 contribution, $C_{hh}(t)$, is shown in red, the contribution from the rest of the system, $C_{rr}(t)$, is shown in green, and the sum of the cross-correlation functions, $C_{rh}(t) + C_{hr}(t)$, is shown in blue. Figure 13a shows that for the A_3 substate, $C(t)$ is essentially identical to $C_{hh}(t)$ over the range of time scales investigated. $C(t)$ follows $C_{hh}(t)$ in the initial rapid decay on the time scale of hundreds of femtoseconds, in the slower decay on the picosecond time scale, and in the ringing on the picosecond time scale. This plot shows that the autocorrelation function of frequency fluctuations induced by the rest of the system, solvent and protein, $C_{rr}(t)$, is not itself negligible, but is nearly canceled by the negative cross-correlation functions, $C_{rh}(t) + C_{hr}(t)$. The ringing characterizing $C(t)$ and $C_{hh}(t)$ for A_3 is absent for A_1, as shown in Figure 13b. For the A_1 substate, $C_{hh}(t)$ decays on the time scale of hundreds of femtoseconds to an offset that is static on the time scale of the simulations. The initial decay of $C(t)$ follows that of $C_{hh}(t)$ for A_1. The slower decay of $C(t)$ on the picosecond time scale follows the contribution from the rest of the system, $C_{rr}(t)$.

These MD simulation results indicate that interactions with His64 play a central role in the dephasing dynamics of CO in MbCO. For the A_3 state, interactions with His64 are the dominant dephasing contributor, and these interactions control CO dephasing on both the femtosecond and picosecond time scales. In the A_1 state, His64 interactions control the initial rapid dephasing on the femtosecond time scale, but dephasing on slower time scales results from the combined effect of interactions with the rest of the protein and solvent. The differences in His64 contributions to the FFCF shown in Figure 13a and 13b might lead one to the conclusion that His64 dynamics are significantly different for the A_1 and A_3 substates, that is, that the imidazole ring undergoes larger amplitude motions in A_3 than in A_1. In fact, this is *not* the case. We have examined the motions of His64 in the A_1 and A_3 substates, and our analysis indicates that the dynamics of the imidazole ring of His64 are

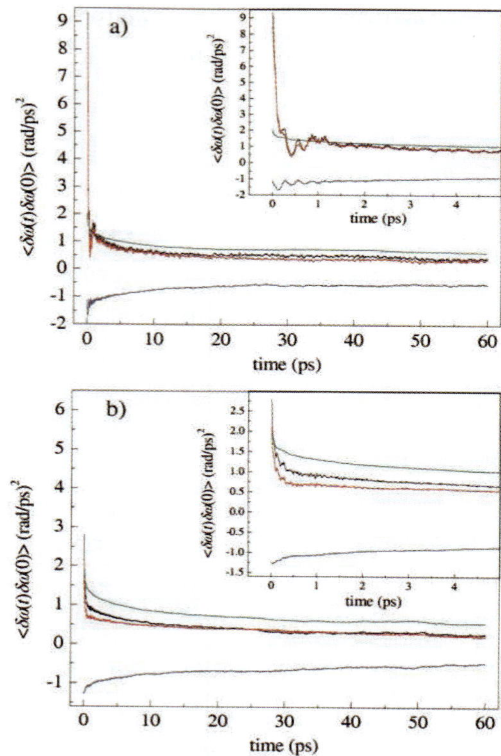

Figure 12. Decomposition of the FFCFs into contributions from the protein and from the solvent for the (a) A_3 and (b) A_1 states seen in MD simulations of MbCO. The total contribution (black), contribution from only the protein (red), contribution from only the solvent (green), and the cross-correlations between the protein and solvent (blue) are shown. For both the A_1 and the A_3 states, most of the dephasing is due to the protein. The protein and solvent dephasing contributions are anticorrelated.

very similar in the two substates. A quantitative calculation of the autocorrelation functions of the C_β−C_γ dihedral angle of His64 for the two states, not shown here, indicates that the dynamics of this angle are comparable for A_1 and A_3. In addition, the mean center-of-mass position of the imidazole ring is unchanged by transitions between states, and the mean-squared fluctuations of this position are comparable for the two states. The essential difference for His64 between the A_1 and A_3 substates is the average dihedral angle. Of course, the magnitude and direction of the electric field of His64 felt at the CO ligand will be dramatically affected by the orientation of the imidazole ring. The different dephasing autocorrelation functions of His64 therefore appear to be reflections of structural differences between the A_1 and A_3 substates, rather than dynamical differences between these states. The larger value of $C_{hh}(0)$ for A_3, as well as the presence of ringing on the picosecond time scale for A_3, reflect these structural differences.

V. Concluding Remarks

Multidimensional vibrational echo signals were measured for sperm whale MbCO, and the dephasing dynamics were found to be in agreement with dephasing dynamics calculated from

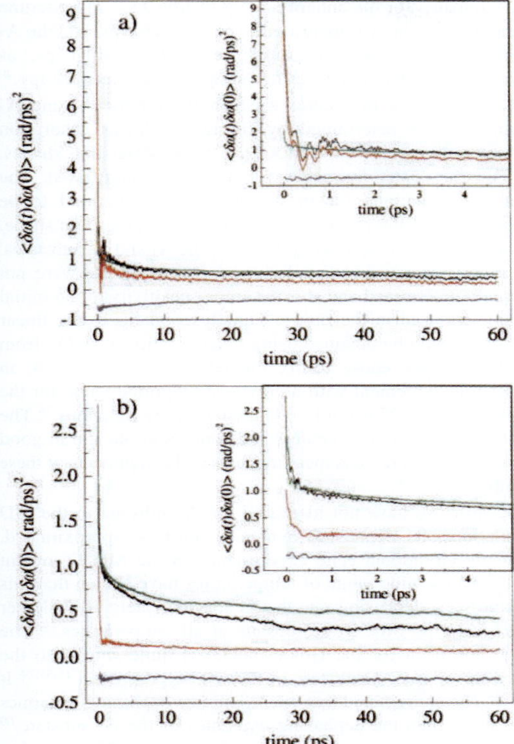

Figure 13. Decomposition of the FFCFs into contributions from His64 and from the rest of the system for the (a) A_3 and (b) A_1 states observed in MD simulations of MbCO. The total contribution (black), contribution from only His64 (red), contributions from all other atoms (green), and the cross-correlations between the His64 and the rest of the system (blue) are all shown. His64 is the dominant dephasing source for the A_3 state. The A_1 state has a substantial portion of the short time dephasing contribution from His64, but the rest of the system contributes appreciably to the dephasing at longer times. The dephasing contribution for His64 is more strongly anticorrelated with its surroundings in the A_3 state.

MD simulations for sperm whale MbCO with the N_ϵ–H tautomer of His64. The two conformational substates B_ϵ and R_ϵ observed in the MD simulations were assigned to the A_1 and A_3 substates of MbCO. Dynamics of His64 constitute a major source of dephasing for the CO ligand, and orientational changes of this residue give rise to the A_1 and A_3 substates.

Our previous vibrational echo studies on MbCO[40,45] treated horse heart MbCO. We have also recently made detailed spectrally resolved stimulated vibrational echo measurements on horse heart MbCO and find the vibrational echo data to be essentially the same as those for sperm whale MbCO,[72] despite minor differences in the peak shapes and positions previously noted in the IR spectra.[15,31] Additionally, we are able to fit the horse heart vibrational echo data using the identical FFCFs derived from the MD simulations on sperm whale MbCO with very similar values of the Stark coupling constant λ, although with different values of the substate concentrations appropriate for horse heart MbCO. The primary structures of sperm whale

MbCO and horse heart MbCO differ at 20 residues.[73] The substitutions are generally conservative, and none of them is near the distal heme pocket. The ligand binding kinetics and heme affinity[73] (the dominant factor in determining folding stability[74]) for the two holoproteins are very similar. These facts strongly suggest that the structural assignments and dephasing dynamics for the A_1 and A_3 substates of sperm whale MbCO are also valid for horse heart MbCO, and also possibly for MbCO from other species.

Ultimately, the goal of understanding protein dynamics is to relate these dynamics to the biological function of the protein. The structural assignments of the A substates of MbCO and the identification of the sources of the vibrational dephasing of CO presented here represent a step toward that goal. Interpreting the stimulated vibrational echo experimental results with MD simulations has provided considerable insight into the structural and dynamical nature of the A substates of myoglobin. Further work toward identifying the functional consequences of these structural differences is needed. The combination of MD simulations and time-resolved spectroscopy can provide some of the atomic level details necessary to help understand the connection between protein structure and function.

Acknowledgment. We thank Ileana Stoica and Professor Ron Elber at Cornell University for helpful discussions regarding MOIL. K.A.M., I.J.F., A.G., B.L.M., and M.D.F. acknowledge the National Institutes of Health (1R01-GM61137) for support of this research. W.G.N., R.A., and R.F.L. acknowledge support from the National Science Foundation (CHE-0105623) and the Petroleum Research Fund of the American Chemical Society. The molecular dynamics portion of this research was carried out using the resources of the Cornell Theory Center, which receives funding from Cornell University, New York State, federal agencies, and corporate partners. K.A.M. was partially supported by an Abbott Laboratories Stanford Graduate Fellowship. W.G.N. acknowledges fellowship support from Cornell's IGERT program in nonlinear systems, funded by NSF Grant DGE-9870681. R.A. is a Research Fellow of the Japan Society for the Promotion of Science (2000).

Appendix A – Calculation of the Vibrational Echo

We calculate the vibrational echo signal from third-order perturbation theory in the radiation–matter interaction.[47,52] The third-order nonlinear polarization can be expressed as the sum of eight relevant terms after application of the rotating wave approximation and the phase-matching condition $\vec{k}_s = \vec{k}_2 + \vec{k}_3 - \vec{k}_1$.[47,52] The material system is treated as a quantum-mechanical three-level system, coupled to a classical solvent that is treated within a second-order cumulant expansion. Within this approximation, the dynamics of the system can be described with a two-time autocorrelation function. Autocorrelation functions of fluctuations in the frequencies of one-quantum transitions are set equal to the classical mechanical autocorrelation function of frequency fluctuations in eq 2: $C(t) = \langle \delta\omega_{10}(t)\delta\omega_{10}(0)\rangle = \langle \delta\omega_{21}(t)\delta\omega_{21}(0)\rangle$. The autocorrelation function of fluctuations in a two-quantum frequency is given by $\langle \delta\omega_{20}(t)\delta\omega_{20}(0)\rangle = 4\langle \delta\omega_{10}(t)\delta\omega_{10}(0)\rangle$, the lifetime of the second excited state is taken to be one-half that of the first excited

(72) Merchant, K. A. Ph.D. Thesis, Stanford University, 2003.

(73) Scott, E. E.; Paster, E. V.; Olson, J. S. *J. Biol. Chem.* **2000**, *275*, 27129–27136.

(74) Hargrove, M. S.; Olson, J. S. *Biochemistry* **1996**, *35*, 11310–11318.

state,[75] and the transition dipole moments are related by $\mu_{21} = \sqrt{2}\mu_{10}$, as appropriate for a nearly harmonic vibration.[76] Within this model, the eight response functions are given by[47,52]

$$R_1 = R_2 = |\vec{\mu}_{10}|^4 e^{-i\omega_0(t_3-t_1)} e^{-(t_1+2t_2+t_3)/2T_1} \exp(-g(t_1) + g(t_2) - g(t_3) - g(t_2+t_1) - g(t_3+t_2) + g(t_1+t_2+t_3)) \quad (A1)$$

$$R_3 = - |\vec{\mu}_{10}|^2|\vec{\mu}_{21}|^2 e^{-i\omega_0(t_3-t_1)} e^{i\Delta t_3} e^{-(t_1+2t_2+3t_3)/2T_1} \exp(-g(t_1) + g(t_2) - g(t_3) - g(t_2+t_1) - g(t_3+t_2) + g(t_1+t_2+t_3)) \quad (A2)$$

$$R_4 = R_5 = |\vec{\mu}_{10}|^4 e^{-i\omega_0(t_1+t_3)} e^{-(t_1+2t_2+3t_3)/2T_1} \exp(-g(t_1) - g(t_2) - g(t_3) + g(t_2+t_1) + g(t_3+t_2) - g(t_1+t_2+t_3)) \quad (A3)$$

$$R_6 = - |\vec{\mu}_{10}|^2|\vec{\mu}_{21}|^2 e^{-i\omega_0(t_3+t_1)} e^{i\Delta t_3} e^{-(t_1+2t_2+3t_3)/2T_1} \exp(-g(t_1) - g(t_2) - g(t_3) + g(t_2+t_1) + g(t_3+t_2) - g(t_1+t_2+t_3)) \quad (A4)$$

$$R_7 = |\vec{\mu}_{10}|^2|\vec{\mu}_{21}|^2 e^{-i\omega_0(t_1+2t_2+t_3)} e^{i\Delta t_2} e^{-(t_1+2t_2+t_3)/2T_1} \exp(g(t_1) - g(t_2) + g(t_3) - g(t_2+t_1) - g(t_3+t_2) - g(t_1+t_2+t_3)) \quad (A5)$$

$$R_8 = - |\vec{\mu}_{10}|^2|\vec{\mu}_{21}|^2 e^{-i\omega_0(t_1+2t_2+t_3)} e^{i\Delta(t_2+t_3)} e^{-(t_1+2t_2+3t_3)/2T_1} \exp(g(t_1) - g(t_2) + g(t_3) - g(t_2+t_1) - g(t_3+t_2) - g(t_1+t_2+t_3)) \quad (A6)$$

where ω_0 is the fundamental transition frequency, Δ is the anharmonicity, and T_1 is the lifetime of the first vibrational excited state. Dynamics of the environment of the CO represented by the FFCFs enter these response functions through the line-broadening function, $g(t)$:

$$g(t) = \int_0^t d\tau_1 \int_0^{\tau_1} d\tau_2 C(\tau_2) \quad (A7)$$

The linear absorption spectrum of the CO vibration is calculated within the same approximations used to calculate the nonlinear polarization and is given as the Fourier transform of $I(t)$, a weighted sum of contributions from the conformational substates:

$$I(t) = \sum_{\alpha} c_{\alpha} I_{\alpha}(t) \quad (A8)$$

$$I_{\alpha}(t) = |\vec{\mu}_{10}|^2 e^{-i\omega_{\alpha}t} e^{-t/2T_1} e^{-g_{\alpha}(t)} \quad (A9)$$

where the subscript α is used to denote the different A substates of the protein.

Appendix B − Fitting the Vibrational Echo for the MbCO A Substates

The A_1 line center determined from Figure 7 is 1945 cm^{-1}, and the anharmonicity is 25.3 cm^{-1}, consistent with previous

measurements for the anharmonicity of MbCO.[77] The lifetime of the A_1 substate was measured to be $T_1 = 16.5$ ps, and the A_3 substate lifetime was measured to be $T_1 = 14.7$ ps, as independently determined using pump−probe spectroscopy.[40]

The values for the center wavelengths for the A_0 and A_3 substates were determined by fitting the linear absorption spectrum of sperm whale MbCO to three Voigt line shapes, holding the center frequency for the A_1 substate fixed. The relative populations for the substates were assumed to be proportional to the relative areas under each Voigt line shape. Because of the strong overlap between the A_1 and A_3 substates, the relative concentrations derived from the fitting were not uniquely determined and showed some sensitivity to the initial parameters used in the fitting. Equally good fits to the linear spectrum could be obtained using relative ratios of A_1/A_3 from 1.2 to 1.6. We chose to fix the ratio of A_1/A_3 to 1.4, in reasonable agreement with a previously reported value for the A_1/A_3 ratio of 1.25 using four Gaussian fitting functions.[78] The final values of the line centers and concentrations are in good agreement with previous measurements and assignments of these quantities.[6,7,12,78]

Because we have not identified the A_0 substate in the MD simulations, its dephasing dynamics must be approximated. Previous vibrational echo experiments on the MbCO mutant H64V over a wide range of temperatures have shown that this mutant has a dephasing rate that is approximately 20% slower than wild-type MbCO A_1 substate at all temperatures.[79] The linear spectrum for the H64V mutant is quite similar to the spectrum of the A_0 substate in the wild-type protein.[15,31,79] It has also been suggested that the A_0 substate dephasing dynamics are slower than the dephasing dynamics of the A_1 substate.[60] We therefore chose to scale the amplitude of the FFCF used to model the A_1 state by 0.64 to give a FFCF for the A_0 substate. The factor 0.64 is the empirically determined scaling constant that gives a vibrational echo decay that is slower by ∼20%. The lifetime of this substate was assumed to be equal to the lifetime of the A_1 substate. While the dephasing function and lifetime for the A_0 substate are only approximate, the overall effect of the error in these approximations on the calculated vibrational echo data is small, because these substates represent ∼5% of the total substate population and contribute even less to the vibrational echo signal.

The values for the A substate line centers and relative populations, along with the FFCFs derived from the MD simulations, were used as input parameters for the calculation of the spectrally resolved stimulated vibrational echo for both His64 tautomers. For the calculated vibrational echo signals in Figure 8, the FFCF for B_{ϵ} (Figure 6a, blue curve) was used for A_1, the FFCF for R_{ϵ} (Figure 6a, red curve) was used for A_3, and the FFCF for B_{ϵ} multiplied by 0.64 was used for A_0. For the calculated vibrational echo signals in Figure 10, the FFCF for B_{δ} (Figure 6b, blue curve) was used for A_1, the FFCF for

(75) The total dephasing is overwhelmingly dominated by pure dephasing processes in this system and would be essentially unaffected by small errors in the second excited-state lifetime.

(76) Fourkas, J. T.; Kawashima, H.; Nelson, K. A. *J. Chem. Phys.* **1995**, *103*, 4393−4407.

(77) Rector, K. D.; Kwok, A. S.; Ferrante, C.; Tokmakoff, A.; Rella, C. W.; Fayer, M. D. *J. Chem. Phys.* **1997**, *106*, 10027−10036.

(78) Potter, W. T.; Hazzard, J. H.; Choc, M. G.; Tucker, M. P.; Caughey, W. S. *Biochemistry* **1990**, *29*, 6283−6295.

(79) Rector, K. D.; Rella, C. W.; Kwok, A. S.; Hill, J. R.; Sligar, S. G.; Chien, E. Y. P.; Dlott, D. D.; Fayer, M. D. *J. Phys. Chem. B* **1997**, *101*, 1468−1475.

(80) Unno, M.; Christian, J. F.; Olson, J. S.; Sage, J. T.; Champion, P. M. *J. Am. Chem. Soc.* **1998**, *120*, 2670−2671.

(81) Merchant, K. A.; Thompson, D. E.; Fayer, M. D. *Phys. Rev. A* **2002**, *65*, 023817.

R$_\delta$ (Figure 6b, red curve) was used for A$_3$, and the FFCF for B$_\delta$ multiplied by 0.64 was used for A$_0$. In each of the vibrational echo calculations, all three FFCFs used in the calculation were scaled by the square of the Stark effect parameter λ which was allowed to vary freely until the best possible agreement between the calculated and measured vibrational echo signals was obtained.

Supporting Information Available: Descriptions of the modifications to the MOIL force-field for the N$_\epsilon$−H tautomer of His64, and contour plots of multidimensional vibrational echo data at various T_w values (PDF). This material is available free of charge via the Internet at http://pubs.acs.org.

JA035654X

JOURNAL OF CHEMICAL PHYSICS   VOLUME 121, NUMBER 24   22 DECEMBER 2004

Dynamics of water probed with vibrational echo correlation spectroscopy

John B. Asbury, Tobias Steinel, and Kyungwon Kwak
Department of Chemistry, Stanford University, Stanford, California 94305

S. A. Corcelli, C. P. Lawrence, and J. L. Skinner
Department of Chemistry, University of Wisconsin, Madison, Wisconsin 53706

M. D. Fayer
Department of Chemistry, Stanford University, Stanford, California 94305

(Received 17 August 2004; accepted 22 September 2004)

Vibrational echo correlation spectroscopy experiments on the OD stretch of dilute HOD in H_2O are used to probe the structural dynamics of water. A method is demonstrated for combining correlation spectra taken with different infrared pulse bandwidths (pulse durations), making it possible to use data collected from many experiments in which the laser pulse properties are not identical. Accurate measurements of the OD stretch anharmonicity (162 cm^{-1}) are presented and used in the data analysis. In addition, the recent accurate determination of the OD vibrational lifetime (1.45 ps) and the time scale for the production of vibrational relaxation induced broken hydrogen bond "photoproducts" ($\sim$2 ps) aid in the data analysis. The data are analyzed using time dependent diagrammatic perturbation theory to obtain the frequency time correlation function (FTCF). The results are an improved FTCF compared to that obtained previously with vibrational echo correlation spectroscopy. The experimental data and the experimentally determined FTCF are compared to calculations that employ a polarizable water model (SPC-FQ) to calculate the FTCF. The SPC-FQ derived FTCF is much closer to the experimental results than previously tested nonpolarizable water models which are also presented for comparison. © *2004 American Institute of Physics.* [DOI: 10.1063/1.1818107]

I. INTRODUCTION

The dynamics and structure of water continue to be a problem of both fundamental interest and practical significance. Water is important as a solvent in a vast number of chemical and biological settings. To a great extent, the unique properties of water derive from its ability to form complex hydrogen bond networks.[1] Water can form up to four hydrogen bonds, but the number and strength of the hydrogen bonds continually fluctuate.[2–5] Therefore, understanding the dynamics of the hydrogen bond networks is central to understanding the nature of water. Characterization of the full dynamics of the hydrogen bond networks is essential to test water models which find application in simulations of protein folding,[6,7] ion hydration,[8] and polymerization reactions.[9]

While there have been a wide variety of experimental methods directed at understanding water, ultrafast infrared experiments that examine the water hydroxyl stretching mode are particularly well suited for the study of the hydrogen bond network dynamics. The frequency of the hydroxyl stretch is sensitive to the strength of the hydrogen bonds and the number of hydrogen bonds.[2,3,10–13] The distribution in the strengths and number of hydrogen bonds gives rise to the very broad hydroxyl stretch absorption line observed in IR spectra of water.[2,3] However, the different strengths and numbers of hydrogen bonds do not give rise to spectroscopically resolvable features in the water spectrum. For that reason, a number of ultrafast infrared experimental methods have been applied to extract information about hydrogen

bond dynamics from under the inhomogeneously broadened hydroxyl stretch absorption band.[5,14–35]

The development of the ultrafast infrared vibrational echo technique,[36–39] and the application of vibrational echo experiments to the study of water,[5,15–19,40] provide a direct method for accessing hydrogen bond dynamics through the hydroxyl stretch frequency evolution. Vibrational echo correlation spectroscopy experiments with full phase information applied to water[5,19] can provide more detailed examination of water dynamics than two pulse vibrational echoes[18] or vibrational echo peak shift measurements.[15–17] The multidimensional stimulated vibrational echo correlation spectroscopy technique measures the population and vibrational dephasing dynamics in two frequency dimensions ω_τ and ω_m.[5,41–46] (In NMR, ω_τ and ω_m are usually called ω_1 and ω_3, respectively.[47]) The ω_m axis is similar to the frequency axis in frequency resolved pump-probe spectroscopy. The ω_τ axis does not have an analog in the pump-probe experiment; it provides an additional dimension of information that is contained only in the correlation spectrum. The time evolution of the shapes of the bands in the correlation spectra provides information on the dynamics of the system produced by structural evolution of the hydrogen bond network.

In the experiments, ultrashort mid-IR pulses, ($\sim$50 fs or $<$4 cycles of light), were employed, making it possible to perform experiments on the entire broad hydroxyl stretching band despite its $>$400 cm^{-1} width. Employing pulses that are transform limited and controlling path lengths with accuracy of a small fraction of a wavelength of light, along with

0021-9606/2004/121(24)/12431/16/$22.00

12432   J. Chem. Phys., Vol. 121, No. 24, 22 December 2004

proper analysis, data are obtained with correct phase relationships across the entire spectrum. The proper phase relationships permit accurate separation of the absorptive and dispersive contributions to the spectra.[5,42,44,48] As a result, the two-dimensional (2D) IR correlation spectra are obtained in a manner akin to 2D NMR spectroscopy.[47]

The OD hydroxyl stretch of dilute HOD in H_2O was investigated previously with vibrational echo correlation spectroscopy experiments on water.[5] The results were compared to theoretically calculated results[3,5,49] obtained from two water simulation models TIP4P (Ref. 50) and SPC/E.[51] Below we present supplementary vibrational echo correlation spectroscopy experiments on water and compare them to additional simulations.

The vibrational echo correlation spectra are measured as a function of the time delay T_w between the second and third pulses in the stimulated vibrational echo pulse sequence. The resulting T_w dependent data are fit by varying the frequency time correlation function (FTCF), which is input into a time dependent diagrammatic perturbation theory calculation of the data.[41,52] In addition, the FTCF obtained from the experiments is compared to calculations based on the SPC-FQ (Ref. 53) water model. A quantum oscillator is embedded in the classical molecular dynamics simulation to calculate the FTCF.[3,5,49,54] The resulting FTCF is then input into a diagrammatic perturbation theory calculation to obtain simulated data in the same form as the experimental data. The previous calculations using the TIP4P and SPC/E models showed relatively poor agreement with the experiments, particularly at long time. In contrast, calculations of the FTCF using the SPC-FQ model[54] are in much better agreement with the data. The better agreement may occur because the polarizable term in the SPC-FQ model introduced collective effects into the molecular dynamics simulation. These collective effects are absent in the TIP4P and SPC/E models because they are do not contain the polarizable term. The experiments have improved accuracy and data analysis procedures. In addition, complementary experiments have been performed to obtain data that are necessary for the analysis of the vibrational echo experiments. Using spectrally resolved ultrafast IR pump-probe experiments, the vibrational anharmonicity of the OD stretch of HOD in water is accurately determined (162 cm^{-1}). The anharmonicity is a necessary input parameter in the diagrammatic perturbation theory calculations used to extract the FTCF from the data and to compare the data to the simulations. Furthermore, an accurate value of the OD stretch lifetime (1.45 ps) (Ref. 55) was employed. The determination of the lifetime took into account the generation of "photoproducts" (broken hydrogen bonds) that are formed following vibrational relaxation.[55] The production of photoproducts changes the hydroxyl stretch spectrum and influences the determination of the vibrational spectral diffusion. Detailed analysis of the growth of the photoproduct spectrum demonstrates that the photoproducts do not significantly impact the vibrational echo correlation spectra because of the relatively long lifetime of the OD stretch in H_2O. This is one of the reasons that OD in H_2O is studied rather than OH in D_2O.[15–18] OH in D_2O has a substantially shorter lifetime ($\sim$0.7 ps).[56,57] The

generation of photoproducts limits the time over which OH in D_2O results can be analyzed in a straightforward manner, a factor that has not been taken into account in previous reports.[17,18]

II. EXPERIMENTAL PROCEDURES

The ultrashort IR pulses employed in the experiments are generated using a Ti:sapphire regeneratively amplified laser/OPA system. The output of the modified Spectra Physics regen is 26 fs transform limited 2/3 mJ pulses at 1 kHz rep rate. These are used to pump an optical parametric amplifier (BBO) employing difference frequency generation ($AgGaS_2$). The output of the OPA is compressed to produce $\sim$50 fs transform limited IR pulses as measured by collinear autocorrelation. For the experiments, the compression was readjusted to give transform limited pulses in the sample as measured by a sample that gave a purely nonresonant signal. The pulses were always transform limited in the sample during the course of an experiment, that is, over the full range of T_ws. However, over the several months during which data were collected, there was some variation in the bandwidth (pulse duration) of the pulse. A change in bandwidth has a substantial influence on the experimental correlation spectra, as discussed below. As detailed in the Appendix, a method was developed to combine data sets taken with different pulse bandwidths. Therefore, it is possible to average together data sets taken under appropriate but not identical conditions.

The IR beam is split into five beams. Three of the beams are the excitation beams for the stimulated vibrational echo. A fourth beam is the local oscillator (LO) used to heterodyne detect the vibrational echo signal. One of the excitation beams is also used for pump-probe experiments, with the fifth beam as the probe beam in the pump-probe experiments. All of the beams that pass through the sample are optically identical and are compensated for group velocity dispersion simultaneously. The vibrational echo signal combined with the LO is passed through a monochromator and detected by a 32 element MCT array. At each monochromator setting, the array detects 32 individual wavelengths.

The sample, 5% HOD in H_2O, was held in a sample cell of CaF_2 flats with a spacing of 6 μm. The peak absorbance of the samples was 0.2. Such a low absorbance is necessary to prevent serious distortions of the pulses as they propagate through the sample. The OD stretch of HOD is used as a probe of water for three reasons. The relatively dilute OD stretch reduces the rate of vibrational excitation transport to the point where transport has a negligible influence on the dynamics.[58] Excitation transport would be a source of spectral diffusion that is not related to structural dynamics of water. Furthermore, it is very difficult to make a pure water sample that is thin enough to have the necessary low optical density for the experiments, and such a thin sample is subject to heating artifacts. Finally, as discussed below and mentioned in the Introduction, the vibrational lifetime of the OD stretch of HOD in H_2O and the time for the onset of the photoproduct spectrum caused by hydrogen bond breaking determine the longest time for which data can be taken and

J. Chem. Phys., Vol. 121, No. 24, 22 December 2004

unambiguously interpreted. The OD stretch of HOD in H_2O lifetime is significantly longer than the OH stretch lifetime in D_2O.

The phase-resolved, heterodyne detected, stimulated vibrational echo was measured as a function of one frequency variable ω_m and two time variables τ and T_w that are defined as the time between the first and second sample-radiation field interactions and the second and third sample-radiation field interactions, respectively. The measured signal is the absolute value squared of the sum of the vibrational echo signal electric field S and the local oscillator electric field L

$$|L+S|^2 = L^2 + 2LS + S^2. \tag{1}$$

The L^2 term is time independent and the S^2 is very small; hence neither contributes to the time dependence of the signal. The $2LS$ term is the heterodyne amplified signal. The monochromator performs an experimental Fourier transform on the radiation. Through data processing that involves chopping to measure L^2 and normalization, the spectrum yields the ω_m frequency axis. As the τ variable is scanned in 2 fs steps, the phase of the vibrational echo signal electric field is scanned relative to the fixed local oscillator electric field, resulting in an interferogram measured as a function of the τ variable. The interferogram contains the amplitude, sign, frequency, and phase of the vibrational echo signal electric field as it varies with τ. By numerical Fourier transformation, this interferogram is converted into the frequency variable ω_τ, providing the ω_τ axis.

The interferogram measured as a function of τ contains both the absorptive and dispersive components of the vibrational echo signal. However, two sets of quantum pathways can be measured independently by appropriate time ordering of the pulses in the experiment.[43,44,46] With pulses 1 and 2 at the time origin, pathway 1 or 2 is obtained by scanning pulse 1 or 2 to negative time, respectively. Adding the Fourier transforms of the interferograms from the two pathways, the dispersive component cancels leaving only the absorptive component. The 2D vibrational echo correlation spectra are constructed by plotting the amplitude of the absorptive component as a function of both ω_m and ω_τ.

Lack of perfect knowledge of the timing of the pulses and consideration of chirp on the vibrational echo pulse requires a "phasing" procedure to be used.[5,40] The projection slice theorem[47] with an auxiliary condition is employed to generate the absorptive 2D correlation spectrum. The projection of the absorptive 2D correlation spectrum onto the ω_m axis is equivalent to the IR pump-probe spectrum recorded at the same T_w, as long as all the contributions to the stimulated vibrational echo are absorptive. Consequently, comparison of the projected 2D stimulated vibrational echo spectrum to the pump-probe spectrum permits the correct isolation of the absorptive vibrational echo correlation spectrum from the 2D spectrum obtained from the addition of the two quantum pathways.

It is possible to come relatively close to the correct correlation spectrum prior to the phasing procedure because the very short pulses permit their time origins to be known within a few femtoseconds. However, the correlation spectrum is very sensitive to small errors in the time origin, even

on the order of 1 fs, and to chirp. Therefore, a well defined phasing procedure based on various potential errors was developed.[45,59] The frequency dependent phasing factor used to correct the 2D spectra has the form

$$S_C(\omega_m,\omega_\tau) = S_1(\omega_m,\omega_\tau)\Phi_1(\omega_m,\omega_\tau)$$
$$+ S_2(\omega_m,\omega_\tau)\Phi_2(\omega_m,\omega_\tau),$$

$$\Phi_1(\omega_m,\omega_\tau) = \exp[i(\omega_m\Delta\tau_{LO,E} + \omega_\tau\Delta\tau_{1,2} + \omega_m\omega_\tau C$$
$$+ \omega_m^2 Q)], \tag{2}$$

$$\Phi_2(\omega_m,\omega_\tau) = \exp[i(\omega_m\Delta\tau_{LO,E} - \omega_\tau\Delta\tau_{1,2} + \omega_m\omega_\tau C$$
$$+ \omega_m^2 Q)].$$

S_C is the correlation spectrum. S_1 and S_2 are the spectra recorded for pathways 1 and 2, respectively. $\Delta\tau_{LO,E}$ accounts for the lack of perfect knowledge of the time separation of the LO pulse and the vibrational echo pulse; $\Delta\tau_{1,2}$ accounts for the lack of perfect knowledge of the time origins of excitation pulses 1 and 2; C accounts for linear chirp caused by the echo propagating through the sample; and Q accounts for the linear chirp caused by propagation of the vibrational echo through the back window of the sample cell. $\Delta\tau_{1,2}$ comes in with opposite sign for pathways 1 and 2.

An additional procedure was developed and employed to produce the very high quality correlation spectra shown below.[5,40] The projection slice theorem reduces a two-dimensional entity to a one-dimensional entity. However, in order to unambiguously assign a correct correlation spectrum it is preferable to use information from both dimensions. Therefore, in the phasing procedure an additional constraint is applied for the ω_τ dimension. We can use information from the absolute value correlation spectrum, which is the sum of the absolute value spectra of pathway 1 (rephasing) and pathway 2 (nonrephasing) discussed above. The absolute value spectrum is independent of the phase factor and peaks at the same frequency along ω_τ as the purely absorptive spectrum. Consequently, the difference in peak positions of a trial phased absorptive spectrum and the absolute value spectrum for each ω_m gives an additional criterion on the quality of the correlation spectrum. The correct correlation spectrum is the one that provides the best fit to the pump-probe spectrum and minimizes the difference in peak positions of the absorptive and absolute value spectra. This procedure holds for symmetric line shapes and hence can be applied to the data discussed here, where the dynamical lines can be well fit to a Gaussian line shape and the deviations are symmetric.

Finally, to test the procedures described above and to make final very small corrections, a method call PEARL for "phasing employing absorptive resonant loci" was used. This procedure will be discussed in detail subsequently.[60] PEARL is the multidimensional, multipeak absorptive equivalent of a Cole-Cole procedure used in dielectric relaxation experiments.[60] Following phasing, the errors in the time origins are $< 100 \times 10^{-16}$ s and the time shift across the entire spectrum due to chirp $\omega_m C$ is $< 100 \times 10^{-16}$ s.

12434   J. Chem. Phys., Vol. 121, No. 24, 22 December 2004

Asbury et al.

III. RESULTS AND DISCUSSION

The hydroxyl stretch frequency provides a sensitive probe of the hydrogen bond dynamics of water through its correlation with the strength and number of hydrogen bonds associated with a water molecule.[2,3,10–13] Ultrafast time-resolved nonlinear vibrational spectroscopy probes the hydroxyl stretch directly, and so provides access to the hydrogen bond dynamics. Vibrational echo spectroscopy[36–39] and, in particular, multidimensional vibrational echo correlation spectroscopy[5,19,40,61] provides superior time resolution and sensitivity to the dynamics of water compared to IR pump-probe techniques.[20–27,33–35] Multidimensional vibrational echo correlation spectroscopy provides experimental observables that permit the parametrization or selection of the most accurate water models.[5,54]

Significant disparities exist between recent measurements of water dynamics made by several groups using different vibrational echo techniques.[5,17–19] The water dynamics are described by the FTCF that each group reported. The FTCF is defined as

$$C(t) = \langle \delta\omega(t)\,\delta\omega(0)\rangle, \tag{3}$$

where $\delta\omega$ is the change in the hydroxyl stretch frequency from its initial value, and the bracket is an ensemble average. The FTCF describes the loss of correlation of the frequency of an ensemble of OD stretches as time progresses. Yeremenko, Pshenichnikov, and Wiersma heterodyne detected a 2-pulse vibrational echo of the OH stretch of HOD in D_2O[18] and reported a biexponential decaying FTCF with 130 and 900 fs time constants having ~60% and ~40% amplitudes, respectively. Fecko *et al.* also studied the OH stretch of HOD in D_2O using 3-pulse vibrational echo peak shift measurements.[17] They reported a FTCF that decayed in ~200 fs (~85% amplitude) with a large oscillation followed by a 1.2 ps exponential decay (~15% amplitude). Both aforementioned experiments utilized frequency integrated techniques which did not discriminate between the 0-1 and 1-2 transitions that are emitted by the hydroxyl stretch and so included both ground and excited state dynamics in their measurements. Most recently, the OD stretch of HOD in H_2O was studied using vibrational echo correlation spectroscopy. The experiments yielded a FTCF described by a triexponential function having 32 fs, 400 fs, and 1.8 ps time constants with corresponding 43%, 16%, and 41% amplitudes.[5,19] These experiments selectively measured the dynamics of the 0-1 transition because the vibrational echo correlation spectroscopy technique spectrally resolves the vibrational echo and facilitates complete discrimination between ground and excited state contributions. A review of other reports of the FTCF of water has been provided recently.[5]

The differences in the dynamics of water reported in these three experimental investigations[5,17–19] suggest that the experimental execution and the experimental method can influence the results. To improve the experimental accuracy of the vibrational echo correlation spectroscopy measurements on water, we have performed the type of experiments reported above[5,19] with improved technique five times over the course of many months and under many different laser conditions. Additionally, we have performed complementary experiments to obtain properties of the hydroxyl stretch that are more accurate than those available in the literature. These auxiliary experiments are necessary for accurate analysis of the vibrational echo experiments and accurate determination of the FTCF using molecular dynamics simulations of three water models.

This report is broken into two main sections: (A) measurement of water dynamics and (B) comparison of experimental and theoretical dynamics predicted from three models of water. Within Sec. III A, we describe our measurement of the vibrational anharmonicity and the vibrational lifetime of the hydroxyl stretch using spectrally resolved infrared pump-probe spectroscopy.[55] We then present the results from five sets of vibrational echo correlation spectra and discuss data analysis procedures that enable us to combine multiple data sets to reduce the error bars in the measurement. Section III B compares the experimental data with predictions from three water models and discusses the differences between the models that affect their predicted dynamics.

A. Accurate measurement of water dynamics

1. OD stretch anharmonicity

The goal of studying the vibrational dynamics of water is to access the ground state equilibration dynamics, which are displayed in the 0-1 transition. The 1-2 transition reports the dynamics of the excited state, which is not the principle objective of the study. The 1-2 transition of the OD stretch vibration of HOD in liquid water significantly overlaps the 0-1 transition. The contribution of the 1-2 transition in the experimental data can be accounted for as long as the anharmonicity (shift in frequency between the 0-1 and 1-2 transitions) is known accurately. The ground state dynamics of water can be measured from the 0-1 transition once the contribution from the 1-2 transition is known. The dynamics can be examined at wavelengths that do not have interference from the 1-2 transition in the frequency resolved correlation spectroscopy experiments. In addition, the contribution of the 1-2 transition can be calculated using diagrammatic perturbation theory.[5,41,52,62,63] The calculations describe the 0-1 and 1-2 transitions and require the anharmonicity as input. By comparing the experimental data with the calculated results, the ground state dynamics can be obtained without any contamination from the excited state.[5,19]

The anharmonicity of the OD stretch vibration of HOD in liquid water has been reported several times with values ranging from 127 (Ref. 64) to ~200 cm^{-1}.[33] The uncertainty in the values from the literature is far greater than can be tolerated in the analysis of the experiments. Consequently, we determined the anharmonicity of the OD stretch of HOD in H_2O using frequency resolved pump-probe spectroscopy.

The pump-probe spectra at delay times T_w shorter than the vibrational lifetime are composed of the bleach of the 0-1 transition and the excited state absorption of the 1-2 transition. A representative difference spectrum collected at $T_w = 200$ fs is displayed in Fig. 1. The bleach of the ground state is a positive signal centered at 2510 cm^{-1}, while the negative excited state absorption signal is centered around 2350 cm^{-1}. To extract a precise anharmonicity constant, we fit the pump-

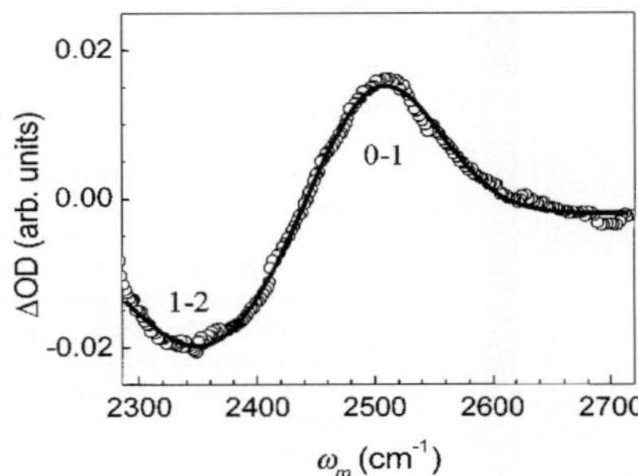

FIG. 1. Transient absorption spectrum of the OD hydroxyl stretch of HOD in water collected at pump-probe delay $T_w = 0.2$ ps. The 0-1 transition is positive going at $\sim$2510 cm^{-1} and the 1-2 transition is negative going at $\sim$2350 cm^{-1}. The solid line through the data is a fit comprised of two Gaussians, one for the 0-1 and another for the 1-2 transition that yields the anharmonicity 162 cm^{-1}.

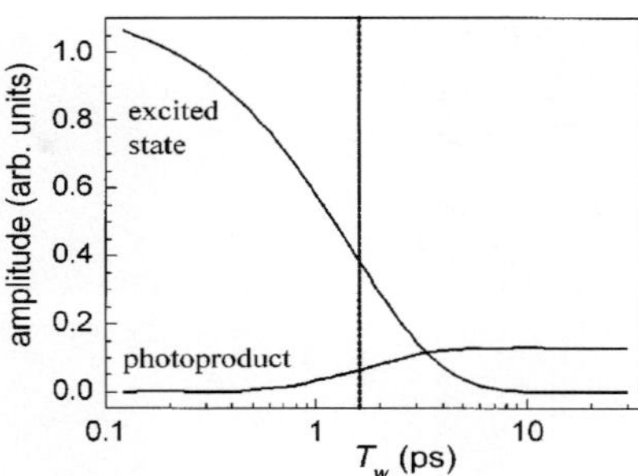

FIG. 2. Time dependent contributions to the signal of the initially excited (upper curve) and photoproduct (lower curve) populations of water species measured from transient absorption experiments. The photoproduct forms following vibrational relaxation of the hydroxyl stretch, which breaks hydrogen bonds.

probe spectrum to a sum of two Gaussian line shapes, one corresponding to the ground state and another to the excited state. The fit is displayed as the solid line through the data in Fig. 1. We obtained a more constrained fit by fixing the center of the ground state peak to be the same as the peak of the OD-hydroxyl stretch absorption in the Fourier transform infrared absorption spectrum (2510 cm^{-1}) (Ref. 5) while leaving the center of the 1-2 transition and the widths of the Gaussians as fitting parameters. We performed the fit described above on hundreds of distinct pump-probe spectra at delay times ranging from 0.12 to 1.0 ps. The difference between the peak positions of the 0-1 and 1-2 transitions that were determined from the fits provides the anharmonicity constant, which we determined to be 162±4 cm^{-1}.

2. OD stretch lifetime and photoproduct formation

The vibrational lifetime of the hydroxyl stretch determines the time duration over which the equilibrium dynamics of water can be unambiguously measured. Beyond this time window, photoproducts that result from hydrogen bond breaking contribute to the signal.[55] If the dynamics of water are measured on the time scale of and beyond the vibrational lifetime, the measured dynamics do not solely reflect the equilibrium fluctuations of water. The photoproducts of hydrogen bond breaking that form following vibrational relaxation contaminate the signal.[55] To determine the time window during which the ground state fluctuations of water may be interrogated and to access the contribution to the signal from the production of photoproducts as a function of time, the vibrational lifetime was determined using broadband transient absorption experiments to monitor the hydroxyl stretch population dynamics.[55] The ultrashort mid-IR pulses ($\sim$50 fs or <4 cycles of light) uniformly excited the 0-1 transition of the OD hydroxyl stretch of HOD in water, which completely eliminated the influence of spectral diffusion from the dynamics. Such precautions have not always been taken into account previously.[20–27,33–35] The ultrafast transient absorption experiments have been described in detail elsewhere.[55] The necessary details are recounted here to

establish the T_w delays, for which the generation of photoproducts can influence the measurements of the dynamics of water. The magnitude of the influence of the photoproducts on the vibrational echo results are determined quantitatively below.

The transient absorption spectra were taken in a magic angle configuration to eliminate orientational dynamics.[55] The transient spectra were decomposed into reactant and photoproduct contributions using singular value decomposition.[55,65] Separation of the vibrational relaxation dynamics from the hydrogen bond breaking and formation processes revealed an exponential vibrational lifetime of $\tau_{vib} = 1.42 \pm 0.05$ ps across the entire hydroxyl stretch band including 0-1 and 1-2 transitions. As a consistency check we analyzed the data by fitting the transient absorption decay at the red side of the 1-2 transition, where the decay is not perturbed by the photoproduct dynamics. The results were virtually the same and the vibrational lifetime was found to be frequency independent with a time constant $\tau_{vib} = 1.45 \pm 0.05$ ps. This value for the lifetime[55] is slightly shorter and much more accurate than previous measurements.[33,35]

From the measurement of the population dynamics, which is briefly discussed here, it was demonstrated that excitation of the hydroxyl stretch and subsequent vibrational relaxation breaks hydrogen bonds and produces spectrally distinct photoproducts.[55] Figure 2 displays the spectrally resolved dynamics of the initially excited and the photoproduct populations at 2500 cm^{-1} (note the time scale is logarithmic). The excited state contribution to the signal (upper curve) decays exponentially (1.45 ps), while the photoproduct (lower curve) grows in nonexponentially as a result of the vibrational relaxation rate and the hydrogen bond breaking rate (810 fs).[55] The dashed line indicates the maximum T_w delay at which a correlation spectrum is reported in this work. It is clear that the measured dynamics will not be substantially affected by the photoproducts for $T_w < 1$ ps, while at times larger than 5 ps the photoproduct prevails. At intermediate times both components contribute to the transient absorption signal and neither of them can be neglected. From the analysis, the photoproduct contributes to the tran-

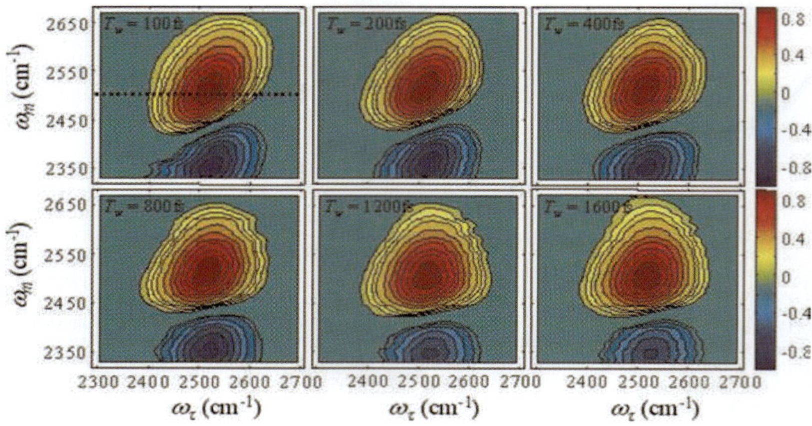

FIG. 3. (Color) The experimental vibrational echo correlation spectra of the OD stretch of HOD in H_2O. The data have been normalized to the maximum positive value. Each contour represents a 10% change. The positive going peak arises from the 0-1 transition. The negative going peak arises from the 1-2 transition. The 0-1 and 1-2 peaks are elongated along the diagonal, indicating inhomogeneity persists in the hydroxyl stretch frequency distribution. Spectral diffusion broadens the widths of the peaks as T_w increases. The dashed line represents a cut at $\omega_m = 2500$ cm^{-1}.

sient absorption signal $\sim$3%, 8%, and 15% of the total at T_w delays of 800, 1200, and 1600 fs, respectively.

The influence of the photoproduct on the vibrational echo correlation spectra is discussed in detail below. From the analysis, the photoproduct signal contributes only $\sim$0.5 cm^{-1} to the measured dynamic line width at the longer T_w delays, (800–1600 fs) which is negligible given that the error bars in the measurement are $\pm$2.5 cm^{-1} (see below). Consequently, it is unnecessary to correct the measured dynamic linewidths for the photoproduct contribution. As has been discussed in detail elsewhere,[55] the growth of the photoproducts does have a significant influence on the determination of the vibrational lifetime and the orientational relaxation.

3. Vibrational echo correlation spectra

Since the first report of vibrational echo correlation spectroscopy experiments investigating the dynamics of water,[5] we have refined the data collection and analysis methods to enhance the accuracy of the measurements of water dynamics. The results of Secs. III A 1 and 2 are used below in the analysis presented in this section. Five distinct sets of time dependent correlation spectra were collected to improve the accuracy. The data sets were collected over the span of several months. Each data set comprised an independent collection of correlation spectra measured for six T_w points at 100, 200, 400, 800, 1200, and 1600 fs. A typical set of correlation spectra from a single experiment is displayed in Fig. 3. The correlation spectra are two-dimensional frequency maps of the frequency fluctuations of the OD stretch. The ω_τ axis (horizontal axis) displays the frequency of the interaction of the sample with the first laser pulse, which occurs at the 0-1 transition frequency in the region around 2500 cm^{-1}. All the features in the correlation spectra appear centered at the 0-1 transition frequency along the ω_τ axis. The ω_m axis displays the frequency at which the vibrational echo is emitted from the sample. Vibrational echo correlation spectroscopy accesses the ground (0), first (1), and second (2) vibrational states. Consequently, two peaks are observed along the ω_m axis corresponding to the 0-1 and the 1-2 transition frequencies. The positive going 0-1 peak appears on the diagonal

(represented by the line $\omega_m = \omega_\tau$) because the vibrational echo is emitted with the same frequency as the first interaction with the radiation field (first pulse). The negative going peak corresponds to the vibrational echo that was emitted on the 1-2 transition. The peak appears redshifted along the ω_m axis by the anharmonicity $\Delta_{an} = 162$ cm^{-1} (Sec. III A 1) because the vibrational echo was emitted from the 1-2 transition, but the first interaction was with the 0-1 transition. Similar to IR transient absorption spectroscopy, the 0-1 peak is positive and the 1-2 peak is negative because they result from a decrease in the ground state population and an increase in the excited state population, respectively. The phase resolution afforded by heterodyne detected vibrational echo correlation spectroscopy displays the sign parity between the 0-1 and 1-2 transitions. The sign parity combined with the measurement of purely absorptive line shapes facilitates the complete removal of the excited state contribution from the ground state dynamics, unlike previous investigations of water dynamics.[17,18]

Looking at the earliest T_w delay ($T_w = 100$ fs) displayed in Fig. 3, the 0-1 and 1-2 peaks appear elongated along the diagonal because their widths along the ω_τ axis are smaller than their total widths (observed by looking at the projection of the peak on the ω_m axis). The width of the projection of the 0-1 peak at a particular ω_m onto the ω_τ axis is the "dynamical linewidth."[5,66,67] Because the hydroxyl stretch of HOD is inhomogeneously broadened for small T_w, the resulting band shapes are asymmetric.[5,66,67] As T_w increases (see the T_w panels of Fig. 3), the dynamic linewidths of the 0-1 and 1-2 peaks (projection of ω_m onto the ω_τ axis) broaden. By $T_w = 1600$ fs, the width of the 0-1 peak along the ω_τ axis has grown almost to the asymptotic value, that is, the width when all possible frequencies have been sampled because of complete structural randomization. Changes in the dynamical linewidth report the spectral diffusion dynamics. The dynamical linewidth can be measured by selecting a cross sectional cut of the 0-1 peak at a particular ω_m value and projecting it onto the ω_τ axis. The horizontal dashed line shown in the $T_w = 100$ fs panel of Fig. 3 illustrates such a cut at the approximate center of the hydroxyl stretch peak (2500

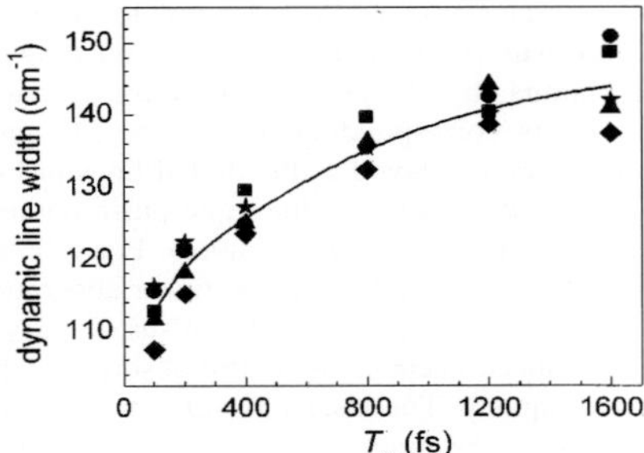

FIG. 4. The dynamic linewidths of 2500 cm^{-1} slices that were extracted from five independent sets of vibrational echo correlation spectra (see text for details). Five distinct experiments were performed over a period of several months. Each experiment produced one correlation spectrum at each T_w point. The triangles report the dynamic linewidths obtained from the correlation spectra that are presented in Fig. 3. The black curve is a guide to the eye. Different bandwidths (pulse durations) were used in the experiments, which prevents the five sets of data from being averaged together without first correcting for the different pulse durations.

cm^{-1}). The dynamical line measured in the correlation spectra of HOD is narrow when spectral diffusion has not randomized the distribution of hydroxyl stretch frequencies. As spectral diffusion occurs, the dynamical linewidth broadens, which is captured by the increased width of the cut at 2500 cm^{-1}.[5,19]

Excited state relaxation occurs with a wavelength independent 1.45 ps time constant as T_w increases,[55] which reduces the amplitude of the emitted vibrational echo and produces photoproducts that are spectrally distinct in about 2 ps (see above). Consequently, the signal amplitudes of the data displayed in Fig. 3 decay by a factor of ~3 from 100 to 1600 fs. The data in Fig. 3 have been normalized to the amplitude of the 0-1 peak, which obscures the decay. Because we want to study the hydrogen bond dynamics of equilibrium water, not the photoproducts caused by hydrogen bond breaking following vibrational relaxation, we restrict our observation window to <2 ps. Studying OD in H_2O enables us to examine the equilibrium hydrogen bond dynamics of water over a time window that is approximately twice to that of investigations of OH in D_2O.[17,18] OH in D_2O has a substantially shorter lifetime (~0.7 ps).[56,57] Therefore, photoproducts will be generated on a shorter time scale. Furthermore, the vibrational quantum of energy deposited upon vibrational relaxation is substantially larger for a OH stretch and may lead to a larger photoproduct contribution. Photoproduct generation has not been taken into account in the analysis of vibrational echo data for the OH in D_2O system.[17,18]

The cuts at 2500 cm^{-1} are well described by a Gaussian line shape. We fit the cuts at $\omega_m = 2500$ cm^{-1} from the correlation spectra presented in Fig. 3 to extract the full width at half maximum (FWHM) and to quantify the change in the dynamical line as a function of T_w delay. The dynamical linewidths extracted from the data in Fig. 3 are the triangles in Fig. 4. The dynamic linewidths display the same trend as the correlation spectra, small width at $T_w = 100$ fs (~112 cm^{-1}), which increases with T_w delay (~143 cm^{-1} by 1600

fs). Four other sets of dynamic linewidths are presented in Fig. 4 (circles, squares, diamonds, and stars) that were extracted from additional sets of correlation spectra. The solid black line is a guide to the eye.

The method we used to extract the dynamics of water from the correlation spectra, selection of cuts at 2500 cm^{-1}, has two distinct advantages that enable the most accurate measurement of the structural evolution of water. First, the method can extract the dynamics of water without interference from the 1-2 transition. Because we accurately determined the anharmonicity (Sec. III A 1), it was possible to assess the contribution of the 1-2 transition at 2500 cm^{-1}. It was determined that the contribution is negligible. Consequently, it is unnecessary to remove its influence[5] when fitting the slice at 2500 cm^{-1} to obtain the dynamic linewidth of the 0-1 transition. (Nonetheless, the 1-2 transition is included in the theoretical calculations presented below.) Recent investigations of water dynamics[17,18] were unable to selectively measure the dynamics of the 0-1 transition because the investigators used frequency integrated two[18] and three[17] pulse vibrational echo measurements. The inclusion in data of the 1-2 transition, which can have different dynamics, complicates the analysis and may influence the FTCF determined from such data.

The second advantage of the method is that it selects the central frequency of the 0-1 transition, which facilitates comparison with theoretical simulations of water (Sec. III B). In a recent vibrational echo correlation spectroscopy study of water, it was found that the dynamical line shapes measured at different wavelengths across the OD hydroxyl stretch band of HOD in water vary in width at very short T_w delays.[19] The wavelength dependence indicated that different water molecules experience fluctuations that are specific to their hydrogen bonded configuration. By $T_w \sim 400$ fs, the dynamical linewidths become wavelength independent because the very local structures of different hydrogen bonded configurations are randomized.[19] The frequency dependence of the dynamical linewidth cannot be described by the diagrammatic perturbation theory treatment used to simulate the vibrational echo correlation spectra because of the Gaussian approximation that is implicit in the theory.[41,52,62,63] The method we use mitigates this short coming by selecting the central frequency of the 0-1 transition to represent the dynamics of the entire distribution. The method is rigorously correct for the longer time scale dynamics ($T_w > 400$ fs), and the oppositely signed errors at the earlier T_w delays cancel because we use the central frequency.

Part of the spread in the different data sets shown in Fig. 4 is not random error. Differences arise if the data sets are collected with different pulse bandwidths. To reduce the error bars in the determination of the time evolution of the dynamic linewidths, a method is needed to combine the results of the five experiments into a single composite set of data. However, the groups of data presented in Fig. 4 cannot be averaged together without first correcting the dynamic linewidths from each experiment for the influence of the corresponding laser pulse bandwidths. During each measurement, the laser pulses were transform limited and did not vary while the data were collected. However, different ex-

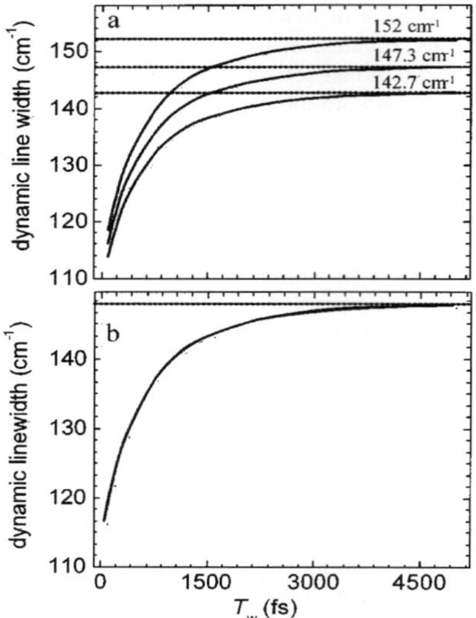

FIG. 5. (a) Dynamic linewidths from calculated vibrational echo correlation spectra that used different pulse durations (45, 51, and 57 fs, top to bottom). The linewidths rise to different limiting values because the bandwidth affects the asymptotic limit. The spread in asymptotic limits is the reason that the data displayed in Fig. 4 cannot simply be averaged together. (b) The three sets of dynamic linewidths displayed in Fig. 5(a) but recast to have the average pulse duration (51 fs) using the correction procedure developed in the Appendix. The exact superposition of all three curves demonstrates that the procedure corrects the experimental data for the influence of finite pulse durations.

periments employed laser pulses that varied in duration from 45 to 57 fs. The pulse durations were precisely determined by measuring the cross correlation of the three laser pulses in a sample that gave a purely nonresonant signal.[40]

To illustrate the importance of correcting the data sets, we calculated correlation spectra with 45, 51, and 57 fs pulse durations to simulate the effect of the actual laser pulses. We employed time dependent diagrammatic perturbation theory to obtain the full third-order nonlinear material response including finite pulse durations.[41,52,62,63] The implementation has been described in detail elsewhere.[5] Briefly, the diagrammatic treatment uses a FTCF as input to obtain the response functions employed in the calculations of the correlation spectra. The FTCF used as input to the calculations was the experimentally determined FTCF that describes the equilibrium fluctuations of water (the method by which the FTCF was obtained from the data is described below). The FTCF was scaled to reproduce the linear linewidth observed from the linear absorption spectrum (FWHM=170 cm^{-1}).[5] Three groups of correlation spectra were calculated using the same FTCF. Only the pulse durations varied between the sets. The top, middle, and bottom curves in Fig. 5(a) display the dynamic linewidths obtained from the correlation spectra calculated with 45, 51, and 57 fs pulse durations, respectively. The dynamic linewidths were acquired from Gaussian fits to the 2500 cm^{-1} cuts out of the correlation spectra. T_w delays were calculated from 0.1 to 5 ps, which is sufficient time

delay for spectral diffusion to be complete. The dynamic linewidths are narrow at small T_w delays and grow to their asymptotic limits by 5 ps for all three groups. The asymptotic limits corresponding to the 45, 51, and 57 fs pulse durations are displayed as the dotted lines in Fig. 5(a). The asymptotic line widths for the three pulse durations are 152, 147.3, and 142.7 cm^{-1}, respectively. For infinite pulse bandwidth, the asymptotic limit is the linear absorption linewidth. However, for finite bandwidth pulses, the asymptotic linewidth is reduced because the entire absorption spectrum is not excited equally. The laser pulse durations used in the simulations reflect the range of pulse durations used in the experiment. Consequently, the asymptotic limit of the data presented in Fig. 4 also varies between 152.0 and 142.7 cm^{-1}. The dynamic linewidth in the experiment does not reach its asymptotic limit within our time window of observation (T_w = 1600 fs), which is limited by the appearance of the broken hydrogen bond photoproducts (see above discussion). Consequently, we rely on precise knowledge of the limit *a priori* in fitting the data. Given that the dynamic linewidth varies by 35 cm^{-1} over the T_w range in the data sets, the variance in the asymptotic limit due to different pulse durations produces substantial shifts not only in the asymptotic limit, but in the entire T_w dependent curve. The variation is $\pm15\%$ for the range of pulse durations used. This produces unacceptable apparent errors in the data.

To address this issue, we developed a procedure to correct the dynamic linewidths for the influence of finite pulse durations. Expressions are provided in the Appendix describing the dependence of the correlation spectra on the laser pulse bandwidth. From this dependence, the correction procedure is developed. The correction procedure was applied to the simulated dynamic linewidths displayed in Fig. 5(a). The linewidths calculated with 45 and 57 fs pulses are recast into the median pulse duration 51 fs. Figure 5(b) displays the three sets of calculated dynamic linewidths after the correction procedure was applied [see Eqs. (A12) and (A13)]. The sets are indistinguishable. What appears as a single curve in Fig. 5(b) is the superposition of three identical curves. The comparison demonstrates that the effects of finite-bandwidth pulses on the dynamic line widths are precisely described by Eqs. (A12) and (A13), and that the procedure described in the Appendix can be used to combine data sets taken with different bandwidths.

Having verified the accuracy of the correction procedure, we applied the method to the five sets of experimental dynamical linewidths presented in Fig. 4. After correction for the differences in bandwidth, the averaged data give the points with error bars shown in Fig. 6. The scatter in the data at each T_w point yields error bars, defined as one standard deviation ±2.5 cm^{-1}. The dynamic linewidth data climb towards a well-defined asymptotic limit of 147.3 cm^{-1} (determined from the pulse duration and the linear linewidth). The dynamic width in Fig. 6 approaches, but does not reach the asymptotic limit by T_w = 1600 fs.

Before analyzing the data presented in Fig. 6 in detail, it is necessary to determine the influence of photoproduct generation on the experimentally determined dynamic linewidths. The real dynamic line shape $G_{\mathrm{dyn}}(\omega_\tau)$, which would

J. Chem. Phys., Vol. 121, No. 24, 22 December 2004

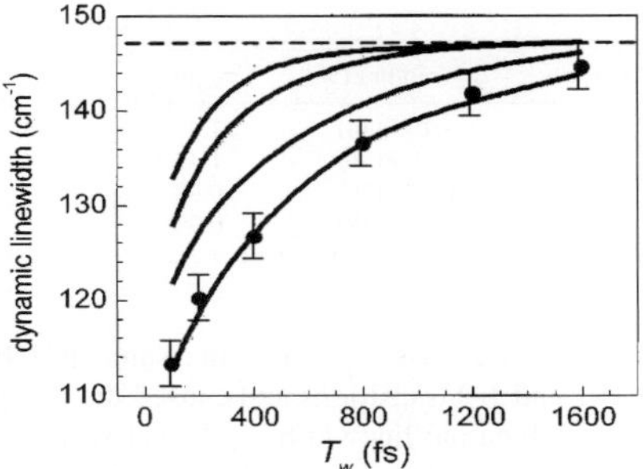

FIG. 6. Comparison of the dynamical linewidths of the simulated and experimental correlation spectra obtained from the $\omega_m = 2500$ cm^{-1} slices. The experimental data appear as circles with error bars. The horizontal line at 147.3 cm^{-1} is the long time asymptotic linewidth. The top curve is from the TIP4P model and the next lower curve is from the SPC/E model. Both models have reached the asymptotic limit by 1.6 ps. Compared to the data, both models have too much amplitude at short time and the long time components are too fast. The third from top curve is from the SPC-FQ model, which is a polarizable water model. The SPC-FQ model displays considerably better agreement with the experimental dynamics that the TIP4P or SPC/E models. Most importantly, the SPC-FQ model has not reached the asymptotic limit by 1.6 ps. The curve through the data is from a fit to the data obtained by varying the parameters of a triexponential trial FTCF until the dynamical linewidths best fit the data. The experimental FTCF provides a fit to the data within experimental error.

be measured in the absence of the photoproduct, can be calculated from the measured dynamic line shape $G_m(\omega_\tau)$ according to the following expression:

$$G_{\text{dyn}}(\omega_\tau) = (1-A)^{-1}[G_m(\omega_\tau) - AG_{\text{pp}}(\omega_\tau)]. \qquad (4)$$

$G_{\text{pp}}(\omega_\tau)$ is the ω_τ dependent line shape corresponding to the photoproduct and A is the fractional contribution to the measured signal from the photoproduct. The influence of photoproducts on the correlation spectra has been discussed in detail in a recent vibrational echo correlation spectroscopy study of methanol oligomers.[45,48,68,69] It was demonstrated that the photoproduct line shape along the ω_τ axis is the asymptotic line shape when it is formed without fine frequency correlation with the initially excited hydroxyl distribution.[69] When fine frequency correlation exists between a photoproduct and its parent molecule that was initially excited, the photoproduct retains a detailed memory of the parent frequency within the inhomogeneous hydroxyl stretch distribution. The dynamical linewidth of the photoproduct is narrow, just like the parent dynamical linewidth was prior to hydrogen bond breaking. This fine frequency correlation, which was observed in methanol oligomers in CCl$_4$,[69] does not exist in water. Consequently, the dynamic linewidth of the photoproduct is the asymptotic linewidth for the total hydroxyl distribution (147.3 cm^{-1}).

The contribution of the photoproduct to the vibrational echo correlation spectrum signal amplitude as a function of T_w delay was determined independently from ultrafast infrared transient absorption experiments (3%, 8%, and 15% at $T_w = 800$, 1200, and 1600 fs, see Fig. 2). Solving Eq. (4) for each T_w point displayed in Fig. 6, we find that the photoproduct produces an increase of ≤ 0.5 cm^{-1} to the true dy-

namic linewidth at all T_w delays. This quantity is insignificant compared to the measured linewidths that have error bars of ± 2.5 cm^{-1}. The photoproduct adds negligibly to the measured widths because the vibrational lifetime (1.45 ps) is comparable to the longest decay component in the frequency fluctuations (1.4 ps, see discussion that follows). The frequency fluctuations have nearly randomized the distribution of hydroxyl stretches before the photoproduct contributes significantly to the signal. Consequently, the dynamic linewidths reported in Fig. 6 do not need to be corrected for the contribution from the photoproduct.

We note that the time dependent contribution from hydrogen bond breaking photoproducts was not considered in other recent reports of water FTCFs,[17,18] which interrogated the OH stretch of HOD in D$_2$O. Recent mixed electronic structure and molecular dynamics simulations of water have demonstrated that the frequency fluctuations in D$_2$O are nearly identical to those in water.[54] Yet, the vibrational lifetime of the OH stretch of HOD in D$_2$O is half the duration compared to the OD stretch of HOD in water. Consequently, the photoproducts will contribute much earlier in the studies involving the OH stretch.[17,18] In addition, a larger quantum of energy is deposited upon vibrational relaxation, which could result in increased photoproduct generation through hydrogen bond breaking. Therefore, photoproduct generation might influence the reported FTCFs.

B. Comparisons of experimental results and theoretical models

The data presented above reflect the equilibrium dynamics of water. They are free of artifacts resulting from excited state (1-2 transition) contributions, photoproduct contamination, and finite pulse duration effects. We now compare the water dynamics obtained experimentally with the dynamics that are predicted by three models of water: TIP4P,[50] SPC/E,[51] and SPC-FQ.[53] The comparisons provide a context for discussing the data and rigorously test the accuracy with which these three models predict the dynamics of water.

The following four step sequence[5] is used to compare the theoretically derived dynamics with the experimental results.

(1) A particular water model simulation is implemented.

(2) The FTCF for a given model of water is generated from methods that involve electronic structure (ES) calculations and classical molecular dynamic (MD) simulations.

(3) We calculate the vibrational echo correlation spectra from the FTCF using time dependent diagrammatic perturbation theory.[5,41,52,62,63]

(4) We extract the T_w dependence of the dynamical linewidth and compare to the experimental data.

The mixed ES/MD method used to calculate the normalized FTCF for the OD stretch of HOD in H$_2$O (Refs. 49 and 54) is based on *ab initio* electronic structure calculations of clusters of water molecules. First, configurations of molecules are generated from a given classical molecular dynamics simulation of the HOD/H$_2$O system. Then a representative set of HOD(H$_2$O)$_n$ clusters are extracted from the simulation. For each cluster a series of *ab initio* calculations (using density functional theory) are performed for different

TABLE I. TFCF parameters [see Eq. (5)].

TFCF	Δ_0 (rad/ps) (%)	τ_0 (ps)	Δ_1 (rad/ps) (%)	τ_1 (ps)	Δ_2 (rad/ps) (%)	τ_2 (ps)
Fit	10.1 (41)	0.048	3.8(15)	0.4	10.9(44)	1.4
SPC-FQ	12.0 (44)	0.048	5.5(20)	0.35	9.8(36)	1.45
SPC/E	12.1 (42)	0.031	9.0(32)	0.28	7.4(26)	0.98
TIP4P	15.2 (55)	0.032	10.2(37)	0.34	2.2 (8)	0.90

values of the OD stretch coordinate, and in doing so an anharmonic potential curve that leads to a 0-1 OD transition frequency for the HOD molecule in that cluster is generated. From looking at a hundred or so clusters, it was determined that there is a linear correlation between the OD frequency and the component of the electric field from the H_2O molecules on the D atom in the direction of the OD bond vector (correlation coefficient=0.9). Assuming this same linear correlation holds for the full liquid, the normalized FTCF simply becomes the normalized electric field time correlation function. The latter can be generated easily from a completely classical molecular dynamics simulation. Using this ES/MD method,[49,54] the simulation was actually performed on the neat H_2O system, which provides much better statistics by averaging over all molecules. It was determined that this change makes a negligible difference.[54] With this method, it is very easy to examine various models of water. In this paper we report comparisons to the data for the OD stretch of TIP4P,[50] SPC/E,[51] and SPC-FQ[53] HOD in H_2O.

The FTCFs obtained from the ES/MD simulations are fit with exponential functions to obtain an analytical representation. The numerical FTCFs can be represented exceedingly well by the sum of three exponentials.[5] Calculations with the numerical FTCFs generated by the simulations produced indistinguishable correlation spectra from those generated with the exponential fits. However, considerable calculation time was saved using the analytical forms. These functions have the form

$$C(t) = \Delta_0^2 \exp(-t/\tau_0) + \Delta_1^2 \exp(-t/\tau_1)$$
$$+ \Delta_2^2 \exp(-t/\tau_2). \tag{5}$$

While the time constants in Eq. (5) were obtained directly from the simulations, the magnitudes of the prefactors were multiplied by a constant to reproduce the full width at half maximum of the experimentally measured linear absorption spectrum (170 cm^{-1}).[5] We will henceforth refer to the triexponential fits as the FTCFs. The parameters of the exponential functions describing the FTCFs for each water model are displayed in Table I. Additionally, parameters describing the FTCF obtained from fitting the data are provided in the table. The method by which this FTCF was obtained is discussed below.

We calculated theoretical vibrational echo correlation spectra from the appropriately scaled FTCFs (Step 3, see above) using diagrammatic perturbation theory to obtain the third-order nonlinear material response, including finite pulse durations.[5,41,52,62,63] The calculations included all contributing pathways to the 0-1 and 1-2 peaks.[5,41,52,62,63] We approximated the frequency fluctuations of the 1-2 transition as being the same as the 0-1 transition (harmonic approximation).

This assumption allowed us to use the same FTCF to describe the 0-1 and 1-2 transitions and caused the two peaks to have the same dynamic linewidths and total widths.

Having input the experimentally determined anharmonicity (162 cm^{-1}, see above), the calculations demonstrated that within the harmonic approximation, the 1-2 peak does not contribute significantly at 2500 cm^{-1}. Therefore, the T_w dependent dynamical linewidths of the 0-1 transition in the theoretical correlation spectra are accurately measured by fitting slices at 2500 cm^{-1} with Gaussian line shapes. The same analysis scheme was used for the experimental correlation spectra. Although the harmonic approximation may not be strictly valid for water, the contribution of the 1-2 peak at 2500 cm^{-1} is so small that employing the harmonic approximation should have a negligible influence on the dynamic linewidth. In any event, any contribution from the 1-2 transition is included in both the measurement and the calculations.

Correlation spectra were calculated for each water model at six T_w delays coinciding with the experimental data (T_w = 100, 200, 400, 800, 1200, and 1600 fs) and at T_w=5 ps (to insure that the calculation reached the correct asymptotic limit). Representative theoretical correlation spectra for the SPC/E model of water have been presented previously.[5] Slices of the correlation spectra were selected at 2500 cm^{-1} and fit with Gaussian line shapes to quantify their widths. The Gaussian function accurately fit the slice and provided a measure of the dynamic linewidth.

The dynamic linewidths calculated from the TIP4P (top curve),[50] SPC/E (second from top),[51] and SPC-FQ (third from top)[53] models of water are displayed as a function of T_w delay in Fig. 6. The dynamic linewidths calculated for the TIP4P and SPC/E models were reported previously[5] and are included here for comparison. The data are displayed in Fig. 6 as the filled circles with error bars. The dashed line at 147.3 cm^{-1} represents the asymptotic dynamic linewidth for the 51 fs pulse duration (see Sec. III A 3). The same pulse duration was used in the diagrammatic perturbation theory calculations. The line through the composite data is a fit, which yielded the experimental FTCF (see discussion that follows). The dynamic linewidths calculated from the water model FTCFs and the experimental FTCF, all ascend (at sufficiently long T_w delay) to the same asymptotic value because the FTCFs were scaled to reproduce the linear linewidth (170 cm^{-1}).[5]

The dynamic linewidths extracted from the calculated correlation spectra display the same trend as the experimental data. The widths are narrow at early T_w and increase toward the asymptotic linewidth with increasing T_w delay because spectral diffusion randomizes the frequencies in the

J. Chem. Phys., Vol. 121, No. 24, 22 December 2004

hydroxyl stretch peak. The rate of increase in the dynamic linewidth for each model is determined by the corresponding FTCF. The triexponential fit parameters corresponding to each water model are displayed in Table I. The comparison in Fig. 6 demonstrates that the TIP4P and SPC/E water models describe the dynamics of water that are reflected by the frequency fluctuations of the hydroxyl (OD) stretch relatively poorly.[5] The dynamic linewidths predicted by both models are much too large at $T_w = 100$ fs and increase to the asymptotic linewidth by 1600 fs, considerably faster than the composite data. These models overemphasize the fast fluctuations and do not contain a slow enough component to account for the slowest fluctuations.

The third water model simulated in this work SPC-FQ,[53] better describes the dynamics of water compared to the TIP4P and SPC/E models. The dynamic linewidth calculated from the SPC-FQ model at $T_w = 100$ fs deviates only ~ 8 cm^{-1} from the experimental dynamic linewidth measured at the same T_w delay. In contrast, the dynamic linewidth calculated from the TIP4P and SPC/E models deviated ~ 20 and ~ 14 cm^{-1}, respectively, at $T_w = 100$ fs. Of greater significance is the fact that the dynamic linewidth calculated from the SPC-FQ model increases more slowly towards and does not reach the asymptotic limit by $T_w = 1600$ fs, in contrast to the TIP4P and SPC/E models. These differences indicate that the SPC-FQ model more accurately describes the frequency fluctuations of water.

Based on the results from the water model calculated FTCFs, a triexponential form for the FTCF was used to fit the experimental data. We begin by entering a triexponential function as the FTCF into the full third-order material response calculation.[5,41,52,62,63] As output, simulated correlation spectra are generated. A slice at 2500 cm^{-1} is used to obtain the time dependent dynamic linewidths for each T_w. This output is compared with the experimental data. The process is iteratively repeated, and the parameters in the FTCF are varied until the dynamical linewidths obtained from the simulated correlation spectra produce the best least squares fit to the experimental data.

The triexponential function describing the experimental FTCF was constrained to be consistent with other complementary experiments and calculations to obtain the greatest confidence in the parameters. Constraint of the initial magnitude of the FTCF was achieved by fitting the linear absorption line shape calculated from the FTCF to the absorption spectrum.[5] The FTCF reproduces the long time asymptotic value of the dynamic line width 147.3 cm^{-1}, which results when 51 fs pulses are used. This procedure also causes the pre-exponential factor Δ_i^2 of one the exponential components to be dependent on the other two. As a result, only two of the three Δ's were varied independently in the fitting procedure. The amplitudes of the Δ's in Table I reflect the appropriate initial magnitude to reproduce the linear line shape. In all three simulations of the FTCF, the fastest component is motionally narrowed, that is, $\Delta_0 \tau_0 < 1$. In this case, this component contributes an exponential decay in the time domain to the total time domain vibrational echo decay with decay rate $\Delta_0^2 \tau_0$. Therefore, τ_0 and Δ_0 are not independent. The SPC-FQ provided the FTCF that produced the T_w dependent

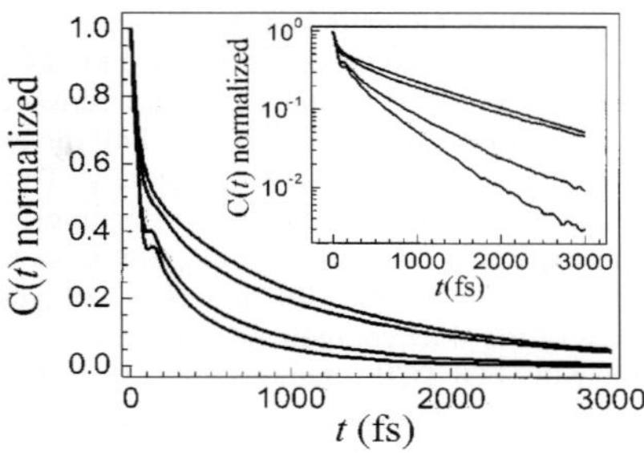

FIG. 7. Comparison of the FTCF corresponding to the TIP4P, (bottom curve) SPC/E, (second from bottom) SPC-FQ (third from bottom) models and the experimental FTCF (top curve). The inset displays the same FTCFs on a semi-long plot with the curves in the same order as in the main portion of the figure. The SPC-FQ model gives an FTCF that is remarkably close to the experimental FTCF, and it is far superior to the other two.

linewidths that are closest to the data. Therefore, $\tau_0 = 48$ fs was fixed and only Δ_0 was varied. Three pulse photon echo peak shift measurements were made on the intermediate time scale and gave $\tau_1 = 400$ fs.[5] This measurement provided the intermediate time constant because once the FTCF has decayed sufficiently, the vibrational echo peak shift as a function of T_w gives the time dependence of the FTCF.[70–73] Therefore, in the fitting procedure, three parameters were allowed to vary independently; the preexponential factors of two of the components and the time constant of the slowest component.

The line through the data points in Fig. 6 is the result of the fitting procedure. Because the error bars on the measured points are small, the fit is highly constrained. In contrast to the dynamical linewidths produced from the water models, the dynamic linewidth obtained from the experimental FTCF has the correct value at $T_w = 100$ fs. The calculated widths pass through the data points within experimental error. The dynamical linewidth at $T_w = 1600$ fs has not reached the asymptotic value and approaches it relatively slowly.

The normalized FTCF that provided the best fit to the experimental dynamical linewidths is displayed in Fig. 7 (top curve) in addition to the FTCFs for the TIP4P (bottom), the SPC/E (second from bottom), and the SPC-FQ (third from bottom) water models. The triexponential parameters for each FTCF are displayed in Table I. The fit through the experimental data has a 1.4 ps time constant for the slowest component that comprises 44% of the decay. We estimate upper and lower bounds of ± 0.2 ps based on a factor of 2 increase in the sum of the squares of the residuals.

The inset in Fig. 7 displays the FTCFs in the same order as the main portion of the figure but on a semilog scale to emphasize the behavior at long time. It is clear from both the main portion of the figure and the inset that the SPC-FQ model comes much closer to reproducing the data than the other two models. In fact, it is remarkable how close the experimentally derived FTCF and the SPC-FQ FTCFs are. The fast components of the theoretical FTCFs comprise approximately half of the total decay amplitude. This feature is reproduced by the experimentally derived FTCF, even

12442   J. Chem. Phys., Vol. 121, No. 24, 22 December 2004

though this amplitude was allowed to vary independently in the fit. The long time behavior of the TIP4P and SPC/E models compared to the experimental FTCF are consistent with their behavior reported previously.[5] The TIP4P model essentially misses the slow component completely. The SPC/E model does better but overestimates the magnitude of the intermediate and underestimates the magnitude of the slow components. In addition, the SPC/E model underestimates the time scale of the slow component. Comparing the experimental and SPC-FQFTCFs demonstrates that this model more accurately describes the fluctuations of water than the TIP4P or SPC/E models. In fact, the time scale of SPC-FQs slow component (1.45 ps) is identical, within experimental error, to the measured value (1.4 ps, see above discussion). The only difference between the experimentally derived FTCF and the SPC-FQFTCF are relatively small differences in the amplitudes of the components (see Table I). It is interesting to note that we scaled the FTCFs derived from the three water models and from the experiment to fit the OD stretch linear absorption spectrum so that the dynamic linewidths (see Fig. 6) would ascend to the same asymptotic value. However, the FTCF generated from the SPC-FQ model accurately predicted the linear linewidth within a few percent without scaling, which further supports the conclusion that the SPC-FQ model more accurately describes the dynamics of water.

The time dependent structural evolution that is obtained from simulations, enables us to correlate specific motions of water with components of the FTCF. The decay components of the TIP4P FTCF have been assigned for the OH stretch of the HOD/D_2O system.[2–4,74,75] The fast component was assigned to hindered translational motion of the hydrogen bonds (stretching of the hydrogen bond length coordinate). There is also a small contribution from hydrogen bond bending.[2–4,76] This assignment was based on the observation of a shallow oscillation of about one cycle in the FTCF with a period that corresponds to the estimated period of hindered translational motion.[3] The oscillation is also present with varying amplitudes in the FTCFs obtained from simulations for the HOD/H_2O system (see Fig. 7).[49] The SPC-FQFTCF does not display the oscillation but exhibits inflections at the period of the hindered translational motion. It is safe to assume that the same physical interpretation holds for the HOD/H_2O system. Therefore, the fast component of the FTCFs mainly reflects the time scale of fluctuations in the hydrogen bond length coordinate. The fact that the oscillation is damped nearly to zero in the SPC-FQ FTCF indicates that it is essentially overdamped in this model while it is just barely underdamped in the other two models. It should be noted that we simulated vibrational echo correlation spectroscopy and vibrational echo peak shift experiments with 50 fs pulse durations using the TIP4P FTCF, which shows the most prominent indication of an oscillation and determined that the oscillation cannot be observed experimentally. The shallowness of the oscillation and the convolution with the pulses eliminate any trace of the oscillation in the data. Using the actual TIP4P FTCF or the triexponential fit that does not have an oscillation produces identical calculated correlation spectra.

The slower portion of the FTCF calculated for the HOD/D_2O system has been assigned to the influence of hydrogen bond breaking and formation. This assignment was based on the similarity in the time constants between the slow component of the FTCF and that of the hydrogen bond population correlation function and from examining the frequency distributions for hydrogen bonded and nonhydrogen bonded oscillators.[3,4,75,76] This correlation function depends on the rate of hydrogen bond formation *and* breaking. In the simulations, the definition of the existence of a hydrogen bond is in terms of the oxygen-oxygen distance, the oxygen-hydrogen distance, and the angles involved.[77] Again, this identification of the slow component with hydrogen bond equilibration should also hold for the HOD in H_2O system. The experimental results demonstrate the time scales obtained from the TIP4P and SPC/E water models are too fast and contain too little amplitude to describe the dynamics that we observe in water. However, the SPC-FQ model provides a much more accurate description of the dynamics of hydrogen bond equilibration (making and breaking). Recent broadband transient absorption experiments that examine hydrogen bond equilibration, which is observed to be complete in 5 ps,[55] support the assignment of the slowest component of the FTCF as the hydrogen bond equilibration time.

It seems likely that the SPC-FQ model more accurately captures the structural fluctuations of water because it contains a method for including water polarizability through the influence of water molecules on each other.[53] Unlike the TIP4P and SPC/E models, the partial charges in the SPC-FQ model are not fixed but rather change position in response to the electric fields produced by the other water molecules. This feature is absent in the TIP4P and SPC/E models.[50,51] The polarizability term introduces collective effects into the intermolecular pair potential between neighboring water molecules. The partial charges assigned to a particular pair of water molecules are influenced by the collective electric field that results from the water molecules surrounding the pair. The partial charges in part determine the intermolecular pair potential that governs the behavior of the water pair. The intermolecular distance and bond angles are affected by the intermolecular pair potential. This collective effect produces the ~50% slower dynamics in the SPC-FQ model compared with the non-polarizable models.[54,78] The collective effect also enhances the magnitude of the slower fluctuations in the SPC-FQ model (see Table I). The improved description of the dynamics in water that resulted from the inclusion of the polarizability term suggests that collective effects dominate the slower time scale motions of water. It will be important to determine if other polarizable water models also do a superior job of reproducing the experimental data.

IV. CONCLUDING REMARKS

Water dynamics were studied using ultrafast heterodyne detected multidimensional stimulated vibrational echo correlation spectroscopy with full phase information. Five completely independent sets of experiments were performed over a period of several months to establish the reproducibility and reduce the error bars in the measurement of the water dynamics. A procedure was described for accounting for the

J. Chem. Phys., Vol. 121, No. 24, 22 December 2004

influence of finite pulse durations on the vibrational echo correlation spectra. Additional experiments were performed to provide highly accurate measurement of the vibrational anharmonicity (162 cm^{-1}) and excited state lifetime for the OD hydroxyl stretch (1.45 ps). Following excited state relaxation, hydrogen bonds break in water and photoproducts form. Infrared pump-probe experiments were employed to determine the buildup time for the photoproducts (~ 2 ps). Therefore, the dynamic linewidths were collected for times shorter than 2 ps to avoid significant contamination from the photoproduct. Careful measurement of the photoproduct contribution at all T_ws proved that the photoproduct contribution was insignificant at all T_w delays that were reported. Therefore, the vibrational echo correlation spectra provide information free from the influence of photoproducts and without complication from the dynamics of the 1-2 transition. From these data, an accurate picture of water dynamics emerged.

The dynamic linewidth, measured as a horizontal cut at 2500 cm^{-1} across the two-dimensional 0-1 peak of the correlation spectra (see Fig. 3), displays the time evolving frequency distribution sampled by water molecules as a function of T_w delay. Analysis of the dynamic linewidths demonstrates that the frequency fluctuations of water are not complete by 2 ps. Fitting the T_w dependent dynamic linewidths using time dependent diagrammatic perturbation theory provided the experimental FTCF.

Electronic structure calculations and molecular dynamics simulations gave the FTCF for the polarizable SPC-FQ water model as well as two nonpolarizable models (TIP4P and SPC/E). These FTCFs were used to calculate the experimental observables, which were compared to the data in Fig. 6. In Fig. 7, the calculated FTCFs and the experimental FTCF are compared. The SPC-FQ FTCF is very close to the experimental FTCF (see Fig. 7 and Table I), while the other two calculated FTCFs have substantial deviations from experiment. The SPC-FQ water model is "polarizable" while TIP4P and SPC/E models are not. The polarizability aspect of the SPC-FQ model may be responsible for its accuracy.

The simulations provide qualitative insights into the physical meaning of the dynamics measured on different time scales. The very short time dynamics are associated with hydrogen bond length fluctuations. The longest time scale observed in the measurements is associated with hydrogen bond equilibration, the making and breaking of hydrogen bonds. The experimental FTCFs slow component comprized $\sim 44\%$ of the decay with a time constant of 1.4 ps. The slow component of the FTCF derived from the SPC-FQ model contained a 1.45 ps time constant that comprised $\sim 36\%$ of the decay. The accurate description of hydrogen bond equilibration is most important for describing water as a chemical and biological agent because it is on this time scale that the complex hydrogen bond network structure undergoes global evolution.

ACKNOWLEDGMENTS

J.B.A., T.S., K.K., and M.D.F. would like to thank AFOSR (Grant No. F49620-01-1-0018) and NIH (Grant No. 2R01GM061137-05) for support of this research and additional support provided by NSF DMR (Grant No. DMR-0332692). C.P.L. and J.L.S. would like to thank NSF (Grant Nos. CHE-9816235 and CHE-0132538) for support of this research. S.A.C. acknowledges the support of a Ruth L. Kirschstein National Research Service Award administered through NIH.

APPENDIX: INFLUENCE OF FINITE PULSE DURATIONS AND COMBINING DATA SETS

Finite-duration laser pulses influence the data collected in a vibrational echo experiment through a time domain (time ordered) convolution between the ideal signal (which would be observed with infinitely short pulses) and the pulses' temporal envelopes. Vibrational echo correlation spectroscopy is a time-domain experiment. The two frequency axes in the correlation spectra, denoted as $S_{\mathrm{VECS}}(\omega_\tau, T_w, \omega_m)$, result from Fourier transformation of the τ and τ_s time dimensions, respectively,

$$S_{\mathrm{VECS}}(\omega_\tau, T_w, \omega_m) = 2\,\mathrm{Re}\int_{-\infty}^{\infty} d\tau \exp(i\omega_\tau \tau)$$

$$\times \int_{-\infty}^{\infty} d\tau_s \exp(i\omega_m \tau_s) P^{(3)}(\tau, T_w, \tau_s)$$

$$\times \int_{-\infty}^{\infty} d\tau_s \exp(i\omega_m \tau_s) E_{\mathrm{LO}}^{*}(\tau_s). \quad \text{(A1)}$$

The product of Fourier transforms along the τ_s dimension results from the hererodyne amplified signal [Eq. (1)] and occurs experimentally in the monochromator. The τ dimension is Fourier transformed numerically after the data are collected. $P^{(3)}(\tau, T_w, \tau_s)$ is the third-order polarization in the sample that emits the vibrational echo, which is induced by interaction with the three laser pulses. τ is the time separation between the first and second laser pulses, and τ_s is the time following the third laser pulse, during which the echo is emitted. By resolving the phase of the vibrational echo, we selectively measure the absorptive (real) part of the polarization. $E_{\mathrm{LO}}^{*}(\tau_s)$ is the electric field of the local oscillator,

$$E_{\mathrm{LO}}(t) = e^{i\vec{k}_{\mathrm{LO}}\cdot\vec{r} - i\omega_{\mathrm{LO}}\tau_s} e^{-\tau_s^2/2\sigma^2}, \quad \text{(A2)}$$

which determines the frequency range along the ω_m axis over which the heterodyne cross term can be detected. The influence of finite laser pulse duration on the polarization in the sample is described by the following time ordered integral:[5,41,52,62,63]

$$P^{(3)}(\tau, T_w, \tau_s) = \int_0^{\infty}\int_0^{\infty}\int_0^{\infty} \sum_i R_i(t_3, t_2, t_1)$$

$$\times E_1^{*}(\tau_s + \tau + T_w - t_3 - t_2 - t_1)$$

$$\times E_2(\tau_s + T_w - t_3 - t_2)$$

$$\times E_3(\tau_s - t_3)dt_1 dt_2 dt_3, \quad \text{(A3)}$$

where the E_i are the electric fields of the three laser pulses

$$E_i(t) = e^{i\vec{k}_i\cdot\vec{r} - i\omega_i t} e^{-t^2/2\sigma^2}, \quad \text{(A4)}$$

and the R_i describe the material response of the sample.[5,41,52,62,63]

Because the correlation spectra are displayed in the frequency domain, we re-express Equations (A1) through (A4) purely in the frequency domain as

$$S(\omega_\tau, \omega_{T_w}, \omega_m)$$

$$= \mathrm{IRF}^+(\omega_\tau, \omega_{T_w}, \omega_m) \sum_i \widetilde{R}_i^+(\omega_\tau, \omega_{T_w}, \omega_m)$$

$$+ \mathrm{IRF}^-(\omega_\tau, \omega_{T_w}, \omega_m) \sum_i \widetilde{R}_i^-(\omega_\tau, \omega_{T_w}, \omega_m). \qquad (A5)$$

The instrument response function,

$$\mathrm{IRF}^+(\omega_\tau, \omega_{T_w}, \omega_m)$$

$$= \exp\!\left(-\frac{(\omega_\tau + \omega_0)^2}{\sigma_\omega^2} - \frac{(-\omega_\tau + \omega_{T_w} - \omega_0)^2}{\sigma_\omega^2} \right.$$

$$\left. -\frac{(-\omega_{T_w} + \omega_m - \omega_0)^2}{\sigma_\omega^2} - \frac{(-\omega_m - \omega_0)^2}{\sigma_\omega^2} \right), \qquad (A6)$$

applies to the rephasing quantum pathways $\widetilde{R}_i^+$ that result when pulse 1 precedes pulses 2 and 3. The converse instrument response function,

$$\mathrm{IRF}^-(\omega_\tau, \omega_{T_w}, \omega_m)$$

$$= \exp\!\left(-\frac{(\omega_\tau + \omega_{T_w} - \omega_0)^2}{\sigma_\omega^2} - \frac{(-\omega_\tau - \omega_0)^2}{\sigma_\omega^2} \right.$$

$$\left. -\frac{(-\omega_{T_w} + \omega_m - \omega_0)^2}{\sigma_\omega^2} - \frac{(-\omega_m - \omega_0)^2}{\sigma_\omega^2} \right), \qquad (A7)$$

applies to the nonrephasing quantum pathways $\widetilde{R}_i^-$ that result when pulse 2 precedes pulses 1 and 3. Here, the frequencies ω_τ, ω_{T_w}, and ω_m are Fourier conjugates of the time variables τ, T_w, and τ_s. σ_ω is the standard deviation of the Gaussian fit to the electric field level spectrum of the laser pulses, and ω_0 is the center frequency of the laser pulses. The instrument response function in the frequency domain [Eqs. (A6) and (A7)] contains the influence of the time-ordered integration ever the pulses expressed in Eqs. (A1) and (A3). The vibrational echo response in the frequency domain, expressed as $\Sigma_i \widetilde{R}_i^\pm(\omega_\tau, \omega_{T_w}, \omega_m)$, is the three-dimensional Fourier transform of its time domain equivalent $\Sigma_i R_i(t_3, t_2, t_1)$.[5,41,52,62,63]

From Eq. (A5), the influence of finite pulse duration can simply be divided out of the vibrational echo signal expressed in the frequency domain. However, the vibrational echo signal represented in Eq. (A5) represents the T_w period in the frequency domain, which is counter intuitive. We want the T_w period expressed in the time domain. Unfortunately, the simplicity of the influence of finite pulse duration is lost when the T_w period is expressed in the time domain. For the purpose of correcting the experimental correlation spectra, we address this problem with the assumption that the dynamics in the T_w period are slow compared to the pulse duration. This assumption removes the dependence of the instrument response function on ω_{T_w}. The instrument response function

is reduced to a two-dimensional Gaussian line shape depending only on ω_τ and ω_m. Following the above assumption, the dependence on pulse spectrum of the signal in the correlation spectra can be described by the expression

$$S_{\mathrm{VECS}}(\omega_\tau, T_w, \omega_m)$$

$$= \exp[-(\omega_\tau - \omega_0)^2/2\sigma_i^2]$$

$$\times \exp[-(\omega_m - \omega_0)^2/2\sigma_i^2] \sum_i \widetilde{R}_i^\pm(\omega_\tau, T_w, \omega_m), \quad (A8)$$

where σ_i is the standard deviation of the intensity level laser spectrum. Therefore, dividing the correlation spectra by the intensity level laser spectrum along both frequency axes deconvolves the correlation spectra from the laser pulses and produces the correlation spectrum that would be measured with infinite bandwidth (delta function) pulses.

In the analysis of the correlation spectra presented in the main body, we selected cross sectional cuts of the 0-1 peaks in the correlation spectra at $\omega_m = 2500 \, \mathrm{cm}^{-1}$ to describe the dynamic line shapes as a function of T_w delay. Looking at a single ω_m value, the effect on the data of the instrument response function along the ω_m axis reduces to an amplitude factor that does not impact the analysis. Consequently, we can neglect the ω_m dependence of the instrument response function in the following discussion. By fitting the cross sectional cuts of the correlation spectra with Gaussian line shapes, the vibrational echo signals at $\omega_m = 2500 \, \mathrm{cm}^{-1}$ are described by the expression

$$S_{\mathrm{VECS}}(\omega_\tau, T_w) = \exp[-(\omega_\tau - \omega_0)^2/2\sigma_{\mathrm{ob}}^2(T_w)], \qquad (A9)$$

where $2.35\sigma_{\mathrm{ob}}(T_w)$ are the FWHM of the Gaussian fits to the dynamic line shapes that are measured as a function of T_w delay (see Fig. 4). Combining Eqs. (A8) and (A9) and evaluating at $\omega_m = 2500 \, \mathrm{cm}^{-1}$, the following expression relates the FWHM of the Gaussian fits to the instrument response function and vibrational echo response, $\widetilde{R}_{0\text{-}1}^\pm(\omega_\tau, T_w)$,

$$\exp[-(\omega_\tau - \omega_0)^2/2\sigma_{\mathrm{ob}}^2(T_w)]$$

$$= \exp[-(\omega_\tau - \omega_0)^2/2\sigma_i^2]\widetilde{R}_{0\text{-}1}^\pm(\omega_\tau, T_w). \qquad (A10)$$

Approximating the vibrational echo response as a Gaussian and taking the natural logarithm of Eq. (A10) produces the equality (assuming the laser is centered on the 0-1 transition)

$$1/2\sigma_{\mathrm{ob}}^2(T_w) = 1/2\sigma_i^2 + 1/2\sigma_R^2(T_w), \qquad (A11)$$

where $2.35\sigma_R(T_w)$ are the FWHM of the Gaussian fits to the vibrational echo response that would be observed in the limit of infinite bandwidth laser pulses ($\sigma_i \rightarrow \infty$). Solving Eq. (A11) for $\sigma_R(T_w)$, we obtain an expression for the width of the vibrational echo response as a function of T_w delay in terms of the experimentally observed widths, $\sigma_{\mathrm{ob}}(T_w)$, and the width of the intensity level laser spectrum σ_i

$$\sigma_R(T_w) = \sigma_i \sigma_{\mathrm{ob}}(T_w)/\sqrt{\sigma_i^2 - \sigma_{\mathrm{ob}}^2(T_w)}. \qquad (A12)$$

Equation (A12) removes the effect of finite pulse bandwidth on the dynamic linewidth measured from $\omega_m = 2500 \, \mathrm{cm}^{-1}$

J. Chem. Phys., Vol. 121, No. 24, 22 December 2004

slices that were selected from the correlation spectra. To be consistent with the data presented in Fig. 4, we solve Eq. (A11) for $\sigma_{\mathrm{ob}}(T_w)$,

$$\sigma_{\mathrm{ob}}(T_w) = \sigma_R(T_w)\sigma_i / \sqrt{\sigma_R^2(T_w) + \sigma_i^2}. \tag{A13}$$

Substitution of the solutions for $\sigma_R(T_w)$ from Eq. (A12) into Eq. (A13) enables the data to be recast with the average pulse duration used in the experiment (51 fs). It should be noted that the correction scheme presented above is general for all ω_m frequency slices, not just the 2500 cm^{-1} slice.

We note that we used full third-order diagrammatic perturbation theory to calculate the dynamic linewidths displayed in Fig. 5(a), which rigorously calculated the effect of finite-bandwidth pulses. The calculations did not use the simplifying assumption that the dynamics during the T_w period are slow compared to the pulse duration, which we made in the development of Eqs. (A12) and (A13). The fact that the three sets of dynamic line widths displayed in Fig. 5(b) superimpose exactly onto each other demonstrates that our treatment precisely corrected the effects of finite-bandwidth pulses and validates the assumption of slow dynamics compared to the pulse duration during the T_w period. If considerably longer pulses were used, the assumption would no longer be valid and the dynamic line widths would report the convolution between the spectral diffusion dynamics during the T_w period and the pulse duration.

[1] P. Schuster, G. Zundel, and C. Sandorfy, *The Hydrogen Bond Recent developments in theory and experiments* (North-Holland, Amsterdam, 1976).

[2] C. P. Lawrence and J. L. Skinner, J. Chem. Phys. **117**, 8847 (2002).

[3] C. P. Lawrence and J. L. Skinner, J. Chem. Phys. **118**, 264 (2003).

[4] R. Rey, K. B. Møller, and J. T. Hynes, J. Phys. Chem. A **106**, 11993 (2002).

[5] J. B. Asbury, T. Steinel, C. Stromberg, S. A. Corcelli, C. P. Lawrence, J. L. Skinner, and M. D. Fayer, J. Phys. Chem. A **108**, 1107 (2004).

[6] V. Daggett and A. Fersht, Nat. Rev. Mol. Cell Biol. **4**, 497 (2003).

[7] A. Almond and J. K. Sheehan, Glycobiology **10**, 329 (2000).

[8] D. Spangberg, R. Rey, J. T. Hynes, and K. Hermansson, J. Phys. Chem. B **107**, 4470 (2003).

[9] G. D. Smith, D. Bedrov, and O. Borodin, Phys. Rev. Lett. **85**, 5583 (2000).

[10] G. C. Pimentel and A. L. McClellan, *The Hydrogen Bond* (Freeman, San Francisco, 1960).

[11] U. Liddel and E. D. Becker, Spectrochim. Acta **10**, 70 (1957).

[12] W. Mikenda, J. Mol. Struct. **147**, 1 (1986).

[13] A. Novak, in *Structure and Bonding*, edited by J. D. Dunitz (Springer, Berlin, 1974), Vol. 18, p. 177.

[14] E. T. J. Nibbering and T. Elsaesser, Chem. Rev. (Washington, D.C.) **104**, 1887 (2004).

[15] J. Stenger, D. Madsen, P. Hamm, E. T. J. Nibbering, and T. Elsaesser, J. Phys. Chem. A **106**, 2341 (2002).

[16] J. Stenger, D. Madsen, P. Hamm, E. T. J. Nibbering, and T. Elsaesser, Phys. Rev. Lett. **87**, 027401 (2001).

[17] C. J. Fecko, J. D. Eaves, J. J. Loparo, A. Tokmakoff, and P. L. Geissler, Science **301**, 1698 (2003).

[18] S. Yeremenko, M. S. Pshenichnikov, and D. A. Wiersma, Chem. Phys. Lett. **369**, 107 (2003).

[19] T. Steinel, J. B. Asbury, S. A. Corcelli, C. P. Lawrence, J. L. Skinner, and M. D. Fayer, Chem. Phys. Lett. **386**, 295 (2004).

[20] H. Graener, G. Seifert, and A. Laubereau, Phys. Rev. Lett. **66**, 2092 (1991).

[21] S. Woutersen, U. Emmerichs, and H. J. Bakker, Science **278**, 658 (1997).

[22] R. Laenen, C. Rausch, and A. Laubereau, J. Phys. Chem. B **102**, 9304 (1998).

[23] R. Laenen, C. Rausch, and A. Laubereau, Phys. Rev. Lett. **80**, 2622 (1998).

[24] G. M. Gale, G. Gallot, F. Hache, N. Lascoux, S. Bratos, and J. C. Leicknam, Phys. Rev. Lett. **82**, 1068 (1999).

[25] H. J. Bakker, S. Woutersen, and H. K. Nienhuys, Chem. Phys. **258**, 233 (2000).

[26] S. Woutersen and H. J. Bakker, Phys. Rev. Lett. **83**, 2077 (1999).

[27] S. Bratos, G. M. Gale, G. Gallot, F. Hache, N. Lascoux, and J. C. Leicknam, Phys. Rev. E **61**, 5211 (2000).

[28] J. C. Deak, S. T. Rhea, L. K. Iwaki, and D. D. Dlott, J. Phys. Chem. A **104**, 4866 (2000).

[29] A. Pakoulev, Z. Wang, Y. Pang, and D. D. Dlott, Chem. Phys. Lett. **380**, 404 (2003).

[30] A. Pakoulev, Z. Wang, and D. D. Dlott, Chem. Phys. Lett. **371**, 594 (2003).

[31] T. Wang, D. Du, and F. Gai, Chem. Phys. Lett. **370**, 842 (2003).

[32] Z. Wang, Y. Pang, and D. D. Dlott, J. Chem. Phys. **120**, 8345 (2004).

[33] M. F. Kropman, H.-K. Nienhuys, S. Woutersen, and H. J. Bakker, J. Phys. Chem. A **105**, 4622 (2001).

[34] G. Gallot, N. Lascoux, G. M. Gale, J. C. Leicknam, S. Bratos, and S. Pommeret, Chem. Phys. Lett. **341**, 535 (2001).

[35] R. Laenen, K. Simeonidis, and A. Laubereau, J. Phys. Chem. B **106**, 408 (2002).

[36] D. Zimdars, A. Tokmakoff, S. Chen, S. R. Greenfield, M. D. Fayer, T. I. Smith, and H. A. Schwettman, Phys. Rev. Lett. **70**, 2718 (1993).

[37] A. Tokmakoff and M. D. Fayer, J. Chem. Phys. **103**, 2810 (1995).

[38] K. D. Rector and M. D. Fayer, Laser Chem. **19**, 19 (1999).

[39] P. Hamm, M. Lim, and R. M. Hochstrasser, Phys. Rev. Lett. **81**, 5326 (1998).

[40] J. B. Asbury, T. Steinel, and M. D. Fayer, J. Lumin. **107**, 271 (2004).

[41] S. Mukamel, Annu. Rev. Phys. Chem. **51**, 691 (2000).

[42] D. M. Jonas, Annu. Rev. Phys. Chem. **54**, 425 (2003).

[43] M. Khalil, N. Demirdoven, and A. Tokmakoff, J. Phys. Chem. A **107**, 5258 (2003).

[44] M. Khalil, N. Demirdoven, and A. Tokmakoff, Phys. Rev. Lett. **90**, 047401(4) (2003).

[45] J. B. Asbury, T. Steinel, C. Stromberg, K. J. Gaffney, I. R. Piletic, A. Goun, and M. D. Fayer, Chem. Phys. Lett. **374**, 362 (2003).

[46] J. D. Hybl, A. A. Ferro, and D. M. Jonas, J. Chem. Phys. **115**, 6606 (2001).

[47] R. R. Ernst, G. Bodenhausen, and A. Wokaun, *Nuclear Magnetic Resonance in One and Two Dimensions* (Oxford University Press, Oxford, 1987).

[48] J. B. Asbury, T. Steinel, C. Stromberg, K. J. Gaffney, I. R. Piletic, and M. D. Fayer, J. Chem. Phys. **119**, 12981 (2003).

[49] S. Corcelli, C. P. Lawrence, and J. L. Skinner, J. Chem. Phys. **120**, 8107 (2004).

[50] W. L. Jorgensen, J. Chandrasekhar, J. D. Madura, R. W. Impey, and M. L. Klein, J. Chem. Phys. **79**, 926 (1983).

[51] H. J. C. Berendsen, J. R. Grigera, and T. P. Straatsma, J. Phys. Chem. **91**, 6269 (1987).

[52] S. Mukamel, *Principles of Nonlinear Optical Spectroscopy* (Oxford University Press, New York, 1995).

[53] S. W. Rick, S. J. Stuart, and B. J. Berne, J. Chem. Phys. **101**, 6141 (1994).

[54] S. Corcelli, C. P. Lawrence, J. B. Asbury, T. Steinel, M. D. Fayer, and J. L. Skinner, J. Chem. Phys. **121**, 8897 (2004).

[55] T. Steinel, J. B. Asbury, and M. D. Fayer, J. Phys. Chem. B (in press).

[56] H.-K. Nienhuys, S. Woutersen, R. A. van Santen, and H. J. Bakker, J. Chem. Phys. **111**, 1494 (1999).

[57] D. Cringus, S. Yeremenko, M. S. Pshenichnikov, and D. A. Wiersma, J. Phys. Chem. B **108**, 10376 (2004).

[58] K. J. Gaffney, I. R. Piletic, and M. D. Fayer, J. Chem. Phys. **118**, 2270 (2003).

[59] T. Steinel, J. B. Asbury, and M. D. Fayer, presented at the 51th Annual Western Spectroscopy Association Conference, Asilomar, California, USA, 2004 (unpublished).

[60] T. Steinel, J. B. Asbury, and M. D. Fayer (unpublished).

[61] J. B. Asbury, T. Steinel, and M. D. Fayer, in *Femtosecond Laser Spectroscopy: Progress in Lasers*, edited by P. Hannaford (Kluwer, Brussels, 2004).

[62] W. M. Zhang, V. Chernyak, and S. Mukamel, J. Chem. Phys. **110**, 5011 (1999).

[63] S. Mukamel and R. F. Loring, J. Opt. Soc. Am. B **3**, 595 (1986).

[64] D. Kroh and A. Ron, Chem. Phys. Lett. **36**, 527 (1975).

[65] M. R. J. Healy, *Matrices for Statisticians* (Clarendon, Oxford, 1986).

[66] A. Tokmakoff, J. Phys. Chem. A **104**, 4247 (2000).

[67] J. D. Hybl, Y. Christophe, and D. M. Jonas, Chem. Phys. **266**, 295 (2001).

[68] J. B. Asbury, T. Steinel, C. Stromberg, K. J. Gaffney, I. R. Piletic, A. Goun, and M. D. Fayer, Phys. Rev. Lett. **91**, 237402 (2003).

[69] J. B. Asbury, T. Steinel, and M. D. Fayer, J. Phys. Chem. B **108**, 6544 (2004).

[70] W. de Boeij, M. S. Pshenichnikov, and D. A. Wiersma, Chem. Phys. Lett. **253**, 53 (1996).

[71] T. H. Joo, Y. W. Jia, J. Y. Yu, D. M. Jonas, and G. R. Fleming, J. Chem. Phys. **104**, 6089 (1996).

[72] K. F. Everitt, E. Geva, and J. L. Skinner, J. Chem. Phys. **114**, 1326 (2001).

[73] A. Piryatinski and J. L. Skinner, J. Phys. Chem. B **106**, 8055 (2002).

[74] C. P. Lawrence and J. L. Skinner, J. Chem. Phys. **117**, 5827 (2002).

[75] K. B. Møller, R. Rey, and J. T. Hynes, J. Phys. Chem. A **108**, 1275 (2004).

[76] C. P. Lawrence and J. L. Skinner, Chem. Phys. Lett. **369**, 472 (2003).

[77] M. Mezei and D. L. Beveridge, J. Chem. Phys. **74**, 622 (1981).

[78] H. Xu, H. A. Stern, and B. J. Berne, J. Phys. Chem. B **106**, 2054 (2002).

PRL **94**, 057405 (2005)

PHYSICAL REVIEW LETTERS

week ending
11 FEBRUARY 2005

Dynamics of Water Confined on a Nanometer Length Scale in Reverse Micelles: Ultrafast Infrared Vibrational Echo Spectroscopy

Howe-Siang Tan,[1] Ivan R. Piletic,[1] Ruth E. Riter,[2,*] Nancy E. Levinger,[2] and M. D. Fayer[1]

[1]*Department of Chemistry, Stanford University, Stanford, California 94305, USA*
[2]*Department of Chemistry, Colorado State University, Fort Collins, Colorado 80523, USA*
(Received 18 June 2004; published 9 February 2005)

The dynamics of water, confined on a nanometer length scale (1.7 to 4.0 nm) in sodium bis-(2-ethylhexyl) sulfosuccinate reverse micelles, is directly investigated using frequency resolved infrared vibrational echo experiments. The data are compared to bulk water and salt solution data. The experimentally determined frequency-frequency correlation functions show that the confined water dynamics is substantially slower than bulk water dynamics and is size dependent. The fastest dynamics ($\sim$50 fs) is more similar to bulk water, while the slowest time scale dynamics is much slower than water, and, in analogy to bulk water, reflects the making and breaking of hydrogen bonds.

DOI: 10.1103/PhysRevLett.94.057405

PACS numbers: 78.30.Cp, 33.15.Vb, 68.65.−k, 78.47.+p

Water plays a major role in a multitude of physical and biological systems. The key feature of liquid water is its formation of dynamic hydrogen bond networks that are responsible for water's unique properties [1]. Hydrogen bonds are constantly being formed, broken, lengthened (weakened), and shortened (strengthened). Hydrogen bond network dynamics in bulk water occurs over a range of time scales from tens of fs to ps [2–7]. However, in many processes, water does not exist in its bulk form, but in confined environments on a nanometer length scale. Such "nanoscopic water" is important in biology [8], geology [9], and chemistry [10].

Reverse micelles (RMs) are widely used as model systems for studying water confined on a nanometer length scale [11]. The water molecules are trapped in nanometer size cavities (one to tens of nm) created by surfactant molecules oriented so that their ionic or polar head groups point inward toward the aqueous phase. Sodium bis(2-ethylhexyl) sulfosuccinate (AOT) is a common surfactant used to make RMs. Optical experiments performed on AOT RMs include linear infrared spectroscopy of the hydroxyl stretch [12], librational motions [13], and terahertz frequency spectroscopy of surface modes [14]. Ultrafast time-resolved experiments have been performed on probe molecules in RMs. These experiments include visible fluorescence spectroscopy [15] and mid-IR pump-probe spectroscopy [16]. The interpretations of such experiments require deciphering the influence of the water dynamics on the observables associated with the probe molecule. Nonetheless, these important results indicate that there are changes in the water dynamics as the size of the water "nanopool" decreases.

Recent vibrational echo experiments [3,5,17] and molecular dynamics simulations [5,6,17] on bulk water show that the structural dynamics associated with the hydrogen bond network can be roughly divided into two classes, very fast local motions involving mainly small changes in hydrogen bond lengths with some angular changes ($<$ 400 fs)

and a longer time scale with more global changes that involve hydrogen bond formation and breaking ($\sim$1.5 ps). In this Letter, we use ultrafast infrared spectrally resolved stimulated vibrational echo (SVE) decays and spectrally resolved vibrational echo peak shift (VEPS) measurements to directly examine the differences between bulk water dynamics and the dynamics of water confined in AOT RMs. In bulk water, the hydrogen bond network is effectively infinite in extent. In the confined water studied here, the nanopools of water range from $\sim$1000 water molecules down to $\sim$50 water molecules. The hydrogen bond networks are not infinite in extent, and they have a rough spherical surface boundary (the RM head groups). The important questions that are directly addressed are the following: how does confinement influence the hydrogen bond network dynamics of water; is the influence of confinement sensitive to the size of the nanopools; and does confinement influence dynamics on all time scales to the same extent?

Using SVE experiments, the frequency evolution (spectral diffusion) of the hydroxyl stretch is monitored. Because the frequency of the hydroxyl stretch is very sensitive to the number of hydrogen bonds a water molecule makes and the strengths of the hydrogen bonds [6,7,18], measuring the spectral diffusion provides direct information on the structural evolution of hydrogen bond networks.

SVE experiments are used to study hydrogen bond networks of water confined in nanometer size pools as a function of pool size, and the results are compared to those obtained on bulk water and NaCl solutions. The ensemble averaged measurement of the vibrational frequency evolution is characterized by the frequency-frequency correlation function (FFCF), $C(t) = \langle \delta\omega(t)\delta\omega(0)\rangle$ [19]. Using the vibrational echo observables, the FFCFs for the various samples are calculated and compared. In the experiments, the dilute OD hydroxyl stretch of HOD in H_2O is investigated to eliminate vibrational excitation transport and to

0031-9007/05/94(5)/057405(4)$23.00

PRL **94**, 057405 (2005)

week ending
11 FEBRUARY 2005

TABLE I. FFCF parameters [Eq. (1)] and T_1's

Sample	Δ_1 (cm^{-1})	τ_1 (ps)	$1/\pi T_2^*$ (cm^{-1})	Δ_2 (cm^{-1})	τ_2 (ps)	Δ_3 (cm^{-1})	τ_3 (ps)	T_1 (ps)
Bulk water[a]	60	0.045	61	20	0.41	55	1.5	1.5
6 M NaCl	74	0.027	56	22	0.32	50	2.1	3.4
10	59	0.033	42	35	0.46	61	5[b]	2.5
5	38	0.054	30	28	0.27	47	8[b]	4.4
2	35	0.065	30	15	0.16	51	22[b]	6.3
2 (peak shift)	28	0.11	32	22	0.15	44	16[b]	6.3

[a]References [5,17].
[b]Error bars: Random error equals ± 1 ps. For $w_0 = 2$, two values are obtained depending on which experimental observable is fit (see text). The difference is not a random error.

assure that the absorption of the sample is not too high. Detailed simulations of water show that experiments on the OD stretch provide an accurate measurement of water dynamics [20]. The FFCF for the OD stretch in bulk water has been determined using SVE experiments and compared to FFCFs obtained from molecular dynamics simulations [3,5,17].

Monodispersed AOT RMs were prepared in CCl$_4$ solvent, with $w_0 = [H_2O]/[AOT] = 10$, 5, and 2. The size of the water pool at the center of the RM can be assigned from w_0 [21]. The nanopool sizes are approximately 4.0, 2.6, and 1.7 nm, containing ~ 1000, ~ 300, and ~ 50 water molecules, respectively [21]. The absorbances were ~ 0.3 for the OD stretch in all samples. Transform limited 50 fs tunable ~ 4 μm pulses were produced using a Ti:Sapphire regenerative amplifier pumped optical parametric amplifier (OPA) system. The OPA was tuned to match the peak of the absorption spectrum of the sample under investigation. The pulses have sufficient bandwidth (> 300 cm^{-1}) to span the entire hydroxyl stretch spectrum. The laser pulse is split into three beams and focused into the sample. The pulse delays between the first-second and second-third pulses are τ and T_w, respectively. The SVE signal was passed through a monochromator prior to detection. The frequency resolved SVE signal was measured at the absorption peak frequency of each sample for the majority of the experiments, but some frequency dependent results are briefly discussed. Frequency resolution avoids the ambiguity introduced when signals from both the 0-1 and 1-2 vibrational transitions are observed simultaneously in a nonfrequency resolved experiment. In addition, the presence or lack of a detectable wavelength dependence provides useful information. SVE traces were collected as a function of τ for a series of T_w's. The decay traces were used to obtain both the time dependence of the SVE signal and the VEPS as a function of T_w. Peak shift data from $T_w = 0$ fs to many picoseconds are reported. At $T_w = 0$ fs, the peak shifts for all of the RMs are >100 fs. There is a narrow nonresonant spike centered at $T_w = 0$ fs, which is small but still present by $\tau = 100$ fs. The nonresonant signal precludes analyzing the shapes of the decays at $T_w = 0$ fs, but, within a small

uncertainty, the $T_w = 0$ fs peak shift can be recovered. By $T_w = 200$ fs, there is no nonresonant signal in the data. IR pump-probe experiments were used to obtain the vibrational lifetimes, T_1 (see Table I, last column) used in the data analysis.

The measured spectral peak positions (cm^{-1}) and widths (FWHM, cm^{-1}) of the OD stretching mode for bulk water, a 6 M NaCl solution, and $w_0 = 10$, 5, and 2 RMs are (2506, 170); (2526, 149); (2539, 174); (2558, 160); (2566, 156), respectively. The spectra shift to higher frequency (blueshift) and become narrower with decreasing w_0 [12]. If the trends in bulk water [1,6,7] are followed in RMs, the blueshift with decreasing nanopool size suggests a decrease in the equilibrium number of hydrogen bonds and possibly a decrease in the average hydrogen bond strength.

Figure 1 displays VEPS data for five samples. The curves through the RM data are empirical biexponential fits to aid the initial qualitative discussion. The peak shifts as a function of T_w are related to the decay of the FFCF [22]. The data in this form enable the qualitative trends to be seen prior to the detailed analysis given below. It can be

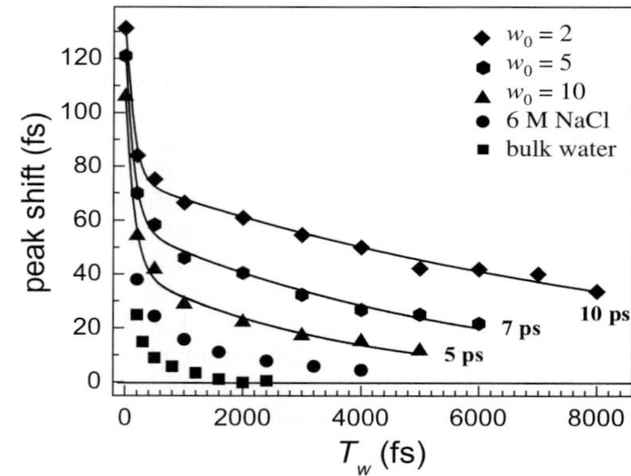

FIG. 1. Spectrally resolved vibrational echo peak shift data for bulk water (squares), 6 M NaCl water solution (circles), $w_0 = 10$ (triangles), 5 (hexagons), and 2 (diamonds) reverse micelles. The lines through the RM data are empirical fits (see text).

PRL **94**, 057405 (2005)

PHYSICAL REVIEW LETTERS

week ending
11 FEBRUARY 2005

seen that the dynamics slow as the nanopool size is decreased. The long components of the empirical fits to the decays in Fig. 1 are 5, 7, and 10 ps for $w_0 = 10$, 5, and 2, respectively. Theory shows that the peak shift data decay in the same manner as the FFCF at sufficiently long time [22]. Therefore, the long component of the empirical fit to the peak shift data should be similar to the FFCF determined by theoretical calculations.

Experiments have shown that the presence of ions can affect water dynamics [23]. If the ionic head groups and the Na^+ counter ions were free ions in bulk water with the same number of ions per water molecule, the ionic strength for water in a $w_0 = 10$ RM would be equivalent to an ~ 5.6 M NaCl solution. The data show that the high ionic strength of the 6 M NaCl solution does have an influence on the peak shift data, slowing the decay compared to bulk water, and there is a significant effect on both the FFCF and T_1 (see Table I). However, the $w_0 = 10$ data decay is substantially slower. The slowest component of the peak shift data for $w_0 = 10$ is 5 ps compared to 2.1 ps for the NaCl solution (see Table I). In addition, simulations show that the Na^+ are mainly associated with the head groups, not free ions [24]. Therefore, the ionic strength is low in the RM water nanopools, which is very different from a 6 M NaCl solution. Furthermore, infrared pump-probe experiments performed on the stretching mode of the probe azide ion in nanopools of water in *nonionic* RMs show that both the vibrational relaxation rates and orientational anisotropy decays differ from bulk water and become increasingly slow as the nanopool size is reduced [16]. Therefore, the

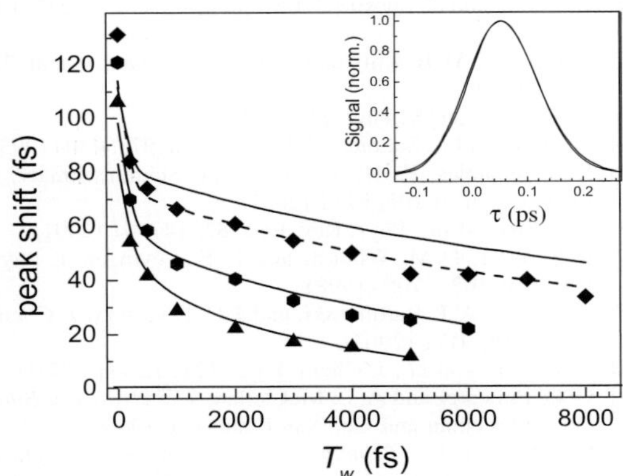

FIG. 2. Inset: $T_w = 200$ fs, $w_0 = 5$ RM vibrational echo data and the calculated curve obtained from the FFCF (Table I) determined by the fit to multiple T_w's and the linear line shape. Main figure: Spectrally resolved vibrational echo peak shift data for $w_0 = 10$ (triangles), 5 (hexagons), and 2 (diamonds) RMs. The solid lines are no adjustable parameter calculations using FFCFs obtained from fitting the vibrational echo decay curves (Table I). The dashed line is a fit to the peak shift data.

differences in RM nanoscopic water dynamics from those of bulk water displayed by the peak shift data are a result mainly of confinement and not significantly a result of ionic strength.

Calculations were performed using diagrammatic perturbation theory [19]. To obtain the FFCF, the SVE decays over the range of T_w's and the linear absorption spectrum were fit simultaneously for each w_0 using an iterative algorithm that included the laser pulse duration. The resulting FFCF was then used to calculate the VEPS with no adjustable parameters. In accord with recent theoretical analysis and simulations of SVE experiments of bulk water [5], a triexponential form for the FFCF,

$$C(t) = \Delta_1^2 e^{-t/\tau_1} + \Delta_2^2 e^{-t/\tau_2} + \Delta_3^2 e^{-t/\tau_3}, \quad (1)$$

was employed.

The inset in Fig. 2 shows one SVE decay curve for $w_0 = 5$ and $T_w = 200$ fs and the theoretically calculated curve obtained once the fit to the T_w data sets and the linear line shape had converged. The main portion of Fig. 2 shows the VEPS data (data and symbols are the same as in Fig. 1) and the calculated peak shifts obtained using the FFCF determined by fitting the decay curves (solid lines). There are no adjustable parameters in the peak shift calculations. The determination of the FFCF did not include the $T_w = 0$ fs data because the shape of the decay curves is obscured by the nonresonant signal. The FFCFs that give rise to the solid lines are given in Table I.

First, consider $w_0 = 10$ and 5. The calculations based on the experimentally determined FFCFs for these two RMs do a very good job of reproducing the peak shift data. The longest component is the same or close to that determined by the empirical fits. The slowest component for $w_0 = 10$ and 5 are 5 and 8 ps, respectively. Both are much slower than the 1.5 ps slowest component for bulk water (see Table I and Fig. 1). The calculated curves miss the $T_w = 0$ fs points by ~ 20 fs, but these points were not in the fits, and their experimental values are uncertain to some extent. The results demonstrate that the calculations are able to quite accurately reproduce the shapes of the decay curves at all T_w's, and thereby obtain an FFCF that reproduces the peak shifts. It is unlikely that there is a slower component of any substantial size that was missed. The $w_0 = 10$ peak shift begins at ~ 108 fs, and by the last data point, it is down to ~ 10 fs. The data and the calculated curve both have a downward slope. Therefore if there is a longer time scale component not contained in the three terms of Eq. (1), its amplitude is very small. The same argument holds true for $w_0 = 5$.

The solid line for $w_0 = 2$ calculated by fitting the shapes of the decay curves and the linear absorption spectrum reproduces the form of the time dependence of the peak shift data but misses the values by ~ 10 fs at the longer times. This is because the FFCF form given in Eq. (1) does not reproduce the shape of the decay as well as the results

PRL **94**, 057405 (2005)   PHYSICAL REVIEW LETTERS   week ending
11 FEBRUARY 2005

shown in the inset of Fig. 2. The fit compromises the shape and misses the peak shifts somewhat. The dashed line is a fit to the peak shifts only. The FFCF parameters for both fits to the $w_0 = 2$ data are given in Table I. The values obtained from the two fits are very similar except for the longest component. The first component is motionally narrowed. Both sets of Δ_1 and τ_1 give almost the same motionally narrowed linewidth $(1/\pi T_2^*)$, ~ 30 cm^{-1}. All of the values reported in Table I are from data taken at the peak of the absorption spectrum for each RM, the bulk water, and the NaCl solution. We have preliminary data on the wavelength dependence of the SVE data. $w_0 = 10$ and 5 show at most a small change in the SVE data at the different wavelengths. However, $w_0 = 2$ shows a very substantial change.

In the theoretical calculations, we have treated the water molecules as belonging to a single ensemble. However, "bound" water molecules, i.e., those directly associated with the surfactant, may have dynamics that are distinct from molecules in the nanopool that are not in direct contact [12,13,24]. The proportion of bound water molecules will increase with decreasing w_0. The different subensembles are not resolvable in the absorption spectra. Theoretical simulations have indicated that the exchange between bound and free water is slower than the time scale of SVE experiments [24]. The wavelength dependence of the $w_0 = 2$ SVE data suggests that the data for these micelles taken at line center contain two subensembles of water molecules that can have different dynamics. This may account for the failure of the fits using the FFCF of Eq. (1) to reproduce the shape of the echo decay curves accurately, resulting in missing the peak shifts. We are currently investigating the wavelength dependence of the data in detail.

Bulk water simulations provide insight into the changes in dynamics observed in RM nanopools. The water FFCF is given in Table I [5,17]. In simulations of water, the fastest component is from the very local fluctuations of hydrogen bond lengths [3,5–7]. This component is motionally narrowed; the motionally narrowed linewidths, $1/\pi T_2^*$, should be compared. The middle component represents the crossover from the very fast to the slow dynamics. The slowest component has been associated with more global structural evolution of the hydrogen bonding network, i.e., the making and breaking of hydrogen bonds [3,5–7]. Simulations of the FFCFs of water in RMs have not yet been performed. We can compare to simulations of bulk water by assuming that the nature of the dynamics is the same. The slowest components of the RM water FFCFs have the largest differences compared to bulk water. While the amplitudes of the slowest components, Δ_3, are similar, the decay times are much longer in the nanopools than in bulk water, and τ_3 increases by more than a factor of 10 as the size of the nanopool is reduced. The increasing τ_3 with decreasing w_0 may be associated with water molecules' inability to move

relative to each other. This is consistent with a reduction in translational and rotational mobility both from confinement as well as the increasing presence of water molecules interacting with the micelle head group boundary [15,24]. The direct measurements of the nanoscopic water dynamics presented here are in accord with the general trends observed in previous time-resolved experiments on probe molecules in RMs [15,16]. A wider range of nanopool sizes and reverse micelles with nonionic head groups are being studied; orientational relaxation measurements will be reported elsewhere.

We thank Ilya Finkelstein for his considerable help with computer programming. This work was supported by the Department of Energy (DE-FG03-84ER13251), NIH (2R01GM061137-05), and NSF DMR (DMR-0332692).

---

*Permanent address: Department of Chemistry, Agnes Scott College, Decatur, GA 30030, USA.

[1] P. Schuster, G. Zundel, and C. Sandorfy, *The Hydrogen Bond. Recent Developments in Theory and Experiments* (North-Holland, Amsterdam, 1976).

[2] H. J. Bakker *et al.*, J. Chem. Phys. **116**, 2592 (2002).

[3] C. J. Fecko *et al.*, Science **301**, 1698 (2003).

[4] E. T. J. Nibbering and T. Elsaesser, Chem. Rev. **104**, 1887 (2004).

[5] J. B. Asbury *et al.*, J. Phys. Chem. A **108**, 1107 (2004).

[6] C. P. Lawrence and J. L. Skinner, J. Chem. Phys. **118**, 264 (2003).

[7] K. B. Møller, R. Rey, and J. T. Hynes, J. Phys. Chem. A **108**, 1275 (2004).

[8] S. K. Pal and A. H. Zewail, Chem. Rev. **104**, 2099 (2004).

[9] R. Sutton and G. Sposito, J. Colloid Interface Sci. **237**, 174 (2001).

[10] S. Habuchi, H. B. Kim, and N. Kitamura, Anal. Chem. **73**, 366 (2001).

[11] N. E. Levinger, Science **298**, 1722 (2002).

[12] G. Onori and A. Santucci, J. Phys. Chem. **97**, 5430 (1993).

[13] D. S. Venables, K. Huang, and C. A. Schmuttenmaer, J. Phys. Chem. B **105**, 9132 (2001).

[14] J. E. Boyd *et al.*, Phys. Rev. Lett. **87**, 147401 (2001).

[15] R. E. Riter, D. M. Willard, and N. E. Levinger, J. Phys. Chem. B **102**, 2705 (1998).

[16] Q. Zhong, A. P. Baronavski, and J. C. Owrutsky, J. Chem. Phys. **119**, 9171 (2003).

[17] J. B. Asbury *et al.*, J. Chem. Phys. **121**, 12431 (2004).

[18] G. C. Pimentel and A. L. McClellan, *The Hydrogen Bond* (W. H. Freeman and Co., San Francisco, 1960).

[19] S. Mukamel, *Principles of Nonlinear Optical Spectroscopy* (Oxford University Press, New York, 1995).

[20] S. Corcelli, C. P. Lawrence, and J. L. Skinner, J. Chem. Phys. **120**, 8107 (2004).

[21] T. Kinugasa *et al.*, Colloids Surf. A **204**, 193 (2002).

[22] M. Cho *et al.*, J. Phys. Chem. **100**, 11944 (1996).

[23] M. F. Kropman and H. J. Bakker, Science **291**, 2118 (2001).

[24] J. Faeder and B. M. Ladanyi, J. Phys. Chem. B **104**, 1033 (2000).

J. Phys. Chem. A **2006**, *110*, 4985−4999

Testing the Core/Shell Model of Nanoconfined Water in Reverse Micelles Using Linear and Nonlinear IR Spectroscopy

Ivan R. Piletic, David E. Moilanen, D. B. Spry, Nancy E. Levinger,[†] and M. D. Fayer*

Department of Chemistry, Stanford University, Stanford, California 94305

Received: February 19, 2006; In Final Form: March 11, 2006

A core/shell model has often been used to describe water confined to the interior of reverse micelles. The validity of this model for water encapsulated in AOT/isooctane reverse micelles ranging in diameter from 1.7 to 28 nm ($w_0 = 2-60$) and bulk water is investigated using four experimental observables: the hydroxyl stretch absorption spectra, vibrational population relaxation times, orientational relaxation rates, and spectral diffusion dynamics. The time dependent observables are measured with ultrafast infrared spectrally resolved pump−probe and vibrational echo spectroscopies. Major progressive changes appear in all observables as the system moves from bulk water to the smallest water nanopool, $w_0 = 2$. The dynamics are readily distinguishable for reverse micelle sizes smaller than 7 nm in diameter ($w_0 = 20$) compared to the response of bulk water. The results also demonstrate that the size dependent absorption spectra and population relaxation times can be quantitatively predicted using a core−shell model in which the properties of the core (interior of the nanopool) are taken to be those of bulk water and the properties of the shell (water associated with the headgroups) are taken to be those of $w_0 = 2$. A weighted sum of the core and shell components reproduces the size dependent spectra and the nonexponential population relaxation dynamics. However, the same model does not reproduce the spectral diffusion and the orientational relaxation experiments. It is proposed that, when hydrogen bond structural rearrangement is involved (orientational relaxation and spectral diffusion), dynamical coupling between the shell and the core cause the water nanopool to display more homogeneous dynamics. Therefore, the absorption spectra and vibrational lifetime decays can discern different hydrogen bonding environments whereas orientational and spectral diffusion correlation functions predict that the dynamics are size dependent but not as strongly spatially dependent within a reverse micelle.

I. Introduction

Water has unique properties that are intimately related to its molecular level structure. Water's properties are strongly influenced by its dynamic hydrogen bond network. The hydrogen bond network and, therefore, the properties of water are sensitive to structural and chemical perturbations. An interface with another phase,[1,2] solutes ranging from ions to large biological molecules,[3,4] pH,[5] temperature and pressure[6,7] or any combination of these impacts the hydrogen bond network structure and dynamics. The modification of hydrogen bond dynamics from their bulk characteristics has important ramifications for protein folding, charge-transfer reactions, and reactions that occur at interfaces or in confined environments. A body of experimental and theoretical work has examined the dynamics of water in a variety of environments, for example, water around large biological molecules,[8] around micelles,[9] in reverse micelles,[10−14] in cyclodextrins,[15] and zeolites.[16] The modified dynamics of water in complex chemical and biological systems probed by dielectric relaxation and solvation dynamics experiments have been reviewed recently.[4]

Recent theories[17] and molecular dynamic simulations[18−20] indicate that at least two types of water molecules are present in complex systems, free and bound. Free suggests water molecules with essentially bulk characteristics separated from the interface in the core of the system. Bound water molecules

are associated with particular groups in proteins, the interface in reverse micelles, etc. and can reside in pockets separated from the bulk water phase. The long time components observed in dielectric relaxation and solvation dynamics experiments have been attributed to exchange between the associated and core populations.[4]

To properly account for the behavior of water in complex systems requires consideration of the complexity of water itself. Bulk water contains many different hydrogen bonding environments determined by the strength and number of local hydrogen bonds.[21] There are faster and slower relaxation time scales that reflect inertial and diffusive dynamics.[22,23] The presence of a solute or interface will modify the hydrogen bond network and new interactions and environments need to be accounted for. As a result, the response of water in complex systems is expected to exhibit even more complexity than bulk water.

Reverse micelles are small molecular assemblies that encapsulate nanoscopic pools of water. The anionic surfactant aerosol-OT (AOT with Na^+ counterion) is known to form spherical reverse micelles in a variety of nonpolar phases. The size of the reverse micelles is readily tuneable by adjusting the w_0 value.

$$w_0 = \frac{[H_2O]}{[AOT]} \qquad (1)$$

Varying the w_0 value allows the dynamics of the nanoscopically confined water to be explored as a function of the water nanopool size. Deviations from bulk behavior can be observed and the size at which bulk-like characteristics appear for a

---

* Corresponding author. E-mail: fayer@stanford.edu.
† Permanent address: Department of Chemistry, Colorado State University, Fort Collins, CO 80523-1872.

10.1021/jp061065c CCC: $33.50   © 2006 American Chemical Society
Published on Web 03/29/2006

4986 *J. Phys. Chem. A, Vol. 110, No. 15, 2006*

particular observable can be determined. The size for the onset of bulk properties is an important question that has been addressed by several groups although different techniques yield different results.[24]

Ultrafast infrared (IR) vibrational spectroscopy[25] is ideally suited to probe the dynamics of water in reverse micelles. Time dependent vibrational measurements on the hydroxyl stretch of water can resolve different dynamical time scales and relate the dynamics to hydrogen bond network structural evolution. In addition to linear IR absorption spectroscopy, two ultrafast vibrational methods are employed to explicate the dynamics of nanoscopic pools of water in reverse micelles, and the results are compared to those for bulk water. Spectrally resolved pump−probe experiments are used to determine the vibrational population relaxation and the orientational relaxation of the OD stretch of HOD in H_2O as a function of reverse micelle size. Spectrally resolved stimulated vibrational echo spectroscopy[26] measures the different contributions to the spectroscopic line shape and thus the underlying system dynamics through spectral diffusion.

The utility of ultrafast vibrational spectroscopy[25] in probing the dynamics of water arises from the intimate connection between the hydrogen bond network and the water hydroxyl stretch. Hydrogen bonding affects the frequency of the hydroxyl stretch significantly, causing the hydroxyl stretch absorption frequency to red shift, as well as the integrated intensity and the spectral width of the hydroxyl band to increase.[27] The broad hydroxyl stretch absorption arises from a broad distribution of hydrogen bonding environments.[28−30] Many studies have analyzed this band for bulk water as well as water in different environments.[12,31−34] Simulations of bulk water have shown the variety of hydrogen bonding environments and their relationship to the absorption spectrum and to dynamical measurements.[22,35] Addressing water dynamics and structure through experiments on the H_2O hydroxyl stretch is complicated because the molecule possesses coupled symmetric and antisymmetric stretch modes that absorb in the same frequency range and overlap substantially. The bend overtone also absorbs within the OH stretch frequency range. The experiments presented below were performed on the OD stretch of dilute HOD in H_2O. This system simplifies the absorption spectrum because the OD stretch mode of HOD is essentially a local mode simplifying the system and making it easier to interpret the dynamics.[12] Furthermore, experiments conducted on the OD stretch of dilute HOD in H_2O eliminate vibrational excitation transfer that interferes with both orientational relaxation and spectral diffusion measurements.

Below we present a comprehensive investigation exploring the effect of confinement on water dynamics in the AOT/water/isooctane reverse micelle system for a large range of sizes. Previous work using these methods exclusively targeted small reverse micelles ($w_0 = 2, 5, 10$) for the AOT/water/CCl_4 system,[12,14,36] demonstrating that confinement substantially affects the hydrogen bond network evolution of water. Here we examine and contrast the results of four OD hydroxyl stretch experimental observables: the IR absorption spectra, vibrational population relaxation times, orientational relaxation rates, and spectral diffusion dynamics. The measurements are conducted on samples from bulk water down to water nanopools with diameter 1.7 nm. All four observables show changes as the size of the nanopool is decreased with major deviations from bulk water occurring for $w_0 = 20$ and smaller.

The experimental results from the four observables will be analyzed in detail using a core−shell model that divides the nanoscopic water pool into two subensembles. Those water molecules that are associated with the headgroups form an outer shell, and those water molecules some distance in from the headgroups form the core. The core will be modeled by the properties of bulk water, and the shell will be modeled by the properties of the $w_0 = 2$ reverse micelle, in which essentially all waters are associated with the headgroups. It will be demonstrated that the size dependent absorption spectra and population relaxation times can be reproduced quantitatively as properly weighted combinations of the bulk water and $w_0 = 2$ spectra and vibrational lifetimes. From the results, the shell thickness is determined. However, the same model fails to reproduce either the spectral diffusion or the orientational relaxation data. It is proposed that strong dynamical coupling between the shell and the core causes the water nanopool to have distinct dynamics from the weighted sum of shell and core dynamics when hydrogen bond structural rearrangement is involved (orientational relaxation and spectral diffusion). However, the absorption spectra and population relaxation times, which significantly depend on the local structure, do reflect the weighted sum of the shell and core properties. Therefore, the separation into core and shell is appropriate for what might be termed static observables but not dynamic observables.

For the core−shell model to be successful in describing the static observables, the time scale for the core and shell populations to exchange must be longer than the time scale of the observables. Thus the exchange must occur slower than ∼10 ps, which encompasses the time scales of time dependent observables presented below. This exchange time scale is supported by simulations in various confined environements[37,38] as well as quasi-elastic neutron scattering experiments[10] that suggest translational mobility of water is severely restricted in these systems. The data and simulations suggest that dynamical exchange takes place on the time scale of tens of picoseconds or longer. Data presented in this paper also illustrate that fast (<10 ps) core−shell population exchange cannot be happening as will be discussed further below.

The paper is organized as follows. Section II describes experimental procedures. Section III presents the results of the four types of experiments. Section IV describes the methods used to analyze the results of each observable in terms of the core−shell model and presents the results of the analysis. The reasons for the successes and failures of the analysis are discussed. Section V presents some concluding comments.

II. Experimental Procedures

Aerosol OT, isooctane (Aldrich Inc.) and water were used without further purification. A 1.0 M stock solution of AOT in isooctane was prepared. This concentration allowed for viable cell path lengths for experiments conducted on the smaller reverse micelles. Karl-Fisher titration was used to determine the residual water content in the surfactant due to its hygroscopic nature (∼0.5 water molecules per headgroup). To prepare solutions of desired w_0, precise volumes of water (5% HOD in H_2O) were added to measured quantities of the stock solution.

The size of the water nanopool is determined by the value of w_0. An empirical formula relating w_0 to the diameter of the water pool may be used for small reverse micelles. The linear relationship $d_{wp} = 0.29w_0 + 1.1$ (nm) applies to reverse micelles in the range $w_0 = 2-20$.[39] For larger reverse micelles, dynamic light scattering measurements provide the diameter of the water pool.[40] In the experiments described in this paper, the sizes ranged from 1.7 to 28 nm. The sizes of the nanopools as well as the approximate number of encapsulated water molecules are listed in Table 1. The numbers were estimated by assuming

J. Phys. Chem. A, Vol. 110, No. 15, 2006 **4987**

TABLE 1: Reverse Micelle Water Pool Sizes and Number of Water Molecules

reverse micelle	diameter, nm	no. of water molecules
$w_0 = 2$	1.7	~40
$w_0 = 5$	2.3	~300
$w_0 = 10$	4.0	~1000
$w_0 = 20$	7.0	~5400
$w_0 = 40$	17	~77000
$w_0 = 60$	28	~350000

that the density of water is preserved within the reverse micelles indicating the level of approximation involved. All samples were housed in cells containing two CaF_2 windows (3 mm thick) and a Teflon spacer with a thickness adjusted to obtain optical densities in the range 0.2–0.4 for all solutions studied. All experiments described in this paper were conducted at 25°C.

The laser system employed in these experiments has been described in detail elsewhere.[12] Briefly, 800 nm pulses are generated by a home-built Ti:sapphire/regenerative amplifier system at 1 kHz. These pulses are used to pump an optical parametric amplifier (OPA) to produce ~4 μJ mid-IR pulses at ~4 μm. The power spectrum of the IR pulses has a Gaussian envelope with a 230 cm^{-1} bandwidth (full width at half-maximum (fwhm)). After the mid-IR pulses are generated, the experimental system is purged with dry CO_2 scrubbed air to remove both atmospheric water absorptions and the strong CO_2 absorption. To characterize the pulses in the time domain, a nonresonant third-order signal is generated that provides a three-pulse correlation from which the pulse duration is determined. Pulse chirp is determined by doing a FROG measurement.[41] Ge and CaF_2 flats, which have opposing effects on the second-order dispersion of the IR pulses, are placed in the beam path at Brewster's angle, and their relative amounts are adjusted to minimize the chirp. The chirp was minimized to be less than a 2 fs across the laser spectrum.

Two types of third-order measurement techniques were employed: pump–probe and stimulated vibrational echo experiments. For the pump–probe experiments, a pump pulse excites chromophores to the first vibrational excited state of the OD hydroxyl stretch of HOD in the H_2O nanopools and the subsequent time evolution is measured with a variably delayed probe pulse. Because the pump and probe pulses are polarized, the decay of the anisotropy (orientational relaxation) as well as the population relaxation can·be measured. In the experiments, the pump beam is polarized horizontally, and the probe is set to ~45° polarization. A polarizer in front of the monochromator is set either horizontal (∥) or vertical (⊥), to record the two components of the anisotropy. To produce directly comparable ∥ and ⊥ datasets needed to extract the decay of the anisotropy and the population, two polarizers, a half-wave plate, and a computer controlled mechanical device for interchanging the polarizers are employed. One polarizer is set ∥. The other polarizer is set ⊥ and then followed by a half wave plate that returns the beam to ∥ following the polarization analysis by the polarizer. Therefore, the light that propagates through the monochromator and measured by an MCT detector is always "∥". A small adjustment of the probe angle around the nominal 45° is made so that in the absence of the pump beam, the MCT detector measured identical intensities for the ∥ and ⊥ polarizations. Scans of the delay line alternated between ∥ and ⊥ every scan. Thus, the magnitude of the ∥ and ⊥ components have the correct relative amplitudes. The correct amplitudes are required because the decay of the anisotropy is so slow in the smaller reverse micelles that tail matching of the data cannot be used. This procedure avoids potential pitfalls in making anisotropy

measurements.[42] The pump beam is chopped and the modulated intensity of the shot-to-shot reference detector normalized probe beam is measured with a lock-in amplifier.

For the stimulated vibrational echo experiments, three variably delayed pulses are crossed in the sample, and the vibrational echo signal emerges in a unique direction. The vibrational echo pulse is dispersed in a monochromator and a particular wavelength is detected with an InSb detector. The monochromator wavelength is set at the peak of the absorption spectrum for each reverse micelle size and bulk water for a majority of the work described in this paper. Spectrally resolving the vibrational echo eliminates contributions from the 1−2 vibrational transition and permits the data to be collected at well-defined wavelengths in the 0−1 absorption spectrum.

In the three-pulse sequence, the time between pulses 1 and 2 is τ, and the time between pulses 2 and 3 is T_w. For fixed T_w, τ is scanned, and a vibrational echo decay curve is recorded. This decay curve reflects the "coherence decay". Its Fourier transform is a dynamic line shape with dynamics on longer time scales eliminated. The dynamics that occur on time scales longer than a single vibrational echo decay can be measured by varying the time delay between the second and third pulses, T_w. An entire decay curve is measured at each T_w, and the full curves are used in a quantitative analysis of the data. As T_w increases, a broader distribution of water structural conformations, and therefore a wider range of frequencies, is sampled through a process termed spectral diffusion. The increase in the sampled frequencies can be viewed as an increase in the dynamic line width that is buried under the absorption spectrum. As the dynamic line width increases, the vibrational echo data decay more rapidly, and the peak of the data move toward $\tau = 0$. For each water nanopool size (reverse micelle size), a set of T_w dependent decay curves is recorded and analyzed using time dependent diagrammatic perturbation theory.[12,43−47] The vibrational lifetime limits the maximum time for which data can be collected. A nonresonant signal is observed for $T_w < 200$ fs. This signal, which is caused by the nonresonant polarizability of the solvent, is centered at $\tau = 0$, and interferes with data for very short T_w. However, because the resonant vibrational echo data are displaced from $\tau = 0$, it is possible to obtain some information even at very short T_w values despite the nonresonant signal. In addition to analyzing the complete decay curves, vibrational echo peak shifts are also employed for comparisons of different reverse micelles and bulk water.[12,14] The peak shift[48] is the difference in the peak of the vibrational echo curve from $\tau = 0$ and has been shown to track the decay of the frequency-frequency correlation function (FFCF) at long times.[48] The entire vibrational echo decay curve, and therefore the peak of the curve is related to the FFCF. The magnitudes of the peak shifts in the experiments are on the order of tens of femtoseconds. Temporal drift between the first two pulses affects the peak shift values. To eliminate the effect of this drift, the two-pulse vibrational echo emitted in a distinct direction was simultaneously detected. The two-pulse echo curve involves only the first two pulses in the sequence; it is independent of T_w. Therefore, its position provides a method to correct for any timing drifts in the system.

In the pump–probe and vibrational echo experiments, vibrational relaxation of the excited hydroxyl stretch can cause local heating of the sample that may be observed on the time scale of 5−100's of ps. The magnitude of the temperature change is small (fraction of 1 K). The signals and local heating effects also scale equivalently with laser intensity. The time scale for the onset of these effects is long enough not to significantly impact the vibrational echo signals because a large fraction of

4988   *J. Phys. Chem. A, Vol. 110, No. 15, 2006*

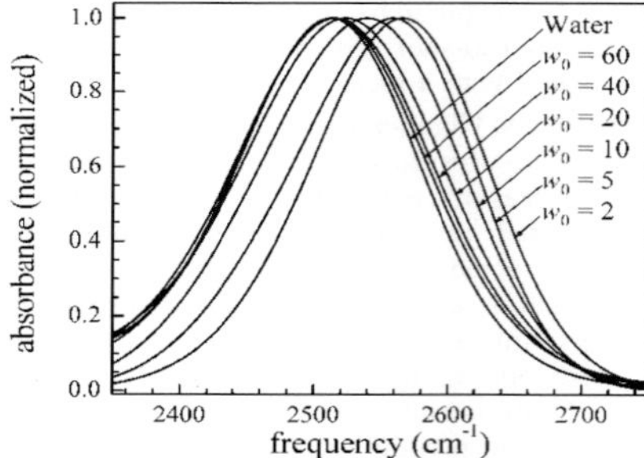

Figure 1. Background-subtracted FTIR absorption spectra of the OD stretch mode of 5% HOD in H$_2$O in bulk water and different size reverse micelles.

the dynamics occur within the first 5 ps for all samples. As a result, its effect is negligibly small for these experiments. The pump–probe experiments do measure dynamics occurring on the time scale of the local heating. The heating effects are well understood,[32,49–52] and are subtracted from the data as described in detail in the following section.

III. Results

A. FTIR Spectroscopy. The linear spectroscopy of the hydroxyl stretch of water in AOT reverse micelles has been studied extensively.[12,33,34,36,53,54] Figure 1 displays the linear absorption spectra for the OD hydroxyl stretch of HOD in H$_2$O for the bulk water and $w_0 = 60, 40, 20, 10, 5$ and 2 AOT reverse micelle samples. These spectra display an increasing blue shift as the reverse micelle size decreases. This shift has been observed by others and has been attributed to a weakening of the hydrogen bond network[33,34,53] because it is well documented that weaker hydrogen bonds absorb on the blue side of the bulk water hydroxyl stretch absorption.[35,55,56] However, attributing a blue shift of the spectrum to weaker hydrogen bonds based on the nature of bulk water can be problematic, and the interpretation of the shift will be discussed in detail below.

In contrast to the IR spectra of pure H$_2$O in reverse micelles which contains structure in the OH stretch region,[33] the spectra of HOD in H$_2$O exhibit no inherent structure, as can be seen in Figure 1. The bend overtone as well as symmetric and antisymmetric stretches that are coupled to the hydrogen bond network complicate the interpretation of the spectrum of pure H$_2$O in terms of different hydrogen bonding environments. The featureless isotopically mixed absorption spectrum of the OD stretch (local mode) of HOD in reverse micelles has led to the conclusion that distinct water environments cannot be resolved by looking at the line shape of the OH or OD stretch.[53] The IR spectrum of low-frequency libration modes of water in reverse micelles suggests the possibility of two environments of water in these systems (associated and core).[57] However, the spectra are exceedingly broad, and it has been argued that the presence of an isosbestic point may not necessarily imply the existence of more than one type of chemical species in a sample.[58] There is the additional complication that the asymmetry of the absorption spectra depicted in Figure 1 may be related to a frequency dependent transition dipole of water.[59] The issue of the difference in transition dipole between bulk water and water associated with headgroups will be discussed as part of the quantitative analysis of the spectrum.

An important feature of Figure 1 is that the peak positions of the $w_0 = 60$ and 40 spectra are only slightly shifted to the

blue from the peak position of bulk water. The spectra of bulk water, $w_0 = 60$, and 40 peak at 2511, 2513, and 2516 cm⁻¹, respectively. Even though the peak shifts are small, there is noticeable broadening on the blue side of the absorption spectra. The FTIR spectra also show a distinct change in the peak position with reverse micelles $w_0 = 20$ and smaller. The $w_0 = 20$ reverse micelles correspond to a water pool diameter of ∼7 nm. Large reverse micelles contain many tens of thousands of water molecules (see Table 1), which might lead one to believe that bulk-like characteristics are to be expected. However, the broadening on the blue side of the $w_0 = 60$ and 40 spectra suggest the existence of a modified hydrogen bond network due to the presence of the interface. As shown below, all of the features in the size dependent spectra can be reproduced by the core–shell model outlined in the Introduction and applied below.

B. Pump–Probe Spectroscopy. The population dynamics of the OD and OH stretch have been studied extensively in different hydrogen bonded systems.[32,49–52] In all of these studies, the signals do not promptly decay to zero but persist for tens of picoseconds. The residual signals are caused by the deposition of heat following vibrational relaxation, which shifts the OD absorption spectrum.[32] These signals need to be accounted for when constructing the isotropic and anisotropic signals. This offset arises from the long-term limit of $G(t)$, a contribution to the signal that grows in as vibrational relaxation occurs. As discussed in the Experimental Section, both the parallel and perpendicular components of the pump–probe signal were measured. Measuring the signal at parallel ($S_{\parallel}$) and perpendicular ($S_{\perp}$) polarizations permits the determination of both the isotropic and anisotropic decays. For a dipole transition such as the OD stretch under study here,

$$S_{\parallel}(t) = P(t)[1 + 0.8C_2(t)] + G(t) \tag{2}$$

$$S_{\perp}(t) = P(t)[1 - 0.4C_2(t)] + G(t) \tag{3}$$

$P(t)$ represents population relaxation and $C_2(t)$ is the second Legendre polynomial correlation function describing the orientational relaxation.[60] $G(t)$ takes into account the thermal heating that results after vibrational relaxation and will be referred to as the growth term. The fact that $G(t)$ is simply added to the signals in eqs 2 and 3 assumes that the thermal heating contribution to the signal is isotropic.[32,61,62] The population relaxation is determined by constructing the isotropic signal

$$I(t) = S_{\parallel} + 2S_{\perp} = 3(P(t) + G(t)) \tag{4}$$

The growth term needs to be accounted for to obtain the vibrational population dynamics $P(t)$. In the simplest situation the population decay is a single exponential given by

$$P(t) = e^{-t/T_1} \tag{5}$$

where T_1 is the vibrational lifetime. $G(t)$ involves the vibrational relaxation of the hydroxyl stretch that decays to one or more intermediate states before energy finally dissipates into the ground-state bath and produces a shift of the absorption spectrum.[32] This model gives the following expression for the growth term

$$G(t) = \frac{\alpha}{k_r + k_b}[k_r(1 - \exp(-k_b t)) + k_b(-1 + \exp(-k_r t))] \tag{6}$$

$k_r = 1/T_1$ and k_b is the rate constant for relaxation of intermediate

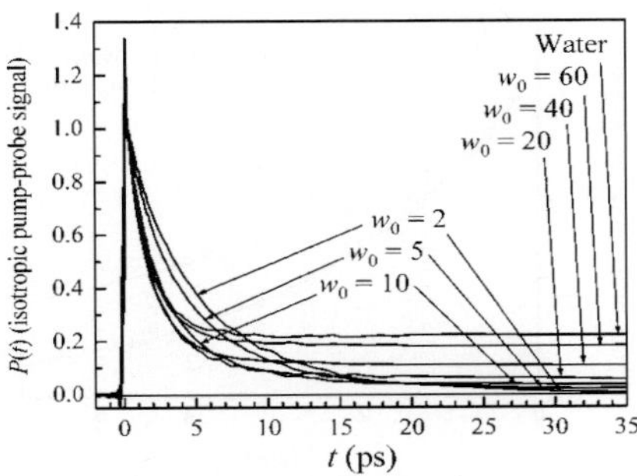

Figure 2. Pump–probe data taken at the center frequency of the absorption spectrum for bulk water and different size reverse micelle samples. A nonresonant spike occurs at $t = 0$, and as a result, the data are normalized at $t = 200$ fs for comparison. The long time offsets in the data are caused by a shift in the equilibrium distribution of hydrogen bonds following vibrational relaxation (local heating).

TABLE 2: Vibrational Lifetimes and Orientational Relaxation Parameters

sample	T_1 (ps)	A_1	τ_1 (ps)	A_2	τ_2 (ps)	θ_c	D_c (ps^{-1})	D_m (ps^{-1})
bulk water	1.7	0.34	2.7					0.062
$w_0 = 60$	1.8	0.33	2.7					0.062
$w_0 = 40$	1.7	0.33	2.8					0.060
$w_0 = 20$	2.1	0.33	3.0					0.056
$w_0 = 10$	2.7	0.17	18	0.16	1.5	42	0.06	0.009
$w_0 = 5$	4.4	0.25	30	0.08	1.0	31	0.05	0.006
$w_0 = 2$	5.2	0.27	50	0.03	0.9	29	0.02	0.003

states leading to the spectral shift responsible for the long time offset, α.[32] To extract the rotational correlation function, the following ratio is calculated,

$$r(t) = \frac{S_{\parallel} - S_{\perp}}{S_{\parallel} + 2S_{\perp}(t) - 3G(t)} = 0.4C_2(t) \qquad (7)$$

where the denominator divides out the vibrational lifetime when the growth term is subtracted. The growth term needs to be subtracted in the denominator to give the orientational dynamics exclusively. In the absence of detectable heating induced spectral changes after vibrational relaxation (no growth term), eqs 4 and 7 yield the typical formulas that represent the isotropic and anisotropic responses.

1. Population Relaxation. Figure 2 presents isotropic pump–probe data obtained using the middle term of eq 4. The data curves do not decay as single exponentials. At very short time, there is a nonresonant spike that tracks the pulse duration and is negligible for the pump–probe delay time, $t > 200$ fs. The data have been normalized to 1 at $t = 200$ fs because the decays are relatively slow, the fastest being 1.7 ps (see Table 2). Two important features can be seen easily in the figure. The first feature is that the decays reflecting vibrational relaxation become increasingly fast with increasing reverse micelle size. The second feature involves the long-lived component, which increases in amplitude as the size of the water nanopool increases and attains a maximum value for bulk water.

To explore the different vibrational relaxation times, the data in Figure 2 were fit using eqs 4 and 6. As discussed in detail below, the intermediate size reverse micelles give $P(t)$'s that are not strictly single exponentials once the growth term is removed. However, initially to see the trends with nanopool size, the decays will be fit as single exponentials. The T_1 values

from the best single-exponential fits are given in Table 2. When bulk water is compared with water confined in a $w_0 = 2$ reverse micelle, T_1 increases by a factor of $\sim$3. T_1 for bulk water and $w_0 = 60$ and 40 are identical within experimental error estimated to be ± 0.1 ps. By $w_0 = 20$, the lifetime has lengthened, becoming substantially longer than that of bulk water for the smaller reverse micelles. The vibrational lifetime trends we observe here are consistent with previous measurements of T_1 for HOD in small AOT/CCl$_4$ reverse micelles ($w_0 \leq 10$),[12] as well as measurements of T_1 for the OH stretch of HOD in D$_2$O.[52] Similarly, experiments probing vibrational relaxation of the azide ion in reverse micelles also show increased lifetimes for small reverse micelles.[63] The wide range of reverse micelle sizes studied here show that bulk like T_1 values are achieved for $w_0 > 20$ ($d_{wp} \sim 7$ nm), and that T_1 differs significantly for smaller reverse micelles.

The increasing long-lived component observed in the pump–probe transients can be attributed to local heating in the system. Thermally induced shifts in the absorption spectra have been observed previously for the AOT/CCl$_4$ reverse micelle system[12] as well as other hydrogen bonded systems.[50,51] The magnitude of this long-lived signal is largest for bulk water and decreases as the reverse micelle size decreases, as shown in Figure 2. In addition, the heat deposition signals decay on a shorter time scale for the smaller reverse micelles, 100 ps, increasing in time (nanoseconds) for the larger reverse micelles. The amplitude of the long-lived component with decreasing size may occur because the interface provides an avenue for some of the vibrational energy to escape the water pool during the vibrational relaxation cascade.[64]

2. Orientational Relaxation. The orientational dynamics were extracted from the pump–probe data using eq 7.[32,61,62] $G(t)$ is obtained from fits of the isotropic data described in the previous section. Small variations in $G(t)$ have very little effect on $r(t)$. As the nanopool size decreases, the offset also decreases, and any error introduced by the choice of $G(t)$ is further reduced. Furthermore, because the data ranges that were fit do not extend further than 35 ps, modifications to $G(t)$ involving a long time decay as heat diffuses out of the smaller reverse micelles was not introduced because the time scale associated with this (tens of picoseconds) is longer than the time range over which the data were examined. In a previous study of small AOT reverse micelles in CCl$_4$, the growth term was not included in the analysis, resulting in some errors in the values reported for the lifetimes and the orientational relaxations.[36] However, the trends presented previously are the same as those discussed below. A reanalysis of the data for the small reverse micelles studied previously[36] using the current method gave essentially identical results as those presented below for the same size micelles. Therefore, changing the solvent from CCl$_4$ to isooctane does not appear to change the dynamics of the nanoscopic water contained in the reverse micelles.

Parts A and B of Figure 3 display the anisotropy decays of bulk water and water in the different reverse micelles. There is the possibility of OD stretch excitation transfer, which would contribute to the decay of the anisotropy.[65] Faster anisotropy decays are measured for higher concentrations of HOD. Because of its strong distance dependence excitation transfer can be eliminated by lowering the concentration.[66–68] A concentration study showed evidence of excitation transfer at 10% HOD but not at 5% or 2.5%. Therefore, 5% HOD in H$_2$O samples were used in the experiments.

In all of the data it is not possible to determine the time dependence of $r(t)$ for $t < 200$ fs because of the large

4990 *J. Phys. Chem. A, Vol. 110, No. 15, 2006*

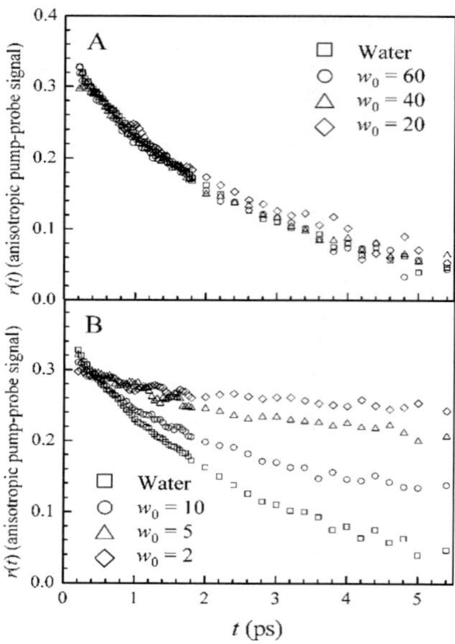

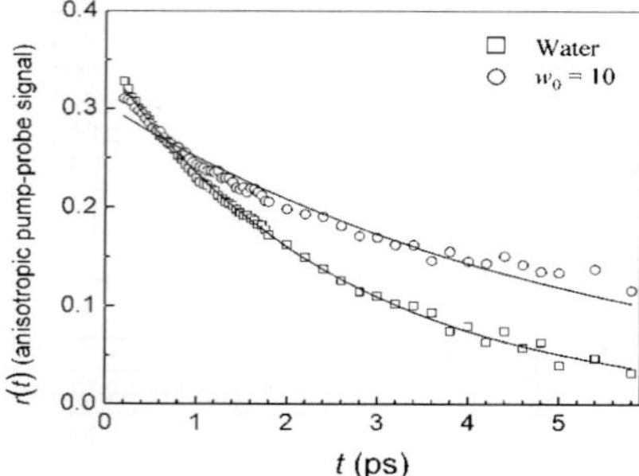

Figure 4. Single-exponential fits to the anisotropy decays of bulk water and the $w_0 = 10$ reverse micelle. Bulk water fits well to a single exponential whereas the $w_0 = 10$ does not.

Figure 3. Orientational anisotropy decays of the OD stretch mode in the various samples (see Table 2). (A) Bulk water, $w_0 = 60$, 40 and 20 show little variability in the anisotropies of these samples. The decays are single exponential. (B) Bulk water, $w_0 = 10$, 5 and 2 anisotropies reveal the progressively longer orientational relaxation time scales in the smaller reverse micelles. The anisotropy decays of the small reverse micelles are biexponential.

nonresonant signal that dominates the data. Therefore, analysis of all the data begins at 200 fs. Extrapolation of the curves back to $t = 0$ never yields $r(t) = 0.4$ because there is a very fast ($\sim$50 fs) inertial component to the orientational relaxation that is complete by 200 fs.[10,37] The inertial component reduces the apparent initial anisotropy obtained by extrapolation to $t = 0$ of the measured curves.[36,62]

Figure 3A shows the data for bulk water and the three largest reverse micelles, $w_0 = 20$, 40 and 60. All of the curves are fit exceptionally well to single-exponential decays given as τ_1 values in Table 2. Like the lifetime (T_1) results, bulk water, $w_0 = 60$ and 40, display the same orientational relaxation decay time within experimental error. The value of τ_1 for $w_0 = 20$ is slightly longer and the difference from the larger reverse micelles is just outside of experimental error.

Figure 3B displays the anisotropy decays for $w_0 = 10$, 5 and 2 reverse micelles as well as bulk water. In contrast to Figure 3A, the anisotropy decays for the small reverse micelles differ substantially from bulk water. In addition, the anisotropy decays of the smaller reverse micelles are not single exponential. Figure 4, which displays anisotropy data for bulk water and $w_0 = 10$, clearly shows the difference in the functional forms of the decays. For clarity, the $w_0 = 10$ data and its single-exponential fit have been offset by $+0.05$ along the vertical axis. Although the water data fit very well to a single exponential, $w_0 = 10$ data are not correctly described as a single exponential. (A biexponential fit to the $w_0 = 10$ data over a much wider range of times is shown in Figure 10 and discussed below.) For the three smallest reverse micelles, Table 2 lists decay constants, τ_1 and τ_2 resulting from biexponential fits. The fast component, τ_2, is substantially faster than bulk water and the slow component, τ_1, is much slower than bulk water. Models for the

biexponential behavior will be discussed below. It is evident that a divergence of dynamical behavior occurs for water nanopool diameters of <7 nm ($w_0 = 20$). For the larger reverse micelles the anisotropy decay is single exponential and very similar to bulk water although the change may begin at $w_0 = 20$. For $w_0 \leq 10$, a $\sim$1 ps component is followed by a component that decays in tens of picoseconds. The most dramatic affects are observed in the long time dynamics of the smallest reverse micelles. This drastic slowing down demonstrates the restrictions of structural dynamics of water molecules in small reverse micelles.

C. Vibrational Echo Spectroscopy. Recently, vibrational echo experiments on the hydroxyl stretch have explicated the dynamics of hydrogen bond networks in bulk water. Vibrational echo peak shift measurements[23,45] and 2D IR vibrational echo correlation spectroscopy experiments[22,44,69,70] combined with molecular dynamics simulations[35,56,59,71−76] have provided detailed information on the nature of structural evolution of bulk water's hydrogen bond network. The results demonstrate that the dynamics of water responsible for spectral diffusion observed in the experiments occur over time scales ranging from <100 fs to >1 ps. The fastest dynamics involve very local hydrogen bond fluctuations, whereas the longest time scale dynamics reflect global hydrogen bond structural rearrangement that complete the randomization of structures and the sampling of all frequencies in the hydroxyl stretch absorption spectrum.

Vibrational echo and vibrational echo peak shift measurements have been extended to small AOT reverse micelles in CCl_4 solvent.[12,14] Large AOT reverse micelles cannot be prepared in CCl_4. Preparing the reverse micelles in isooctane allowed us to revisit the small reverse micelles and extend the examination to large ones as well. Figure 5 displays time dependent vibrational echo traces for $T_w = 1.2$ ps. The water $w_0 = 60$, and 40 decays are virtually identical. The first decay that departs from bulk water behavior is for $w_0 = 20$. For water nanopools $w_0 < 20$ the differences in the decays become large. The substantial change in the nature of the dynamics at $w_0 = 20$ is consistent with the observations made of the vibrational lifetimes and orientational relaxation dynamics, where significant departure from bulk water dynamics first appear for reverse micelles smaller than $w_0 = 20$.

Data like those displayed in Figure 5 were collected for all 7 samples over a range of T_w values, each T_w yielding a distinct curve. The curves can be characterized by their shapes and their peak shifts, that is, the displacement of the peak of the decay curve from the absolute position of $\tau = 0$. First the data on the 7 samples will be discussed qualitatively in terms of the peak

J. Phys. Chem. A, Vol. 110, No. 15, 2006 **4991**

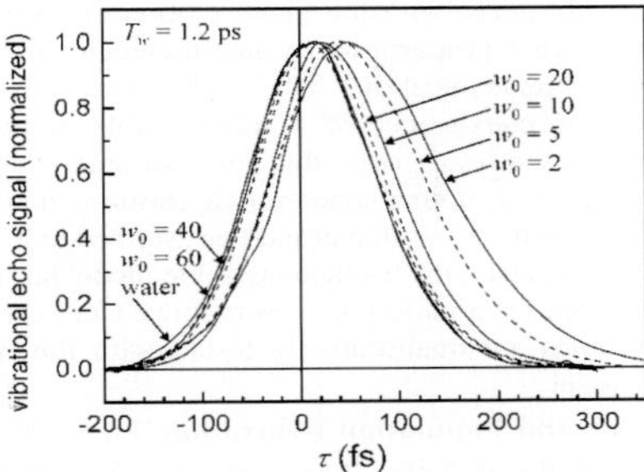

Figure 5. Vibrational echo decays (τ scan) at $T_w = 1.2$ ps for bulk water and different reverse micelles detected at the frequency of the center of the absorption line for each sample. Bulk water (solid line), and $w_0 = 60$, 40 and 20 (dashed lines) are not easily discernible whereas the smaller reverse micelles clearly exhibit slower decays.

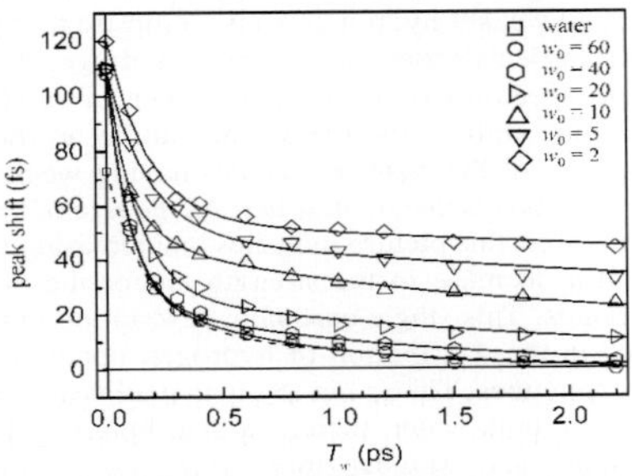

Figure 6. Vibrational echo peak shifts for bulk water and different reverse micelles. The lines through the data are phenomenological biexponential fits for reverse micelles (solid lines) and bulk water (dashed line). As the size of the water nanopools decrease, the spectral diffusion dynamics slow as manifested by the longer decays in the peak shifts with increasing T_w.

shifts. As discussed further below, the peak shifts as well as the detailed shape of the decay curves can be calculated using a model FFCF. The peak shifts provide a convenient representation of the trends in the data.

Figure 6 displays peak shift data for bulk water and the six different reverse micelles. The solid curves represent phenomenological biexponential fits rather than diagrammatic perturbation theory[43] analysis of the data,[12,14] which will be presented below. The vibrational echo peak shifts as a function of T_w are related to the decay of the FFCF.[48,77] These phenomenological fits allow assessment of qualitative trends in the data prior to the detailed analysis. As structural fluctuations over a longer period of time (increased T_w) cause more and more of the spectroscopic line to be sampled through spectral diffusion, the peak shift goes to zero. The peak shift data at short T_w (sub 500 fs) are related to the extent of the fastest structural fluctuations. As the water nanopool becomes smaller, the decay of the peak shift slows, demonstrating that the rate of structural fluctuations decrease with the nanopool size. Peak shifts for bulk water and the two largest reverse micelles, $w_0 = 60$ and 40, though quite similar, include some significant differences. Peak shifts for bulk water and $w_0 = 60$ are virtually identical after a few hundred femtoseconds. However, the initial peak shift value for water is much smaller than for $w_0 = 60$. Therefore, the very fast fluctuations in bulk water are not the same as in a water nanopool with a diameter of ~28 nm. The slower fluctuations are almost indistinguishable. Peak shifts for

TABLE 3: FFCF Parameters

sample	T_2^* (ps)	Δ_2 (cm^{-1})	τ_2 (ps)
bulk water	0.12	54	1.1
$w_0 = 60$	0.12	57	1.2
$w_0 = 40$	0.10	53	1.2
$w_0 = 20$	0.10	55	2.5
$w_0 = 10$	0.14	66	3.8
$w_0 = 5$	0.16	54	6
$w_0 = 2$	0.23	56	15

$w_0 = 60$ and $w_0 = 40$ are virtually identical at very short times but differ slightly at longer times. Nonetheless, except for the initial peak shift difference between bulk water and $w_0 = 60$ and $w_0 = 40$, these three samples have very similar dynamics. The significant changes in the structural fluctuations as manifested through the hydroxyl stretch frequency evolution begin at $w_0 = 20$ and become increasingly pronounced as the water nanopool size continues to decrease.

Once the FFCF has decayed significantly, further decay of the peak shift is directly related to the decay of the FFCF.[48,77] The slowest components of the phenomenological biexponential fits for the peak shifts for water, $w_0 = 60$, 40, 20, 10, 5, and 2, are approximately 1, 1, 1, 2, 4, 5, and 11 ps, respectively. These reflect the trend in the slowing of the longest component of the dynamics (complete global structural randomization of the hydrogen bond network), and in fact, are not far from the values obtained using time dependent diagrammatic perturbation theory (see Table 3).[43]

The time dependent diagrammatic perturbation theory method was used to calculate the third-order polarization generated in the sample.[43,78] The calculations employed finite pulse durations with the associated spectra. The vibrational echo curves taken at each T_w and the linear line shape were fit simultaneously. The parameters used in the functional form of the FFCF were varied in fitting the data. Previous studies using 2D IR vibrational echo measurements performed with heterodyne detection (polarization level measurements) used a triexponential form for the FFCF. The full 2D correlation spectra (measured and calculated)[44,69] revealed a very fast component (<100 fs) that gave rise to a motionally narrowed contribution, a low amplitude intermediate component (~400 fs), and the slowest component, 1.4 ps. The slowest component arises from global structural rearrangement of the hydrogen bond network. In the bulk water experiments, the nonresonant contribution to the signal at short time is much smaller than in the current experiments. Furthermore, the polarization level measurements afford better short time resolution compared to the intensity level measurements presented here. Given the nature of the data in the current experiments, we found a biexponential form for the FFCF was sufficient to fit the data. This reduces the number of parameters and makes comparisons between systems easier. Of particular interest here are comparisons of the longest time scale component for each system. The number obtained for bulk water is slightly faster than that obtained in the analysis of the more detailed data because the intermediate time scale component was in part subsumed into the long time scale component. However, for the purposes of understanding trends, the difference is not significant.

The functional form of the FFCF used in fitting the vibrational echo data is

$$C(t) = \Delta_1 e^{-t/\tau_1} + \Delta_2 e^{-t/\tau_2} \qquad (8)$$

where Δ_1 and Δ_2 are amplitude factors and τ_1 and τ_2 are the corresponding decay times. It is important to note that the FFCF

4992 *J. Phys. Chem. A, Vol. 110, No. 15, 2006*

is the input into a detailed calculation of the observable. A biexponential FFCF does not produce a biexponential decay, but rather a nonexponential decay with a functional form that depends on the relative values of the parameters.

Table 3 contains the best fit parameters for all samples studied. These will be used below in the context of the core−shell model. The root-mean-squared amplitude Δ_1 and time scale τ_1 cannot be independently determined because the fastest time scale dephasing dynamics are motionally narrowed, that is, $\Delta_1\tau_1 < 1$,[79,80] giving a Lorentzian contribution to the dynamic spectrum. The line width of the motionally narrowed component is $\Gamma = 1/\pi T_2^*$, with

$$1/T_2^* = \Delta_1^2\tau_1 \qquad (9)$$

(Δ_1 in rad/s, τ_1 in s, and T_2^* in s) Although Δ_1 and τ_1 can be varied, they must simultaneously satisfy $\Delta_1\tau_1 < 1$ and eq 9, which results in a well-defined value for T_2^*. Therefore, Table 3 lists T_2^* rather than Δ_1 and τ_1. The T_2^* values are similar until the reverse micelles become small, and then there is some increase. The Δ_2 values are all very similar. The major trend with decreasing water nanopool size is a substantial increase in the long time scale time constant, τ_2, for $w_0 \leq 20$. The results of these detailed fits show the same trend as the peak shift results (Figure 6), and the simple fits to the long component of the peak shifts are almost in quantitative agreement with the results given in Table 3. The change in dynamical behavior at $w_0 = 20$ is in accord with the observations of the orientational relaxation and population relaxation results.

The time dependent perturbation theory method[43] used to obtain the FFCFs involves approximations that assume the dynamics are identical across the absorption spectrum. Experiments on bulk water have shown that on short time scales (<400 fs) the rate of spectral diffusion, and therefore hydrogen bond structural dynamics, depends somewhat on wavelength because the water hydrogen bond network reflects water molecules that have different numbers and strengths of hydrogen bonds.[70] The short time scales involve very local dynamics that are sensitive to these differences. The longer time scale global hydrogen bond network rearrangements[22,44,75,76,81] involve many water molecules that are not sensitive to local details that determine the instantaneous position of a particular hydroxyl stretch in the absorption line.[70] The fact that these wavelength dependent differences are observed shows that the Gaussian approximation that underlies the diagrammatic perturbation theory determination of the FFCF is not strictly valid. Furthermore, it has been shown recently that the hydroxyl stretch transition dipole matrix element varies across the bulk water absorption line, and this variation has an impact on the vibrational echo observables.[59] For these reasons, vibrational echo data presented in Table 3 were all taken at the equivalent frequency for each sample, the peak of the absorption spectrum. Therefore, the trends observed in Table 3 will be valid although the approximations in the theory used to obtain the FFCF introduce some error.

IV. Quantitative Test of the Core−Shell Model

Four distinct size dependent observables have been presented and discussed above. They are the IR absorption spectrum, the hydroxyl stretch population relaxation, the water orientational relaxation, and the hydroxyl stretch spectral diffusion. The discussion given above, particularly regarding small reverse micelles where deviations from bulk water are pronounced, did not directly address the issue of treating nanoscopic water as a single ensemble or as two subensembles. A single ensemble

picture does not agree with the many reports in the literature that assign distinct properties to water molecules in separate regions in the reverse micelles.[33,34,54,57,82−84] A core−shell model is based on the coexistence of a subensemble of molecules associated with the headgroups that form an outer shell of the water nanopool, and another subensemble forming the nanopool core, which has bulk water like properties ascribed to it. Though presented ubiquitously, the two subensemble model has not been rigorously tested for a variety of observables. Here each of the four observables are quantitatively tested with the two subensemble model.

A. Spectra and Population Relaxation. First consider the absorption spectra as a function of reverse micelle size. As shown in Figure 1, as the reverse micelle decreases in size, the spectrum shifts to the blue. One proposition to explain the shift is based on the spectrum of bulk water.[85] Water has a distribution of the number and strengths of hydrogen bonds.[21,31] On the blue side of the bulk water spectrum, water molecules with fewer and weaker hydrogen bonds comprise the spectrum; the red side reflects more and stronger hydrogen bonds. To describe the spectrum of nanoscopic water in AOT reverse micelles, a blue shift of the entire line can be attributed to a single ensemble of the water molecules having weaker and/or fewer hydrogen bonds than bulk water. As the size of the reverse micelle decreases, this picture indicates an increasing blue shift results from a decrease in the strength and/or the number of hydrogen bonds. This single ensemble description implies that the changes in the distribution of hydrogen bonding environments are continuous. This is not to say that all water molecules are alike. As in bulk water, there is a distribution of hydrogen bond strengths and number that gives rise to the broad spectroscopic line, which is inhomogeneously broadened on short time scales, as demonstrated by vibrational echo experiments.[70] However, this is a single ensemble description in the sense that it does not associate the spectroscopic changes with changes in the proportion of shell and core subensembles.

In a two subensemble model of the spectrum, overlapping contributions from a shell subensemble and a core subensemble comprise the spectrum. To test this idea, we will take the core spectrum to be the bulk water spectrum and the shell spectrum to be the $w_0 = 2$ spectrum. The $w_0 = 2$ reverse micelles have ~40 water molecules. A very simple model in which the thickness of the shell is the diameter of one water molecule, places essentially all of the $w_0 = 2$ water molecules in the shell. Figure 7A shows the bulk water and $w_0 = 2$ spectra that were used as basis spectra. All of the spectra for $w_0 = 60$ through $w_0 = 5$ were fit with the single adjustable parameter, the relative proportion of the two basis spectra. Neither the peak positions nor the widths of the basis spectra were changed in the fitting procedure. Figure 7B shows the result of this fit for $w_0 = 10$. The total $w_0 = 10$ spectrum is composed of 54% of the core (bulk water) spectrum and 46% of the shell ($w_0 = 2$) spectrum. (These are not the fractions of water molecules in the core and shell, as discussed below.) Overall, the agreement is very good with only some error in the red tail. The shape and peak position are reproduced well. Figure 7C shows the result for $w_0 = 60$. Spectra shown in Figure 1 reveal the small but noticeable differences between the bulk water spectrum and the $w_0 = 60$ spectrum; that is, the $w_0 = 60$ spectrum displays a small shift in the peak position and a broadening on the blue side of the line. The total $w_0 = 60$ spectrum is composed of 93% of the core (bulk water) spectrum and 7% of the shell ($w_0 = 2$) spectrum. These changes are reproduced quite well by the single adjustable parameter fit. The fits to all of the spectra, $w_0 = 60$

J. Phys. Chem. A, Vol. 110, No. 15, 2006 **4993**

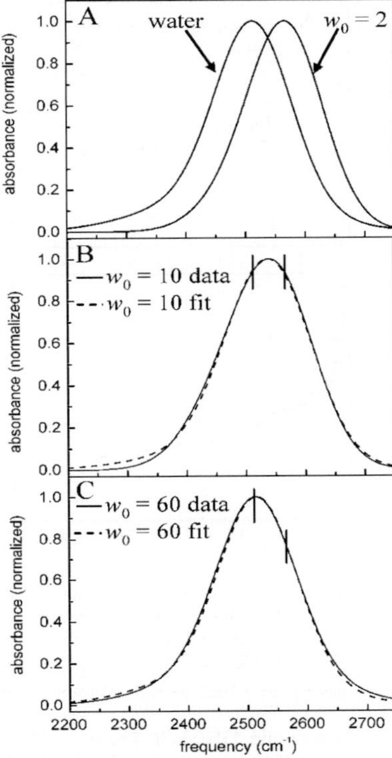

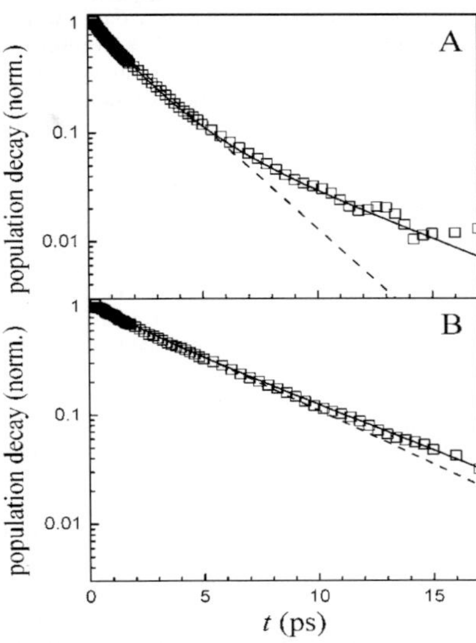

Figure 7. Comparison of the core−shell model to the reverse micelle size dependence of the absorption spectrum. (A) OD stretch spectra of bulk water (model of the core) and $w_0 = 2$ spectrum (model of the shell). (B) $w_0 = 10$ spectrum (solid line) and fit (dashed line). The fit is the weighted sum of the spectra in A. (C) $w_0 = 60$ spectrum and fit. The fit is the weighted sum of the spectra in (A). The small vertical lines show the peak positions of the spectra in (A).

Figure 8. Population decay data (squares), best single-exponential fit (dashed line), and best fit with the core−shell model, that is, the weighted sum of bulk water (core model) $w_0 = 2$ (shell model) population decays: (A) $w_0 = 20$; (B) $w_0 = 5$.

to $w_0 = 5$, using the bulk water and the $w_0 = 2$ basis spectra gave results similar to those shown in Figure 7, although the fit for $w_0 = 5$ is not quite as good.

Not only are the changes in peak position reproduced by the core−shell model, but the changes in line widths are also reproduced. Adding an increasing amplitude of the $w_0 = 2$ spectrum (shell) to the water (core) will shift the peak to the blue and broaden the spectrum. But for the smallest reverse micelles, the spectrum narrows again as they become predominantly shell and lose amplitude on the red (core) side. In the single ensemble model of the spectrum, it is possible to explain the peak shift qualitatively as a shift of the distribution of hydrogen bonds to weaker and/or fewer, but it is less clear how to explain, even qualitatively, the broadening and then narrowing of the spectra as the size of the reverse micelles decreases. In contrast, the core−shell model reproduces that spectral feature quantitatively with a single adjustable parameter.

Next we consider the population relaxation data. A single ensemble will produce a single-exponential decay of the excited vibrational population. The results of single-exponential fits to the population data obtained after removing the growth term as discussed in connection with eq 4 are presented in Table 2 as T_1 values. As mentioned above, not all of the decays fit well to single exponentials. Following the very short time (~200 fs), which is masked by the nonresonant contribution to the signal,

bulk water and $w_0 = 2$ give single-exponential decays, within experimental error. Figure 8A shows the $w_0 = 20$ data and the best single-exponential fit (dashed line) on a semilog plot. In the fit, there are three adjustable parameters, T_1, k_r (the rate constant in the growth term in eq 6), and an overall amplitude scaling factor. The long time limit of the growth term was fixed to the value of the data at ~30 ps for all samples. On a longer time scale this term actually decays for the smaller reverse micelles but the dynamics at later times do not affect the data analysis presented here.[12] At short time, the single-exponential fits relatively well, but at longer times, the fit is poor. Figure 8B shows similar data for $w_0 = 5$ and the best single-exponential fit (dashed line). Again, a single exponential does a reasonable but not perfect job of reproducing the data, missing at long time. In addition to the single-ensemble model (single-exponential decay) not reproducing the data, there is no quantitative explanation for the trend of the lifetime with size, as given in Table 2.

The vibrational population relaxation data can also be analyzed quantitatively with the core−shell model. The two subensemble model can be applied by using the population dynamics measured for bulk water and $w_0 = 2$. For both of these samples, the T_1, and k_r values are known. Therefore, rather than three parameters, these fits include only two adjustable parameters, the relative proportion of core (bulk water) and shell ($w_0 = 2$) population dynamics to be added together and an overall scaling factor. Initially, it may seem that the relative proportion should be the same as that used to fit the spectral data. However, the linear absorbance is proportional to $|\mu|^2 C$, where μ is the transition dipole bracket[86] and C is the concentration, whereas the pump−probe signal is proportional to $|\mu|^4 C$ because it depends on the coupling of the oscillators to both the pump and probe pulses. The solid curves in Figure 8A,B are the best fits using the core−shell model. The results show that the two ensemble fit is substantially better with no systematic deviation between the fit and the data. The fit is better than the single-exponential fit despite the fact that there is one fewer adjustable parameter. The $w_0 = 5$ data are closer to a single exponential than the $w_0 = 20$ data because the signal from the smaller reverse micelle is dominated by the shell contribution ($w_0 = 2$), which is single exponential. Fits to all of the different size reverse micelles, $w_0 = 60$ to 5, are of very similar quality.

For each reverse micelle size, fitting the spectra gives the relative proportion of $|\mu|^2C$ for bulk water and $w_0 = 2$, whereas fitting the population dynamics gives the relative proportion of $|\mu|^4C$. Using the two results for each reverse micelle size permits the independent determination of $|\mu|^2$ and C. From the results it is found that $|\mu|^2$ for bulk water is 1.4 ± 0.2 times larger than $|\mu|^2$ for $w_0 = 2$. This is an important result that can be independently tested by experiment. Samples of bulk water and $w_0 = 2$ (both 5% HOD) were prepared with known path lengths. The concentration of the $w_0 = 2$ sample was determined from the quantitative preparation. FT-IR spectra of the OD stretch in each sample and on the same samples but without the 5% HOD were collected. The samples without HOD were used for background subtraction. The areas of the background subtracted spectra were integrated. The ratio of the areas after taking into account the differences in path length and concentration yield the ratio of $|\mu|^2$ between the two samples. These FT-IR experiments yield a ratio of $|\mu|^2$ for bulk water to $w_0 = 2$ of 1.6 ± 0.1. This result provides very strong support of the core—shell model used to fit the spectra and the pump—probe population data.

With the ratio of $|\mu|^2$ determined, the ratio of C for the core to shell is also determined from the spectra. The fraction of each sample that is shell for bulk water and $w_0 = 60$, 40, 20, 10, 5, and 2, is 0.0, 0.10, 0.14, 0.26, 0.53, 0.80, and 1.0, respectively. Using the diameters of the reverse micelles given in Table 1, and a model of a spherical core surrounded by a spherical shell, the fraction that is shell for each reverse micelle can be used to determine the thickness of the shell. The shell thickness is ~ 0.4 nm. This value is consistent with calculations performed for similar systems.[20] As the reverse micelle increases in size, the fraction of the water molecules found in the shell region rapidly decreases. In a pump—probe experiment, the core is emphasized further by the dependence of the signal on $|\mu|^4C$. The fact that $|\mu|^2$ is ~ 1.6 times larger for the core than for the shell, combined with very large fraction of the water molecules in the core, accounts for the virtually identical T_1 values measured for bulk water, $w_0 = 60$ and 40 (Table 2).

Finally, the core—shell model for the vibrational lifetime predicts a well-defined wavelength dependence for the observed nonexponential decay. As the wavelength is shifted to the blue, more water molecules in the shell contribute, and the shape of the decay contains more amplitude of the slow component. When the wavelength is shifted to the red, more water molecules in the core contribute, although there is an additional complication because detection wavelengths on the red side of the spectra also contain contributions from the first vibrationally excited state to the second vibrationally excited-state (1−2) transition. In a preliminary study of the wavelength dependence of $w_0 = 10$, we were able to reproduce the both blue side and red side data using the core—shell model. A full wavelength study will be presented subsequently.

B. Spectral Diffusion and Orientational Relaxation. To further test the core—shell model, we applied it to observables that are strongly related to hydrogen bond network structural evolution. Wavelength dependent vibrational echo data were collected for $w_0 = 10$ as a good test case because our linear-IR absorption and pump—probe measurements indicate it is ap-

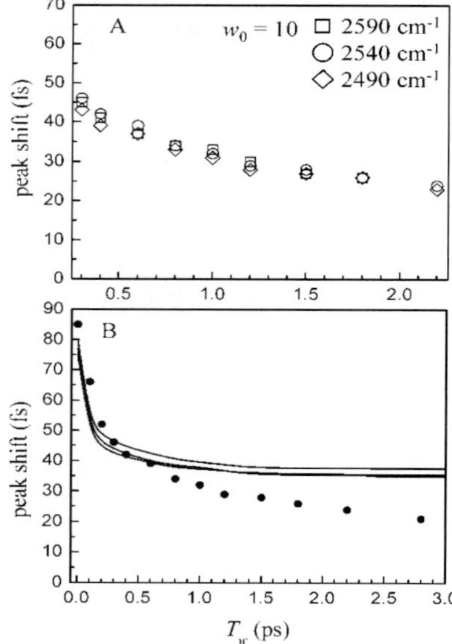

Figure 9. (A) Wavelength dependent vibrational echo peak shift data for $w_0 = 10$. There is no detectable wavelength dependence. (B) $w_0 = 10$ peak shift data for the center frequency (circles) and the no adjustable parameter core—shell model calculation of the three wavelengths shown in (A). In contrast to Figures 7 and 8, the core—shell model does not provide a proper description of the spectral diffusion.

proximately 47% core and 53% shell. The $w_0 = 10$ absorption spectrum peaks at ~ 2540 cm^{-1} with a fwhm of ~ 175 cm^{-1} (see Figures 1 and 7B). Experiments were conducted at points 50 cm^{-1} to higher frequency (2590 cm^{-1}) and 50 cm^{-1} to lower frequency (2490 cm^{-1}). Full vibrational echo curves were collected as a function of T_w. The results, shown as the T_w dependent peak shifts in Figure 9A, display almost no wavelength dependence. Although these results pertain to AOT/isooctane, similar wavelength independent data were observed previously for $w_0 = 10$ in AOT/CCl$_4$.[12]

The surprising lack of wavelength dependence would seem to indicate that the two subensemble model cannot apply because signals measured at the blue and red wavelengths were expected to have substantially different contributions from the shell and core. However, in the two subensemble model, each subensemble contributes to the polarization. In the intensity level vibrational echo experiments conducted here, the signal reflects the absolute value squared of the sum of the contributions from the core and the shell subensembles. The signal is proportional to $|P_c + P_s|^2 = P_c^2 + 2P_cP_s + P_s^2$, where P_i is the core and shell contributions to the total polarization. This leads to a nonnegligible cross term, $2P_cP_s$, that can produce substantial mixing and potentially reduce the wavelength dependence.

The core—shell model can be tested quantitatively for the vibrational echo observables. The FFCFs were determined for both bulk water and $w_0 = 2$ (see Table 3). Therefore, a detailed diagrammatic perturbation theory calculation can be performed for the core—shell model. For $w_0 = 10$ or any of the reverse micelles, the appropriately weighted FFCFs for the core and the shell are used in the calculation. The vibrational echo signal at the intensity level measured in these experiments is proportional to $|\mu|^8C^2$. To obtain the observables, the calculations are performed at the polarization level prior to taking the absolute

J. Phys. Chem. A, Vol. 110, No. 15, 2006 **4995**

value square of the results. The FFCF for each subensemble is weighted by the their relative values of $|\mu|^4 C$. From the experiments described in section IV.A, the relative values of $|\mu|^4 C$ are known. Therefore, the wavelength dependent vibrational echo calculations can be performed *using no adjustable parameters*.

Figure 9B displays the vibrational echo peak shift data for $w_0 = 10$ at the center frequency. Only the data set obtained at the center frequency is displayed because the data for all three wavelengths are so similar. The solid curves are the no adjustable parameter calculations for the three wavelengths displayed in Figure 9A. Several features of these data and calculations are interesting. First, like the data in Figure 9A, the calculations show little wavelength dependence, but the calculated wavelength dependence is still greater than that displayed by the data. Second, the calculated curves fundamentally have the wrong shape. Using an adjustable parameter to change the weighting of the core and shell contributions moves the calculated curves up and down but does not change their shape. At short times, the calculated curves decay too fast, and at long times, they decay far too slowly. On the time scale of the figure, the calculated curves are essentially horizontal after ∼2 ps.

The form of the calculated curves can be understood qualitatively. At short times, the decay is dominated by the very fast core (bulk water) decay. For $w_0 = 10$, the core would dominate the signal except for the cross term, which essentially heterodyne amplifies the shell contribution. Very qualitatively, after ∼2 ps the core peak shift has decayed to zero and is no longer changing. However, the shell peak shift is still significant, and it is hardly changing because its decay is very slow, $\tau_2 = 15$ ps (see Table 3). Combining a zero peak shift that is not changing with a significant peak shift that is barely changing gives a moderate peak shift that does not decay on the time scale of the plot. The vibrational echo results demonstrate that a rigid separation in core and shell subensembles does not apply, although it works very well for both the absorption spectra and the vibrational lifetimes.

The core–shell model can also be used to calculate the anisotropic pump–probe signal discussed in section III.B2. The results of the measurements are given in Table 2. The experiments illustrate that $w_0 = 10$ or smaller reverse micelles give biexponential anisotropy decays whereas the larger reverse micelles gave single-exponential decays. Like the vibrational echo calculations, the anisotropic pump–probe calculations can be done without adjustable parameters. However, to implement the core–shell model, the relations given in section III.B need to be rewritten to account for the two subensembles contributing to the signal.

The anisotropic pump–probe signal is constructed in the usual manner by combining two different measurements of the probe intensity, parallel and perpendicular, as in eq 7. As discussed in connection with eqs 4 and 6, the growth term, $G(t)$, can be removed even when two ensembles are present. However, unlike the population relaxation, $P(t)$, after eliminating $G(t)$, the resulting anisotropy equation does not necessarily isolate the orientational relaxation dynamics. To see this, eq 7 may be extended to involve two subensembles as

$$r(t) = \frac{2}{5}\left(\frac{f_C e^{-(t/T_{1C})}C_2^C(t) + f_S e^{-(t/T_{1S})}C_2^S(t)}{f_C e^{-(t/T_{1C})} + f_S e^{-(t/T_{1S})}}\right) \quad (10)$$

The prefactors f_C and f_S are proportional to $\mu^4 C$, which, as discussed above, are not the same for the core and the shell.

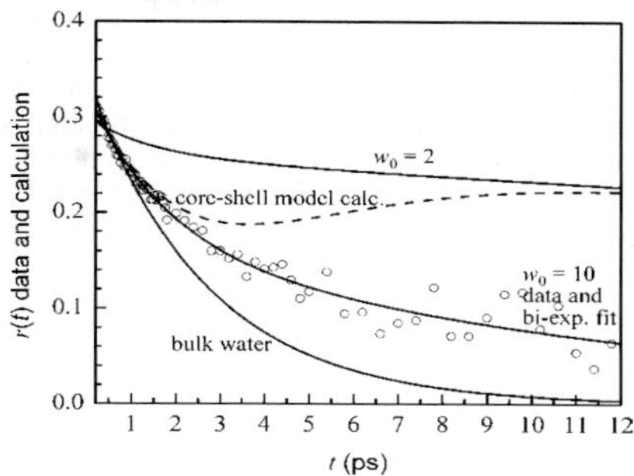

Figure 10. $w_0 = 10$ orientational relaxation data (circles) and biexponential fit (solid curve through circles). The upper and lower curves are the results of fits to the $w_0 = 2$ and bulk water orientational relaxation data, respectively. See Table 2 for fit parameters. The dashed curve is the core–shell model no adjustable parameter calculation of the data. See text for explanation of the shape. The core–shell model does not properly describe the orientational relaxation in reverse micelles.

This equation cannot be factored to give the pure orientational dynamics when the vibrational lifetimes and orientational correlation times of the two subensembles differ. Some interesting limits may be examined. If $T_{1C} = T_{1S}$, the anisotropy reduces to

$$r(T_w) = \frac{2}{5}\left[\left(\frac{f_C}{f_C + f_S}\right)C_2^C(t) + \left(\frac{f_S}{f_C + f_S}\right)C_2^S(t)\right] \quad (11)$$

This implies that the anisotropy will be a weighted sum of the orientational dynamics from each subensemble scaled by the transition dipoles and concentrations. If $C_2^C(t) = C_2^S(t)$, then the anisotropy reduces to

$$r(t) = \frac{2}{5}C_2(t) \quad (12)$$

which is also a result that only contains the purely orientational response. In the case of the core–shell model for AOT reverse micelles, neither of these limits are appropriate. The vibrational lifetimes and orientational correlation times are 1.7 and 2.6 ps, respectively, for bulk water and they are 5.2 and ∼50 ps (major component), respectively, in a $w_0 = 2$ reverse micelle. Using the core–shell model, these differing time scales produce a calculated anisotropy that does not purely reflect the orientational dynamics.

To apply the core–shell model to the anisotropy decays using eq 10 requires knowledge of all of the parameters with reasonable accuracy. Again, we use our data for $w_0 = 10$ as a test case. Figure 10 displays the measured anisotropy of the reverse micelle and the best biexponential fit (solid line) through the data. The parameters of the fit are given in Table 2. Also shown are the anisotropy decays of $w_0 = 2$ and bulk water obtained from fits to the data. The curves begin at 200 fs because the short time inertial dynamics are obscured by the nonresonant signal centered at $t = 0$.

The dashed curve shown in Figure 10 is the result of the no adjustable parameter calculation using the core–shell model given by eq 10. Initially, the match is relatively good, but at longer times the curve grows and approaches the $w_0 = 2$ curve. At still longer times, the calculated curve decays as the $w_0 = 2$ curve. The recurrence in the calculated curve does not mean that the model predicts an increase in the orientational anisot-

4996 *J. Phys. Chem. A, Vol. 110, No. 15, 2006*

ropy. As can be seen from eq 10, $r(t)$ is not the anisotropy decay as it would be for a single ensemble. Rather, this curve reflects the interplay of the four terms in eq 10. To understand the shape of the curve, it is only necessary to consider the time scale of the decays of the four terms. A fast lifetime multiplied by a fast anisotropy decay comprises the first term (core term, C) in the numerator, yielding the fastest decay that looks like bulk water. The second term in the numerator (shell term, S) decays very slowly because it is a product of a slow lifetime and an even slower anisotropy decay. The second fastest term is the core term in the denominator. It is the decay of this term that is fast compared to either of the shell terms that causes the calculated curve to increase after $\sim$3.5 ps. After the two core terms have fully decayed, the remaining shell terms in the numerator and denominator are the anisotropy decay of $w_0 = 2$. Even if the relative amounts of core and shell are adjusted, it is not possible to reproduce the data. The core$-$shell model of the anisotropy decay fails to fit or explain the data.

The core$-$shell model cannot reproduce the orientational relaxation data even qualitatively. However, a model suggested previously for orientational relaxation of small reverse micelles can produce biexponential decays for small reverse micelles[36] and provides an explanation for the transition to single-exponential decays for the larger reverse micelles. For a single ensemble of molecules undergoing unrestricted orientational diffusion $C_2(t)$ (eq 7) decays as a single-exponential decay. This model can describe the dynamics observed for water and the larger reverse micelles and yields an orientational diffusion constant[87]

$$\tau_1 = \frac{1}{6D} \tag{13}$$

The orientational dynamics of water in the smaller reverse micelles, $w_0 \leq 10$ shown in Figure 3B warrants the use of a more detailed model that can account for the restrictive orientational motions produced by the hydrogen bond network. The wobbling-in-a-cone model, which describes relaxation of a single ensemble of molecules that are orientationally restricted, can be applied. This model has been described in detail elsewhere[88] and was used to examine the orientational dynamics of water in small AOT/CCl$_4$ reverse micelles.[12] The model describes the relaxation when the orientational diffusion is restricted initially to a range of angles within a cone. In the absence of other relaxation processes, the wobbling-in-a-cone model orientational correlation function decays exponentially to a constant value. However, if there are slower time scale processes that result in complete orientational relaxation, then the wobbling-in-a-cone model yields a biexponential decay that reflects the limited diffusion within the cone as well as slower complete orientational diffusion that causes the anisotropy to decay to zero. In the standard model,[88,89] the two processes occur simultaneously. The shorter correlation time describes the restricted motion that samples a cone with semi-angle θ. The longer correlation time accounts for slower unrestricted motion. The correlation function is given by

$$C_2(t) = [Q^2 + (1 - Q^2) \exp(-t/\tau_w)] \exp(-t/\tau_m) \tag{14}$$

where Q^2, the generalized order parameter that describes the degree of restriction on the wobbling-in-the-cone motion, is the amplitude of the slow component of the biexponential decays, and $0 \leq Q^2 \leq 1$. $Q^2 = 0$ corresponds to unrestricted reorientation, and $Q^2 = 1$ reflects no orientational relaxation. Q^2 is obtained from data by taking the amplitude of the slow

decay of $r(t)$ and multiplying it by 2.5 (see eq 7). τ_w is the time constant that arises from the wobbling-in-the-cone diffusion, and τ_m is the final diffusive full orientational relaxation time. For the studies reported here, τ_m is given by the long time constant τ_1 in Table 2 and directly gives the orientational diffusion coefficient, D_m, of the angularly unrestricted orientation relaxation using eq 13. In contrast, τ_w is not directly related to experimental observables. Rather,

$$\tau_w = (\tau_2^{-1} - \tau_1^{-1})^{-1} \tag{15}$$

The cone semi-angle θ is obtained from Q^2 using[89]

$$Q^2 = \left[\frac{1}{2}(\cos \theta)(1 + \cos \theta)\right]^2 \tag{16}$$

If $\theta \leq 30°$, the wobbling-in-a-cone diffusion constant, D_w, is approximately[89]

$$D_w \cong 7\theta^2/24\tau_w \tag{17}$$

with θ in radians. From eq 17 it is clear that the wobbling diffusion constant is not determined solely by τ_w. The more accurate expression for any value of θ_w leading to values reported in Table 2 is[89]

$$D_w = \frac{x_w^2(1 + x_w)^2\{\ln[(1 + x_w)/2] + (1 - x_w)/2\}}{\tau_w(1 - Q^2)[2(x_w - 1)]} + \frac{(1 - x_w)(6 + 8x_w - x_w^2 - 12x_w^3 - 7x_w^4)}{24\tau_w(1 - Q^2)} \tag{18}$$

where $x_w = \cos \theta_w$. In the limit of $\theta_w = 180°$; i.e., there is no restriction to the orientational diffusion, $Q^2 = 0$ and $D_w = 1/6\tau_w$, as expected.

Values for θ, D_w and D_m are given in Table 2 for $w_0 = 10$, 5, and 2. Because τ_1 is so much longer than τ_2, τ_2 and τ_w differ very little (see eq 15). These values reveal a pronounced change between $w_0 = 10$ and 20. In addition to requiring the wobbling-in-a-cone model for $w_0 = 10$ and small reverse micelles but not above $w_0 = 20$, D_m drops by a factor of 6 and continues to drop with decreasing w_0. Furthermore, the cone angle decreases with decreasing w_0, indicating increasing restriction in the reverse micelle environment.

Our observation of single-exponential decay for reorientation in large reverse micelles and biexponential decay in smaller reverse micelles is consistent with the wobbling-in-a-cone model. The wobbling occurs because a restricted range of orientations is sampled on a time scale short compared to complete orientational relaxation that requires complete randomization of the hydrogen bond network (breaking and making new hydrogen bonds). If the cone angular limit is reached rapidly compared to the hydrogen bond network randomization, then a separation of time scales exists, and a biexponential decay will be observed. For $w_0 = 10$, wobbling (1.5 ps) is still fast compared to complete randomization (18 ps). By $w_0 = 20$, τ_1 has dropped a factor of 6 to 3 ps, and the trend in τ_2 is that it is becoming longer. There is no longer a separation of time scales. Complete randomization is occurring prior to the cone angular limit be reached and the result is a single-exponential decay. Thus, it is possible to explain the orientational relaxation data without the core$-$shell model, whereas the core$-$shell model is both quantitatively and qualitatively incorrect.

C. Reconciling the Successes and Failures of the Strict Core$-$Shell Model. In section IV.A, a core$-$shell model was shown to have remarkable success in reproducing the reverse

J. Phys. Chem. A, Vol. 110, No. 15, 2006 **4997**

micelle size dependent trends in the absorption spectra and the population relaxation. If the reverse micelle core is modeled by bulk water data and the shell by the $w_0 = 2$ data, the appropriate addition of a core component and a shell component could reproduce all of the data with remarkable fidelity (see Figures 7 and 8). The results define a strict core−shell model in the sense that the observables are the weighted sum of the observables associated with each subensemble. However, when the identical strict core−shell model was applied in the appropriate manner to spectral diffusion and orientational relaxation in section IV.B, the calculations based on the two subensembles produced results that were fundamentally at odds with the data. The reason for success with two observables and the failure with two others lies in the differences in the nature of the properties measured in the four types of experiments.

The steady-state spectra and the population relaxation are very sensitive to the quantized energy levels of the system, whereas the hydrogen bond structural rearrangement dominates the spectral diffusion and the orientational relaxation. The absorption spectra measure the distribution of the frequencies of the OD hydroxyl stretch and report the instantaneous distribution of frequencies of the system. OD oscillators in different environments will have different stretch frequencies. Indeed, ions change both the water structure and dynamics.[3,90] It is not surprising that shell water molecules in close proximity to the charged surfactant headgroups with associated Na^+ counterions will have a different distribution of frequencies than those in the core, which are removed from the immediate vicinity of the headgroups.[17] Although the spectrum depends only on the frequency of the OD stretch, vibrational relaxation depends on the frequency both of the OD stretch and of other quantized modes of the system. The density of states, in part, determines the vibrational relaxation rate and is expected to differ for water in reverse micelles compared with bulk water.[91] Vibrational relaxation can be described theoretically in terms of the Fourier component of the force-force correlation function at the oscillator frequency.[92−94] However, the process is quantized. A polyatomic molecule can relax into a distribution of internal and bath modes. A variety of pathways that conserve energy can be found, but not all are equally rapid.[94] For example, the OD stretch can relax into a bend and one or more quanta of the continuum of low frequency bath modes, such as torsional modes, hydrogen bond modes, and collective coupled orientational and translational motions. Because the environment affects its frequency, the OD stretch differs in the core and shell regions. However, the environment may not affect all modes in the same way. If the bend, for example, shifts differently than the stretch, then the energy that must be transferred into the bath for vibrational relaxation to occur will be different in the core and shell regions. The result will be that different bath modes with different couplings and density of states and possibly a different number of bath modes will be required to conserve energy during the relaxation process. In addition, the bath modes themselves can be expected to be different in the core and shell regions. The relaxation pathways may differ because of the proximity of the reverse micelle headgroups. Vibrational energy can transfer to modes of the surfactant in very small reverse micelles within several ps after the vibrational excitation of the water hydroxyl stretch.[64] The net result is that, on average, the two regions produce distinct population relaxation times that are independent of each other. The fact that the population decay is composed of two distinct components is strong support for previous work[10,37,38] that suggests that the core−shell water

exchange is slow. The results presented here demonstrate that exchange must occur on a time scale >10 ps.

Spectral diffusion and orientational relaxation depend on the structural dynamics of the hydrogen bonded network rather than on the location of energy levels per se. The core and shell regions, though physically distinct, are intimately connected through the hydrogen bond network. Simulations[10,19,37] and physical considerations[14,36] show that the Na^+ counterions are associated with the sulfonate anionic headgroups. These structures will have very large dipoles. Although the detailed structure of water associated with the headgroup/counterion is unknown, it is safe to say that the arrangement of water molecules in a shell of waters at the interface will differ substantially from bulk water. Consider a hypothetical system possessing an infinite extent of water molecules with the structure found near the interface. Such a system will have hydrogen bond dynamics that lead to global hydrogen bond rearrangement and randomization of the network that is very different from bulk water. If this infinite extent shell structure were contacted with an infinite bulk water system, at the boundary between the two there would have to be accommodation akin to the interface of two incommensurate crystal lattices. When two incommensurate crystal lattices are joined, the boundary region structure differs from both pure crystalline materials because of accommodation.[95] However, if the two types of crystals are macroscopic in extent, some distance from the boundary, their unperturbed lattice structures will exist. In the reverse micelle, a thin shell layer meets a relatively small core region. The structures of these regions differ, but their structures and particularly their long time scale global hydrogen bond dynamics are strongly coupled. Because core waters are hydrogen bonded to shell waters, the rearrangements responsible for complete spectral diffusion and full orientational relaxation in the small reverse micelles require motions that are not independent in the two regions. Therefore, the strict core−shell model does not describe the dynamics. That is, adding the data of only the shell ($w_0 = 2$) to the data that are only the core (bulk water) does not reproduce the spectral diffusion data or the orientational relaxation data for other size reverse micelles as shown in Figures 9 and 10.

The results shown in Figures 9 and 10 demonstrate that, in contrast to the spectrum and the population relaxation, a strict core−shell model does not apply to observables that depend on global hydrogen bond rearrangements. However, differences in the hydrogen bond dynamics in the core and shell regions may still be observable. As discussed in connection with Figure 9, intensity level vibrational echo measurements of spectral diffusion will not reveal core/shell differences because of the cross term between different subensembles accounts for a large part of the observed signal. However, 2D IR vibrational echo correlation spectroscopy, which heterodyne detects the vibrational echo pulse at the polarization level eliminates the cross term.[22,44,69] This method can detect wavelength dependent differences in spectral diffusion in bulk water.[70]

Although the cross term that influences the vibrational echo measurements does not influence the orientational relaxation measurements, the core/shell model still does not adequately describe orientational observables. Molecular dynamics (MD) simulations of small AOT reverse micelles that calculate orientational relaxation show differences between associated and core water molecules.[10] Both subensembles give nonexponential decays that can be described with a wobbling-in-a-cone model that we have applied in section IV.B.[36] The simulations are consistent with the results presented here that the biexponential orientational relaxation decays cannot be assigned to two distinct

4998 *J. Phys. Chem. A, Vol. 110, No. 15, 2006*

subensembles each with single-exponential orientation relaxation. The simulations also suggest that there may be two subensembles, each displaying nonexponential orientational relaxation. Preliminary wavelength dependent measurements of orientational relaxation in $w_0 = 10$ reverse micelles show changes in the orientational relaxation with wavelength which indicate that some difference in dynamics exist in different parts of the reverse micelles, but not in the manner described by the *strict* core shell model. Detailed wavelength dependent studies will be presented subsequently.

V. Concluding Remarks

A detailed study of the nature and dynamics of water in the nanoscopically confined environment of AOT reverse micelles has been presented. Four observables were studied as a function of the water nanopool size: the hydroxyl stretch absorption spectrum, vibrational population relaxation, orientational relaxation, and spectral diffusion. All four show distinct trends with size. Figure 1 and the parameters in Tables 2 and 3 show that the major changes from bulk water behavior appear to occur when the reverse micelles become smaller than $w_0 = 20$, which corresponds to a water nanopool diameter of <7 Å. In the small reverse micelles, the orientational relaxation slows dramatically as does the global hydrogen bond randomization determined by measurements of spectral diffusion using vibrational echoes.

A strict core−shell model was applied to all four observables. In the model, bulk water and $w_0 = 2$ properties were used as surrogates of the core and shell (headgroup region) of the reverse micelles, respectively. It was found that the size dependent spectra and the size dependent population relaxation measured with pump−probe spectroscopy could be reproduced by the appropriately weighted sum of the core and shell surrogates. The combined fits to the spectra and population relaxation permitted the relative core−shell concentrations of water and the core−shell ratio of the water transition dipole moments to be determined. The transition dipole ratio was confirmed by independent spectroscopic measurements. The results permitted the shell thickness to be estimated at ∼0.4 nm. The excellent agreement between the core−shell model and the spectra and the population relaxation data (Figures 7 and 8) lend tremendous credence to viewing the reverse micelles as being composed of distinct environments *for these observables*, which depend mainly on the immediate local structure and not on the hydrogen bond network structural evolution.

In contrast to the spectra and population relaxation, detailed calculations using the strict core−shell model could not reproduce the orientational relaxation or the spectral diffusion. These observables depend strongly on global hydrogen bond network rearrangement dynamics. It was proposed the continuous hydrogen bond network couples the core and shell regions producing network dynamics that are intimately linked in the two regions and cannot be separated into distinct independent subensembles modeled as a superposition of a core region (bulk water) and a shell region ($w_0 = 2$). There still may be differences in these regions in the dynamics. Initial wavelength dependent anisotropy experiments show changes in the orientational relaxation with the detection wavelength. Detailed wavelength dependent studies and studies of other reverse micelles are in progress. These additional experiments will clarify issues that have been brought to the fore here.

Acknowledgment. This work was supported by the Department of Energy (DE-FG03-84ER13251), and the National Science Foundation (DMR-0332692).

Note Added after Print Publication. This paper was published on the Web on March 29, 2006, and in the April 20, 2006, issue with an incorrect received date at the top of p 4985. The correct received date is February 19, 2006. The electronic version was corrected and reposted to the Web issue on August 07, 2006. An Addition and Correction appears in the August 31, 2006, issue (Vol. 110, No. 34).

References and Notes

(1) Scatena, L. F.; Brown, M. G.; Richmond, G. L. *Science* **2001**, *292*, 908.

(2) Eisenthal, K. B. *Acc. Chem. Res.* **1993**, *26*, 636.

(3) Kropman, M. F.; Bakker, H. J. *Science* **2001**, *291*, 2118.

(4) Nandi, N.; Bhattacharyya, K.; Bagchi, B. *Chem. Rev.* **2000**, *100*, 2013

(5) Kim, J.; Schmitt, U. W.; Gruetzmacher, J. A.; Voth, G. A.; Scherer, N. F. *J. Chem. Phys.* **2002**, *116*, 737.

(6) Tassaing, T.; Danten, Y.; Besnard, M. *J. Mol. Liq.* **2002**, *101*, 149.

(7) Gorbaty, Y. E.; Bondarenko, G. V.; Kalinichev, A. G.; Okhulkov, A. V. *Mol. Phys.* **1999**, *96*, 1659.

(8) Pal, S. K. P., J.; Bagchi, B.; Zewail, A. H. *J. Phys. Chem. B.* **2002**, *106*, 12376.

(9) Balasubramanian, S.; Pal, S.; Bagchi, B. *Phys. Rev. Lett.* **2002**, *89*, 115505/1.

(10) Harpham, M. R.; Ladanyi, B. M.; Levinger, N. E.; Herwig, K. W. *J. Chem. Phys.* **2004**, *121*, 7855.

(11) Willard, D. M.; Riter, R. E.; Levinger, N. E. *J. Am. Chem. Soc.* **1998**, *120*, 4151.

(12) Piletic, I. R.; Tan, H.-S.; Fayer, M. D. *J. Phys. Chem. B.* **2005**, *109*, 21273.

(13) Sarkar, N.; Das, K.; Datta, A.; Das, S.; Bhattacharyya, K. *J. Phys. Chem.* **1996**, *100*, 10523.

(14) Tan, H.-S.; Piletic, I. R.; Riter, R. E.; Levinger, N. E.; Fayer, M. D. *Phys. Rev. Lett.* **2005**, *94*, 057405(4).

(15) Vajda, S.; Jimenez, R.; Rosenthal, S. J.; Fidler, V.; Fleming, G. R.; Castner, E. W. *J. Chem. Soc., Faraday Trans.* **1995**, *91*, 867.

(16) Beta, I. A.; Bohlig, H.; Hunger, B. *Phys. Chem. Chem. Phys.* **2004**, *6*, 1975.

(17) Nandi, N.; Bagchi, B. *J. Phys. Chem. B* **1997**, *101*, 10954.

(18) Faeder, J.; Ladanyi, B. M. *J. Phys. Chem. B* **2001**, *105*, 11148.

(19) Abel, S.; Sterpone, F.; Bandyopadhyay, S.; Marchi, M. *J. Phys. Chem. B* **2004**, *108*, 19458.

(20) Senapati, S.; Berkowitz, M. L. *J. Chem. Phys.* **2003**, *118*, p 1937.

(21) Stillinger, F. H. *Science* **1980**, *209*, 451.

(22) Asbury, J. B.; Steinel, T.; Stromberg, C.; Corcelli, S. A.; Lawrence, C. P.; Skinner, J. L.; Fayer, M. D. *J. Phys. Chem. A* **2004**, *108*, 1107.

(23) Fecko, C. J.; Eaves, J. D.; Loparo, J. J.; Tokmakoff, A.; Geissler, P. L. *Science* **2003**, *301*, 1698.

(24) De, T. K.; Maitra, A. *Adv. Colloid Interface Sci.* **1995**, *59*, 95.

(25) Tokmakoff, A.; Fayer, M. D. *Acc. Chem. Res.* **1995**, *28*, 437.

(26) Merchant, K. A.; Thompson, D. E.; Fayer, M. D. *Phys. Rev. Lett.* **2001**, *86*, 3899.

(27) Henri-Rousseau, O.; Blaise, P. *Adv. Chem. Phys.* **1998**, *103*, 1.

(28) Laenen, R.; Rausch, C.; Laubereau, A. *J. Phys. Chem. B* **1998**, *102*, 9304.

(29) Woutersen, S.; Emmerichs, U.; Bakker, H. J. *Science (Washington, D.C.)* **1997**, *278*, 658.

(30) Pimentel, G. C.; McClella. Al. *Annu. Rev. Phys. Chem.* **1971**, *22*, 347.

(31) Marechal, Y. *J. Chem. Phys.* **1991**, *95*, 5565.

(32) Steinel, T.; Asbury, J. B.; Fayer, M. D. *J. Phys. Chem. A* **2004**, *108*, 10957.

(33) Onori, G.; Santucci, A. *J. Phys. Chem.* **1993**, *97*, 5430.

(34) Nucci, N. V.; Vanderkooi, J. M. *J. Phys. Chem. B* **2005**, *109*, 18301.

(35) Rey, R.; Møller, K. B.; Hynes, J. T. *J. Phys. Chem. A* **2002**, *106*, 11993.

(36) Tan, H.-S.; Piletic, I. R.; Fayer, M. D. *J. Chem. Phys.* **2005**, *122*, 174501(9).

(37) Faeder, J.; Ladanyi, B. M. *J. Phys. Chem. B* **2000**, *104*, 1033.

(38) Thompson, W. H. *J. Chem. Phys.* **2004**, *120*, 8125.

(39) Kinugasa, T.; Kondo, A.; Nishimura, S.; Miyauchi, Y.; Nishii, Y.; Watanabe, K.; Takeuchi, H. *Colloids Surf. A−Physicochem. Eng. Aspects* **2002**, *204*, 193.

(40) Zulauf, M.; Eicke, H.-F. *J. Phys. Chem.* **1979**, *83*, 480.

(41) Kane, D. J.; Trebino, R. *IEEE J. Quantum Electron.* **1993**, *29*, 571.

(42) Tan, H.-S.; Piletic, I. R.; Fayer, M. D. *J. Opt. Soc. Am. B: Opt. Phys.* **2005**, *22*, 2009.

(43) Mukamel, S. *Principles of Nonlinear Optical Spectroscopy*; Oxford University Press: New York, 1995.

J. Phys. Chem. A, Vol. 110, No. 15, 2006 **4999**

(44) Asbury, J. B.; Steinel, T.; Kwak, K.; Corcelli, S.; Lawrence, C. P.; Skinner, J. L.; Fayer, M. D. *J. Chem. Phys.* **2004**, *121*, 12431.

(45) Stenger, J.; Madsen, D.; Hamm, P.; Nibbering, E. T. J.; Elsaesser, T. *J. Phys. Chem. A.* **2002**, *106*, 2341.

(46) Zanni, M. T.; Asplund, M. C.; Hochstrasser, R. M. *J. Chem. Phys.* **2001**, *114*, 4579.

(47) Merchant, K. A.; Xu, Q.-H.; Thompson, D. E.; Fayer, M. D. *J. Phys. Chem. A* **2002**, *106*, 8839.

(48) Cho, M. H.; Yu, J. Y.; Joo, T. H.; Nagasawa, Y.; Passino, S. A.; Fleming, G. R. *J. Chem. Phys.* **1996**, *100*, 11944.

(49) Woutersen, S.; Emmerichs, U.; Bakker, H. J. *J. Chem. Phys.* **1997**, *107*, 1483.

(50) Gaffney, K.; Piletic, I.; Fayer, M. D. *J. Phys. Chem. A* **2002**, *106*, 9428.

(51) Gaffney, K. J.; Davis, P. H.; Piletic, I. R.; Levinger, N. E.; Fayer, M. D. *J. Phys. Chem. A* **2002**, *106*, 12012.

(52) Dokter, A. M.; Woutersen, S.; Bakker, H. J. *Phys. Rev. Lett.* **2005**, *94*, 178301.

(53) Novaki, L. P.; El Seoud, O. A. *J. Colloid Interface Sci.* **1998**, *202*, 391.

(54) Christopher, D. J.; Yarwood, J.; Belton, P. S.; Hills, B. P. *J. Colloid Interface Sci.* **1992**, *152*, 465.

(55) Lawrence, C. P.; Skinner, J. L. *J. Chem. Phys.* **2002**, *117*, 8847.

(56) Møller, K. B.; Rey, R.; Hynes, J. T. *J. Phys. Chem. A.* **2004**, *108*, 1275.

(57) Venables, D. S.; Huang, K.; Schmuttenmaer, C. A. *J. Phys. Chem. B* **2001**, *105*, 9132.

(58) Geissler, P. L. *J. Am. Chem. Soc.* **2005**, *127*, 14930.

(59) Schmidt, J. R.; Corcelli, S. A.; Skinner, J. L. *J. Chem. Phys.* **2005**, *123*, 044513.

(60) Tokmakoff, A. *J. Chem. Phys.* **1996**, *105*, 1.

(61) Loparo, J. J.; Fecko, C. J.; Eaves, J. D.; Roberts, S. T.; Tokmakoff, A. *Phys. Rev. B* **2004**, *70*, 180201.

(62) Rezus, Y. L. A.; Bakker, H. J. *J. Chem. Phys.* **2005**, *123*, 114502.

(63) Zhong, Q.; Baronavski, A. P.; Owrutsky, J. C. *J. Chem. Phys.* **2003**, *118*, 7074.

(64) Deak, J. C.; Pang, Y.; Sechler, T. D.; Wang, Z.; Dlott, D. D. *Science* **2004**, *306*, 473.

(65) Szabo, A. *J. Chem. Phys.* **1984**, *81*, 150.

(66) Forster, T. *Discuss. Faraday Soc.* **1959**, *27*, 7.

(67) Gochanour, C. R.; Andersen, H. C.; Fayer, M. D. *J. Chem. Phys.* **1979**, *70*, 4254.

(68) Gochanour, C. R.; Fayer, M. D. *J. Phys. Chem.* **1981**, *85*, 1989.

(69) Asbury, J. B.; Steinel, T.; Fayer, M. D. *J. Luminescence* **2004**, *107*, 271.

(70) Steinel, T.; Asbury, J. B.; Corcelli, S. A.; Lawrence, C. P.; Skinner, J. L.; Fayer, M. D. *Chem. Phys. Lett.* **2004**, *386*, 295.

(71) Lawrence, C. P.; Skinner, J. L. *Chem. Phys. Lett.* **2003**, *369*, 472.

(72) Piryatinski, A.; Lawrence, C. P.; Skinner, J. L. *J. Chem. Phys.* **2003**, *118*, 9672.

(73) Rey, R.; Moller, K. B.; Hynes, J. T. *Chem. Rev.* **2004**, *104*, 1915.

(74) Rey, R.; Hynes, J. T. *J. Chem. Phys.* **1996**, *104*, 2356.

(75) Corcelli, S.; Lawrence, C. P.; Skinner, J. L. *J. Chem. Phys.* **2004**, *120*, 8107.

(76) Corcelli, S.; Lawrence, C. P.; Asbury, J. B.; Steinel, T.; Fayer, M. D.; Skinner, J. L. *J. Chem. Phys.* **2004**, *121*, 8897.

(77) Joo, T. H.; Jia, Y. W.; Yu, J. Y.; Jonas, D. M.; Fleming, G. R. *J. Chem. Phys.* **1996**, *104*, 6089.

(78) Mukamel, S. *Annu. Rev. Phys. Chem.* **2000**, *51*, 691.

(79) Kubo, R. A Stochastic Theory of Line-Shape and Relaxation. In *Fluctuation, Relaxation and Resonance in Magnetic Systems*; Ter Haar, D., Ed.; Oliver and Boyd: London, 1961.

(80) Joo, T.; Albrecht, A. W. *J. Chem. Phys.* **1993**, *99*, 3244.

(81) Lawrence, C. P.; Skinner, J. L. *J. Chem. Phys.* **2003**, *118*, 264.

(82) Wong, M.; Thomas, J. K.; Nowak, T. *J. Am. Chem. Soc.* **1977**, *99*, 4730.

(83) Pileni, M. P. *J. Phys. Chem.* **1993**, *97*, 6961.

(84) Maitra, A.; Jain, K. T.; Shervani, Z. *Colloids Surf.* **1990**, *47*, 255.

(85) Bertie, J.; Lan, Z. *Appl. Spectrosc.* **1996**, *50*, 1047.

(86) Fayer, M. D. *Elements of Quantum Mechanics*; Oxford University Press: New York, 2001.

(87) Berry, R. S.; Rice, S. A.; Ross, J. *Physical Chemistry*, 2nd ed.; Oxford University Press: New York, 2000.

(88) Wang, C. C.; Pecora, R. *J. Chem. Phys.* **1980**, *72*, 5333.

(89) Lipari, G.; Szabo, A. *Biophys. J.* **1980**, *30*, 489.

(90) Kropman, M. F.; Bakker, H. J. *J. Chem. Phys.* **2001**, *115*, 8942.

(91) Boyd, J. E.; Briskman, A.; Colvin, V. L.; Mittleman, D. M. *Phys. Rev. Lett.* **2001**, *87*, 147401.

(92) Oxtoby, D. W. *Adv. Chem. Phys.* **1981**, *47*, 487.

(93) Oxtoby, D. W. *Annu. Rev. Phys. Chem.* **1981**, *32*, 77.

(94) Kenkre, V. M.; Tokmakoff, A.; Fayer, M. D. *J. Chem. Phys.* **1994**, *101*, 10618.

(95) Theodorou, G.; Rice, T. M. *Phys. Rev. B* **1978**, *18*, 2840.

Substrate binding and protein conformational dynamics measured by 2D-IR vibrational echo spectroscopy

Ilya J. Finkelstein, Haruto Ishikawa, Seongheun Kim, Aaron M. Massari[†], and M. D. Fayer[‡]

Department of Chemistry, Stanford University, Stanford, CA 94305

Edited by Robin M. Hochstrasser, University of Pennsylvania, Philadelphia, PA, and approved December 27, 2006 (received for review November 10, 2006)

Enzyme structural dynamics play a pivotal role in substrate binding and biological function, but the influence of substrate binding on enzyme dynamics has not been examined on fast time scales. In this work, picosecond dynamics of horseradish peroxidase (HRP) isoenzyme C in the free form and when ligated to a variety of small organic molecule substrates is studied by using 2D-IR vibrational echo spectroscopy. Carbon monoxide bound at the heme active site of HRP serves as a spectroscopic marker that is sensitive to the structural dynamics of the protein. In the free form, HRP assumes two distinct spectroscopic conformations that undergo fluctuations on a tens-of-picoseconds time scale. After substrate binding, HRP is locked into a single conformation that exhibits reduced amplitudes and slower time-scale structural dynamics. The decrease in carbon monoxide frequency fluctuations is attributed to reduced dynamic freedom of the distal histidine and the distal arginine, which are key residues in modulating substrate binding affinity. It is suggested that dynamic quenching caused by substrate binding can cause the protein to be locked into a conformation suitable for downstream steps in the enzymatic cycle of HRP.

horseradish peroxidase | ultrafast

Enzyme-substrate binding is a dynamic process that is intimately coupled to protein structural fluctuations (1, 2). High-throughput screening methods have identified enzymes that bind structurally diverse inhibitors within their active sites, often with binding affinities exceeding those of the biologically derived substrate (3, 4). Such experimental observations illuminate our understanding of protein–substrate interactions, underscoring that static protein structure information alone is insufficient to describe the diverse influences of ligand binding at an active site (2). A complete description of protein–ligand interactions requires information on the modification of protein dynamics, if any, when a ligand binds.

Proteins rapidly interconvert within an ensemble of similar conformations (5–10). A subset of these rapidly interconverting states may be favorable for binding a given substrate. A different subset of conformations may accommodate a structurally heterologous ligand. Although conceptually appealing, this mechanism is difficult to probe experimentally because it requires sensitivity to protein structural dynamics on fast time scales. The questions addressed here are, does substrate binding influence protein dynamics, and do different substrates binding to the same protein produce distinct changes in protein dynamics?

Ultrafast 2D-IR vibrational echo spectroscopy can probe protein conformational fluctuations under thermal equilibrium conditions on time scales ranging from subpicoseconds to $\approx$100 picoseconds or longer. 2D-IR vibrational echo spectroscopy is somewhat akin to 2D-NMR techniques but reports on protein dynamics that occur on fast time scales. The methods have recently been applied to study model enzymes (11, 12), protein unfolding (10), peptide dynamics in membranes (9), and protein equilibrium fluctuations in aqueous and confined environments (7, 8, 13, 14). In this work, we use 2D-IR vibrational echo

spectroscopy to examine the equilibrium structural fluctuations of horseradish peroxidase (HRP) in the absence and presence of small molecule substrates with dissociation constants spanning three orders of magnitude. HRP is a type III peroxidase-family glycoprotein that oxidizes a variety of organic molecules in the presence of hydrogen peroxide as the oxidizing agent (15). HRP has proven to be amenable to protein engineering and has been of intense interest in bioindustrial and enantiospecific catalysis applications (4, 16).

The active site of HRP comprises a solvent-exposed iron-heme prosthetic group that participates in the enzymatic catalysis cycle (4, 16). The heme can bind carbon monoxide (CO), which has frequently been exploited as a site-specific reporter of protein structure (8, 13, 17, 18) and dynamics (8, 19–21). The time dependence of the CO transition frequency is a spectroscopic reporter of protein structural fluctuations (8, 13, 21). Within the dynamic Stark-effect model, structural fluctuations of groups within the protein generate a time-dependent electric field at the heme active site (8, 13, 21). The CO transition frequency is exquisitely sensitive to electric fields (20, 22). Therefore, CO-frequency fluctuations are sensitive to global and local protein structural evolution.

The small molecule substrates used in this study are benzhydroxamic acid (BHA) analogs that have been investigated as a general class of tightly binding inhibitors for HRP and other peroxidases (4, 23) (see Fig. 1). A wealth of structural (24–27), biochemical (16, 28, 29), and spectroscopic evidence (19, 20, 30, 31) has identified His-42 and Arg-38 as the key HRP residues in modulating substrate binding and enzymatic activity. These distal heme residues are highly conserved in the peroxidase family (4). Substrates and intermediates in the catalytic reaction of HRP interact with the distal residues through an extensive hydrogen-bonding network within the active site (24, 32) (see Fig. 2). Substrates such as BHA can participate strongly with this intrinsic hydrogen-bond network, showing the strongest binding affinities. The substrates selected for this study incorporate key structural modifications that tune their propensity for making hydrogen bonds within the active site and thus exhibit dissociation constants between $K_d = 4300 - 0.16\ \mu M$ from ferric HRP (16).

Aqueous HRP in the free form (without substrate) adopts two distinct spectroscopic states and, as shown below, has structural

Author contributions: I.J.F., A.M.M., and M.D.F. designed research; I.J.F., H.I., S.K., A.M.M., and M.D.F. performed research; I.J.F. and M.D.F. analyzed data; and I.J.F. and M.D.F. wrote the paper.

The authors declare no conflict of interest.

This article is a PNAS direct submission.

Abbreviations: BHA, benzhydroxamic acid; MbCO, myoglobin-CO; 2-NHA, 2-napthohydroxamic acid; NMBZA, N-methylbenzamide; BZA, benzamide; BZH, benzhydrazide; FFCF, frequency–frequency correlation function.

[†]Present address: Department of Chemistry, University of Minnesota, Minneapolis, MN 55455.

[‡]To whom correspondence should be addressed. E-mail: fayer@stanford.edu.

© 2007 by The National Academy of Sciences of the USA

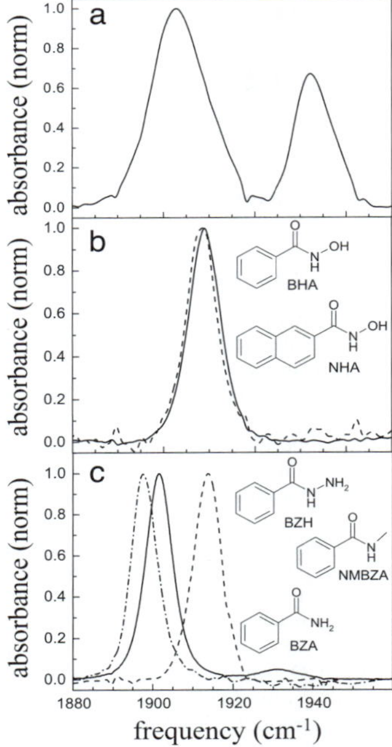

Fig. 1. Normalized FTIR spectra of the CO-stretching mode bound to HRP in the free form (*a*) and when complexed to 2-NHA (dashed line) and BHA (solid line) (*b*) and NMBZA (dash-dot line), BZH (dashed line), and BZA (solid line) (*c*). Structures and abbreviations of the substrates used in this study are shown in the appropriate places.

dynamics on the <1- and 20-ps time scales. After substrate binding, HRP occupies a single structural state. The protein dynamics, as reported by the CO transition-frequency fluctuations measured with the 2D-IR experiments, decrease in ampli-

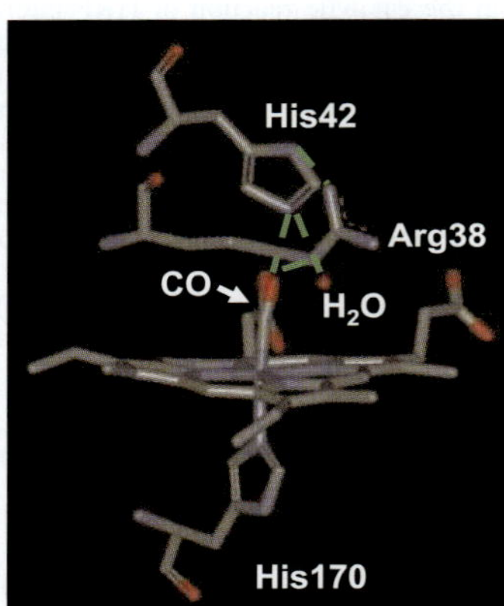

Fig. 2. Crystal structure of the active site of free HRP taken from the Protein Data Bank (ID code 1W4Y). His-170 tethers the heme group in the active site. The residues Arg-38 and His-42 and a disordered water molecule define the distal active-site structure and hydrogen-bonding network. Possible hydrogen-bonding interactions are shown as dashed green lines.

Table 1. Linear spectra of free and substrate-bound HRP

	ν_{CO}, cm^{-1}	FWHM, cm^{-1}
HRP blue	1,932.7	9.0
HRP red	1,903.7	13.0
2-NHA	1,908.3	7.3
BHA	1,909.0	7.3
BZH	1,913.3	9.0
NMBZA	1,898.0	6.9
BZA	1,901.5	8.6

tude and slow down significantly. The protein dynamics of HRP are compared with myoglobin-CO (MbCO) and a mutant H64V (the distal histidine is replace by valine) that exhibits a similar reduction in observed dynamics relative to MbCO. The results are used to comment on the possible role of the biologically derived substrates in reordering the protein active site to facilitate downstream events in the enzymatic pathway.

Results and Discussion

Linear Spectra. The background-subtracted and normalized linear absorption spectra of free and inhibitor-bound HRP–CO in ^2H_2O are presented in Fig. 1. At pD 7.4, free HRP occupies two spectroscopically distinct structural conformations, with CO absorption bands centered at 1,903 and 1,934 cm^{-1} (33, 34) (Fig. 1*a*). The x-ray crystal structure of HRP–CO shows an extensive hydrogen-bond network at the active site that incorporates the CO, His-42, Arg-38, and a disordered water molecule in the heme cavity (24) (see Fig. 2). Resonance Raman spectra of the distal histidine imidazolium (35) and other experimental data (34) suggest that in the red state the CO is nearly normal to the heme plane and oriented such that it has a strong interaction with the distal histidine and a weaker one with the distal arginine. In the blue state the Fe–C–O linkage, although linear, is somewhat bent relative to the heme normal, allowing the CO to have a strong interaction with the distal arginine and a weaker one with the distal histidine. Resonance Raman evidence indicates that the crystal structure of the heme pocket shown in Fig. 2 resembles the blue spectroscopic state (35). This structural assignment of the blue state is reinforced by pH titration studies of HRP. Deprotonation of His-42 at pH 8.7 is correlated with a total disappearance of the red state in HRP, indicating that CO in the red state participates in hydrogen bonding with a hydrogen on the distal histidine (35).

Fig. 1*b* shows the background-subtracted linear spectra of HRP ligated with BHA (solid curve) and 2-napthohydroxamic acid (2-NHA) (dashed curve). Fig. 1*c* displays the linear spectra of HRP–*N*-methylbenzamide (NMBZA), HRP–benzamide (BZA), and HRP–benzhydrazide (BZH) complexes. After substrate binding, the CO band becomes a single, essentially Gaussian peak centered between 1,898 and 1,915 cm^{-1} (33, 34). The ligated HRP spectra are spread around the frequency of the unligated HRP red state and are significantly lower in frequency than the blue unligated state. The CO-band peak positions and widths are summarized in Table 1. A change in the transition frequency of the type observed here is almost certainly caused by a rearrangement of the distal cavity structure (8, 20). For example, in MbCO it is well documented that the three spectroscopically distinct substates, A_0, A_1, and A_3, arise from distinct configurations of the distal histidine (8, 36).

Substrate binding in HRP is enthalpically stabilized, largely by the strong substrate carbonyl–Arg-38 hydrogen bond (16). The HRP mutant R38L does not bind BHA (37). BHA and 2-NHA are well matched to form hydrogen bonds with Arg-38, His-42, and Pro-139 and have very high binding affinities (K_d = 2.5 and 0.16 μM, respectively) to the ferric protein (16, 32). The sub-

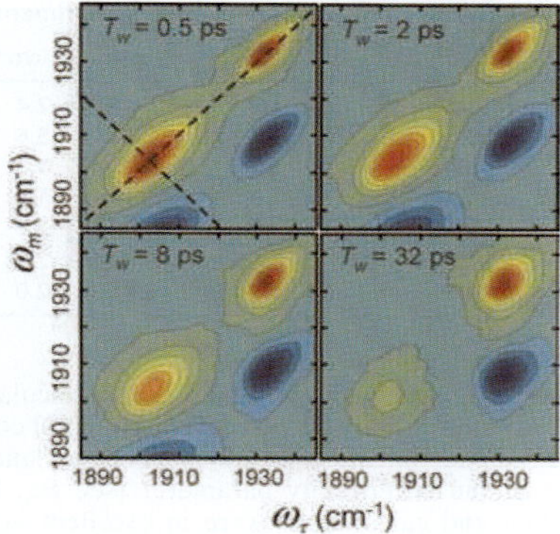

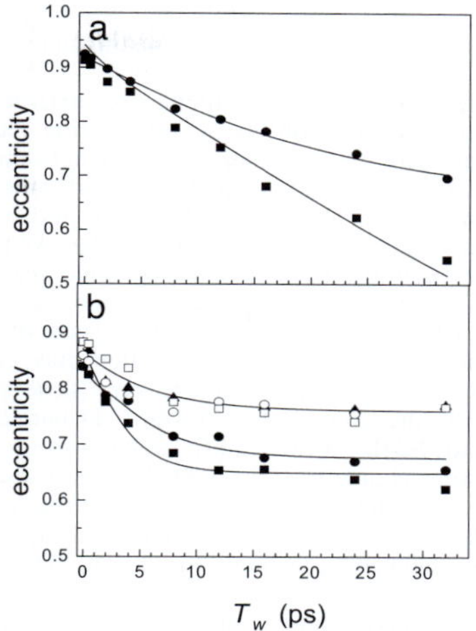

Fig. 3. 2D-IR spectra of free HRP as a function of increasing T_w. The dashed lines illustrate the diagonal and antidiagonal slices through the data for calculating the eccentricity parameter.

Fig. 4. A comparison of HRP dynamics in the free and substrate-bound states. (*a*) T_w-dependent eccentricities of the blue (circles) and red (squares) states of free HRP. (*b*) Eccentricities of HRP ligated with 2-NHA (filled squares), BHA (filled circles), BZA (open squares), BZH (open circles), and NMBZA (triangles). The solid lines are obtained by simultaneously fitting the linear and 2D-IR data to determine the FFCFs.

strates BZA, NMBZA, and BZH are less well suited to making hydrogen bonds within the active site and have K_d values of 4,300, 360, and 110 μM, respectively.

2D-IR Spectroscopy. The absorption frequencies of substrate-bound HRP do not report the time-dependent protein structural fluctuations (20). 2D-IR vibrational echo spectroscopy provides a dynamic spectrum that is sensitive to structural fluctuations of HRP and can be used to investigate how these dynamics are influenced by the binding of different substrates. Vibrational correlation spectra of HRP in the free form for a series of increasing "waiting" times, T_w, are presented in Fig. 3 as contour plots with 10% intervals. (The data have much finer contours used in analysis.) The ω_τ frequency axis is associated with the first interaction with the radiation field (first IR pulse), and the ω_m frequency axis is associated with the third interaction with the radiation field (third IR pulse), which is the emission frequency of the vibrational echo pulse. (ω_τ and ω_m are the equivalent of ω_1 and ω_3 in 2D NMR; see *Materials and Methods* below.) Two positive bands (red) along the diagonal at (ω_τ, ω_m) = (1,903 cm^{-1}, 1,903 cm^{-1}) and (1,932 cm^{-1}, 1,932 cm^{-1}) arise from the 0-to-1 vibrational transitions of the CO stretch of the red and blue states of HRP, respectively. The negative peaks (blue) arise from the 1-to-2 transitions and are displaced from the diagonal bands by a CO vibrational anharmonicity of $\approx$25 cm^{-1} (38). These negative-going bands are directly below their corresponding positive-going 0-to-1 bands. The dynamical information obtained from the 1-to-2 bands is the same as that obtained from the 0-to-1 bands; therefore, only the 0-to-1 bands are analyzed below.

As can be seen in Fig. 3, both the relative amplitudes and the shapes of the 0-to-1 bands (bands on the diagonal) change with increasing T_w. At early T_w, the amplitudes of the two states are determined by the strengths of their transition dipole moments and concentrations. Vibrational energy relaxation to the ground vibrational state reduces the amplitudes of both bands at longer T_w. IR pump-probe experiments were used to measure the vibrational lifetimes of the red and blue states, which were determined to be 8 and 12 ps, respectively. Therefore, although the red state has a higher equilibrium concentration, the band at (ω_τ, ω_m) = (1,903 cm^{-1}, 1,903 cm^{-1}) decays significantly faster than the blue band, and the ratio of the peak amplitudes changes with increasing T_w.

The change in shape of the bands as T_w is increased reflects protein structural dynamics. The vibrational echo pulse se-

quence used to collect 2D-IR spectra displays inhomogeneous broadening along the diagonal and dynamic broadening along the antidiagonal (39) (shown as dashed lines in Fig. 3). At short T_w, the 2D dynamic line shape has significant inhomogeneous broadening, which manifests itself as elongation along the diagonal. As T_w is increased, the experiment picks up longer time-scale protein dynamics that increases the antidiagonal width and decreases the diagonal width but to a lesser extent than the antidiagonal. In the long time limit, all protein fluctuations contribute to the 2D spectrum, which would lead to a 2D-IR shape with equal diagonal and antidiagonal linewidths.

To compare the 2D line-broadening behavior of the different states of HRP, we define the eccentricity, ε, of a 2D band according to

$$\varepsilon(T_w) = [1 - \sigma_{AD}^2(T_w)/\sigma_D^2(T_w)]^{1/2}, \qquad [1]$$

where σ_{AD} and σ_D are the bandwidths along the antidiagonal and diagonal slices, respectively (see dashed lines in Fig. 3 *Upper Left*). This definition for the eccentricity displays several convenient limits. In the case of large inhomogeneous broadening, $\sigma_{AD} \ll \sigma_D$, and $\varepsilon \to 1$. At longer T_w, slower time-scale protein dynamics contributes to the antidiagonal width, causing the eccentricity to decay to zero. In the long time limit in which all protein structures have been sampled, $\varepsilon \to 0$. The eccentricity is a convenient way to succinctly summarize protein dynamics contained in the 2D-IR spectra. The full 2D spectral shapes are used in fitting the data.

The T_w-dependent eccentricities of free and substrate-bound HRP are presented in Fig. 4. Curves through the data are obtained from 2D-IR calculations using a dynamic Stark-effect model of protein dynamics that is described below. In Fig. 4*a*, the eccentricities of the red and blue spectroscopic states of free HRP are shown as squares and circles, respectively. Both states have nearly identical initial eccentricities, indicating a similar ratio of dynamic to inhomogeneous broadening at very short T_w. The eccentricity of the red states decays faster than that of the blue state and reaches $\approx$0.5 at the longest experimentally accessible T_w. Neither the red states nor the blue states have sampled all structural configurations on the time scale of the experiment.

Fig. 4b presents the T_w-dependent eccentricities of HRP complexed with the five substrates. Introducing the substrate into the protein active site fundamentally changes the nature of the dynamics, as is evident in the qualitative differences between the T_w-dependent eccentricities in Fig. 4, a and b. Substrate binding shifts dynamics to longer time scales. With a bound substrate, the eccentricity decays on the time scale of a few picoseconds at early T_w but plateaus after $T_w = \approx 10$ ps, becoming nearly T_w-independent. In the cases of BHA and 2-NHA bound to HRP, the eccentricities may continue to decrease but on time scales that are very slow, particularly compared with the free-HRP red state. Dynamics in HRP complexed with NMBZA are significantly constrained, and the eccentricity stops changing at a relatively large value of ≈ 0.75. A large T_w-independent eccentricity indicates that a significant fraction of the total CO linewidth undergoes frequency fluctuations on time scales that are much slower than the experimental observation window. Fig. 4 demonstrates qualitatively that substrate binding to HRP significantly slows the short-time-scale structural dynamics of the protein and shifts a significant fraction of the fluctuations to much longer time scales.

Underlying Protein Dynamics. A quantitative description of the amplitudes and time scales of CO-frequency fluctuations is provided by the frequency–frequency correlation function (FFCF) (8, 13, 21). Within standard approximations, both the linear absorption spectrum and the 2D-IR vibrational echo spectra at all T_w are simultaneously described by the FFCF (40). A multiexponential form of the FFCF, $C(t)$, conveniently organizes a distribution of protein-fluctuation time scales and has been found to reproduce the influence of structural dynamics on the CO frequency in other heme proteins (7, 8, 41). A biexponential plus a constant FFCF was sufficient to describe most of the observed protein dynamics in HRP:

$$C(t) = \Delta_1^2 e^{-t/\tau_1} + \Delta_2^2 e^{-t/\tau_2} + \Delta_3^2. \qquad [2]$$

The static component in $C(t)$, Δ_3, is the contribution to the CO-frequency distribution that occurs from protein structures that interconvert slower than the observable experimental time scales. The 2D-IR experiment is sensitive to dynamics that occur out to several T_w (42), which corresponds to ≈ 100 ps in the experiments presented below. Structural fluctuations slower than ≈ 100 ps appear static to the experiment and are described by Δ_3. The Δ_3 term can be viewed as multiplied by another exponential, but as far as the experiment is concerned in all but one case discussed below, $\tau_3 = \infty$. Δ_i is the amplitude of frequency fluctuations (in units of angular frequency) resulting from structural evolution with a characteristic time τ_i. For $\Delta_i^{-1}\tau_i \ll 1$ for a given exponential term, this component of the FFCF is motionally narrowed (41, 43). A motionally narrowed term in $C(t)$ contributes a T_w-independent symmetric 2D Lorentzian line with a width $\Gamma^* = \pi\Delta^2\tau$ to the total 2D-IR spectrum. For a motionally narrowed component, Δ and τ cannot be determined independently; rather, Γ^* can be obtained. The fastest protein fluctuations can contribute a motionally narrowed exponential to the FFCF and arise from very fast local motions of small groups, analogous to low-frequency vibrations, that do not significantly modify the enzyme topology. These fluctuations have been observed in experiments and simulations of several similar heme-CO proteins and will not be discussed further here (7, 41).

FFCFs were extracted from the experimental data by using an iterative fitting algorithm. The FFCF obtained from analysis of the data using response-theory calculations (40) was deemed

Table 2. FFCFs derived from linear and 2D-IR experimental data

	Γ^*, cm^{-1}	Δ_2, cm^{-1}	τ_2, ps	Δ_3, cm^{-1}	τ_3, ps
HRP blue	1.40	3.2	15.0	2.4	∞
HRP red	0.76	3.1	1.5	5.6	21
2-NHA	0.62	3.2	2.6	1.8	∞
BHA	1.40	2.3	4.4	1.9	∞
BZA	1.40	2.8	4.5	2.7	∞
BZH	1.60	2.6	2.6	2.7	∞
NMBZA	1.30	1.7	5.4	2.0	∞

correct when it could be used to simultaneously calculate 2D-IR spectra that accurately reproduced the experimental correlation plots at all T_w, the linear absorption spectrum, and the T_w dependence of the eccentricity parameter (see Eq. 1). In all cases, the data and calculations were in excellent agreement. Notably, the dynamics of the red state of free HRP could not be modeled adequately by an FFCF of the form shown in Eq. 2. The static term in the FFCF was replaced with a third exponential that has a 21-ps^{-1} time constant.

The parameters defining the experimentally derived FFCFs for free HRP and HRP with the five substrates bound are presented in Table 2. Each of the FFCFs has a motionally narrowed component characterized by the Lorentzian linewidth (Γ^*) that contributes to the total 2D-IR linewidth but does not contribute to the T_w dependence. The FFCF of the free-HRP red state is different from all of the other spectroscopic lines. In addition to the motionally narrowed component, two additional time scales, $\tau_2 = 1.5$ ps and $\tau_3 = 21$ ps, are observed for the red state. This τ_3 is the decay-time constant in the exponential factor that formally multiplies the Δ_3^2 term in Eq. 2 but is not shown because τ_3 is infinite for all of the other systems. The results indicate that the full range of structures that give rise to the red-state absorption spectrum have been sampled by ≈ 100 ps. The blue state of free HRP is well described by a single 15-ps exponential but lacks a short-time-scale component (aside from the motionally narrowed term that is common to all of the systems). In the blue state a relatively large set of CO-frequency fluctuations occur on time scales slower than ≈ 100 ps (42) and necessitate a static term in the FFCF ($\tau_3 = \infty$).

In comparing the behavior of free vs. substrate-bound HRP, the question arises as to the nature of the structural fluctuations that give rise to the observed dynamics. The influence of BHA binding on HRP–CO dynamics has been studied by Kaposi and coworkers both experimentally and by using molecular dynamics simulations (18, 20). The two spectroscopic states of free HRP were modeled by changing the protonation state of the distal histidine. In the protonated state (taken to be the red peak in the simulations), Arg-38 was relatively far from the CO and exhibited large RMSD fluctuations. In the deprotonated case (blue state), Arg-38 was significantly closer to the CO. Introducing BHA into the active site reoriented Arg-38 toward the substrate, shortened the Arg-38–CO distance slightly, and reduced the RMSD fluctuations of both the distal histidine and arginine.

Our results are in good qualitative agreement with the simulations. It is clear from the FFCF parameters given in Table 2 that the dynamics of HRP with a bound substrate are very different from the red state of free HRP. The free blue and red states have a 15- and 21-ps component, respectively. In contrast, the FFCFs for HRP with bound substrates have a slowest component of ≈ 2 to ≈ 5 ps in the observation time window. The linear absorption spectra of HRP with bound substrates are more similar to the red state of free HRP than to the blue state. The red-state 1.5-ps-fast component is followed by a slow component of 21 ps that takes the FFCF essentially to zero by ≈ 100 ps, completing the dynamics. None of the HRP–substrate systems

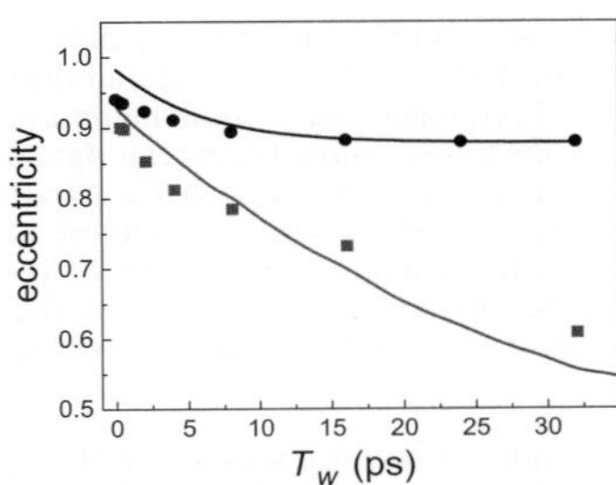

Fig. 5. The eccentricities of the A_1 state of wild-type MbCO (blue squares) and the mutant H64V (black circles). The curves are calculated eccentricities of HRP–NMBZA (black) and red-state free HRP (blue), offset to overlap with the myoglobin data.

show a component of their FFCFs that reflects full structural sampling by 100 ps. All of the HRP–substrate systems have substantial structural dynamics too slow to be in the experimentally accessible window, which are manifested as the constant term in the FFCF (see Table 2). In comparison with the free-HRP red state, substrate binding significantly reduces the protein fluctuations. The red-state 1.5-ps component becomes two or three times as long, and the 21-ps component becomes too slow to measure. When the protein binds a substrate, the active-site structure exhibits a single spectroscopic state in which fluctuations of the distal arginine are strongly constrained. The small fraction of the dynamics that occur within the experimental window for HRP–substrate systems indicates that the dynamics of the distal histidine are also severely constrained. Our results are consistent with the suggestion that loss of dynamic freedom by Arg-38 is responsible for much of the entropic loss after substrate binding (16).

Further insights into the changes in HRP dynamics after ligand binding can be obtained by comparison to MbCO. The active site of HRP is relatively large, can bind a diverse array of substrate molecules, and contains a distal histidine and arginine. The active site of myoglobin cannot accommodate substrates and does not have a distal arginine, but it also contains a distal histidine and a heme that binds CO. A wealth of experiments and simulations has underscored the pivotal role of the distal histidine in modulating the CO frequency in myoglobin. The H64V mutant (the distal histidine is replaced by a nonpolar valine) displays significantly decreased vibrational dephasing because of the elimination of this polar group (8, 13). It is instructive to compare the dynamics of free HRP to that of wild-type myoglobin and substrate-bound HRP with H64V.

The eccentricities of wild-type horse heart MbCO (blue squares) and H64V (black circles) are presented in Fig. 5 as a function of T_w. MbCO can adopt three distinct spectroscopic states that arise from different structural configurations of the distal histidine relative to the CO (8, 17). The data shown in Fig. 5 are for the A_1 band of MbCO, the main band in the absorption spectrum. In the A_1 state, it has been shown that the distal histidine does not form a hydrogen bond with the CO (8). The eccentricities of H64V are obtained from an FFCF derived from previous homodyne vibrational echo experiments (13). The lines are fits to the HRP–NMBZA (black curve) and free-HRP (blue curve) experimental eccentricities that were discussed above but have been offset along the vertical axis to account for different initial values of the HRP and myoglobin data (different motionally narrowed components). It is apparent from Fig. 4 that despite differences in the structure and function of HRP and Mb, the proteins display qualitatively similar trends in how the CO frequency is modulated by protein structural fluctuations. The

dynamics of the red state of free HRP are very similar to those of the wild-type MbCO and are still evolving even at the longest experimentally accessible time scales. Introducing NMBZA into the active site of HRP reduces the rate and amplitude of protein dynamics in a fashion analogous to that of removing the distal histidine in MbCO.

In H64V, the distal histidine has been removed, and the rate and magnitude of the vibrational dephasing on the experimental time scales are reduced accordingly. On the basis of the similarity between the MbCO data and the HRP red-state data shown in Fig. 4, it is reasonable to assume that a substantial contribution to the vibrational dephasing comes from the fluctuations of the distal histidine and the distal arginine. The 2D-IR experiment is sensitive to time-dependent electric fields around the CO, so the striking similarity in HRP–substrate and H64V dynamics highlights that substrate binding in HRP renders the distal residues nearly static on the 2D-IR experimental time scale. The clear conclusion to be drawn is that substrate binding locks up the distal ligands, constraining the structural fluctuations in the active site. The result is that the time scale of the fluctuations is pushed out to long times (>100 ps).

Concluding Remarks. The 2D-IR vibrational echo experiments demonstrate that unligated HRP has a dynamically "loose" red state in which motions of the distal ligands contribute substantially to vibrational dephasing. After substrate binding, both the amplitudes and time scales of CO-frequency fluctuations decrease markedly.

It is striking that the protein dynamics are significantly decreased when additional potential sources of CO-frequency perturbations are introduced into the active site. The distal residues participate in every step of the enzymatic cycle of HRP, and the results presented here indicate that substrate binding reorganizes and dynamically constrains these residues. On the basis of the observation that substrates in the HRP active site significantly decreases structural fluctuations of the distal residues, we suggest that the protein may exploit this feature of substrate binding to catalyze further steps in the enzymatic pathway. Thus, after recognition of biologically occurring substrates, the protein active site is not only reorganized structurally but also dynamically, priming the enzyme to sample the portion of the conformational energy landscape that may lead to subsequent steps in the reaction. In the enzymatic cycle of HRP, substrate binding occurs after H_2O_2 enters the active site and binds at the heme. The distal histidine serves as a general proton source, and the distal arginine stabilizes reactive intermediates (4). Thus, both residues must be favorably positioned to facilitate downstream events in the enzymatic cycle, a process that may be afforded by the dynamic quenching observed after substrate binding (20).

Materials and Methods

Materials and Sample Preparation. HRP type VIA (isoenzyme C) and horse heart myoglobin were purchased as lyophilized powder from Sigma–Aldrich (St. Louis, MO), BZH, BHA, BZA, NMBZA, and ^2H_2O were purchased from Aldrich. All reagents were of the highest available quality and used without further purification. 2-NHA was synthesized according to published protocols (23), recrystallized twice, and deemed pure by ^{1}H-NMR.

CO forms of HRP and myoglobin were prepared according to previously published protocols (8). Typical absorption at the peak of the CO-stretch band was ≈ 0.1 absorbance units above the protein and ^2H_2O background, which is typically ≈ 0.4.

2D-IR Spectroscopy. The 2D-IR spectrometer is similar to an experimental setup described previously (44, 45). Three transform-limited (110 fs, 150 cm^{-1} FWHM) mid-IR laser pulses are sequen-

274

tially time-delayed before they are brought into the sample. The vibrational echo pulse generated in the sample is emitted in the phase-matched direction, made colinear with a local oscillator pulse, and dispersed through a monochromator onto a 32-element mercury-cadmium-telluride array detector. The signal is mixed with the local oscillator to obtain full-time, frequency, and phase information about the vibrational echo wave packet. The laser frequency was tuned to coincide with the maximum of the absorption frequency for each protein. For studies on free HRP, the laser frequency was set at $1,920\ cm^{-1}$. In all samples, the broadband laser pulse afforded sufficient bandwidth to observe both 0-to-1 and 1-to-2 vibrational transitions.

For each protein, a family of vibrational echo 2D spectra is obtained as a function of three variables: the emitted vibrational echo frequencies, ω_m, and the variable time delays between the first and second pulses (τ) and the second and third pulses (T_w, "waiting" time). Frequency correlation maps are created by numerically Fourier transforming the τ scan data as a function of the emission frequencies, ω_m, at each T_w. Absorptive 2D-IR spectra are created via the dual-scan method (44, 46). The amplitude of the 2D-IR data at each fixed T_w is plotted as a function of the Fourier-transformed τ decay, ω_τ (horizontal axis in the 2D plots), and emission frequency, ω_m (vertical axis of the 2D plots).

This work was supported by National Institutes of Health Grant 2 R01 GM-061137-05. A.M.M. was supported by the National Institutes of Health Postdoctoral Fellowship, and H.I. is a Fellow of the Human Frontier Science Program.

1. Jimenez R, Salazar G, Yin J, Joo T, Romesberg FE (2004) *Proc Natl Acad Sci USA* 101:3803–3808.
2. Ma B, Shatsky M, Wolfson HJ, Nussinov R (2002) *Protein Sci* 11:184–197.
3. Vazquez-Laslop N, Zheleznova EE, Markham PN, Brennan RG, Neyfakh AA (2000) *Biochem Soc Trans* 28:517–520.
4. Veitch NC, Smith AT (2000) *Adv Inorg Chem* 51:107–162.
5. Fenimore PW, Frauenfelder H, McMahon BH, Parak FG (2002) *Proc Natl Acad Sci USA* 99:16047–16051.
6. Frauenfelder H, McMahon BH, Austin RH, Chu K, Groves JT (2001) *Proc Natl Acad Sci USA* 98:2370–2374.
7. Massari AM, Finkelstein IJ, Fayer MD (2006) *J Am Chem Soc* 128:3990–3997.
8. Merchant KA, Noid WG, Akiyama R, Finkelstein IJ, Goun A, McClain BL, Loring RF, Fayer MD (2003) *J Am Chem Soc* 125:13804–13818.
9. Mukherjee P, Kass I, Arkin IT, Zanni MT (2006) *Proc Natl Acad Sci USA* 103:3528–3533.
10. Chung HS, Khalil M, Smith AW, Ganim Z, Tokmakoff A (2005) *Proc Natl Acad Sci USA* 102:612–617.
11. Wang J, Chen J, Hochstrasser RM (2006) *J Phys Chem B Condens Matter Mater Surf Interfaces Biophys* 110:7545–7555.
12. Bredenbeck J, Helbing J, Kumita JR, Woolley GA, Hamm P (2005) *Proc Natl Acad Sci USA* 102:2379–2384.
13. Finkelstein IJ, Goj A, McClain BL, Massari AM, Merchant KA, Loring RF, Fayer MD (2005) *J Phys Chem B Condens Matter Mater Surf Interfaces Biophys* 109:16959–16966.
14. Massari AM, McClain BL, Finkelstein IJ, Lee AP, Reynolds HL, Bren KL, Fayer MD (2006) *J Phys Chem B Condens Matter Mater Surf Interfaces Biophys* 110:18803–18810.
15. Veitch NC (2004) *Phytochemistry* 65:249–259.
16. Aitken SM, Turnbull JL, Percival MD, English AM (2001) *Biochemistry* 40:13980–13989.
17. Spiro TG, Wasbotten IH (2005) *J Inorg Biochem* 99:34–44.
18. Kaposi AD, Vanderkooi JM, Stavrov SS (2006) *Biophys J* 91:4191–4200.
19. Khajehpour M, Troxler T, Vanderkooi JM (2003) *Biochemistry* 42:2672–2679.
20. Kaposi AD, Prabhu NV, Dalosto SD, Sharp KA, Wright WW, Stavrov SS, Vanderkooi JM (2003) *Biophys Chem* 106:1–14.
21. Williams RB, Loring RF, Fayer MD (2001) *J Phys Chem B Condens Matter Mater Surf Interfaces Biophys* 105:4068–4071.
22. Park ES, Andrews SS, Hu RB, Boxer SG (1999) *J Phys Chem B Condens Matter Mater Surf Interfaces Biophys* 103:9813–9817.
23. Summers JB, Mazdiyasni H, Holms JH, Ratajczyk JD, Dyer RD, Carter GW (1987) *J Med Chem* 30:574–580.
24. Carlsson GH, Nicholls P, Svistunenko D, Berglund GI, Hajdu J (2005) *Biochemistry* 44:635–642.
25. Meno K, Jennings S, Smith AT, Henriksen A, Gajhede M (2002) *Acta Crystallogr D* 58:1803–1812.
26. Gajhede M, Schuller DJ, Henriksen A, Smith AT, Poulos TL (1997) *Nat Struct Biol* 4:1032–1038.
27. Henriksen A, Gajhede M, Baker P, Smith AT, Burke JF (1995) *Acta Crystallogr D* 51:121–123.
28. Howes BD, Heering HA, Roberts TO, Schneider-Belhadadd F, Smith AT, Smulevich G (2001) *Biopolymers* 62:261–267.
29. Howes BD, Rodriguez-Lopez JN, Smith AT, Smulevich G (1997) *Biochemistry* 36:1532–1543.
30. Smulevich G, Feis A, Indiani C, Becucci M, Marzocchi MP (1999) *J Biol Inorg Chem* 4:39–47.
31. Kaposi AD, Fidy J, Manas ES, Vanderkooi JM, Wright WW (1999) *Biochim Biophys Acta* 1435:41–50.
32. Henriksen A, Schuller DJ, Meno K, Welinder KG, Smith AT, Gajhede M (1998) *Biochemistry* 37:8054–8060.
33. Holzbaur IE, English AM, Ismail AA (1996) *J Am Chem Soc* 118:3354–3359.
34. Ingledew WJ, Rich PR (2005) *Biochem Soc Trans* 33:886–889.
35. Hashimoto S, Takeuchi H (2006) *Biochemistry* 45:9660–9667.
36. Johnson JB, Lamb DC, Frauenfelder H, Müller JD, McMahon B, Nienhaus GU, Young RD (1996) *Biophys J* 71:1563–1573.
37. Smulevich G, Paoli M, Burke JF, Sanders SA, Thorneley RN, Smith AT (1994) *Biochemistry* 33:7398–7407.
38. Rector KD, Kwok AS, Ferrante C, Tokmakoff A, Rella CW, Fayer MD (1997) *J Chem Phys* 106:10027–10036.
39. Roberts ST, Loparo JJ, Tokmakoff A (2006) *J Chem Phys* 125:084502.
40. Mukamel S (1995) *Principles of Nonlinear Optical Spectroscopy* (Oxford Univ Press, New York).
41. Massari AM, Finkelstein IJ, McClain BL, Goj A, Wen X, Bren KL, Loring RF, Fayer MD (2005) *J Am Chem Soc* 127:14279–14289.
42. Bai YS, Fayer MD (1989) *Phys Rev B Condens Matter* 39:11066.
43. Schmidt J, Sundlass N, Skinner J (2003) *Chem Phys Lett* 378:559–566.
44. Asbury JB, Steinel T, Fayer MD (2004) *J Luminescence* 107:271–286.
45. Zanni MT, Gnanakaran S, Stenger J, Hochstrasser RM (2001) *J Phys Chem B Condens Matter Mater Surf Interfaces Biophys* 105:6520–6535.
46. Khalil M, Demirdoven N, Tokmakoff A (2003) *Phys Rev Lett* 90:047401–047404.

Hydrogen bond dynamics in aqueous NaBr solutions

Sungnam Park[†‡] and M. D. Fayer[†§]

[†]Department of Chemistry, Stanford University, Stanford, CA 94305; and [‡]Stanford Synchrotron Radiation Laboratory, Stanford Linear Accelerator Center, Menlo Park, CA 94025

This contribution is part of the special series of Inaugural Articles by members of the National Academy of Sciences elected on May 1, 2007.

Contributed by M. D. Fayer, August 19, 2007 (sent for review July 27, 2007)

Hydrogen bond dynamics of water in NaBr solutions are studied by using ultrafast 2D IR vibrational echo spectroscopy and polarization-selective IR pump–probe experiments. The hydrogen bond structural dynamics are observed by measuring spectral diffusion of the OD stretching mode of dilute HOD in H_2O in a series of high concentration aqueous NaBr solutions with 2D IR vibrational echo spectroscopy. The time evolution of the 2D IR spectra yields frequency–frequency correlation functions, which permit quantitative comparisons of the influence of NaBr concentration on the hydrogen bond dynamics. The results show that the global rearrangement of the hydrogen bond structure, which is represented by the slowest component of the spectral diffusion, slows, and its time constant increases from 1.7 to 4.8 ps as the NaBr concentration increases from pure water to $\approx$6 M NaBr. Orientational relaxation is analyzed with a wobbling-in-a-cone model describing restricted orientational diffusion that is followed by complete orientational randomization described as jump reorientation. The slowest component of the orientational relaxation increases from 2.6 ps (pure water) to 6.7 ps ($\approx$6 M NaBr). Vibrational population relaxation of the OD stretch also slows significantly as the NaBr concentration increases.

ultrafast 2D IR spectroscopy | water dynamics in ionic solutions

Water plays an important role in chemical and biological processes. In aqueous solutions, water molecules dissolve ionic compounds, charged chemical species, and biomolecules by forming hydration shells (layers) around them. Pure water undergoes rapid structural evolution of the hydrogen bond network that is responsible for water's unique properties (1). A question of fundamental importance is how the dynamics of water in the immediate vicinity of an ion or ionic group differ from those of pure water. For monatomic ions, molecular ions, charged groups of large molecules, or charged amino acids on the surfaces of proteins, the basic structure of hydration shells is determined by ion–dipole interactions between water molecules and the charged group (2, 3). Such ion–dipole interactions will influence both the structure and dynamics of water in the proximity of ions. Water dynamics in ion hydration shells play a significant role in the nature of systems such as proteins and micelles (3) and in processes such as ion transport through transmembrane proteins (4).

Over the last several years, the application of ultrafast IR vibrational echo spectroscopy (5, 6), particularly 2D IR vibrational echo experiments, (7, 8) combined with molecular dynamics (MD) simulations (9–12) have greatly enhanced understanding of the hydrogen bond dynamics in pure water. These experiments directly examine the dynamics of water rather than study the indirect influence of water dynamics on a probe molecule (13). The ultrafast 2D IR vibrational echo experiments on water (7, 8) and other hydrogen-bonded systems (14) build upon earlier IR pump–probe experiments that have been extensively used to study the vibrational population relaxation and orientational relaxation dynamics of pure water (15, 16), water in nanoscopic environments (17–19), and ionic solutions (20). Pump–probe experiments have also been used to investigate the

hydrogen bond dynamics in aqueous solutions (20) and other hydrogen-bonded liquids (21).

Water molecules around monatomic ions provide the simplest system for the investigation of the dynamics of water in the presence of charges. Therefore, examining the dynamics of water in simple ionic solutions, such as sodium bromide (NaBr), can provide valuable insights that will increase understanding of chemical processes and biological systems that involve charged species in aqueous solutions. In pure water, water molecules are hydrogen-bonded to neighboring water molecules in a more or less tetrahedral geometry making an extended hydrogen bond network. Hydrogen bonds are continually breaking and reforming, and hydrogen bond lengths (strengths) are continually changing (9). The structure of water fluctuates on femtosecond to picosecond time scales (6, 9). The slowest component of the fluctuations is associated with the global structural rearrangement of the hydrogen bond network (7). As salt is dissolved in water, water molecules form hydration shells around ions, and consequently the local hydrogen bond network is perturbed.

In this paper, a very detailed study of the hydrogen bond dynamics of water in highly concentrated NaBr solutions is presented. Four experimental observables, which are based on measurements on the OD stretch mode of HOD in H_2O, were used. Ultrafast 2D IR vibrational echo experiments were used to measure spectral diffusion, which is directly related to the time evolution of the hydrogen bond structure. Polarization-selective IR pump–probe experiments were used to measure orientational relaxation dynamics of water, which is affected by the interactions of water molecules with their surroundings. The IR pump–probe experiments also were used to measure vibrational population relaxation dynamics of the OD stretch of HOD, which is very sensitive to the local environment. Finally, Fourier transform (FT) IR absorption spectra report on the overall effect of NaBr concentration on the OD stretch frequency.

In the experiments presented below, highly concentrated NaBr solutions are investigated. The number of water molecules per NaBr molecule ranges from 8 to 32. In low-concentration NaBr solutions, these numbers of water molecules would correspond to less than one hydration shell for each ion to approximately two hydration shells, respectively. By examining such highly concentrated NaBr solutions, all of water molecules are placed in close proximity to ions, and therefore, the influence of water–ion interactions is not diluted out by large quantities of bulk water that are not affected by the ions. Some limited initial vibrational echo experiments on highly concentrated salt solutions have been performed (22, 23). These experiments showed that the dynamics of water in concentrated salt solutions differ

Author contributions: S.P. and M.D.F. designed research; S.P. performed research; S.P. and M.D.F. analyzed data; and S.P. and M.D.F. wrote the paper.

The authors declare no conflict of interest.

Abbreviations: MD, molecular dynamics; FT, Fourier transform; AOT, sodium bis(2-ethylhexyl)sulfosuccinate; FFCF, frequency–frequency correlation function; CLS, center line slope.

[§]To whom correspondence should be addressed. E-mail: fayer@stanford.edu.

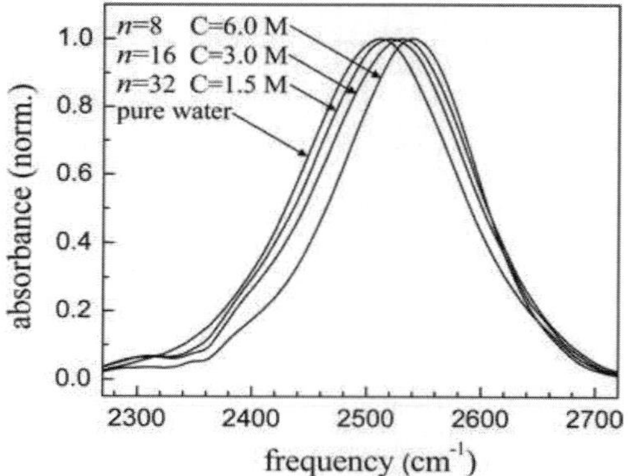

Fig. 1. Linear FT IR spectra of the OD stretch of HOD in pure water and aqueous NaBr solutions (H_2O background subtracted). $n = 8$, 16, and 32 are the numbers of water molecules per NaBr, and correspond to $\approx$6 M, 3 M, and 1.5 M NaBr solutions, respectively. The peak positions (widths) of the OD stretch of HOD in pure water, $n = 32$, 16, and 8 are 2,509 cm^{-1} (160 cm^{-1}), 2,519 cm^{-1} (163 cm^{-1}), 2,529 cm^{-1} (159 cm^{-1}), and 2,539 cm^{-1} (141 cm^{-1}), respectively.

significantly from those of pure water, but not to the extent indicated earlier by less direct two-color pump–probe measurements (20). The results presented here confirm and extend earlier vibrational echo experiments. The surprising result is that even at the highest concentration with eight water molecules per NaBr ($\approx$6 M solution) the time scale associated with the compete rearrangement of the hydrogen bond structure is slower than in pure water by only approximately a factor of three.

Results and Discussion

Linear FT IR Absorption Spectra. NaBr solutions were chosen for study among aqueous halide salt solutions because the OD stretch band of HOD in aqueous NaBr solutions shows a reasonably large blue-shift in the FT IR spectrum compared with that of HOD in pure water, which suggests a large change in the hydrogen bond structure. The OD stretching band in halide salt solutions shifts more to the blue as the halide ion goes from F^- to I^-. Subsequent experiments will examine other salt solutions.

All of the experiments were conducted on the OD stretch of 5% HOD in pure H_2O and aqueous NaBr solutions. The OD stretch of HOD in H_2O was examined for two reasons. First, it is a single local mode, which simplifies the absorption spectrum (24, 25). Second, of considerable importance for the time resolved experiments is the fact that measurements on the OD stretch of dilute HOD in H_2O eliminate vibrational excitation transfer (26). We are interested in the ground-state equilibrium properties of the aqueous solutions. H_2O or D_2O vibrational excitation transport interferes with measurements of spectral diffusion and orientational relaxation. MD simulations have shown that measurements on dilute HOD in H_2O yield the dynamics of water (7).

The H_2O background-subtracted FT IR spectra of the OD stretch are shown in Fig. 1. The curves are labeled by the number of water molecules per NaBr molecule, n. $n = 8$, 16, and 32 correspond to $\approx$6 M, 3 M, and 1.5 M NaBr solutions, respectively. The spectra are characteristically broad (25), and their peak positions are increasingly blue-shifted as the NaBr concentration increases. It is known for pure water that the broad spectrum is associated with different numbers and a wide range of strengths of the hydrogen bonds (27). For pure water, subensembles of water molecules with more and/or stronger hydrogen bonds have red-shifted absorptions, whereas subensembles with fewer and/or weaker hydrogen bonds are blueshifted. In some other water systems, the blue-shift of the entire water spectrum has been interpreted as reflecting an overall weaker distribution of the hydrogen bonds (25, 28). However, this is not necessarily true. For example, it has been shown recently that the blue-shift of the water spectrum in small sodium bis(2-ethylhexyl)sulfosuccinate (AOT) reverse micelles is actually caused by the superposition of two distinct spectra of water molecules (18). One of the spectra is the bulk-like spectrum of water molecules in the core of the reverse micelle. The other is the substantially blue-shifted spectrum of water molecules associated with the ionic sulfonate head groups with sodium counter ions. Furthermore, a recent study showed that four distinct hydrogen-bonded systems containing water molecules could have essentially identical blue-shifted spectra of water molecules, although they had very different hydrogen bond dynamics (29). In the type of systems being considered here, in which the hydroxyl groups are as likely or more likely to be interacting with ions as with other water molecules, the blue-shift should not be interpreted as an indication of weakening of the hydrogen bond structure in aqueous solutions.

The width of the OD stretch band is relatively narrow in the $n = 8$ NaBr solution compared with pure water. In pure water, the hydroxyl group (OH and OD) is hydrogen-bonded only to the oxygen atoms in neighboring water molecules, making an extended hydrogen bond network leading to a broad distribution of the OD stretch frequencies as discussed above. On the other hand, in the $n = 8$ NaBr solution, there are four water molecules per ion on average. Given the fact that in dilute solutions the hydration numbers of Na^+ and Br^- are 4 and 6, respectively (30, 31), Na^+ and Br^- ions in the $n = 8$ NaBr solution are separated by a few water molecules. The detailed structure of such concentrated salt solutions is not known, but it is reasonable to assume that water molecules are bound to ions and are mainly interacting with ions rather than with a network of neighboring water molecules. At this high concentration, the hydrogen bond network of water is not only substantially disrupted but is unlikely to exist. There is nothing like bulk water at the highest concentration and even at the lower concentrations investigated here.

Water molecules in NaBr solutions can be hydrogen-bonded to ions by three different ion–dipole interactions in HOD–Br^-, DOH–Br^-, and HDO–Na^+. The last two configurations may have less influence on the OD stretch frequency than the direct interaction of the OD groups with bromide ions. The broad line width for the $n = 8$ NaBr solution reflects the distribution of water–ion interactions rather than water–water interactions found in pure water. In the $n = 16$ NaBr solution with eight water molecules per ion, there are several extra water molecules beyond what would be the first hydration shells in dilute solution. The hydrogen bond structure of water in the $n = 16$ solution will still be substantially different from that of pure water. In the $n = 32$ solution, where there are 16 water molecules per ion, there are a significant number of water molecules beyond the first hydration shells. Nonetheless, the high concentration of ions should still have a substantial influence on many of water molecules. FT IR spectrum of the $n = 32$ solution is blue-shifted and broader than that of pure water (see Fig. 1). As discussed below, not only is the spectrum of the $n = 32$ NaBr solution different from that of pure water, but also all of the dynamic observables are different.

2D IR Vibrational Echo Spectroscopy. 2D IR vibrational echo spectroscopy is an ultrafast IR analog of 2D NMR, but it operates on time scales many orders of magnitude shorter than NMR. The 2D IR vibrational echo experiments measure the hydrogen bond dynamics of water through the time evolution of the hydroxyl stretch 2D IR spectra. In the experiment, there are three excitation pulses, $\approx$60 fs in duration, tuned to the OD

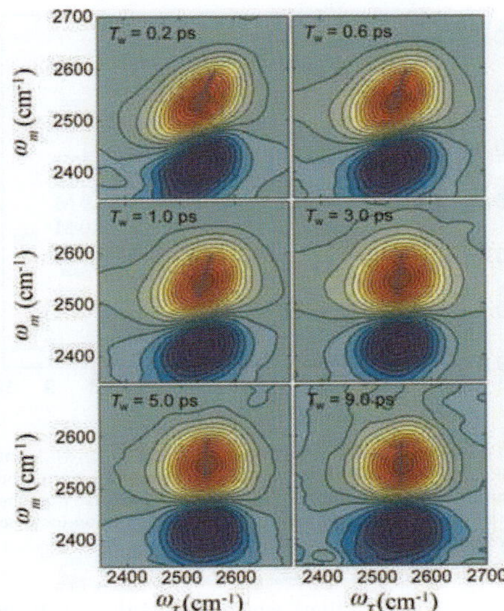

Fig. 2. 2D IR spectra measured with the $n = 8$ ($\approx$6 M) NaBr solution. Red peaks (0–1 transition) are positive-going. Blue peaks (1–2 transition) are negative-going. The peak shape changes as T_w increases. The blue lines are the center lines.

stretch frequency. The short IR pulses have sufficient spectral bandwidth to span the entire OD stretch spectrum. The time between pulses 1 and 2 and between pulses 2 and 3 is called τ and T_w, respectively. The vibrational echo signal radiates from the sample at a time $\leq \tau$ after the third pulse. The vibrational echo signals are recorded by scanning τ at fixed T_w. The signal is spatially and temporally overlapped with a local oscillator for heterodyne detection, and combined pulse is dispersed although a monochromator onto an array detector. Heterodyne detection provides both amplitude and phase information. Taking the spectrum of the heterodyne-detected echo signal performs one of the two FTs and gives the ω_m axis (where subscript m stands for monochromator) in the 2D IR spectrum. When τ is scanned, temporal interferograms are obtained at each ω_m. The temporal interferograms are numerically Fourier transformed to give the other axis, the ω_τ axis. 2D IR spectra are obtained for a range of T_w values.

As T_w increases, the peak shapes in the 2D IR spectra change. These changes are directly related to the structural evolution of the water–ion system through the influence of the structural evolution on the OD stretch frequencies. Very qualitatively, the 2D IR experiment can be viewed as follows. The first and second pulses act to label the initial frequencies of the molecular oscillators. Between the second and third pulses, structural evolution causes the initially labeled frequencies to change (spectral diffusion). The third pulse ends the evolution period, and the vibrational echo signal reads out the final frequencies of the initially frequency-labeled molecular oscillators. As T_w increases, there is more time for structural evolution, and, therefore, larger frequency changes. The structural evolution (frequency changes) is reflected in the changes of peak shape of the 2D IR spectra. The 2D IR vibrational echo experiments have been reviewed recently (23, 32, 33).

2D IR spectra of the OD stretch of HOD in the $n = 8$ NaBr solution are shown in Fig. 2 as a function of T_w. The red bands (on the diagonal) are positive-going and result from the transi-

tion between the ground and first excited vibrational states (0–1). The blue bands (off-diagonal) are negative-going and are associated with emission of the vibrational echo signal at the first overtone (1–2) transition frequency, which is shifted along the ω_m axis by the OD stretching mode anharmonicity. Focusing on the 0–1 transition (red positive band), a noticeable feature of 2D IR spectra is the change in peak shape as T_w increases. The red band is substantially elongated along the diagonal at $T_w = 0.2$ ps. As T_w increases, the peak shape becomes more symmetrical. This change is the signature of spectral diffusion. In the long time limit, all structural configurations are sampled and, consequently, all frequencies are sampled. In this limit, spectral diffusion is complete and the peak shape becomes completely symmetrical. However, because of the overlap with the negative-going peak that arises from emission at the 1–2 transition frequency, the bottom of the 0–1 transition band is eaten away. Spectral diffusion dynamics are obtained by analyzing the change of peak shapes as a function of T_w.

The change of peak shapes in the 2D IR spectra with increasing T_w can be used to determine the time scales and amplitudes of various contributions to the structural evolution of the water–ion systems using methods based on diagrammatic perturbation theory (34). The frequency–frequency correlation function (FFCF) connects the experimental observables to the underlying dynamics. The FFCF is the probability that an oscillator with a given initial frequency still has the same frequency at time t later, averaged over all starting frequencies. Once the FFCF is known, all linear and nonlinear optical experimental observables can be calculated by time-dependent diagrammatic perturbation theory (34). Conversely, the FFCF can be extracted from 2D IR spectra with additional input from linear FT IR absorption spectra. In general, to determine the FFCF from 2D IR and linear FT IR spectra, full calculations of linear and nonlinear third-order response functions are performed iteratively until the calculation results converge to the experimental results (7).

Here, we used the center line slope (CLS) method (23, 35). The CLS method is a new approach for extracting the FFCF from the T_w dependence of the 2D IR spectra that is accurate and much simpler to implement numerically than iterative fitting methods with calculations of all response functions based on time-dependent diagrammatic perturbation theory. Furthermore, the CLS provides a more useful quantity to plot for visualizing differences in the dynamics as a function of the NaBr concentration than a series of full 2D IR spectra (23, 35). In the CLS method, frequency slices through the 2D IR spectrum parallel to the ω_τ axis at various ω_m values are projected onto the ω_τ axis (35). These projections are a set of spectra with peak positions, ω_τ^{max}, on the ω_τ axis. The plot of ω_τ^{max} vs. ω_m form a line called the center line. Center lines are shown in Fig. 2. In the absence of a homogenous contribution to the 2D IR spectrum, the peak shape in the 2D IR spectrum at $T_w = 0$ ps would be essentially a line along the diagonal at 45°. The slope of this center line would be 1. At very long time, the peak shape in the 2D IR spectrum becomes symmetrical, and the center line is vertical (slope is infinite). The change in the slope of the center line with increasing T_w can be seen in Fig. 2. It has been shown theoretically that the normalized FFCF is equal to the T_w dependence of the inverse of the slope of the center line (35). For simplicity, the inverse of the slope of the center line is called the CLS. Therefore, the CLS will vary from a maximum value of 1 at $T_w = 0$ ps to a minimum value of 0 at a sufficiently long time. It has also been shown theoretically that combining the analysis of the CLS with the linear FT IR absorption spectrum, the full FFCF can be obtained, including the T_w independent homogenous component (35).

A homogenous contribution to the peak shape in the 2D IR spectra and to the line shape of the linear FT IR absorption spectra can arise from three sources: very fast structural fluctuations that produce a motionally narrowed contribution to the FFCF (7, 36), vibrational population relaxation, and orientational relaxation. Both vibrational population relaxation and orientational relaxation times were measured independently using polarization-selective IR pump–probe experiments. However, these contributions are quite small compared with the main homogeneous contribution for the water systems studied in this paper, which is the motionally narrowed component. Motional narrowing occurs when there is some portion of the structural fluctuations that are extremely fast, such that $\Delta\tau \ll 1$, where τ is the time scale of the fast fluctuations and Δ is the amplitude of the associated frequency fluctuations. Motionally narrowed fluctuations produce a Lorentzian contribution to both the 2D IR spectrum and the linear FT IR absorption spectrum. The T_w-independent homogeneous contribution manifests itself by broadening the 2D IR spectrum along the ω_τ axis even at $T_w = 0$ ps. This broadening reduces the initial value of the CLS to a number <1.

A multiexponential form of the FFCF, $C(t)$, was used to model the multitime scale dynamics of the structural evolution of the water systems. This form has been used previously for the analysis of pure water (7) and nanoscopic water pools in AOT reverse micelles (22, 37). The FFCF has the form

$$C(t) = \sum_{i=1}^{n} \Delta_i^2 e^{-t/\tau_i}. \quad [1]$$

The Δ_i and τ_i terms are the amplitudes and correlation times, respectively, of the frequency fluctuations induced by water–ion structural fluctuations. τ_i reflects the time scale of a set of structural fluctuations and the Δ_i is the range of the OD stretch frequency sampled due to the structural fluctuations. In the motionally narrowed limit (fast modulation limit) with $\Delta\tau \ll 1$, Δ and τ cannot be determined independently but $\Delta^2\tau = 1/T_2^*$ is obtained as a single parameter describing the motionally narrowed Lorentzian contribution. T_2^* is called the pure dephasing time. The Lorentzian contribution to the 2D IR spectrum also has contributions from lifetime (T_1) and orientational relaxation (T_{or}). The total homogeneous (Lorentzian) contribution, T_2, is given by $1/T_2 = 1/T_2^* + 1/(2T_1) + 1/(3T_{or})$, where T_2 is determined from the CLS combined with the linear FT IR absorption spectrum (23, 35), and T_1 and T_{or} are measured by polarization-selective IR pump–probe experiments. The total homogeneous line width is $\Gamma = 1/(\pi T_2)$. Γ is dominated by the pure dephasing, T_2^*.

The CLS procedure is well defined and reliable. However, for water systems, it has been shown that there are non-Condon effects; that is, the transition dipole moments are frequency dependent, which influence the vibrational echo observables to some extent (11, 38). The non-Condon effects have been included in MD simulations of pure water, and they have some influence on the calculation of experimental observables (11, 38). However, there is no direct method for going from the experimental observables to the FFCF that includes non-Condon effects. Therefore, as with all current methods when applied to water, the FFCFs obtained using the CLS must be considered approximate to some extent. Nonetheless, the FFCFs are very useful for comparing the influence of NaBr concentration on the dynamics of water. Furthermore, the FFCFs obtained with the CLS can be used in calculations to reproduce the experimental results, and then any theoretical method can be used for additional analysis.

Fig. 3 shows the T_w-dependent CLS, $C_{CLS}(T_w)$, and the normalized FFCF, $C_N(T_w)$. The points are the experimental data obtained from 2D IR spectra. As the NaBr concentration increases, the vibrational lifetime of the OD stretch also in-

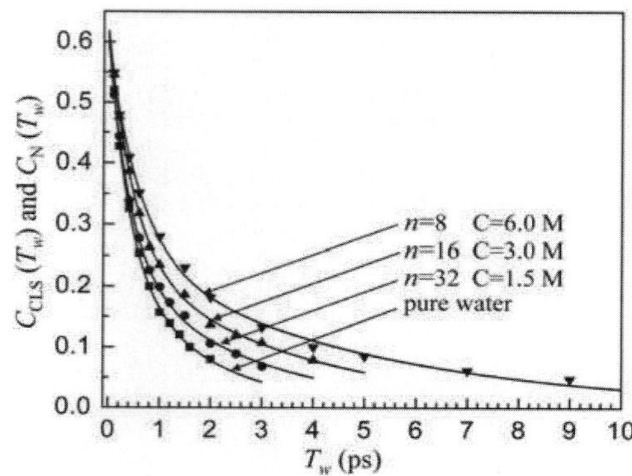

Fig. 3. The inverse of the slope of the center line (CLS), $C_{CLS}(T_w)$, obtained from 2D IR spectra of pure water and aqueous NaBr solutions. The lines through the data are the normalized FFCF, $C_N(T_w)$, determined from the 2D IR spectra and linear FT IR spectra. As the NaBr concentration increases, the dynamics of water slow.

creases (see the next section), which permits the 2D IR spectra to be taken out to longer times. The solid lines are the normalized FFCF, $C_N(T_w)$, determined from the CLS and the FT IR absorption spectrum. As the NaBr concentration increases, the decays of $C_N(T_w)$ get slower. The hydrogen bond dynamics slow as the NaBr concentration increases. The decrease in value of $C_N(T_w = 0$ ps) from 1 is indicative of a homogeneous contribution to the FFCF.

Quantitative information on the differences in the hydrogen bond dynamics in NaBr solutions is given in Table 1. To fit the data and obtain the lines through the data shown in Fig. 3, three terms are used in Eq. **1**. One of the components of the FFCF is motionally narrowed for each sample. For this component, the homogeneous line width, $\Gamma = 1/(\pi T_2)$ is reported because the Δ and τ values are not independently determined. For the other two exponential components, both the Δ and τ values are given. The values for pure water are almost identical to those reported previously, although the methodology has been improved, and therefore, the current values are used for comparison with the NaBr solutions.

Examining Table 1, it is seen that the major change with increasing NaBr concentration is in the slowest time constant, τ_2, of the FFCF. The homogeneous line widths, Γ, of pure water and the $n = 32$ and $n = 16$ NaBr solutions are identical within experimental error. Only for the $n = 8$ NaBr solution is Γ slightly smaller, and this value is within the error bars of the others. The Δ values, which give the amplitudes of the fluctuations on different time scales, are again very similar, with the $n = 8$ NaBr solution being slightly smaller, but on the edge of the error bars. The time constants, τ_1 and τ_2, show a systematic trend, both become longer as the NaBr concentration increases. Although the error bars for successive values overlap, the trend is clear. τ_1 increases by a factor of ≈ 1.66, whereas τ_2 increases by ≈ 3.

Table 1. FFCF parameters

Sample	Γ, cm^{-1}	Δ_1, cm^{-1}	τ_1, ps	Δ_2, cm^{-1}	τ_2, ps
Water	76 ± 14	41 ± 8	0.38 ± 0.08	34 ± 11	1.7 ± 0.5
n = 32	78 ± 9	43 ± 5	0.43 ± 0.06	34 ± 7	2.6 ± 0.5
n = 16	74 ± 6	41 ± 3	0.49 ± 0.05	34 ± 4	3.5 ± 0.5
n = 8	67 ± 6	35 ± 3	0.63 ± 0.09	30 ± 4	4.8 ± 0.6

To understand the trends in the FFCF, we can draw on MD simulations of pure water that have been used to model previous 2D IR vibrational echo experiments (7). The simulations used a number of standard water models, such as SPC/E and SPC-FQ, to calculate the FFCF of a quantum OD oscillator of HOD embedded in the simulations. The FFCFs were then used to calculate the 2D IR spectra, which were compared with the experimental results (7). Although the agreement between the simulation results and 2D IR experimental results was not quantitative and varied depending on which water model was used, the simulation results were sufficiently close to the experimental results to provide understanding of the nature of the fluctuations that occur on different time scales. The FFCFs produced from the simulations are functions that can be well approximated as a triexponential function like that given by Eq. 1. The simulations produce a motionally narrowed component on a subpicosecond time scale and then a T_w-dependent spectral diffusion on a slower time scale, which is captured by the two terms in our model FFCFs that are not motionally narrowed.

The simulations of pure water show that the motionally narrowed component arises from extremely fast (tens of femtoseconds), very local fluctuations mainly in the hydrogen bond length. These exceedingly fast fluctuations amount to overdamped oscillations of single hydrogen bonds. There is some experimental evidence that the strongest hydrogen bonds (red side of the absorption spectrum) are slightly underdamped (6, 38). However, these ultrafast motions do not change the global structure of hydrogen bond networks. The slowest time (1.7 ps) is associated with the final complete reorganization of hydrogen bond structure. This is the time scale for global rearrangement of hydrogen bond networks, and it requires breaking and re-forming hydrogen bonds as well as reorienting and switching hydrogen bond partners. The complete structural reorganization means that the OD stretch oscillators will have sampled all structures and, therefore, all vibrational frequencies. This slowest decay time is the time scale on which spectral diffusion is complete, and all frequencies under the inhomogeneously broadened absorption spectra have been sampled. The $\approx$400-fs time scale is a transition between the strictly local hydrogen bond length fluctuations and the final complete structural reorganization. The intermediate and slowest time scale components in the FFCF should not necessarily be viewed as distinct processes.

It is useful to take the results from the MD simulations of pure water and carry them over to understanding the FFCFs obtained from highly concentrated NaBr solutions. It is reasonable to assume that ultrafast motionally narrowed contribution to the FFCF still arises from very local hydrogen bond length fluctuations. However, in highly concentrated NaBr solutions, the hydrogen bonds are no longer among water molecules but rather between water molecules and ions. At the highest concentration, $n = 8$, most of the OD groups will be hydrogen bond donors to Br^- ions. This change in the nature of the OD hydrogen bond length fluctuations may be manifested in the reduction in Γ seen in Table 1. As mentioned above, the slowest component of the FFCF displays the largest change with increasing NaBr concentration. The slowest component slows by almost a factor of three going from pure water to the $n = 8$ NaBr solution. Therefore, as more and more hydrogen bonds are made to Br^- ions rather than among water molecules, the rate of global structural rearrangement slows. The structural randomization is only a factor of three slower than in pure water even when the number of water molecules is less than would be required to make single hydration shells in dilute aqueous solutions.

IR Pump–Probe Experiments. Polarization-selective pump–probe experiments were used to measure the vibrational population relaxation and the orientational relaxation of the OD stretch in aqueous NaBr solutions. The parallel ($S_\parallel$) and perpendicular

($S_\perp$) components of the pump–probe signal measured with two beam polarization configurations are given by

$$S_\parallel = P(t)[1 + 0.8C_2(t)] \qquad [2]$$

and

$$S_\perp = P(t)[1 - 0.4C_2(t)], \qquad [3]$$

where $P(t)$ is the population relaxation, and $C_2(t) = \langle P_2[\boldsymbol{\mu}(t)\cdot\boldsymbol{\mu}(0)]\rangle$ is the second-order Legendre polynomial describing transition dipole orientational correlation function (orientational relaxation).

Vibrational population relaxation, $P(t)$, is obtained from

$$P(t) = S_\parallel + 2S_\perp. \qquad [4]$$

Orientational relaxation dynamics are determined from the anisotropy decay, $r(t)$, which is calculated from the parallel and perpendicular components of the pump–probe signals (18).

$$r(t) = \frac{S_\parallel - S_\perp}{S_\parallel + 2S_\perp} = 0.4C_2(t). \qquad [5]$$

In water systems, vibrational population relaxation is followed by the deposition of heat into the solution, which produces a long-lived orientationally isotropic signal (16). This long-lived isotropic signal, which decays on thermal diffusion time scales, is caused by the temperature dependence of the water spectrum. The magnitude of the temperature change in these experiments is small (a fraction of 1 K). The pump–probe signals and local heating effects scale equivalently with laser intensity. The heating effects are well understood and are subtracted from pump–probe signals using the standard procedure (16, 39).

Orientational relaxation. Fig. 4A displays the anisotropy decays, $r(t)$, for pure water and aqueous NaBr solutions. As the NaBr concentration increases, the orientational relaxation slows. The anisotropy decays do not begin at the maximum value of 0.4 (see Eq. 5) because of an ultrafast inertial component that occurs on a tens-of-femtoseconds time scale (17, 40, 41). For times <100 fs, when the pump and probe pulses are overlapped in time, there is a very large nonresonant signal that obscures the inertial component. The difference in the amplitude of the measured anisotropy decay and 0.4 gives a measure of the amplitude of the inertial component. In the following text, only the anisotropy decay after the inertial component is discussed.

The anisotropy decay of HOD in pure water is a single exponential with a time constant of 2.6 ps (18, 41). However, the anisotropy decays in aqueous NaBr solutions are clearly not a single exponential and are fit very well by a biexponential function of the form

$$r(t) = a_1\exp(-t/\tau_{or1}) + a_2\exp(-t/\tau_{or2}), \qquad [6]$$

where $\tau_{or2} > \tau_{or1}$. Experiments on HOD in water nanopools in small AOT reverse micelles (18) and in nanoscopic water channels in Nafion fuel cell membranes (19) also displayed biexponential decays of the anisotropy. Both of these systems have interfacial layers composed of sulfonate head groups with associated sodium counter ions. The nanoscopic environment of water modifies the nature of the hydrogen bond network and alters both spectral diffusion (18) and orientational relaxation (18, 19). In the nanoscopic water systems, the fast component of the biexponential decay of the anisotropy was analyzed by using a wobbling-in-a-cone model based on restricted orientational diffusion (18, 19, 42).

The orientational relaxation is modeled as two processes (17). The motions at short times are restricted by the boundary condition of a cone, and orientational diffusion occurs by the

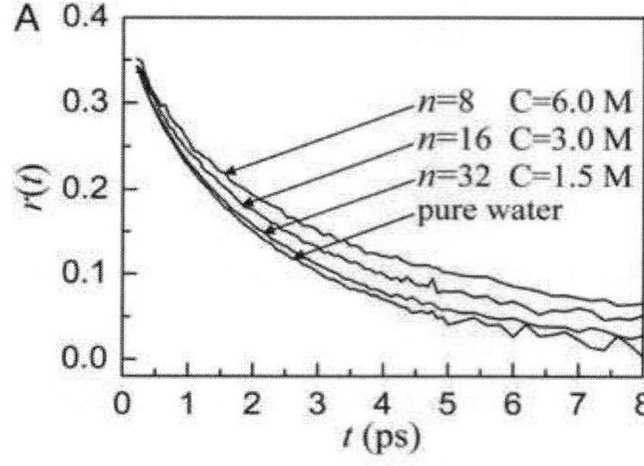

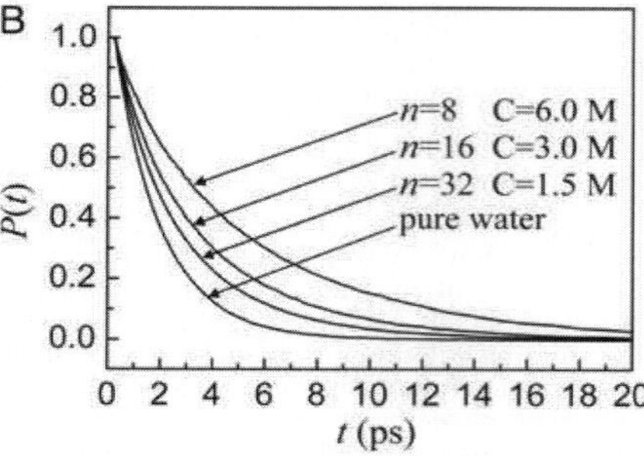

Fig. 4. Results of pump–probe measurements. (*A*) Anisotropy decays, *r*(*t*), of the OD stretch of HOD in pure water and aqueous NaBr solutions. As the NaBr concentration increases, the orientational relaxation slows. (*B*) Population relaxation, *P*(*t*), of the OD stretch of HOD in pure water and aqueous NaBr solutions. As the NaBr concentration increases, the population relaxation slows.

wobbling motion of water within a cone of semiangle of θ_c. After the initial wobbling period, a longer time decay of the anisotropy is associated with complete orientational randomization. The essential idea is that, on a time scale short compared with the rearrangement of hydrogen bond structure (breaking and re-forming hydrogen bonds), the OD bond vector, which is essentially the transition dipole vector, can sample a restricted range of orientations but cannot be completely randomized. In the wobbling-in-a-cone model, the orientational correlation function is given by

$$C_2(t) = [S^2 + (1 - S^2)\exp(-t/\tau_w)]\exp(-t/\tau_l), \qquad [7]$$

where S ($0 \le S \le 1$) is the generalized order parameter that describes the degree of restriction of the orientational motion, τ_w is the time constant of the wobbling-in-a-cone motion, and τ_l is the time constant for the complete orientational relaxation (18, 19, 42). By comparing Eqs. **6** and **7**, $\tau_w = (\tau_{or1}^{-1} - \tau_{or2}^{-1})^{-1}$, $\tau_l = \tau_{or2}$, and $S = \sqrt{a_2}$. The cone semiangle θ_c is obtained from the generalized order parameter $S = 0.5\cos\theta_c(1 + \cos\theta_c)$.

The parameters from fitting the anisotropy decays with Eqs. **6** and **7** are given in Table 2. In the absence of the ultrafast inertial component, $a_1 + a_2 = 0.4$. The decrease from 0.4 gives the amplitude of the inertial component. For the three NaBr solutions, the wobbling time constant τ_w and cone semiangle θ_c are identical within experimental error. However, the long time constant, τ_l, increases as the NaBr concentration increases. In the $n = 8$ NaBr solution, τ_l is almost three times longer than in pure water. Like the slowest component of the spectral diffusion (see Table 1), the complete randomization of the OD orientation requires a global reorganization of the hydrogen bond structure.

Table 2. Orientational relaxation parameters

Sample	a_1	τ_w, ps	θ_c, °	a_2	τ_l, ps
Water				0.35 ± 0.01	2.6 ± 0.1
$n = 32$	0.16 ± 0.02	1.7 ± 0.2	38 ± 4	0.21 ± 0.03	3.9 ± 0.3
$n = 16$	0.17 ± 0.03	1.6 ± 0.1	38 ± 2	0.21 ± 0.01	5.1 ± 0.2
$n = 8$	0.17 ± 0.01	1.6 ± 0.1	38 ± 2	0.21 ± 0.01	6.7 ± 0.3

Although the FFCF and the orientational correlation function cannot be directly compared, the slowest component of each requires rearrangements of the hydrogen bond structure. It is interesting to compare the effect of NaBr concentration on the slowest components of the FFCF and orientational correlation function shown in Tables 1 and 2, respectively. This can be done by taking the ratios, R, of the time constants of the slowest components for NaBr solutions with respect to the value for pure water. For the FFCF, $R_{FFCF} = 1.5$, 2.1, and 2.8 going from low concentration to high concentration. For the orientational correlation function, $R_{or} = 1.5$, 2.0, and 2.6. These trends support the picture that the slowest components of both experimental observables are closely related to the global rearrangement of the hydrogen bond structure by breaking and reforming of the hydrogen bonds.

Recently, Laage and Hynes (12, 43) have shown that the orientational relaxation of water molecules in pure water and ionic solutions is described by a jump reorientation model rather than Gaussian orientational diffusion. This model was recently applied to nanoscopic water in Nafion fuel cell membranes (19). Water molecules undergo complete randomization of their orientations by switching their hydrogen bond partners in a concerted manner, which involves large angular jumps. The decay time τ_l is related to the jump reorientation time constant, τ_j by Ivanov's jump reorientation model (44)

$$\tau_l = \tau_j\left(1 - \frac{1}{5}\frac{\sin(5\,\theta_j/2)}{\sin(\theta_j/2)}\right)^{-1}, \qquad [8]$$

where τ_j is the average delay time between jumps, and θ_j is the jump angle. On a time scale that is short compared with τ_j, the orientational motion is restricted by the intact hydrogen bonds and the angular motion is described by the wobbling-in-a-cone model. In MD simulations of water, one of the characteristics used to define the existence of a hydrogen bond is the angle between the O–D bond and the O–O axis being <30° (12). This definition permits a significant amount of orientational space to be sampled without breaking hydrogen bonds. Here, there are both water–water and water–ion hydrogen bonds. In aqueous NaBr solutions, it is reasonable to assume that, over some range of angles, the orientational motion of the OD bond will be caused by fluctuations of the intact hydrogen bond network and will be diffusive in nature within a cone.

From the measurement of τ_l, it is not possible to independently determine τ_j and θ_j. It has been found from MD simulations of an aqueous 1 M NaCl solution that the jump angles of water molecules are 63° and 70° for switching their hydrogen bond partners from water to water and from Cl⁻ to water, respectively (12, 43). For $\theta_j = 63°$, $\tau_l = 1.17\tau_j$, and for $\theta_j = 70°$, $\tau_l = 1.03\tau_j$. If $\theta_j = 72°$, $\tau_l = \tau_j$. Therefore, the measured τ_l is closely related to τ_j.

Vibrational population relaxation. Fig. 4*B* displays $P(t)$, vibrational population relaxation of the OD stretch in pure water and aqueous NaBr solutions at the peak of their absorptions. As the NaBr concentration increases, population relaxation becomes significantly slower. In pure water, $P(t)$ is a single exponential decay with a lifetime $T_1 = 1.8$ ps. However, $P(t)$ for the NaBr solutions are biexponential decays:

Table 3. Population relaxation parameters

Sample	A	T_{1f}, ps	$1 - A$	T_{1s}, ps
Water	1.0	1.81 ± 0.003		
$n = 32$	0.32 ± 0.01	1.19 ± 0.02	0.68	3.21 ± 0.02
$n = 16$	0.27 ± 0.01	1.20 ± 0.03	0.73	3.99 ± 0.03
$n = 8$	0.19 ± 0.01	1.21 ± 0.04	0.81	5.74 ± 0.02

$$P(t) = A \exp(-t/T_{1f}) + (1 - A)\exp(-t/T_{1s}),$$

where T_{1f} is the fast decay time constant and T_{1s} is the slow decay time constant. Table 3 lists the population relaxation parameters.

The biexponential decays indicate that there are two distinct OD environments. Biexponential decays have been observed for vibrational population relaxation of water in nanoscopic environments of AOT reverse micelles (18) and Nafion membranes (19). Detailed analysis of linear FT IR absorption spectra, spectral diffusion, orientational relaxation, and vibrational population relaxation of water molecules in small AOT reverse micelles have demonstrated that the vibrational population relaxation is very sensitive to the immediate local environment (18). In contrast, spectral diffusion and orientational relaxation depend on the dynamics of the hydrogen bond network. The hydrogen bond network couples the different environments so that the structural dynamics of the hydrogen bond network are not separable despite the system having different local structures (18). It has been demonstrated that a system with two distinct orientational relaxation times associated with two independent environments does not give rise to a biexponential decay of the orientational correlation function (18). Therefore, the vibrational population relaxation reports on local structural differences, whereas the spectral diffusion and orientational relaxation measure global hydrogen bond network dynamics.

It has been shown that in HOD–D$_2$O systems, the HOD bend overtone ($\approx$2,900 cm^{-1}) is coupled to the OH stretch mode ($\approx$3,400 cm^{-1}) and provides an efficient relaxation pathway for the OH stretch (45). The broad bandwidths of these transitions leads to substantial overlap of the spectra, which results in efficient wavelength-dependent relaxation of the OH stretch. As a result, the lifetime of the OH stretch of HOD in D$_2$O ($T_1 = 0.7$ ps) is shorter than that of the OD stretch of HOD in H$_2$O ($T_1 = 1.8$ ps). The HOD bend overtone is not an efficient relaxation pathway for the OD stretch ($\approx$2,500 cm^{-1}), because it is uphill and the bands do not overlap. For the OD stretch of HOD in H$_2$O, the OD stretch lifetime is not wavelength-dependent (16). It seems unlikely that the relatively small shift of the OD spectrum with NaBr concentration is by itself responsible for the change in the vibrational population relaxation with concentration.

The rate of vibrational population relaxation will depend on the strength of the coupling to intramolecular molecular modes, the continuum of low-frequency modes of the environment, and a density of states factor (46). In Table 3, it is seen that the fast decay time (T_{1f}) for the NaBr solutions is shorter than that of pure water and is constant with NaBr concentration. However, its amplitude decreases as the NaBr concentration increases. The slow decay time (T_{1s}) is significantly longer than that of pure water and becomes increasingly long as the NaBr concentration increases. The implication is that there is one environment that does not change with NaBr concentration, but the probability that the OD oscillator is in that environment decreases with increasing NaBr concentration. The other environment increases in probability, and its nature also changes as the NaBr concentration increases. Vibrational population relaxation is exceedingly sensitive to small changes in the energy levels of accepting modes and the continuum density of states. In contrast to orientational relaxation and spectral diffusion, it is much more difficult to ascribe changes in vibrational population relaxation to specific variations in the systems. One possibility is that the fast vibrational population relaxation (T_{1f}) comes from the OD being hydrogen-bonded to the oxygen atom of another water molecule, but the coupling and/or density of states are modified by the ionic environment, resulting in a decay time that is somewhat faster than in pure water. The slow vibrational population relaxation (T_{1s}) could arise from the OD hydrogen-bonded to Br$^-$ ions. As the NaBr concentration increases, the probability of the OD being hydrogen-bonded to Br$^-$ ions increases, and there is an increasing probability that the oxygen atom and the OH group are interacting with ions as DOH–Br$^-$ and HDO–Na$^+$. If the oxygen atom and the OH group are interacting with ions rather than with water molecules, the change in these interactions could lengthen the lifetime of the OD stretch.

Concluding Remarks

In liquid water, water molecules can form up to four hydrogen bonds in an approximately tetrahedral geometry forming an extended hydrogen bond network. The hydrogen bond network evolves continuously, breaking and reforming hydrogen bonds as well as reorienting and switching hydrogen bond partners. In aqueous NaBr solutions, the hydrogen bond network of water is partially or completely disrupted depending on the concentration of NaBr salt. The HOD molecules in highly concentrated NaBr solutions studied here are hydrogen-bonded to ions as HOD–Br$^-$, DOH–Br$^-$, and HDO–Na$^+$ and form partial or complete hydration shells around the ions. As a consequence, the structural evolution of the water–ion network is very different from the evolution of the hydrogen bond network in pure water.

The FFCFs extracted from 2D IR vibrational echo experiments and the anisotropy decays measured with polarization-selective IR pump–probe experiments as a function of ion concentration in aqueous NaBr solutions were compared with those for pure water (see Tables 1 and 2). The slowest time components of both the FFCFs and the anisotropy decays demonstrate that the time scale for complete rearrangement of the hydrogen bond structure increases with NaBr concentration. At the highest concentration with eight water molecules per NaBr molecule ($\approx$6 M NaBr), the evolution of the hydrogen bond structure is slower than in pure water by a factor of three.

Previously, two-color IR pump–probe experiments were performed by pumping the OH stretch of HOD in D$_2$O salt solutions (6 M NaCl, NaBr, or NaI) near the center of the OH stretching band and measuring vibrational population relaxation at different frequencies (20). The results were interpreted as spectral diffusion and were analyzed in terms of a correlation time constants τ_c. The correlation times were reported to be 20 to 50 times longer than those of pure water. MD simulations on halide–water systems suggested that the very long correlation times measured by two-color pump–probe experiments were similar to residence times of water molecules in the first hydration shells of halide anions in dilute solutions (30, 47–49). However, there is no bulk water in the highly concentrated salt solutions, and, as shown in previous studies of various water systems (5–7, 18), the FFCFs of water are nonexponential in nature and decays on multiple time scales. A model with two subsets of water molecules, water bound to ions and bulk water, seems to be insufficient to describe the hydrogen bond dynamics of water in the very highly concentrated salt solutions.

The very long correlations times obtained from two-color pump–probe experiments are in contrast to the results obtained here from direct measurements of spectral diffusion dynamics using the 2D IR vibrational echo experiments for the 6 M NaBr and are supported by the anisotropy decays occurring on similar time scales. Recently, MD simulations performed on ionic-solution, hydrogen-bond lifetime correlation functions were calculated (48, 49). These simulations show that the lifetime of water–Br$^-$ hydrogen bonds is a few picoseconds and is somewhat longer than that of water–water hydrogen bonds (49). The results presented in this paper indicate that the global rearrangement of the hydrogen bond structure in aqueous NaBr solutions is associated with breaking and reforming hydrogen bonds between water molecules and ions in a manner analogous to that of pure water. The time scales of hydrogen-bond lifetime correlation functions are qualitatively in accord with the results presented here.

Perhaps the most significant result to emerge from this work is that the time scale for the structural evolution of the hydrogen bonds in the presence of ions is longer, but not vastly longer, than that of pure water. In the $\approx$6 M NaBr solution, there are only eight water molecules per NaBr. In this solution, water molecules must be closely associated with ions, but the time scale for spectral diffusion and orientational relaxation is only a factor of three longer than that of bulk water. These results have implication for the nature of water dynamics in other aqueous systems with charged species, such as at the surfaces of biological membranes or in the vicinity of charged amino acids on the surfaces of proteins.

Materials and Methods

Samples. n = 8, 16, and 32 NaBr solutions were prepared by mixing 0.1, 0.05, and 0.025 mol of NaBr with 0.8 mol of 5% HOD

in H_2O, respectively. The samples were placed in a cell with two 3-mm-thick CaF_2 windows and a 6-μm path length to give a maximum absorbance of the OD stretch of $\approx$0.3 including the H_2O background absorption.

2D IR Vibrational Echo Experiments. The experimental details of 2D IR vibrational echo spectroscopy are described in ref. 23 and briefly in the text. The vibrational echo signal emitted in a unique phase-matched direction is spatially and temporally overlapped with a local oscillator and is dispersed through a monochromator onto a 32 MCT array detector with frequency resolution of $\approx$6 cm^{-1}. A range of 2,340–2,700 cm^{-1} is measured in the experiments.

Pump–Probe Experiments. The pump–probe setup is similar to an experimental setup described previously (19). The pump and probe beams with a relative intensity of 9:1 are focused onto the sample. The probe beam is dispersed through a monochromator onto the array detector. The measurements are performed with two polarization configurations, i.e., the probe beam is parallel or perpendicular to the pump beam polarization.

We thank Kelly J. Gaffney and Kyungwon Kwak for helpful discussions and David E. Moilanen for assistance in making the IR pump–probe measurements. This work was supported by Air Force Office of Scientific Research Grant F49620-01-1-0018, National Science Foundation Grant DMR 0652232, Department of Energy Grant DE-FG03-84ER13251, and National Institutes of Health Grant 2 R01 GM-061137-05.

1. Schuster P, Zundel G, Sandorfy C (1976) *The Hydrogen Bond: Recent Developments in Theory and Experiments* (North–Holland, Amsterdam).
2. Marcus Y (1985) *Ion Solvation* (Wiley, New York).
3. Bagchi B (2005) *Chem Rev* 105:3197–3219.
4. Gouaux E, MacKinnon R (2005) *Science* 310:347–349.
5. Stenger J, Madsen D, Hamm P, Nibbering ETJ, Elsaesser T (2002) *J Phys Chem A* 106:2341–2350.
6. Fecko CJ, Eaves JD, Loparo JJ, Tokmakoff A, Geissler PL (2003) *Science* 301:1698–1702.
7. Asbury JB, Steinel T, Kwak K, Corcelli SA, Lawrence CP, Skinner JL, Fayer MD (2004) *J Chem Phys* 121:12431–12446.
8. Loparo JJ, Roberts ST, Tokmakoff A (2006) *J Chem Phys* 125:194521.
9. Lawrence CP, Skinner JL (2003) *J Chem Phys* 118:264–272.
10. Moller K, Rey R, Hynes J (2004) *J Phys Chem A* 108:1275–1289.
11. Schmidt JR, Corcelli SA, Skinner JL (2005) *J Chem Phys* 123:044513.
12. Laage D, Hynes JT (2006) *Science* 311:832–835.
13. Pal SK, Peon J, Zewail AH (2002) *Proc Natl Acad Sci USA* 99:1763–1768.
14. Asbury JB, Steinel T, Stromberg C, Gaffney KJ, Piletic IR, Goun A, Fayer MD (2003) *Phys Rev Lett* 91:237402.
15. Woutersen S, Emmerichs U, Bakker HJ (1997) *Science* 278:658–660.
16. Steinel T, Asbury JB, Zheng JR, Fayer MD (2004) *J Phys Chem A* 108:10957–10964.
17. Tan H-S, Piletic IR, Fayer MD (2005) *J Chem Phys* 122:174501.
18. Piletic I, Moilanen DE, Spry DB, Levinger NE, Fayer MD (2006) *J Phys Chem A* 110:4985–4999.
19. Moilanen DE, Piletic IR, Fayer MD (2007) *J Phys Chem C* 111:8884–8891.
20. Kropman MF, Bakker HJ (2001) *Science* 291:2118–2120.
21. Piletic IR, Gaffney KJ, Fayer MD (2003) *J Chem Phys* 119:423–434.
22. Tan H-S, Piletic IR, Riter RE, Levinger NE, Fayer MD (2005) *Phys Rev Lett* 94:057405.
23. Park S, Kwak K, Fayer MD (2007) *Laser Phys Lett* 4:704–718.
24. Rey R, Moller KB, Hynes JT (2002) *J Phys Chem A* 106:11993–11996.
25. Lawrence CP, Skinner JL (2002) *J Chem Phys* 117:8847–8854.
26. Gaffney KJ, Piletic IR, Fayer MD (2003) *J Chem Phys* 118:2270–2278.
27. Pimentel GC, McClellan AL (1960) *The Hydrogen Bond* (Freeman, San Francisco).
28. Onori G, Santucci A (1993) *J Phys Chem* 97:5430–5434.
29. Piletic IR, Moilanen DE, Levinger NE, Fayer MD (2006) *J Am Chem Soc* 128:10366–10367.
30. Raugei S, Klein ML (2002) *J Chem Phys* 116:196–202.
31. Krekeler C, Hess B, Site LD (2006) *J Chem Phys* 125:054305.
32. Finkelstein IJ, Zheng J, Ishikawa H, Kim S, Kwak K, Fayer MD (2007) *Phys Chem Chem Phys* 9:1533–1549.
33. Zheng J, Kwak K, Fayer MD (2007) *Acc Chem Res* 40:75–83.
34. Mukamel S (1995) *Principles of Nonlinear Optical Spectroscopy* (Oxford Univ Press, New York).
35. Kwak K, Park S, Finkelstein IJ, Fayer MD (2007) *J Chem Phys* 127:124503.
36. Woutersen S, Pfister R, Hamm P, Mu Y, Kosov DS, Stock G (2002) *J Chem Phys* 117:6833–6840.
37. Piletic IR, Tan H-S, Fayer MD (2005) *J Phys Chem B* 109:21273–21284.
38. Schmidt JR, Roberts ST, Loparo JJ, Tokmakoff A, Fayer MD, Skinner JL (2007) *Chem Phys*, in press.
39. Dokter AM, Woutersen S, Bakker HJ (2005) *Phys Rev Lett* 94:178301.
40. Harpham MR, Ladanyi BM, Levinger NE, Herwig KW (2004) *J Chem Phys* 121:7855.
41. Rezus YLA, Bakker HJ (2005) *J Chem Phys* 123:114502.
42. Lipari G, Szabo A (1982) *J Am Chem Soc* 104:4546–4559.
43. Laage D, Hynes JT (2007) *Proc Natl Acad Sci USA* 104:11167–11172.
44. Ivanov EN (1964) *Sov Phys JETP* 18:1041–1045.
45. Deak JC, Rhea ST, Iwaki LK, Dlott DD (2000) *J Phys Chem A* 104:4866–4875.
46. Kenkre VM, Tokmakoff A, Fayer MD (1994) *J Chem Phys* 101:10618.
47. Nigro B, Re S, Laage D, Rey R, Hynes JT (2006) *J Phys Chem A* 110:11237–11243.
48. Guardia E, Laria D, Marti J (2006) *J Phys Chem B* 110:6332–6338.
49. Chowdhuri S, Chandra A (2006) *J Phys Chem B* 110:9674–9680.

Neuroglobin dynamics observed with ultrafast 2D-IR vibrational echo spectroscopy

Haruto Ishikawa[†], Ilya J. Finkelstein[†‡], Seongheun Kim[†], Kyungwon Kwak[†], Jean K. Chung[†], Keisuke Wakasugi[§], Aaron M. Massari[†¶], and Michael D. Fayer[†‖]

[†]Department of Chemistry, Stanford University, Stanford, CA 94305; and [§]Department of Life Sciences, Graduate School of Arts and Sciences, University of Tokyo, Tokyo 153-8902, Japan

Contributed by Michael D. Fayer, August 15, 2007 (sent for review July 25, 2007)

Neuroglobin (Ngb), a protein in the globin family, is found in vertebrate brains. It binds oxygen reversibly. Compared with myoglobin (Mb), the amino acid sequence has limited similarity, but key residues around the heme and the classical globin fold are conserved in Ngb. The CO adduct of Ngb displays two CO absorption bands in the IR spectrum, referred to as N_3 (distal histidine in the pocket) and N_0 (distal histidine swung out of the pocket), which have absorption spectra that are almost identical with the Mb mutants L29F and H64V, respectively. The Mb mutants mimic the heme pocket structures of the corresponding Ngb conformers. The equilibrium protein dynamics for the CO adduct of Ngb are investigated by using ultrafast 2D-IR vibrational echo spectroscopy by observing the CO vibration's spectral diffusion (2D-IR spectra time dependence) and comparing the results with those for the Mb mutants. Although the heme pocket structure and the CO FTIR peak positions of Ngb are similar to those of the mutant Mb proteins, the 2D-IR results demonstrate that the fast structural fluctuations of Ngb are significantly slower than those of the mutant Mbs. The results may also provide some insights into the nature of the energy landscape in the vicinity of the folded protein free energy minimum.

myoglobin mutants | protein dynamics | energy landscape

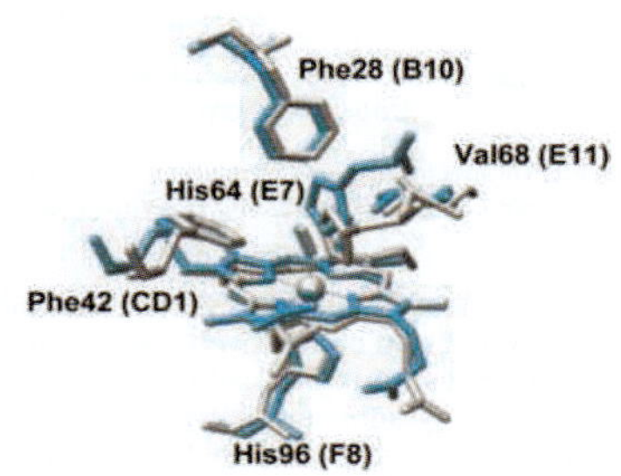

Fig. 1. Crystal structure of the active site of CO-bound ferrous mouse Ngb (gray) and L29F Mb (blue) mutant protein taken from the Protein Data Bank. The heme and some selected amino acid residues are shown.

Neuroglobin (Ngb) is a recently discovered family of vertebrate globin proteins. Ngb is expressed predominantly in nerve tissue. It has been hypothesized that Ngb facilitates O_2 diffusion to protect neuronal cells from hypoxia and ischemia (1). Ngb concentration in the brain is fairly low, too low to play the role that myoglobin (Mb) plays in the red muscles. However, because the expression level of Ngb is enhanced under hypoxic conditions *in vitro* as well as ischemia *in vivo* (2, 3), Ngb may be involved in neuronal responses to ischemia. A much higher concentration of Ngb is found in the retina. The concentration is high enough to deliver O_2 to mitochondria (4) and may facilitate O_2 diffusion in a manner similar to Mb. Another possibility is that Ngb is involved in redox-coupled sensor regulation for signal transduction in the brain. Ngb binds to the GDP-bound form of the α-subunit of heterotrimeric G protein under oxidative stress (5).

The 3D structures of Ngb from human and mouse, both having 151 amino acids, have been published (6, 7). Comparisons of Ngb with vertebrate Mb and Hb sequences show only minor similarities at the amino acid level (1), but Ngb features the conserved classical globin fold and contains heme (6). The heme iron in both the ferrous and ferric forms of Ngb is hexacoordinated (8), in contrast to mammalian Mb and Hb, which contain pentacoordinated heme iron. Although hexacoordinated heme has been reported in plant, bacteria, and invertebrate globins, its physiological significance is not yet understood (9). In Ngb, an external gaseous ligand must compete with the sixth ligand, the distal histidine (E7 in helix notation), for binding. Several key residues in the distal pocket for Mb, such as His-64 (E7), Val-68 (E11), and Phe-42 (CD1), are conserved in Ngb, although Leu-29 (B10) is replaced by Phe in Ngb (see Fig. 1). Phenylal-

anine is found at position B10 in many nonsymbiotic plant Hbs that have hexacoordinated binding to the heme (9). Previous studies revealed that Phe at position B10 influences the distal histidine and facilitates stable oxygen binding (10). Another feature of Ngb is that the structural analysis has shown the presence in human Ngb of an intramolecular disulfide bond, which affects its oxygen affinity (11).

Here we report measurements of the fast equilibrium structural dynamics of Ngb and comparisons of the Ngb dynamics with that of Mb mutant proteins. The Mb mutants mimic the structure of the active site of Ngb. The measurements are made by using 2D-IR vibrational echo spectroscopy, which is an ultrafast IR analog of 2D-NMR but operates on time scales many orders of magnitude shorter than NMR. 2D-IR vibrational echo spectroscopy can directly characterize protein motions on time scales from subpicosecond to 100 ps or longer (12–22).

A folded protein at equilibrium resides in a state that is in a free energy local minimum. A protein can have several such local minima, substates, separated by relatively high barriers. The substates are distinguished by differences in some well defined aspects of the protein's structure. Ngb and Mb are examples of proteins with several substates. As discussed below, the substates differ by the configuration of the distal histidine (19, 23–25). One way to view the local minimum of a substate is that it is relatively

---

Author contributions: H.I., I.J.F., and M.D.F. designed research; H.I., I.J.F., S.K., J.K.C., A.M.M., and M.D.F. performed research; H.I. and K.W. contributed new reagents/analytic tools; H.I., I.J.F., K.K., and M.D.F. analyzed data; and H.I. and M.D.F. wrote the paper.

The authors declare no conflict of interest.

Abbreviations: CLS, center line slope; FFCF, frequency–frequency correlation function; Mb, myoglobin; Ngb, neuroglobin.

[‡]Present address: Department of Biochemistry and Biophysics, Columbia University, New York, NY 10032.

[¶]Permanent address: Department of Chemistry, University of Minnesota, Minneapolis, MN 55455.

[‖]To whom correspondence should be addressed. E-mail: fayer@stanford.edu.

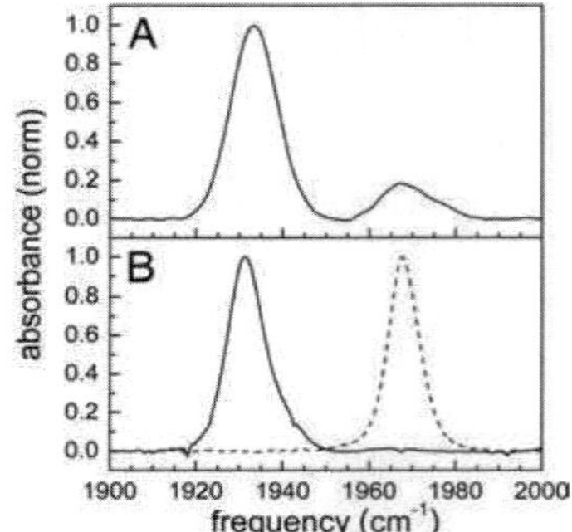

Fig. 2. Normalized FTIR spectra of the CO stretching mode bound to Ngb (*A*) and Mb mutant proteins L29F (solid curve) and H64V (dashed curve) (*B*).

broad with a rough energy landscape (26). This landscape has numerous minima separated by relatively low barriers. Changes in protein structure are caused by transitions from one landscape minimum to another. A substantial fraction or all of these transitions occur on subpicosecond to 100-ps time scales, depending on the protein. These structural fluctuations can be observed by using 2D-IR vibrational echo spectroscopy.

For heme proteins, 2D-IR vibrational echo experiments use the heme-ligated CO vibration as a direct sensor of protein dynamics (15, 16). The IR liner absorption spectrum of the CO stretching mode of heme proteins generally displays several bands. The different bands reflect structural differences, i.e., distinct structural substates, and the width of the bands reflects the range of CO vibrational energies that are associated with the distribution of protein structural configurations of each substate (15, 23, 24). The linear IR absorption spectrum cannot provide information on a protein's structural dynamics. The dynamical information is obtained from the time evolution of the 2D-IR line shapes that are acquired with the vibrational echo experiments (16).

The peak positions of the stretching bands of CO bound to Ngb in the absorption spectrum are almost identical with those of particular Mb mutant proteins. However, it will be shown below that, despite the great similarity in the absorption spectra, heme pocket structures, and backbone folds, the structural dynamics of Ngb and Mb are distinct. Ligand rebinding studies of Ngb show that CO rebinding kinetics after optical excitation-induced dissociation of the CO from the heme is different from that of Mb (27, 28). The results presented below demonstrate that the equilibrium structural fluctuations of Ngb are considerably slower than those of Mb. The results also provide insights into the nature of the energy landscape near the folded protein's free energy minimum.

Results and Discussion

Linear FTIR Spectroscopy. The background-subtracted linear FTIR spectrum of CO bound to human Ngb is shown in Fig. 2*A*. There are two CO absorption bands at 1,932 cm^{-1} and 1,968 cm^{-1}. These bands have been called the N_3 (lower frequency) and N_0 (higher frequency) conformers, respectively, in analogy to the A_3 and A_0 bands of Mb (27, 29). The significant width of the CO stretching bands of Ngb and their Gaussian shapes imply structural heterogeneity associated with each protein substate that is sensed by the CO bound to the active site. Although the range of frequencies that make up a substate's absorption band is sensitive to the details of the heme pocket, it is also determined by the global structural variations of the protein (19, 20, 30, 31). The frequency of N_3 corresponds closely to the MbCO A_3

conformation, which has the distal histidine localized in the heme pocket (19), whereas the frequency of N_0 is almost identical with the MbCO A_0 conformer, which has the distal histidine rotated out of the pocket (25). Substitution in Ngb of the distal histidine by the nonpolar amino acid residues, valine or alanine, causes a substantial increase in the population of N_0 conformer (27).

The A_1 substate of wild-type MbCO is predominately populated and dominates the absorption spectrum. It does not correspond to either the N_3 or N_0 states of Ngb in vibrational frequency. The A_1 and A_3 bands of MbCO arise from different conformations of the distal histidine in the heme pocket (19, 20). The A_3 conformer places the hydrogen on the protonated ε-nitrogen of the distal histidine, much closer to the CO than the A_1 conformer (19). Because of the similarities of the N_3 and A_3 spectra and the N_0 and A_0 spectra, it is desirable to compare their dynamics by using 2D-IR vibrational echo spectroscopy. However, both the A_3 and A_0 bands are too small in wild-type MbCO to be useful for the 2D-IR experiments.

To compare the dynamics of Ngb and Mb, the mutant Mb proteins L29F and H64V were used. The dynamics of MbCO and MbCO mutants have been measured by using vibrational echo spectroscopy (see refs. 16, 18–21, 32–34). The background-subtracted normalized FTIR spectra of CO bound to the mutants are shown in Fig. 2*B*. The L29F mutant has one major band at 1,932 cm^{-1} corresponding closely to the N_3 spectrum of Ngb. Leu-29 (B10) in Mb is substituted by Phe in Ngb, although other key residues in the distal pocket of Mb are preserved in Ngb. Therefore, the L29F mutant mimics the heme pocket structure of Ngb (Fig. 1). Experiments, theory, and molecular dynamics simulations have shown that both the local and global coupling between protein structure and the CO vibrational transition frequency is mainly through the electric field along the CO bond (19, 20, 24, 29, 30, 35–37). The protein residues range from charged, to polar, to nonpolar. All of these, as well as, to some extent, the solvent, contribute to the electric field. The vibrational states couple to the electric field via the Stark effect (37). Differences in structure produce changes in the electric field at the CO and, therefore, changes in frequency. The range of frequencies associated with the inhomogeneous line widths of the CO bands is determined by the range of structures that produce the electric field at the CO. 2D-IR vibrational echoes monitor the interconversion of the structures via the time evolution of the CO frequency.

The spectrum of the Mb A_0 band has a frequency almost the same as the Mb mutant, H64V, which has the distal histidine replaced by a valine. H64V mimics the situation in which the distal histidine is rotated out of the heme pocket. Both H64V and the N_0 conformation for Ngb absorb at 1,968 cm^{-1}. The spectral positions and line widths of the Ngb bands and the mutant bands are given in Table 1.

2D-IR Spectroscopy. Ultrafast 2D-IR vibrational echo spectroscopy can measure fast protein dynamics under thermal equilibrium conditions through the changes in the 2D-IR line shapes with time (13, 15–17, 19). In the vibrational echo experiments, three IR excitation pulses, ≈110 fs in duration, tuned to the vibrational absorption frequency are used. The times between pulses 1 and 2 and pulses 2 and 3 are called τ and T_w, respectively. The vibrational echo pulse emerges from the sample in a unique direction at a time $\leq \tau$ after the third pulse. A 2D spectrum is recorded by scanning τ at fixed T_w. The vibrational echo pulse is heterodyne detected through a monochromator. Heterodyne detection provides phase information in addition to signal amplitudes. Taking the spectrum of the heterodyned echo pulse performs one of the two Fourier transforms and gives the ω_m-axis (m for monochromator) in the 2D frequency domain spectrum. When τ is scanned, time interferograms are generated

Table 1. Parameters for CO bound to Ngb and Mb and mutants

Protein	Line center, cm^{-1}	FWHM, cm^{-1}	T_1, ps
Ngb N$_3$	1,932.7	12.1	19.3
Ngb N$_0$	1,968.1	9.8	18.4
Mb L29F	1,931.8	9.6	15.8
Mb H64V	1,968.0	8.8	24.1

at each ω_m. The interferograms are numerically Fourier transformed to give the ω_τ-axis of the 2D spectrum. Two-dimensional vibrational echo spectra are recorded as a function of T_w.

As T_w increases, the shape of the 2D spectral bands change. These changes are directly related to the structural evolution of the protein through the influence of the structural changes on the frequency of the CO vibrational mode. The experiment can be viewed qualitatively as follows. The first and second pulses act to label the initial frequencies of the molecules. Between the second and third pulses, structural evolution of the proteins causes the initially labeled frequencies to change (spectral diffusion). The third pulse ends the evolution period, and the vibrational echo pulse reads out the final frequencies. As T_w increases, there is more time for structural evolution and, therefore, larger frequency changes. The frequency changes are reflected in the change in shapes of the 2D-IR vibrational echo spectra. Full descriptions of the method have been presented (16, 38–40).

Fig. 3A shows 2D-IR spectra of CO bound to Ngb at several values of T_w. The red bands are positive going and correspond to the 0–1 vibrational transition. The blue bands are negative going. They arise from vibrational echo emission at the frequency of the 1–2 vibration transition (41) and are shifted along the ω_m-axis by the CO stretching mode's anharmonicity. As T_w increases, all of the peaks decay at a rate determined by the

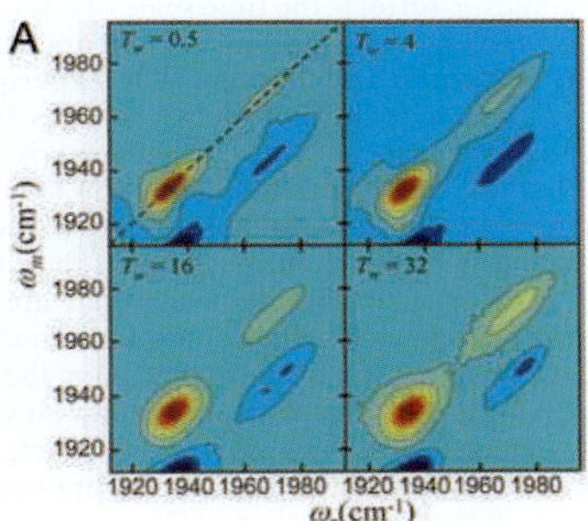

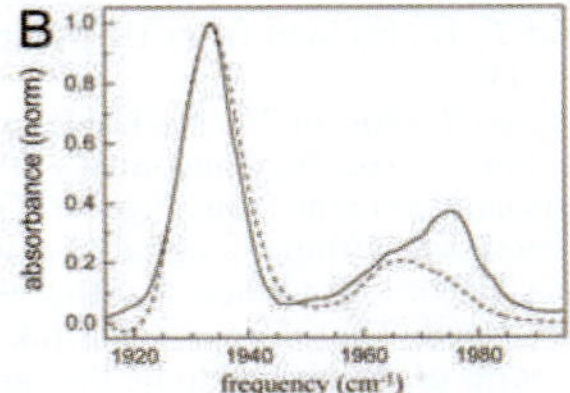

Fig. 3. 2D-IR vibrational echo spectra of NgbCO. (*A*) 2D-IR spectra at various times, T_w. Each contour corresponds to a 10% signal change. The red bands (positive going) correspond to the 0–1 vibrational transition. The blue bands (negative going) arise from vibrational echo emission at the 1–2 transition frequency. The dashed line in *Upper Left* is the diagonal. (*B*) Diagonal slice spectra of CO-bound Ngb at T_w = 0.25 ps (dashed curve) and 50 ps (solid curve).

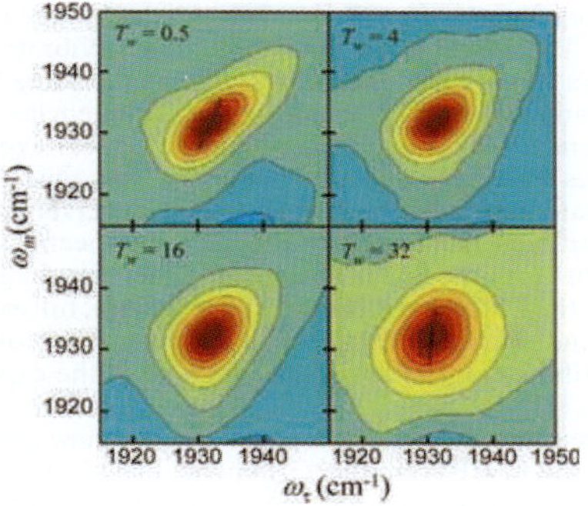

Fig. 4. 2D-IR spectra of CO bound to L29F Mb mutant protein. Each contour corresponds to a 10% signal change. The dashed lines are the center lines.

vibrational lifetime T_1. In Fig. 3A, the peaks are normalized to the largest peak in each panel. Therefore, the decay of the vibrationally excited population manifests itself as a relative reduction in the amplitude compared with the largest peak. The dashed line from upper right to lower left in Fig. 3A *Upper Left* is the diagonal. The two bands on the diagonal correspond to the 0–1 transitions of the two peaks in the absorption spectrum shown in Fig. 2A. These two bands are located at $(\omega_\tau, \omega_\mathrm{m})$ = (1,932 cm^{-1}, 1,932 cm^{-1}) and $(\omega_\tau, \omega_\mathrm{m})$ = (1,968 cm^{-1}, 1,968 cm^{-1}) for the N$_3$ and N$_0$ conformers, respectively.

Fig. 3A illustrates two important features. As T_w increases, the bands go from highly elongated along the diagonal to less elongated and increasingly broad along the ω_τ-axis. In the long time limit, they would become round. The change in shape is a manifestation of spectral diffusion caused by protein structural evolution. Detailed analysis presented below of the time dependence of the shapes quantifies the time evolution of the protein structures.

As T_w increases, the shape of the N$_3$ changes but the peak position remains fixed. However, not only does the N$_0$ band change shape, but the center frequency also changes, as can be seen in Fig. 3B, which displays two diagonal cuts through the 2D spectrum. The dotted-dashed curve is the cut at T_w = 0.25 ps, and the solid curve is the cut at T_w = 50 ps. The spectra have been normalized to the N$_3$ band maximum. The N$_0$ line clearly shifts. The shift occurs because the N$_0$ line is produced by two distinct protein conformers with overlapping spectra and different vibrational lifetimes. The lower-frequency conformation in the N$_0$ band has a shorter CO lifetime than the higher-frequency band, so at long time, the high-frequency band is revealed. In Fig. 2A, the N$_0$ absorption is somewhat elongated to the high-frequency side, suggesting that there may be two peaks. In Fig. 3B it is clear that there are two peaks.

To compare the dynamics of Ngb and Mb, the 2D-IR spectra of the Mb mutants, L29F and H64V, were also measured. Fig. 4 shows 2D-IR spectra of the CO stretch for the L29F mutant at several T_w points. Only the 0–1 transition region is shown. The band is located on the diagonal at $(\omega_\tau, \omega_\mathrm{m})$ = (1,933 cm^{-1}, 1,933 cm^{-1}) for L29F and at $(\omega_\tau, \omega_\mathrm{m})$ = (1,968 cm^{-1}, 1,968 cm^{-1}) for H64V (not shown). The peak positions of the Mb mutants are time-independent. The structural evolution of L29F is manifested by the spectral-diffusion-induced change in shape of the 2D-IR band from elongated to close to round as T_w increases.

The changes in the shapes of the 2D bands with increasing T_w can be used to determine the time scales and amplitudes of various contributions to the fast structural dynamics of the proteins by using methods based on diagrammatic perturbation theory (42). The frequency–frequency correlation function

(FFCF) connects the experimental observables to the underlying dynamics. The FFCF is the probability that a vibrational oscillator with a given initial frequency still has the same frequency at time t later, averaged over all starting frequencies. Once the FFCF is known, all linear and nonlinear optical experimental observables can be calculated by using the time-dependent diagrammatic perturbation theory (42). Conversely, the FFCF can be extracted from 2D vibrational echo spectra with additional input from the linear absorption spectrum. In general, to determine the FFCF from 2D and linear spectra, full calculations of linear and nonlinear response functions are performed iteratively until the calculated results converge to the experimental results (40, 43, 44).

Here the center line slope (CLS) method, a new approach for extracting the FFCF from the T_w dependence of the 2D spectra, is used (39, 45). It is accurate and much simpler numerically to implement than the full diagrammatic perturbation theory. Furthermore, it provides a more useful quantity to plot for visualizing differences in the dynamics of the various proteins than a series of 2D line shapes (45). A slice through the 2D spectrum at a particular ω_m is a spectrum along the ω_τ-axis. The peak of this spectrum is at a particular ω_τ-value. So the peak is a point with ω_m-, ω_τ-coordinates. Slices are taken over a range of ω_m values, and the resulting set of points forms the center line. Two such center lines are shown in Fig. 4 for $T_w = 0.5$ ps and 32 ps. In the absence of a homogeneous contribution to the spectrum (see below), at $T_w = 0$ the 2D line shape would be a line along the diagonal at 45°. Then, the slope of the center line would be 1. At very long time, the 2D line shape becomes symmetrical, and the center line is vertical (the slope is infinite). It has been shown theoretically that the T_w-dependent part of the FFCF equals the T_w dependence of the inverse of the slope of the center line (45), which is referred to as the CLS. Therefore, the CLS will vary from a maximum of 1 at $T_w = 0$ to a minimum of 0 at sufficiently long time. In Fig. 4, it can be seen that for $T_w = 32$ ps the slope of the center line is steeper than at $T_w = 0.5$ ps. It has also been shown theoretically that combining the analysis of the CLS with the linear absorption spectrum, the full FFCF can be obtained, including the T_w-independent homogeneous component (45).

A homogeneous contribution to the 2D line shape and the linear absorption line shape can arise from three sources: very fast structural fluctuations that produce a motionally narrowed contribution (16, 43, 44, 46); the vibration lifetime, T_1; and orientational relaxation. Because the proteins are so large, the orientational relaxation contribution is negligible. The T_1 contribution was measured independently by using IR pump–probe experiments. The T_1 contribution to the homogeneous line width $(1/\pi 2T_1)$ is small. It varies for the different bands but is ≈ 0.3 cm^{-1}. The T_1 values are listed in Table 1. The homogeneous contribution mainly comes from the motionally narrowed component. Motional narrowing occurs when some portion of the structural fluctuations are extremely fast, such that $\Delta\tau < 1$, where τ is the time scale of the fluctuations and Δ is the range (amplitude) of the frequency fluctuations. Motionally narrowed fluctuations produce a Lorentzian contribution to both the linear absorption spectrum and the 2D-IR spectrum. The T_w-independent homogeneous contribution manifests itself by broadening the 2D spectrum along the ω_τ-axis even at $T_w = 0$. This broadening reduces the initial value of the CLS to a number <1, which permits the determination of the homogeneous component (45).

A multiexponential form of the FFCF, $C(t)$, was used to model the multi-time scale dynamics of the protein structural fluctuations. This form has been widely used, and in particular, it has been found applicable in studies of the structural dynamics of heme-CO proteins (see refs. 15, 16, 18–20, 31–33, 47). The FFCF has the form

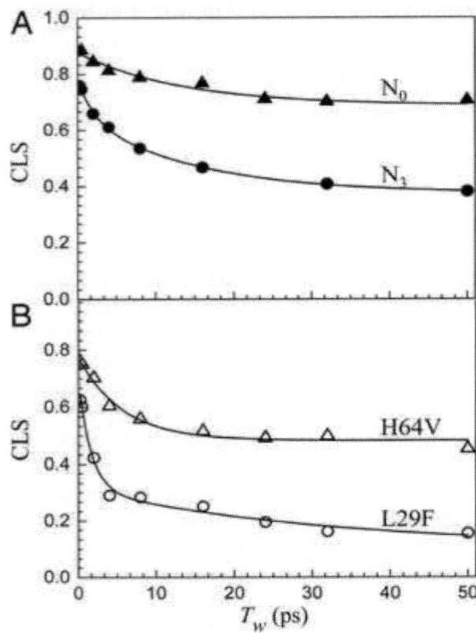

Fig. 5. T_w-dependent CLS for N_0 conformer of Ngb (*A*, filled triangles) and N_3 conformer of Ngb (*A*, filled circles) and for Mb mutant proteins H64V (*B*, open triangles) and L29F (*B*, open circles).

$$C(t) = \sum_{i=1}^{n} \Delta_i^2 e^{-t/\tau_i} + \Delta_s^2. \qquad [1]$$

The Δ_i and τ_i terms are the amplitudes and correlation times, respectively, of the frequency fluctuations induced by protein structural dynamics. τ_i reflects the time scale of a set of structural fluctuations, and Δ_i is the range of CO frequencies sampled because of the structural fluctuations. The experimental time window is $\approx 5T_1$, because lifetime decay reduces the signal to zero. The 2D-IR vibrational echo experiment is sensitive to fluctuations a few times longer than this window (48), i.e., several hundred picoseconds, because some portion of slower fluctuations will occur in the experimental window if their time scale is not too slow. Protein structural dynamics that are sufficiently slow will appear as static inhomogeneous broadening, which is reflected in $C(t)$ by Δ_s, a static term. In obtaining the FFCF from the data, the Δ_i and the τ_i are determined. However, for a motionally narrow term ($\Delta\tau < 1$), only the product, $\Delta^2\tau = 1/T_2^*$, can be obtained. T_2^* is called the pure dephasing time. The pure dephasing time is determined by using $1/T_2 = 1/T_2^* + 1/T_2$, where T_2 is determined from the CLS with use of the linear absorption spectrum (45) and T_1 is obtained from IR pump–probe experiments (see Table 1).

The T_w-dependent CLS for the N_3, N_0, L29F, and H64V bands are presented in Fig. 5. The N_0 band arises from two highly overlapped signals as is evident from Fig. 3B. Given the small amplitude of the N_0 signal, within the signal-to-noise ratio we can see no difference in the dynamics for the two underlying substates of the N_0 band. Therefore, we will take the dynamics of the two components of the N_0 line to be the same and present the data for N_0 as if they are a single entity.

Fig. 5A shows the CLS data for the Ngb N_3 and N_0 bands. It is clear from inspection of the data that the dynamics of the two Ngb substates are very different. The FFCF parameters are given in Table 2. The deviation from the maximum possible CLS value at $T_w = 0$ is indicative of a motionally narrowed term. T_2^* is more than a factor of 2 faster for N_3 than N_0. The N_0 band has an

Table 2. FFCF parameters obtained from 2D-IR and absorption spectra

Protein	T_2^*, ps	Δ_1, cm^{-1}	τ_1, ps	Δ_2, cm^{-1}	τ_2, ps	Δ_s, cm^{-1}
Ngb N_3	5.0	1.9	2.0	2.7	14	3.1
Ngb N_0	11.9	1.8	11.5			3.5
Mb L29F	4.8	2.8	1.7	2.2	66	
Mb H64V	7.7	2.1	5.2			2.7

intermediate decay, 11.5 ps, followed by a static component. As discussed above, the static component shows that there are protein structural fluctuations occurring on time scales longer than several hundred picoseconds, which are outside of the experimental window. In contrast, the N_3 line has a fast (2 ps) decay followed by another, intermediate decay of 14 ps, and then a static term. Although the intermediate decay times are similar for the two Ngb substates, N_3 has a fast-decay component, reflecting fast structural fluctuations, which is not present for the N_0 substate.

The CLS data for the Mb mutants L29F and H64V are displayed in Fig. 5B. The difference between the CLS decays of L29F and H64V is qualitatively similar to the difference between the N_3 and N_0 decays shown in Fig. 5A. The decay of L29F is significantly faster than that of H64V in the same manner that the decay of N_3 is faster than the decay of N_0. The fundamental difference between L29F and H64V is the distal histidine. L29F has the distal histidine in the heme pocket. Its spectrum is almost identical with the spectrum of the A_3 band of wild-type Mb. By combining vibrational echo experiments and molecular dynamics simulations, it has been shown that the A_3 configuration of Mb has the protonated ε-nitrogen of the imidazole side group of the distal histidine directed toward the CO (19). The conformation of the distal histidine results in a strong interaction between it and the CO bound at the active site. In contrast, H64V has the distal histidine replaced with a nonpolar valine, almost eliminating the interaction between residue 64 and the active site. Vibrational echoes and molecular dynamics simulations have shown that the reduction of CO vibrational spectral diffusion in H64V compared with the A_3 (and A_1) band of Mb is caused mainly by the elimination of the distal histidine (32). In N_3 the distal histidine is in the heme pocket but in N_0 it is swung out of the pocket. The IR CO vibrational absorption spectra of the N_0 and H64V bands, as well as the A_0 band of Mb, are almost identical. Elimination of the distal histidine (H64V) has the same effect on the absorption spectrum as its being swung out of the pocket. Therefore, it is reasonable to ascribe the difference between the N_3 and N_0 spectral diffusion to the difference in the configuration of the distal histidine. As with Mb, the fast component of the dynamics of N_3 can be substantially ascribed to motions of the distal histidine strongly interacting with the active site, which are absent for N_0.

Although the nature of the relationship of the dynamics of N_3/N_0 to L29F/H64V is the same, the dynamics of Ngb is fundamentally different from the dynamics of Mb. The dynamics of N_3 is substantially slower than that of L29F, and the dynamics of N_0 is significantly slower than that of H64V. The fastest dynamics values of N_3 and L29F are similar. They have essentially identical T_2^* values. L29F has a somewhat faster τ_1 with somewhat greater amplitude, Δ_1, but overall the fast dynamics values are similar. However, there is a major difference on the slower time scales. N_3 has an intermediate time scale component and a static component of significant amplitude. L29F does not have a static component. Instead, it has a single 66-ps decay. Thus, the equilibrium structural fluctuations of L29F have a slowest component that is relatively fast, only 66 ps. In L29F, all possible structures that give rise to the inhomogeneous broad-

ened CO absorption line are sampled in a few times at 66 ps. Although the heme pockets of N_3 and L29F are similar and they are both globin proteins, their amino acid sequences are substantially different. The 2D-IR vibrational echo experiment is sensitive to both local and global structural fluctuation. The similarities in the heme pockets indicate that the differences in dynamics arise from the different global structural fluctuations of the two proteins.

The dynamics of the N_0 substate of Ngb is also slower than the dynamics of H64V. Both have static components. The major difference between the two is in the moderately fast fluctuations. The τ_1 values for N_0 and H64V are 11.5 ps and 5.2 ps, respectively. Neither of these species has a distal histidine in the heme pocket, supporting the proposition that the differences in dynamics between Ngb and Mb involve differences in global structural fluctuations rather than differences in very local interactions at the active site.

Concluding Remarks

Ultrafast 2D-IR vibrational echo experiments have determined the time scales and relative magnitudes of the fast equilibrium structural fluctuations of the two substates, N_3 and N_0, of the protein neuroglobin by observing the vibrational dephasing dynamics of the stretching mode of CO bound at the heme active site. It is found that the structural fluctuations of the N_0 substate are much slower than those of the N_3 substate. The heme pocket structures of Ngb and Mb are similar despite the lack of similarity in their amino acid sequences. The linear IR spectra of CO bound at the active sites of the N_3 and N_0 conformers of Ngb are almost identical with the liner IR spectra of L29F and H64V Mb mutant proteins, respectively. N_3 and L29F have a distal histidine in the active site pocket, whereas N_0 and H64V do not. H64V displays substantially slower vibrational spectral diffusion than L29F. Therefore, the reduction in structural fluctuations detected by CO bound at the active site of Ngb in the N_0 substate relative to the N_3 substate is substantially due to the distal histidine being swung out of the heme pocket.

The vibrational echo experiments demonstrate that the equilibrium structural fluctuations of Ngb are significantly different from those of Mb despite the globin structure of the proteins, the near identity of their heme pocket structures, and the almost identical CO FTIR spectra. However, away from the heme pocket, the primary structures of Ngb and Mb are quite different. Ngb has a disulfide bond not present in Mb. Even in the absence of an exogenous bound ligand, Ngb is hexacoordinated with the distal histidine in the sixth ligand, whereas Mb is pentacoordinated. As revealed by the 2D-IR vibrational echo experiments, the net result of all structural differences is that the Ngb structure is substantially more constrained on fast time scales than Mb.

As shown in Fig. 3B, the N_0 band is actually composed of two substates, N_0' and N_0'', each of which has an absorption line width that is narrower than the total line width reported in Table 1. Thus, the range of structural variations of the N_0' and N_0'' substates is small, and as can be seen in Fig. 5A, there is not a great deal of structural dynamics out to ≈ 100 ps. The reduced structural sampling on fast time scales in Ngb, particularly the N_0' and N_0'' substates, may slow ligand migration through the protein to the active site compared with Mb.

A folded protein in a particular substate occupies a minimum on the free energy landscape (26). However, the minimum is broad and rough with many local minima separated by low barriers. Transitions among these minima give rise to the structural fluctuations that cause the spectral diffusion measured by the 2D-IR vibrational echo experiment. In L29F, all of these local minima are sampled rapidly. Spectral diffusion is complete, and therefore all structures within the broad free energy minimum are sampled, in a few hundred picoseconds. In contrast, N_3 has a static component in the FFCF that indicates that some of the dynamics occur on time scales that are long compared with several hundreds of picoseconds. The implication is that the rough landscape of N_3 around the folding minimum has higher barriers than L29F.

The FFCF obtained for L29F has three terms (see Table 2). The motionally narrowed term does not permit the determination of the time constant, τ. However, simulations of Mb show that the motionally narrowed τ is tens of femtoseconds (32). With this value, L29F has structural fluctuations on three time scales: tens of femtoseconds, 1.7 ps, and 66 ps. The tens-of-femtosecond dynamics values are too fast for true structural changes. They reflect thermally excited oscillatory motions of small groups that are local and highly damped. The slower time constants are the time scales for true structural changes, such as changes in conformations of side chains. Although other explanations may exist, the fact that L29F has two well separated time constants suggests a two-tier energy landscape near the free energy minimum of the folded state. There is a set of low barriers that are crossed relatively rapidly and a set of higher barriers that are crossed more slowly. For the Ngb substates and H64V, the static terms in the FFCF imply that there is an even higher tier or tiers of barriers that push the time scale for part of the structural evolution out of the ≈ 100-ps time window of the experiments.

Materials and Methods

Expression and purification of His_6-tagged human Ngb was performed as described in ref. 5. Purity of protein and contamination of disulfide-dependent formation of dimmers were checked by SDS/PAGE under reduced and nonreduced conditions. The mutant sperm whale Mb proteins L29F and H64V were expressed and purified as described in ref. 49.

The CO forms of Ngb and mutant Mb proteins were prepared according to protocols published in refs. 19, 20, and 31. For both the linear FTIR and vibrational echo measurements, ≈ 20 μl of the sample solution was placed in a sample cell with CaF_2 windows and a 50-μm Teflon spacer.

We thank Prof. R. Kopito (Stanford University) for the use of protein expression and purification equipment; Dr. J. Christianson and Dr. C. Patel (Stanford University) for invaluable assistance with sample preparation; and Prof. John S. Olson (Rice University, Houston, TX) for providing the myoglobin mutant proteins. This work was supported by National Institutes of Health Grant 2 R01 GM-061137-05. H.I. was supported by the Human Frontier Science Program. S.K. was supported by a fellowship from the Korea Research Foundation funded by the Korean government.

1. Burmester T, Weich B, Reinhardt S, Hankeln T (2000) *Nature* 407:520–523.
2. Sun Y, Jin K, Peel A, Mao XO, Xie L, Greenberg DA (2003) *Proc Natl Acad Sci USA* 100:3497–3500.
3. Sun Y, Jin K, Mao XO, Zhu Y, Greenberg DA (2001) *Proc Natl Acad Sci USA* 98:15306–15311.
4. Schmidt M, Giessl A, Laufs T, Hankeln T, Wolfrum U, Burmester T (2003) *J Biol Chem* 278:1932–1935.
5. Wakasugi K, Nakano T, Morishima I (2003) *J Biol Chem* 278:36505–36512.
6. Pesce A, Dewilde S, Nardini M, Moens L, Ascenzi P, Hankeln T, Burmester T, Bolognesi M (2003) *Structure (London)* 11:1087–1095.
7. Vallone B, Nienhaus K, Matthes A, Brunori M, Nienhaus GU (2004) *Proc Natl Acad Sci USA* 101:17351–17356.
8. Dewilde S, Kiger L, Burmester T, Hankeln T, Baudin-Creuza V, Aerts T, Marden MC, Caubergs R, Moens L (2001) *J Biol Chem* 276:38949–38955.
9. Kundu S, Trent JT, III, Hargrove MS (2003) *Trends Plants Sci* 8:387–393.
10. Smagghe BJ, Kundu S, Hoy JA, Halder P, Weiland TR, Savage A, Venugopal A, Goodman M, Premer S, Hargrove MS (2006) *Biochemistry* 45:9735–9745.
11. Hamdane D, Kiger L, Dewilde S, Green BN, Pesce A, Uzan J, Burmester T, Hankeln T, Bolognesi M, Moens L, *et al.* (2003) *J Biol Chem* 278:51713–51721.
12. Volkov V, Hamm P (2004) *Biophys J* 87:4213–4225.
13. Chung HS, Khalil M, Smith AW, Ganim Z, Tokmakoff A (2005) *Proc Natl Acad Sci USA* 102:612–617.
14. Fang C, Hochstrasser RM (2005) *J Phys Chem B* 109:18652–18663.
15. Finkelstein IJ, Ishikawa H, Kim S, Massari AM, Fayer MD (2007) *Proc Natl Acad Sci USA* 104:2637–2642.
16. Finkelstein IJ, Zheng J, Ishikawa H, Kim S, Kwak K, Fayer MD (2007) *Phys Chem Chem Phys* 9:1533–1549.
17. Mukherjee P, Kass I, Arkin IT, Zanni MT (2006) *Proc Natl Acad Sci USA* 103:3528–3533.
18. Massari AM, Finkelstein IJ, Fayer MD (2006) *J Am Chem Soc* 128:3990–3997.
19. Merchant KA, Noid WG, Akiyama R, Finkelstein I, Goun A, McClain BL, Loring RF, Fayer MD (2003) *J Am Chem Soc* 125:13804–13818.
20. Merchant KA, Noid WG, Thompson DE, Akiyama R, Loring RF, Fayer MD (2003) *J Phys Chem B* 107:4–7.
21. Rector KD, Rella CW, Kwok AS, Hill JR, Sligar SG, Chien EYP, Dlott DD, Fayer MD (1997) *J Phys Chem B* 101:1468–1475.
22. Rella CW, Rector KD, Kwok AS, Hill JR, Schwettman HA, Dlott DD, Fayer MD (1996) *J Phys Chem* 100:15620–15629.
23. Hartmann H, Parak F, Steigemann W, Petsko GA, Ponzi DR, Frauenfelder H (1982) *Proc Natl Acad Sci USA* 79:4967–4971.
24. Phillips GN, Jr, Teodoro ML, Li T, Smith B, Olson JS (1999) *J Phys Chem B* 103:8817–8829.
25. Augspurger JD, Dykstra CE, Oldfield E (1991) *J Am Chem Soc* 113:2447–2451.
26. Frauenfelder H (1995) *Nat Struct Biol* 2:821–823.
27. Sawai H, Makino M, Mizutani Y, Ohta T, Sugimoto H, Uno T, Kawada N, Yoshizato K, Kitagawa T, Shiro Y (2005) *Biochemistry* 44:13257–13265.
28. Kriegl JM, Bhattacharyya AJ, Nienhaus K, Deng P, Minkow O, Nienhaus GU (2002) *Proc Natl Acad Sci USA* 99:7992–7997.
29. Morikis D, Champion PM, Springer BA, Sligar SG (1989) *Biochemistry* 28:4791–4800.
30. Williams RB, Loring RF, Fayer MD (2001) *J Phys Chem B* 105:4068–4071.
31. Merchant KA, Thompson DE, Xu Q-H, Williams RB, Loring RF, Fayer MD (2002) *Biophys J* 82:3277–3288.
32. Finkelstein IJ, Goj A, McClain BL, Massari AM, Merchant KA, Loring RF, Fayer MD (2005) *J Phys Chem B* 109:16959–16966.
33. Massari AM, Finkelstein IJ, McClain BL, Goj A, Wen X, Bren KL, Loring RF, Fayer MD (2005) *J Am Chem Soc* 127:14279–14289.
34. Hill JR, Dlott DD, Rella CW, Peterson KA, Decatur SM, Boxer SG, Fayer MD (1996) *J Phys Chem* 100:12100–12107.
35. Oldfield E, Guo K, Augspurger JD, Dykstra CE (1991) *J Am Chem Soc* 113:7537–7541.
36. Spiro TG, Wasbotten IH (2005) *J Inorg Biochem* 99:34–44.
37. Park ES, Boxer SG (2002) *J Phys Chem B* 106:5800–5806.
38. Zheng J, Kwak K, Fayer MD (2007) *Acc Chem Res* 40:75–83.
39. Park S, Kwak K, Fayer MD (2007) *Laser Phys Lett* 4:704–718.
40. Asbury JB, Steinel T, Fayer MD (2004) *J Lumin* 107:271–286.
41. Rector KD, Kwok AS, Ferrante C, Tokmakoff A, Rella CW, Fayer MD (1997) *J Chem Phys* 106:10027–10036.
42. Mukamel S (2000) *Annu Rev Phys Chem* 51:691–729.
43. Asbury JB, Steinel T, Stromberg C, Corcelli SA, Lawrence CP, Skinner JL, Fayer MD (2004) *J Phys Chem A* 108:1107–1119.
44. Asbury JB, Steinel T, Kwak K, Corcelli SA, Lawrence CP, Skinner JL, Fayer MD (2004) *J Chem Phys* 121:12431–12446.
45. Kwak K, Park S, Finkelstein IJ, Fayer MD (2007) *J Chem Phys*, in press.
46. Woutersen S, Pfister R, Hamm P, Mu Y, Kosov DS, Stock G (2002) *J Chem Phys* 117:6833–6840.
47. Finkelstein IJ, Massari AM, Fayer MD (2006) *Biophys J* 92:3652–3662.
48. Bai YS, Fayer MD (1989) *Phys Rev B* 39:11066–11084.
49. Varadarajan R, Szabo A, Boxer SG (1985) *Proc Natl Acad Sci USA* 82:5681–5684.

10054   *J. Phys. Chem. B* **2008,** *112,* 10054–10063

Native and Unfolded Cytochrome c—Comparison of Dynamics using 2D-IR Vibrational Echo Spectroscopy

Seongheun Kim,[†] **Jean K. Chung,**[†] **Kyungwon Kwak,**[†] **Sarah E. J. Bowman,**[‡] **Kara L. Bren,**[‡] **Biman Bagchi,**[†,§] **and M. D. Fayer*,**[†]

Department of Chemistry, Stanford University, Stanford, California 94305 and University of Rochester, Rochester, New York 14627-0216

Received: March 14, 2008; Revised Manuscript Received: May 16, 2008

Unfolded vs native CO-coordinated horse heart cytochrome c (h-cyt c) and a heme axial methionine mutant cyt c_{552} from *Hydrogenobacter thermophilus* (*Ht*-M61A) are studied by IR absorption spectroscopy and ultrafast 2D-IR vibrational echo spectroscopy of the CO stretching mode. The unfolding is induced by guanidinium hydrochloride (GuHCl). The CO IR absorption spectra for both h-cyt c and *Ht*-M61A shift to the red as the GuHCl concentration is increased through the concentration region over which unfolding occurs. The spectra for the unfolded state are substantially broader than the spectra for the native proteins. A plot of the CO peak position vs GuHCl concentration produces a sigmoidal curve that overlays the concentration-dependent circular dichroism (CD) data of the CO-coordinated forms of both *Ht*-M61A and h-cyt c within experimental error. The coincidence of the CO peak shift curve with the CD curves demonstrates that the CO vibrational frequency is sensitive to the structural changes induced by the denaturant. 2D-IR vibrational echo experiments are performed on native *Ht*-M61A and on the protein in low- and high-concentration GuHCl solutions. The 2D-IR vibrational echo is sensitive to the global protein structural dynamics on time scales from subpicosecond to greater than 100 ps through the change in the shape of the 2D spectrum with time (spectral diffusion). At the high GuHCl concentration (5.1 M), at which *Ht*-M61A is essentially fully denatured as judged by CD, a very large reduction in dynamics is observed compared to the native protein within the ~100 ps time window of the experiment. The results suggest the denatured protein may be in a glassy-like state involving hydrophobic collapse around the heme.

I. Introduction

Protein folding is a fundamentally important problem that has been intensely investigated for decades.[1–10] Although a great deal is known about protein folding, some of the most important issues remain unresolved. The theoretical models of protein folding can be broadly divided into two categories. The first model is that protein folding proceeds through a unique predetermined sequence of metastable intermediate states of increasingly lower energy until the native state is reached.[2–5] The native state is assumed to be the state of lowest free energy. The alternative folding funnel model[7,8,11,12] based on energy landscape considerations advocates a multitude of pathways and no unique sequence of intermediate states. The driving force in the funnel model is also the free energy bias toward the native state, but entropy considerations are included in the description of stabilization. While experimental results have been proposed in favor of both the models,[6,10] it is fair to say that it is not clear which model is correct or whether either model is uniquely correct. Indeed, treatments of protein folding data benefit from consideration of both of these models which are not mutually exclusive.

A major difficulty in experimentally addressing the protein folding problem is the proper identification and characterization

of non-native conformational states on the folding energy landscape. These conformations can be short lived, which can inhibit thorough examination. To improve characterization of folding intermediates, considerable research has been directed toward unfolding of proteins from their native states. The utility of this approach is that a non-native state can be stabilized by use of an appropriate amount of a denaturant such as guanidinium hydrochloride (GuHCl) or urea. These denaturants are known to destabilize proteins in different ways.[13–17] The guanidinium cation may disrupt salt bridges and other Coulombic interactions within the protein.[18,19] In contrast, urea is uncharged and generally denatures a protein by disrupting its secondary structures and operates at higher concentrations than GuHCl. By using a particular denaturant it is possible to create an ensemble of non-native states of nearly identical characteristics which can then be probed experimentally.

Because of their relative simplicity and high solubility, proteins in the cytochrome c (cyt c) family have served as important model systems in protein folding studies using a host of optical and magnetic resonance techniques.[13,17,20–33] Class I cyts c proteins are soluble and monomeric with a single heme group covalently bound to a Cys-X-X-Cys-His motif in which the two Cys form thioether linkages to the heme and the His binds the heme iron.[34] Figure 1 displays the active site with the Cys linkages shown. A range of experiments on cyts c have been directed toward distinguishing between the sequential model and the folding funnel model and characterizing the non-native states. A series of papers by Englander and co-workers (who used a hydrogen exchange method) has indicated the

---

* To whom correspondence should be addressed. E-mail: fayer@stanford.edu.
† Stanford University.
‡ University of Rochester.
§ Permanent address: Solid State and Structural Chemistry Unit, Indian Institute of Science, Bangalore 560 012, India.

10.1021/jp802246h CCC: $40.75   © 2008 American Chemical Society
Published on Web 07/23/2008

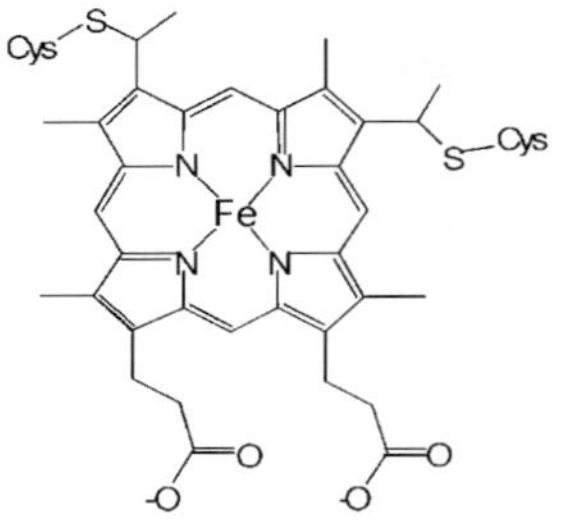

Figure 1. Active site of Class I cyt *c* with a single heme group covalently bound to a Cys-X-X-Cys-His motif in which the two Cys form thioether linkages to the heme.

existence of a definite pathway of unfolding through a well-defined sequence of intermediate states.[10,35] These studies seem to run counter to the folding funnel picture. Subsequent theoretical studies by Wolynes and co-workers suggested that the hydrogen exchange experiments can be explained within the folding funnel picture.[6] Comparative studies of the folding of Class I cyts *c* from different subfamilies suggest that proteins in this class share a common folding nucleus, which translates to a common folding mechanism.[26,36]

Class I cyts *c* as a group are among the most actively studied proteins in the context of chemical and thermal denaturation, in which the heme plays an important role and serves as a valuable probe.[13,16,17,21–30,37–41] The thioether bonds ensure that the heme prosthetic group is closely associated with the protein in the native and unfolded configurations. The heme is axially ligated by side chains of a histidine and a methionine. The methionine−iron bond may be disrupted by binding of exogenous ligands which serve as probes of the physical properties of the heme and its environment.[42] Alternatively, mutation of the axial methionine to nonligating alanine creates a site for ligand probes to bind the heme iron.[43]

In this paper we report a new approach to studying the denaturation of two different Class I cyts *c*. The stretching mode of CO bound to the reduced Fe of these cyts *c* is studied as a function of GuHCl concentration using infrared absorption spectroscopy and ultrafast two-dimensional infrared (2D-IR) vibrational echo spectroscopy. Experiments are conducted on CO-coordinated ferrous cyt c_{552} mutant from *Hydrogenobacter thermophilus* (in which the heme axial ligand Met61 is replaced with Ala, *Ht*-M61A) and on the CO derivative of ferrous horse heart cyt *c* (h-cyt *c*). Both are Class I cyts *c* proteins, but h-cyt *c* is from the mitochondrial cyt *c* subfamily, and *Ht* cyt c_{552} is from the cyt c_8 subfamily (sometimes called the cyts c_{551}).[44] The folding of these proteins has been proposed to involve a common folding nucleus and mechanism resulting from their similar folding topology.[26,36]

The linear absorption spectra display a single CO band for the native proteins that shifts to lower frequency as the GuHCl concentration is increased. For h-cyt *c*, a comparison of the CO absorption peak shift with GuHCl concentration is almost identical to the concentration dependence of the CD data, demonstrating that the CO vibration is sensitive to the structural changes induced by the denaturant. In addition to the absorption band shift with increasing concentration, the CO band develops a shoulder and then becomes a much broader band at high (>5 M) GuHCl concentration. The CO band is inhomogeneously broadened, and the width reflects the range of structural configurations available to the protein. The increase in width

when the protein is unfolded indicates the existence of a broader range of structures in the denatured state than in the native state.

The 2D-IR vibrational echo experiments are conducted on the CO bound to the *Ht*-M61A ferrous heme. 2D-IR vibrational echo spectroscopy[45–47] can probe protein conformational fluctuations under thermal equilibrium conditions on timescales ranging from subpicosecond to ~100 ps or longer.[48–53] 2D-IR vibrational echo spectroscopy reports on protein dynamics that occur on fast timescales. The method has recently been applied to biological problems[54,55] including the study of model enzymes,[56,57] protein unfolding,[50] peptide dynamics in membranes,[51] and protein equilibrium fluctuations in aqueous and confined environments.[48,49,52,53,58]

Here we employ 2D-IR vibrational echo spectroscopy to examine the equilibrium structural fluctuations of different states of *Ht*-M61A by observing the change in the 2D lineshapes with time for three samples: the native protein and the protein in low- and high-concentration GuHCl. Within the time window of the experiments (subpicosecond to ~100 ps), the native protein displays considerable fast structural dynamics with approximately >50% of the accessible structures being sampled in <100 ps. However, the unfolded protein formed at high GuHCl concentration (5.1 M) shows a dramatic change in the structural dynamics with only ~15% of the protein configurations being sampled in <100 ps. There is a significant decrease in the homogeneous (motionally narrowed) contribution to the 2D spectra and a substantial decrease in the fraction of structures sampled on the longer time scale (100 ps). The changes indicates that the unfolded state may be in a compact, hydrophobicity driven collapsed glassy state.

II. Experimental Procedures

A. Sample Preparation. Preparation of *Ht*-M61A utilized an *E. coli*-based expression system.[59,60] Molecular biology procedures and materials and preparation of *Ht*-M61A are described in detail elsewhere.[20,61,62] To prepare aqueous samples of CO-ligated *Ht*-M61A for IR studies, 10 mg of lyophilized protein was dissolved in 1.0 mL of pD 7.4 phosphate buffer (50 mM) in D_2O. The solutions were reduced with a 5-fold excess of dithionite (Aldrich) and stirred under a CO atmosphere for 1 hour. The solutions were centrifuged at 3000 relative centrifugal force for 15 min through a 0.45-μm acetate filter (Pall Nanosep MF) to remove particulates. CO-ligated *Ht*-M61A was then denatured by adding GuHCl phosphate buffer to obtain final concentrations of 3.2 and 5.1 M GuHCl. The samples were further concentrated by repeated centrifugation (Eppendorf 5415D) over modified polyetersulfone membranes (Pall Nanosep 3K Omega) to a final protein concentration of 6−8 mM. The sample was then placed in a sample cell with CaF_2 windows and a 56 μm Teflon spacer.

Horse heart cyt *c* (Type VI from Sigma) in 6 M GuHCl was prepared by dissolving 6 mM protein in deoxygenated GuHCl and 50 mM potassium phosphate (pD 7.4). The protein was reduced and ligated with CO as described above. Samples in the range of 0.0003−5.8 M GuHCl were prepared by combining the h-cyt *c*/6 M GuHCl solution with a 50 mM phosphate buffer solution. To remove light scattering sources such as dust particles, samples were filtered through a 0.45-μm acetate filter (Pall Nanosep MF) and then concentrated by centrifugation to a final protein concentration of 6−8 mM before loading in a gastight 50 μm path length sample cell with CaF_2 windows.

B. 2D-IR Vibrational Echo Spectroscopy. The experimental setup is similar to those described previously.[46,47,63] Briefly, the mid-IR pulses with center frequency adjusted to

10056 *J. Phys. Chem. B, Vol. 112, No. 32, 2008*

the absorption frequency of each protein sample were generated by an optical parametric amplifier pumped with a regeneratively amplified Ti:Sapphire laser. The bandwidth and duration of the mid-IR pulses were 150 cm^{-1} and 110 fs, respectively. Three mid-IR pulses were sequentially time delayed before they were crossed and focused in the sample. The vibrational echo pulse generated in the phase-matched direction was made collinear with a local oscillator pulse, dispersed through a monochromator, and detected with a 32-element HgCdTe array detector. The signal was interfered with the local oscillator to obtain full time, frequency, and phase information from the vibrational echo wave packet.

The 2D-IR spectrum is obtained as a function of three variables: the emitted vibrational echo frequencies, ω_m, and the variable time delays between the first and second pulses (τ) and the second and third pulses (T_w, "waiting" time). The 2D spectrum at each T_w was obtained by numerically Fourier transforming the τ scan data at each emission frequency, ω_m, to give the ω_τ axis. The 2D-IR data presented below are plotted as a function of ω_τ and ω_m.

C. Circular Dichroism Spectroscopy. Samples of CO-ligated *Ht*-M61A for circular dichroism (CD) spectroscopy were prepared as described above, except that $10\ \mu$M protein samples were prepared in 50 mM sodium phosphate buffer in H$_2$O. GuHCl solutions were prepared from a stock solution of 8.2 M GuHCl in 50 mM sodium phosphate buffer, pH 7.0, with a final range of GuHCl concentrations from 0.0 to 7.0 M. Refractive index measurement was used to determine the GuHCl concentration of each sample as described.[64] The pH of each sample was adjusted to 7.0 prior to data collection. CD measurements were performed on an Aviv Instruments model 202 spectropolarimeter using a quartz cell with 0.100 cm path length. CD spectra of samples were recorded every 0.5 nm over a range of 224−220 nm with an averaging time of 5 s and a bandwidth of 1.00 nm. The change in the CD signal at 222 nm was analyzed to follow the unfolding of the protein at different GuHCl concentrations. The concentration used for the CD experiments is very low compared to those used for the vibrational echo experiments. A protein concentration study of the denaturation of h-cyt c with GuHCl measured with CD over the protein concentration range 0.014−0.224 mM showed no dependence on concentration.[29] The highest concentration used in that study is well below the concentration employed in the vibrational echo experiments, but it suggests that the CD results will be unchanged at even higher concentrations. NMR experiments carried out on high concentrations (1−3 mM) of the wild-type *Ht*-cyt c show no evidence of aggregation.[65] Furthermore, as discussed below, the denaturation studies using IR absorption experiments on h-cyt c and *Ht*-M61A as a function of GuHCl concentration show the same denaturation curves as found with CD. These IR absorption studies were performed on samples with the same high concentration used in the vibrational echo experiments. Therefore, the high protein concentrations used in the IR experiments do not appear to influence the results.

III. Results

A. Time-Independent Spectroscopy. A complete set of CO-ligated h-cyt c IR absorption spectra of the CO stretching mode were taken as a function of GuHCl concentration, and a small number of spectra were taken on *Ht*-M61A. Most of the absorption experiments were conducted on h-cyt c because it was readily available.

The heme iron of h-cyt c is axially ligated with His18 and Met80 to yield a six-coordinated form under physiological

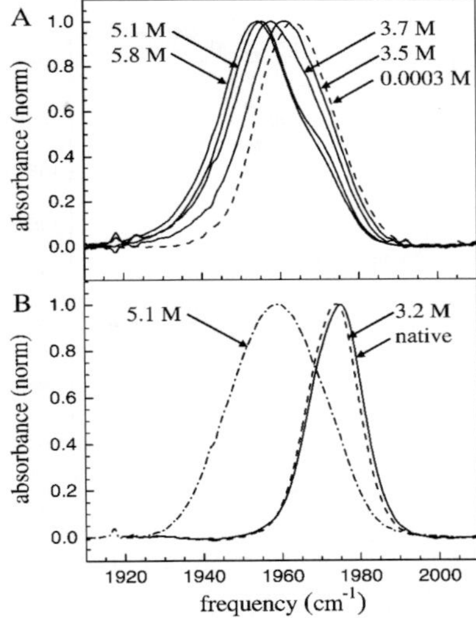

Figure 2. Vibrational absorption spectra of the stretching mode of CO bound to horse heart cyt c (A) and the mutant *Ht*-M61A (B) at room temperature, pD 7.4, for various concentrations of GuHCl.

conditions. When the protein is denatured under high GuHCl or urea concentration, Met80 dissociates from the heme and heme iron (Fe^{2+}) can bind small ligands such as CO and NO tightly in the denatured state.[29] When the denaturant is diluted, the denatured protein with CO bound refolds to form CO-ligated ferrocyt c.[29] Studies have shown that unfolded h-cyt c at neutral pH displays non-native ligation to the heme by His26 and His33.[13,29,66−68] Partly as a result of formation of this improperly ligated structure, the refolding of h-cyt c at neutral pH is quite complex. However, binding CO to heme in the unfolded state eliminates the possibility of forming improperly ligated structures.

Figure 2A displays several spectra of the CO stretch of h-cyt c-CO. As the concentration of GuHCl increases, the band shifts to the red (lower frequency) and broadens. In addition, there is a shoulder on the blue side of the line for the higher concentrations that decreases as the concentration increases further. Figure 2B shows spectra of the mutant protein M61A-CO for the three GuHCl concentrations that were studied with the 2D-IR vibrational echo experiments discussed below.

An earlier study using a native *Ht*-M61A FT-IR spectroscopy at room temperature reported a single CO stretching band at 1974 cm^{-1} with 14.7 cm^{-1} fwhm.[20] In that study, the double mutant *Ht*-M61A/Q64N was prepared to replace Gln64 with Asn in addition to the M61A mutation. According to an NMR study,[20] Gln64 in *Ht*-M61A is expected to be oriented out of the heme pocket to be consistent with a non-hydrogen-bonding interaction with the CO ligand, while Asn64 is oriented into the active site in *Ht*-M61A/Q64N to donate a hydrogen bond. The CO band in a native *Ht*-M61A is attributed to a conformation in which the distal Gln64 is positioned out of heme pocket. Thus, the CO band position in *Ht*-M61A is higher in frequency ($\sim$9 cm^{-1}) than that in *Ht*-M61A/Q64N.

With *Ht*-M61A, again there is a red shift and broadening as the concentration of GuHCl increases. While the trend in peak positions and widths is similar, they are not identical. For h-cyt c the spectrum shifts from 1963 (0.0003 M) to 1955 cm^{-1} (5.8 M), while for *Ht*-M61A the spectrum shifts from 1975 (native)

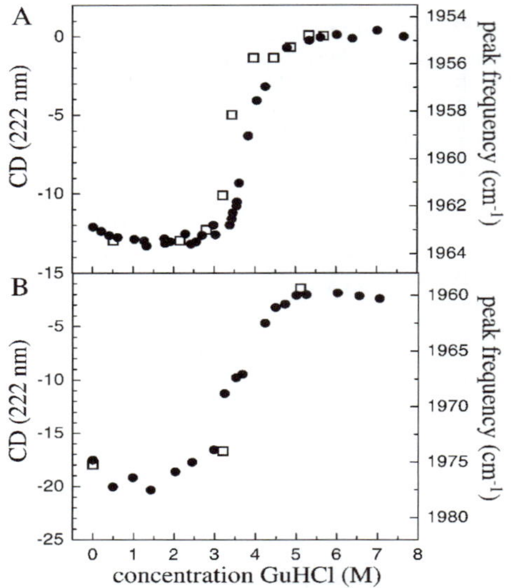

Figure 3. CD data (filled circles) and CO stretching mode absorption band shift for CO bound to (A) horse heart cyt *c* and (B) the mutant *Ht*-M61A as a function of GuHCl concentration. The CD data for horse heart cyt *c* is from ref 29. The close correspondence between the CD data and the vibrational band shift data shows that the CO vibrational frequency is sensitive to the same global structural changes as CD.

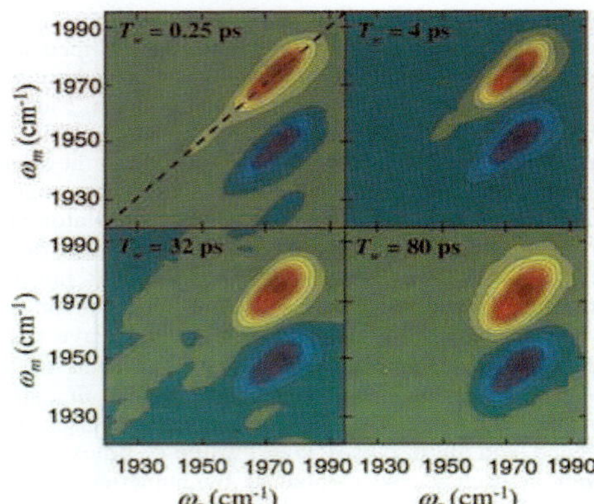

Figure 4. 2D-IR vibrational echo spectra of the stretching mode of CO bound to *Ht*-M61A at several delays, T_w. The diagonal is the dashed line shown in the upper left panel. The bands on the diagonal are positive going, and the off-diagonal bands are negative going. As T_w increases, the shape of the 2D spectrum changes, becoming less elongated. The change in shape with T_w is caused by protein structural evolution-induced spectral diffusion.

to 1959 cm^{-1} (5.1 M). In addition, the highest concentration h-cyt *c* peak has a width of 25 cm^{-1} fwhm, while for *Ht*-M61A it is 28 cm^{-1} fwhm. Using CO as a reporter, *Ht*-M61A has a somewhat greater response to unfolding than does h-cyt *c*. The CO peak for *Ht*-M61A shifts further to the red at high GuHCl concentration and the high-concentration absorption bandwidth is larger.

Figure 3A displays two types of data as a function of GuHCl concentration for h-cyt *c*. The squares are the peak positions of the CO absorption band. The circles are CD data from the literature.[29] Two data sets have been scaled so that the amplitudes match at the highest and lowest GuHCl concentration. The change in the peak position with GuHCl concentration closely mimics the CD data, suggesting that the CO vibrational band frequency is sensitive to global unfolding. Figure 3B shows CD data taken on *Ht*-M61A (circles). The signal-to-noise ratio for the CD of *Ht*-M61A is not as good as for the spectra of h-cyt *c*. The squares are the *Ht*-M61A-CO peak positions for the three concentrations studied below with 2D-IR experiments. While there are only three points, they appear to be consistent with the CD data. In the future this plot will be filled in. For now, we will assume that the vibrational frequency of *Ht*-M61A is sensitive to the extent of denaturation in the same manner as h-cyt *c* and the global changes in the protein structure monitored by CD. It is important to note, however, that protein unfolding is complex; *Ht* cyt *c* unfolding involves a well-populated intermediate non-native state,[26] and although h-cyt *c* unfolding can be approximated with a two-state model, close examination has shown it to be highly heterogeneous.[16,32,69,70] Thus, the interpretation of how denaturation impacts the CO stretch will be revisited as models of folding and our understanding of 2D-IR as a probe of folding evolve.

B. 2D-IR Vibrational Echo Spectroscopy. 2D-IR vibrational echo spectroscopy directly examines structural degrees

of freedom on very fast time scales, ~100 fs to ~100 ps. The vibrational echo signals are recorded by scanning τ at fixed T_w to produce a 2D spectrum. T_w is then increased, and another spectrum is taken. As T_w increases, the peak shapes in the 2D-IR spectra change. These changes are directly related to the structural evolution of the protein through the influence of the changes in structure on the CO stretch frequency. Very qualitatively, the 2D-IR experiment can be viewed as follows. The first and second pulses act to label the initial frequencies of the molecular oscillators. Between the second and third pulses, structural evolution causes the initially labeled frequencies to change (spectral diffusion). The third pulse ends the evolution period, and the vibrational echo signal reads out the final frequencies of the initially frequency-labeled molecular oscillators. As T_w increases, there is more time for structural evolution to occur and, therefore, larger frequency changes. The structural evolution (frequency change) is reflected in the changes of peak shape of the 2D-IR spectra. The 2D-IR vibrational echo experiments have been reviewed recently.[45–47]

Figure 4 displays 2D-IR vibrational echo spectra for native *Ht*-M61A at several times, T_w. In the $T_w = 0.25$ ps panel, the dashed line is the diagonal. There are two bands in each of the spectra. The band on the diagonal (positive going) arises from the 0−1 vibrational transition of the CO stretching mode. The band below it, off-diagonal (negative going), arises from vibrational echo emission at the 1−2 transition frequency, which is shifted to lower frequency by the anharmonicity of the vibrational potential.[71,72] Both types of bands provide the same information. Therefore, we will focus on the 0−1 transition band on the diagonal.

At short T_w, the peaks in the 2D-IR spectra reveal a strong elongation along the diagonal. For the shortest two T_ws shown, there is a long tail going to low frequency. This tail arises from a low-amplitude band in the absorption spectrum that is centered at ~1953 cm^{-1}. This peak has a shorter lifetime than the main band further to the blue. It has decayed and is not visible in the $T_w = 32$ ps spectrum. The important feature for the main band is that as T_w increases the shape becomes less elongated along the diagonal. Elongation along the diagonal is caused by

10058 *J. Phys. Chem. B, Vol. 112, No. 32, 2008*

Kim et al.

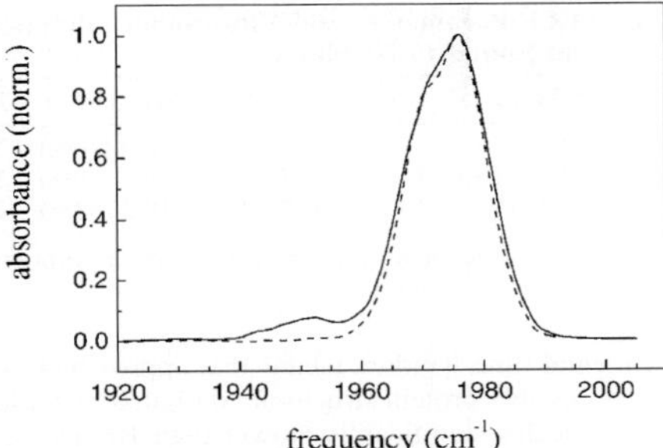

Figure 5. Native *Ht*-M61A-CO spectra produced by cutting the 2D vibrational echo spectrum along the diagonal at two T_ws: (solid curve) $T_w = 0.5$ ps; (dashed curve) $T_w = 32$ ps. The data show that the native protein has three conformational substates (see text).

inhomogeneous broadening. At short time there are many structural configurations of the protein that have not yet been sampled by structural evolution. As time increases, structural evolution causes spectral diffusion, resulting in the change in shape. The longest time for the measurements is limited to a few T_1's, where T_1 is the vibrational lifetime. Here data could be collected out to 120 ps. The spectrum is sensitive to structural fluctuations that are a few times the longest T_w measured because some portion of slower fluctuations will occur in the experimental window if their time scale is not too slow.[73] If there were no time limit on the measurement, then the 2D bands would eventually become round when spectral diffusion is complete, that is, when all possible structural configurations have been sampled.

Figure 5 displays spectra that are slices along the diagonal of the 2D spectra taken at two times: $T_w = 0.25$ (solid curve) and 32 ps (dashed curve). The two spectra have been normalized. At short time (0.25 ps), the small peak at ~ 1953 cm^{-1} is apparent. At long time (32 ps), the 1953 cm^{-1} peak is absent because it has a shorter lifetime. For CO bound to hemes or metalloporphyrins it is well documented that the CO vibrational lifetime decreases essentially linearly as the frequency of the transition decreases.[74-76] In both Figures 2B and 5, the main band for the native protein shows some asymmetry. In Figure 5 the 2D diagonal cut at 32 ps strongly suggests that there are two substates responsible for the absorption band. The difference between the 2D diagonal spectrum at 0.25 and 32 ps arises from a slight difference in the lifetimes of the two substates. The spectra shown in Figure 5 indicate that the native *Ht*-M61A protein exists in three conformational substates.

The change of peak shapes in the 2D-IR spectra with increasing T_w (see Figure 4) can be employed to determine the time scales and amplitudes of various contributions to the structural evolution of the protein using methods based on diagrammatic perturbation theory.[77,78] The frequency–frequency correlation function (FFCF) connects the experimental observables to the underlying dynamics. The FFCF is the joint probability distribution that the frequency has a certain initial value at $t = 0$ and another value at a later time t. As the structure evolves, the initial frequencies of the protein bound COs change and the FFCF decays. Once the FFCF is known, all linear and nonlinear optical experimental observables can be calculated by time-dependent diagrammatic perturbation theory.[77,78] Conversely, the FFCF can be extracted from 2D-IR spectra with additional input from linear FT-IR absorption spectra. In general, to determine the FFCF from 2D-IR and linear FT-IR spectra,

full calculations of linear and nonlinear third-order response functions are performed iteratively until the calculation results converge to the experimental results.[79]

Here, the center line slope (CLS) method is employed.[80,81] The CLS method is an approach for extracting the FFCF from the T_w dependence of the 2D-IR spectra that is accurate and much simpler to implement numerically than iterative fitting methods with calculations of all response functions based on time-dependent diagrammatic perturbation theory. Furthermore, the CLS provides a more useful quantity to plot for visualizing differences in the dynamics as a function of GuHCl concentration than a series of full 2D-IR spectra.[52,53,80,81] In the CLS method employed here, frequency slices through the 2D-IR spectrum parallel to the ω_m axis at various ω_τs are projected onto the ω_m axis.[80,81] These projections are a set of spectra with peak positions, ω_m^{max}, on the ω_m axis. The plot of ω_m^{max} vs ω_τ forms a line called the center line. In the absence of a homogeneous contribution to the 2D-IR spectrum (see below), the peak shape in the 2D-IR spectrum at $T_w = 0$ ps would be essentially a line along the diagonal at 45°. The slope of this center line would be 1. At very long time, the peak shape in the 2D-IR spectrum becomes symmetrical and the center line is horizontal (slope is zero). It has been shown theoretically that the normalized FFCF is equal to the T_w dependence of the slope of the center line (CLS).[80,81] Therefore, the CLS will vary from a maximum value of 1 at $T_w = 0$ ps to a minimum value of 0 at a sufficiently long time. It has also been shown theoretically that by combining the analysis of the CLS with the linear FT-IR absorption spectrum, the full FFCF can be obtained, including the T_w-independent homogeneous component.[80,81]

A homogeneous contribution to the peak shape in the 2D-IR spectra and the line shape of the linear FT-IR absorption spectra can arise from three sources: very fast structural fluctuations that produce a motionally narrowed contribution to the FFCF,[79,82] vibrational population relaxation, and orientational relaxation. Because the rotational diffusion time of the protein is long relative to the vibrational lifetime, the contribution from orientational relaxation is negligible. The vibrational population relaxation times (vibrational lifetimes) were measured independently using IR pump–probe experiments. The lifetime contribution to the homogeneous line width is small. Motional narrowing occurs when there is some portion of the structural fluctuations that are extremely fast, such that $\Delta\tau < 1$, where τ is the time scale of the fast fluctuations and Δ is the amplitude of the associated frequency fluctuations. Motionally narrowed fluctuations produce a Lorentzian contribution to both the 2D-IR spectrum and the linear FT-IR absorption spectrum. The T_w-independent homogeneous contribution manifests itself by broadening the 2D-IR spectrum along the ω_τ axis even at $T_w = 0$ ps. This homogeneous broadening reduces the initial value of the CLS to a number less than 1, which permits its determination.

A multiexponential form of the FFCF, $C(t)$, is used to model the multitime scale dynamics of the structural evolution of the protein systems. This form has been used previously for the analysis of a number of heme proteins.[20,46,48,49,52,53,58,62,83] The FFCF has the form

$$C(t) = \sum_{i=1}^{n} \Delta_i^2 e^{-t/\tau_i} + \Delta_s^2 \tag{1}$$

The Δ_i and τ_i terms are the amplitudes and correlation times, respectively, of the frequency fluctuations induced by protein structural dynamics. τ_i reflects the time scale of a set of structural fluctuations, and the Δ_i is the range of CO frequencies sampled due to those structural fluctuations. The experimental time

J. Phys. Chem. B, Vol. 112, No. 32, 2008 **10059**

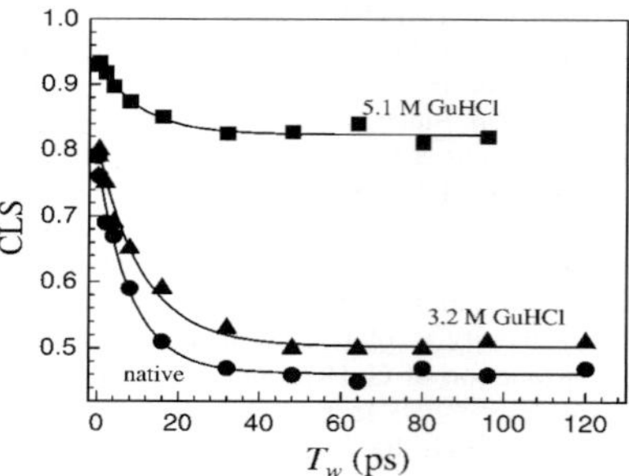

Figure 6. Center line slope (CLS) data for native *Ht*-M61A-CO in aqueous solution and in 3.2 and 5.1 M GuHCl solutions. The dynamics in the ∼100 ps experimental time window of the denatured protein (5.1 M GuHCl) are substantially changed from those of the native protein.

window is a few T_1's because lifetime decay reduces the signal to zero. The 2D-IR vibrational echo experiment is sensitive to fluctuations a few times longer than this window[73] because some portion of slower fluctuations will occur in the experimental window if their time scale is not too slow. Protein structural dynamics that are sufficiently slow will appear as static inhomogeneous broadening, which is reflected in $C(t)$ by Δ_s, a static term. In obtaining the FFCF from the data, Δ_i and τ_i are determined. However, for a motionally narrow term ($\Delta\tau < 1$), only the product, $\Delta^2\tau = 1/T_2^*$, can be obtained. T_2^* is called the pure dephasing time. The pure dephasing time is obtained using $1/T_2 = 1/T_2^* + 1/2T_1$, where T_2 is determined from the CLS with the linear absorption spectrum[80,81] and T_1 is obtained from IR pump–probe experiments. The pure dephasing line width is $\Gamma = 1/\pi T_2^*$.

Figure 6 displays CLS data for native *Ht*-M61A and for the protein in 3.2 and 5.1 M GuHCl solutions. As can be seen in Figure 3, 3.2 M is at the onset of the denaturation while at 5.1 M denaturation is essentially complete. As will be discussed below, the FFCF for all three samples is well described by three terms in eq 1. As discussed above, the difference between 1 and the $T_w = 0$ value of the CLS is related to the homogeneous component of the dynamics. All three curves show a deviation from 1 at $T_w = 0$. At relatively short times, <∼40 ps, there is a decay. Following the decay, the CLS becomes horizontal. The horizontal portion of the data is described by the static term in the FFCF and indicates that there are dynamics that are too slow to be observed within the time frame of the experiment.

First consider the CLS data for the native state qualitatively. During the 120 ps of observation, the CLS has decayed to 0.46. This means that over one-half of the spectroscopic line has been sampled by spectral diffusion. Part of this sampling is described by the homogeneous component. The homogeneous component is caused by exceedingly fast fluctuations on an atomic distance scale that may be thought of as thermally populated low-frequency vibrations.[58,62,83] The dynamics on the tens of picosecond time scale involve much larger distance scales. A reasonable question is: what is the distance scale of motions that occur on these time scales? Incoherent quasielastic neutron scattering experiments on native bovine α-lactalbumin observed collective motions on the tens of picosecond time scale.[84] The correlation length for such fluctuations was reported to be 18 Å.[84] Therefore, the fluctuations that cause the decay of the CLS should represent motions of amino acids and collections of amino acids. The fact that the CLS becomes flat and that 46% of the inhomogeneous absorption line is not sampled within

TABLE 1: FFCF Parameters and Vibrational Lifetimes, T_1, of the CO Band Bound to *Ht*-M61A[a]

GuHCl conc.	Γ (cm⁻¹)	τ_1 (ps)	Δ_1 (cm⁻¹)	Δ_s (cm⁻¹)	T_1 (ps)
0.0 M	2.7 (22%)	8	4.2 (32%)	3.9 (46%)	29
3.2 M	2.5 (21%)	11	4.0 (29%)	3.5 (50%)	27
5.1 M	1.5 (7%)	10	4.7 (11%)	10.3 (82%)	23

[a] Percentages are fractional contributions to the absorption line-widths.

the experimental time window means that approximately one-half of the accessible protein structural configurations interconvert on times scales significantly slower than 100 ps.

At the onset of denaturation (see Figure 6, 3.2 M GuHCl), the CLS changes but not by a great deal. The absorption spectrum (Figure 2B) is also almost the same as that of the native protein. Quantitative analysis of the data (see below) clearly delineates the small differences. However, the fully denatured protein (see Figure 6, 5.1 M GuHCl) shows dramatic changes from the native protein. The absorption spectrum is significantly different (see Figure 2B). In the CLS data, the homogeneous component is substantially reduced and the decay plateaus with <20% of the spectroscopic line sampled. Therefore, on time scales of 100 ps there is a substantial change, the nature of which is discussed below.

The CLS decay results are quantified in Table 1. The table gives the parameters obtained from the CLS determination of the FFCF and the vibrational lifetime, T_1, for each sample. The numbers in parentheses give the fraction of the particular contribution of the FFCF to the entire absorption line width. There are three terms in the FFCF (see eq 1). The first data column has the values of the pure dephasing linewidths, $\Gamma = 1/\pi T_2^*$. This component of the FFCF is determined by the difference between 1 and the $T_w = 0$ value in Figure 6. The next two columns are the decay constant and the amplitude factor that give rise to the decays that are seen in all three curves in Figure 6. The column labeled Δ_s is the amplitude of the static term, which sets the value of the long time plateau in the data. Comparing the results for the native protein (0.0 M GuHCl) to the protein at the onset of denaturation (3.2 M GuHCl), as can be seen from the CD and IR spectral shift data in Figure 3, the FFCF parameters are very similar except for τ_1. The decay time τ_1 is longer for the slightly denatured protein, and Δ_s is a little larger. These differences can be seen in the data in Figure 6.

As can be seen in Figure 6, the large change in the dynamics occurs when the protein is fully denatured in 5.1 M GuHCl. Looking at the parameters in Table 1, the pure dephasing line width, Γ, becomes much narrower (T_2^* becomes longer) by almost a factor of 2. The decay time constant, τ_1, is slower than in the native protein, although it does not change significantly from the value in 3.2 M GuHCl. In addition to the large change in the homogeneous component, the other large change is in the amplitude of the static term, Δ_s. This term goes from <4 cm⁻¹ to >10 cm⁻¹. For the denatured protein, the static term is responsible for 82% of the absorption line width compared to 46% for the native protein. This is seen in Figure 6 as the much higher value for the plateau for the denatured protein. The results demonstrate that denaturation changes the dynamics within the experimental window. The increase in the inhomogeneous line width shows that there are a wider range of structures in the denatured state, and within the time window of the experiment, a smaller fraction of these structures are sampled compared to the native state. Denaturation causes a large fraction of the dynamics to occur on time scales that are long compared to several hundred picoseconds because measurements within the

10060 *J. Phys. Chem. B, Vol. 112, No. 32, 2008*

time window of $\sim$100 ps are sensitive to structural fluctuations that occur on time scales out to several times the experimental window.[73]

IV. Discussion

Could the change in the *Ht*-M61A dynamics be caused by the increase in viscosity of 5.1 M GuHCl compared to water? A detailed study of the viscosity dependence of CO vibrational dephasing for a number of heme proteins including *Ht*-M61A has been reported.[85] For many orders of magnitude increase in viscosity over aqueous protein solutions it was determined that the influence of viscosity on the observed spectral diffusion is very mild. For *Ht*-M61A, the dependence on viscosity goes as $\eta^{0.13}$, where η is the solution viscosity.[85] Furthermore, the homogeneous component is essentially unchanged by viscosity. In going from water to 5.1 M GuHCl, the viscosity only increases by 40%.[86] Given the very weak dependence of the vibrational dephasing on viscosity $\eta^{0.13}$ and the lack of influence of the viscosity on the homogeneous component, the change in viscosity is not responsible for the observed change in protein dynamics at 5.1 M GuHCl.

Insight into the influence of denaturation can be obtained by examining the results of placing *Ht*-M61A and other heme proteins in a glassy trehalose solvent.[62] All of the heme proteins showed the same behavior. The homogeneous component of the FFCF (the extremely fast motions) remained essentially unchanged. However, the slower times scale dynamics within the experimental window, picoseconds to tens of picoseconds, were eliminated in the trehalose glass. Molecular dynamics simulations of the myoglobin-CO mutant H64V-CO (the distal histidine replaced with a valine) were conducted to understand the influence of a trehalose glassy solvent on protein dynamics and the vibrational echo observables.[62] In the simulations, an aqueous protein solution was equilibrated and then the solvent was immobilized. The dynamics of the H64V-CO were investigated, and the FFCFs were determined for the immobilized solvent and a normal aqueous solution. The simulations reproduced the nature of the experimental results semiquantitatively. Like the experiments, the simulations showed that an ultrafast homogeneous component remained with the immobile solvent, but the slower time scale dynamics within the experimental and simulation windows were shut down. In addition, an analytical theory has been used to understand the viscosity dependence of protein dynamics using a breathing sphere model.[87,88] The understanding that emerged from the simulations and theory showed that elimination of dynamics on the tens of picoseconds time scale is caused by locking up the surface topology of the protein. The internal motions of the protein structure depend on the ability of the surface topology to change.

The experiments in trehalose show how slowing or eliminating protein motions manifest themselves in the experimental observables. However, there are major differences between the trehalose experimental results and the results for denatured *Ht*-M61A shown in Figure 6. The key features of the trehalose experiments are that the homogeneous component is unchanged while the observable fluctuations on slower time scale are eliminated.[62] In contrast, the denatured *Ht*-M61A data show a substantial change in the homogeneous component and the decay over the first few tens of picoseconds is not eliminated. The comparison between the linear spectra is significant. When *Ht*-M61A is put into glassy trehalose, the line width (fwhm) of the linear IR spectrum does not change and the peak position is only shifted by 3 cm^{-1} to higher frequency in trehalose compared to aqueous solution.[62] This demonstrates that the basic structure of the protein and the variations in structure about the average are almost unchanged when the surface topology is locked up. The $\sim$10 ps time scale dynamics are pushed out to a long time beyond the experimental time window, but the structure is unchanged. This behavior is very different from the results for *Ht*-M61A in 5.1 M GuHCl. The *Ht*-M61A band shifts substantially to lower frequency and broadens considerably in going from the native to denatured protein. The shift in the position and change in width demonstrate that major structural changes have taken place in contrast to *Ht*-M61A in trehalose. Therefore, the changes in the FFCF observed for denatured *Ht*-M61A are caused by a change in structure, not by locking up the original native structure.

Looking at Table 1, the denatured protein's FFCF has a much smaller homogeneous line width (Γ), somewhat slower intermediate time scale dynamics (τ_1), and a much larger static component (Δ_s) compared to the native protein. One possible view of the denatured state is that it is comprised of a large number of structures that are each only slightly different from the native protein. Such a range of structures could account for the large increase in the line width of the denatured protein. However, this picture is inconsistent with the data. First, in addition to the increase in line width, the absorption spectrum of the fully denatured protein is shifted substantially to the red; thus, the average structure has changed. Furthermore, there is a large change in Γ, discussed in detail below, which demonstrates a substantial change from the native structure. The change in the intermediate time scale dynamics, $\sim$10 ps, occurs at the onset of denaturation (see Table 1), indicating a change in structure. However, the large change in Γ, the change in the absorption peak position, and the increase in absorption line width are very different for the fully denatured protein than the mildly denatured and native proteins, demonstrating additional substantial changes in structure for the fully denatured protein.

Γ is reduced dramatically, by almost a factor of 2, in going from the native to the denatured protein. Γ is the motionally narrow component of the FFCF. For a motionally narrowed contribution, the homogeneous line width is $\Gamma = \Delta^2\tau/\pi$, so it is not possible to determine from the experiments whether Δ or τ has changed or both. The 2D-IR vibrational echo experiments only determine the product, $\Delta^2\tau$.

Although it is not possible to determine Δ and τ from experiment, it is useful to examine a possible scenario using the results of molecular dynamics simulations. MD simulations have determined the FFCF for myoglobin-CO (MbCO).[89] In the MD simulations, Δ and τ for the motionally narrowed component were determined. From MD simulations of MbCO we know that the homogeneous component is caused by the very fast localized fluctuations of very small groups of atoms. These motions can be viewed as very low frequency vibrations, such as torsions of side chains. The MbCO MD simulation gave the motionally narrowed $\tau = 0.14$ ps. Note that k_BT corresponds to 0.16 ps, where k_B is Boltzmann's constant and T is the absolute temperature. Thus, these very fast thermal fluctuations of small local modes have $\tau \approx h/k_BT$. It is reasonable to assume that τ is approximately the same for Mb and cyt c. Both Mb and cyt c are small globular heme proteins, and their very local motions might be expected to be similar.

Because the motionally narrowed component is due to very local motions of small groups of atoms, for heuristic purposes we will assume that τ does not change when the protein dentures. This is an assumption that will permit the discussion of a plausible explanation for the reduction in Γ when the protein unfolds. With the assumption that $\tau = 0.14$ ps, we can obtain

Native and Unfolded Cytochrome c

J. Phys. Chem. B, Vol. 112, No. 32, 2008 **10061**

Δ for the native and denatured states. The values are 42 and 31 cm^{-1}, respectively. This decrease in Δ could be consistent with a hydrophobically collapsed state (see below) for the denatured protein in which the packing is more dense. The denser packing of the collapsed state could limit the amplitude of the local motions, thereby reducing Δ. Of course this argument is not rigorous, but it is indicative of a possible explanation for the large decrease in Γ with denaturation.

A hydrophobically collapsed denatured state is consistent with the funnel model of protein folding.[7,8,11,12] As discussed in the Introduction, the opposite view of protein folding from the funnel model is folding or unfolding through a well-defined set of intermediates.[2–5,10] In the sequential unfolding model, cyt c unfolds with as many as five intermediates through a sequence of unravelings of particular structural motifs.[10,35] The denatured protein is unbundled with all regions exposed to solvent. Could this model be consistent with the IR experiments reported above? It is reasonable to assume that if the protein denatures in this manner, then there would be a wide variety of conformations that could result in the observed increase in the line width. The change in structure could result in an overall shift in the average frequency, although a red shift is generally associated with an increase electric field at the CO or increased back bonding from the iron−heme to the CO π^* antibonding molecular orbital.[58,76,83,90,91] In MbCO, when the distal His moves away from the CO, the result is a blue shift.

The change in the 2D-IR results may not be consistent with a protein that is denatured through a sequence of unraveling intermediates. The denatured system should be more extended and floppy than the native protein. It is unlikely that such a system would have slower dynamics on the 10 ps time scale than the native protein. In a recent experiment, it was shown that cleaving a single disulfide bond in a heme protein led to a somewhat faster decay of the FFCF.[53] If large dissociative structural changes occur, then it would seem unlikely that the dynamics would slow. Furthermore, such a denatured state could have a wide range of configurations, but they might be expected to be sampled readily, which is inconsistent with the very large static component, Δ_s, of the FFCF found for the denatured state. Finally, consider the heuristic argument given above for the homogeneous component, Γ, and again assume that τ is unchanged. If this is the case, a less compact and less dense structure would yield a larger Δ and therefore a larger Γ, in contrast to the experimentally observed decrease in Γ for the denatured protein.

The arguments presented above, while qualitative, indicate that the state of the denatured protein is more consistent with a compact collapsed state suggested by funnel models than an unraveled expanded state suggested by the hierarchical unfolding models. The *Ht*-M61A results show that the denatured protein dynamics within the experimental window sample only a small portion of the total structural space. The implication is that motions over distances from a few Ångstroms to less than or equal to $\sim$20 Å[84] are slower compared to the native protein. However, the dynamics are not slowed to the extent that occurs when the protein's surface topology is completely locked up by a glassy solvent. The results shown in Figure 6 do not seem to be consistent with a picture in which the denatured protein has lost its native structure by unraveling, significantly expanding in size[92] and becoming floppy, i.e., taking on a "random coil" form. The initial experiments presented in Figure 6 may indicate that the unfolded state of *Ht*-M61A is a compact, hydrophobicity driven collapsed state.

In the context of the funnel model, theoretical studies[7,8] have suggested the existence of a collapsed and glassy unfolded state in close proximity of the native state, particularly with respect to the size of the protein. In the case of proteins in the cyt c family, theory[6] and experiments[9] indicate that the initial stage of folding is formation of a hydrophobicity induced collapsed state with most of the native contacts not formed. These contacts form subsequently along the folding pathway. For unfolding the predicted scenario is reversed. It has been suggested that for cyt c the hydrophobic heme pocket plays an important role in stabilization of the hydrophobic collapsed state.[6]

The theoretical study shows that because the heme-induced hydrophobic collapse occurs before a large fraction of native contacts are formed, the collapsed state can accommodate many non-native, possibly hydrophobic contacts.[6] The hydrophobic collapsed state is predicted to be an important intermediate state. A similar conclusion was reached earlier by Takahashi and co-workers, who used small-angle X-ray scattering and resonance Raman to study folding of cyt c.[9,66] It was found that along the folding pathway, the size of the protein decreases first without much change in the helical content. Subsequent to the collapse, the secondary structure is acquired.

The large static component of the FFCF for time scales longer than a few hundred picoseconds for *Ht*-M61A was observed for strongly denaturing conditions. Traditionally, denatured proteins were considered to exist in an extended conformation. A number of studies, however, have revealed residual structure in denatured proteins (for review, see ref 70). Gray, Winkler, and co-workers characterized distributions of dansyl fluorophore−heme distances in *Saccharomyces cerevisiae* iso-1-cyt c, a Class I cyt c, through a range of denaturing conditions and during folding. Their data reveal that denatured cyt c at equilibrium exists in a range of compact and extended conformations. In addition, both compact and extended conformations are observed in kinetic folding studies. These data indicate that even under strongly denaturing conditions a significant portion of the ensemble of cyt c structures exists in a collapsed state.[32,69,70] It has been conjectured that the collapsed state can be glassy with slow collective dynamics.[8] The large static component observed by the 2D-IR vibrational echoes when the denatured state is formed demonstrates that there is only a small amount of structural dynamics in the experimental time window (intermediate distance scale). The reduction in dynamics may be a signature of the hydrophobicity induced collapsed state.

V. Concluding Remarks

Here we examined the dynamics of denatured cyt c mutant *Ht*-M61A in comparison to the protein in its native form using ultrafast 2D-IR vibrational echo spectroscopy and linear IR absorption experiments. The similarity in trend between the CO peak shift curve and the CD curve in cyt c demonstrates that the vibrational stretching frequency of CO bound to cyt c is sensitive to the same types of changes in structure as a CD experiment.

The vibrational absorption lines are inhomogeneously broadened. The inhomogeneous broadening means that the ensemble of proteins has a heterogeneous distribution of structures. The 2D-IR vibrational echo experiment measures spectral diffusion, which occurs because of the interconversion of one protein structure to another. In terms of an energy landscape picture, under a particular set of equilibrium conditions, the protein occupies a broad rough minimum on the energy landscape. (There may be more than a single minimum that gives rise to structural substates[93,94] as discussed for the native protein in connection with Figure 5.) The roughness is caused by a collection of local minima between barriers of a range of heights. Some of the barriers are low enough that transitions are constantly taking place between the minima that correspond to variations in the protein structure within the experimental time window. Spectral diffusion is caused by transitions among the local minima. The time dependence of the spectral diffusion yields the time evolution of the protein structure.

In the time window of the 2D-IR vibrational echo experiments ($\sim$100 fs to $\sim$100 ps), native *Ht*-M61A displays significant structural dynamics. Somewhat greater than 50% of the accessible structures that give rise to the inhomogeneous CO absorption line are sampled. When the protein is denatured in 5.1 M GuHCl, the inhomogeneous line is substantially broader than that for the native protein, which indicates a broader range of structures. The dynamics of the denatured protein in the experimental time window sample only a small fraction, less than 20%, of the increased range of structures. Therefore, a large fraction of the structural dynamics is occurring on time scales much greater than 100 ps.

In terms of the energy landscape picture, the results on the denatured protein in 5.1 M GuHCl indicate that the free energy landscape has become rather rugged or rough due to the appearance of deep minima, accompanied by barriers, in the region of the free energy landscape that contains the denatured state. Glassy behavior is a consequence of this ruggedness. Note that if ϵ is the variance that characterizes the ruggedness of the landscape, then configuration diffusion in the landscape varies as $\exp(-(\epsilon/k_\mathrm{B}T)^2)$.[95] Thus, ruggedness can slow down configurational fluctuations drastically, giving rise to the much slower spectral diffusion in the denatured protein. The observed vibrational dynamics could be consistent with a hydrophobicity driven collapsed glassy state with such a rugged free energy landscape. Because glassy dynamics are predicted by landscape theory as being due to the appearance of ruggedness, it may be possible to estimate the ruggedness with inherent structure analysis[96] and explore the possible role of salt concentration in augmenting the ruggedness.

Clearly it will be important to fill in the 2D-IR vibrational echo data for concentrations of GuHCl between 3.2 and 5.1 M and concentrations > 5.1 M. The only limitations on obtaining additional data is that the experiments are complex and the mutant protein must be prepared in substantial quantity. While *Ht*-M61A under denaturing conditions may be in a hydrophobic collapsed state, there may be well defined non-native structures between those of the native state and the fully denatured state.[97] This is already suggested by the data shown in Figure 6 for the mildly denatured protein. These could be manifested as steps in the changes in dynamics as a function of GuHCl concentration. In addition, it will be important and interesting to perform the 2D-IR vibrational echo experiments on *Ht*-M61A using urea as a denaturant. Urea is believed to have a different mechanism for denaturation,[13–17,98] and it is possible that the denatured state will be distinct and display different dynamics from the protein denatured with GuHCl.

The results presented in this paper are an important target for simulations. The time scales of the experiments, $\sim$100 ps, are readily amenable to simulation. The relationship between simulations and the calculation of the 2D-IR vibrational echo spectra has been developed and applied to MbCO.[58,62,83,99,100] Comparison of the simulation results for the native and denatured protein would provide a bench mark for simulations that are employed to understand protein folding and folding intermediates. If the simulations can reproduce the observed behavior, then they should be able to provide a molecular level view of the structures responsible for the change in the dynamics observed for the denatured state.

Acknowledgment. We thank Professor Abani K. Bhuyan, School of Chemistry, University of Hyderabad, India, for providing the CD data shown in Figure 3A. This work was supported by the National Institutes of Health (2 R01 GM-061137-05), the Air Force Office of Scientific Research (F49620-01-1-0018), and the National Science Foundation (DMR 0652232). S.K. was supported by the Korea Research Foundation Grant funded by the Korean Government (MOE-HRD, basic Research Promotion Fund) (KRF-2006-C00038). K.L.B. acknowledges support from the National Institutes of Health (R01- GM63170).

References and Notes

(1) Daggett, V.; Fersht, A. *Nat. Rev. Mol. Cell Biol.* **2003**, *4*, 497.

(2) Daggett, V.; Fersht, A. R. *Trends Biochem. Sci.* **2003**, *28*, 18.

(3) Laurents, D. V. a. B., R. L. *Biophys. J.* **1998**, *75*, 428.

(4) Dill, K. A.; Shortle, D. *Annu. Rev. Biochem.* **1991**, *60*, 795.

(5) Wallace, L. A.; Matthews, C. R. *BioPhys. Chem.* **2002**, *101–102*, 113.

(6) Weinkam, P.; Zong, C. H.; Wolynes, P. G. *Proc. Natl. Acad. Sci. U.S.A.* **2005**, *102*, 12401.

(7) Sasai, M.; Wolynes, P. G. *Phys. Rev. Lett.* **1990**, *65*, 2740.

(8) Chahine, J.; Nymeyer, H.; Leite, V. B. P.; Socci, N. D.; Onuchic, J. N. *Phys. Rev. Lett.* **2002**, *88*, 168101.

(9) Akiyama, S.; Takahashi, S.; Kimura, T.; Ishimori, K.; Morishima, I.; Nishikawa, Y.; Fujisawa, T. *Proc. Natl. Acad. Sci. U.S.A.* **2002**, *99*, 1329.

(10) Hoang, L.; Bedard, S.; Krishna, M. M. G.; Lin, Y.; Englander, S. W. *Proc. Natl. Acad. Sci. U.S.A.* **2002**, *99*, 12173.

(11) Onuchic, J. N.; LutheySchulten, Z.; Wolynes, P. G. *Annu. Rev. Phys. Chem.* **1997**, *48*, 545.

(12) Bryngelson, J. D.; Onuchic, J. N.; Socci, N. D.; Wolynes, P. G. *Proteins-Struct. Funct. Genet.* **1995**, *21*, 167.

(13) Elove, G. A.; Bhuyan, A. K.; Roder, H. *Biochemistry* **1994**, *33*, 6925.

(14) Fedurco, M.; Augustynski, J.; Indiani, C.; Smulevich, G.; Antalik, M.; Bano, M.; Sedlak, E.; Glascock, M. C.; Dawson, J. H. *Biochim. Biophys. Acta, Proteins Proteomics* **2004**, *1703*, 31.

(15) Gupta, R.; Yadav, S.; Ahmad, F. *Biochemistry* **1996**, *35*, 11925.

(16) Latypov R. F.; Cheng, H.; Roder, N. A.; Zhang, J.; Roder, H. *J. Mol. Biol.* **2006**, *357*, 1009.

(17) Russell, B. S.; Melenkivitz, R.; Bren, K. L. *Proc. Natl. Acad. Sci. U.S.A.* **2000**, *97*, 8312.

(18) Smith, J. S.; Scholtz, J. M. *Biochemistry* **1996**, *35*, 7292.

(19) Monera, O. D.; Cyril, M. K.; Hodges, R. S. *Protein Sci.* **1994**, *3*, 1984.

(20) Massari, A. M.; McClain, B. L.; Finkelstein, I. J.; Lee, A. P.; Reynolds, H. L.; Bren, K. L.; Fayer, M. D. *J. Phys. Chem. B* **2006**, *110*, 18803.

(21) Bren, K. L.; Kellogg, J. A.; Kaur, R.; Wen, X. *Inorg. Chem.* **2004**, *43*, 7934.

(22) Jones, C. M.; Henry, E. R.; Hu, Y.; Chan, C. K.; Luck, S. D.; Bhuyan, A.; Roder, H.; Hofrichter, J.; Eaton, W. A. *Proc. Natl. Acad. Sci. U.S.A.* **1993**, *90*, 11860.

(23) Sinibaldi, F.; Howes, B. D.; Piro, M. C.; Caroppi, P.; Mei, G.; Ascoli, F.; Smulevich, G.; Santucci, R. *J. Biol. Inorg. Chem.* **2006**, *11*, 52.

(24) Kumar, R.; Prabhu, N. P.; Bhuyan, A. K. *Biochemistry* **2005**, *44*, 9359.

(25) Droghetti, E.; Smulevich, G. *J. Biol. Inorg. Chem.* **2005**, *10*, 696.

(26) Travaglini-Allocatelli, C.; Gianni, S.; Brunori, M. *Trends Biochem. Sci.* **2004**, *29*, 535.

(27) Rischel, C.; Jorgensen, L. E.; Foldes-Papp, Z. *J. Phys. Condens. Matter* **2003**, *15*, 1725.

(28) Hagen, S. J.; Latypov, R. F.; Dolgikh, D. A.; Roder, H. *Biochemistry* **2002**, *41*, 1372.

(29) Bhuyan, A. K.; Udgaonkar, J. B. *J. Mol. Biol.* **2001**, *312*, 1135.

(30) Panda, M.; Benavides-Garcia, M. G.; Pierce, M. M.; Nall, B. T. *Protein Sci.* **2000**, *9*, 536.

(31) Jimenez, R.; Romesberg, F. E. *J. Phys. Chem. B* **2002**, *106*, 9172.

(32) Lyubovitsky, J. G.; Gray, H. B.; Winkler, J. R. *J. Am. Chem. Soc.* **2002**, *124*, 5481.

(33) Winkler, J. R. *Curr. Opin. Chem. Biol.* **2004**, *8*, 169.

(34) Scott, R. A.; Mauk, A. G. *Cytochrome C: A Multidisciplinary Approach*; University Science Books: Sausalito, CA, 1996.

(35) Krishna, M. M. G.; Lin, Y.; Rumbley, J. N.; Englander, S. W. *J. Mol. Biol.* **2003**, *331*, 29.

(36) Ptitsyn, O. B. *J. Mol. Biol.* **1998**, *278*, 655.

(37) Varhac, R.; Antalik, M.; Bano, M. *J. Biol. Inorg. Chem.* **2004**, *9*, 12.

(38) Russell, B. S.; Bren, K. L. *J. Biol. Inorg. Chem.* **2002**, *7*, 909.

(39) Roccatano, D.; Daidone, I.; Ceruso, M. A.; Bossa, C.; Di Nola, A *Biophys. J.* **2003**, *84*, 1876.

(40) Droghetti, E.; Oellerich, S.; Hildebrandt, P.; Smulevich, G. *Biophys. J.* **2006**, *91*, 3022.

(41) Dobson, C. M.; Sali, A.; Karplus, M. *Angew. Chem., Int. Ed.* **1998**, *37*, 868.

(42) Anderson, J. L. R.; Chapman, S. K. *Dalton Trans.* **2005**, *13*, 9999.

(43) Bren, K. L.; Gray, H. B. *J. Am. Chem. Soc.* **1993**, *115*, 10382.

(44) Ambler, R. P. *Biochim. Biophys. Acta* **1991**, *1058*, 42.

(45) Zheng, J.; Kwak, K.; Fayer, M. D. *Acc. Chem. Res.* **2007**, *40*, 75.

(46) Finkelstein, I. J.; Zheng, J.; Ishikawa, H.; Kim, S.; Kwak, K.; Fayer, M. D. *Phys. Chem. Chem. Phys.* **2007**, *9*, 1533.

(47) Park, S.; Kwak, K.; Fayer, M. D. *Laser Phys. Lett.* **2007**, *4*, 704.

(48) Finkelstein, I. J.; Ishikawa, H.; Kim, S.; Massari, A. M.; Fayer, M. D. *Proc. Natl. Acad. Sci. U.S.A.* **2007**, *104*, 2637.

(49) Massari, A. M.; Finkelstein, I. J.; Fayer, M. D. *J. Am. Chem. Soc.* **2006**, *128*, 3990.

(50) Chung, H. S.; Khalil, M.; Smith, A. W.; Ganim, Z.; Tokmakoff, A. *Proc. Natl. Acad. Sci. U.S.A.* **2005**, *102*, 612.

(51) Mukherjee, P.; Kass, I.; Arkin, I. T.; Zanni, M. T. *Proc. Natl. Acad. Sci. U.S.A.* **2006**, *103*, 3528.

(52) Ishikawa, H.; Finkelstein, I. J.; Kim, S.; Kwak, K.; Chung, J. K.; Wakasugi, K.; Massari, A. M.; Fayer, M. D. *Proc. Natl. Acad. Sci. U.S.A.* **2007**, *104*, 16116.

(53) Ishikawa, H.; Kim, S.; Kwak, K.; Wakasugi, K.; Fayer, M. D. *Proc. Natl. Acad. Sci. U.S.A.* **2007**, *104*, 19309.

(54) Rella, C. W.; Rector, K. D.; Kwok, A. S.; Hill, J. R.; Schwettman, H. A.; Dlott, D. D.; Fayer, M. D. *J. Phys. Chem.* **1996**, *100*, 15620.

(55) Fayer, M. D. *Annu. Rev. Phys. Chem.* **2001**, *52*, 315.

(56) Wang, J.; Chen, J.; Hochstrasser, R. M. *J. Phys. Chem. B* **2006**, *110*, 7545.

(57) Bredenbeck, J.; Helbing, J.; Kumita, J. R.; Woolley, G. A.; Hamm, P. *Proc. Natl. Acad. Sci. U.S.A.* **2005**, *102*, 2379.

(58) Merchant, K. A.; Noid, W. G.; Akiyama, R.; Finkelstein, I.; Goun, A.; McClain, B. L.; Loring, R. F.; Fayer, M. D. *J. Am. Chem. Soc.* **2003**, *125*, 13804.

(59) Karan, E. F.; Russell, B. S.; Bren, K. L. *J. Biol. Inorg. Chem.* **2002**, *7*, 260.

(60) Fee, J. A.; Chen, Y.; Todaro, T. R.; Bren, K. L.; Patel, K. M.; Hill, M. G.; Gomez-Moran, E.; Loehr, T. M.; Ai, J.; Thöny-Meyer, L.; Williams, P. A.; Stura, E.; Sridhar, V.; McRee, D. E. *Protein Sci.* **2000**, *9*, 2074.

(61) Wen, X.; Bren, K. L. *Biochemistry* **2005**, *44*, 5225.

(62) Massari, A. M.; Finkelstein, I. J.; McClain, B. L.; Goj, A.; Wen, X.; Bren, K. L.; Loring, R. F.; Fayer, M. D. *J. Am. Chem. Soc.* **2005**, *127*, 14279.

(63) Asbury, J. B.; Steinel, T.; Fayer, M. D. *J. Lumin.* **2004**, *107*, 271.

(64) Pace, C. N.; Scholtz, J. M. Measuring the Conformational Stability of a Protein. In *Protein Structure: A Practical Approach*; Creighton, T. E., Ed.; Oxford University Press: New York, 1997; pp 299.

(65) Karan, E. F.; Russell, B. S.; Bren, K. L. *J. Biol. Inorg. Chem.* **2002**, *7*, 260.

(66) Takahashi, S.; Yeh, S. R.; Das, T. K.; Chi-Kin, C.; Gottfried, D. S.; Rousseau, D. L. *Nat. Struct. Biol.* **1997**, *4*, 44.

(67) Elove, G. A.; Chaffote, A. F.; Roder, H.; Goldberg, M. E. *Biochemistry* **1992**, *31*, 6876.

(68) Sosnick, T. R.; Mayne, L.; Hiller, R.; Englander, S. W. *Nat. Struct. Biol.* **1994**, *1*, 149.

(69) Pletneva, E. V.; Gray, H. B.; Winkler, J. R. *J. Mol. Biol.* **2005**, *345*, 855.

(70) Pletneva, E. V.; Gray, H. B.; Winkler, J. R *Proc. Natl. Acad. Sci. U.S.A.* **2005**, *102*, 18397.

(71) Rector, K. D.; Kwok, A. S.; Ferrante, C.; Tokmakoff, A.; Rella, C. W.; Fayer, M. D. *J. Chem. Phys.* **1997**, *106*, 10027.

(72) Golonzka, O.; Khalil, M.; Demirdoven, N.; Tokmakoff, A. *Phys. Rev. Lett.* **2001**, *86*, 2154.

(73) Bai, Y. S.; Fayer, M. D. *Phys. Rev. B.* **1989**, *39*, 11066.

(74) Dlott, D. D.; Fayer, M. D.; Hill, J. R.; Rella, C. W.; Suslick, K. S.; Ziegler, C. J. *J. Am. Chem. Soc.* **1996**, *118*, 7853.

(75) Hill, J. R.; Dlott, D. D.; Rella, C. W.; Peterson, K. A.; Decatur, S. M.; Boxer, S. G.; Fayer, M. D. *J. Phys. Chem.* **1996**, *100*, 12100.

(76) Hill, J. R.; Rosenblatt, M. M.; Ziegler, C. J.; Suslick, K. S.; Dlott, D. D.; Rella, C. W.; Fayer, M. D. *J. Phys. Chem.* **1996**, *100*, 18023.

(77) Mukamel, S. *Principles of Nonlinear Optical Spectroscopy*; Oxford University Press: New York, 1995.

(78) Mukamel, S. *Annu. Rev. Phys. Chem.* **2000**, *51*, 691.

(79) Asbury, J. B.; Steinel, T.; Kwak, K.; Corcelli, S. A.; Lawrence, C. P.; Skinner, J. L.; Fayer, M. D. *J. Chem. Phys.* **2004**, *121*, 12431.

(80) Kwak, K.; Park, S.; Finkelstein, I. J.; Fayer, M. D. *J. Chem. Phys.* **2007**, *127*, 124503.

(81) Kwak, K.; Rosenthal, D. E.; Fayer, M. D. *J. Chem. Phys.* **2008**, accepted.

(82) Woutersen, S.; Pfister, R.; Hamm, P.; Mu, Y.; Kosov, D. S.; Stock, G. *J. Chem. Phys.* **2002**, *117*, 6833.

(83) Merchant, K. A.; Noid, W. G.; Thompson, D. E.; Akiyama, R.; Loring, R. F.; Fayer, M. D. *J. Phys. Chem. B* **2003**, *107*, 4.

(84) Bu, Z.; Neumann, D. A.; Lee, S. H.; Brown, C. M.; Engelman, D. M.; Han, C. C. *J. Mol. Biol.* **2000**, *301*, 525.

(85) Finkelstein, I. J.; Massari, A. M.; Fayer, M. D. *Biophys. J.* **2007**, *92*, 3652.

(86) Kawahara, K.; Tanford, C. *J. Biol. Chem.* **1966**, *241*, 3228.

(87) Rector, K. D.; Jiang, J.; Berg, M.; Fayer, M. D. *J. Phys. Chem. B* **2001**, *105*, 1081.

(88) Rector, K. D.; Engholm, J. R.; Rella, C. W.; Hill, J. R.; Dlott, D. D.; Fayer, M. D. *J. Phys. Chem. A* **1999**, *103*, 2381.

(89) Merchant, K. A.; Thompson, D. E.; Xu, Q.-H.; Williams, R. B.; Loring, R. F.; Fayer, M. D. *Biophys. J.* **2002**, *82*, 3277.

(90) Augspurger, J. D.; Dykstra, C. E.; Oldfield, E. *J. Am. Chem. Soc.* **1991**, *113*, 2447.

(91) Hill, J. R.; Dlott, D. D.; Fayer, M. D.; Rella, C. W.; Rosenblatt, M. M.; Suslick, K. S.; Ziegler, C. J. *J. Phys. Chem.* **1996**, *100*, 218.

(92) Tsong, T. Y. *J. Biol. Chem.* **1974**, *249*, 1988.

(93) Frauenfelder, H.; Parak, F.; Young, R. D. *Annu. Rev. Biophys. Biophys. Chem.* **1988**, *17*, 471.

(94) Frauenfelder, H.; Sligar, S. G.; Wolynes, P. G. *Science* **1991**, *254*, 1598.

(95) Zwanzig, R. *Proc. Natl. Acad. Sci. U.S.A.* **1988**, *85*, 2029.

(96) Debenedetti, P. G.; Stillinger, F. H. *Nature* **2001**, *410*, 259.

(97) Russell, B. S.; Melenkivitz, R.; Bren, K. L. *Proc. Natl. Acad. Sci. U.S.A.* **2000**, *97*, 8312.

(98) Gianni, S.; Brunori, M.; Travaglini-Allocatelli, C. *Protein Sci.* **2001**, *10*, 1685.

(99) Williams, R. B.; Loring, R. F.; Fayer, M. D. *J. Phys. Chem. B* **2001**, *105*, 4068.

(100) Finkelstein, I. J.; Goj, A.; McClain, B. L.; Massari, A. M.; Merchant, K. A.; Loring, R. F.; Fayer, M. D. *J. Phys. Chem. B* **2005**, *109*, 16959.

JP802246H

THE JOURNAL OF CHEMICAL PHYSICS **127**, 124503 (2007)

Frequency-frequency correlation functions and apodization in two-dimensional infrared vibrational echo spectroscopy: A new approach

Kyungwon Kwak, Sungnam Park, Ilya J. Finkelstein, and M. D. Fayer
Department of Chemistry, Stanford University, Stanford, California 94305, USA

(Received 6 June 2007; accepted 23 July 2007; published online 26 September 2007)

Ultrafast two-dimensional infrared (2D-IR) vibrational echo spectroscopy can probe structural dynamics under thermal equilibrium conditions on time scales ranging from femtoseconds to $\sim$100 ps and longer. One of the important uses of 2D-IR spectroscopy is to monitor the dynamical evolution of a molecular system by reporting the time dependent frequency fluctuations of an ensemble of vibrational probes. The vibrational frequency-frequency correlation function (FFCF) is the connection between the experimental observables and the microscopic molecular dynamics and is thus the central object of interest in studying dynamics with 2D-IR vibrational echo spectroscopy. A new observable is presented that greatly simplifies the extraction of the FFCF from experimental data. The observable is the inverse of the center line slope (CLS) of the 2D spectrum. The CLS is the inverse of the slope of the line that connects the maxima of the peaks of a series of cuts through the 2D spectrum that are parallel to the frequency axis associated with the first electric field-matter interaction. The CLS varies from a maximum of 1 to 0 as spectral diffusion proceeds. It is shown analytically to second order in time that the CLS is the T_w (time between pulses 2 and 3) dependent part of the FFCF. The procedure to extract the FFCF from the CLS is described, and it is shown that the T_w independent homogeneous contribution to the FFCF can also be recovered to yield the full FFCF. The method is demonstrated by extracting FFCFs from families of calculated 2D-IR spectra and the linear absorption spectra produced from known FFCFs. Sources and magnitudes of errors in the procedure are quantified, and it is shown that in most circumstances, they are negligible. It is also demonstrated that the CLS is essentially unaffected by Fourier filtering methods (apodization), which can significantly increase the efficiency of data acquisition and spectral resolution, when the apodization is applied along the axis used for obtaining the CLS and is symmetrical about $\tau=0$. The CLS is also unchanged by finite pulse durations that broaden 2D spectra. © *2007 American Institute of Physics.* [DOI: 10.1063/1.2772269]

I. INTRODUCTION

Ultrafast two-dimensional infrared (2D-IR) vibrational echo experiments probe fast dynamics in condensed matter systems with exceptional detail. They have recently been applied to study the hydrogen bond network of water,[1–3] the equilibrium dynamics of aqueous and membrane bound proteins,[4–6] ultrafast exchange and isomerization dynamics,[7–10] and bath mediated solute structure fluctuations.[11,12] 2D-IR vibrational echo spectra are acquired by heterodyne detection of the stimulated vibrational echo wave packet. They report the time dependent frequency evolution of an ensemble of chromophores as the molecule-bath system undergoes equilibrium structural fluctuations. In a 2D-IR vibrational echo experiment, three ultrafast mid-IR pulses with experimentally controlled delay times generate and manipulate a coherent superposition of the probe's ground and first two excited vibrational states. The time between pulses 1 and 2 is τ (the first coherence period), and the time between pulses 2 and 3 is T_w (the population period). The vibrational echo pulse is generated after pulse 3 at a time $\leqslant \tau$ (the second coherence period). 2D vibrational echo spectra are obtained by scanning τ at fixed T_w.

During the first coherence period, the molecules are frequency labeled. During the population period, the frequency-labeled molecules can evolve to different frequencies (spectral diffusion) because of microscopic molecular events. During the second coherence period, the final frequencies of the frequency-labeled molecules are read out. A 2D spectrum is obtained with the initial labeled frequencies as one axis and the final frequencies of the molecules as the other axis. A set of such 2D spectra is measured as a function of T_w. By analyzing the amplitude, position, and peak shapes of the 2D spectra, detailed information on structure and dynamics of the molecular system is determined. Spectral diffusion results in changes in peak shapes as a function of T_w.[1,13] Appearance of off-diagonal peaks results from incoherent and coherent population transfers by anharmonic interactions[14,15] or chemical exchange.[8] Off-diagonal peaks occurring at $T_w=0$ can arise from coupling of different vibrational modes.[16] Vibrational population relaxation and molecular reorientation lead to decay of the amplitudes of all peaks.[4,8,17]

A key link between experimental observables and the underlying molecular and intermolecular structural fluctuations is the frequency-frequency correlation function (FFCF), also known as the vibrational solvation correlation function.

124503-2   Kwak *et al.*   J. Chem. Phys. **127**, 124503 (2007)

Within conventional approximations,[18] the FFCF captures the frequency response of a vibrational mode to the bath dynamics, where the bath can be a solvent or, for systems such as proteins, the protein itself. In addition, the FFCF provides a key connection between 2D-IR vibrational echo experiments and molecular dynamics simulations.[13,19,20] However, the highly nonlinear relationship between the FFCF and spectroscopic observables significantly complicates the extraction of the FFCFs from 2D-IR spectra. To obtain the FFCF from experimental data, a trial FFCF is generally parameterized as a combination of decaying functions, and spectroscopic observables are calculated from a response function formalism that was developed by Mukamel and co-workers.[18,21–23] A nonlinear fitting routine is employed to vary the multiple FFCF parameters to obtain agreement between the calculated spectra and the experimental spectra. The numerical problem is greatly increased when finite pulse durations need to be included as a set of three time ordered integrals. The computational complexity and questionable convergence of multiparameter nonlinear fitting routines has spurred the development of simpler methods that try to obtain the FFCF directly from experimental data.[24,25]

Increasing interest in heterodyne detected 2D-IR vibrational echo spectroscopy has led to various approaches for obtaining the 2D-IR line shape equations analytically.[26–28] Among these, Kwac *et al.* included spectral diffusion effects in their line shape equation for the narrow band pump-broadband probe IR experiments. In addition to the standard cumulant expansion and Condon approximations, a short time approximation was assumed for the two coherence periods. Using this line shape function, the time dependent slopes of the nodal plane of 2D-IR spectra were proven to be proportional to the normalized FFCF.[28] More recently, Roberts *et al.* showed that, in 2D-IR vibrational echo experiments, the ellipticity of the band shape is also proportional to T_w dependent portion of the FFCF.[29] Both methods independently derived the same line shape function that is the product of two Gaussians whose widths change with increasing spectral diffusion. 2D spectra invariably have a motionally narrowed component that is T_w independent. Neither approached dealt with extraction of the motionally narrowed component.

The characteristic T_w dependence of an inhomogeneously broadened 2D-IR band caused by spectral diffusion is a change in shape from elongation along the diagonal axis at short T_w (waiting time) toward a symmetric band at long waiting time, as shown in Figs. 1(a) and 1(b). The ω_τ axis is the axis of the first radiation field-matter interaction, and the ω_m axis is the axis of the third interaction and vibrational echo emission. Besides fitting a set of 2D-IR spectra and listing the parameters that define the FFCF, the change in the 2D spectral band shape can be presented in other ways. For example, the change in the band shape can be described in terms of one-dimensional cuts through the data parallel to the ω_τ axis. Projection of this cut onto the ω_τ axis has a line shape with a width that is called the dynamic linewidth.[1,13] Another method is to obtain linewidth from cuts taken perpendicular to the diagonal (antidiagonal) and along the diag-

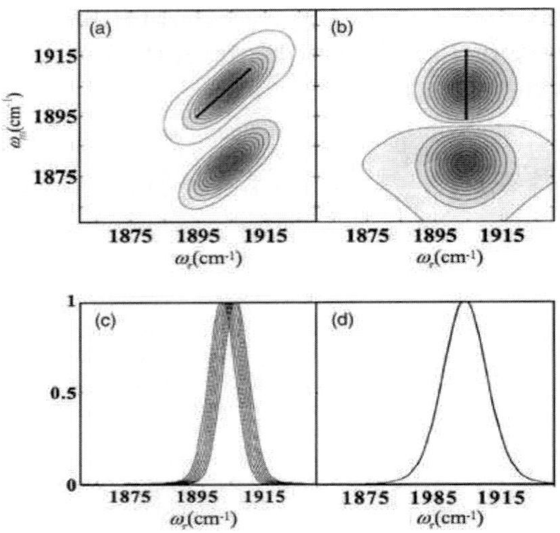

FIG. 1. (a) Calculated 2D-IR vibrational echo spectrum of HRP-CO red state at $T_w=0.2$ ps. The parameters used in the calculation are listed in the Table I as HRP red experiment. (b) Calculated HRP-CO 2D-IR spectrum at $T_w=60$ ps. The heavy lines in (a) and (b) are the center lines, which connect the peak position (maximum value of the projection onto the ω_τ axis) for cuts at each ω_m. (c) and (d) are the normalized projected spectra for (a) and (b) at several ω_m. (c) shows the distribution of peak positions at short T_w. In (d), all of the peak positions are identical corresponding to the vertical line in (b).

onal to form the closely related functions, the eccentricity[30] or ellipticity.[29] In practice, the determination of these linewidths may be difficult because linewidths are very sensitive to experimental noise and errors that may result from inadequate sampling or Fourier transform truncation artifacts.

Rather than attempting to quantify changes in peak widths, we propose a new method that reports on both spectral diffusion and the FFCF by tracking changes in the frequency dependent positions of the peak maxima of slices through the 2D-IR data. The observable is the inverse of the center line slope (CLS) of the 2D spectrum, which varies from a maximum of 1 to 0 as spectral diffusion proceeds. The CLS is the inverse of the slope of the line that connects the maxima of the peaks of a series of cuts through the 2D spectrum that are parallel to the ω_τ frequency axis. A key feature of the proposed method is that it eliminates the need for line shape analysis and the possible practical artifacts inherent therein.

The necessity of performing numerical Fourier transforms to obtain 2D-IR spectra imposes conditions on the data acquisition of the time-domain interferograms. To avoid frequency aliasing in 2D frequency space, a minimum sampling interval, the Nyquist interval, is required for the maximum frequency which is to be resolved.[31] Therefore, points are taken with a few femtosecond intervals, and the time to collect an interferogram can be relatively long. However, truncation of the interferogram is not an option if accurate line shapes are required because truncation artifacts can make

J. Chem. Phys. **127**, 124503 (2007)

peaks broader or add side lobes, which can interfere with neighboring peaks. To avoid truncation artifacts, it is necessary to take data until the interferogram has decayed to zero, necessitating long scans and good signal-to-noise ratios at long times when the signal has decayed almost to zero. In 2D-IR vibrational echo experiments with more than one peak, the scanning time is always determined by the narrowest peak.

In analogy to the wide array of data processing techniques developed for handling complex 2D-NMR data, acquisition of 2D-IR vibrational echo spectra can be enhanced by applying data processing techniques to the time-domain interferograms.[32] Apodization, or windowing, is one of the most important procedures and is routinely employed for slowly decaying interferograms and to spectrally resolve overlapping transitions in NMR.[33,34] Apodization usually involves multiplication of an interferogram by a simple windowing function before numerical Fourier transformation. The primary goal of this procedure is to improve the signal-to-noise ratio or spectral resolution in the resulting 2D spectrum.[35] Finite data acquisition time and the resulting truncation of the interferogram effectively define an apodization window that may lead to spectral artifacts and line broadening in the frequency domain. Numerical apodization performed with a decaying window function can smooth the abruptly truncated edge of the interferogram and reduce spectral leakage around the main peak.

Apodization can reduce data acquisition time and improve the overall signal-to-noise ratio by focusing the data acquisition effort on those portions of the interferogram where signal is still relatively strong. However, these substantial advantages are mitigated by the fact that a rapidly decaying window function will significantly alter the observed 2D-IR vibrational echo band shape. Numerical deconvolution of the true band shape from the effect of the windowing function is frequently difficult and is generally numerically unstable.[36] However, the CLS method does not depend on line shapes to obtain the time dependent spectral diffusion, but rather the inverse slope of the center line. The center line is determined by the peak maxima of cuts through the 2D band. Although the shape of the band is changed by apodization, it will be demonstrated below that the positions of the peak maxima and, therefore, the CLS is not affected by apodization provided that the apodization is performed along the frequency axis used to obtain the CLS and the apodization function is symmetrical about $\tau=0$. Furthermore, the CLS is not influenced by finite pulse durations.

2D-IR vibrational echo spectra need to be "phased" correctly to obtain an essential absorptive spectrum. Obtaining the absorption part of the 2D-IR spectrum using the dual-scan method[37] combined with proper phasing using the pump-probe projection theorem,[38] corresponds to the phase correction in NMR. It is demonstrated below that apodization does not affect the procedures used to obtain a properly phased absorptive 2D-IR spectrum.

II. THEORETICAL DEVELOPMENT

A. Response functions for 2D-IR spectra in the short time approximation

Here, we will derive the line shape function for the 2D-IR vibrational echo experiments using the short time approximation. A similar approach has already been employed by other groups in related contexts.[25,28,29] It is included here so that the derivation of the important results is complete and to include the effects of lifetime and orientational relaxation, which have not been treated previously. The linear and third order response functions using diagrammatic perturbation theory have been presented.[18] The linear IR absorption spectrum can be expressed as a Fourier transform of the linear response function, $R^1(t)$,

$$R^1(t) = |\mu_{0,1}|^2 e^{-i\langle\omega_{0,1}\rangle t}\exp[-g_1(t)]\exp(-t/3T_{\text{or}})$$
$$\times\exp(-t/2T_1),\tag{1}$$

where $\mu_{0,1}$ is the transition dipole for the ground vibrational state, 0, to the first vibrationally excited state, 1. $\langle\omega_{0,1}\rangle$ is the ensemble average 0–1 transition frequency, and the vibrational lifetime and orientational relaxation are included phenomenologically via T_1 and T_{or}, respectively. The line shape function $g_1(t)$ is

$$g_1(t) = \int_0^t d\tau_2 \int_0^{\tau_2} d\tau_1 \langle\delta\omega_{1,0}(t)\delta\omega_{1,0}(0)\rangle,\tag{2}$$

where $\langle\delta\omega_{1,0}(\tau_1)\delta\omega_{1,0}(0)\rangle$ is the frequency-frequency correlation function (FFCF) for the 0-1 transition frequency. An FFCF that is a sum of exponential terms has been used to describe a wide variety of experimental systems.[1,4,13,39,40] It has also been found that the vibrational systems that have been studied contain a motionally narrowed component in addition to dynamics that are not motionally narrowed.[1,4,13,40,41] Therefore, we will consider the form of the FFCF to contain a motionally narrowed term as well as a sum of exponential terms. Motional narrowing can be represented as delta function in the FFCF. Then the FFCF has the form

$$C_1(t) = \langle\delta\omega_{1,0}(\tau_1)\delta\omega_{1,0}(0)\rangle = \frac{\delta(t)}{T_2^*} + \sum_i \Delta_i^2 \exp(-t/\tau_i),\tag{3}$$

where T_2^* is the pure-dephasing time, which is homogeneous at all times. Δ_i is the frequency fluctuation amplitude and τ_i is the correlation time of the ith component. Because the contribution to line broadening from the finite vibrational lifetime and orientational relaxation are also purely homogeneous, these can be combined with the pure dephasing into a single homogeneous dephasing term. Then the FFCF is

$$C_1(t) = \langle\delta\omega_{1,0}(\tau_1)\delta\omega_{1,0}(0)\rangle = \frac{\delta(t)}{T_2} + \sum_i \Delta_i^2 \exp(-t/\tau_i),\tag{4}$$

where

J. Chem. Phys. **127**, 124503 (2007)

$$\frac{1}{T_2} = \frac{1}{T_2^*} + \frac{1}{2T_1} + \frac{1}{3T_{\text{or}}}. \tag{5}$$

This substitution significantly simplifies the subsequent treatment of the linear and third-order response functions while simultaneously including the effects of the finite lifetime and orientational relaxation in the overall treatment.

Within the form of the FFCF given in Eq. (4), the first-order response function is given by

$$R^1(t) = |\mu_{0,1}|^2 e^{-i\langle\omega_{0,1}\rangle t} \exp[-g_1(t)]. \tag{6}$$

The third-order response function for the quantum pathways responsible for the stimulated vibrational echo signal are given by

$$R_1^3(t_3,T_w,t_1) = R_2^3(t_3,T_w,t_1)$$
$$= |\mu_{0,1}|^4 e^{-i\langle\omega_{0,1}\rangle(-t_1+t_3)}$$
$$\times\exp[-g_1(t_1) + g_1(T_w) - g_1(t_3) - g_1$$
$$\times(t_1 + T_w) - g_1(T_w + t_3) + g_1(t_1 + T_w + t_3)]$$
$$\times \exp(-T_w/T_1)(1 + 0.8\exp(-T_w/T_{\text{or}})),$$

$$R_3^3(t_3,T_w,t_1) = -|\mu_{0,1}|^2|\mu_{1,2}|^2 e^{-i[\langle\omega_{0,1}\rangle(-t_1+t_3)-\Delta t_3]}$$
$$\times\exp[-g_1^*(t_1) + g_2(T_w) - g_3(t_3)$$
$$- g_2(t_1 + T_w) - g_2(T_w + t_3)$$
$$+ g_2(t_1 + T_w - t_3)]$$
$$\times\exp(-T_w/T_1)(1 + 0.8\exp(-T_w/T_{\text{or}})),$$

$$R_4^3(t_3,T_w,t_1) = R_5^3(t_3,T_w,t_1) \tag{7}$$
$$= |\mu_{0,1}|^4 e^{-i\langle\omega_{0,1}\rangle(t_1+t_3)} \exp[-g_1(t_1) - g_1(T_w)$$
$$- g_1(t_3) + g_1(t_1 + T_w) + g_1(T_w + t_3)$$
$$- g_1(t_1 + T_w + t_3)]$$
$$\times\exp(-T_w/T_1)(1 + 0.8\exp(-T_w/T_{\text{or}})),$$

$$R_6^3(t_3,T_w,t_1) = -|\mu_{0,1}|^2|\mu_{1,2}|^2 e^{-i[\langle\omega_{0,1}\rangle(t_1+t_3)-\Delta t_3]}$$
$$\times\exp[-g_1(t_1) - g_2^*(T_w) - g_3(t_3)$$
$$+ g_2^*(t_1 + T_w) + g_2^*(T_w + t_3) - g_2^*(t_1 + T_w + t_3)]$$
$$\times\exp(-T_w/T_1)(1 + 0.8\exp(-T_w/T_{\text{or}})).$$

In the above, $\mu_{1,2}$ is the transition dipole matrix elements for the 1-2 vibrational transitions and Δ is the vibrational anharmonicity. $g_2(t)$ represents cross correlation between the fundamental and excited transition frequency,

$$g_2(t) = \int_0^t d\tau_2 \int_0^{\tau_2} d\tau_1 \langle\delta\omega_{2,1}(t)\delta\omega_{1,0}(0)\rangle. \tag{8}$$

$g_3(t)$ is the autocorrelation of the excited transition frequency,

$$g_3(t) = \int_0^t d\tau_2 \int_0^{\tau_2} d\tau_1 \langle\delta\omega_{2,1}(t)\delta\omega_{2,1}(0)\rangle. \tag{9}$$

These two functions can be different from $g_1(t)$ and also from each other in a three level vibrational system. The quantum correction to the time-correlation function[42] is not considered here. Therefore, the FFCF is a real quantity and $g_i(t)$ are also real.

The first three response functions represent rephasing pathways (R) and the last three are nonrephasing pathways (NR). There are actually two more response functions (reverse echoes) that occur only when the time ordering is such that T_w is negative, that is, pulse 3 comes before pulses 1 and 2. In the dual-scan method used to obtain absorptive 2D-IR vibrational echo spectra,[37] this never occurs. Then the additional pathways can only contribute for T_w's that are approximately equal to or less than the pulse duration, and all three pulses overlap in time. Generally in this situation, the sample will produce a nonresonant contribution that arises from the electronic polarizability of all of the molecules in the sample, solutes and solvent. The nonresonant signal usually obscures or distorts the resonant single. For these reasons, the two reverse echo response functions are not included in the analysis.

An absorptive 2D-IR signal, S_{2D}, is obtained via the dual-scan method[43] according to

$$S_{2D}(\omega_\tau,\omega_m,T_w) \propto \text{Re}[\tilde{R}_R(\omega_\tau,\omega_m,T_w) + \tilde{R}_{NR}(\omega_\tau,\omega_m,T_w)], \tag{10}$$

where $\tilde{R}_R$ and $\tilde{R}_{NR}$ are defined as

$$\tilde{R}_R(\omega_\tau,\omega_m,T_w) = \int_0^\infty dt_1 \int_0^\infty dt_3 \exp(i\omega_m t_3 - i\omega_\tau t_1)$$
$$\times R_R(t_1,T_w,t_3),$$

$$\tilde{R}_{NR}(\omega_\tau,\omega_m,T_w) = \int_0^\infty dt_1 \int_0^\infty dt_3 \exp(i\omega_m t_3 + i\omega_\tau t_1) \tag{11}$$
$$\times R_{NR}(t_1,T_w,t_3).$$

From Eq. (7), it is evident that the lifetime and orientational relaxation terms cause the intensity of the various response functions to decay as T_w is increased. These decay terms can be effectively removed by normalizing the individual T_w dependent 2D-IR spectra. In addition, these terms are independent of the Fourier transformation along t_1 and t_3 [see Eq. (11)]. Thus, the effect of the lifetime and orientational relaxation terms can be dropped to further simplify the response functions, but it should be emphasized that this normalization does not affect the overall 2D-IR line shape. As shown in Eq. (5), the combination of a motionally narrowed term [see Eq. (4)] with the lifetime and orientational relaxation produces a single Lorentzian contribution to the line shape.

Usually, an analytical form of the frequency domain response functions cannot be obtained because the line shape functions $g_i(t)$ are a complicated set of nested integrals of exponential functions. Instead, numerical calculations are

124503-5   2D-IR vibrational echo spectroscopy

J. Chem. Phys. **127**, 124503 (2007)

used to obtain the frequency domain response functions. During the numerical calculations, the direct relation between the signal and FFCF is lost. Multiparameter nonlinear fitting methods are generally used to obtain the FFCF from frequency domain spectra.

Using a short time approximation for the two coherence periods,[28] the $g_i(t)$ can be expanded with a Taylor expansion to second order in time, and the line shape functions and third-order response functions become analytically tractable. For example, the first two third-order response functions become

$$R_1^3(t_3, T_w, t_1) = R_2^3(t_3, T_w, t_1)$$

$$= |\mu_{0,1}|^4 e^{-i\langle\omega_{0,1}\rangle(-t_1+t_3)} \exp\left[-\frac{C_1(0)}{2}t_1^2 - \frac{t_1}{T_2}\right.$$

$$\left. + C_1(T_w)t_1t_3 - \frac{C_1(0)}{2}t_3^2 - \frac{t_3}{T_2}\right], \quad (12)$$

where $C_1(t)$ is given in Eq. (4). However, analytical solutions for the frequency domain response still cannot be derived from equations of this form. Therefore, we temporarily take

$1/T_2=0$. This approximation and the property of Dirac delta function guarantee that $C(t)$ no longer has a motionally narrowed component. Below, we will introduce a procedure for recovering the motionally narrowed component from experimental data. The resulting response functions have been presented elsewhere,[25,28,29] so only final result will be summarized here.

For the pathways that involve only the 0 and 1 vibrational levels (equivalent to ground state bleaching and stimulated emission in a pump-probe experiment),

$$\widetilde{R}_{0\to1}^3(\omega_\tau, \omega_m, T_w)$$

$$= \frac{4\pi}{(C_1(0)^2 - C_1(T_w)^2)^{1/2}}$$

$$\times \exp\left(-\frac{C_1(0)\omega_m^2 - 2C_1(T_w)\omega_m\omega_\tau + C_1(0)\omega_\tau^2}{2\{C_1(0)^2 - C_1(T_w)^2\}}\right). \quad (13)$$

For the pathways that result in a 1-2 coherence following the third interaction (excited state absorption), three different FFCFs are involved so the equation becomes more complex.

$$\widetilde{R}_{1\to2}^3(\omega_\tau, \omega_m, T_w) = \frac{-2\pi\kappa^2}{(C_1(0)C_3(0) - C_2(T_w)^2)^{1/2}} \exp\left(-\frac{C_1(0)(\omega_m + \Delta)^2 - 2C_2(T_w)(\omega_m + \Delta)\omega_\tau + C_3(0)\omega_\tau^2}{2\{C_1(0)C_3(0) - C_2(T_w)^2\}}\right), \quad (14)$$

where κ is defined as μ_{01}/μ_{12} which is $\sqrt{2}$ under harmonic approximation. Also, to reduce the complexity of the equation, the average transition frequency $\langle\omega_{01}\rangle$ is taken as 0. In addition to $C_1(t)$ defined in Eq. (3), two other correlation functions, $C_2(t)$ and $C_3(t)$, are needed for the three level system. The former is the cross correlation function between the 0-1 and 1-2 transition frequencies. The latter is the auto-correlation of the 1-2 transition frequency. The $0\to1$ bands in the 2D-IR vibrational echo spectra only depend on $C_1(t)$. Hence, the FFCF of the fundamental frequency can be obtained by analyzing the $0\to1$ transition even though $C_2(t)$ and $C_3(t)$ may be different from $C_1(t)$.

Using the above form of the line shape function the analytical relationship between the FFCF and all 2D-IR experimental observables can be examined. Earlier studies have proposed the dynamic linewidth,[13] the eccentricity,[30] and the slope of nodal plane[41] as simplified experimental observables related to the FFCF. These observables conveniently summarize 2D-IR spectra in a reduced one-dimensional form and can increase the accuracy and efficiency of nonlinear fitting routines. However, these fitting routines still require response function calculations with a parameterized model FFCF treated as a multivariable fitting parameter. To avoid these difficulties, methods for extracting the FFCF directly from 2D-IR spectra have begun to emerge. Recently, using the short time approximation, the ellipticity of the 2D-IR line shape was analyzed.[29] Like the eccentricity, the ellipticity is

obtained from the widths of the diagonal and antidiagonal cuts through the 2D-IR band. It was shown that, within the short time approximation, the T_w dependent portion of the FFCF could be recovered directly from the experimental data. However, the method requires two linewidths at each T_w, which can be subject to errors associated with determining line shapes. Also, a method for obtaining the motionally narrowed contribution to the FFCF was not developed. The CLS method is in the same spirit as the ellipticity approach but, as discussed in Sec. I, has advantages of not requiring linewidths and not being susceptible to influences on the line shapes, such as apodization. In addition, a method for obtaining the motionally narrowed contribution to the FFCF using the linear spectrum and a relatively simple calculation has been developed.

B. The center line slope

The change in shape of the 2D-IR spectrum caused by spectral diffusion can be described in terms of the center line slope. Figures 1(a) and 1(b) show model calculations for $T_w=0.2$ ps and for $T_w=60$ ps, at which time spectral diffusion is almost complete. The calculations are based on the FFCF determined for the CO stretching mode of CO bound at the active site of the enzyme horseradish peroxidase (HRP).[30] The FFCF for HRP has the form given in Eq. (4), and will be used in detailed model calculations presented below. The heavy lines are the center lines. At a given ω_m, a

124503-6   Kwak *et al.*

J. Chem. Phys. **127**, 124503 (2007)

slice through the 2D spectrum parallel to the ω_τ axis when projected onto the ω_τ axis is a spectrum. The peak of this spectrum is one point on the center line. Taking many such slices and determining the peak for each produces a set of points. The line connecting the resulting points is the center line. At short T_w, the center line has a significant slope. As T_w increases, the 2D spectrum becomes more symmetrical. At sufficiently long time, when spectral diffusion has sampled all frequencies, the 2D band is symmetrical, and all cuts have the same peak frequency, which is the frequency of the peak of the linear IR absorption spectrum. The center line is vertical (infinite slope). Figures 1(c) and 1(d) show the spectra (normalized) projected onto to the ω_τ axis for several ω_m slices. At short T_w [200 fs, Fig. 1(c)], there is a range of peak positions, yielding a center line with a slope. At very long T_w [60 ps, Fig. 1(d)], all of the peak frequencies are identical, giving the vertical center line.

In the limit of complete spectral diffusion, the long time limit, the 2D spectrum is symmetrical and the center line is vertical. In the other limit, $T_w=0$, and in the absence of a homogeneous component, the 2D spectrum is a thin line along the diagonal. The center line would be at 45°. As discussed below, the FFCF is related to the inverse of the center line slope. The inverse of the CLS has a maximum value of 1 at $T_w=0$ and goes to 0 in the long time limit. The maximum value of 1 can only occur in the absence of a homogeneous component. As the size of the homogeneous contribution increases, the initial value of the inverse of the CLS decreases. (The inverse of the CLS will also be referred to as the CLS.) The change in the center line as a function of T_w is shown in Fig. 2(a). The line with the smallest slope is for $T_w=0.2$ ps, and the line with the largest slope is for $T_w=60$ ps.

The relationship between the CLS (inverse of the center line slope) and the FFCF can be derived using the approximate 2D-IR line shape functions given in Eqs. (13) and (14). Here, we will concentrate only on the 0-1 band in the 2D-IR spectrum. The same procedure can be applied to the band involving vibrational echo emission at the 1-2 transition frequency. First, to define the slope of the line connecting the peak positions, at least the peak maxima for two ω_m slices are needed. One point is selected as the center frequency of the 2D spectrum. This is the slice along ω_τ at the ω_m which corresponds to the peak frequency of the linear IR absorption spectrum. The center frequency slice spectrum can be expressed as

$$\widetilde{R}^3_{0\rightarrow1}(\omega_\tau,0,T_w) = \frac{4\pi}{(C_1(0)^2 - C_1(T_w)^2)^{1/2}}$$
$$\times \exp\left(-\frac{C_1(0)\omega_\tau^2}{2\{C_1(0)^2 - C_1(T_w)^2\}}\right). \quad (15)$$

Clearly, this slice spectrum has a maximum at $(\omega_\tau,\omega_m)=(0,0)$. The other cut at $\omega_m=\delta$ has the spectrum projected onto the ω_τ axis of

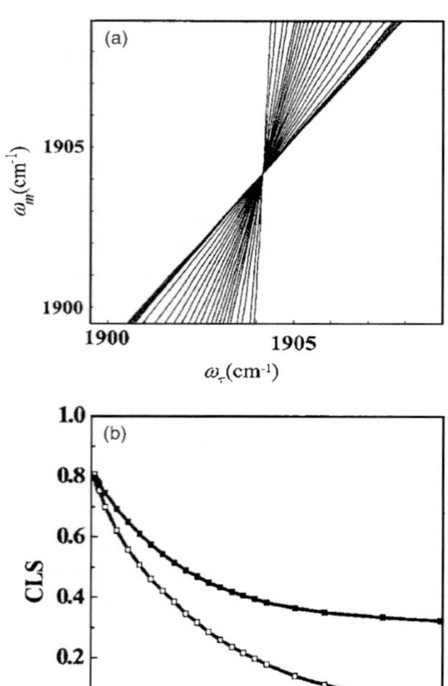

FIG. 2. (a) The progression of center lines showing the change in slope toward the vertical spectral diffusion proceeds for increasing values of T_w. (b) The CLS (inverse of the center line slopes) as a function of T_w for the HRP-CO blue state (upper curve, closed squares) and the red state (lower curve, open squares).

$$\widetilde{R}^3_{0\rightarrow1}(\omega_\tau,\delta,T_w)$$
$$= \frac{4\pi}{(C_1(0)^2 - C_1(T_w)^2)^{1/2}}$$
$$\times \exp\left(-\frac{C_1(0)\delta^2 - 2C_1(T_w)\delta\omega_\tau + C_1(0)\omega_\tau^2}{2\{C_1(0)^2 - C_1(T_w)^2\}}\right).$$
$$(16)$$

Using the first derivative of this spectrum, $\partial\widetilde{R}^3_{0\rightarrow1}(\omega_\tau^{Max},\delta,T_w)/\partial\omega_\tau=0$, the peak position of this slice is $(\omega_\tau,\omega_m)=(C_1(T_w)/C_1(0)\delta,\delta)=(C_1^N(T_w)\delta,\delta)$, where $C_1^N(T_w)=C_1(T_w)/C_1(0)$ is the normalized FFCF. Therefore, the slope of the line is

$$S(T_w) = \frac{1}{C_1^N(T_w)}. \quad (17)$$

Equation (17) is the important result. Within the short time approximation, the normalized FFCF, $C_1^N(T_w)$, is directly proportional to the inverse of the center line slope, which we refer to as the CLS. It should be emphasized that $C_1^N(T_w)$ does not include a motionally narrowed component. The change in slope reflects the T_w dependent spectral diffusion. The $1/T_2$ contribution to the FFCF and 2D line shape is T_w

TABLE I. HRP FFCF input parameters from Ref. 40, and parameters determined from the CLS method as discussed in the text.

		T_2 (ps)	Δ_2 (rad/ps)	τ_2 (ps)	Δ_3 (rad/ps)	τ_3 (ps)	T_1 (ps)
HRP red	Experiment	7.5	0.58	1.5	1.06	21	8
	CLS (norm)	NA	0.07[a]	1.6	0.75[a]	21	
	CLS and linewidth	3.9	0.32	1.6	1.05	21	
	CLS and line shape	7.3	0.58	1.6	1.05	21	
HRP blue	Experiment	5.8	0.60	15	0.45	∞	12
	CLS (norm)	NA	0.5[a]	15	0.31[a]	∞	
	CLS and linewidth	5.3	0.58	15	0.46	∞	
	CLS and line shape	5.7	0.6	15	0.46	∞	

[a]Normalized amplitude from normalized CLS, unitless, not rad/ps.

independent. In the next section, numerically calculated 2D-IR spectra using known FFCF input parameters are used to verify this relationship, access its accuracy, and demonstrate the procedure for recovering the T_w independent $1/T_2$ contribution to the FFCF.

III. TESTING THE CLS METHOD

To check the validity of the CLS method, numerical calculations of response functions were performed with known FFCFs that were obtained from experiments by fitting the experimental data with the full response function time dependent diagrammatic perturbation theory method.[21,22,44] First, the specific procedures to extract the FFCF will be shown using the FFCFs from 2D-IR vibrational echo measurements on the CO stretching mode of HRP.[30] HRP is an enzyme that can bind a variety of substrates. Without a bound substrate, HRP displays two CO peaks in the FT-IR spectrum because it exists in two conformational substates related to the configuration of the distal residues. The FFCFs for the two CO lines will be extracted using CLS from the 2D spectra. Second, the method for recovering the homogeneous contribution T_2 and the absolute amplitudes of inhomogeneous components will be demonstrated by also utilizing linear IR absorption spectra. The HRP-CO system was chosen because the FFCFs of the CO stretch of this protein contain the various components discussed in connection with Eq. (4), including a motionally narrowed component, a relatively slow spectral diffusion component, and a static component. The HRP-CO line shapes are almost Gaussian, but they are quite narrow with bandwidths of 10 and 15 cm^{-1}. A narrow peak gives rise to a slow decay time of the time-domain signal as τ is scanned, which makes this system a stringent test of the short time approximation. Also the 2D-IR experimentally obtained FFCFs from the deuterated hydroxyl stretching bands of phenol-OD in two solvents, pure CCl_4 and mesitylene, were used to test the CLS method for systems that have almost homogeneously broadened Lorentzian absorption bands.

The FFCFs for HRP-CO that we will try to duplicate with CLS were obtained by iterative fitting of the 2D-IR vibrational echo experiments with response function calcula-tions of the T_w dependent 2D-IR spectra and the linear line shapes.[30] The protein is so large that orientational relaxation can be neglected. The population relaxation times, T_1, were measured with IR pump-probe experiments.[30] The two CO absorption bands are referred to as the red state (lower absorption frequency, 1903.7 cm^{-1}) and the blue state (higher absorption frequency, 1932.7 cm^{-1}).[30] Both FFCFs have the form given in Eq. (4). The slow exponential component for the blue state is so slow that it appears as a constant on the accessible time scale of the experiments ($\sim 5T_1$). Therefore, for the blue state, the last term in Eq. (4) is just Δ_2^2. The parameters obtained from the experiments are given in Table I (labeled as experiment) and are used to calculate the 2D-IR spectra. Because the lines are so narrow compared to the bandwidth of the pulses used in the experiments, finite pulse durations were not included in obtaining the FFCFs.[30]

As discussed above, a plot of the peak frequency (ω_τ^{max}) at each ω_m point forms a line in two-dimensional frequency space such as those shown in Fig. 2(a) for the HRP red state. The slopes of such lines are determined and the inverse, the CLS, is plotted versus T_w in Fig. 2(b) for both the blue state (top curve) and the red state (bottom curve). As can be seen in Fig. 2(b), the FFCFs for the two states are quite different. The blue state decays to a constant, which shows that there is a static component to the FFCF on the accessible time scale of the experiment. The red state is decaying to zero, indicating that on the time scale of ~ 100 ps all protein structural configurations associated with the red conformational substrate are sampled.

Another important feature of Fig. 2(b) is that neither of the curves begins at 1. This immediately suggests that there is a homogeneous term composed of a motionally narrowed component, a lifetime term, and an orientational relaxation term. As discussed in Sec. II A, the homogeneous component was dropped, that is, $1/T_2=0$, from the FFCF to derive the analytical equations relating the CLS to the FFCF. The CLS gives only the T_w dependent portion of the FFCF. The homogeneous contribution to the 2D-IR line shape does not depend on T_w. CLS plots are the normalized FFCF without homogeneous contributions. If there is no homogeneous contribution, in general, the initial value of the CLS can still be somewhat smaller than 1, which is a result of the short time approximation (see Appendix 1). The error inherent in FFCF caused by the short time approximation, which produces a total amplitude for the T_w dependent portion of the FFCF being a floor, is tested in the examples given below. Numerical simulations of CLS from FFCFs with various homogeneous contributions were used to see the effect of the inclusion of homogeneous components. Procedures and results of the numerical simulation are given in Appendix 2. Here the procedures that are validated in Appendix 3 are applied.

Time constants and relative amplitudes of the FFCF components are obtained by fitting the CLS to a trial function for the FFCF. A multiexponential decay function was used. The CLS for the red state of HRP, open squares in Fig. 2(b), is fit best by a biexponential function without an offset. The relative amplitudes and decay time constants obtained from fitting are listed as CLS (norm) in the Table I. The time constants are reproduced essentially perfectly for both fast and slow inhomogeneous components. In many tests, we have determined that the time constants are always accurate. The initial value of the CLS is 0.82, which is the sum of the relative amplitudes of the two inhomogeneous components.

A homogeneous component decreases the initial value of the CLS from 1. The difference between CLS at $T_w=0$ and 1 is related to the homogeneous contribution, to the FFCF [see equation (4)] and the Lorentzian contribution, $1/\pi T_2$, and to the linewidth, full width at half maximum (FWHM), of the IR absorption spectrum. The initial value of CLS represents the inhomogeneous contribution in line broadening of IR spectrum. Within the short time approximation used to drive the relationship between the FFCF and the CLS, the amplitude of an inhomogeneous component with a very fast time constant is decreased. Therefore, the CLS method cannot tell the difference between the magnitude of a homogeneous component and error introduced into the amplitude from a very fast inhomogeneous component by the short time approximation. The extent of this error is tested in the examples presented here, and it is small.

For the red state of HRP, 82% of IR line is ascribed to inhomogeneous broadening and the remaining 18% to the homogeneous contribution (see Table I). Using the procedure that is shown to be a good approximation in Appendix 3, the homogeneous line broadening, $1/\pi T_2$, is obtained by the product $0.18 \times$ FWHM of IR absorption spectrum, which gives T_2 in the FFCF [Eq. (4)]. If the CLS can be fit as a single exponential, that is, the inhomogeneous part of FFCF is a single component, the amplitude of this factor is obtained as $\Delta_i = \sqrt{0.82} \times (\text{FWHM})/(2\sqrt{2} \ln 2)$. In this formula, $2(2 \ln 2)^{1/2}$ is required to change the FWHM of the IR absorption line into the standard deviation and the overall square root is needed because relative amplitude from the CLS involves the squares of absolute amplitudes.[39] When inhomogeneous part has multiple components, the entire inhomogeneous broadening will be divided following the ratio between the relative amplitudes, for example, for the red state of HRP [see Table I, CLS (norm)], $0.07/(0.07+0.75)$ and $0.75/(0.07+0.75)$. Again the relative amplitudes from the CLS are the ratio between the squares of amplitudes. Therefore, the square of the amplitude of the 1.5 ps compo-

nent is $\Delta_i^2 = (0.07/0.82) \times (\text{FWHM}/2\sqrt{2 \ln 2})^2$. To sum up this procedure, the amplitude of an inhomogeneous component can be estimated as $\Delta_i = \sqrt{a_i} \times (\text{FWHM})/(2\sqrt{2} \ln 2)$. a_i is the relative amplitude obtained from fitting the T_w dependence of the CLS. The results of this estimation are listed in Table I as CLS and linewidth. The amplitude of the slow component is accurate. The amplitude of the fast component and T_2 is about a factor of 2 off. The time constants are correct, the amplitude of the slow component is correct, and the other two factors are somewhat off. Given the simplicity and ease of this procedure, which involves no response function calculations, the results are reasonable. This procedure is not rigorously correct because FWHM of the IR spectrum is the result of convolutions between the homogeneous and inhomogeneous contributions. Simple division of FWHM will lead to some error. However, as shown in Appendix 3, the error is small in all cases from almost purely Lorentzian to purely Gaussian lines. As shown in other examples below, if a very fast inhomogeneous component does not exist, this simple procedure is virtually quantitative.

With the simple procedure just presented, CLS cannot completely distinguish the homogeneous broadening and the initial part of inhomogeneous broadening, leading to the errors in Table I CLS with linewidth. An inhomogeneous component with a very fast time component will push this initial decay into the homogenous component. The result is a decrease of the amplitude of the fast inhomogeneous component and an increase of the homogenous dephasing time. The true T_2 is no less than the estimation using only the absorption linewidth, and the amplitude of a very fast inhomogeneous component cannot be smaller than the estimated amplitude. Very fast means that the decay constant is comparable to the free induction decay time (see Appendix 1 for details).

More accurate results can be obtained by employing a more complicated but not difficult procedure. This procedure fits the linear absorption spectrum rather than using percentages of FWHM. The absorption spectrum is the Fourier transform of the linear response function given in Eq. (6). The linear response function is found using the known parameters obtained from the CLS and the FWHM of the absorption spectrum. For the red state of HRP, there are three exponential terms in the FFCF. Of these, the two time constants and the amplitude of the slow component are known. Only the homogenous component and the amplitude of the fast decay component were treated as fitting variables in FFCF for calculating the IR spectrum. The response function was numerically Fourier transformed and compared to the absorption spectrum obtained from the calculation using the reported FFCF.[30] Because there is no noise on the calculated spectrum, the fit only used the line shape down to 20% of the maximum amplitude. This cutoff prevented the possibility that the fitting was determined by the low amplitude wings of the spectrum that would not be accessible from a real spectrum with noise. The two fitting variables are constrained to be larger than or equal to the values obtained from the FWHM method. The upper limit is set using Eq. (5) as $T_2 \leqslant 1/(1/2T_1+1/3T_{\text{or}})$. With these constraints, the linear response function calculation is iterated to obtain the best fit

TABLE II. FFCF input parameters from phenol in CCl_4, and parameters determined from the CLS method as discussed in the text.

		T_2 (ps)	Δ_2 (rad/ps)	τ_2 (ps)	T_2^{a}	T_1	T_{or}
Phenol-OD in CCl_4	Experiment	0.9	0.55	5	1.04	12.5	2.9
	CLS (norm)	NA	0.19[a]	5			
	CLS and linewidth	0.88	0.52	5			
	CLS and line shape	0.9	0.55	5			
Phenol-OD in mesitylene	Experiment	0.45	1.2	13	0.48	7.6	5.5
	CLS (norm)	NS	0.27[a]	13			
	CLS and linewidth	0.47	1.3	13			
	CLS and line shape	0.47	1.3	13			

[a]Normalized amplitude from normalized CLS, unitless, not rad/ps.

to the experimental spectrum. The results are given in Table I as CLS and line shape. The agreement between the experimental values and the parameters obtained using CLS and the line shape fitting is essentially perfect. The information lost because of the short time approximation was recovered using the IR spectrum and linear response function calculation even though the experimental spectrum was only fit down the 20% of the peak value.

Table I also includes analysis of the HRP blue state. Fitting the CLS shows that the FFCF has a slow component and a constant component. The fact that the normalized CLS amplitudes do not sum to 1 indicates that there is also a homogeneous component. The decay times match the experimental values. Because both of the inhomogeneous components are slow, the simple FWHM linewidth method should work well. As can be seen in Table I CLS and linewidth, the results are actually quite close to the experimental values. These are obtained without any complicated analysis. When the line shape method is used, fitting the linear absorption spectrum as described above produces virtually perfect agreement with the experimental values, as shown in Table I CLS and line shape.

The HRP absorption line shapes are narrow and almost Gaussian with substantial inhomogeneous broadening. The method was also tested using the experimental FFCF for the OD stretch of HOD in pure water H_2O.[1,13] The absorption spectrum is very broad, almost Gaussian with significant inhomogeneous broadening.[1,13,45] The CLS method works very well, with agreement comparable to that displayed in Table I. The method was also applied to a concentrated NaBr solution.[46] As another test approaching the opposite limit, experimentally determined FFCFs of the OD stretch of phenol-OD (the hydroxyl H replaced with D) in both CCl_4 and mesitylene were used for the 2D-IR calculation.[47] Phenol-OD displays very narrow and almost Lorentzian IR spectra, implying that the absorption lines are almost homogeneously broadened.[47] Both systems display 2D-IR vibrational echo spectra with the characteristic starlike shape associated with nearly homogeneously broadened line.[48] The inhomogeneous contribution to the absorption line is small and therefore the spectral diffusion does not have a great impact on the 2D-IR line shapes.

The FFCFs were derived from the iterative fitting of the 2D-IR spectra and the linear IR line shapes to third-order and linear response function calculations.[47] The resulting FFCFs show a large homogenous component and small inhomogeneous component.[47] The homogenous component was ascribed to very fast density fluctuations in the first solvation of the phenol, and the spectral diffusion to diffusive motions of solvent molecules in the first solvent shell.[47] The same procedure used for the HRP protein was applied to extract the FFCFs using the CLS method from the 2D-IR spectra. All the input parameters for calculating 2D-IR spectrum are listed in Table II experiment. For completeness, the lifetimes and orientational decay times, measured using polarization selective pump-probe experiments, are also given.[47] These do not come into the calculations but show that the homogeneous component is mainly composed of a motionally narrowed contribution to the dynamic line shape, rather than arising from the lifetime or orientational relaxation. The CLS for both samples could be well fit with a single exponential decay with an initial value of ~ 0.2. The initial values show that there is a very large homogenous contribution to the lines. The CLS time constants for the two samples are accurate. Both the simple FWHM linewidth method and the more detailed line shape method produce the amplitudes and T_2 values that are in excellent agreement with the experimentally determined numbers.

Additional details relating to the simple FWHM method are given in Appendix 3 and errors introduced by the short time approximation are discussed in Appendix 1. The amplitude factor is reasonably accurate using the FWHM method if the decay time constant, $\tau > \sim 5 \times \text{FID}$, where FID is the free induction decay, and its duration is taken to be the FID half-width.

IV. APODIZATION AND THE CLS

In NMR, a variety of numerical methods is used to improve signal-to-noise ratios, resolutions, or data acquisition times. One that is very useful is apodization or windowing.[33,35] Apodization involves multiplying an interferogram by a known simple function. A decaying function is used to reduce data acquisition time and improve signal-to-

124503-10   Kwak *et al.*

J. Chem. Phys. **127**, 124503 (2007)

noise ratio. A growing function can be employed to simplify highly overlapping spectra by narrowing the line shapes. Of particular interest here is apodization with a decaying function along the ω_τ axis. In many 2D-IR vibrational echo experiments, the ω_m axis is obtained by detecting the heterodyned vibrational echo wave packet through a monochromator using an IR array detector.[46,49] Taking the spectrum of the wave packet experimentally performs the necessary Fourier transform to give the ω_m axis. There is no interferogram. This axis is referred to as the ω_m axis because it is obtained with the monochromator. It corresponds to the ω_3 axis in 2D-NMR. At each frequency along the ω_m axis where there is signal, an interferogram is recorded by scanning the time delay τ between the first and second pulses. Therefore, the numerical Fourier transforms are applied to multiple one-dimensional interferograms, corresponding to the same type of data processing used in one-dimensional NMR.

The interferograms for the ω_τ axis need to be scanned to sufficiently long τ so that they decay to zero to avoid Fourier transform artifacts. A good deal of time consuming data collection is required to obtain good signal-to-noise ratios at long τ. If the interferograms are simply truncated at short time, the Fourier transforms will contain high frequency artifacts.

If the ω_τ axis interferogram is multiplied by a decaying function, it can be numerically taken to zero smoothly at a τ that is short compared to the complete decay of the interferogram. This avoids Fourier transform artifacts, but it also distorts the 2D-IR spectrum. Multiplying by a decaying function will produce an artificially broadened spectrum, while multiplying by an increasing function will produce an artificially narrowed spectrum along the ω_τ axis. If accurate line shapes are important for extraction of the FFCF, then apodization can only be used with a deconvolution procedure to try to recover the true line shapes. However, as we will show here, apodization along the ω_τ axis does not change the CLS even though the line shapes change a good deal. Therefore, the FFCF can be obtained in the same manner as described above even if ω_τ axis apodization is employed.

To test the influence of apodization on the CLS, response function calculations are performed to obtain the T_w dependence of the CLS with and without apodization. In the response function calculations, two Fourier transforms are performed to obtain the 2D frequency domain spectrum. The Fourier transform for t_1 (time between the first and second pulses, τ) gives the ω_τ axis, and the Fourier transform for t_3 (time after the third pulse) gives the ω_m axis. In the calculations, the t_3 Fourier transform is performed, which is the equivalent in the experiment of using the monochromator to obtain the ω_m axis. However, the apodization function is applied to the t_1 interferograms, and then the t_1 Fourier transforms are performed. This is the equivalent to experimentally collecting the interferograms at each ω_m, applying an apodization function to the experimental interferograms, and then Fourier transforming.

As a first example, a two sided exponential decay centered at $\tau=0$ is used as the apodization function. The FFCF is that of the HRP red state used to produce the 2D spectra

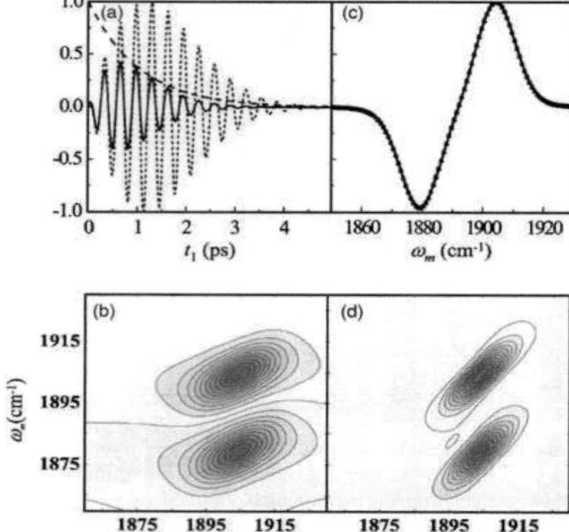

FIG. 3. (a) HRP-CO red state interferogram at a single ω_m (short dashes). Apodization function—decaying exponential (long dashes). Apodized interferogram (solid curve). (b) 2D-IR spectrum after apodization along the ω_τ axis with a decaying biexponential. The spectrum is broadened compared to Fig. 1(a). (d) 2D-IR spectrum after apodization along the ω_τ axis with an increasing biexponential. The spectrum is narrowed compared to Fig. 1(a). (c) Demonstration that apodization does not interfere with phasing using the projection theorem, see text.

shown in Figs. 1(a) and 1(b). The parameters are given in Table I HRP red experiment. In Fig. 3(a), the positive time portion interferogram (rephasing scan) at one ω_m is plotted before (small dashes) and after (solid) applying apodization. The exponential function, $\exp(-t_1/\gamma)$, with $\gamma=1$, is also plotted (large dashes). (The negative time portion of the interferogram, the nonrephasing scan, which is not shown, is apodized by the other side of the two sided exponential.) The same function is applied to the rephasing and nonrephasing interferograms to avoid possible distortion of the absorptive line shape because the ratio of the rephasing and nonrephasing signal determines the shape of 2D-IR spectrum. As can be seen in the figure, apodization changes the interferogram a great deal although there is no change in the frequency of the oscillations. Figure 3(b) shows the resulting 2D spectrum with apodization. The spectrum is substantially broadened, as can be seen by comparison to Fig. 1(a), which is calculated with the same FFCF at the same T_w but without apodization. The broadening only occurs along the ω_τ axis because apodization was only applied to the interferograms that arise from scanning t_1, the time in the first coherence period (time between pulses 1 and 2, τ). The ω_m axis is unaffected. Figure 3(d) shows the result of using an increasing function for apodization, $\gamma=-3$. The spectrum is narrowed compared to the spectrum without apodization [Fig. 1(a)].

Before calculating the CLS of the apodized 2D-IR spectra, the effect of apodization on the phasing process needs to

124503-11   2D-IR vibrational echo spectroscopy

J. Chem. Phys. **127**, 124503 (2007)

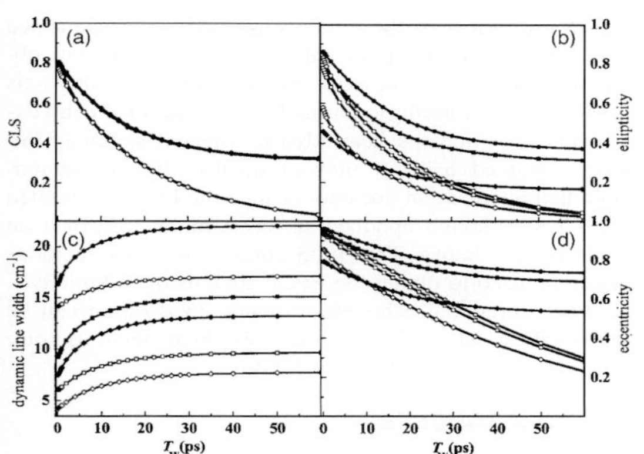

FIG. 4. Calculations using the FFCFs of the HRP red and blue states. Two apodization functions are used, $\exp(-t_1/\gamma)$, with $\gamma=1$ (decaying exponential) and $\gamma=-3$ (increasing exponential) along the ω_τ axis. In all parts, squares are without apodization ($\blacksquare$,$\square$), circles are with decaying apodization ($\bullet$,$\circ$), and diamonds are with increasing apodization ($\blacklozenge$,$\lozenge$). The open symbols are for the HRP red state, and the closed symbols are for the HRP blue state. (a) The CLS is unaffected by apodization. (b) Dynamic line widths. (c) Ellipticity. (d) Eccentricity. The dynamic line width, the ellipticity, and the eccentricity are greatly affected by apodization along the ω_τ axis.

be addressed. In real experiments, there are distortion to the 2D spectra caused by errors in knowing the exact $t_1=0$ position, chirp, and the exact time between pulse 3 and the local oscillator pulse used for heterodyne detection.[46,49] The dual-scan method is used to produce 2D line shapes that are mainly absorptive by adding the rephasing and nonrephasing spectra.[37] To determine the correct phase correction factors, frequency resolved pump-probe spectra are used. The projection theorem[38] states that the frequency resolved pump-probe spectrum should be equal to the projection of the 2D-IR vibrational echo spectrum onto the ω_m axis. The projection is obtained by integrating the 2D-IR spectrum along the ω_τ axis. Apodization is applied to the interferograms prior to Fourier transformation and phase correction. Therefore, it is important that apodization does not change the projection of the 2D-IR spectrum onto the ω_m axis. In fact, apodization does not change the projection because it only affects the spectrum along ω_τ. An example is given in Fig. 3(c). The solid line is the projection of the data in Fig. 1(a) (no apodization) and the two sets of points (square and triangles) are the projections of the apodized spectra given in Figs. 3(b) and 3(d). The three projections are indistinguishable even though the three 2D-IR spectra are very different.

To demonstrate the influence of apodization on the CLS, calculations were performed using the FFCFs of the red and blue states of HRP (see Table I). In addition to obtaining the T_w dependent CLS, three other methods used to characterize the time evolution of 2D-IR spectra are obtained with and without apodization. These are the dynamic linewidth,[1,13] the ellipticity,[29] and the eccentricity,[30] which are defined below.

Figure 4(a) shows CLS calculations using the FFCFs of the HRP red and blue states. The upper curve is for the blue state, and the lower curve is for the red state. Two apodization functions are used, $\exp(-t_1/\gamma)$, with $\gamma=1$ (decaying ex-

ponential) and $\gamma=-3$ (increasing exponential). In all the parts of Fig. 4, squares are without apodization, circles are with decaying apodization, and diamonds are with increasing apodization. The open symbols are for the HRP red state, and the closed symbols are for the HRP blue state. As can be clearly seen in Fig. 4(a), apodization does not influence the T_w dependence of the CLS. Although the differences are very small, the squares, diamonds, and circles do not overlap exactly. This difference is not caused by the apodization but by the short time approximation because applying an apodization function changes the length of the interferogram. The accelerated decay of interferogram by apodization reduces the error caused by short time approximation and the prolonged interferogram increases the error (see Appendix for details). This change in amplitude in CLS is corrected using the method involving the IR spectrum as described in Sec. II. Therefore, the FFCF can be extracted even though apodization is applied to the data. Apodization along the ω_τ axis changes the shapes of the 2D spectra, but it does not change the position of the center point (maximum value) at each ω_m. The increasing apodization narrows peaks [see Fig. 3(d)]. This type of apodization may be useful in congested spectra with overlapping off-diagonal peaks along the ω_τ axis. The results show that it may be possible to separate peaks and determine their FFCFs with apodization and CLS.

Figure 4(b) shows plots of the dynamic linewidths.[1,13] The dynamic linewidth is a cut through the data at the center of the 0-1 portion of the spectrum, parallel to the ω_τ axis, and projected onto the ω_τ axis. The FWHMs of the projected line shapes are plotted. It is clear from the curves that apodization has a dramatic influence on the dynamic linewidth. This is to be expected because apodization changes the shapes of the 2D spectra along the ω_τ axis. Therefore, it changes the FWHM of the projection on to the ω_τ axis.

The ellipticity[29] and the eccentricity[30] are two other observables that are sensitive to spectral diffusion. Both of them use diagonal and antidiagonal widths of the 2D-IR spectra. The diagonal width is the standard deviation of the cut through the 2D spectrum along the diagonal. The antidiagonal width is the standard deviation of the cut perpendicular to the diagonal through the center of the 0-1 portion of the spectrum. Using the same procedure applied above to show the relationship between the CLS and the FFCF, the direct relationships between the ellipticity and the eccentricity can be derived. The derivation has been published for the ellipticity[29] but the procedure for obtaining the full FFCF including the homogeneous contribution and the true amplitudes of each component was not developed. The ellipticity[29] (El) and the eccentricity[30] (Ec) are given by

$$\mathrm{El}(T_w) = \frac{\sigma_\mathrm{D}^2(T_w) - \sigma_\mathrm{AD}^2(T_w)}{\sigma_\mathrm{D}^2(T_w) - \sigma_\mathrm{AD}^2(T_w)}, \tag{18}$$

$$\mathrm{Ec}(T_w) = \sqrt{1 - \frac{\sigma_\mathrm{AD}^2(T_w)}{\sigma_\mathrm{D}^2(T_w)}}. \tag{19}$$

Figures 4(c) and 4(d) show the results of calculating the ellipticity and the eccentricity without apodization and with the two apodization functions. Like the dynamic linewidth,

124503-12   Kwak *et al.*

J. Chem. Phys. **127**, 124503 (2007)

apodization has a substantial affect on both the ellipticity and the eccentricity.

The important result is that only the CLS is immune to apodization along the ω_τ axis. Therefore, it is possible to improve signal-to-noise ratios and reduce data collection time using a decaying apodization function, or increase resolution and peak separation using an increasing apodization and still extract the FFCF in a simple manner using the CLS. There are several limitations on apodization that need to be kept in mind if the CLS is not going to be distorted. First, the apodization function should be symmetrical around $\tau=0$ so that its effect is the same on the rephasing and nonrephasing scans. Second, apodization along the ω_m will change the CLS. However, if apodization along only ω_m is performed, and if the cuts through the 2D spectra are taken parallel to ω_m rather than parallel to ω_τ, then the equivalent of the CLS is obtained, and it is not distorted by ω_m apodization. The ω_m apodization procedure is presented in Appendix 5. However, apodization along both axes will significantly change the CLS and prevent the FFCF from being obtained. In the examples given above, only exponential functions were employed. In Appendix 4, several other functions are used, and the generality of using the CLS method to obtain the FFCF with apodization is demonstrated.

A related issue is the influence of pulse duration on the CLS. In the examples given above, the experiments were conducted with pulses that were sufficiently short that their bandwidths were much wider than the absorption spectra including the 1-2 transitions. Therefore, the 2D-IR spectra are not affected by the finite pulse duration. However, for vibrations with broad spectra, such as water, the 2D-IR vibrational echo spectra can be changed by the finite bandwidth of the pulses.[13,50] Provided that the pulses are reasonably short, that is, the band width is sufficient to span the spectrum even if it is not vastly wider than the spectrum, the finite pulse duration (bandwidth) has a negligible effect on the CLS. This is also true and has been demonstrated for the ellipticity.[29]

V. CONCLUDING REMARKS

We have presented a new approach for extracting the frequency-frequency correlation function from 2D-IR vibrational echo spectra. The direct relationship between the CLS and the T_w dependent portion of the normalized FFCF was derived analytically using a short time approximation. A detailed procedure to obtain the full FFCF from the 2D-IR vibrational echo and absorption spectra, including the homogeneous contribution and the absolute rather than relative amplitudes of the inhomogeneous components, was delineated. Tests of the procedures using known FFCFs were given that show that the CLS method works very well in cases in which the lines are substantially inhomogeneously broadened and in cases in which the lines are almost homogeneously broadened. The usefulness of the method is that the full FFCF can be obtained without using complex response function calculations to fit the 2D-IR vibrational echo line shapes. The CLS method has recently been applied to water and concentrated salt solutions.[46]

The usefulness of the CLS method is further enhanced by its insensitivity to apodization of the interferogram obtained for the ω_τ axis (ω_1 axis). The ω_τ axis is the only axis that produces an interferogram in 2D-IR vibrational echo experiments in which the heterodyned detected signal is frequency resolved using a monochromator. It was demonstrated that apodization does not change the FFCFs extracted using CLS, although apodization has a major influence on the 2D-IR lineshapes. This is in contrast to other methods that can be used to obtain the FFCF such as the ellipticity.[29] In 2D-IR vibrational echo experiments, apodization can be used to reduce data collection times, improve signal-to-noise ratios, and increase spectral resolution.

ACKNOWLEDGMENTS

This work was supported by grants from AFOSR (F49620-01-1-0018) and NSF (DMR-0652232).

APPENDIX: DETAILS OF THE CLS METHOD

The decrease of the initial CLS value from 1 is caused by a homogeneous contribution to the 2D spectra. However, such a decrease can also be caused by errors introduced by the short time approximation. The short time approximation effectively results in the transfer of part of the inhomogeneous contribution that undergoes fast spectral diffusion into the homogeneous component. Three methods were used to analyze the CLS with the results presented in Tables I and II. The third method, which includes a response function analysis of the absorption line, was shown to be quite accurate. The second method is also accurate if there are slow inhomogeneous components, but no fast inhomogeneous component. Below, numerical simulations are used to separately delineate the effect of a homogeneous contribution and the errors induced by the short time approximation. The linear relationship between the $T_w=0$ reduction of the CLS from 1 and the homogeneous contribution is examined numerically. It is found that the initial value of the CLS is related to the inhomogeneous contribution to the IR absorption linewidth. The extraction of the FFCF amplitudes and the homogeneous contribution using the simple division of the IR linewidth into homogeneous and inhomogeneous parts is shown to be approximately correct by comparison to the rigorous convolutions that give a Voight function. Furthermore, CLS are with a variety of different apodization functions, and it is demonstrated that any function can be used for apodization if the same function is applied to the rephasing and nonrephasing scans. While apodization along the ω_τ axis was discussed in the body of the paper, it is shown that apodization along ω_m axis produces the same results as that from the apodization along the ω_τ axis when the CLS is determined using cuts parallel to the ω_m axis.

1. Influence of the short time approximation

The short time approximation or fast dephasing time approximation has usually been applied to broad absorption lines.[24,29] A broadband in the frequency domain corresponds to fast decay in the time-domain signal. Therefore, including only the first or second order terms of a Taylor expansion

311

may be sufficient to describe the dephasing during the coherence periods. For example, the hydroxyl stretch of water has a very wide absorption line. However, many infrared transitions have narrow peaks that are nonetheless inhomogeneously broadened and have Gaussian line shapes. The CO stretch of the HRP protein, which was analyzed in detail above, has a narrow but inhomogeneously broadened absorption spectrum. Even for broad lines, it is not clear to what extent the short time approximation mixes a very fast inhomogeneous component with a homogeneous contribution. Therefore, it is important to examine the application of the short time approximation.

Here deviations of the CLS from an input normalized FFCF are determined for various cases. For simplicity, a single exponential function with one time constant and one amplitude is used as the FFCF, and homogeneous broadening is not included. Because there is no homogeneous broadening, reductions in the initial value from 1 are only caused by the short time approximation. A 2D-IR spectrum is calculated from a given FFCF. Then, the CLS obtained from the calculated 2D-IR spectrum is used to determine the time constant and relative amplitude. For this study, the absolute amplitude is not needed because, with a single inhomogeneous term in the FFCF, the relative amplitude can be directly compared to 1, which is the correct value of the relative amplitude for all FFCFs with a single component. A time standard is required to compare the results from different FFCFs to assign the τ value in the exponential decay as fast or slow. The free induction decay (FID) is a good time standard for systems with different dynamics. The FID time is defined as the time to decay to the half maximum of the envelope of interferogram. This envelope can be obtained by Fourier transforming the IR absorption spectrum. Therefore, in a real experiment, it is not necessary to know the FFCF. A time constant obtained by fitting the CLS can be compared to the Fourier transform of the IR absorption spectrum.

To compare the results from systems with dynamically different FFCFs, the ratio of the time constant to the FID time is used as the horizontal axis. The amplitude is fixed at $\Delta = 5$ rad/ps in the FFCF, and a range of τ values is used. For each τ value, the spectrum is calculated and the FID determined. The ratio τ/FID was varied from 0.13 to 10. For each ratio (a particular τ), more than 20 2D-IR spectra with various T_w points were calculated from the FFCF. The $T_w = 0$ values obtained by fitting the resulting CLS obtained from the 2D spectra are plotted in Fig. 5(a). Motionally narrowed cases with $\Delta\tau < 1$ are not considered as discussed above. The smallest τ/FID=0.49 plotted corresponds to $\Delta\tau = 1$. The inhomogeneous component at τ/FID=1 shows a 30% reduction from the correct value of 1. As the ratio increases, the deviation from 1 decreases. At τ/FID=5, the error is only 10%. A 10% error in the amplitude in many cases is within experimental error. When the ratio is very large compared to 1, the error becomes negligibly small. It is for this reason that the slowest components of the FFCFs discussed in the body of the paper were taken to be accurate and used in determining other parameters.

For a given τ/FID ratio, it may be possible to know and correct for the error introduced by the short time approxima-

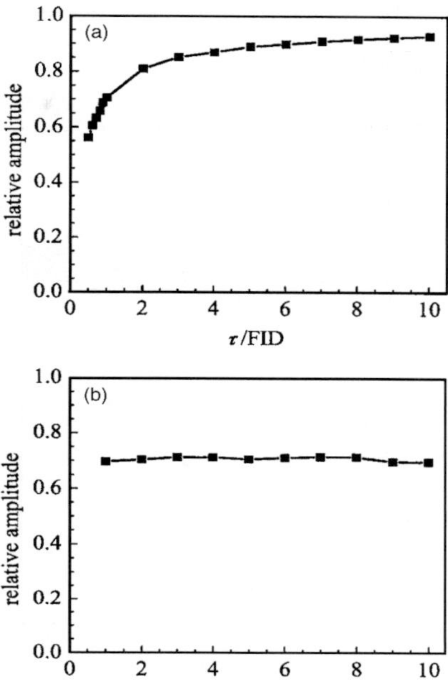

FIG. 5. (a) The relative amplitude vs τ/FID points, for a strictly inhomogeneously broadened FFCF. τ is the exponential decay constant in the FFCF and FID is the free induction decay time (see text). At smaller values of τ/FID than shown, the line is homogeneously broadened. For τ/FID>5, the error is $<10\%$. (b) The relative amplitude is plotted for various absolute amplitudes with τ/FID$=1$. The constant value at various amplitudes shows that the error in amplitude caused by the short time approximation is determined only by the ratio τ/FID.

tion in the amplitude. In Fig. 5(b), the ratio is fixed at τ/FID$=1$, and the amplitude is varied from 1 to 10. The results show that the error, $\sim 30\%$, is independent of the amplitude. Because the FID is known from the absorption spectrum and τ is known from fitting the CLS, Fig. 5(b) can be used to correct the relative amplitude. However, as discussed in the body of the paper and shown below, the homogenous component also causes a decrease of the initial value of the CLS.

2. Influence of a homogeneous component

CLS is inversely proportional to the normalized FFCF, $C_1^N(T_w)$, in the absence of a homogeneous component. Here, the effect of a homogeneous component is examined through numerical calculations. An FFCF with a homogeneous component and one inhomogeneous component, $C(t) = \delta(t)/T_2 + \Delta_{\text{In}}^2 \exp(-t/\tau_{\text{In}})$, is used. The homogeneous line width is $W_h = 1/\pi T_2$. The inhomogeneous component is fixed with $\Delta_{\text{In}} = 1.9$ rad/ps and $\tau_{\text{In}} = 25$ ps. This τ_{In} is sufficiently long that the error in the amplitude of the inhomogeneous contribution introduced by the short time approximation is very small. T_2 was varied from 25 to 0.15 ps to generate FFCFs that gave rise to W_h/FWHM with values ranging from 0.05 to 0.9. FWHM is the full width at half maximum of the total

124503-14   Kwak *et al.*

J. Chem. Phys. **127**, 124503 (2007)

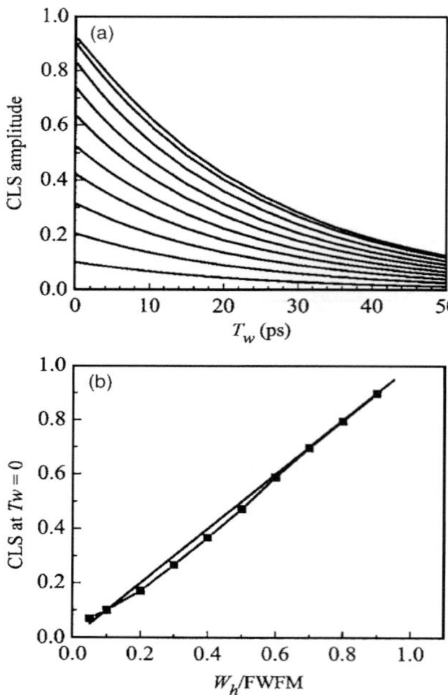

FIG. 6. (a) T_w dependent CLS curves with increasing homogeneous contribution. The fraction of homogeneous linewidth to FWHM was increased from 0.05 to 0.9, with the fraction as (top to bottom) 0.05, 0.1, 0.2, 0.3, 0.4, 0.5, 0.6, 0.7, 0.8, and 0.9. The initial values of these curves show that a linear relationship holds between initial amplitude and the fraction of homogeneous width in the total linewidth. (b) The initial amplitudes (squares) of the CLS (T_w=0) are plotted vs the fraction of the homogeneous contribution to the absorption linewidth. In addition, a line of slope of 1 is shown to demonstrate the virtually linear relationship between the initial value of the CLS and the homogeneous fraction of the line.

absorption line. The T_w dependent CLS were determined from the 2D spectra calculated from the different FFCFs (different homogeneous contributions) and are plotted in Fig. 6(a). As shown in the figure, the effect of the homogeneous component is to decrease the initial amplitude from 1. The top curve in Fig. 6(a) has the smallest homogeneous contribution, T_2=25 ps, and the bottom curve is for T_2=0.15 ps. In the 2D-IR vibrational echo spectra, the increasing amount of the homogeneous contribution results in less elongation along the diagonal.

To check the relationship between decrease in the T_w=0 value of the CLS from 1 and the magnitude of the homogeneous contribution, the decrease from 1 is plotted as a function of the ratio W_h/FWHM in Fig. 6(b). The points are calculated from the CLS and the homogeneous and total linewidths using the FFCFs. The line with slope 1 was drawn to emphasize the almost linear one-to-one relationship between these quantities throughout the entire change of homogeneous contributions. This relationship demonstrates that, for systems in which the inhomogeneous terms in the FFCF decay slowly (τ/FID > ~5) so that there is little error in the relative amplitude of the inhomogeneous terms, the homogeneous and inhomogeneous components of the FFCF can be

obtained from the CLS and the absorption linewidth without using response function calculations of the absorption line shape.

3. Determination of inhomogeneous component from the CLS

In addition to the homogeneous contribution to the absorption linewidth discussed in Appendix 2, there is also the inhomogeneous portion of the total linewidth, W_{In}. In the simple methods, CLS and linewidth, for determining the absolute amplitudes and the homogeneous T_2, we employed the relationship,

$$\left(\frac{W_G}{\text{FWHM}}\right)^2 + \frac{W_L}{\text{FWHM}} \cong 1, \tag{A1}$$

where FWHM is the full width at half maximum of the absorption line and W_G and W_L are the FWHM of the Gaussian (inhomogeneous) and Lorentzian (homogeneous) components of the line. W_G is 2.35 times the standard deviation of the Gaussian and $W_L = 1/\pi T_2$. The relation given in Eq. (A1) is not strictly correct because the total linewidth is determined by the convolution of the Gaussian and Lorentzian contributions to give a Voight line shape.

Because the Voight function has no exact analytical form, a very accurate approximation for the Voight function is used[51] to demonstrate that the relationship in Eq. (A1) is quite accurate. The half-width at half maximum of Voight function can be expressed with very good accuracy as[51]

$$b_{1/2}(a) = a + (\ln 2)^{1/2} \exp[-0.6055a + 0.718a^2$$
$$- 0.0049a^3 + 0.000\,136a^4], \tag{A2}$$

with $a \equiv (\ln 2)^{1/2} W_L/W_G$. $b_{1/2}(a)$, in normalized units, is related to the FWHM of the Lorentizian and Gaussian components of the Voight function by

$$b_{1/2}(a) = a\frac{\text{FWHM}}{W_L} = (\ln 2)^{1/2}\frac{\text{FWHM}}{W_G}. \tag{A3}$$

Then, Eq. (A1) can be rewritten as

$$\frac{a}{b_{1/2}(a)} + \left(\frac{\sqrt{\ln 2}}{b_{1/2}(a)}\right)^2 \cong 1. \tag{A4}$$

To test the quality of Eq. (A1), Eq. (A4) is used to evaluate Eq. (A1) for a range of W_L/W_G. The results are plotted in Fig. 7. $W_L/W_G \cong 0$ is the almost inhomogeneous case for which the line shape is approximately Gaussian. The $W_L/W_G \cong 20$ is the almost homogeneously broadened case for which the line shape is approximately Lorentzian. As shown in the figure, Eq. (A1) is satisfied very well for quite a wide range of W_L/W_G.

4. CLS is independent of the apodization function

In the body of the paper, biexponential functions were used to demonstrate the effects of apodization along the ω_τ axis. The results from CLS are not changed by the different apodization functions and extents of apodization. To show this, numerical simulations were used.

J. Chem. Phys. **127**, 124503 (2007)

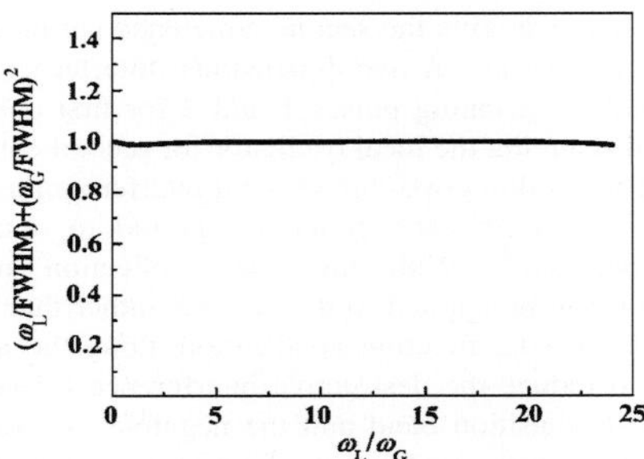

FIG. 7. Equation (A1) is plotted as a function of W_L/W_G. The almost constant value of 1 demonstrates the validity of the approximation used to obtain the homogeneous contribution and the absolute amplitudes of the FFCF using the simple CLS and linewidth methods.

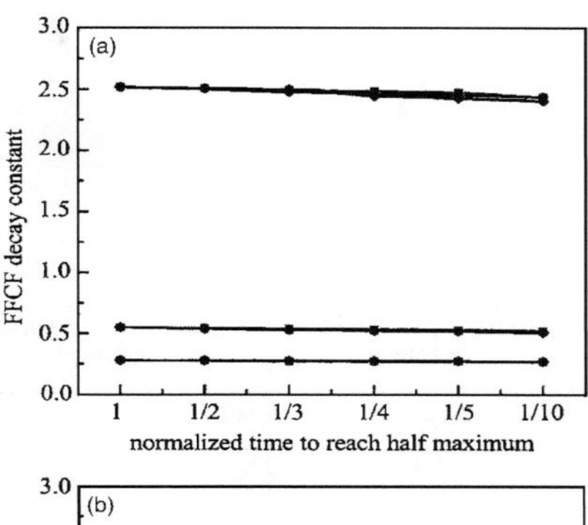

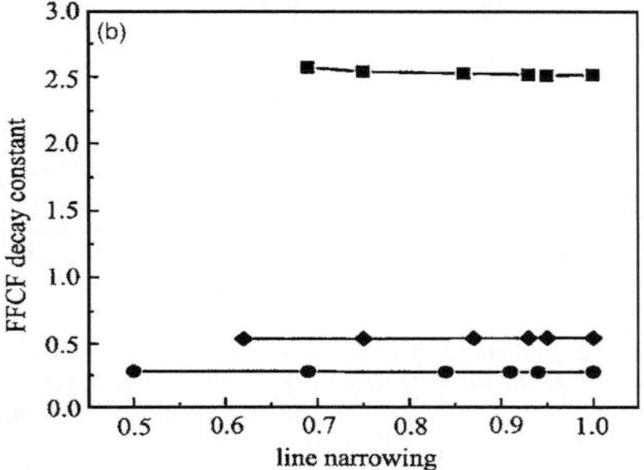

FIG. 8. Time constants calculated with CLS before and after apodization along the ω_τ axis. Three different FFCFs with $\tau/\text{FID}=0.5$, 1, and 5, with time constants of 0.27, 0.51, and 2.5, bottom to top, respectively, were used. (a) Three different decay functions centered on $\tau=0$, two-sided exponential decay (squares), Gaussian (diamonds), and hat function (circles), were used. The time for the interferogram to decay to half maximum is varied from 1 (no apodization) to 1/10. The decay constants extracted with CLS are virtually unchanged by apodization along the ω_τ axis. (b) Time constants obtained with CLS after increasing apodization with a two-sided exponential function (exponential rising function) are unchanged. The horizontal axis is the ratio between the dynamic linewidth before and after apodization. 0.5 is a reduction in the width along the ω_τ axis by a factor of 2.

The influence of changing the apodization function and the extent of apodization is tested using three apodization functions, each centered at $\tau=0$. They are a two-sided exponential decay, a Gaussian and a "hat" function. The hat function is a constant that extends from zero to the peak of the interferogram envelope and then decays to zero as a Gaussian at positive times, with the identical shape at negative times. The FFCF used to calculate the 2D response is the single exponential function used in Appendix 1. Among the various FFCFs presented in Fig. 5, three FFCFs were selected as a model for fast ($\tau/\text{FID}=0.5$), intermediate ($\tau/\text{FID}=1$), and slow ($\tau/\text{FID}=5$) dynamics. The extent of apodization was varied from 1/2 to 1/10 of the initial interferogram. The time to decay to half maximum of the interferogram was calculated before the second Fourier transform along the ω_τ axis. Then the time constant of each apodization function was set such that the interferogram multiplied by the apodization function reached half maximum at 1/2, 1/3, 1/4, 1/5, and 1/10 of the time for the interferogram without apodization to reach half maximum. 2D spectra with different apodization functions and different decays were calculated, and the FFCFs were extracted from these spectra using CLS.

Figure 8(a) shows the results of the calculations for the three FFCFs with three apodization functions [two-sided exponential decay (squares), Gaussian (diamonds), and hat function (circles)] and six different extents of apodization. On the horizontal axis, 1 is no apodization. As shown in the figure, there is almost no changes in the time constants extracted with CLS from the apodized 2D spectra even through there are extensive changes in the shapes of the spectra. Short time approximation causes some change in the initial amplitude, but as discussed in the body of the paper, the amplitudes can be accurately obtained using the CLS plus fitting the linear line shape. Even when the interferogram is forced to decay ten times faster than in the absence of apodization, the CLS gives an accurate determination of the FFCF. The 1/10 apodization can reduce the data collection time in an experiment tremendously. Apodization of the CLS along the ω_τ axis is insensitive to the function form of the apodization function and the extent of apodization.

Enhancement of resolution along the ω_τ axis is also demonstrated. Two apodization functions are used, a rising two-sided exponential and a shifted Gaussian function, $\exp(a\tau/T_2^*-b(\tau/T_2^*)^2)$. The shifted Gaussian function is composed of the rising exponential function, $\exp(a\tau/T_2^*)$, to compensate the decay of interferogram and the Gaussian decay function, $\exp(-b(\tau/T_2^*)^2)$, to make the interferogram decay to zero. T_2^* is the FID of unprocessed interferogram. However, a, b, and T_2^* are used as adjustable parameter to achieve the desired resolution.[52] This function is used on both the rephasing and nonrephasing interferograms.

First, the simple rising exponential functions with various time constants are applied for resolution enhancement. Because the main purpose for applying a rising exponential function is to reduce the linewidth, the ratio between the linewidth before and after the apodization was used as a measure of the extent of apodization. This ratio is calculated from the dynamic linewidth at the shortest T_w. Figure 8(b) shows the results of the calculations. The horizontal axis is the reduction of the linewidth along the ω_τ axis. The calculations are for the three FFCFs used in Fig. 8(a). The results show that the FFCF decay constant can be extracted with the

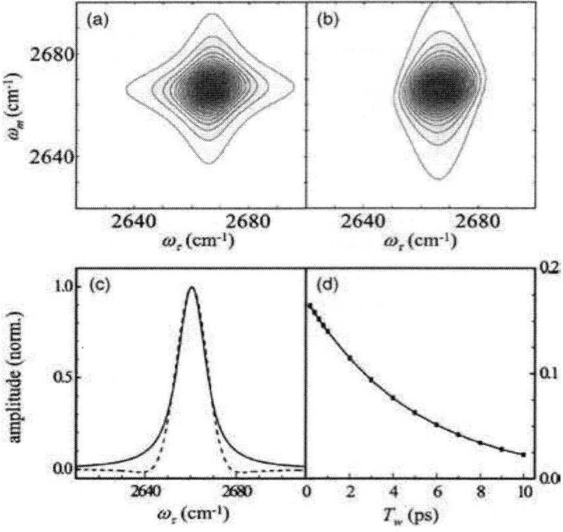

FIG. 9. The effects of the shifted Gaussian apodization function on the 2D IR spectrum and CLS. (a) Calculated 2D-IR vibrational echo spectrum of phenol-OD in CCl_4 at T_w=0.2 ps. The parameters for calculation are listed in the Table II. (b) Calculated 2D-IR spectrum with apodization using the shifted Gaussian function, $\exp(a\tau/T_2^* - b(\tau/T_2^*)^2)$, at T_w=0.2 ps. Here 1, 0.4, and 0.72 ps are used for a, b, and T_2^*. (c) Slices cut through the ω_m center frequency projected onto the ω_τ axis. Original spectrum—solid curve. Apodized spectrum—dashed curve. Apodization of this form eliminates the Lorentzian wings of the spectrum. (d) CLS obtained from the 2D-IR spectra with (●) and without (■) apodization.

CLS. For each decay constant, the line stops at the point where further narrowing by increased apodization produces side lobes along the ω_τ axis on either side of the 2D band. These side lobes do not change the CLS, but they could interfere with increased resolution.

Second, the shifted Gaussian function is used for resolution enhancement by reducing the extended wing of a spectrum without changing the FWHM. Similar results can be achieved by the sine bell function.[32] To clearly show the effect of the shifted Gaussian function, the 2D-IR spectrum of phenol in CCl_4 was used because it has an almost Lorentzian line shape. Parameters for calculating 2D-IR spectrum are listed in Table II. In Fig. 9(a), the 2D-IR spectrum at T_w=0.2 ps without apodization is shown. It has long wings along both axes. 2D-IR spectrum after apodization along the ω_τ axis is presented in Fig. 9(b). The long wing along the ω_τ axis is reduced. To clearly show the reduction of the wing along the ω_τ axis, slices cut through the ω_m center frequency projected onto the ω_τ axis are shown in Fig. 9(c). The solid curve is the spectrum without apodization and the dashed curve is that with apodization. The shifted Gaussian function eliminates the extended Lorentzian wing without changing the FWHM of spectrum. Finally, Fig. 9(d) shows that the CLS before (squares) and after (circles) apodization has not change.

5. Apodization along the ω_m axis

In the body of the paper, we discuss apodization along the ω_τ axis because, frequently, data are collected through a

monochromator to give the second coherence period Fourier transformed axis, ω_m. A two-dimensional interferogram can be obtained by scanning pulses 1 and 2 for first coherence period and scanning the local oscillator for second coherence periods. The resulting two-dimensional interferogram is Fourier transformed for each coherence period to obtain the 2D-IR spectrum.[32] With this data collection method, apodization can be applied to the ω_m axis rather than the ω_τ axis. An increasing function apodization along the ω_m axis can help to reduce the destructive interference between the positive 0-1 transition band and the negative 1-2 transition band. If necessary, apodization along the ω_m axis can be accomplished even though a monochromator is used for the ω_m axis. The time-frequency interferogram obtained using the monochromator can be converted into a time-time interferogram using inverse Fourier transformation, and an apodization function can be applied along the ω_m axis. Then 2D-IR spectrum with apodization along the ω_m axis can be obtained through double Fourier transformation. Response function with apodization along the ω_m axis was calculated to see the effect of apodization along the ω_m axis. The calculated 2D-IR spectra show the same effects as displayed in Figs. 3(b) and 3(d) but the changes are along the ω_m axis rather than along the ω_τ axis.

When apodization is applied along the ω_m axis, CLS is obtained by taking cuts through the 2D spectrum parallel to the ω_m axis rather than the ω_τ axis. Resulting CLS shows inverted behavior relative to the ω_τ axis CLS. The ω_m CLS has a center line slope close to 45° at short T_w and the line becomes horizontal at long T_w. As can be proved easily using Eq. (16), the CLS obtained through cuts parallel to the ω_m axis is directly proportional to normalized FFCF. The resulting CLS apodized along the ω_m axis shows the same behavior as the ω_τ axis CLS apodized along ω_τ axis. Apodization along the axis used to calculate the CLS does not change the CLS. Apodization along the other axis does change it. Therefore, it is possible to apodize along either axis, but not both.

[1] J. B. Asbury, T. Steinel, C. Stromberg, S. A. Corcelli, C. P. Lawrence, J. L. Skinner, and M. D. Fayer, J. Phys. Chem. A **108**, 1107 (2004).

[2] C. J. Fecko, J. J. Loparo, S. T. Roberts, and A. Tokmakoff, J. Chem. Phys. **122**, 054506 (2005).

[3] M. L. Cowan, B. D. Bruner, N. Huse, J. R. Dwyer, B. Chugh, E. T. J. Nibbering, T. Elsasser, and R. J. D. Miller, Nature (London) **434**, 199 (2005).

[4] I. J. Finkelstein, J. Zheng, H. Ishikawa, S. Kim, K. Kwak, and M. D. Fayer, Phys. Chem. Chem. Phys. **9**, 1533 (2007).

[5] L. P. DeFlores and A. Tokmakoff, J. Am. Chem. Soc. **128**, 16520 (2006).

[6] P. Mukherjee, I. Kass, I. T. Arkin, and M. T. Zanni, Proc. Natl. Acad. Sci. U.S.A. **103**, 3528 (2006).

[7] Y. S. Kim and R. M. Hochstrasser, Proc. Natl. Acad. Sci. U.S.A. **102**, 11185 (2005).

[8] J. Zheng, K. Kwak, J. B. Asbury, X. Chen, I. Piletic, and M. D. Fayer, Science **309**, 1338 (2005).

[9] K. Kwak, J. Zheng, H. Cang, and M. D. Fayer, J. Phys. Chem. B **110**, 10384 (2006).

[10] J. Zheng, K. Kwak, J. Xie, and M. D. Fayer, Science **313**, 1951 (2006).

[11] J. Wang, J. Chen, and R. M. Hochstrasser, J. Phys. Chem. B **110**, 7545 (2006).

[12] E. C. Fulmer, F. Ding, and M. T. Zanni, J. Chem. Phys. **122**, 034302 (2005).

[13] J. B. Asbury, T. Steinel, K. Kwak, S. A. Corcelli, C. P. Lawrence, J. L. Skinner, and M. D. Fayer, J. Chem. Phys. **121**, 12431 (2004).

J. Chem. Phys. **127**, 124503 (2007)

[14] O. Golonzka, M. Khalil, N. Demirdoven, and A. Tokmakoff, Phys. Rev. Lett. **86**, 2154 (2001).

[15] M. Khalil, N. Demirdoven, and A. Tokmakoff, J. Chem. Phys. **121**, 362 (2004).

[16] O. Golonzka, M. Khalil, N. Demirdoven, and A. Tokmakoff, J. Chem. Phys. **115**, 10814 (2001).

[17] M. Khalil, N. Demirdoven, and A. Tokmakoff, J. Phys. Chem. A **107**, 5258 (2003).

[18] S. Mukamel, *Principles of Nonlinear Optical Spectroscopy* (Oxford University Press, New York, 1995).

[19] J. D. Eaves, J. J. Loparo, C. J. Fecko, S. T. Roberts, A. Tokmakoff, and P. L. Geissler, Proc. Natl. Acad. Sci. U.S.A. **102**, 13019 (2005).

[20] S. A. Corcelli, C. P. Lawrence, and J. L. Skinner, J. Chem. Phys. **120**, 8107 (2004).

[21] S. Mukamel and R. F. Loring, J. Opt. Soc. Am. B **3**, 595 (1986).

[22] S. Mukamel, Annu. Rev. Phys. Chem. **51**, 691 (2000).

[23] W. M. Zhang, V. Chernyak, and S. Mukamel, J. Chem. Phys. **110**, 5011 (1999).

[24] M. H. Cho, J. Y. Yu, T. H. Joo, Y. Nagasawa, S. A. Passino, and G. R. Fleming, J. Chem. Phys. **100**, 11944 (1996).

[25] A. Piryatinski and J. L. Skinner, J. Phys. Chem. B **106**, 8055 (2002).

[26] A. Tokmakoff, J. Phys. Chem. A **104**, 4247 (2000).

[27] K. Okumura, A. Tokmakoff, and Y. Tanimura, Chem. Phys. Lett. **314**, 488 (1999).

[28] K. Kwac and M. Cho, J. Phys. Chem. A **107**, 5903 (2003).

[29] S. T. Roberts, J. J. Loparo, and A. Tokmakoff, J. Chem. Phys. **125**, 084502 (2006).

[30] I. J. Finkelstein, H. Ishikawa, S. Kim, A. M. Massari, and M. D. Fayer, Proc. Natl. Acad. Sci. U.S.A. **104**, 2637 (2007).

[31] R. N. Bracewell, *The Fourier Transform and Its Application* (McGraw Hill, Boston, 2000).

[32] M. T. Zanni, M. C. Asplund, and R. M. Hochstrasser, J. Chem. Phys. **114**, 4579 (2001).

[33] E. G. Codding and G. Horlick, Appl. Spectrosc. **27**, 85 (1973).

[34] A. G. Marshall and F. R. Verdun, *Fourier Transforms in NMR, Optical, and Mass Spectrometry* (Elsevier, New York, 1990).

[35] J. C. Hoch and A. S. Stern, *NMR Data Processing* (Wiley-Liss, New York, 1996).

[36] W. H. Press, S. A. Teukolsky, W. T. Vetterling, and B. P. Flannery, *Numerical Recipes in C*, 2nd ed. (Cambridge University Press, New York, 1999).

[37] M. Khalil, N. Demirdoven, and A. Tokmakoff, Phys. Rev. Lett. **90**, 047401 (2003).

[38] M. Sarah, G. Faeder, and D. M. Jonas, J. Phys. Chem. A **103**, 10489 (1999).

[39] J. R. Schmidt, N. Sundlass, and J. L. Skinner, Chem. Phys. Lett. **378**, 559 (2003).

[40] I. J. Finkelstein, H. Ishikawa, S. Kim, A. M. Massari, and M. D. Fayer, Proc. Natl. Acad. Sci. U.S.A. **104**, 2637 (2007).

[41] S. Woutersen, R. Pfister, P. Hamm, Y. Mu, D. S. Kosov, and G. Stock, J. Chem. Phys. **117**, 6833 (2002).

[42] C. P. Lawrence and J. L. Skinner, Proc. Natl. Acad. Sci. U.S.A. **102**, 6720 (2005).

[43] M. Khalil, N. Demirdoven, and A. Tokmakoff, Phys. Rev. Lett. **90**, 047401 (2003).

[44] S. Mukamel, *Principles of Nonlinear Optical Spectroscopy* (Oxford University Press, New York, 1999).

[45] T. Steinel, J. B. Asbury, J. R. Zheng, and M. D. Fayer, J. Phys. Chem. A **108**, 10957 (2004).

[46] S. Park, K. Kwak, and M. D. Fayer, Laser Phys. Lett. **4**, 704 (2007).

[47] K. Kwak, S. Park, and M. D. Fayer, Proc. Natl. Acad. Sci. U.S.A. (to be published).

[48] R. R. Ernst, G. Bodenhausen, and A. Wokaun, *Nuclear Magnetic Resonance in One and Two Dimensions* (Oxford University Press, Oxford, 1987).

[49] J. B. Asbury, T. Steinel, and M. D. Fayer, J. Lumin. **107**, 271 (2004).

[50] P. Kjellberg, B. Bruggemann, and T. Pullerists, Phys. Rev. B **74**, 24303 (2006).

[51] H. O. Di Rocco, D. I. Iriarte, and J. Pomarico, Appl. Spectrosc. **55**, 822 (2001).

[52] R. Freeman, *A Handbook of Nuclear Magnetic Resonance* (Addison-Wesley-Longman, Singapore, 1997).

THE JOURNAL OF CHEMICAL PHYSICS **128**, 204505 (2008)

Taking apart the two-dimensional infrared vibrational echo spectra: More information and elimination of distortions

Kyungwon Kwak, Daniel E. Rosenfeld, and M. D. Fayer[a]
Department of Chemistry, Stanford University, Stanford, California 94305, USA

(Received 19 March 2008; accepted 22 April 2008; published online 23 May 2008)

Ultrafast two-dimensional infrared (2D-IR) vibrational echo spectroscopy can probe the fast structural evolution of molecular systems under thermal equilibrium conditions. Structural dynamics are tracked by observing the time evolution of the 2D-IR spectrum, which is caused by frequency fluctuations of vibrational mode(s) excited during the experiment. However, there are a variety of effects that can produce line shape distortions and prevent the correct determination of the frequency-frequency correlation function (FFCF), which describes the frequency fluctuations and connects the experimental observables to a molecular level depiction of dynamics. In addition, it can be useful to analyze different parts of the 2D spectrum to determine if dynamics are different for subensembles of molecules that have different initial absorption frequencies in the inhomogeneously broadened absorption line. Here, an important extension to a theoretical method for extraction of the FFCF from 2D-IR spectra is described. The experimental observable is the center line slope (CLSω_m) of the 2D-IR spectrum. The CLSω_m is obtained by taking slices through the 2D spectrum parallel to the detection frequency axis (ω_m). Each slice is a spectrum. The slope of the line connecting the frequencies of the maxima of the sliced spectra is the CLSω_m. The change in slope of the CLSω_m as a function of time is directly related to the FFCF and can be used to obtain the complete FFCF. CLSω_m is immune to line shape distortions caused by destructive interference between bands arising from vibrational echo emission, from the 0-1 vibrational transition (positive), and from the 1-2 vibrational transition (negative) in the 2D-IR spectrum. The immunity to the destructive interference enables the CLSω_m method to compare different parts of the bands as well as comparing the 0-1 and 1-2 bands. Also, line shape distortions caused by solvent background absorption and finite pulse durations do not affect the determination of the FFCF with the CLSω_m method. The CLSω_m can also provide information on the cross correlation between frequency fluctuations of the 0-1 and 1-2 vibrational transitions. © *2008 American Institute of Physics.*
[DOI: 10.1063/1.2927906]

I. INTRODUCTION

Ultrafast two-dimensional infrared (2D-IR) vibrational echo spectroscopy is a rapidly developing tool for studying fast structural dynamics under thermal equilibrium conditions in condensed matter systems.[1,2] 2D-IR vibrational echo experiments have been successfully applied to study the hydrogen bonding network evolution in water[3–5] and alcohol oligomers,[6] the equilibrium dynamics of aqueous and membrane bound proteins,[7–11] fast chemical exchange and isomerization,[12–14] intramolecular vibrational energy relaxation,[15] and dynamical solute-solvent interactions.[16–18]

In a 2D-IR vibrational echo experiment, three ultrafast mid-IR pulses tuned to the vibrational frequency of interest impinge on the sample. The delay times of the pulses are controlled, and their interactions with the vibrational oscillators generate a third-order polarization in the sample, the vibrational echo, which is emitted in the phase matched direction. The vibrational echo pulse is combined with another mid-IR pulse [the local oscillator (LO)] for heterodyne detection, which provides phase information and amplifies the

signal. The combined vibrational echo and LO pulse is frequency resolved using a monochromator. The time between the first pulse and second pulse τ is called the first coherence period. The time between the second and third pulse T_w is called the population time because the second pulse takes molecules from the coherent superposition states produced by the first pulse to a population state, either in the 0 or in the 1 vibrational levels. The third pulse again excites the molecules into a coherent superposition state of either the 0-1 vibrational levels or the 1-2 vibrational levels. The oscillating electric dipole generated during the final coherence period emits the vibrational echo in the phase matched direction.[19]

Qualitatively, the measurement of dynamics with 2D-IR vibrational echo spectroscopy can be understood in the following manner. During the first coherence period, the molecules are labeled with their initial frequencies. Microscopic molecular events can cause the frequency-labeled molecules to evolve to different frequencies during T_w. In the second coherence period, the final frequencies are read out. The 2D spectrum is composed of the initial frequencies of molecules along a horizontal axis, the ω_τ axis, and the final frequencies along a vertical axis, the ω_m axis. A set of 2D spectra is

---

[a]Electronic mail: fayer@stanford.edu.

204505-2   Kwak, Rosenfeld, and Fayer

J. Chem. Phys. **128**, 204505 (2008)

measured as a function of T_w. The frequency evolution of the molecular oscillators (spectral diffusion) during the T_w period causes the shape of the 2D spectrum to change as T_w is increased. The amplitude, position, and peak shape of the 2D spectra provide detailed information on molecular dynamics and structure.[2,7] Line shape analysis is difficult to perform. Even in simple cases, like the determination of the linewidth of a linear IR spectrum, different fitting functions can produce different linewidths.

Spectral diffusion dynamics of the molecular systems can be quantified with the frequency-frequency correlation function (FFCF).[20–26] The FFCF is a key to understanding the structural evolution of molecular systems in terms of amplitudes and time scales of the dynamics. The FFCF is the joint probability distribution that the frequency has a certain initial value at $t=0$ and another value at a later time t. The FFCF connects the experimental observables to the underlying dynamics. Once the FFCF is known, all linear and nonlinear optical experimental observables can be calculated by time-dependent diagrammatic perturbation theory.[21]

The approach most directly related to the underlying theory for the analysis 2D-IR vibrational echo spectra is to apply a global fitting routine to extract the FFCF. The global fitting is done by fitting the multiple T_w dependent 2D-IR spectra and the linear spectrum using the nonlinear and linear response functions. The functional form of the FFCF is assumed. Fluctuation amplitudes and time constants are used as fitting variables. A full response function calculation is performed to obtain the 2D and linear spectra. These are compared to the data, and the parameters in the FFCF are iterated to obtain the best fit. The complexity of this method and the questionable convergence to the global minimum of such multivariable nonlinear fitting has led to searches for observables that are sensitive to the FFCF but do not require the full fitting procedure. A variety of methods have been used with limited success to characterize and extract the FFCF of 2D spectra. These include nodal plane slope,[27–29] ellipticity,[30–33] and dynamic linewidth.[24] None of these methods is wholly satisfactory. The dynamic linewidth and ellipticity can be strongly influenced by the destructive interference between the positive going 0-1 band and the negative going 1-2 band. Both of these methods require determining the linewidths of cuts through the 2D spectrum. Both methods can be used, in principle, to examine dynamics in different portions of the absorption line, but the methods abilities are limited by interference between the two bands. The nodal plane slope uses the interference of the 0-1 and 1-2 bands in the 2D spectrum. However, this nodal plane cannot be defined when the anharmonic shift is too large, and it cannot be used to analyze dynamics that may be distinct in different parts of the spectrum. The slope of the phase spectrum has been shown to be proportional to the FFCF.[30,34] However, the relationship between the observable and the FFCF is satisfied only at the center frequency, making it difficult to compare the dynamics of unresolved subensembles.

Recently, a new method for determining the FFCF from 2D-IR vibrational echo spectra was introduced[35] and applied to a number of systems.[10,11,26] It was shown that the FFCF is directly related to the *inverse* of the slope of the line that

connects the peak frequencies of spectra obtained by taking slices parallel to ω_τ for a range of ω_m. We will refer to this method as CLSω_τ because the slices are parallel to ω_τ. The peak of a slice spectrum at one ω_m in a 2D spectrum gives one point on the 2D frequency map $(\omega_m, \omega_\tau^*)$, where ω_τ^* is the frequency of the maximum of the slice spectrum for the cut at ω_m. Connecting these maximum points for a range of ω_m generates a line. At short T_w, the line is tilted close to the diagonal line and has a slope of ~ 1. At sufficiently long T_w, spectral diffusion is complete, and the line is vertical. Therefore, the inverse of the slopes of the lines change from ~ 1 to 0 as T_w increases. The details of how to recover the full FFCF including the homogeneous component from 2D and 1D IR spectra have been provided previously.[35] The method only requires the determination of peak positions, which are much easier to measure than linewidths. Furthermore, the CLSω_τ method is not affected by finite pulse durations and other complications that make other methods prone to errors. However, the direct relation between the FFCF and CLSω_τ cannot be guaranteed when destructive interference between the 0-1 and 1-2 bands causes peak position changes as well as line shape distortions in a 2D-IR spectrum.

In this paper, a modification of the CLSω_m method is presented. This method is called the CLSω_m method because slices through the 2D spectrum are taken parallel to the ω_m axis for a range of ω_τ. The points $(\omega_m^*, \omega_\tau)$ are plotted, and they give a center line. The slope of this center line (not the inverse of the slope) varies from ~ 1 to 0 as T_w is increased, and the change in slope is directly related to the FFCF. It is proven that the CLSω_m method is not affected by line shape distortion, which can be caused by peak intensity differences or by the destructive interference between the 0-1 and 1-2 bands. It is numerically demonstrated that peak shape distortion caused by background absorption does not affect the determination of the FFCF. The CLSω_m method can also be used to investigate possible differences in the dephasing of in the 0-1 and 1-2 bands. Piryatinski and Skinner have shown that the initial slope of the frequency resolved echo peak shift is proportional to the cross correlation function between 0-1 and 1-2 transition frequencies when the 1-2 signal is analyzed.[36] However, there has been no report of how this cross correlation function can be directly obtained from the 2D-IR vibrational echo spectra. Here, we prove that the CLSω_m of the 1-2 band is proportional to the cross correlation function when the anharmonicity is sufficiently large relative to the IR linewidth.

II. THEORETICAL DEVELOPMENT

A. The 2D line shape function in the short time approximation

A 2D-IR vibrational echo line shape function can be derived using a two dimensional Fourier transform of the third order response functions about the two coherence periods. Within the Condon and cumulant approximation, the line shape function of the linear IR and 2D-IR spectra can be described using one common line shape function, $g(t)$.[21] A linear IR spectrum includes only the autocorrelation function between the fundamental transition frequencies because only

J. Chem. Phys. **128**, 204505 (2008)

the transition between the ground state and first excited state is probed. This is also true for the positive going band in the 2D-IR spectrum. The equivalent of ground state bleaching and stimulated emission contribute to this band and only involve the 0-1 transition between the ground state and the first excited state. As a result, the line shape of this positive band can be described using the autocorrelation function of the 0-1 transition frequencies. In contrast, the negative going band that arises from vibrational echo emission at the 1-2 transition frequency (equivalent to excited state absorption) involves the transition between the first excited state and second excited state (1-2) as well as that between the ground state and first excited state (0-1).[37]

To develop a tractable analytical equation for a 2D-IR

spectrum, the short time approximation for each coherence period was applied and a simplified 2D-IR line shape equation including spectral diffusion is employed. This approach has been used previously.[27,30,35] Here, we will use the short time approximate 2D-IR line shape equation to prove the direct proportionality between the $CLS\omega_m$ and the FFCF. Also, we will show that the $CLS\omega_m$ is not affected by cancellation of the positive 0-1 transition band by the negative 1-2 transition band within the harmonic approximation. The method for dealing with this problem is very similar to that used to develop the $CLS\omega_\tau$ method. Many of the details have been presented previously.[35] Following the previous approach, the 2D-IR line shape function is

$$R(\omega_m,\omega_\tau) = \frac{4\pi}{\sqrt{C_1(0)^2 - C_1(T_w)^2}} \exp\left(-\frac{C_1(0)\omega_m^2 - 2C_1(T_w)\omega_m\omega_\tau + C_1(0)\omega_\tau^2}{2\sqrt{C_1(0)^2 - C_1(T_w)^2}}\right) - \frac{2\pi s^2}{\sqrt{C_1(0)C_3(0) - C_2(T_w)^2}}$$

$$\times \exp\left(-\frac{C_1(0)(\omega_m + \Delta)^2 - 2C_2(T_w)(\omega_m + \Delta)\omega_\tau + C_3(0)\omega_\tau^2}{2\sqrt{C_1(0)C_3(0) - C_2(T_w)^2}}\right). \tag{1}$$

Here, Δ is the anharmonicity (the difference in frequency between the 0-1 and 1-2 transitions) and s is μ_{21}/μ_{10}, where μ_{ij} is the transition dipole matrix element for the two transitions. s is $\sqrt{2}$ in the harmonic approximation (harmonic oscillator) and is generally $\sim\sqrt{2}$. Three different correlation functions are defined as

$$C_1(t) = \langle \delta\omega_{1,0}(\tau_1)\delta\omega_{1,0}(0)\rangle, \tag{2a}$$

$$C_2(t) = \langle \delta\omega_{2,1}(\tau_1)\delta\omega_{1,0}(0)\rangle, \tag{2b}$$

$$C_3(t) = \langle \delta\omega_{2,1}(\tau_1)\delta\omega_{2,1}(0)\rangle. \tag{2c}$$

The two terms for the 0-1 and 1-2 transitions in the response functions cannot be separated because destructive interference must be included to provide an actual description of the 2D-IR signal. In the previous development of $CLS\omega_\tau$, only one of the three terms was considered because we assumed there was little effect from interference.

B. The ω_m center line slope for a weakly anharmonic system

In the previous work, two points in the 2D frequency space were chosen to calculate the slope of the line that they define. The points $(\omega_m, \omega_\tau^*)$ were calculated by setting the partial derivative of response function to zero at two fixed ω_m points. Here, we derive the equivalent relationship more generally by not assuming that a straight line connects the maxima. The goal of the derivation is to calculate a general form for the slope of the curve connecting the maxima of the spectral slices, $d\omega_m^*/d\omega_\tau$. This slope can be directly calculated by first setting the partial derivative of the response function to zero, $\partial R(\omega_m,\omega_\tau)/\partial\omega_m = F(\omega_m,\omega_\tau) = 0$ at $\omega_m = \omega_m^*$. The equation $F(\omega_m,\omega_\tau) = 0$ defines ω_m^* as an implicit function of ω_τ. Taking the total derivative of $F(\omega_m,\omega_\tau)$ gives the equation for $d\omega_m^*/d\omega_\tau$ through the relation between the total derivative and partial derivative $dF(\omega_m^*,\omega_\tau)/d\omega_\tau = [\partial F(\omega_m^*,\omega_\tau)/\partial\omega_m^*](d\omega_m^*/d\omega_\tau) + \partial F(\omega_m^*,\omega_\tau)/\partial\omega_\tau = 0$.

After some simplification,

$$\frac{d\omega_m^*}{d\omega_\tau} = \frac{\dfrac{C_1(0)\{C_1(T_w) - C_2(T_w)\}\omega_\tau + C_1(0)C_1(T_w)\Delta}{(C_1(T_w)\omega_\tau - C_1(0)\omega_m)^2} - D(\omega_\tau,\omega_m,T_w,s)\left[\dfrac{(C_1(T_w)\omega_m - C_1(0)\omega_\tau)}{\sqrt{C_1(0)^2 - C_1(T_w)^2}} - \dfrac{(C_2(T_w)(\omega_m + \Delta) - C_3(0)\omega_\tau)}{\sqrt{C_1(0)C_3(0) - C_2(T_w)^2}}\right]}{\dfrac{C_1(0)\{C_1(T_w) - C_2(T_w)\}\omega_m + C_1(0)^2\Delta}{(C_1(T_w)\omega_\tau - C_1(0)\omega_m)^2} + D(\omega_\tau,\omega_m,T_w,s)\left[\dfrac{(C_1(T_w)\omega_\tau - C_1(0)\omega_m)}{\sqrt{C_1(0)^2 - C_1(T_w)^2}} - \dfrac{(C_2(T_w)\omega_\tau - C_3(0)(\omega_m + \Delta))}{\sqrt{C_1(0)C_3(0) - C_2(T_w)^2}}\right]}, \tag{3}$$

where $D(\omega_\tau,\omega_m,T_w,s)$ is defined as

204505-4   Kwak, Rosenfeld, and Fayer

J. Chem. Phys. **128**, 204505 (2008)

$$D(\omega_\tau, \omega_m, T_w, s) = \left(\frac{2\{C_1(0)C_3(0) - C_2(T_w)^2\}}{s^2 C_1(0)^2 - C_1(T_w)^2}\right)\exp\left(-\frac{C_1(0)\omega_m^2 - 2C_1(T_w)\omega_m\omega_\tau + C_1(0)\omega_\tau^2}{2\sqrt{C_1(0)^2 - C_1(T_w)^2}}\right.$$

$$\left. + \frac{C_1(0)(\omega_m + \Delta)^2 - 2C_2(T_w)(\omega_m + \Delta)\omega_\tau + C_3(0)\omega_\tau^2}{2\sqrt{C_1(0)C_3(0) - C_2(T_w)^2}}\right). \tag{4}$$

This expression cannot be simplified further, which illustrates that the slope of the maximum points is a complicated function of both frequencies and all variables appearing in the response function. Making the harmonic approximation for the three level system reduces all three correlation functions to a single function.

When all three correlation functions are equal, $C(t) = C_1(t) = C_2(t) = C_3(t)$, the slope becomes the normalized FFCF

$$\frac{\partial \omega_m}{\partial \omega_\tau} = \frac{C(T_w)}{C(0)}. \tag{5}$$

It should be emphasized that this relationship is true after making the single assumption for the correlation functions. The harmonic approximation implies that the dephasing of the 0-1 transition is the same as that for the 1-2 transition, and that $\mu_{21}/\mu_{10} = \sqrt{2}$. In general, this is a good approximation, and all three correlation functions can be taken to be the same.

C. The CLS in nonoverlapping case

For an oscillator with a large anharmonicity, the 0-1 and 1-2 transition bands in a 2D-IR spectrum are well separated. Each peak can be analyzed independently without considering the effect of band overlap. This greatly simplifies the equations for $\text{CLS}\omega_\tau$ and $\text{CLS}\omega_m$ because each peak can be treated separately. For $\text{CLS}\omega_\tau$, which uses the slices parallel to ω_τ at specific ω_m, the equation that defines the set of $(\omega_m, \omega_\tau^*)$ will be given for the 0-1 and 1-2 transition peaks. When only the 0-1 or the 1-2 peak is included in the response function, the equation for $\text{CLS}\omega_\tau$ using the relation $\partial R(\omega_m, \omega_\tau)/\partial \omega_\tau = 0$ is

$$C_1(T)\omega_m - C_1(0)\omega_\tau^* = 0$$

or

$$C_2(T_w)(\omega_m + \Delta) - C_3(0)\omega_\tau^* = 0, \tag{6}$$

$$d\omega_m/d\omega_\tau^* = \frac{C_1(0)}{C_1(T_w)} \quad (0-1 \text{ transition}), \tag{7}$$

$$d\omega_m/d\omega_\tau^* = \frac{C_3(0)}{C_2(T_w)} \quad (1-2 \text{ transition}). \tag{8}$$

Equation (7) shows that $\text{CLS}\omega_\tau$ is proportional to the *inverse* of the normalized correlation function. The same result was shown previously, but here we used a more general method to obtain the $\text{CLS}\omega_\tau$ equation. The result also proves that within the cumulant and short time approximation the maxi-

mum points are connected by a straight line. The second equation for the 1-2 transition gives new information about the cross correlation function $C_2(T_w)$ normalized by the initial amplitude of autocorrelation function $C_3(0)$. If we neglect any differences in the three correlation functions (harmonic approximation), the 0-1 and 1-2 transitions will have the same information and the CLS for each peak will be identical. However, in the general case, the cross correlation function $C_2(T_w)$ can only be obtained from the vibrational echo spectroscopy. $C_3(0)$ can be estimated from the linear IR linewidth of the 1-2 transitions although this can be experimentally.

$\text{CLS}\omega_m$, which uses the slices parallel to ω_m at specific ω_τ, has the equivalent equations for the case of nonoverlapping 0-1 and 1-2 bands. Using the relation $\partial R(\omega_m, \omega_\tau)/\partial \omega_m = 0$, it is readily shown that,

$$d\omega_m^*/d\omega_\tau = \frac{C_1(T_w)}{C_1(0)} \quad (0-1 \text{ transition}), \tag{9}$$

$$d\omega_m^*/d\omega_\tau = \frac{C_2(T_w)}{C_1(0)} \quad (1-2 \text{ transition}). \tag{10}$$

The $\text{CLS}\omega_m$ for the 0-1 transition is directly proportional to the FFCF. The $\text{CLS}\omega_m$ for the 1-2 transition displays a different relation to the FFCF than $\text{CLS}\omega_\tau$. It is the cross correlation function normalized by the fluctuation amplitude of the autocorrelation function, $C_1(0)$. $C_1(0)$ can be obtained by analyzing the 0-1 transition. Therefore, in principle, $C_2(T_w)$ can be obtained from the $\text{CLS}\omega_m$. In the original derivation of $\text{CLS}\omega_\tau$, it was shown that the homogeneous component can be determined and errors arising from the short time approximation can be eliminated by employing the linear absorption spectrum in addition to the $\text{CLS}\omega_\tau$ data obtained from the 2D spectra.[35] This is also true for the 0-1 transition $\text{CLS}\omega_m$. We have not found a similar method for the determination of a $C_2(T_w)$. The time constants for the decay of $C_2(T_w)$ can be easily obtained and the relative amplitudes of the components can be found, but corrections for the short time approximation's effect on the amplitude of fast decay components cannot be made. However, as shown below in a model calculation, the errors are small.

III. APPLICATIONS

A. $\text{CLS}\omega_m$ independent of line shape distortion from destructive interference

If the anharmonicity is not large compared to the 2D spectral bandwidths, the 0-1 and 1-2 transitions overlap. Because the two bands have opposite signs, destructive inter-

204505-5   2D-IR vibrational echo spectra   J. Chem. Phys. **128**, 204505 (2008)

TABLE I. Input parameter (superscript *in*) for generating 2D and 1D spectra and the results obtained from the CLS ω_m calculations (superscript *ob*).

	T_2 (ps)	Δ_2 (cm^{-1})	τ_2 (ps)	Δ_3 (cm^{-1})	τ_3 (ps)
$C_1^{in}(t)$	0.14	41	0.38	34	1.7
$C_2^{in}(t)$	0.14 (0.45)[a]	43 (0.36)[a]	0.43	34 (0.22)[a]	2.6
$C_1^{ob}(t)$	0.14	42	0.37	33	1.7
$C_2^{ob}(t)$	0.46[a]	0.31[a]	0.41	0.23[a]	2.5

[a]Fraction of normalized amplitude.

ference produces distortions of the band shapes. Numerical simulations were performed to see the effects of destructive interference on the CLSω_m. Two cases will be considered. The first case is when the harmonic approximation applies, that is, the correlation functions in Eqs. (2a)–(2c) are equal. For this example, the FFCF determined for the OD hydroxyl stretching mode of dilute HOD in water will be used.[26] The parameters are listed in Table I. The lifetime and orientational relaxation time were included in the calculations although they have little effect. In the second case treated in Sec. III C below, we will consider the consequences of the dynamics of the 1-2 transition being different from those of the 0-1 transition.

The calculated spectra are shown in Fig. 1. At short T_w (A, 200 fs), the center line connecting the maxima lies close to the diagonal. In the absence of homogeneous broadening, the line would be at 45° and have a slope of 1. At 200 fs, the combination of homogeneous dephasing and spectral diffusion makes the slope less than 1. Panel B is the long time

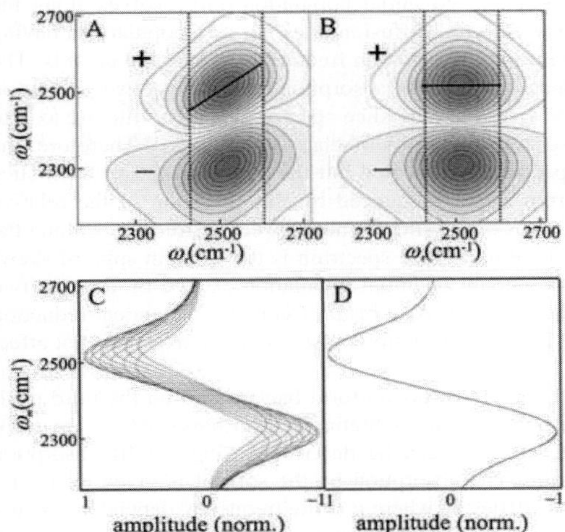

FIG. 1. (A) Calculated 2D-IR vibrational echo spectrum of the OD stretch of HOD/H$_2$O at T_w=0.2 ps. The parameters used in the calculation are listed in the Table I. (B) Calculated HOD/H$_2$O 2D-IR spectrum at T_w =5 ps. The heavy lines in (A) and (B) are the center lines. The dotted lines show the range of ω_τ used. (C) and (D) are the projected spectra for (A) and (B) at several ω_m. (C) shows the distribution of peak positions at short T_w. In (D), all of the peak positions are identical which correspond to the horizontal line in (B). To clearly show that all of the slice spectra in (D) have the same maximum position and line shape, each slice spectrum is normalized.

limit. Spectral diffusion is complete, and the center line is parallel to the ω_τ axis. The spectra of the slices at various ω_τ points are shown for short and long T_w in Figs. 1(c) and 1(d). The maximum frequencies of the slice spectra ω_m^* at various ω_τ form the set of points in 2D frequency space as $(\omega_m^*, \omega_\tau)$. Connecting these maximum frequencies, points at multiple ω_τ points forms a line. The solid line shown in Fig. 1(a) is a plot of the frequencies of the positive going peaks (0-1 transition) in Fig. 1(c). Figure 1(d) shows that at long time the peak frequencies of all of the slice spectra are the same, which yields the horizontal line in Fig. 1(b). At long time, all frequencies in the absorption line have been sampled by spectral diffusion, and, therefore, the slice spectra are the same for all ω_τ.

To examine the effect of the overlap of the 0-1 and 1-2 bands on the CLSω_m, the anharmonicity and the transition dipole of the 1-2 transition were varied separately without changing other parameters in the FFCF for the OD stretch of HOD in water used in the calculations. The water 2D-IR spectra at five different anharmonicities and four different s values (ratios of the 0-1 and 1-2 transition dipoles) were calculated. To clearly show the line shape distortion from the change in the anharmonicity and transition dipole, slice spectra along the ω_m axis at T_w=0.2 ps are shown in Figs. 2(a) and 2(b), respectively. As the extent of overlap increases [Fig. 2(a)], the slice spectrum shapes change. In Figs. 2(a) and 2(b), the slice spectra were taken at center frequency of the 0-1 transition, ω_τ=2508 cm^{-1}. If the slice spectra are taken at other ω_τ positions, the spectra will shift to a lower or a higher frequency, depending on the choice of ω_τ. However, the shape changes caused by the overlap display the same trend presented in the figures. The calculated linear IR spectrum with the same parameters has a 160 cm^{-1} full width at half maximum (FWHM). As the anharmonicity becomes small compared to the FWHM, the overlap region becomes increasingly steep, but the line shape on the high frequency side of the 0-1 band (positive going) and the low frequency side of the 1-2 band (negative going) are basically unchanged. s^2 was varied from 1 to 3 with a fixed anharmonicity of 200 cm^{-1}. The slice spectra are shown in Fig. 2(b). The amplitudes have been normalized by the peak intensity of the 0-1 transition. In the harmonic approximation, s^2 is 2, which results in the 0-1 and 1-2 peaks having equal intensities because there are twice as many quantum pathways that contribute to the 0-1 peak than that contribute to the 1-2 peak. In addition to the change in the relative size of the 0-1 and 1-2 bands, the line shape within the overlap region depends on s.

Figures 2(c) and 2(d) show that CLS$s\omega_m$ curves for the

J. Chem. Phys. **128**, 204505 (2008)

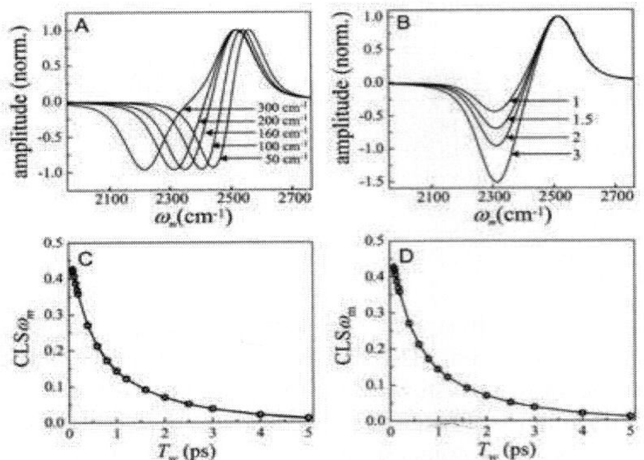

FIG. 2. The effects of the destructive interference between the 0-1 transition positive region of the 2D spectrum and the 1-2 negative region on the $CLS\omega_m$ determination. (A) The slice spectra at $T_w=0.2$ ps cut at the ω_τ center frequency of the transition (2508 cm^{-1}) are presented for various anharmonicities: 300, 200, 160, 100, and 50 cm^{-1}. (B) The slice spectra at $T_w=0.2$ ps cut at the center frequency are presented for various ratios of the transition dipoles squared, s^2: 1, 1.5, 2, 3. All slice spectra are normalized at the maximum intensity of 0-1 peak. (C) $CLS\omega_m$ curves from each 2D-IR spectrum with different anharmonicities used for A are plotted: 300 cm^{-1} ($\square$), 200 cm^{-1} ($\bigcirc$), 160 cm^{-1} ($\triangle$), 100 cm^{-1} ($\triangledown$), and 50 cm^{-1} ($\Diamond$). The points and the curves overlap so well that they cannot be distinguished. (D) $CLS\omega_m$ curves for various ratios of the transition dipoles squared, s^2: 1 ($\square$), 1.5 ($\bigcirc$), 2 ($\triangle$), 3 ($\triangledown$). The points and curves overlap so well that they cannot be distinguished.

various anharmonicities (C) and the transition dipole ratios (D). The curves are identical. Although there are five curves in C and four curves in D, they cannot be distinguished from each other. At each point, there are multiple symbols. As proven analytically above, line shape distortions caused by destructive interference do not change the value of the $CLS\omega_m$. The $CLS\omega_m$ curves are directly related to the FFCF. The method for extracting the entire FFCF, including time constants, amplitudes, and the homogeneous component, has been described previously for the $CLS\omega_\tau$ method in detail.[35] However, the $CLS\omega_\tau$ method is susceptible to line shape distortions. Therefore, for systems in which the anharmonicity is not large compared to the FWHM, when using the $CLS\omega_\tau$ method, it is necessary to limit the center lines to the high frequency side of the 0-1 transition where the line shape is not affected by overlap. In contrast, line shape distortions do not affect the $CLS\omega_m$ and, therefore, do not interfere with the determination of the FFCF.

B. $CLS\omega_m$ not affected by background absorption

Studies of the effects of high optical density in nonlinear optical experiment have focused on the absorbance of the chromophore under investigation, usually a solute in a liquid or solid solvent.[38–40] The effect of solvent background absorption is generally not considered because the small transition dipole μ of solvent molecules at the wavelength of interest produces a negligible contribution to the third order polarization. The solvent can have a significant absorption due to its high concentration, but because the third order nonlinear signal depends on μ^8 in an intensity level experiment or μ^4 in a heterodyne detected polarization level experiment, the solvent does not contribute to the signal.[41] However, the solvent absorption will reduce the amplitudes of the excitation pulses and reduce the amplitude of the signal pulse as it propagates through the sample. If the solvent background absorption is uniform across the spectrum of the solute peaks of interest, it will have little effect other than reducing the amplitude of the signal. However, in general, the background absorption will not be flat, and it can cause line shape distortions and changes in the relative peak intensities. For example, consider a sample in which the solvent absorbs more strongly on the blue side of the solute spectrum than on the red side. As the vibrational echo propagates through the sample, the blue side of its frequency distribution will be attenuated more than the red side. The excitation pulses will also be attenuated more on the blue side than on the red side, resulting in the generation of more signal on the red side of the spectrum.

The influence of solvent background absorption on the $CLS\omega_m$ can be separated into the first two excitation pulses, which are associated with the ω_τ axis and the third excitation pulse and the vibrational echo pulse, which are associated with the ω_m axis. Here, we are considering the experimental situation in which the LO pulse that combines with the vibrational echo pulse does not pass through the sample. Therefore, the LO is not changed by the sample absorption.

The first two excitation pulses produce the ω_τ dependent initial population. A nonuniform solvent absorption will distort the initial population along the ω_τ axis. Spectral diffusion will cause this ω_τ frequency-labeled initial population to broaden and move its center frequency. However, the frequency-labeled initial population only evolves along the ω_m axis. Spectral diffusion does not mix populations having different initial excitation frequencies along the ω_τ axis. The frequency dependent absorption of the first two excitation pulses will cause the slice spectra parallel to the ω_m axis to have distortions on their relative amplitudes. Therefore, the 2D spectrum is distorted but the *peak position* of each slice spectrum is not influenced by the distortion of the relative amplitudes of the slices. Thus, the center frequency along the ω_m axis of each slice spectrum is the same in spite of skewing the amount of initial population excited by the first two excitation pulses. As a result, frequency dependent reduction in the first two excitation pulses along ω_τ axis does not affect the $CLS\omega_m$.

The effect of nonuniform background on the third excitation pulse and the vibrational echo pulse, which are associated with ω_m, can be described using the IR absorption spectrum. The absorption by the solvent is given by Beer's Law, $I=I(0)e^{-\varepsilon lc}$, where ε is the frequency dependent extinction coefficient, l is the sample thickness, and c is the solvent concentration. The electric field of the third laser pulse becomes $E_3 \propto E_3(0)e^{-\varepsilon lc/2}$. Also, the reduction in the generated echo field can be described in the same way as $E_{echo} \propto E_{echo}(0)e^{-\varepsilon lc/2}$. Combining these two frequency dependent absorptions and considering only the ω_m axis, the signal is $S \propto E_L(E_{echo}(0)E_3(0)e^{-\varepsilon lc})$. E_L is the LO field, which does not pass through the sample. E_1 and E_2 have been omitted because they do not influence the slice spectra along the ω_m

J. Chem. Phys. **128**, 204505 (2008)

axis. This is an approximate treatment to determine if frequency dependent distortions of the signal caused by the solvent absorption spectrum will influence the determination of the FFCF using the CLSω_m method. The more correct way to do the calculation is to divide the sample into slices and calculate the signal generated in each slice, with the excitation pulses and signal pulse attenuated as they pass through the sample.[42] If the polarization generated by an excitation pulse is sufficiently strong, it needs to be added to the excitation field, which is attenuated in each slice of the sample.[43] In 2D-IR experiments, the polarization generated in the sample is small. Here, we are not attempting to obtain an accurate description of the 2D spectrum as it is distorted by the nonuniform solvent absorption, but rather to see if a distortion influences the CLSω_m method. The simplified approach employed here is sufficient for this purpose.

To determine the effect of nonuniform solvent absorption on the CLSω_m method, 2D-IR vibrational echo spectra with background absorption were calculated. First, the response function without background absorption was calculated using the FFCF of the OD stretch of HOD in water as above. This absorption is centered at 2508 cm^{-1}. As a simple model, the solvent absorption was taken to be a Gaussian function centered at 3000 cm^{-1} with a 1177 cm^{-1} FWHM. This places a large sloping wing under the OD stretch absorption. To simulate the linear IR absorption spectrum with the solvent absorption, the IR spectrum was calculated from the Fourier transform of the linear response function with the same FFCF parameters as used for the 2D-IR calculations discussed above. Then, the model solvent absorption was added to the calculated IR spectrum. As discussed above, the absorption of the first two excitation pulses does not affect the CLSω_m results, so the background absorption was only taken into consideration for the third pulse and the vibrational echo pulse.

To calculate the signal as modified by the background absorption, the spectrum for the vibrational echo emission at the 0-1 and 1-2 transitions was calculated separately. The signal for the 0-1 and 1-2 transitions was reduced according to the amount of absorption in their corresponding frequency ranges, and then the two peaks are combined. The results are shown in Figs. 3(a) and 3(b) for 0.2 and 5 ps, respectively. These two spectra can be compared to Figs. 1(a) and 1(b), which show the 2D-IR spectra without solvent absorption. The shapes are changed and because the solvent absorption is larger in the 0-1 peak region, the 0-1 peak is reduced more than the 1-2 peak, which is evident from number of contours.

Figure 3(c) shows slice spectra along the ω_m axis. The slices are for ω_τ at the center of the spectrum, 2508 cm^{-1}. The absolute value of peaks height for the 0-1 and 1-2 transitions are equal when solvent absorption is not included (solid curve). The solvent absorption is the line in the upper portion of the panel. The inset shows the solvent absorption (line) and the resulting absorption spectrum for the OD stretch and the solvent absorption. The slice spectrum calculated with the solvent absorption is shown in the main part of Fig. 3(c) as the dash and dot curve. Because the background absorption is different across the spectrum, it has a nonuniform effect on the slice spectrum. Both the 0-1 and 1-2 peak

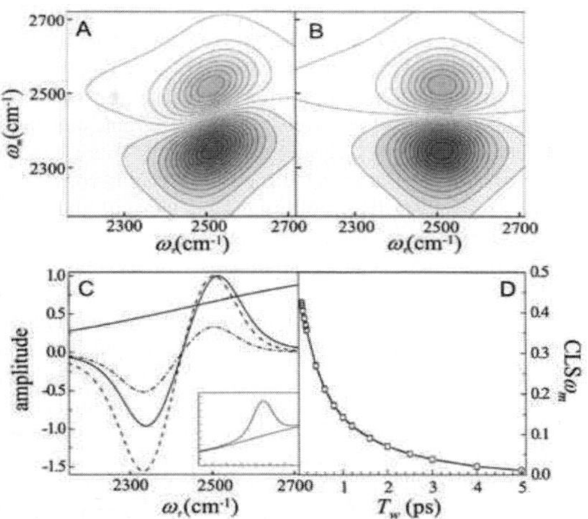

FIG. 3. The effect of solvent background absorption on 2D-IR spectra and CLSω_m determination. (A) Calculated 2D-IR vibrational echo spectrum of the OD stretch of HOD/H$_2$O at T_w=0.2 ps with solvent absorption. All the parameters for calculating the 2D-IR spectrum are the same as ones used in Fig. 1(a). The model solvent absorption spectrum is shown in Fig. 3(c). (B) Calculated HOD/H$_2$O 2D-IR spectrum at T_w=5 ps with solvent absorption. The absolute magnitude of 0-1 band is smaller than that of 1-2 band because the solvent spectrum absorption is greater at high frequency. (C) Slice spectra at T_w=0.2 without solvent absorption (solid curve) and with solvent absorption effect (dash and dot curve). For comparison, the 0-1 peak of the slice spectrum with solvent absorption is normalized (dashed curve). The line in the upper part of the panel is the solvent absorption, and the inset shows the solvent absorption (line) and the calculated linear absorption spectrum with the solvent absorption added. (D) CLSω_m curves for the model with solvent absorption (circles) and without solvent absorption (squares). The CLSω_m curves are identical.

amplitudes are reduced, but that of the 1-2 peak is not as reduced as much. To see the difference between the slice spectra with and without the solvent absorption more clearly, the slice spectrum with solvent absorption has been normalized at the peak of the 0-1 transition (dashed curve). The strong solvent absorption at higher frequency pushes the 0-1 peak to a lower frequency and decreases the linewidth. Also, the amplitudes of the 0-1 and 1-2 peaks are very different. CLSω_m were calculated using the simulated spectra with and without solvent absorption and are shown in Fig. 3(d). As is clear from Fig. 3(d), the background solvent absorption does not change the CLSω_m curve and, therefore, will not change the determination of the FFCF in spite of the fact that the 2D spectral shapes are distorted. Fitting of the 2D spectra by using a full response function calculation gives an incorrect determination of the FFCF. The background absorption would have to be included in the calculation. However, using the CLSω_m method, the correct. FFCF will be obtained without considering the spectral distortion induced by solvent absorption.

C. Determining the cross correlation function

As shown in Sec. II A, the 2D-IR signal consists of three different correlation functions [see Eqs. (2a)–(2c)]. Femtosecond mid-IR pulses can have enough bandwidth to excite a vibrational oscillator from the first to second vibrational excited state; this transition is lower in frequency than the fundamental because of the vibrational anharmonicity. There are open questions as to how the different correlation functions can be studied, if they are significantly different, and theoretical reasons why they should be similar.[36] There has not been a method for extracting the three different correlation functions from a 2D-IR spectrum. Here, we will show that the cross correlation function, $C_2(t)$, between 0-1 and 1-2 transition frequencies can be obtained by analyzing the 1-2 band.

It is possible for the vibrational oscillator dephasing dynamics to be different when it is in the first excited state than in the ground state because of differences in solute-solvent interactions caused by the different bond lengths and dipole moments of the vibrationally excited state. Here, we calculate the 2D-IR spectra including different dynamics for the 0-1 and 1-2 transitions. Therefore, $C_1(t) \neq C_2(t)$ and $C_3(t)$. For these calculations, we use the OD/water FFCF parameters for the 0-1 transition, which determines $C_1(t)$. Different dynamics for the 1-2 transition results in a different $C_3(t)$, the FFCF for the 1-2 transition. There is no simple way to obtain the cross correlation, $C_2(t)$ from the knowledge of $C_1(t)$ and $C_3(t)$. In the calculations, $C_3(t)$ has little effect on the 2D line shapes. It only comes into play during the very brief second coherence period. Therefore, we will simply take $C_2(t) = C_3(t) \neq C_1(t)$. For $C_2(t) = C_3(t)$, we will use the FFCF parameters for the OD stretched of HOD in a 1.5M NaBr aqueous, which are different but not greatly different from those of pure water.[26] These input parameters are given in Table I as $C_1^{in}(t)$ and $C_2^{in}(t)$. We also need the lifetime of the second excited state. We use the simple harmonic oscillator result that the second excited state is a factor of 2 faster than the first excited state lifetime.[37] The orientational relaxation is taken to be the same in the 0 and 1 states, and the lifetime and orientational relaxation parameters for pure water are used.[26] The lifetime and the orientational relaxation make very little contribution to the linear or 2D line shape because the homogeneous dephasing T_2 is dominated by the pure dephasing T_2^*.

As a first illustration, the anharmonicity was set to 300 cm^{-1}, which is large compared to the line width of 160 cm^{-1}, to eliminate line shape distortion from overlapping peaks. Below, the influence of the size of the anharmonicity compared to the line width will be considered. The 2D spectra were calculated (see Table I for input parameters), and the linear IR spectrum was also calculated using $C_1^{in}(t)$. The linear spectrum is used in determining T_2. First, all three correlation functions were set equal, and the 0-1 and 1-2 bands were analyzed separately to check that the analysis of the 0-1 band and the 1-2 band give the same results. The calculated CLSω_m are plotted in Fig. 4(a). The circles show the CLSω_m obtained from the 0-1 transition and the squares are the CLSω_m from the 1-2 band. They overlap well, but close examination reveals that there is a very small difference in the magnitude. This difference comes from the difference in lifetime used for the 2-1 relaxation and the 1-0 relaxation. The lifetime makes a small contribution to the

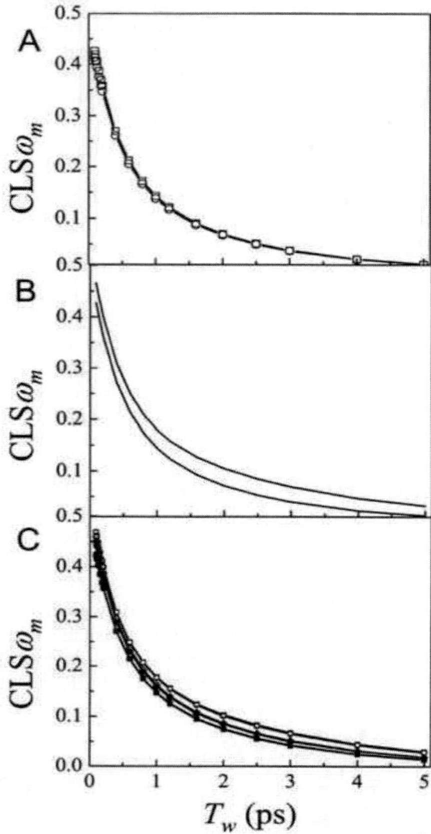

FIG. 4. (A) CLSω_m curves obtained from the 0-1 band (□) and the 1-2 band (○) using the same FFCF for both transitions (OD of HOD in H$_2$O). The two curves are virtually identical but have a very small difference caused by the faster vibrational decay of second excited state. (B) CLSω_m curves obtained from the 0-1 band (lower curve) and the 1-2 band (upper curve) from the 2D spectra calculated with response function theory using as inputs $C_1(t) \neq C_2(t)$ and $C_3(t)$ (see text). The difference in the curves show that it is in principle possible to obtain $C_2(t)$ as well as $C_1(t)$ using the CLSω_m method. (C) The effect of overlapping peaks in determination of cross correlation function is examined for 2D spectra with $C_1(t) \neq C_2(t)$ and $C_3(t)$ as in (B). The anharmonicity is varied from 200 to 50 cm^{-1} [200 cm^{-1} (□), 160 cm^{-1} (○), and 50 cm^{-1} (◇)]. The closed symbols represent CLSω_m from the 0-1 band and open symbols are for the CLSω_m from 1-2 band. The FWHM of the absorption line width is 160 cm^{-1}. $C_2(t)$ can be obtained as long as the anharmonicity is not small compared to the linewidth.

homogeneous component of the FFCF.[35] The homogeneous component manifests itself as the initial difference between the plotted CLSω_m and 1.

Figure 4(b) shows the results of the calculations in which $C_1(t) \neq C_2(t)$ and $C_3(t)$. For Fig. 4(b), the anharmonicity is still taken to be 300 cm^{-1}. It is clear from Fig. 4(b) that the CLSω_m obtained from the 0-1 band (lower curve) is distinguishable from that obtained from the 1-2 band (upper curve). The differences in CLSω_ms show that, in principle, both $C_1(t)$ and $C_2(t)$ can be measured.

In Table I, $C_1^{ob}(t)$ and $C_2^{ob}(t)$ obtained from the CLSω_m analysis are compared to the inputs to the 2D response function calculations using $C_1^{in}(t)$ and $C_2^{in}(t)$. In the determination of the 0-1 FFCF from CLSω_m, it is necessary to use the absorption line shape to obtain the absolute FFCF and to determine the homogeneous component.[35] Without use of the absorption line, the FFCF time constants can be determined and the relative fractional amplitudes of the components can be determined, but the absolute values in wave numbers and an accurate T_2 in picoseconds cannot be determined. There is no equivalent of the absorption line to use in conjunction with the determination of $C_2(t)$. Therefore, in Table I for $C_2^{ob}(t)$, the time constants and fractional amplitudes are listed. As can be seen in Table I, the agreement between the input parameters and those obtained from the CLSω_m analysis is very good.

In the calculations presented in Fig. 4(b), the anharmonicity (300 cm^{-1}) is large compared to the linewidth (160 cm^{-1}). Figure 4(c) shows the results of the calculations in which the anharmonicity is varied with all other parameters identical to those used in Fig. 4(b). Three sets of calculations were performed with anharmonicities of 200, 160, and 50 cm^{-1}. The results for 200 and 160 cm^{-1} (as well as 300 cm^{-1}, not shown) are identical. These results are the upper curves obtained from the 1-2 band and the lower curves obtained from the 0-1 band. The two curves in Fig. 4(c) that virtually overlap and are between the upper and lower curves are for 50 cm^{-1} anharmonicity. These cannot be distinguished. The 50 cm^{-1} anharmonicity is less that a third of the line width. The 0-1 and 1-2 bands overlap almost completely. Whether information on $C_2(t)$ can be obtained is dependent on the magnitude of the anharmonicity relative to the line width, not the absolute linewidth. To determine $C_2(t)$, there must be a reasonable degree of separation between the 0-1 and 1-2 bands.

IV. CONCLUDING REMARKS

We have presented a new approach, the CLSω_m method, for extracting the FFCF from 2D-IR vibrational echo spectra. The CLSω_m method is an extension and improvement on the previously described CLSω_τ method.[35] Details of the practical aspects of using the CLSω_m method are the same as in CLSω_τ method and are described in Ref. 35. Here, it was demonstrated that distortions of the 2D line shapes caused by overlap of the positive going 0-1 band and the negative going 1-2 band do not affect the determination of the FFCF using the CLSω_m. This is not the case for the previously detailed CLSω_τ method. Because of the lack of distortion caused by overlapping bands, it is possible to use the CLSω_m method to examine different frequency ranges of a 2D spectrum to determine if the dynamics vary with frequency.[44] Investigation of a frequency dependence can be accomplished by using slice spectra on the blue side of the line and on the red side of the line, and comparing the CLSω_m. If the CLSω_m are different on the two sides of the line, then frequency dependent subensembles that have different dynamics that exist in the molecular system.

It was also demonstrated that the CLSω_m is not sensitive to distortions in the 2D spectra produced by nonuniform background absorption. Frequently, a very weak absorption of interest is riding on top of a large sloping solvent absorption. The solvent does not produce a signal because it has a small transition dipole, but it has significant absorption owing to its high concentration. Large background absorption often occurs in biological samples, where the vibration of interest produces a small absorption on top of very significant water and protein absorptions. Using the CLSω_m method, the FFCF can be accurately obtained in spite of distortions to the 2D spectrum produced by the solvent absorption. It was shown previously that the CLS method, whether CLSω_m or CLSω_τ, yields the correct FFCF even when the 2D spectra are distorted by finite pulse durations or apodization along one axis.[35] In addition, the CLSω_m was shown to be capable of obtaining more information than the 0-1 transition FFCF. The method is also able to obtain the cross correlation function between fluctuations of the 0-1 and 1-2 vibrational transitions.

In addition to the usefulness of the CLSω_m to obtain various types of information without distortion, it is also important for its speed in data analysis and its lack of sensitivity to systematic errors. In contrast to doing full response theory calculations of the 2D and linear spectra to iteratively fit the experimental spectra to obtain the FFCF, the CLSω_m yields the FFCF in a simple and fast set of calculations.[35] The reduction in fitting time can be a factor greater than 100. In the CLSω_m method, it is necessary to determine peak positions rather than linewidths or line shapes. Peak positions are much easier to obtain accurately than linewidths or line shapes. The net result is that the CLSω_m method can provide more information in a manner that is less susceptible to error than other approaches. Therefore, the CLSω_m method increases the utility of ultrafast 2D-IR vibrational echo spectroscopy for the extraction of dynamical information from complex molecular systems.

ACKNOWLEDGMENTS

The authors would like to thank the Air Force Office of Scientific Research (F49620-01-1-0018) for support of this work. Daniel E. Rosenfeld was supported by the Fannie and John Hertz Foundation and a Stanford Graduate Fellowship.

[1] J. Zheng, K. Kwak, and M. D. Fayer, Acc. Chem. Res. **40**, 75 (2007).
[2] R. M. Hochstrasser, Adv. Chem. Phys. **132**, 1 (2006).
[3] J. B. Asbury, T. Steinel, K. Kwak, S. Corcelli, C. P. Lawrence, J. L. Skinner, and M. D. Fayer, J. Chem. Phys. **121**, 12431–12446 (2004).
[4] M. L. Cowan, B. D. Bruner, N. Huse, J. R. Dwyer, B. Chugh, E. T. J. Nibbering, T. Elsaesser, and R. J. D. Miller, Nature (London) **434**, 199 (2005).
[5] J. J. Loparo, S. T. Roberts, and A. Tokmakoff, J. Chem. Phys. **125**, 194521 (2006).
[6] J. B. Asbury, T. Steinel, C. Stromberg, K. J. Gaffney, I. R. Piletic, and M. D. Fayer, J. Chem. Phys. **119**, 12981 (2003).
[7] I. J. Finkelstein, J. Zheng, H. Ishikawa, S. Kim, K. Kwak, and M. D. Fayer, Phys. Chem. Chem. Phys. **9**, 1533 (2007).
[8] L. P. DeFlores, Z. Ganim, S. F. Ackley, H. S. Chung, and A. Tokmakoff, J. Phys. Chem. B **110**, 18973 (2006).
[9] P. Mukherjee, I. Kass, I. T. Arkin, and M. T. Zanni, Proc. Natl. Acad. Sci. U.S.A. **103**, 3528 (2006).
[10] H. Ishikawa, I. J. Finkelstein, S. Kim, K. Kwak, J. K. Chung, K. Wakasugi, A. M. Massari, and M. D. Fayer, Proc. Natl. Acad. Sci. U.S.A. **104**, 16116 (2007).

J. Chem. Phys. **128**, 204505 (2008)

[11] H. Ishikawa, S. Kim, K. Kwak, K. Wakasugi, and M. D. Fayer, Proc. Natl. Acad. Sci. U.S.A. **104**, 19309 (2007).

[12] J. Zheng, K. Kwak, J. B. Asbury, X. Chen, I. Piletic, and M. D. Fayer, Science **309**, 1338 (2005).

[13] Y. S. Kim and R. M. Hochstrasser, Proc. Natl. Acad. Sci. U.S.A. **102**, 11185 (2005).

[14] J. Zheng, K. Kwak, J. Xie, and M. D. Fayer, Science **309**, 1951 (2006).

[15] M. Khalil, N. Demirdoven, and A. Tokmakoff, J. Chem. Phys. **121**, 362 (2004).

[16] J. Wang, J. Chen, and R. M. Hochstrasser, J. Phys. Chem. B **110**, 7545 (2006).

[17] E. C. Fulmer, F. Ding, and M. T. Zanni, J. Chem. Phys. **122**, 034302 (2005).

[18] K. Kwak, S. Park, and M. D. Fayer, Proc. Natl. Acad. Sci. U.S.A. **104**, 14221 (2007).

[19] S. Park, K. Kwak, and M. D. Fayer, Laser Phys. Lett. **4**, 704 (2007).

[20] S. Mukamel, Annu. Rev. Phys. Chem. **51**, 691 (2000).

[21] S. Mukamel, *Principles of Nonlinear Optical Spectroscopy* (Oxford University Press, New York, 1995).

[22] M. Khalil, N. Demirdoven, and A. Tokmakoff, J. Phys. Chem. A **107**, 5258 (2003).

[23] C. J. Fecko, J. J. Loparo, S. T. Roberts, and A. Tokmakoff, J. Chem. Phys. **122**, 054506 (2005).

[24] J. B. Asbury, T. Steinel, K. Kwak, S. A. Corcelli, C. P. Lawrence, J. L. Skinner, and M. D. Fayer, J. Chem. Phys. **121**, 12431 (2004).

[25] J. B. Asbury, T. Steinel, C. Stromberg, S. A. Corcelli, C. P. Lawrence, J. L. Skinner, and M. D. Fayer, J. Phys. Chem. A **108**, 1107 (2004).

[26] S. Park and M. D. Fayer, Proc. Natl. Acad. Sci. U.S.A. **104**, 16731 (2007).

[27] K. Kwac and M. Cho, J. Chem. Phys. **119**, 2256 (2003).

[28] S. Woutersen, R. Pfister, P. Hamm, Y. Mu, D. S. Kosov, and G. Stock, J. Chem. Phys. **117**, 6833 (2002).

[29] J. D. Eaves, J. J. Loparo, C. J. Fecko, S. T. Roberts, A. Tokmakoff, and P. L. Geissler, Annu. Rev. Phys. Chem. **102**, 13019 (2005).

[30] S. T. Roberts, J. J. Loparo, and A. Tokmakoff, J. Chem. Phys. **125**, 084502 (2006).

[31] I. J. Finkelstein, H. Ishikawa, S. Kim, A. M. Massari, and M. D. Fayer, Proc. Natl. Acad. Sci. U.S.A. **104**, 2637 (2007).

[32] C. Fang, J. D. Bauman, K. Das, A. Remorino, E. Arnold, and R. M. Hochstrasser, Proc. Natl. Acad. Sci. U.S.A. **105**, 1472 (2008).

[33] D. Kraemer, M. L. Cowan, A. Paarmann, N. Huse, E. T. J. Nibbering, T. Elsaesser, and R. J. D. Miller, Proc. Natl. Acad. Sci. U.S.A. **105**, 437 (2008).

[34] J. J. Loparo, S. T. Roberts, and A. Tokmakoff, J. Chem. Phys. **125**, 194522 (2006).

[35] K. Kwak, S. Park, I. J. Finkelstein, and M. D. Fayer, J. Chem. Phys. **127**, 1245031 (2007).

[36] A. Piryatinski and J. L. Skinner, J. Phys. Chem. B **106**, 8055 (2002).

[37] P. Hamm and R. M. Hochstrasser, in *Ultrafast Infrared and Raman Spectroscopy*, edited by M. D. Fayer (Dekker, New York, 2001), Vol. 26, pp. 273–347.

[38] O. Kinrot and Y. Prior, Phys. Rev. A **51**, 4996 (1995).

[39] N. Belabas and D. M. Jonas, J. Opt. Soc. Am. B **22**, 655 (2005).

[40] D. Keusters and W. S. Warren, J. Chem. Phys. **119**, 4478 (2003).

[41] *Ultrafast Infrared and Raman Spectroscopy*, edited by M. D. Fayer (Dekker, New York, 2001), Vol. 26.

[42] K. D. Rector, D. A. Zimdars, and M. D. Fayer, J. Chem. Phys. **109**, 5455 (1998).

[43] R. W. Olson, H. W. H. Lee, F. G. Patterson, and M. D. Fayer, J. Chem. Phys. **76**, 31 (1982).

[44] T. Steinel, J. B. Asbury, S. A. Corcelli, C. P. Lawrence, J. L. Skinner, and M. D. Fayer, Chem. Phys. Lett. **386**, 295 (2004).

Section 4. Recent Review Articles

Recent Review Articles

The following three papers are recent review articles of current ultrafast 2D IR vibrational echo experiments from the Fayer group. The papers have significant overlap but cover the experiments at different levels. The papers treat a variety of topics, but are not comprehensive. The first article is in Accounts of Chemical Research. This is a journal for the general scientific reader who is a non-expert. It is a very basic introduction to the technique with examples of what can be measured. The second paper is published in the journal Physical Chemistry Chemical Physics. It is a much higher level introduction to the method. It presents examples using both biological and chemical systems. The third paper is published in Laser Physics Letters. This paper contains a great deal of detail on how we actually perform the experiments. The experimental setup is presented with a detailed diagram showing the multiple beam paths. Important issues of how the data are processed are addressed. The 2D IR vibrational echo experiments provide both amplitude and phase information. However, obtaining the correct spectrum without phase distortions is tricky. Some of the approaches used to do this are given. Also, this paper gives the first application of the new approach for extracting the frequency-frequency correlation function. The two detailed papers on the theory of this method for obtaining the FFCF are included in the previous section. For those who want to get some idea of how we actually do the experiments, this paper lays it out in some detail.

Since these review articles were written, we have made some major advances in the methodology that have not been published. The timing between the three pulses that comprise the vibrational echo pulse sequence and the timing between the vibrational echo and the local oscillator that is used to heterodyne detect the echo to provide phase information must be set and maintained to an accuracy of less than one femtosecond. In normal ultrafast experiments — even ones operating on very fast time scales — a dirft in time of a few femtoseconds is not important. However, in these experiments, a few femtoseconds are a good fraction of an IR optical cycle, and therefore, produce significant phase error. We now have a method that continually monitors the critical timings and makes corrections to maintain the critical phase relationships. This is a major step towards making these experiments hands-off and robust.

Acc. Chem. Res. **2007**, *40*, 75–83

Ultrafast 2D IR Vibrational Echo Spectroscopy

JUNRONG ZHENG, KYUNGWON KWAK, AND
M. D. FAYER*

*Department of Chemistry, Stanford University,
Stanford, California 94305*

Received May 10, 2006

ABSTRACT
The experimental technique and applications of ultrafast two-dimensional infrared (2D IR) vibrational echo spectroscopy are presented. Using ultrashort infrared pulses and optical heterodyne detection to provide phase information, unique information can be obtained about the dynamics, interactions, and structures of molecular systems. The form and time evolution of the 2D IR spectrum permits examination of processes that cannot be studied with linear infrared absorption experiments. Three examples are given: organic solute–solvent complex chemical exchange, dynamics of the hydrogen-bond network of water, and assigning peaks in an IR spectrum of a mixture of species.

I. Introduction

Ultrafast 2D IR vibrational echo spectroscopy[1–11] is an ultrafast IR analog of 2D NMR[12] that directly probes the structural degrees of freedom of molecules.[13] The 2D IR vibrational echo technique involves a three-pulse sequence that induces and then probes the coherent evolution of excitations (vibrations) of a molecular system. The first pulse in the sequence causes vibrational modes of a molecular ensemble to initially "oscillate" with the identical phase. The later pulses generate observable signals that are sensitive to changes in environments of individual molecules during the experiment, even if the aggregate populations in distinct environments do not change. For example, formation and dissociation of molecular complexes under thermal equilibrium conditions can be observed. The 2D IR vibrational echo spectrum can also display intramolecular interactions and dynamics that are

not observable in linear vibrational absorption experiments. A critical difference between the 2D IR and NMR variants is that the IR pulse sequence is sensitive to dynamics on timescales 6–10 orders of magnitude faster than the NMR pulse sequence.

2D IR vibrational echo spectroscopy has several characteristics that make it a useful tool for the study of problems involving rapid dynamics under thermal equilibrium in condensed phases, e.g., fast chemical exchange reactions, solute/solvent interactions, water dynamics, and intramolecular interactions. Such problems are important and ubiquitous in nature and difficult to study by other means. 2D IR vibrational echo experiments have temporal resolution < 50 fs, which is sufficiently fast to study the fastest chemical processes. In contrast to electronic excitation, the vibration excitation associated with 2D IR experiments produces a negligible perturbation of a molecular system that does not change the chemical properties of the samples. 2D IR experiments can also be useful as a tool for chemical structural analysis by revealing the relationship among different mechanical degrees of freedom of a molecular[14,15] or biomolecular system.[3,10,16] 2D IR vibrational echo experiments have been successfully applied to study fast chemical exchange reactions and solution dynamics,[9,17,18] water dynamics,[19,20] hydrogen network evolution,[6] intramolecular vibrational energy relaxations,[5] protein structures and dynamics,[3,10,16,21] and mixed chemicals analysis.[15]

In 1995 an *Accounts of Chemical Research* article was published[22] summarizing recent fast condensed phase vibrational echo experiments.[1,23,24] The experiments were 1D with a time resolution of ~1 ps. Since those first experiments, the field has advanced greatly. Vibrational echoes have gone multidimensional with full phase information, significantly increasing the types of information that can be obtained. Furthermore, tabletop laser systems now provide sub 50 fs IR pulses, making 2D IR vibrational echo spectroscopy increasingly available to a wide range of researchers. In this Account we will first introduce the experimental technique and then describe its applications in the study of chemical exchange reactions,[9,18] water dynamics,[19] and chemical analysis.[15] Interesting topics that are not covered here are experiments by Tokmakoff and co-workers on the coupling between vibrational modes,[14] population relaxation,[5] and structure and dynamics of proteins,[10] experiments by Hochstrasser and co-workers on solution dynamics of small molecules,[25] hydrogen-bond chemical exchange,[17] and small peptide models of biological systems,[26] experiments by Zanni and co-workers on fifth-order vibrational echo spectroscopy[8] and membrane peptides,[7] and experiments by Fayer and co-workers on hydrogen-bonded oligomers,[6] nanoscopic water,[27] and protein structure and dynamics.[16,28]

Junrong Zheng was born in Chaozhou, China, in 1973. He obtained a B.S. degree (1997) in Chemistry and M.S. degree in Polymer Chemistry (2000) from Beijing University and another M.S. degree in Polymer Physics (2003) from Rensselaer Polytechnic Institute. Currently he is a doctoral candidate in physical chemistry at Stanford University in the group of Professor Michael D. Fayer. His research concentrates on development and applications of 2D IR vibrational echo spectroscopy.

Kyungwon Kwak was born in Ansung, Korea, in 1973. He received his B.A. degree in Chemistry from Korea University in 1999 and earned his M.S. degree in Physical Chemistry from Korea University with Professor Minhaeng Cho in 2001. Currently he is a doctoral candidate in physical chemistry at Stanford University in the group of Professor Michael D. Fayer. His research concentrates on development and applications of 2D IR vibrational echo spectroscopy.

Michael D. Fayer was born in Los Angeles, CA, in 1947. He obtained a B.S. (1969) and Ph.D. (1974) degrees in Chemistry from the University of California at Berkeley. He joined the faculty at Stanford University in 1974 and is now the David Mulvane Ehrsam and Edward Curtis Franklin Professor of Chemistry. His research interests include dynamics and intermolecular interactions in condensed-matter systems and biological molecules investigated with ultrafast nonlinear visible and infrared optical methods and theory.

II. Experimental Method

In a 2D IR vibrational echo experiment three ultrashort IR pulses tuned to the frequency of the vibrational modes

10.1021/ar068010d CCC: $37.00   © 2007 American Chemical Society
Published on Web 09/27/2006

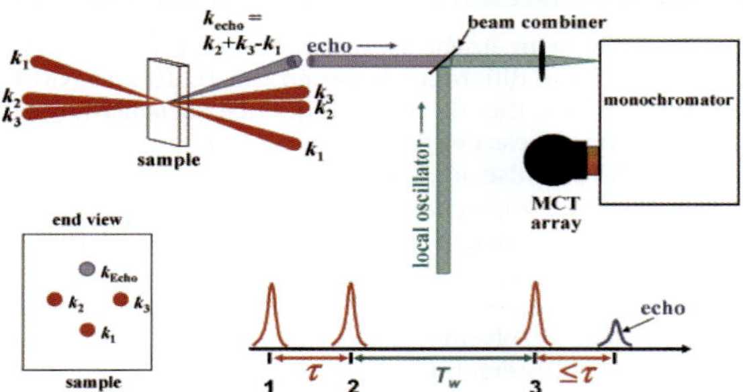

FIGURE 1. Schematic of the ultrafast 2D vibrational echo spectrometer showing the wave vectors (k_i) and the vibrational echo pulse sequence.

of interest are crossed in the sample. Because the pulses are very short, they have a broad bandwidth, which makes it possible to simultaneously excite a number of vibrational modes or a very broad spectral feature such as the hydroxyl stretch band of water. The times between pulses 1 and 2 and pulses 2 and 3 are called τ and T_w, respectively (see Figure 1; a time period refers to the time between the pulse peaks). At a time $\leq \tau$ after the third pulse, a fourth IR pulse is emitted in a unique direction. This is the vibrational echo, the signal in the experiments. The vibrational echo is the infrared vibrational equivalent of the magnetic resonance spin echo[29] and the electronic excitation photon echo.[30] Before going into more quantitative detail, a qualitative description is given of how the vibrational echo pulse arises and how it is possible to obtain phase information, not just the intensity, from the vibrational echo.

The first pulse in the sequence places the vibrational oscillators into a coherent superposition state of the vibrational ground state (0) and vibrational first excited state (1) with all of the oscillators initially oscillating in phase. The initial macroscopic phase relationships among the oscillators decay rapidly. This is called free induction decay. The phase relationships can decay for a number of reasons. Several peaks in the IR spectrum may be excited. These are at different frequencies, so the vibrational oscillators oscillate at different intrinsic frequency. Even for a single mode, molecules in different environments, e.g., different local solvent structures, will have different oscillator frequencies, which is called inhomogeneous broadening. In addition, there are dynamic interactions with the environment, e.g., local solvent structure fluctuations, which cause the frequency of a given oscillator to evolve in time. These frequency fluctuations are called dynamic dephasing and spectral diffusion. The rest of the pulse sequence can preserve information about the phase relationships that are seemingly lost during the free induction decay. Only the spectral diffusion at sufficiently long time can totally destroy the phase relations among the oscillators that cannot be recovered by the rest of the pulse sequence. However, even then, the signal is not zero because the initial two pulses also set up a spatial relationship among the oscillators that contributes to the signal.

The second pulse in the sequence stores the phase (and spatial) information induced by the first pulse as a complex frequency and spatial pattern of differences of the populations of the 0 and 1 vibrational states. After the waiting period, T_w, the third pulse again generates coherent superposition states of the oscillators. Initially the oscillators are not in phase, but the pulse sequence initiates a rephasing process.[29] At a time $\leq \tau$ after the third pulse, the vibrational oscillators are again oscillating in phase. Each vibrational oscillator has associated with it a microscopic oscillating electric dipole. When rephasing has occurred and all of the oscillators are in phase, the sample has a macroscopic oscillating electric dipole. A macroscopic oscillating electric dipole emits radiation, which is the source of the vibrational echo. The vibrational echo pulse is short because the oscillators again get out of phase just as they did after the first pulse.

Both the intensity of the 2D vibrational echo pulse and its time structure contain important information. If the echo pulse is sent directly into an IR detector, its intensity is measured but phase information is lost. Phase information is obtained by allowing the vibrational echo pulse to interfere with another pulse, called the local oscillator. Interference can provide phase information just as in a spatial interference pattern created by two crossed laser beams. In a spatial interference pattern, the fringe pattern is observed as a spatial variation in the intensity, which gives information on the phase relations of the electric fields of the two beams. In the heterodyne-detected vibrational echo experiments, the local oscillator pulse and the vibrational echo pulse are collinear and phase information is obtained by observing the interference pattern (interferogram) as a function of time.

In a dynamic system, the first laser pulse "labels" the initial structures of the species in the sample. The second pulse ends the first time period τ and starts clocking the "reaction time", during which the "labeled" species experience population dynamics. The third pulse ends the population dynamics period of length T_w and begins a third period of length $\leq \tau$, which ends with emission of the echo pulse. The echo signal reads out the information about the final structures of all "labeled" species. There are two types of time periods in the experiment. The periods between the pulses 1 and 2 and between pulse 3 and the echo pulse are called coherence periods. During these periods the vibrations are in coherent superpositions of two vibrational states. Fast vibrational oscillator frequency fluctuations induced by fast structural fluctuation of the system cause dynamic dephasing, which is one contribution to the line shapes in the 2D spectrum. During the period T_w between pulses 2 and 3, called the population period, a vibration is in a particular state not a superposition state. Slower structural fluctuations of the system, spectral diffusion, contribute to the 2D line shapes. Other processes during the population period also produce changes in the 2D spectrum. For example, chemical exchange can occur in which two species in equilibrium are interconverting one to the other without changing the overall number of either species. Chemical exchange causes new peaks to grow in as T_w is increased. In an experiment, τ is scanned for fixed T_w. The recorded signals are converted into a 2D vibrational echo spectrum. Then T_w is increased and another spectrum is obtained. The series of spectra taken as a function of T_w provides information on dephasing, spectral diffusion, and population dynamics.

A 2D IR vibrational echo experimental setup is illustrated schematically in Figure 1. Briefly, three successive ultrashort IR pulses with wave vectors (propagation directions) k_1, k_2, and k_3 are applied to the sample to induce the subsequent emission of the time-delayed vibrational echo in a distinct direction ($k_e = -k_1 + k_2 + k_3$). The IR pulses ($\sim$50 fs) are produced using a regeneratively amplified Ti:Sapphire laser-pumped optical parametric amplifier system.[31] The vibrational echo pulse is detected with frequency and phase resolution by interfering it with a fifth (local oscillator) pulse, and the combined pulses are dispersed in a monochromator and then detected with a 32-element MCT IR array detector. As discussed above, the function of the local oscillator is to phase resolve (through the interferogram between the echo and local oscillator) and optically heterodyne amplify the vibrational echo signal.

A simple example is given in Figure 2 for the hydroxyl stretch (OD) of phenol-OD in CCl_4. Figure 2a shows the linear FT-IR absorption spectrum of the OD stretch. There is a single peak. In the 2D vibrational echo spectrum, there are two frequency axes, which require two Fourier transforms of the time domain data to convert the time structure of the echo into 2D frequency data. As shown in Figure 1, the vibrational echo pulse, which is overlapped with the local oscillator (LO) pulse, is passed through the monochromator. Taking the spectrum of the pulse is an experimental Fourier transform. Thus, one of the two Fourier transforms is performed by the monochromator to provide the vertical axis in the spectrum, ω_m (m for monochromator, see Figure 2c). The other axis is obtained by scanning τ. Scanning τ produces an interferogram (see Figure 2b) as the echo pulse changes its phase relationship relative to the fixed LO pulse. There is one interferogram for each frequency on the ω_m axis for which there is signal. The numerical Fourier transforms of these τ scan interferograms provide the ω_τ axis. The data $S(\omega_\tau, T_w, \omega_m)$ are then plotted for each T_w in three dimensions, the amplitude as a function of both w_t and ω_m. Details of the method including phase error corrections have been presented.[4,31]

In Figure 2c there is a positive going peak (red) on the diagonal (dashed line) and a negative going peak (blue) off diagonal. The frequency of the first interaction of the radiation field (first pulse) with the vibration is the frequency on the ω_τ axis. The frequency of the third interaction (third pulse), which is also the frequency of the echo emission, is the frequency on the ω_m axis. If $\omega_m = \omega_\tau$, the peak is on the diagonal. This is the situation for the red peak in Figure 2c, which corresponds to the 0−1 vibrational transition of the hydroxyl stretch. In Figure 2c each contour is a 10% change in amplitude. The blue off-

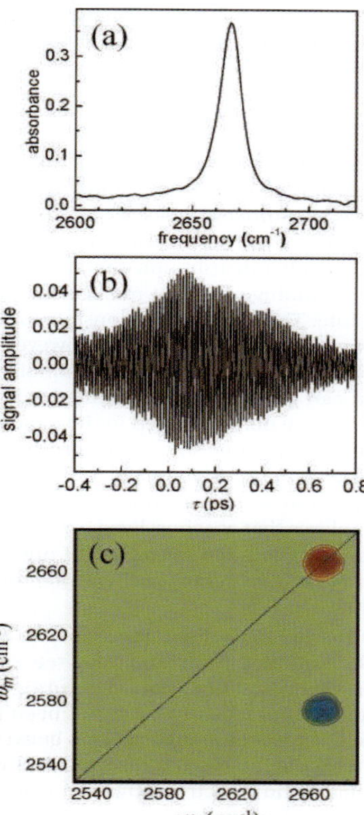

FIGURE 2. (a) FTIR spectrum of the hydroxyl stretch (OD) of phenol in CCl_4. (b) An interferogram obtained by scanning τ for a particular wavelength ω_m. (c) 2D IR spectrum of the hydroxyl stretch of phenol in CCl_4 at $T_w = 16$ ps.

diagonal peak involves the 1−2 transition. The first interaction produces a coherent superposition state of the 0−1 transition. One of the quantum pathways for the second interaction (second pulse) is to produce a population in the 1 state (first vibrationally excited state). Then the third pulse can couple the 1 state to the 2 state and produce a coherent superposition of 1 and 2. This results in the vibrational echo being emitted at the 1−2 transition frequency, which is shifted to lower frequency than the 0−1 transition by the vibrational anharmonicity. Therefore, the ω_τ frequency is the 0−1 frequency (2670 cm^{-1}), but the ω_m frequency is the 1−2 frequency (2570 cm^{-1}). The off-diagonal anharmonicity peak is negative going because there is a 180° phase shift in the echo pulse electric field relative to the 0−1 echo. The spectrum in Figure 2c is for $T_w = 16$ ps, a long time compared to the time scale of solvent fluctuations in this system. Therefore, spectral diffusion is complete, and the peaks in the 2D spectrum are symmetrical, essentially round. At short time, the peaks are elongated along the direction of the diagonal. The elongation is the signature of inhomogeneous broadening. As discussed below in connection with experiments on water, the change in shape with increasing T_w measures spectral diffusion and, thus, the structural evolution of the system.

III. Applications

A. Solute–Solvent Complexes and Chemical Exchange.

The simplest view of a solute in a liquid solvent is to treat the solvent as a featureless dielectric continuum. A much better description involves the use of a radial distribution function. The radial distribution function describes the probability of finding a solvent molecule some distance from the solute. This model is responsible for the picture of solvent shells surrounding a solute. However, solute and solvent molecules have anisotropic intermolecular interactions that can give rise to well-defined solute–solvent complexes that are not accounted for by a radial distribution function. For organic solute–solvent systems, intermolecular interactions are generally relatively weak, a few kcal/mol. Such weak interactions will produce solute–solvent complexes that are short lived.[9,11,18] Although short lived, the dissociation and formation of organic solute–solvent complexes can influence chemical processes such as reactivity.

Organic solute–solvent complexes are in equilibrium between the complex form and the free solute form. Because formation and dissociation occurs on a picosecond time scale, until recently it has not been possible to observe the chemical exchange process between the two forms of the solute. To discuss the method by which 2D IR vibrational echo spectroscopy can study chemical exchange, it is sufficient to focus on the 0–1 region of the spectrum. Identical considerations apply to the 1–2 region, and analysis of the 1–2 region can provide additional information.[9]

Figure 3 illustrates the influence of chemical exchange on the 2D vibrational echo spectrum. Two species, A and B, with vibrational transition frequencies, ω_A and ω_B, are in thermal equilibrium. Figure 3a shows the spectrum at short time prior to chemical exchange. There are two peaks on the diagonal: one for species A and one for species B. Figure 3b shows what would happen if A goes to B. The left part of the spectrum is a schematic representation of the combined quantum pathways that lead to the signal.[32] A dashed arrow represents a coherence (coherent superposition state) produced by a radiation field. The solid arrow represents a population, and the curved arrow represents the vibrational echo emission. The first pulse produces coherence between the states of species A at frequency ω_A. After time τ, the second pulse produces a population. There are several pathways, and a population can be produced in either the ground state (0) or the first excited state (1). During the period T_w some A's turn into B's (A → B). The third pulse again produces coherence, but it is coherence of species B at ω_B followed by echo emission at ω_B. Because the first interaction (frequency on the ω_τ axis) is at ω_A but the last interaction and echo emission (frequency on the ω_m axis) is at ω_B, an

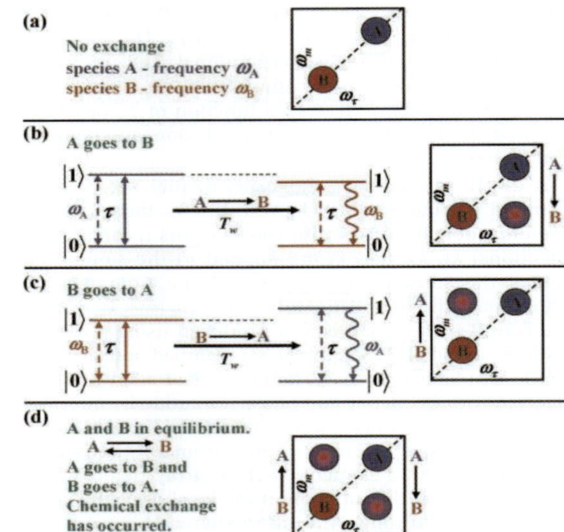

FIGURE 3. Schematic illustrations of the observation of chemical exchange between two species A and B. (a) At short time prior to chemical exchange, the two species give two peaks on the diagonal. (b) If some A turns into B, an off-diagonal peak will develop. (c) If some B turns into A, the opposite off-diagonal peak develops. (d) At equilibrium, as much A turns into B as B turns into A, and two equal amplitude off-diagonal peaks are generated.

off-diagonal peak is generated as shown in the right portion of Figure 3b.

Figure 3c shows what happens if B's turn into A's (B → A). The first pulse makes a coherence between the 0 and 1 states of B at frequency ω_B. The second pulse makes a population on B. If during the period T_w some B's turn into A's, the third pulse produces a coherence of species A and echo emission from A at frequency ω_A. Because the first interaction is at ω_B and the last interaction and emission is at ω_A, a peak is generated off diagonal as shown in the right portion of Figure 3c. In a real system A and B are in equilibrium. Therefore, the number of A's turning into B's in a given time period is equal to the number of B's turning into A's. As shown in Figure 3d, the result is to produce two off-diagonal peaks. Because some A's and B's may not have undergone chemical exchange or may have undergone chemical exchange but reverted back to the original species prior to the third pulse, there are also diagonal peaks. The model spectrum in Figure 3d is the spectrum for a time long compared to the chemical exchange time, while the spectrum in Figure 3a is for a time short compared to the chemical exchange time. The rate of chemical exchange can be determined by observing the growth of the off-diagonal peaks in the 2D vibrational echo spectrum.[9,11,17,18]

Figure 4a displays the spectrum of the hydroxyl OD stretch of phenol in a mixed solvent of benzene (20 mol %) and CCl_4 (80 mol %). The high-frequency peak is the free phenol (see Figure 2a). The low-frequency peak is the phenol benzene complex.[9] The structure of the complex,

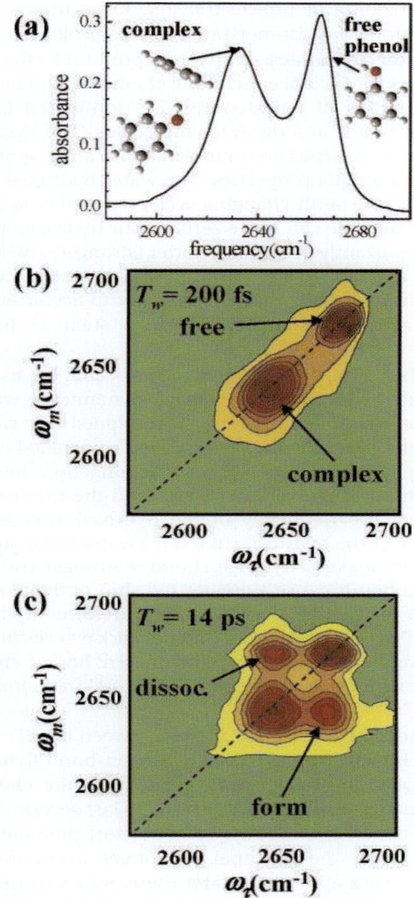

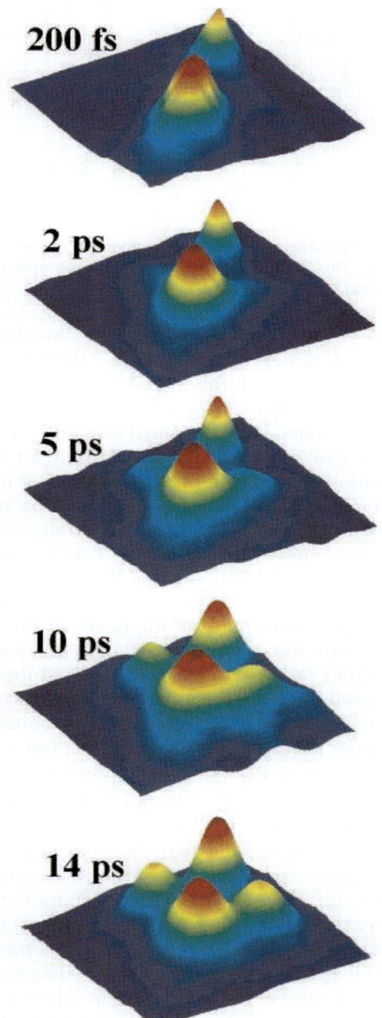

FIGURE 4. (a) FT-IR absorption spectra of the OD stretch of the phenol–benzene complex and free phenol. (b) 2D vibrational echo spectrum prior to significant exchange showing two peaks on the diagonal. (c) 2D spectrum after substantial exchange with two off-diagonal peaks produced by dissociation and formation of the complex.

FIGURE 5. 2D vibrational echo spectra as T_w increases showing growth of off-diagonal peaks because of chemical exchange.

determined from electronic structure calculations, is shown in the figure.[9] The species are in equilibrium. Complexes are constantly dissociating and free phenols are associating with benzene to form complexes. Figure 4b shows the 2D vibrational echo spectrum at short time, $T_w = 200$ fs; no significant exchange has occurred. As in Figure 3a, there are two species in the system with different vibrational transition frequencies. At short time, the two species give two peaks on the diagonal, which correspond to the two peaks in the absorption spectrum (Figure 4a). At long time, 14 ps, extensive chemical exchange has occurred. Off-diagonal peaks have grown in as can be seen clearly in Figure 4c. The two peaks correspond to complex dissociation and formation. Figure 4c has the appearance of the schematic shown in Figure 3d, and the discussion surrounding the origin of the off-

diagonal peaks in Figure 3 applies to the real data shown in Figure 4.

The growth of the off-diagonal peaks with increasing T_w can be used to directly determine the thermal equilibrium chemical exchange rate. Figure 5 displays 2D spectra as three-dimensional representations. At 2 ps, the off-diagonal peaks are just appearing in these plots. By 5 ps they are clearly evident and continue to grow as can be seen in the 10 and 14 ps plots. Data like these are used for detailed analysis of the chemical exchange kinetics.[9,11,18] In addition to chemical exchange, there are other dynamical processes that contribute to the time-dependent changes in the spectrum. Spectral diffusion causes the peaks to change shape. The change in shape can be

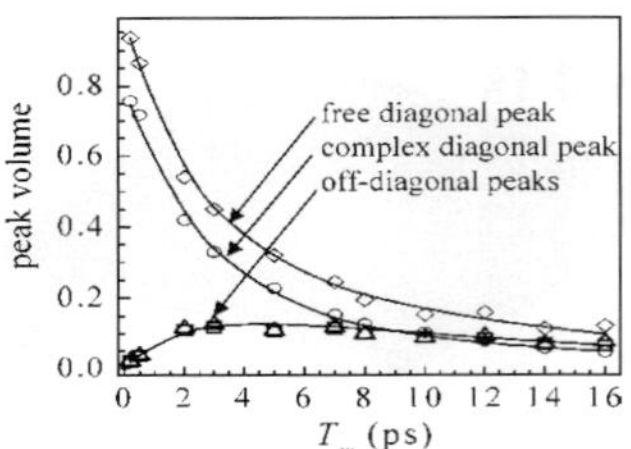

FIGURE 6. Peak volumes (symbols) of the four peaks in the 2D spectra vs T_w. Fits to the data (solid curves) using one adjustable parameter yield the dissociation time, 8 ps.

removed from consideration by fitting the time-dependent peak volumes rather than the peak amplitudes. The vibrational lifetimes of the hydroxyl stretch and the orientational relaxation rates cause all of the peaks to decay in amplitude, while chemical exchange causes the off-diagonal peaks to grow in. The vibrational lifetimes and orientational relaxation rates are measured independently using ultrafast IR pump–probe spectroscopy.[9,11,18] In addition, the equilibrium constant and transition dipole moments of the species are necessary. These are obtained from the FT-IR absorption spectra. With the input constants known, there is only one adjustable parameter to fit the T_w dependence of all of the peaks in the spectra. Because the rate of complex dissociation is equal to the rate of complex formation, we can fit everything using *a single adjustable parameter*, the complex dissociation time, $\tau_d = 1/k_{cf}$, where k_{cf} is the rate constant for dissociation of the complex (c) to the free form (f).

Figure 6 displays the data and fits for the chemical exchange dynamics of the phenol–benzene complex. There are four peaks in the 2D vibrational echo spectra (see Figures 4 and 5): the two diagonal peaks for the complex and free species and the two off-diagonal peak for cf and f → c. As can be seen from Figure 6, the data are fit very well with the single parameter τ_d. (The mathematical details used to fit the data have been presented.[9,11,18]) The data for the two off-diagonal peaks are identical within experimental error, which is one of the tests that shows that the thermal equilibrium between the complex and free species is not perturbed by the experiments. The results of the fitting yields $\tau_d = 8$ ps. Experiments in which benzene is replaced by substituted benzenes with electron-withdrawing (bromo) or electron-donating groups (methyls) change the dissociation times because they decrease or increase the π system electron density.[9] Complexes with phenol by bromobenzene, benzene, and p-xylene give dissociation times of 6, 8, and 21 ps, respectively. The trend in the dissociation times follows the same trend as the measured enthalpies of formation for the complexes, −1.2, −1.7, and −2.2 kcal/mol, respectively.[9]

In addition to complexes with phenol, other organic solute–solvent complexes have been studied,[18] and more studies are in progress. The 2D vibrational echo chemical exchange method can also be applied to problems such as isomerization or proton transfer. Experiments to directly examine fast isomerization are in progress.

B. Water Dynamics. Water has profound effects on diverse fields of science including chemistry and biology. The properties of liquid water are dominated by the hydrogen bonds among water molecules. The hydrogen bonds produce structured networks that are responsible for water's unique properties. The water hydrogen-bond network is constantly changing over a range of time scales. A water molecule can make zero to four hydrogen bonds that are constantly becoming shorter (stronger) and longer (weaker) and formed and broken. The constantly changing hydrogen-bond networks permit water to accommodate a wide array of chemical processes such as protein folding[33] and ion hydration.[34]

Ultrafast 2D IR vibrational echoes can be used to investigate hydrogen-bond network dynamics in water.[19] When the experimental results are combined with simulations of the experimental observables,[19] a detailed understanding of water emerges. The connection between hydrogen-bond network dynamics and the experiments is through the frequency of the hydroxyl stretch. The frequency of the hydroxyl stretch is lowered (red shifted) when water makes a hydrogen bond.[35] Stronger and more hydrogen bonds cause a greater red shift of the hydroxyl oscillator frequency than weaker and fewer hydrogen bonds.[35,36] As the hydrogen-bond network evolves in time, the strength and number of hydrogen bonds change, which in turn causes the water oscillators' frequencies to change.[19,35,36]

To understand the relationship between the 2D vibrational echo spectrum and the hydrogen-bond dynamics qualitatively an analogy can be made to the chemical exchange experiments discussed in the last section. In the phenol–benzene complex system at short time there are two peaks on the diagonal. At longer times the off-diagonal peaks appear. In water there are a vast number of hydrogen-bonded structures. At short time these structures give rise to an elongated band down the diagonal of the 2D spectrum. As time goes on, exchange of structure makes "off-diagonal" peaks grown in. However, these are not resolved. Instead, on either side of the diagonal effectively continuous groups of off-diagonal peaks grow in. The result is a change in shape of the short time elongated band to a more and more symmetrical band rather than the appearance of new peaks.

To obtain a more quantitative view, consider a single water molecule hydroxyl oscillator. At $t = 0$, it will have a particular frequency, $\omega(0)$. As the local H-bond structure changes, the frequency will change. At long time, independent of its starting frequency the oscillator can have any frequency in the entire broad hydroxyl stretch absorption spectrum because it has lost memory of its starting frequency (structure). A measure of the frequency evolution, and therefore the structural evolution, is the frequency–frequency correlation function (FFCF). The FFCF is related to the probability that an oscillator with initial frequency $\omega(0)$ still has the same frequency at time t later, averaged over all starting frequencies. Vibrational echo experiments make the FFCF an experimentally observable quantity.[32] The FFCF is the basic input into time-dependent diagrammatic perturbation theory that is used to calculate nonlinear optical experimental observables in general[32] and vibrational echo signals in particular.[19] The FFCF and diagrammatic perturbation theory should provide an accurate description for systems such as phenol complexes. However, for water recent theoretical work has shown that it is necessary to go beyond diagrammatic perturbation theory because of non-Condon effects that influence experimental observables.[37] Nonetheless, even for water the FFCF provides important qualitative insights and an almost quantitative description of water dynamics.

The experiments[19] and simulations[19,38] look at the time evolution of the hydroxyl stretching frequency (the OD stretch of low concentration HOD in H_2O), which reports on the structural evolution of the water network. HOD is used to avoid vibrational excitation transfer, which would influence the experiments and is not a ground-state equilibrium process. Figure 7 displays 2D vibrational echo spectra of water as a function of T_w. The figure shows both the 0−1 (red, positive going) and 1−2 (blue, negative going) OD hydroxyl stretch transitions. Each contour represents a 10% change in signal. At 100 fs (top) the 0−1 band is substantially elongated along the diagonal. As time increases, the band changes shape and becomes increasingly symmetrical. By 1.6 ps, the 0−1 band is basically symmetrical along the ω axis. It would be essentially round, but the overlap with the negative going 1−2 band eats away the bottom portion.

The change in shape is directly related to the structural evolution of the hydrogen-bond network. When the shape becomes symmetrical, all hydrogen-bond structural configurations have been sampled, which happens by ∼3 ps. Data such as those presented in Figure 7 can be analyzed quantitatively to extract the FFCF. The experimental FFCF is compared to FFCFs obtained from simulations of water to determine the nature of the hydrogen-bond network motions on different time scales.[19,38,39] The results show that the shortest time scale fluctuations (less than a few hundred femtoseconds) involve very local hydrogen-bond motions mainly of the length of a hydrogen bond. The longer time scale dynamics are global hydrogen-bond network rearrangements that lead to complete randomization of the water structure. The 2D vibrational echo results determined this slowest component of the FFCF to be 1.5 ps. This is the time scale on which the water hydrogen-bond network restructures to accommodate processes such as solvation or protein folding.

C. Chemical Structure Analysis. 2D vibrational echoes can also find uses in chemical analysis.[15] Frequently a molecule will have several modes of the same type on a single molecule, for example, the various C−H stretching modes of toluene. In a mixture of two or more unknown chemical species, it can be difficult to determine which peaks in an IR spectrum belong to a particular molecule or even how many molecules are in the mixture by examining the linear IR spectrum.

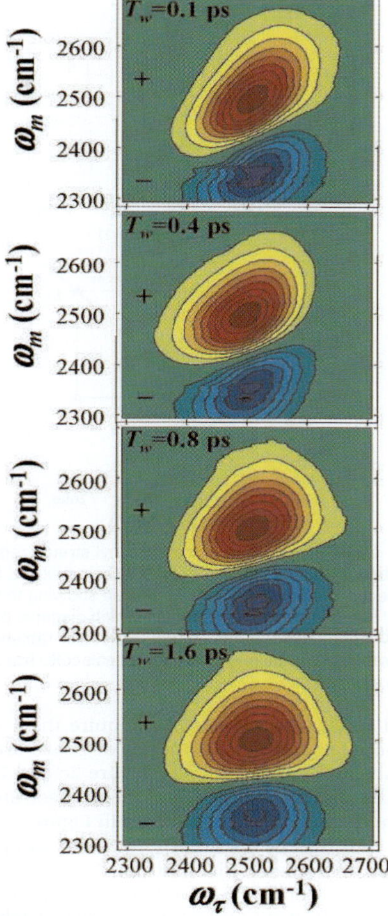

FIGURE 7. 2D vibrational echo spectra of the OD stretch of HOD in water as a function of T_w showing the 0−1 transition (red) and the 1−2 transition (blue). The change in shape of the bands with T_w yields the hydrogen-bond network dynamics.

Vibrational modes on a molecule can be coupled through anharmonic terms in the vibrational potential. In 2D NMR magnetic dipole−magnetic dipole coupling between spins produces off-diagonal peaks. In the 2D IR spectrum coupling between vibrational modes of a molecule also produces off-diagonal peaks. In a mixture of molecules, the off-diagonal peaks can be useful in determining which peaks in an FT-IR absorption spectrum belong to the same molecule.[15]

Figure 8a shows an absorption spectrum with four peaks in the IR region that corresponds to metal carbonyl CO stretching modes. From the absorption spectrum it is not possible to determine if there are four molecules each with one mode, one molecule with four modes, or anything in between. In a 2D IR vibrational echo spectrum there are peaks that only correspond to transitions involv-

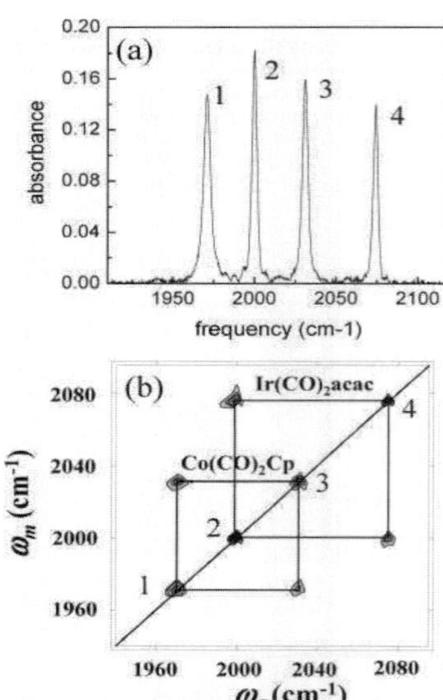

FIGURE 8. (a) FT-IR spectrum of the carbonyl stretch region for a mixed solution of Ir(CO)$_2$(acac) and Co(CO)$_2$Cp in hexane. (b) 2D IR vibrational echo spectrum of the same sample showing the positive peaks only. The off-diagonal peaks show which diagonal peaks are coupled by anharmonic terms in the molecular potential and, therefore, which peaks belong to the same molecule. The lines are aids to the eye.

ing 0 and 1 states and peaks that require the 1 state to first be populated. These two types of peaks are easily distinguishable by their sign (see Figure 2c and 7).[4,15]

Figure 8b shows a 2D vibrational echo spectrum of the same sample that gave the spectrum in Figure 8a.[15] Only the eight positive peaks are shown. The sample is a mixture of two metal dicarbonyls, (acetylacetonato)-dicarbonylIridium (Ir(CO)$_2$(acac)) and (cyclopentadinyl)-dicarbonylcobalt (Co(CO)$_2$Cp) in hexane. Each metal carbonyl has two modes that give rise to two peaks in the absorption spectrum. The result is four peaks in the absorption spectrum and four peaks on the diagonal of the 2D spectrum. However, in addition to the diagonal peaks in the 2D spectrum, there are off-diagonal peaks. Modes that are on the same molecule and coupled via anharmonic terms in the potential generate "coherence transfer" peaks, which form a square with the diagonal peaks that are coupled[4] and therefore on the same molecule.[15] It is clear from the 2D spectrum that there are pairs of coupled peaks. The four peaks in Figure 8a are composed of two pairs of peaks, with peaks 1 and 3 going together and peaks 2 and 4 going together. The peaks corresponding to the two molecules are labeled in Figure 8b. Thus, additional information on mixtures of molecules can be obtained from the 2D spectrum. Because the 2D vibrational echo measurements occur on an ultrafast time scale, it is possible to perform such experiments in reaction mixtures to help identify short-lived intermediates and products.

IV. Concluding Remarks

Ultrafast infrared 2D vibrational echo spectroscopy is an emerging tool for the study of chemical, biological, and materials problems. In the early days of NMR home-built machines were used to do the initial experiments. Now ultrasophisticated multidimensional NMR instruments can be purchased as commercial packages. The laser equipment necessary to perform the ultrafast IR experiments is improving rapidly. The very complex optics, electronics, and computer software necessary to perform the 2D IR experiments are also advancing rapidly. It is easy to envision a future in which ultrafast multidimensional IR vibrational echo spectroscopy and related experiments will be as accessible to the scientific community as NMR.

The authors gratefully acknowledge John B. Asbury, Tobias Steinel, Christopher Stromberg, and Xin Chen, who were involved in many of the experiments that are discussed in this paper. This research was supported by a grant from the AFOSR (F49620-01-1-0018) and NSF (DMR-0332692).

References

(1) Zimdars, D.; Tokmakoff, A.; Chen, S.; Greenfield, S. R.; Fayer, M. D.; Smith, T. I.; Schwettman, H. A. Picosecond Infrared Vibrational Echoes in a Liquid and Glass Using a Free Electron Laser. *Phys. Rev. Lett.* **1993**, *70*, 2718−2721.

(2) Merchant, K. A.; Thompson, D. E.; Fayer, M. D. Two-Dimensional Time-Frequency Ultrafast Infrared Vibrational Echo Spectroscopy. *Phys. Rev. Lett.* **2001**, *86*, 3899−3902.

(3) Zanni, M. T.; Hochstrasser, R. M. Two-Dimensional Infrared Spectroscopy: A Promising New Method for the Time Resolution of Structures. *Curr. Opin. Struct. Biol.* **2001**, *11*, 516−522.

(4) Khalil, M.; Demirdoven, N.; Tokmakoff, A. Obtaining Absorptive Line Shapes in Two-Dimensional Infrared Vibrational Correlation Spectra. *Phys. Rev. Lett.* **2003**, *90*, 047401(4).

(5) Khalil, M.; Demirdoven, N.; Tokmakoff, A. Vibrational Coherence Transfer Characterized with Fourier-Transform 2D IR Spectroscopy. *J. Chem. Phys.* **2004**, *121*, 362−373.

(6) Asbury, J. B.; Steinel, T.; Fayer, M. D. Hydrogen Bond Networks: Structure and Evolution after Hydrogen Bond Breaking. *J. Phys. Chem. B* **2004**, *108*, 6544−6554.

(7) Mukherjee, P.; Krummel, A. T.; Fulmer, E. C.; Kass, I.; Arkin, I. T.; Zanni, M. T. Site-Specific Vibrational Dynamics of the Cd3 Zeta Membrane Peptide Using Heterodyned Two-Dimensional Infrared Photon Echo Spectroscopy. *J. Chem. Phys.* **2004**, *120*, 10215−10224.

(8) Fulmer, E. C.; Ding, F.; Zanni, M. T. Heterodyned Fifth-Order 2D-IR Spectroscopy of the Azide Ion in an Ionic Glass. *J. Chem. Phys.* **2005**, *122*, 034302(12).

(9) Zheng, J.; Kwak, K.; Asbury, J. B.; Chen, X.; Piletic, I.; Fayer, M. D. Ultrafast Dynamics of Solute-Solvent Complexation Observed at Thermal Equilibrium in Real Time. *Science* **2005**, *309*, 1338−1343.

(10) DeCamp, M. F.; DeFlores, L.; McCracken, J. M.; Tokmakoff, A.; Kwac, K.; Cho, M. Amide I Vibrational Dynamics of *N*-Methylacetamide in Polar Solvents: The Role of Electrostatic Interactions. *J. Phys. Chem. B.* **2005**, *109*, 11016−11026.

(11) Kwak, K.; Zheng, J.; Cang, H.; Fayer, M. D. Ultrafast 2D IR Vibrational Echo Chemical Exchange Experiments and Theory. *J. Phys. Chem. B.* **2006**, accepted for publication.

(12) Scheurer, C.; Mukamel, S. Magnetic Resonance Analogies in Multidimensional Vibrational Spectroscopy. *Bull. Chem. Soc. Jpn.* **2002**, *75*, 989−999.

(13) Mukamel, S. Multidimensional Femtosecond Correlation Spectroscopies of Electronic and Vibrational Excitations. *Ann. Rev. Phys. Chem.* **2000**, *51*, 691−729.

(14) Golonzka, O.; Tokmakoff, A. Polarization-Selective Third-Order Spectroscopy of Coupled Vibronic States. *J. Chem. Phys.* **2001**, *115*, 297−309.

(15) Asbury, J. B.; Steinel, T.; Fayer, M. D. Using Ultrafast Infrared Multidimensional Correlation Spectroscopy to Aid in Vibrational Spectral Peak Assignments. *Chem. Phys. Lett.* **2003**, *381*, 139−146.

(16) Merchant, K. A.; Noid, W. G.; Akiyama, R.; Finkelstein, I.; Goun, A.; McClain, B. L.; Loring, R. F.; Fayer, M. D. Myoglobin—CO Substate Structures and Dynamics: Multidimensional Vibrational Echoes and Molecular Dynamics Simulations. *J. Am. Chem. Soc.* **2003**, *125*, 13804—13818.

(17) Kim, Y. S.; Hochstrasser, R. M. Chemical Exchange 2D IR of Hydrogen-Bond Making and Breaking. *Proc. Natl. Acad. Sci.* **2005**, *102*, 11185—11190.

(18) Zheng, J.; Kwak, K.; Chen, X.; Asbury, J. B.; Fayer, M. D. Formation and Dissociation of Intra-Intermolecular Hydrogen Bonded Solute— Solvent Complexes: Chemical Exchange 2D IR Vibrational Echo Spectroscopy. *J. Am. Chem. Soc.* **2006**, *128*, 2977—2987.

(19) Asbury, J. B.; Steinel, T.; Kwak, K.; Corcelli, S.; Lawrence, C. P.; Skinner, J. L.; Fayer, M. D. Dynamics of Water Probed with Vibrational Echo Correlation Spectroscopy. *J. Chem. Phys.* **2004**, *121*, 12431—12446.

(20) Fecko, C. J.; Eaves, J. D.; Loparo, J. J.; Tokmakoff, A.; Geissler, P. L. Local and Collective Hydrogen Bond Dynamics in the Ultrafast Vibrational Spectroscopy of Liquid Water. *Science* **2003**, *301*, 1698—1702.

(21) Merchant, K. A.; Noid, W. G.; Thompson, D. E.; Akiyama, R.; Loring, R. F.; Fayer, M. D. Structural Assignments and Dynamics of the a Substates of Mbco: Spectrally Resolved Vibrational Echo Experiments and Molecular Dynamics Simulations. *J. Phys. Chem. B* **2003**, *107*, 4—7.

(22) Tokmakoff, A.; Fayer, M. D. Infrared Photon Echo Experiments: Exploring Vibrational Dynamics in Liquids and Glasses. *Acc. Chem. Res.* **1995**, *28*, 437—445.

(23) Tokmakoff, A.; Zimdars, D.; Sauter, B.; Francis, R. S.; Kwok, A. S.; Fayer, M. D. Vibrational Photon Echoes in a Liquid and Glass: Room Temperature to 10 K. *J. Chem. Phys.* **1994**, *101*, 1741.

(24) Tokmakoff, A.; Kwok, A. S.; Urdahl, R. S.; Francis, R. S.; Fayer, M. D. Multilevel Vibrational Dephasing and Vibrational Anhar- monicity from Infrared Photon Echo Beats. *Chem. Phys. Lett.* **1995**, *234*, 289.

(25) Asplund, M. C.; Lim, M.; Hochstrasser, R. M. Spectrally Resolved Three Pulse Photon Echoes in the Vibrational Infrared. *Chem. Phys. Lett.* **2000**, *323*, 269—277.

(26) Kim, Y.; Hochstrasser, R. M. Dynamics of Amide-I Modes of the Alanine Dipeptide in D_2O. *J. Phys. Chem. B* **2005**, *109*, 6884—6891.

(27) Piletic, I.; Moilanen, D. E.; Spry, D. B.; Levinger, N. E.; Fayer, M. D. Testing the Core/Shell Model of Nanoconfined Water in Reverse Micelles Using Linear and Nonlinear IR Spectroscopy. *J. Phys. Chem. A* **2006**, *110*, 4985—4999.

(28) Massari, A. M.; Finkelstein, I. J.; Fayer, M. D. Dynamics of Proteins Encapsulated in Silica Sol-Gel Glasses Studied with IR Vibrational Echo Spectroscopy. *J. Am. Chem. Soc.* **2006**, *128*, 3990—3997.

(29) Hahn, E. L. Spin Echoes. *Phys. Rev.* **1950**, *80*, 580—594.

(30) Abella, I. D.; Kurnit, N. A.; Hartmann, S. R. Photon Echoes. *Phys. Rev.* **1966**, *14*, 391—406.

(31) Asbury, J. B.; Steinel, T.; Fayer, M. D. Vibrational Echo Correlation Spectroscopy Probes Hydrogen Bond Dynamics in Water and Methanol. *J. Lumin.* **2004**, *107*, 271—286.

(32) Mukamel, S. *Principles of Nonlinear Optical Spectroscopy*; Oxford University Press: New York, 1995.

(33) Daggett, V.; Fersht, A. The Present View of the Mechanism of Protein Folding. *Nat. Rev. Mol. Cell Biol.* **2003**, *4*, 497—502.

(34) Spangberg, D.; Rey, R.; Hynes, J. T.; Hermansson, K. Rate and Mechanisms for Water Exchange around Li^+(Aq) from Md Simu- lations. *J. Phys. Chem. B* **2003**, *107*, 4470—4477.

(35) Lawrence, C. P.; Skinner, J. L. Vibrational Spectroscopy of HOD in Liquid D_2O. III. Spectral Diffusion, and Hydrogen-Bonding and Rotational Dynamics. *J. Chem. Phys.* **2003**, *118*, 264—272.

(36) Lawrence, C. P.; Skinner, J. L. Vibrational Spectroscopy of HOD in Liquid D_2O. I. Vibrational Energy Relaxation. *J. Chem. Phys.* **2002**, *117*, 5827.

(37) Schmidt, J. R.; Corcelli, S. A.; Skinner, J. L. Pronounced Non- Condon Effects in the Ultrafast Infrared Spectroscopy of Water. *J. Chem. Phys.* **2005**, *123*, 044513(13).

(38) Corcelli, S.; Lawrence, C. P.; Skinner, J. L. Combined Electronic Structure/Molecular Dynamics Approach for Ultrafast Infrared Spectroscopy of Dilute HOD in Liquid H_2O and D_2O. *J. Chem. Phys.* **2004**, *120*, 8107.

(39) Møller, K. B.; Rey, R.; Hynes, J. T. Hydrogen Bond Dynamics in Water and Ultrafast Infrared Spectroscopy: A Theoretical Study. *J. Phys. Chem. A* **2004**, *108*, 1275—1289.

AR068010D

Probing dynamics of complex molecular systems with ultrafast 2D IR vibrational echo spectroscopy

Ilya J. Finkelstein, Junrong Zheng, Haruto Ishikawa, Seongheun Kim, Kyungwon Kwak and Michael D. Fayer*

Received 12th December 2006, Accepted 16th January 2007
First published as an Advance Article on the web 20th February 2007
DOI: 10.1039/b618158a

Ultrafast 2D IR vibrational echo spectroscopy is described and a number of experimental examples are given. Details of the experimental method including the pulse sequence, heterodyne detection, and determination of the absorptive component of the 2D spectrum are outlined. As an initial example, the 2D spectrum of the stretching mode of CO bound to the protein myoglobin (MbCO) is presented. The time dependence of the 2D spectrum of MbCO, which is caused by protein structural evolution, is presented and its relationship to the frequency–frequency correlation function is described and used to make protein structural assignments based on comparisons to molecular dynamics simulations. The 2D vibrational echo experiments on the protein horseradish peroxidase are presented. The time dependence of the 2D spectra of the enzyme in the free form and with a substrate bound at the active site are compared and used to examine the influence of substrate binding on the protein's structural dynamics. The application of 2D vibrational echo spectroscopy to the study of chemical exchange under thermal equilibrium conditions is described. 2D vibrational echo chemical exchange spectroscopy is applied to the study of formation and dissociation of organic solute–solvent complexes and to the isomerization around a carbon–carbon single bond of an ethane derivative.

I. Introduction

In 1950, the NMR "spin echo"[1] ushered in the last half century of the development of coherent spectroscopic methods. The spin echo became the basis for the diverse range of pulsed NMR experiments, including multidimensional experiments, that are in use in fields of research from medicine to geology.[2,3] In 1964, the basic concepts inherent in the spin echo were extended to visible spectroscopy using the earliest pulsed lasers to perform "photon echo" experiments on electronic excited states.[4,5] Photon echoes have been widely used to study electronic excited state dynamics in systems such as low temperature crystals [6–10] and glasses,[11–14] proteins,[15–17] and photosynthetic chromophore clusters.[18,19] The advent of short pulse infrared (IR) sources made it possible to perform "vibrational echo" experiments on the vibrations of condensed matter systems such as liquids, glasses and proteins beginning in the early 1990's.[20–24] These first one-dimensional two-pulse vibrational echo experiments were conducted with an IR free electron laser.[20] However, rapid advances in laser technology now make it possible to conduct far more sophisticated vibrational echo experiments, particularly two-dimensional (2D) vibrational echoes, using table top laser systems.

2D IR vibrational echo spectroscopy[25–35] is an ultrafast IR analog of 2D NMR but it operates on molecular vibrations instead of spins. Vibrations are the structural degrees of freedom of molecules. It is the direct probing of molecular structure and the dynamics of molecular structure and intermolecular interactions on ultrafast time scales that makes 2D vibrational echo spectroscopy a powerful tool that is becoming increasingly useful as the methodology develops.

The 2D IR vibrational echo signal is generated by a sequence of three ultrashort IR pulses tuned to the vibrational transitions of interest. The pulse sequence induces and then probes the coherent evolution of excitations (vibrations) of a molecular system. The first pulse in the sequence causes vibrational modes of an ensemble of molecules to "oscillate" initially all with the identical phase. The later pulses generate observable signals that are sensitive to changes in environments of individual molecules during the experiment, even if the aggregate populations in distinct environments do not change. For example, the structural fluctuations of a protein or the formation and dissociation of molecular complexes under thermal equilibrium conditions can be observed. The 2D IR vibrational echo spectrum can also display intramolecular interactions and dynamics that are not observables in a linear IR vibrational absorption experiment because they are masked by the inhomogeneous broadening that, in general, dominates a linear absorption spectral line shape in complex condensed matter molecular systems. A critical difference between the 2D IR and NMR variants is that the IR pulse sequence is sensitive to dynamics on timescales 6 to 10 orders of magnitude faster than NMR.

2D IR vibrational echo spectroscopy has several characteristics that make it a useful tool for the study of problems involving rapid dynamics under thermal equilibrium conditions in condensed phases. Such problems are ubiquitous in

Department of Chemistry, Stanford University, Stanford, CA 94305, USA. E-mail: fayer@stanford.edu

nature and difficult to study by other means. 2D IR vibrational echo experiments have temporal resolution of <100 fs, which is sufficiently fast to study the fastest chemical processes. In contrast to electronic excitation, the vibrational excitation associated with the 2D IR experiments produces a negligible perturbation of a molecular system, which does not change the chemical properties of the samples. 2D IR experiments can also be useful as a tool for chemical structural analysis by revealing the relationship among different mechanical degrees of freedom of a molecular or biomolecular system. 2D IR vibrational echo experiments have been successfully applied to study fast chemical exchange reactions and solution dynamics,[29,31,36] water dynamics,[37,38] hydrogen network evolution,[28] intramolecular vibrational energy relaxations,[34] protein structures and dynamics,[30,39,40] and mixed chemicals analysis.[41]

Since the first fast condensed matter vibrational echo experiments in 1993,[20] the field has advanced a great deal. The first experiments were 1D with a time resolution of ~ 1 ps. Today, multi-dimensional phase resolved vibrational echoes have become almost routine, greatly increasing the types of information that can be obtained. Furthermore, tabletop laser systems now provide sub 50 fs IR pulses, making 2D IR vibrational echo spectroscopy increasingly available to a wide range of researchers. In this article the 2D vibrational echo will be described and two types of experiments will be used to illustrate the applications of the technique. These are the investigation of protein dynamics and structure through the observation and analysis of spectral diffusion, and the study of chemical process, *i.e.*, solute–solvent complex formation and dissociation and molecular isomerization around a carbon–carbon single bond, by measuring the rate of chemical exchange.

II. The 2D vibrational echo method

In a 2D IR vibrational echo experiment, three ultrashort IR pulses tuned to the frequency of the vibrational modes of interest are crossed in the sample. Because the pulses are very short, they have a broad bandwidth, which makes it possible to simultaneously excite a number of vibrational modes or a very broad spectral feature. The IR pulses (~ 50 fs) are produced using a regeneratively amplified Ti:Sapphire laser pumped optical parametric amplifier system.[42] A 2D IR vibrational echo experimental setup is illustrated schematically in Fig. 1. The pulse sequence is shown at the bottom. The times between pulses 1 and 2 and pulses 2 and 3 are called τ and T_w, respectively. The three successive ultrashort IR pulses with wave vectors (propagation directions) k_1, k_2, and k_3 are applied to the sample to induce the subsequent emission of the time delayed 4th pulse, the vibrational echo. The vibrational echo pulse is emitted from the sample in a distinct direction (wave vector $k_e = -k_1 + k_2 + k_3$).

Both the intensity of the echo pulse and its time structure contain important information in the 2D IR vibrational echo experiment. If the echo pulse is sent directly into an IR detector, its intensity is measured, but phase and sign information is lost. Complete information is obtained by allowing the vibrational echo pulse to interfere with a 5th pulse called the

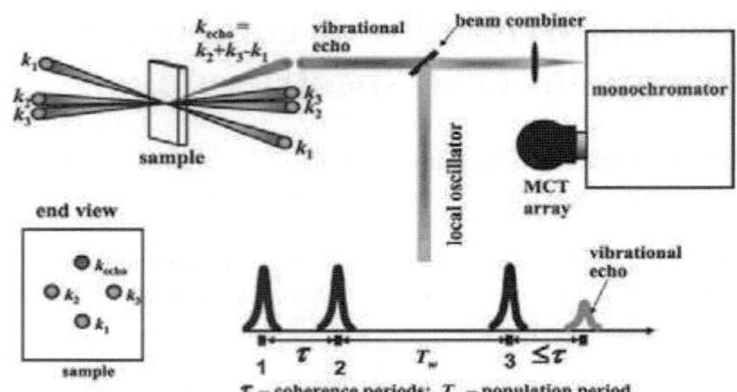

Fig. 1 Schematic layout of the 2D IR experiment. Three mid-IR input beams are labeled as k_1, k_2, and k_3. The emitted vibrational echo pulse and the local oscillator pulse are combined before they are dispersed through a monochromator. The time course of the echo pulse sequence is presented below the experimental layout.

local oscillator. In the heterodyne detected vibrational echo experiments, the local oscillator and vibrational echo pulses are collinear, and the phase information is obtained by observing the interference pattern (interferogram) as a function of time (see below). In addition to providing phase and sign information, combining the vibrational echo with the local oscillator optically heterodyne amplifies the vibrational echo signal. The combined pulses are dispersed in a monochromator and then detected with a 32-element MCT IR array detector, which measures the signal at 32 wavelengths simultaneously.

Qualitatively, the vibrational echo experiment works in the following manner. The first pulse in the sequence places the vibrational oscillators into a coherent superposition state of the vibrational ground state (0) and vibrational first excited state (1), with all of the oscillators initially in phase. The initial phase relationships among the oscillators decay rapidly causing a decay of the initial macroscopic polarization that is generated by an in-phase ensemble of oscillators. The decay of the macroscopic polarization is the free induction decay. The phase relationships among the oscillators decay because of inhomogeneous broadening of the spectral line (a spread of transition frequencies associated with the finite absorption linewidth) with additional contributions from fast fluctuations of the transition frequencies caused by structural dynamics of the system. The rest of the pulse sequence can recover the phase relationships seemingly lost during the free induction decay because of inhomogeneous broadening. Only structural fluctuations that randomize the initial starting frequencies of the ensemble of oscillators (produced by spectral diffusion) and occur over a sufficiently long time can totally destroy the phase relationships among the oscillators. This destruction of phase relations caused by spectral diffusion cannot be recovered by the rest of the pulse sequence. However, even then, the signal is not zero because the initial two pulses also set up a spatial relationship (a transient grating[43]) among the oscillators that contributes to the signal.

The second pulse in the sequence stores the phase (and spatial) information induced by the first pulse as a complex frequency and spatial pattern of differences of the populations of the 0 and 1 vibrational states. To see how the information is stored, first consider the spatial information. The first pulse produces a polarization that will interfere with the electric field of the second pulse if it arrives on a time comparable to the free induction decay. Because the paths of the two beams are crossed, the interference produces a spatial fringe pattern, alternating regions of light and dark. Where it is light, excited state population is produced. Therefore, there are alternating spatial regions of excited vibrations and only ground state vibrations. This pattern can be thought of a population grating. This pattern lasts for the vibrational lifetime. The phase information is stored in a similar manner. Following the first pulse, vibrations at each frequency across the absorption line are in superposition states. Depending on the particular vibrational frequency of a molecule's superposition state, it will have a phase relationship relative to the electric field of the second pulse that will either drive the vibration into the excited state or drive it into the ground state. Because of the spread in frequencies across the absorption band, along the frequency axis, there are alternating bands of excited states and ground states. This frequency pattern can be thought of as a frequency grating. Whether a molecule is in the excited state or ground state following the second pulse is determined by its phase relationship to the electric field of the second pulse. Therefore, the second pulse stores phase information in the form of the frequency grating.

After the waiting period, T_w, (see Fig. 1) the third pulse again generates coherent superposition states of the oscillators. Initially, the oscillators are not in phase, but the pulse sequence initiates a rephasing process.[1] At a time $\leq \tau$ after the third pulse, the vibrational oscillators are again oscillating in phase. Each vibrational oscillator has associated with it a microscopic oscillating electric dipole. When rephasing has occurred, the sample again has a macroscopic polarization that acts as the source of the fourth pulse, the vibrational echo. The pulse is short because the dipoles again get out of phase just as they did after the first pulse.

In a dynamic system, the first laser pulse "labels" the initial structures of the species in the sample and initiates the first coherence period. The second pulse ends the first coherence period, τ, and starts clocking the waiting time during which the labeled species undergo structural dynamics that change their frequencies. The third pulse ends the waiting period of length T_w, and begins the last coherence period of length $\leq \tau$, which ends with the emission of the vibrational echo pulse (see Fig. 1). The echo signal contains information about the final structures of all labeled species.

During the coherence periods fast vibrational frequency fluctuations induced by fast structural fluctuations cause dynamic dephasing, which is one contribution to the line shapes in the 2D spectrum. During the T_w period, called the population period, the oscillators are either in the ground or first excited vibrational state, not a superposition state. Slower structural fluctuations of the system give rise to spectral diffusion (slower time scale evolution of the oscillator frequencies) that contributes to the 2D line shapes. Spectral diffusion can be thought of as a smearing of the frequency grating as molecules wander from frequency grating excited state frequency regions into frequency grating ground state regions

and *vice versa*. Then the third pulse does not produce as large a vibrational echo signal because of destruction of the frequency grating by spectral diffusion during the population period. Spectral diffusion destroys the phase information that was stored in the frequency grating immediately following the second pulse. Other processes during the population period also produce changes in the 2D spectrum. For example, chemical exchange can occur in which two species in equilibrium are interconverting one to the other without changing the overall number of either species. Chemical exchange causes new peaks to grow in as T_w is increased. In an experiment, τ is scanned for fixed T_w. The recorded signals are converted into a 2D vibrational echo spectrum. Then T_w is increased and another spectrum is obtained. The series of spectra taken as a function of T_w provides information on dephasing, spectral diffusion, and population dynamics.

To obtain the well resolved 2D vibrational echo spectra discussed below, it is necessary to acquire essentially pure absorption spectra. The measured interferogram contains both the absorptive and dispersive components of the vibrational echo signal. To isolate the absorptive components, two sets of quantum pathways are measured independently by appropriate time ordering of the pulses in the experiment.[26] With pulses 1 and 2 at the time origin, pathway 1 or 2 is obtained by scanning pulse 1 or 2 to negative time, respectively. By adding the Fourier transforms of the interferograms from the two pathways, the dispersive component can be substantially cancelled leaving only the absorptive component.[26,44] Following a phasing procedure[44,45] based on the projection slice theorem,[26,44,46] the 2D vibrational echo spectra are constructed by plotting the amplitude of the absorptive part of spectrum (see below).

III. 2D vibrational echo experiments

A. Illustration of the technique—myoglobin-CO

To illustrate the nature of the method, experiments on the CO stretch of CO bound to the active site of the protein myoglobin is used as an example. Carbonmonoxy-myoglobin (MbCO) has been extensively studied both experimentally[47–56] and computationally.[57–60] Myoglobin (Mb) is a small globular heme protein weighing approximately 17 kDa that is found in mammalian muscle tissue. The prosthetic heme group can reversibly bind a number of small molecule ligands such as O_2, CO, and NO. The crystal structure of Mb has been known for over 40 years.[61–63] The CO stretching mode of MbCO has a strong transition dipole, making its infrared absorption an easily monitored experimental observable that can be used to track kinetics and dynamics in the protein. Studies on Mb, and in particular on MbCO, serve as tests for many of the ideas on the relationship between structure and function in proteins.

Fig. 2a shows the linear FT-IR absorption spectrum of the CO stretch of MbCO. There are three bands labeled A_0, A_1, and A_3 that correspond to different configurations of the distal histidine (H64). These bands will be discussed below. In the 2D vibrational echo spectrum, there are two frequency axes, which require two Fourier transforms of the time domain data to convert the time structure of the echo observable into 2D

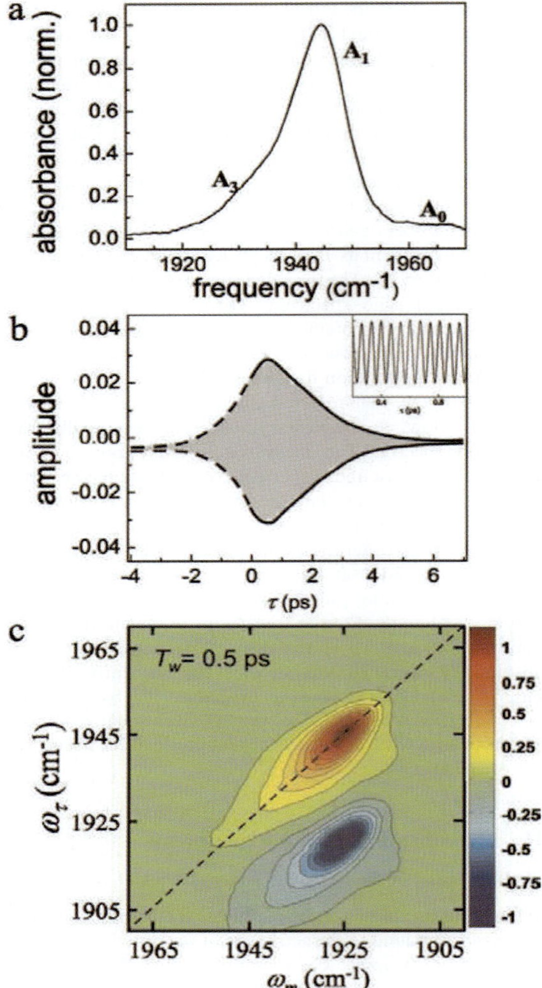

Fig. 2 (a) The FT-IR spectrum of MbCO. (b) A time-domain interference pattern collected in the 2D IR experiment at the center of the A_1 transition in MbCO. The interferogram is shown in gray and envelope as a solid blue or red line. Positive τ (solid envelope) is defined by beam 1 arriving at the sample before beam 2 and negative τ (dashed envelope) indicates that beam 2 arrives before beam 1. A small portion of the interferogram with high time resolution is shown in the inset. (c) 2D IR absorptive spectrum of MbCO at $T_w = 0.5$ ps. The positive going bands correspond to 0–1 transitions of the A_1 and A_3 states and negative going bands are the 1–2 transitions, displaced along ω_m by the anharmonicity. The A_0 state is not visible at this contour level (10%).

frequency data. As shown in Fig. 1, the vibrational echo pulse, which is overlapped with the local oscillator (LO) pulse, is passed through the monochromator. Taking the spectrum of the pulse performs one of the Fourier transforms and provides the vertical axis in the 2D spectrum, ω_m (m for monochro-

mator, see Fig. 2c). The other frequency axis is obtained by scanning τ. Scanning τ produces an interferogram (see Fig. 2b) as the echo pulse changes its phase relationship relative to the fixed LO pulse. There is one interferogram for each frequency on the ω_m axis for which there is signal. The numerical Fourier transforms of the τ scan interferograms provide the ω_τ axis. The data $S(\omega_\tau, \omega_m, T_w)$ are then plotted for each T_w in a three dimensional representation, that is, the amplitude as a function of both ω_τ and ω_m (the ω_1 and ω_3 axes, respectively in 2D NMR). More experimental details of the method including phase error corrections have been presented in detail.[26,42]

In the 2D IR spectrum of MbCO (Fig. 2c) there is a positive going band (red) on the diagonal (dashed line) and a negative going band (blue) off-diagonal. Each contour is a 10% change in amplitude. The positive band on the diagonal reflects the ground state to first vibrationally excited state (0–1 transition) of the CO stretch of MbCO. The negative going off-diagonal band involves the 1–2 transition. The frequencies of the first interaction of the radiation field (first pulse) with the vibrations are the frequencies on the ω_τ axis. The frequencies of the third interaction (third pulse) with the vibrations, which is also the frequency of the echo emission, is the frequency on the ω_m axis. If $\omega_m = \omega_\tau$, the peak is on the diagonal. This is the situation for the red band in Fig. 2c (the 0–1 transition).

The blue off-diagonal 1–2 band arises as follows. The first interaction produces a coherent superposition state of the 0–1 transition. One of the quantum pathways for the second interaction (second pulse) produces a population in the 1 state (first vibrationally excited state). Then the third pulse can couple the 1 state to the 2 state, and produce a coherent superposition of 1 and 2. This results in the vibrational echo being emitted at the 1–2 transition frequency, which is shifted to lower frequency than the 0–1 transition by the vibrational anharmonicity. Therefore, the ω_τ frequency is the 0–1 frequency (1945 cm^{-1}), but the ω_m frequency is the 1–2 frequency (1921 cm^{-1}). The off-diagonal anharmonicity peak is negative going because there is a 180° phase shift in the echo pulse electric field relative to the 0–1 echo. The spectrum in Fig. 2c is for $T_w = 0.5$ ps, a short time compared to the time scale of protein structural fluctuations in this system. Therefore, spectral diffusion is not complete, and the 0–1 band in the 2D spectrum is elongated along the diagonal and the 1–2 band is elongated along a line parallel to the diagonal. The elongation is the signature of inhomogeneous broadening. However, while the band is dominated by the A_1 peak (see Fig. 2a) there is also some contribution along the diagonal from the A_3 band. With the 10% contours shown, the A_0 band is not visible but can be seen when finer contours are used. As discussed below, the change in shape with increasing T_w measures spectral diffusion, and, thus, the structural evolution of the system. The contribution to the band from the A_1 and A_3 peaks can be separated and analyzed to determine their respective dynamics.[40,64]

Fig. 3 displays 2D vibrational echo data for MbCO at four T_ws. The diagonal 0–1 band and the off-diagonal 1–2 band contain essentially the same information. Because the vibrations are nearly harmonic, it can be shown that the 0–1 and 1–2 bands will be nearly identical in their time dependent evolution, which is confirmed by experimental studies.

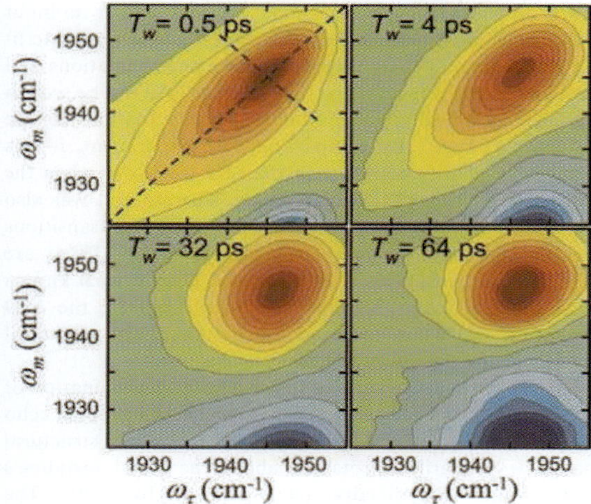

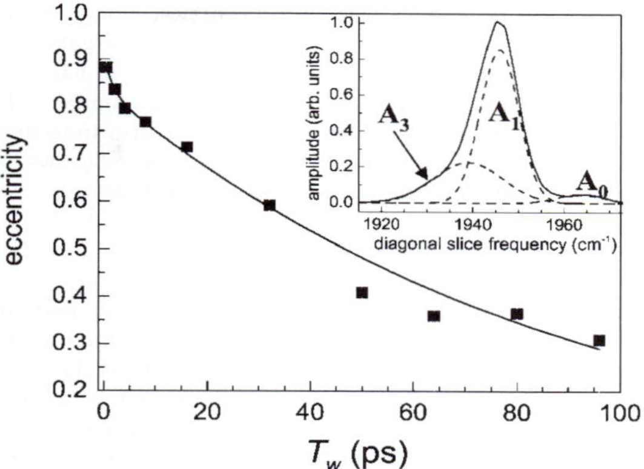

Fig. 3 A series of 2D IR spectra for MbCO as a function of increasing T_w. Mainly the 0–1 transition is shown and the data are presented at 10% contour levels. Dashed lines through the top left panel indicate the diagonal (long line) and anti-diagonal (short line) slice through the center of the A$_1$ peak.

Fig. 4 Eccentricity of the A$_1$ substate of MbCO. The line through the data is a bi-exponential fit. Inset: Normalized diagonal slice through the 2D IR data for MbCO at T_w = 0.5 ps. The spectrum is fit to three Gaussian curves, centered at frequencies of the A$_i$ bands in MbCO.

Therefore, only the 0–1 band will be discussed. As T_w increases, the shape of the spectrum changes, going from elongated along the diagonal to essentially round. The change in shape of the bands as T_w is increased reflects protein structural dynamics. The vibrational echo pulse sequence used to collect 2D IR spectra displays inhomogeneous broadening along the diagonal and dynamic broadening along the anti-diagonal (shown as dashed lines in the upper left panel of Fig. 3).[65] At short T_w the 2D dynamic line shape has significant inhomogeneous broadening, which manifests itself as elongation along the diagonal. As T_w is increased, the experiment picks up longer time scale protein dynamics that increases the anti-diagonal width and decreases the diagonal width, but to a lesser extent than the change in the anti-diagonal. In the long time limit, all protein fluctuations contribute to the 2D spectrum, which would lead to a 2D IR shape with equal diagonal and anti-diagonal linewidths.

To analyze the time evolution of the 2D spectrum, the full 2D band shapes are used. It is convenient to discuss the time evolution of 2D spectral band shapes by defining the eccentricity, ε,

$$\varepsilon(T_w) = \sqrt{1 - \frac{\sigma_{AD}^2(T_w)}{\sigma_D^2(T_w)}}, \qquad (1)$$

where σ_{AD} and σ_D are the band widths along the anti-diagonal and diagonal slices, respectively. This definition for the eccentricity displays several convenient limits. In the case of large inhomogeneous broadening, $\sigma_{AD} \ll \sigma_D$, and $\varepsilon \to 1$. At longer T_ws, slower timescale protein dynamics contribute to the anti-diagonal width, causing the eccentricity to decay to zero. In the long time limit in which all protein structures have been sampled, $\varepsilon \to 0$. The eccentricity is a convenient way to

succinctly summarize protein dynamics contained in the 2D IR spectra. The full 2D spectral shapes are used in fitting the data.

As shown in Fig. 2a, the CO absorption of MbCO consists of three bands, the A$_0$, A$_1$, and A$_3$, that arise from distinct orientations of the distal histidine relative to the CO.[64,66] Fig. 4 inset shows the diagonal of the 2D vibrational echo spectrum at T_w = 0.5 ps with the diagonal spectrum decomposed into the three peaks, which are fit as Gaussians. It has been found in the study of several heme proteins that the band shapes are well described as Gaussians. However, as discussed below, the quantitative analysis does not assume a particular line shape. The eccentricity of the A$_1$ state as a function of T_w is shown in the main part of Fig. 4. It was determined by measuring the antidiagonal width at the peak frequency and subtracting the contribution to the width from the overlapping A$_3$ state, which was measured on the red side of the band. The eccentricity of the A$_1$ state of MbCO decays as a biexponential. As discussed immediately below, there is actually a third, T_w independent component in the 2D signal that arises from fast, motionally narrowed protein dynamics that is revealed in the full analysis of the data.

A quantitative description of the amplitudes and timescales of CO frequency fluctuations is provided by the frequency–frequency correlation function (FFCF).[55,64,67–70] Structural fluctuations give rise to frequency fluctuations. The FFCF is the probability that a molecule with frequency ω at time $t = 0$, still has frequency ω at some later time averaged over the complete ensemble of molecules. As time increases, molecules sample an increasingly wide range of frequencies, and the FFCF decays. At sufficiently long time, all accessible structures are sampled, and all frequencies in the absorption line are sampled as well. Therefore, the FFCF decays to zero.

Within standard approximations, both the linear absorption spectrum and the 2D IR vibrational echo spectra at all T_w can be simultaneously calculated with the appropriate FFCF.[68] The FFCF describes the nature of the underlying dynamics that give rise to the experimental observables. A multi-exponential form of the FFCF, $C(t)$, conveniently organizes a distribution of protein fluctuation time scales and has been found to reproduce the influence of structural dynamics on the CO frequency in other heme proteins.[56,64,71] A tri-exponential FFCF was sufficient to describe the protein dynamics of MbCO and horseradish peroxidase discussed below,

$$C(t) = \Delta_1^2 e^{-t/\tau_1} + \Delta_2^2 e^{-t/\tau_2} + \Delta_3^2 e^{-t/\tau_3} \qquad (2)$$

Δ_i is the amplitude of frequency fluctuations (in units of angular frequency) resulting from structural evolution with a characteristic time τ_i. In some of the experiments discussed below, the last term is a static component in $C(t)$, Δ_3^2, that is, $\tau_3 = \infty$. A static contribution occurs from protein structures that interconvert slower than the observable experimental timescales. The 2D IR experiment is sensitive to dynamics that occur out to several T_w.[72] If $\Delta_i^{-1}\tau_i \ll 1$ for a given exponential term, then this component of the FFCF is motionally narrowed.[71,73] A motionally narrowed term in $C(t)$ contributes a T_w independent symmetric 2D Lorentzian line with a width $\Gamma^* = \pi\Delta^2\tau$ to the total 2D IR spectrum. For a motionally narrowed component, Δ and τ cannot be determined independently, but rather Γ^* is obtained. The motionally narrowed component of the FFCF arises from very fast local motions of small groups that do not significantly modify the protein topology. These fluctuations have been observed in experiments and simulations of several heme-CO proteins.[56,71]

FFCFs were obtained from the data by iterative fitting using response theory calculations.[67,68] The FFCF was deemed correct when it could be used to simultaneously calculate 2D IR spectra at all T_w, and the linear absorption spectrum. Such calculations can also reproduce the T_w dependence of the eccentricity (see eqn (1)). For the experiments on MbCO and on horseradish peroxidase (HRP) discussed below, the data and calculations were in excellent agreement.[64,74]

From studying the dynamics of MbCO and combining the results with molecular dynamics (MD) simulations, it was possible to determine the structures of the A_1 and A_3 substates.[40,64] It has been well documented that the A_0 state has the distal histidine swung out of pocket that contains the active iron heme site.[75,76] It was also known that the A_1 and A_3 states have the distal histidine in the pocket. However, these structures interconvert too fast to be amenable to study using other techniques such as NMR. Using the vibrational echo experiments, the FFCFs of both the A_1 and A_3 states were determined. The results were then compared to MD simulations. However, it was first necessary to develop a method to calculate the ultrafast vibrational echo observables from a classical MD simulation of MbCO.[70] The method uses the MD simulations to determine the fluctuating electric field along the CO transition dipole. The electric field fluctuations are caused by the motions of all of the groups in the proteins, which carry partial charges. The simulation produces the time dependent fluctuating electric field and then through the Stark coupling constant, the fluctuating CO frequency.[77]

Fig. 5a shows a portion of a frequency trajectory obtained from one run of the MD simulations.[40,64] From the time dependent frequency shift, the FFCF is calculated. Then the

FFCF obtained from the MD simulations is used as an input in the same diagrammatic perturbation calculations used to fit the 2D IR data. The agreement between the simulations and the experiments was found to be excellent.[64] As can be seen in Fig. 5a, periodically, the nature of the fluctuations in the frequency changes. By comparison with experiment, it was possible to identify segments of the MD trajectory when the protein is in either the A_1 or A_3 state. Therefore, it was also possible to identify the points in time when the transitions occurred between the two structural substates. These are indicated in Fig. 5a. This is the key point. Once it what known when the transitions between substates occurred, the MD simulation was examined, and the nature of the structural changes at the transition points were identified.

Fig. 5b shows the results of the structural determination of the A_1 and A_3 states based on the combined vibrational echo experiments and the MD simulation.[40,64] The structural change is primarily a rotation about the distal histidine's C_β–C_α bond. The dihedral angle changes by $\sim 40°$. The distance form the protonated nitrogen N_ε–H to the CO is ~ 5 Å in the A_1 state but only ~ 3 Å in the A_3 state. Thus, the vibrational echo experiments not only provided insight into

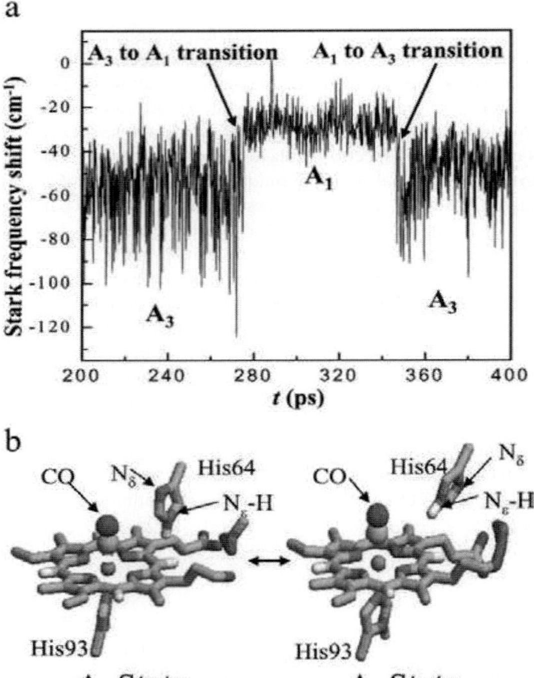

Fig. 5 (a) The time dependent Stark frequency shift taken from a segment of an MD simulation of MbCO. Discontinuous steps in the Stark shift signal structural interconversion between the A_1 and A_3 states of MbCO. (b) Snapshots of the structures of the A_1 and A_3 states of MbCO obtained from MD simulations.

the dynamics of the protein but also provided the necessary observables enabling the solution of a long standing structural problem.

B. The Influence of substrate binding of enzyme dynamics

Enzyme–substrate binding is a dynamic process that is intimately coupled to protein structural fluctuations.[78,79] A complete description of protein–ligand interactions requires information on the modification of protein dynamics when a ligand binds. Recent observations that many proteins can bind structurally heterologous substrates with high affinity in the same active site has further highlighted the importance of protein dynamics in modulating substrate affinity.[79]

A mechanism for multiple substrate recognition within the same binding site has been proposed based on the concept of a protein's distribution of structural states in equilibrium.[79] Proteins are rapidly interconverting within an ensemble of similar conformations.[64,80–82] A subset of these rapidly interconverting states may be favorable for binding a given substrate. A different subset of conformations may accommodate a structurally heterologous ligand. Although conceptually appealing, this mechanism is difficult to probe experimentally as it requires sensitivity to protein structural dynamics on fast timescales.

Using 2D IR vibrational echo spectroscopy we are able to address the questions of whether substrate binding influences protein dynamics and whether different substrates binding to the same protein produce distinct changes in protein structural fluctuations. 2D IR vibrational echo experiments were conducted to examine the equilibrium structural fluctuations of horseradish peroxidase (HRP) in the absence and presence of small molecule substrates with dissociation constants spanning three orders of magnitude.[74] HRP is a type III peroxidase family glycoprotein that oxidizes a variety of organic molecules in the presence of hydrogen peroxide as the oxidizing agent.[83] HRP has proven to be amenable to protein engineering and its reactivity towards a wide variety of organic substrates has made it of intense interest in bio-industrial and enantiospecific catalysis applications.[84–86]

The active site of HRP is comprised of a solvent exposed iron heme prosthetic group (see Fig. 6) that participates in the enzymatic catalysis cycle.[85,86] The heme can bind carbon monoxide (CO), which has been frequently exploited as a site specific reporter of protein structure[55,64,66,87] and dynamics.[64,70,88] As in the MbCO experiments discussed above, the time dependence of the CO transition frequency is a spectroscopic reporter of protein structural fluctuations.[55,64,70] Like MbCO, the CO transition frequency of HRP is highly sensitive to electric fields.[87–89]

Five small molecule substrates that are benzhydroxamic acid (BHA) analogs have been studied with 2D vibrational echoes.[74] Only one of them will be discussed here. These substrates have been investigated as a general class of tightly binding inhibitors for HRP and other peroxidases.[85,90] A wealth of structural,[91,92] biochemical,[84,86,93] and spectroscopic evidence[88,94,95] has identified His42 and Arg38 as the key HRP residues in modulating substrate binding and enzymatic activity. Fig. 6 shows several key residues in the active site of HRP.

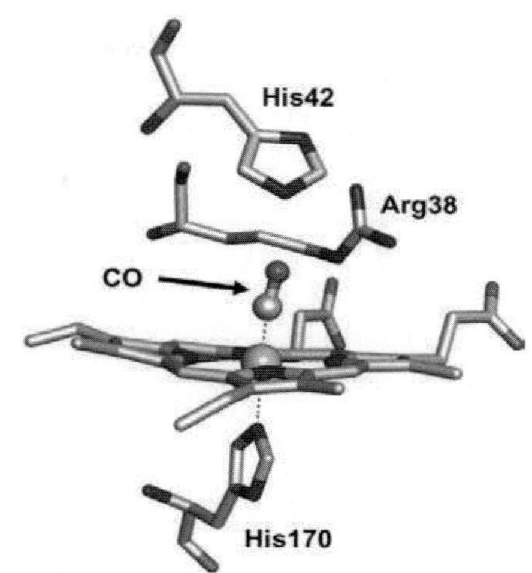

Fig. 6 X-Ray crystal structure of the active site of HRP-CO (PDB ID: 1W4Y). The distal residues His42 and Arg38 are important in the enzymatic cycle of HRP and are key players in modulating the protein's substrate specificity.

These distal heme residues are highly conserved in the peroxidase family.[85] Substrates and intermediates in the catalytic reaction of HRP interact with the distal residues *via* an extensive hydrogen bonding network within the active site.[91,96]

Aqueous HRP-CO in D_2O in the free form (without substrate) pD 7.4 adopts two distinct spectroscopic states[97,98] at 1903 cm^{-1} and 1934 cm^{-1} as shown in the background subtracted and normalized linear absorption spectrum in Fig. 7a. Upon substrate binding, HRP occupies a single structural state resulting in a change in the spectrum to a single peak (Fig. 7b). Resonance Raman spectra of the distal histidine imidazolium[99] and other experimental data[98] suggest that in the red state the CO is nearly normal to the heme plane and oriented such that it has a strong interaction with the distal histidine and a weaker one with the distal arginine. In the blue state, the Fe–C–O linkage, although linear, is somewhat bent relative to the heme normal, allowing the CO to form a strong hydrogen bond with the distal arginine and a weaker one with the distal histidine.

Fig. 7b shows the background subtracted linear spectra of HRP ligated with BHA, a single, essentially Gaussian peak centered at 1909 cm^{-1}.[97,98] Spectra with four other substrates[74] show that the ligated HRP spectra are spread around the frequency of the unligated HRP red state and are significantly lower in frequency than the blue unligated state. The disappearance of the blue CO band and narrowing of the absorption peaks relative to the free HRP red state spectrum suggests decreased protein conformational freedom.

Fig. 8 shows the 2D vibrational echo spectra of the free form of HRP-CO at several T_ws (10% contours). In each panel, the spectra are normalized to the largest peak. The two positive going bands on the diagonal correspond to the two peaks in the linear IR spectrum shown in Fig. 7a. The

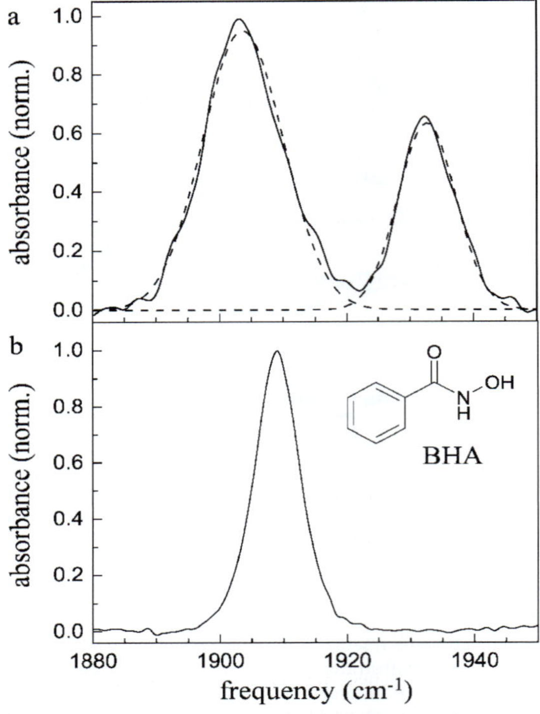

Fig. 7 FT-IR spectra of HRP-CO in the (a) free form with two Gaussian fits and (b) ligated with the substrate BHA. The structure of BHA is shown in the lower panel.

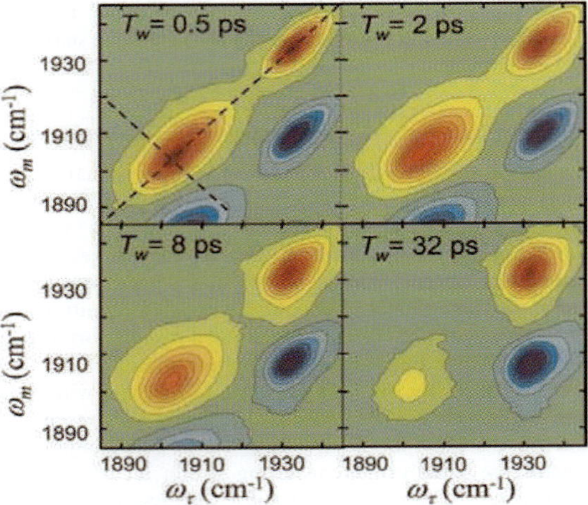

Fig. 8 2D IR spectra of HRP-CO in the free form as a function of increasing T_w.

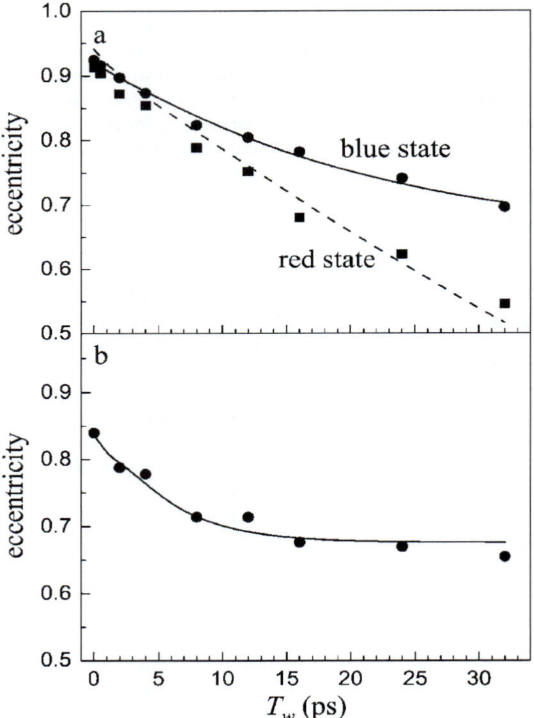

Fig. 9 (a) Eccentricities of the red and blue states of HRP-CO in the free form. (b) Eccentricity of HRP-CO when ligated to BHA. Lines through the data are obtained by simultaneously fitting the linear spectrum and a series of 2D IR spectra as a function of T_w to a multi-exponential FFCF.

two off-diagonal negative going bands arise from the two associated 1–2 transitions. The diagonal and antidiagonal are shown in the upper left panel for the low frequency band. As T_w increases, the peaks become more symmetrical resulting in decreasing eccentricities. The low frequency band has a significantly shorter vibrational life time that gives rise to its decrease in amplitude relative to the higher frequency band by $T_w = 32$ ps. The full data sets contain much finer gradations and the signals can be analyzed at the longest T_ws.

Fig. 9a shows the T_w dependent eccentricities for the blue and red states of the free HRP-CO, and Fig. 9b shows the eccentricities with the BHA substrate bound. Lines through the data are obtained from the full 2D IR calculations that determine the FFCFs as described above. Each of the FFCFs has a motionally narrowed component that contributes to the total 2D IR linewidth but does not contribute to the T_w dependence. The FFCF of the red state of free HRP is different from all of the other spectroscopic bands, those shown here and the results for four other bound substrates discussed elsewhere.[74] In addition to the motionally narrowed component, the red state of free HRP has two additional timescales, $\tau_2 = 1.5$ ps and $\tau_3 = 21$ ps. τ_3 is infinite for all of the other systems. The results demonstrate that the full range of structures that give rise to the red state absorption spectrum have been sampled by ~ 100 ps. The blue state of free HRP is well described by a single 15 ps exponential and lacks a short timescale component (aside from the motionally narrowed term that is common to all of the systems). In the blue state a relatively large set of CO frequency fluctuations occur on timescales slower than ~ 100 ps[72] and necessitate a static term in the FFCF ($\tau_3 = \infty$). The difference between the dynamics of the free states and with BHA bound is clear from the differences in the eccentricities shown in Fig. 9a and b. With BHA bound, the FFCF has a 4.4 ps component (τ_2) and τ_3 is infinite, in addition to a motionally narrowed component.

It is clear from Fig. 9 and the FFCF parameters that the dynamics of HRP with a bound substrate are very different than the red state of free HRP. The free blue state and red state have a 15 ps and 21 ps component, respectively. In contrast, the FFCFs for HRP with bound substrates have a slowest component of ~ 2 ps to ~ 5 ps in the observation time window.[74] The linear absorption spectra of HRP with bound substrates are more similar to the red state of free HRP than to the blue state. The red state 1.5 ps fast component is followed by a slow component of 21 ps that takes the FFCF essentially to zero by ~ 100 ps, completing the dynamics. None of the HRP-substrate systems show a component of their FFCFs that reflects full structural sampling by 100 ps. All of the HRP-substrate systems have substantial structural dynamics too slow to be in the experimentally accessible window, which are manifested as the constant term in the FFCF. In comparison to the red state of free HRP, substrate binding significantly reduces the protein fluctuations. The red state 1.5 ps component becomes two or three times as long and the 21 ps component becomes too slow to measure. When the protein binds a substrate, the active site structure exhibits a single spectroscopic state in which fluctuations of the distal arginine are strongly constrained. The small fraction of the dynamics that occur within the experimental window for HRP-substrate systems indicates that the dynamics of the distal histidine are also severely constrained. Our results are consistent with the suggestion that loss of dynamic freedom by Arg38 is responsible for much of the entropic loss upon substrate binding.[86]

Further insights into the changes in HRP dynamics upon ligand binding can be obtained by comparison to MbCO. The active site of HRP is relatively large, can bind a diverse array of substrate molecules, and contains a distal histidine and arginine. The active site of myoglobin cannot accommodate substrates and does not have a distal arginine, but it also contains a distal histidine and a heme that binds CO. A wealth of experiments and simulations has underscored the pivotal role of the distal histidine in modulating the CO frequency in myoglobin. The H64V mutant (the distal histidine is replaced by a non-polar valine) displays significantly decreased vibrational dephasing because of the elimination of this polar group.[55,64] It is instructive to compare the dynamics of free HRP to that of wild-type myoglobin and substrate-bound HRP with H64V. When this is done, it is found that the dynamics of the A_1 state of MbCO are very similar to the free HRP red state while the HRP-substrate dynamics are very similar to those of H64V.[74]

In H64V, the distal histidine has been removed, and the rate and magnitude of the vibrational dephasing on the experimental time scales are reduced accordingly. Based on the similarity between the MbCO data and the HRP red state data, it is reasonable to assume that a substantial contribution to the vibrational dephasing of the free HRP comes from the fluctuations of the distal histidine and the distal arginine. The striking similarity in HRP-substrate and H64V dynamics highlights that substrate binding in HRP renders the distal residues nearly static on the 2D IR experimental timescale. The clear conclusion to be drawn is that substrate binding locks up the distal ligands, constraining the structural fluctuations in the active site. The result is that the time scale of the fluctuations is pushed out to long times (> 100 ps). It is striking that the protein dynamics are significantly *decreased* when additional potential sources of CO frequency perturbations are introduced into the active site. The distal residues participate in every step of the enzymatic cycle of HRP and the results indicate that substrate binding reorganizes and dynamically constrains these residues.[74] Based on the observation that substrates in the HRP active site significantly decreases structural fluctuations of the distal residues, it is possible that the protein may exploit this feature of substrate binding to catalyze further steps in the enzymatic pathway. Thus, upon recognition of biologically occurring substrates, the protein active site is not only reorganized structurally but also dynamically, priming the enzyme to sample the portion of the conformational energy landscape that may lead to subsequent steps in the reaction.[74]

C. Chemical exchange vibrational echo spectroscopy

1. The effect of chemical exchange on the 2D spectrum. Fig. 10 illustrates the influence of chemical exchange on the 2D vibrational echo spectrum. Two species, A and B, with vibrational transition frequencies, ω_A and ω_B, are in thermal equilibrium. Species A is converting to species B, and *vice versa*, but there is no net change in the populations of A and B because the rate of A's going to B's equals the rate of B's going to A's. Fig. 10a shows the 2D spectrum at very short time prior to chemical exchange. (Only the 0–1 portion of the spectra is shown.) There are two peaks on the diagonal, one for species A and one for species B with frequencies on both the ω_τ and ω_m axes of ω_A and ω_B. Fig. 10b shows what would happen if some of the A's convert to B's. The right part of Fig. 10b is a schematic representation of the combined quantum pathways that lead to the signal.[68] A dashed arrow represents a coherence (coherent superposition state) produced by a radiation field. The solid arrow represents a population, and the curved arrow represents the vibrational echo emission. The first pulse produces a coherence between the states of species A at frequency ω_A. After time τ, the second pulse produces a population. There are several pathways, and a population can be produced in either the ground state (0) or the first excited state (1). During the period T_w, some A's turn into B's (A $\rightarrow$ B). The third pulse again produces a coherence, but it is a coherence of species B at ω_B, followed by echo emission at ω_B. Because the first interaction (frequency on the ω_τ axis) is at ω_A but the last interaction and echo emission (frequency on the ω_m axis) is at ω_B, an off-diagonal peak is generated as shown in the left portion of Fig. 10b. Fig. 10c shows what happens if B's turn into A's (B $\rightarrow$ A). Everything is equivalent to the

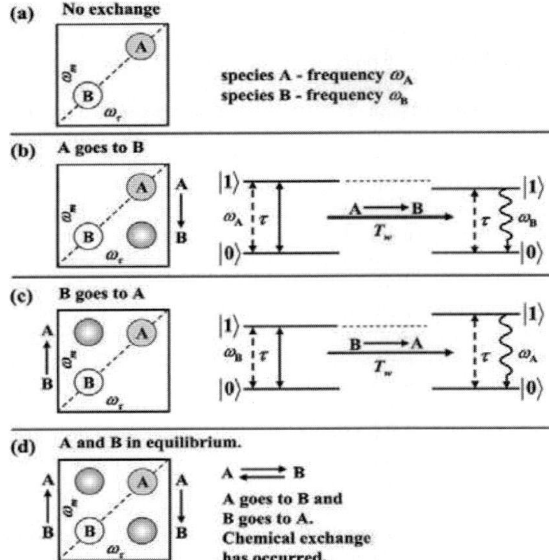

Fig. 10 Schematic illustration of the influence of chemical exchange between two species, A and B, on the 2D vibrational echo spectrum. See text for details.

description of Fig. 10b, but the initial frequency is ω_B and the final frequency is ω_A.

In a real system, A and B are in equilibrium. Therefore, the number of A's turning into B's in a given time period is equal to the number of B's turning into A's. As shown in Fig. 10d, the result is to produce two off-diagonal peaks. Because some A's and B's may not have undergone chemical exchange, or may have undergone chemical exchange but reverted back to the original species prior to the third pulse, there are also diagonal peaks. The model spectrum in Fig. 10d is the spectrum for a time long compared to the chemical exchange time, while the spectrum in Fig. 10a is for a time short compared to the chemical exchange time. The rate of chemical exchange can be determined by observing the growth of the off-diagonal peaks in the 2D vibrational echo spectrum.[29,31,36,100]

2. Organic solute–solvent complex formation and dissociation. Solute and solvent molecules have anisotropic intermolecular interactions that can give rise to well-defined solute–solvent complexes. For organic solute–solvent systems, intermolecular interactions are generally relatively weak, a few kcal mol^{-1}. Such weak interactions will produce solute–solvent complexes that are short lived.[29,36,100] Although short lived, the dissociation and formation of organic solute–solvent complexes can influence chemical processes such as reactivity.

Organic solute–solvent complexes are in equilibrium between the complex form and the free solute form. Because formation and dissociation occurs on a ps time scale, until recently it has not been possible to observe the chemical exchange process between the two forms of the solute.[29,36,100]

To obtain information about solute–solvent complex formation and dissociation using 2D IR vibrational echo chemical exchange spectroscopy, it is sufficient to focus on the 0–1 region of the spectrum. Identical considerations apply to the 1–2 region, but analysis of the 1–2 region can provide additional information.[29] Comparisons of the 0–1 and 1–2 regions of the spectra have been used to demonstrate that vibrational excitation does not perturb the chemical equilibrium.[29]

Fig. 11a displays the spectrum of the hydroxyl OD stretch of phenol in a mixed solvent of benzene (20 mol%) and CCl$_4$ (80 mol%). The high frequency peak is the free phenol, and the low frequency peak is the phenol–benzene complex.[29] The structure of the complex, determined from electronic structure calculations, is shown in the figure.[29] The species are in equilibrium, with complexes constantly dissociating and free phenols associating with benzene to form complexes. Fig. 11b shows the 2D vibrational echo spectrum at short time, $T_w = 200$ fs; no significant exchange has occurred. As in Fig. 10a, there are two species in the system with different vibrational transition frequencies. At short time, the two species give two peaks on the diagonal, which correspond to the 2 peaks in the absorption spectrum. At long time, 14 ps, extensive chemical exchange has occurred. Off-diagonal peaks have grown in as can be seen clearly in Fig. 11c. The two peaks correspond to complex dissociation and formation. Fig. 11c has the appearance of the schematic shown in Fig. 10d, and the discussion surrounding the origins of the off-diagonal peaks in Fig. 10 applies to the experimental data shown in Fig. 11.

The growth of the off-diagonal peaks with increasing T_w can be used to directly determine the thermal equilibrium rate for complex formation and dissociation. Fig. 12 displays 2D spectra as three dimensional representations. At 2 ps, the off-diagonal peaks are just appearing in these plots. By 5 ps they are clearly evident, and continue to grow as can be seen in the 14 ps plot. Data like these are used for detailed analysis of the chemical exchange kinetics.[29,36,100,101] In addition to chemical exchange, there are other dynamical processes that contribute to the time dependent changes in the spectrum. Spectral diffusion causes the peaks to change shape. The rate of chemical exchange without complications arising from the changing peak shapes can be obtained by measuring the time dependent peak volumes.[29,100] The vibrational lifetimes of the hydroxyl stretch and the orientational relaxation rates cause all of the peaks to decay in amplitude while chemical exchange causes the off-diagonal peaks to grow in. The vibrational lifetimes and the orientational relaxation rates are measured independently using ultrafast IR pump–probe spectroscopy.[29,36,100] In addition, the equilibrium constant and the transition dipole moments of the species are necessary inputs in calculations and are obtained from the FT-IR absorption spectra.[29,100]

Because the two off-diagonal peaks grow in at the same rate (see Fig. 13), the rate of complex formation and dissociation are equal, that is, the system is in equilibrium. The fact that vibrational excitation does not take the system out of equilibrium was further tested by comparing time evolution of the 0–1 and 1–2 regions of the spectra.[29] With the system in equilibrium and the other input parameters known,[29,100] there is only one adjustable parameter to fit the T_w dependence of all

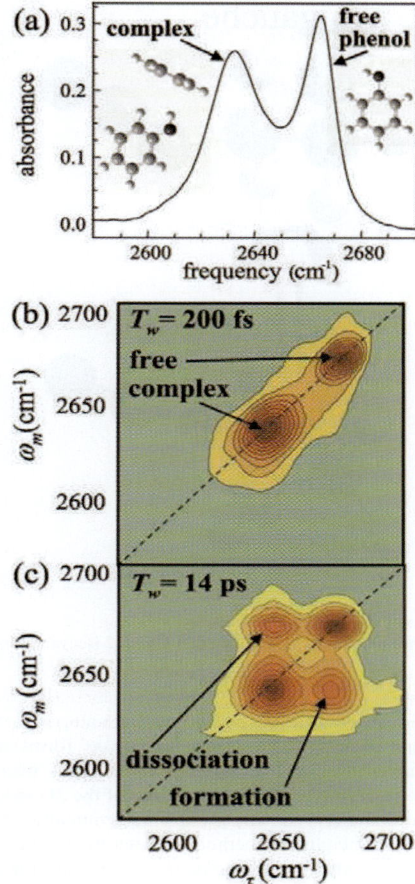

Fig. 11 (a) Absorption spectrum of phenol-OD in the mixed solvent benzene and CCl$_4$. The structure of the phenol–benzene complex is shown. (b) The 2D vibrational echo spectrum at short time. There are two peaks on the diagonal. (c) The 2D vibrational echo spectrum at long time after substantial chemical exchange has occurred (phenol–benzene complex formation and dissociation). Off-diagonal peaks have grown in.

of the peaks in the spectra. Because the rate of complex dissociation is equal to the rate of complex formation, we can fit everything using *a single adjustable parameter*, the complex dissociation time, $\tau_d = 1/k_{cf}$, where k_{cf} is the rate constant for dissociation of the complex (c) to the free form (f).

Fig. 13 displays the data and fits for the chemical exchange dynamics of the phenol–benzene complex. There are four peaks in the 2D vibrational echo spectra (see Fig. 10–12), the two diagonal peaks for the complex (filled diamonds and curve) and free (filled circles and curve) species, and the two off-diagonal peaks (unfilled squares, triangles and dashed curve) for $c \rightarrow f$, and $f \rightarrow c$. As can be seen from Fig. 13, the data are fit very well with the single parameter τ_d. (The

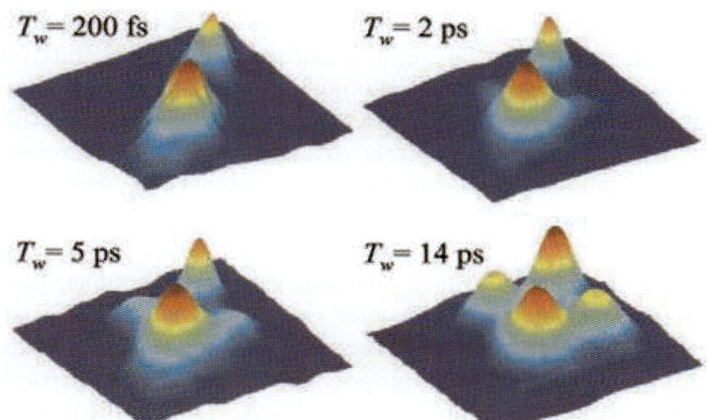

Fig. 12 Three dimensional representation of the T_w dependence of the 2D vibrational echo chemical exchange data. The off-diagonal peaks grow in as time progresses.

mathematical details used to fit the data have been presented.) The data for the two off-diagonal peaks are identical within experimental error, which is one of the tests that shows the thermal equilibrium between the complex and free species is not perturbed by the experiments. The results of the fitting yields $\tau_d = 10$ ps.[101,102]

Eight π hydrogen bond complexes of phenol and phenol derivatives with benzene and benzene derivatives were investigated.[101,102] The series of complexes have a wide range of hydrogen bond strengths. The measured dissociation times for the eight complexes range form 3 to 31 ps. The bond dissociation enthalpies (negative of the formation enthalpies) range from 0.6 to 2.5 kcal mol^{-1}. It was found that the hydrogen bond dissociation times, τ, are correlated with the dissociation enthalpies in a manner akin to an Arrhenius equation.[101] The correlation can be qualitatively understood in terms of

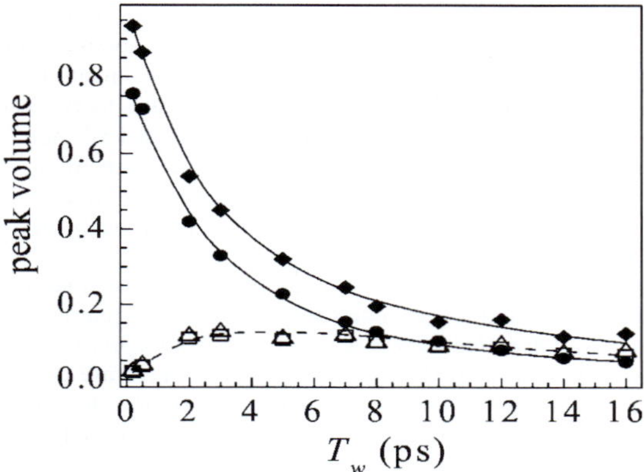

Fig. 13 Data (points) and fits (curves) for the chemical exchange dynamics of the phenol–benzene complex. The complex diagonal peak—filled diamonds and curve. The free diagonal peak—filled circles and curve. The two off-diagonal peaks—unfilled squares, triangles and dashed curve. The fits are with the single adjustable parameter $\tau_d = 10$ ps, the complex dissociation time.

transition state theory. In this model, the activation enthalpy scales linearly with the bond dissociation enthalpy and the activation entropy is essentially independent of the molecular make up of the complex. These experiments are providing the first detailed information on the dynamics of weakly hydrogen bonded organic species and the relationship between dynamics and strengths of the hydrogen bonds.

3. Isomerization around a carbon–carbon single bond. Isomerization of organic molecules is responsible for the vast diversity of their chemical structures and the ability of both small molecules and large biopolymers to undergo structural changes without breaking chemical bonds. During the course of isomerization a molecule is transformed from one relatively stable conformation to another by passing through unfavorable configurations. Ethane and its derivatives are textbook examples of molecules that undergo this type of isomerization.[103] In ethane, as one of the two methyl groups rotates 360^0 around the central carbon–carbon singe bond, it will alternate three times between an unstable eclipsed conformation and the preferred staggered conformation. The transition from one staggered state to another leaves ethane structurally identical. Therefore, the result of ethane isomerization cannot be observed through a change in chemical structure. In a 1,2-disubstituted ethane derivative, the molecule can undergo a similar isomerization. However, a 1,2-disubstituted ethane has two distinct staggered conformations, *gauche* and *trans*. The two conformations have distinguishing characteristics because of the change in the relative positions of the two substituents.[103] Carbon–carbon bond isomerization has been the subject of intense theoretical and experimental study since Bischoff found 100 years ago that rotation about the C–C single bond in ethane is not completely free.[104]

The *trans-gauche* isomerization of 1,2-disubstituted ethane derivatives, *e.g.* *n*-butane, is one of the simplest cases of a first-order chemical reaction. This type of isomerization has served as a basic model for modern chemical reaction kinetic theory and molecular dynamics simulation studies in condensed phases.[105–109] In spite of extensive theoretical investigation, until recently[110] no corresponding kinetic experiments have been performed to test the results, partially due to the low rotational energy barrier of the *n*-butane (~ 3.4 kcal mol^{-1}) and other simple 1,2-disubstituted ethane derivatives.[111] Theoretical studies show that the isomerization time scale ($1/k$, k is the rate constant) is 10–100 ps at room temperature in liquids.[105–109,112]

The room temperature time scale is much shorter than the microsecond and longer time scale measurements that can be made with dynamic nuclear magnetic resonance spectroscopy, a widely used method for studying slow temperature dependent isomerization kinetics at low temperatures for compounds with high barriers.[113] Other methods to study fast isomerization dynamics under thermal equilibrium such as linear IR and Raman line shape analysis,[114,115] are hampered by the contributions from multiple factors in addition to isomerization.[116–118]

The 2D vibrational echo chemical exchange spectroscopic method has been applied to the study of the ultrafast *trans-gauche* isomerization dynamics of a simple 1,2-disubstituted

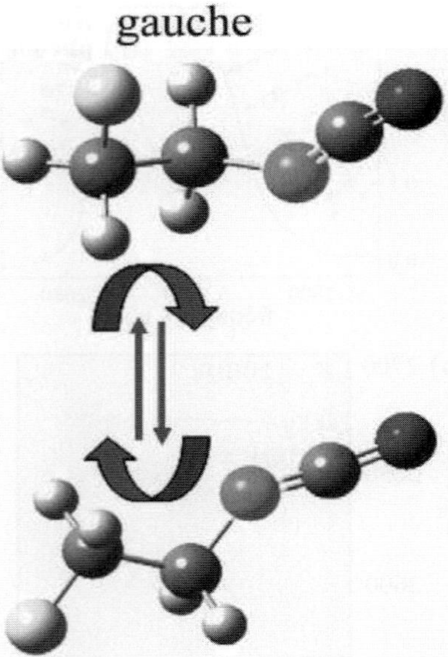

Fig. 14 The *gauche* and *trans* forms of 1-fluoro-2-isocyanato-ethane.

ethane derivative, 1-fluoro-2-isocyanato-ethane, in a room temperature liquid.[110] The *gauche* and *trans* forms of the molecule are shown in Fig. 14. The experiments were performed by observing the time dependence of the 2D spectrum of the isocyanate (N=C=O) group's antisymmetric stretching mode. In principle, the method for determining the rate of isomerization is identical to the method employed for determining the rate of solute–solvent complex formation and dissociation discussed above. In IR absorption spectrum shown in Fig. 15, the *gauche* and *trans* isomers are sufficiently resolved to be able to detect the growth of off-diagonal peaks caused by chemical exchange between the two isomers. The peaks have essentially equal amplitude, which means that half of the molecules are in the *gauche* conformation and half are in the *trans* conformation. However, there is a third peak in the spectrum that is unassigned. It may be an overtone or a combination band. The unassigned transition is coupled through anharmonic terms in the molecular Hamiltonian to the *gauche* transition for molecules in the *gauche* form and to the *trans* transition for molecules in the *trans* form. In the absence of chemical exchange, the influence of coupling between modes on the same molecule is well known.[34,119,120] Coupling between a pair of modes produces off-diagonal coherence transfer peaks (positive) and combination band peaks (negative). The coherence transfer peaks form a square pattern with the diagonal peaks; the off-diagonal peaks are the upper left and lower right corners of the square. The combination band peaks are directly below the coherence transfer peaks, shifted by the combination band shift. If there is a mixture of two species, then there will be two sets of coherence

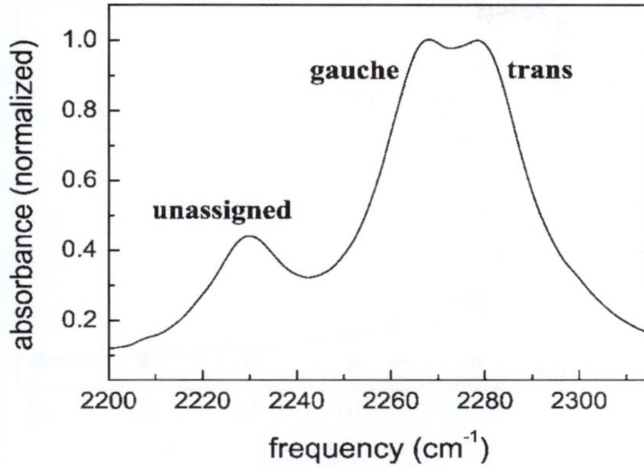

Fig. 15 The IR absorption spectrum of the isocyanate (N=C=O) group's antisymmetric stretching mode. The *gauche* and *trans* isomers are resolved and there is a third peak in the spectrum that is unassigned.

transfer and combination band peaks.[41] Here the sample is a mixture of two species, the *gauche* and *trans* isomers. In addition to the off-diagonal peaks that are time dependent, vibrational population relaxation between modes on the same molecule causes additional peaks to grow in with the vibrational relaxation times.[34]

The process of interest is the chemical exchange, which is caused by *gauche–trans* isomerization. The *gauche* and *trans* peaks are much closer together than either is to the unassigned peaks.[110] The peaks that arise from the coupling to the unassigned peak are much farther off-diagonal than the chemical exchange peaks that arise from isomerization. However, one of the negative going (blue) peaks that grows in because of vibrational relaxation from the *trans* mode to the unassigned peak overlaps with the lower right chemical exchange off-diagonal position that will grow in as isomerization proceeds. The affected chemical exchange peak is shown schematically in Fig. 10b. The details of the vibrational relaxation peak that interferes with the lower right chemical exchange peak can be determined by examining the molecule 1-bromo-2-isocyanato-ethane. The bromo group is very large compared to the fluoro group, and it produces enough steric hindrance to significantly slow down the C–C isomerization. On the time scale of the experiments on 1-fluoro-2-isocyanato-ethane, isomerization of the bromo compound does not occur. The T_w dependence of bromo compound shows that a negative going peak appears below the *trans* diagonal peak in the position in which a positive going off-diagonal peak would grow in if chemical exchange between the *trans* and *gauche* isomer occurred.[110] Therefore, in contrast to the data shown in Fig. 11 and 12, in which the off-diagonal chemical exchange peaks for solute–solvent formation and dissociation are of equal amplitude, in the isomerization experiments on 1-fluoro-2-isocyanato-ethane, the upper left chemical exchange peak will be prominent, but the lower right peak is mainly obscured.

Fig. 16 left column displays five T_w dependent 2D vibrational echo spectra of 1-fluoro-2-isocyanato-ethane in a CCl_4

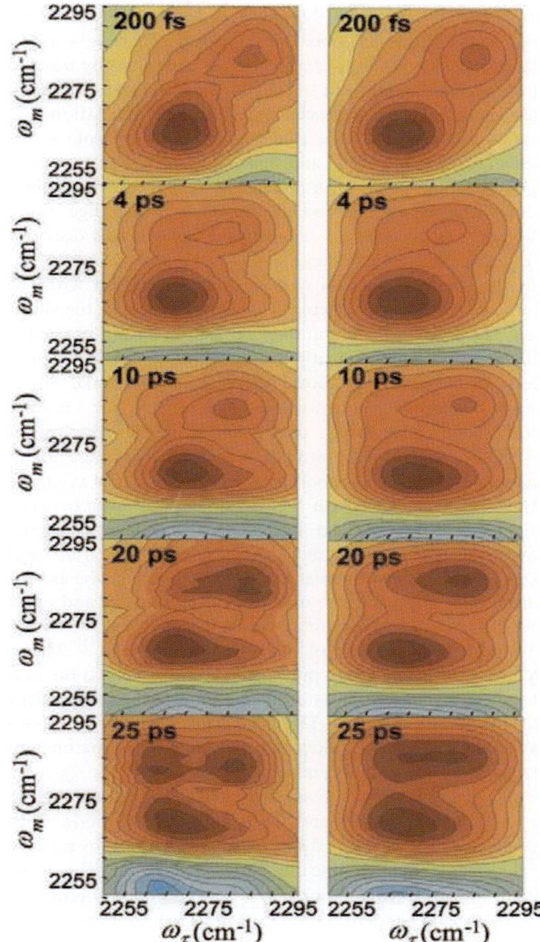

Fig. 16 Left column: five T_w dependent 2D vibrational echo spectra of 1-fluoro-2-isocyanato-ethane in a CCl_4 solution at room temperature. The 200 fs panel, negligible isomerization has occurred. T_w = 25 ps panel, isomerization has proceeded to a substantial degree as evidenced by the growth of off-diagonal peaks, particular to the upper left. Right column: corresponding calculated 2D vibrational echo spectra with one adjustable parameter, the isomerization rate about the carbon–carbon single bond.

solution at room temperature. The 200 fs panel corresponds to a very short T_w at which negligible isomerization has occurred. As discussed in detail above (see Fig. 10a and 11b),[29] when no isomerization has occurred, the initial and final structures of each labeled species in the sample are unchanged. Therefore, the ω_τ and ω_m values of each peak are identical, and the peaks appear only on the diagonal. The two peaks representing the *gauche* (ω_m = 2265 cm^{-1}) and *trans* (ω_m = 2280 cm^{-1}) conformers are clearly visible on the diagonal. For T_w = 25 ps, isomerization has proceeded to a substantial degree. The obvious change is the additional peak that has appeared at the upper left ($\omega_\tau = 2265$ cm^{-1} and $\omega_m = 2280$ cm^{-1}). This peak arises from *gauche* to *trans* isomerization. There is a corresponding peak to the lower right that is generated by *trans* to *gauche* isomerization but it is significantly negated by a negative going peak produced by vibrational relaxation[34,110] as discussed above.[110] The diagonal peaks arise from molecules with the same initial and final structures. The growth of the off-diagonal peaks with increasing T_w permits determination of the isomerization rate (chemical exchange rate), although it is necessary to analyze the growth and decay of all of the peaks for accuracy.[36]

In the same manner as discussed in connection with the solute–solvent complex chemical exchange problem, the vibrational lifetimes and orientational relaxation rates are needed and measured independently using IR pump–probe experiments.[110] As described above, there is an additional vibrational relaxation pathway, distinct from the regular vibrational lifetime decay, that stems from the coupling of the NCO antisymmetric stretch of both conformers to an unassigned vibrational mode at ~ 2230 cm^{-1}. The coupling induces fast back and forth population equilibration between the unassigned mode and the NCO antisymmetric stretch (equilibration time constants are 0.9 ps for the *trans* and 1.9 ps for the *gauche* as measured by IR pump–probe experiments).[110] In analyzing the data, all of the kinetics and other input information were obtained from independent experiments with the exception of the isomerization rate.[110] Therefore, only one unknown parameter, the isomerization rate constant, $k_{TG} = k_{GT}$, was used in fitting the T_w dependent 2D vibrational echo data. The *trans* to *gauche* and *gauche* to *trans* rate constants are taken to be equal within experimental error because the equilibrium constant is 1.

Fig. 16 right column shows calculated 2D spectra using the known input parameters and the results of fitting k_{TG}. Both the measured and calculated spectra are normalized by making the largest peak at each time unity. Given the complexity of the system, the model calculations do a very good job of reproducing the time dependent data. Of particular importance is the growth of the off-diagonal red peaks and the negative going peaks at the bottom of each panel. Fig. 17 shows the T_w dependent peak volume data (points) and the results of the fits (curves). The diagonal peaks for *gauche* (squares) and *trans* (circles) and the off-diagonal peaks (diamonds and stars) are all fit very well (solid lines) using the isomerization rate constant as the only adjustable parameter. The off-diagonal peaks grow in at the same rate, consistent with $k_{TG} = k_{GT}$ within experimental error. The fits yield $1/k_{TG} = 1/k_{GT} = 43 \pm 10$ ps. The error bars arise mainly from the uncertainty in the parameters that go into the calculations.

Based on the experimental results for the 1-fluoro-2-isocyanato-ethane, it is possible to calculate approximately the *gauche–trans* isomerization rate of *n*-butane and the rotational isomerization rate of ethane under the same conditions used in this study (CCl$_4$ solution at room temperature, 297 K).[110] It is interesting to obtain a number for *n*-butane because there have been a large number of theoretical calculations for the isomerization of this molecule.[106–109] Transition state theory[121] was employed with the assumption that the prefactors for all of the systems are the same. This assumption is reasonable

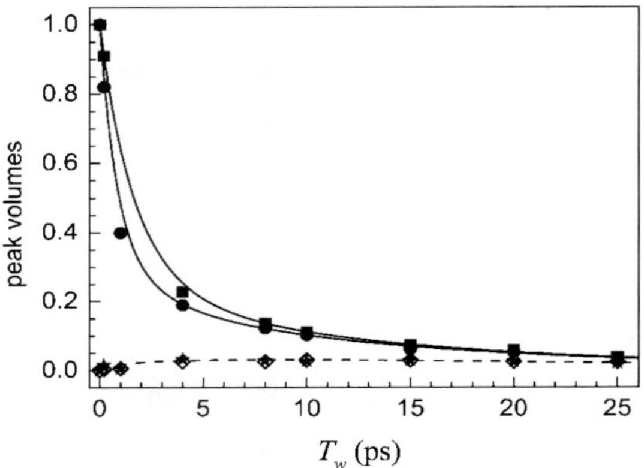

Fig. 17 T_w dependent peak volume data (points) and the results of the fits (curves). The diagonal peaks for *gauche* (squares) and *trans* (circles) and the off-diagonal peaks (diamonds and stars) are all fit (solid lines) using the *gauche–trans* isomerization rate constant as the single adjustable parameter. The fits yield an inverse isomerization rate of 43 ps.

because the transition states and the barrier heights are quite similar for the three systems. DFT calculations (the B3LYP level and 6-31 + G(d,p) basis set) for all the systems were performed to obtain the barrier heights. With the zero point energy correction, the *trans* to *gauche* isomerization of *n*-butane has a calculated barrier of 3.3 kcal mol^{-1}. The barrier for ethane is calculated to be 2.5 kcal mol^{-1}. This value differs from the 2.9 kcal mol^{-1}[122,123] that has been obtained using more extensive electronic structure calculations. However, the 2.5 kcal mol^{-1} value used for comparison with the 3.3 kcal mol^{-1} obtained for 1-fluoro-2-isocyanato-ethane. By calculating the two barriers with the same method, it was hoped that there is some cancellation of errors.

Using the calculated barriers for 1-fluoro-2-isocyanato-ethane and for *n*-butane and the assumption that the prefactors are the same, a ~ 40 ps time constant for the *n*-butane *trans* to *gauche* isomerization time constant ($1/k_{TG}$) was calculated.[110] This value is identical within error to the 43 ps time found for the *n*-butane isomerization in CCl$_4$ at 300 K from MD simulations.[107] Other MD simulations gave isomerization rates in liquid *n*-butane at slightly lower temperatures: 52 ps (292 K),[108] 57 ps (292 K),[106] 50 ps (273 K)[106] and 61 ps (<292 K).[109] All of these values are reasonably close to the value obtained based on the experimental measurements of 1-fluoro-2-isocyanato-ethane. In the same manner, the isomerization time constant for ethane was found to be ~ 12 ps.[110]

IV. Concluding remarks

Here the method and some applications of 2D IR vibrational echo spectroscopy have been illustrated. In the 2D IR vibrational echo experiment, the vibrational echo is heterodyne detected, which provides amplitude and phase information, both of which are required to Fourier transform into the frequency domain and to obtain essentially absorptive 2D spectra. By eliminating a substantial portion of the dispersive contribution, the spectral resolution is greatly enhanced. In addition, in comparison to intensity level measurements, measuring the 2D vibrational echo spectrum at the polarization level *via* heterodyne detection eliminates cross terms between different transitions that can greatly complicate analysis.

Two types of applications of 2D vibrational echo spectroscopy were described, that is, the measurement of spectral diffusion with examples from the study of protein dynamics and the measurement of chemical exchange with examples of solute–solvent complex dynamics and isomerization around a carbon–carbon single bond. Such measurements are useful in an increasingly wide variety of chemical and biophysical problems.

The utility of 2D IR vibrational echo spectroscopy for a wide variety of applications can be seen by contrasting it to NMR and to Vis/UV spectroscopy. NMR provides extremely high resolution and exquisite structural discrimination. In addition, there are a large number of pulse sequences that make NMR applicable to a wide variety of problems. However, NMR intrinsically operates on slow time scales. The RF frequency of <1 GHz limits the direct measurement of dynamics to relatively slow processes. Vis/UV experiments can operate on ultrafast time scales and have been applied to a large number of dynamic phenomena. These are generally limited to photoinduced processes rather than measuring dynamics under thermal equilibrium conditions, in contrast to all of the examples discussed above in which the dynamics were observed for systems in equilibrium at room temperature. Furthermore, for complex molecular systems in liquids, solids, or biological systems at room temperature, Vis/UV spectra are generally extremely broad with few identifiable features. The relationship between the time evolution of the Vis/UV observables and structural evolution of a system can be difficult to sort out.

Vibrational spectroscopy and particularly ultrafast 2D IR spectroscopy occupies a middle ground between NMR and Vis/UV spectroscopies. Relatively well resolved spectroscopic lines with well defined structural identity can be studied on time scales much faster than are accessible to NMR. In contrast to Vis/UV spectroscopies, 2D IR experiments can be conducted without perturbing the equilibrium ground state of a system because exciting a single vibrational quantum is generally a negligibly small perturbation. As demonstrated above, 2D vibrational echo experiments with heterodyne detection are akin to 2D NMR. However, they are intrinsically different. NMR has the great advantage of having all of the spin transitions quantized in the same direction along the magnetic field and all the transitions of a given type of spin, *e.g.*, proton, have the same transition dipole matrix element. These properties make it possible to apply a pulse that will produce the same flip angle for all of the spins. The ability to have the same flip angle makes possible the complex pulse sequences used in NMR. In 2D vibrational echo experiments, the transition dipoles point in random directions. Therefore, the projection of the light's electric field on to the transition dipole varies from some maximum value to zero. Also, different transitions of the same type, *e.g.* a hydroxyl stretch,

have different transition moments. It is not possible to have the same flip angle for all of the vibrations, precluding the more sophisticated pulse sequences used in NMR. Nonetheless, ultrafast heterodyne detected 2D vibrational echo spectroscopy is finding an increasingly diverse range of applications as the methodology advances.

Acknowledgements

Work on protein dynamics was supported by a grant from the NIH (2 R01 GM-061137-05). Work on chemical exchange dynamics was supported by grants from AFOSR (F49620-01-1-0018) and from NSF (DMR-0332692).

References

1 E. L. Hahn, *Phys. Rev.*, 1950, **80**, 580.
2 M. A. Brown and R. C. Semelka, *MRI: Basic Principles and Applications*, Wiley-Liss, New York, 1999.
3 K. J. Dunn, D. J. Bergman and G. A. LaTorraca, *Nuclear Magnetic Resonance (Handbook of Geophysical Exploration: Seismic Exploration)*, Elsevier Science, Oxford, UK, 2002.
4 N. A. Kurnit, I. D. Abella and S. R. Hartmann, *Phys. Rev. Lett.*, 1964, **13**, 567.
5 I. D. Abella, N. A. Kurnit and S. R. Hartmann, *Phys. Rev.*, 1966, **14**, 391.
6 T. J. Aartsma and D. A. Wiersma, *Chem. Phys. Lett.*, 1978, **54**, 415.
7 W. H. Hesselink and D. A. Wiersma, *Phys. Rev. Lett.*, 1979, **43**, 1991.
8 D. E. Cooper, R. W. Olson, R. D. Wieting and M. D. Fayer, *Chem. Phys. Lett.*, 1979, **67**, 41.
9 R. W. Olson, H. W. H. Lee, F. G. Patterson and M. D. Fayer, *J. Chem. Phys.*, 1982, **76**, 31.
10 R. J. Gulotty, C. A. Walsh, F. G. Patterson, W. L. Wilson and M. D. Fayer, *Chem. Phys. Lett.*, 1986, **125**, 507.
11 L. W. Molenkamp and D. A. Wiersma, *J. Chem. Phys.*, 1985, **83**, 1.
12 M. Berg, C. A. Walsh, L. R. Narasimhan, K. A. Littau and M. D. Fayer, *J. Chem. Phys.*, 1988, **88**, 1564.
13 C. A. Walsh, M. Berg, L. R. Narasimhan and M. D. Fayer, *J. Chem. Phys.*, 1987, **86**, 77.
14 C. A. Walsh, M. Berg, L. R. Narasimhan and M. D. Fayer, *Chem. Phys. Lett.*, 1986, **130**, 6.
15 D. T. Leeson and D. A. Wiersma, *Phys. Rev. Lett.*, 1995, **74**, 2138.
16 D. T. Leeson, D. A. Wiersma, K. Fritsch and J. Friedrich, *J. Phys. Chem. B*, 1997, **101**, 6331.
17 X. J. Jordanides, M. J. Lang, X. Song and G. R. Fleming, *J. Phys. Chem. B*, 1999, **103**, 7995.
18 D. M. Jonas, M. J. Lang, Y. Nagasawa, T. Joo and G. R. Fleming, *J. Phys. Chem.*, 1996, **100**, 12660.
19 D. Zigmantas, E. L. Read, T. Mancal, T. Brixner, A. T. Gardiner, R. J. Cogdell and G. R. Fleming, *Proc. Natl. Acad. Sci. U. S. A.*, 2006, **103**, 12672.
20 D. Zimdars, A. Tokmakoff, S. Chen, S. R. Greenfield, M. D. Fayer, T. I. Smith and H. A. Schwettman, *Phys. Rev. Lett.*, 1993, **70**, 2718.
21 C. W. Rella, A. Kwok, K. D. Rector, J. R. Hill, H. A. Schwettmann, D. D. Dlott and M. D. Fayer, *Phys. Rev. Lett.*, 1996, **77**, 1648.
22 C. W. Rella, K. D. Rector, A. S. Kwok, J. R. Hill, H. A. Schwettman, D. D. Dlott and M. D. Fayer, *J. Phys. Chem.*, 1996, **100**, 15620.
23 A. Tokmakoff, D. Zimdars, R. S. Urdahl, R. S. Francis, A. S. Kwok and M. D. Fayer, *J. Phys. Chem.*, 1995, **99**, 13310.
24 A. Tokmakoff and M. D. Fayer, *Acc. Chem. Res.*, 1995, **28**, 437.
25 K. A. Merchant, D. E. Thompson and M. D. Fayer, *Phys. Rev. Lett.*, 2001, **86**, 3899.
26 M. Khalil, N. Demirdoven and A. Tokmakoff, *Phys. Rev. Lett.*, 2003, **90**, 047401(4).

27 J. B. Asbury, T. Steinel, C. Stromberg, S. A. Corcelli, C. P. Lawrence, J. L. Skinner and M. D. Fayer, *J. Phys. Chem. A*, 2004, **108**, 1107.
28 J. B. Asbury, T. Steinel and M. D. Fayer, *J. Phys. Chem. B*, 2004, **108**, 6544.
29 J. Zheng, K. Kwak, J. B. Asbury, X. Chen, I. Piletic and M. D. Fayer, *Science*, 2005, **309**, 1338.
30 M. T. Zanni and R. M. Hochstrasser, *Curr. Opin. Struct. Biol.*, 2001, **11**, 516.
31 Y. S. Kim and R. M. Hochstrasser, *Proc. Natl. Acad. Sci. U. S. A.*, 2005, **102**, 11185.
32 M. C. Asplund, M. T. Zanni and R. M. Hochstrasser, *Proc. Natl. Acad. Sci. U. S. A.*, 2000, **97**, 8219.
33 M. T. Zanni, S. Gnanakaran, J. Stenger and R. M. Hochstrasser, *J. Phys. Chem. B*, 2001, **105**, 6520.
34 M. Khalil, N. Demirdoven and A. Tokmakoff, *J. Chem. Phys.*, 2004, **121**, 362.
35 M. F. DeCamp, L. DeFlores, J. M. McCracken, A. Tokmakoff, K. Kwac and M. Cho, *J. Phys. Chem. B*, 2005, **109**, 11016.
36 J. Zheng, K. Kwak, X. Chen, J. B. Asbury and M. D. Fayer, *J. Am. Chem. Soc.*, 2006, **128**, 2977.
37 J. B. Asbury, T. Steinel, K. Kwak, S. Corcelli, C. P. Lawrence, J. L. Skinner and M. D. Fayer, *J. Chem. Phys.*, 2004, **121**, 12431.
38 C. J. Fecko, J. D. Eaves, J. J. Loparo, A. Tokmakoff and P. L. Geissler, *Science*, 2003, **301**, 1698.
39 K. D. Rector, C. W. Rella, A. S. Kwok, J. R. Hill, S. G. Sligar, E. Y. P. Chien, D. D. Dlott and M. D. Fayer, *J. Phys. Chem. B*, 1997, **101**, 1468.
40 K. A. Merchant, W. G. Noid, D. E. Thompson, R. Akiyama, R. F. Loring and M. D. Fayer, *J. Phys. Chem. B*, 2003, **107**, 4.
41 J. B. Asbury, T. Steinel and M. D. Fayer, *Chem. Phys. Lett.*, 2003, **381**, 139.
42 J. B. Asbury, T. Steinel and M. D. Fayer, *J. Lumin.*, 2004, **107**, 271.
43 M. D. Fayer, *Annu. Rev. Phys. Chem.*, 1982, **33**, 63.
44 J. B. Asbury, T. Steinel, C. Stromberg, K. J. Gaffney, I. R. Piletic, A. Goun and M. D. Fayer, *Chem. Phys. Lett.*, 2003, **374**, 362.
45 J. B. Asbury, T. Steinel and M. D. Fayer, *J. Lumin.*, 2004, **107**, 217.
46 R. R. Ernst, G. Bodenhausen and A. Wokaun, *Nuclear Magnetic Resonance in One and Two Dimensions*, Oxford University Press, Oxford, 1987.
47 A. Ansari, J. Berendzen, D. Braunstein, B. R. Cowen, H. Frauenfelder, M. K. Hong, I. E. T. Iben, J. B. Johnson, P. Ormos, T. B. Sauke, R. Scholl, A. Schulte, P. J. Steinbach, J. Vittitow and R. D. Young, *Biophys. Chem.*, 1987, **26**, 337.
48 W. S. Caughey, H. Shimada, M. G. Choc and M. P. Tucker, *Proc. Natl. Acad. Sci. U. S. A.*, 1981, **78**, 2903.
49 K. D. Rector, J. R. Engholm, C. W. Rella, J. R. Hill, D. D. Dlott and M. D. Fayer, *J. Phys. Chem. A*, 1999, **103**, 2381.
50 W. D. Tian, J. T. Sage, P. M. Champion, E. Chien and S. G. Sligar, *Biochemistry*, 1996, **35**, 3487.
51 G. N. Phillips, Jr, M. L. Teodoro, T. Li, B. Smith and J. S. Olson, *J. Phys. Chem. B*, 1999, **103**, 8817.
52 M. D. Fayer, *Annu. Rev. Phys. Chem.*, 2001, **52**, 315.
53 D. Morikis, P. M. Champion, B. A. Springer and S. G. Sligar, *Biochemistry*, 1989, **28**, 4791.
54 E. Oldfield, K. Guo, J. D. Augspurger and C. E. Dykstra, *J. Am. Chem. Soc.*, 1991, **113**, 7537.
55 I. J. Finkelstein, A. Goj, B. L. McClain, A. M. Massari, K. A. Merchant, R. F. Loring and M. D. Fayer, *J. Phys. Chem. B*, 2005, **109**, 16959.
56 A. M. Massari, I. J. Finkelstein and M. D. Fayer, *J. Am. Chem. Soc.*, 2006, **128**, 3990.
57 C. Rovira, B. Schulze, M. Eichinger, J. D. Evanseck and M. Parrinello, *Biophys. J.*, 2001, **81**, 435.
58 R. Elber and M. Karplus, *Science*, 1987, **235**, 318.
59 D. Vitkup, D. Ringe, G. A. Petsko and M. Karplus, *Nat. Struct. Biol.*, 2000, **7**, 34.
60 B. Kushkuley and S. S. Stavrov, *Biophys. J.*, 1997, **72**, 899.
61 J. C. Kendrew, R. E. Dickerson, B. E. Strandberg, R. G. Hart, D. R. Davies, D. C. Phillips and V. C. Shore, *Nature*, 1960, **185**, 422.
62 G. S. Kachalova, A. N. Popov and H. D. Bartunik, *Science*, 1999, **284**, 473.
63 J. Vojtechovsky, K. Chu, J. Berendzen, R. M. Sweet and I. Schlichting, *Biophys. J.*, 1999, **77**, 2153.
64 K. A. Merchant, W. G. Noid, R. Akiyama, I. Finkelstein, A. Goun, B. L. McClain, R. F. Loring and M. D. Fayer, *J. Am. Chem. Soc.*, 2003, **125**, 13804.
65 S. T. Roberts, J. J. Loparo and A. Tokmakoff, *J. Chem. Phys.*, 2006, **125**, 084502.
66 T. G. Spiro and I. H. Wasbotten, *J. Inorg. Biochem.*, 2005, **99**, 34.
67 S. Mukamel, *Annu. Rev. Phys. Chem.*, 2000, **51**, 691.
68 S. Mukamel, *Principles of Nonlinear Optical Spectroscopy*, Oxford University Press, New York, 1995.
69 P. Hamm and R. M. Hochstrasser, in *Ultrafast Infrared and Raman Spectroscopy*, ed. M. D. Fayer, Marcel Dekker Inc., New York, 2001, vol. 26, p. 273.
70 R. B. Williams, R. F. Loring and M. D. Fayer, *J. Phys. Chem. B*, 2001, **105**, 4068.
71 A. M. Massari, I. J. Finkelstein, B. L. McClain, A. Goj, X. Wen, K. L. Bren, R. F. Loring and M. D. Fayer, *J. Am. Chem. Soc.*, 2005, **127**, 14279.
72 Y. S. Bai and M. D. Fayer, *Phys. Rev. B*, 1989, **39**, 11066.
73 J. Schmidt, N. Sundlass and J. Skinner, *Chem. Phys. Lett.*, 2003, **378**, 559.
74 I. J. Finkelstein, H. Ishikawa, S. Kim, A. M. Massari and M. D. Fayer, *Proc. Natl. Acad. Sci. U. S. A.*, 2007, **104**, 2637.
75 W. D. Tian, J. T. Sage and P. M. Champion, *J. Mol. Biol.*, 1993, **233**, 155.
76 D. P. Braunstein, K. Chu, K. D. Egeberg, H. Frauenfelder, J. R. Mourant, G. U. Nienhaus, P. Ormos, S. G. Sligar, B. A. Springer and R. D. Young, *Biophys. J.*, 1993, **65**, 2447.
77 E. Park, S. Andrews and S. G. Boxer, *J. Phys. Chem.*, 1999, **103**, 9813.
78 R. Jimenez, G. Salazar, J. Yin, T. Joo and F. E. Romesberg, *Proc. Natl. Acad. Sci. U. S. A.*, 2004, **101**, 3803.
79 B. Ma, M. Shatsky, H. J. Wolfson and R. Nussinov, *Protein Sci.*, 2002, **11**, 184.
80 H. Frauenfelder, B. H. McMahon, R. H. Austin, K. Chu and J. T. Groves, *Proc. Natl. Acad. Sci. U. S. A.*, 2001, **98**, 2370.
81 P. Mukherjee, I. Kass, I. T. Arkin and M. T. Zanni, *Proc. Natl. Acad. Sci. U. S. A.*, 2006, **103**, 3528.
82 A. W. Smith, H. S. Chung, Z. Ganim and A. Tokmakoff, *J. Phys. Chem. B*, 2005, **109**, 17025.
83 N. C. Veitch, *Phytochemistry*, 2004, **65**, 249.
84 A. T. Smith and N. C. Veitch, *Curr. Opin. Chem. Biol.*, 1998, **2**, 269.
85 N. C. Veitch and A. T. Smith, *Adv. Inorg. Chem.*, 2000, **51**, 107.
86 S. M. Aitken, J. L. Turnbull, M. D. Percival and A. M. English, *Biochemistry*, 2001, **40**, 13980.
87 S. D. Dalosto, N. V. Prabhu, J. M. Vanderkooi and K. A. Sharp, *J. Phys. Chem. B*, 2003, **107**, 1884.
88 A. D. Kaposi, N. V. Prabhu, S. D. Dalosto, K. A. Sharp, W. W. Wright, S. S. Stavrov and J. M. Vanderkooi, *Biophys. Chem.*, 2003, **106**, 1.
89 E. S. Park, S. S. Andrews, R. B. Hu and S. G. Boxer, *J. Phys. Chem. B*, 1999, **103**, 9813.
90 J. B. Summers, H. Mazdiyasni, J. H. Holms, J. D. Ratajczyk, R. D. Dyer and G. W. Carter, *J. Med. Chem.*, 1987, **30**, 574.
91 G. H. Carlsson, P. Nicholls, D. Svistunenko, G. I. Berglund and J. Hajdu, *Biochemistry*, 2005, **44**, 635.
92 A. Henriksen, M. Gajhede, P. Baker, A. T. Smith and J. F. Burke, *Acta Crystallogr., Sect. D*, 1995, **51**, 121.
93 B. D. Howes, H. A. Heering, T. O. Roberts, F. Schneider-Belhadadd, A. T. Smith and G. Smulevich, *Biopolymers*, 2001, **62**, 261.
94 M. Khajehpour, T. Troxler and J. M. Vanderkooi, *Biochemistry*, 2003, **42**, 2672.
95 G. Smulevich, A. Feis, C. Indiani, M. Becucci and M. P. Marzocchi, *J. Biol. Inorg. Chem.*, 1999, **4**, 39.
96 A. Henriksen, D. J. Schuller, K. Meno, K. G. Welinder, A. T. Smith and M. Gajhede, *Biochemistry*, 1998, **37**, 8054.
97 I. E. Holzbaur, A. M. English and A. A. Ismail, *J. Am. Chem. Soc.*, 1996, **118**, 3354.
98 W. J. Ingledew and P. R. Rich, *Biochem. Soc. Trans.*, 2005, **33**, 886.
99 S. Hashimoto and H. Takeuchi, *Biochemistry*, 2006, **45**, 9660.

100 K. Kwak, J. Zheng, H. Cang and M. D. Fayer, *J. Phys. Chem. B*, 2006, **110**, 19998.
101 J. Zhang and M. D. Fayer, *J. Am. Chem. Soc.*, 2006, accepted.
102 J. R. Zheng, K. Kwak, J. Asbury, X. Chen, I. R. Piletic and M. D. Fayer, *Science*, 2005, **309**, 1338.
103 J. March, *Advanced Organic Chemistry*, John Wiley and Sons, New York, 1985.
104 W. J. Orville-Thomas, *Internal Rotation in Molecules*, John Wiley and Sons, 1974.
105 T. A. Weber, *J. Chem. Phys.*, 1978, **69**, 2347.
106 D. Brown and J. H. R. Clarke, *J. Chem. Phys.*, 1990, **92**, 3062.
107 R. O. Rosenberg, B. J. Berne and D. Chandler, *Chem. Phys. Lett.*, 1980, **75**, 162.
108 R. Edberg, D. J. Evans and G. P. Morris, *J. Chem. Phys.*, 1987, **87**, 5700.
109 J. Ramirez and M. Laso, *J. Chem. Phys.*, 2001, **115**, 7285.
110 J. Zheng, K. Kwak, J. Xie and M. D. Fayer, *Science*, 2006, 1951.
111 A. Streitwieser and R. W. Taft, *Progress in Physical Organic Chemistry*, John Wiley and Sons, New York, 1968.
112 D. Chandler, *J. Chem. Phys.*, 1978, **68**, 2959.
113 L. M. Jackman and F. A. Cotton, *Dynamic Nuclear Magnetic Resonance Spectroscopy*, Academic Press, New York, 1975.
114 B. Cohen and S. Weiss, *J. Phys. Chem.*, 1983, **87**, 3606.
115 J. J. Turner, F. W. Grevels, S. M. Howdle, J. Jacke, M. T. Haward and W. E. Klotzbucher, *J. Am. Chem. Soc.*, 1991, **113**, 8347.
116 R. A. MacPhail and H. L. Strauss, *J. Chem. Phys.*, 1985, **82**, 1156.
117 H. L. Strauss, *J. Am. Chem. Soc.*, 1992, **114**, 905.
118 N. E. Levinger, P. H. Davis, P. Behera, D. J. Myers, C. Stromberg and M. D. Fayer, *J. Chem. Phys.*, 2003, **118**, 1312.
119 M. Khalil and A. Tokmakoff, *Chem. Phys.*, 2001, **266**, 213.
120 N. Demirdoven, M. Khalil, O. Golonzka and A. Tokmakoff, *J. Phys. Chem. A*, 2001, **105**, 8030.
121 I. N. Levine, *Physical Chemistry*, McGraw-Hill Book Company, New York, 1978.
122 V. Pophristic and L. Goodman, *Nature*, 2001, **411**, 565.
123 F. M. Bickelhaupt and E. J. Baerends, *Angew. Chem., Int. Ed.*, 2003, **42**, 4183.

Ultrafast 2D-IR vibrational echo spectroscopy: a probe of molecular dynamics

S. Park, K. Kwak, and M.D. Fayer *

Department of Chemistry, Stanford University, Stanford, California 94305, USA

Received: 19 April 2007, Accepted: 25 April 2007
Published online: 9 May 2007

Key words: two dimensional infrared vibrational echo spectroscopy; frequency-frequency correlation function; third-order four wave mixing; heterodyne detection; spectral diffusion; hydrogen bond

PACS: 78.47.+p, 82.53.Kp, 82.53.Uv

1. Introduction

Over the last decade, ultrafast multidimensional nonlinear spectroscopies with multiple visible [1] and infrared (IR) pulses [2–4] have been developed and have combined with rapid advances in ultrashort and high power laser technology to reveal the dynamics of complex molecular systems that previously could not be studied by one-dimensional (1D) spectroscopic methods. Molecular dynamics on sub-picosecond and picosecond time scales are measured in the multidimensional nonlinear spectroscopy by manipulating the sequences of ultrashort femtosecond pulses and examining the nonlinear response of the system as a function of the timing between the pulses.

Ultrafast two dimensional IR (2D-IR) vibrational echo spectroscopy is an ultrafast IR experimental method that is

* Corresponding author: e-mail: fayer@stanford.edu

Laser Phys. Lett. **4**, No. 10 (2007)

in some respects akin to 2D NMR [5], but it probes molecular vibrations instead of nuclear spins [5]. 2D-IR vibrational echo methods are an important development that follows the first ultrafast 1D-IR vibrational echo experiments conducted over a decade ago [6]. Vibrations are the mechanical degrees of freedom of molecules and thus are sensitive to molecular dynamics of molecular structure and intermolecular interactions. As opposed to electronic excitations, the vibrational excitations in 2D-IR vibrational echo spectroscopy produce a negligible perturbation of a molecular system with less energetic IR photons. In contrast to the electronic excitation, a vibrational excitation does not change the chemical properties of the molecular system under study [7,8]. In general, 2D-IR vibrational echo spectroscopy can be used to study molecular systems under a thermal equilibrium condition and measure molecular dynamics occurring on subpicosecond to picosecond time scales [2,4,7–25]. 2D-IR vibrational echo spectroscopy has been applied to study chemical exchange reactions [2,3,24], molecular internal rotation dynamics [4], water dynamics [17,22,23], hydrogen bond dynamics of alcohols [16,26], vibrational population transfer [27], intramolecular vibrational coupling [28], and mixed chemical analysis [29]. An increasing number of research groups are applying 2D-IR spectroscopy to the study of biological problems [8,11,18,25,30–42]. The applications of 2D-IR vibrational echo spectroscopy are increasing rapidly in areas of physics, chemistry, biology, and materials science.

2D-IR vibrational echo spectroscopy involves three IR pulses that excite molecular vibrations, and because of nonlinear radiation field-matter interactions, the fourth IR pulse is emitted, the vibrational echo, which is the signal in the experiment. The first IR pulse excites the molecules to a coherence state where their wavefunctions are superpositions of the $v = 0$ and $v = 1$ energy levels. In effect, the first interaction causes the molecules to "oscillate" in phase at their initial frequencies. The phase relationships among the molecules decay quickly because of static inhomogeneous and dynamic broadening mechanisms. The decay of the phase relationships produces a vibrational free induction decay (FID). The second IR pulse transfers the coherence state into a population state in either v = 0 or v = 1 energy level, depending on the phase of the particular molecule at the time of the pulse. During the population period, the molecules can undergo spectral diffusion, that is, their frequencies evolve because of dynamic structural evolution of the system. Spectral diffusion causes molecules to lose memory of their initial frequencies. In the water systems discussed below, time dependent evolution of water's hydrogen bond network causes spectral diffusion of the hydroxyl stretching vibration of water because the stretching vibration frequency is sensitive to the strength and number of hydrogen bonds [16,17]. Other processes called chemical exchange, in which the initial frequency makes a sudden jump because a molecule undergoes an abrupt change in its intra- or intermolecular chemical structure, can also occur during the population period [2,4,7,24]. The third IR pulse brings the molecules

to a second coherence state where their wavefunctions are again superpositions of $v = 0$ and $v = 1$ energy levels or between $v = 1$ and $v = 2$ energy levels. As in an NMR spin echo, the phase relationships lost during the first FID can be restored, with the result that a macroscopic oscillating electric dipole is generated. This oscillating dipole gives rise to the emission of the vibrational echo pulse. The phase relationships are again lost via a second FID. The vibrational echo is overlapped spatially and temporally with another pulse, the local oscillator. The combined vibrational echo and local oscillator pulse is detected. The local oscillator, through the heterodyne detection, provides a phase reference so that both intensity and phase information are obtained from the vibrational echo pulse. During the second coherence period (time between the third pulse and the vibrational echo emission), the molecules oscillate at their final frequencies, which can be different from those they had during the first coherence period (time between pulses 1 and 2).

In short, during the first coherence period, the molecules are frequency-labeled. During the population period, the frequency-labeled molecules can evolve to different frequencies (spectral diffusion) because of microscopic molecular events. During the second coherence period, the final frequencies of the frequency-labeled molecules are read out. A 2D correlation map (spectrum) is obtained with the initial labeled frequencies as one axis in the 2D spectrum and the final frequencies of the molecules as the other axis. The time between pulses 1 and 2 is τ and the time between pulses 2 and 3 is T_w. The vibrational echo pulse is generated after pulse 3 at a time $\leq \tau$. 2D vibrational echo spectra are obtained by scanning τ at fixed T_w. A set of such 2D spectra is obtained as a function of T_w. By analyzing the amplitudes, positions, and peak shapes of the 2D spectra, detailed information on structure and dynamics of the molecular system is obtained. Spectral diffusion results in changes in peak shapes as a function of T_w [17,43]. Appearance of off-diagonal peaks results from incoherent and coherent population transfers by anharmonic interactions [27,44] or chemical exchange [2]. Off-diagonal peaks occurring at $T_w = 0$ can arise from state coupling of different vibrational modes [45]. Vibrational population relaxation and molecular reorientation lead to decay of the amplitudes of all peaks [7,8,15].

The paper is organized as follows. A brief theoretical background of the ultrafast 2D-IR vibrational echo experiment is given in Sec. 2. Details of the 2D-IR vibrational echo experiment are presented in Sec. 3 including the experimental setup, data acquisition method, and "phasing" process required to obtain essentially absorptive 2D spectra. In Sec. 4, new 2D-IR vibrational echo experiments comparing the dynamics of water and a concentrated NaBr salt solution are discussed. The inverse of the center line slope (CLS) is defined as a new experimental observable in 2D vibrational spectra that is useful for quantifying spectral diffusion dynamics. Sec. 5 presents some concluding remarks.

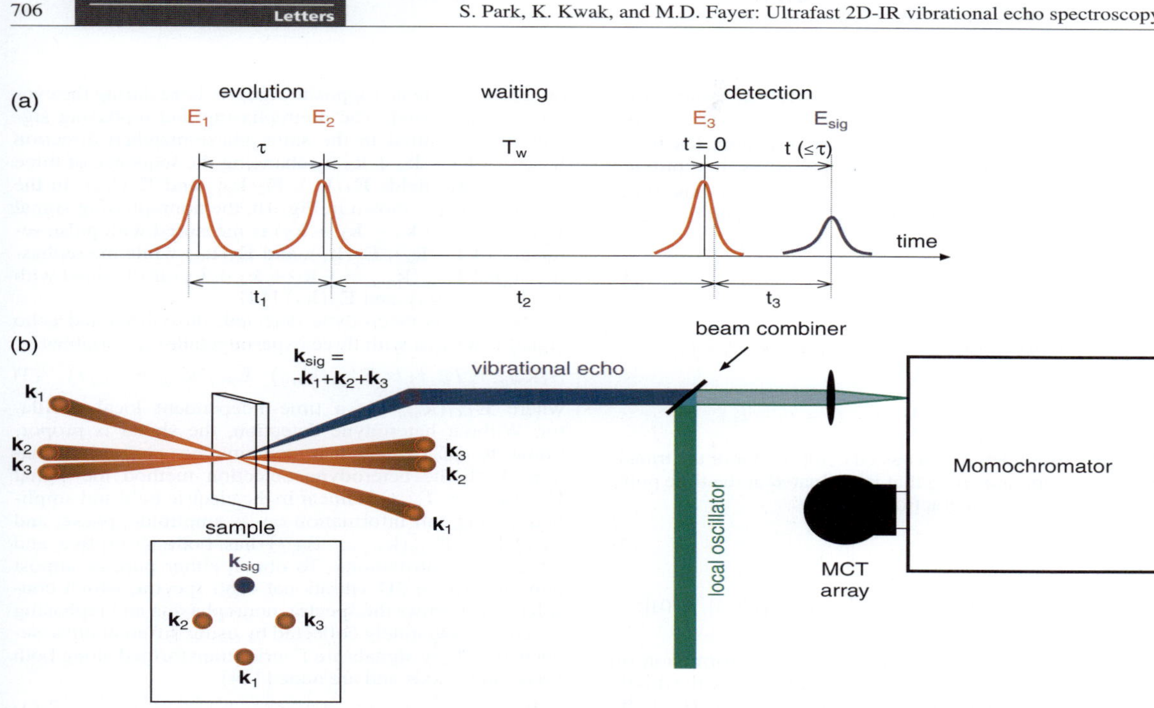

Figure 1 (online color at www.lphys.org) (a) – 2D-IR vibrational echo pulse sequence with the time variables (t_1, t_2, and t_3) of field-matter interactions, pulse sequence ($\mathbf{E}_1$, $\mathbf{E}_2$, and $\mathbf{E}_3$), experimental time variables (τ, T_w, and t), time periods (evolution, waiting, and detection). Time flows from left to right. (b) – schematic of the non-collinear excitation beam geometry, phase-matched direction, and heterodyne detection of the vibrational echo signal in 2D-IR experiments

2. Theoretical background

The 2D-IR vibrational echo experiment is a nonlinear four wave mixing spectroscopy involving three input IR electric fields in a non-collinear beam geometry [15,46–48]. Fig. 1 illustrates the sequence of three field-matter interactions, experimental time variables, time periods, non-collinear beam geometry, and phase-matched condition. The system is initially in thermal equilibrium before any field-matter interactions. The frequency of the IR pulse is tuned to the transition frequency of the vibrational mode or modes of interest. Because the pulse is very short, its band width can span a number of modes of a very broad band such as that associated with the hydroxyl stretching vibrational mode of water. After the first field-matter interaction, the molecular vibrational modes are in coherence states oscillating with initial vibrational transition frequencies (0-1). The second field-matter interaction brings the system to population states, either the ground state (0) or the first vibrationally excited state (1). Following the third field-matter interaction, the molecules are again in a coherence state oscillating with their final transition fre-

quencies. The signal field is radiated with those final transition frequencies in the wave vector phase-matched direction and is combined with a local oscillator (LO) for heterodyne detection. There are three time periods in the experiment: evolution (τ), waiting (T_w), and detection (t) as shown in Fig. 1a. To obtain the 2D-IR spectra in the frequency domain, two Fourier transforms are required. One is done experimentally when the monochromator spectrally resolves the signal-local oscillator pulse. This transform gives the ω_m axis corresponding to the time variable t. When τ is scanned, an interferogram is obtained at each ω_m. These interferograms are numerically Fourier transformed to provide the ω_τ axis. The details of these procedures are discussed below. In the 2D-IR vibrational echo experiment, 2D frequency correlation maps (spectra) between the evolution and detection time periods are obtained as a function of waiting time (T_w), which provides direct information on the time evolution of the molecular system.

In third-order four wave mixing experiments, three input electric fields ($\mathbf{E}_1$, $\mathbf{E}_2$, and $\mathbf{E}_3$ each with a unique wave vector $\mathbf{k}_1$, $\mathbf{k}_2$, and $\mathbf{k}_3$, respectively) interact with

Laser Phys. Lett. **4**, No. 10 (2007)

a system and create a third-order nonlinear polarization $\mathbf{P}^{(3)}$ that subsequently radiates a signal field $\mathbf{E}_{sig}$ in the $\mathbf{k}_{sig} = -\mathbf{k}_1 + \mathbf{k}_2 + \mathbf{k}_3$ phase-matched direction. The third-order nonlinear polarization is expressed as time ordered integrals of a response function $\mathbf{R}^{(3)}(t_3, t_2, t_1)$ and three input fields $\mathbf{E}(\mathbf{k})$ [46]. The radiated signal field is written as [46]

$$\mathbf{E}_{sig}(\mathbf{k}_{sig}, t) \propto i\mathbf{P}^{(3)}(\mathbf{k}_{sig}, t) = \tag{1}$$

$$= \int_0^\infty dt_3 \int_0^\infty dt_2 \int_0^\infty dt_1 \mathbf{R}^{(3)}(t_3, t_2, t_1)\mathbf{E}_3(\mathbf{k}_3, t - t_3) \times$$

$$\times \mathbf{E}_2(\mathbf{k}_2, t - t_3 - t_2)\mathbf{E}_1(\mathbf{k}_1, t - t_3 - t_2 - t_1) .$$

The response function is a nested commutator of the transition dipole operator $\boldsymbol{\mu}(t)$ that is evaluated at the time point of the field-matter interactions [15]

$$\mathbf{R}^{(3)}(t_3, t_2, t_1) = \tag{2}$$

$$= \left(\frac{i}{\hbar}\right)^3 \langle[[[\boldsymbol{\mu}(t_3 + t_2 + t_1), \boldsymbol{\mu}(t_2 + t_1)], \boldsymbol{\mu}(t_1)], \boldsymbol{\mu}(0)]\rangle .$$

The response function contains all of the information on the microscopic dynamics of the system because the dipole operator evolves under the system Hamiltonian H_o in the total Hamiltonian $H = H_o + H_{int}$ where H_{int} is the term for the interaction of the system with the radiation fields. H_{int} is evaluated in the perturbative limit using the rotating wave approximation (RWA). Therefore, only terms in the response function that are resonant with the electric fields are retained. Each term in the response function in the RWA represents a different density matrix pathway that depends on the sequence of field-matter interactions. Only terms that contribute to the signal in the phase-matched direction are included.

A three-level system is considered for the vibrational response function because the $v = 1 \rightarrow v = 2$ transition needs to be included as well as the fundamental transition ($v = 0 \rightarrow v = 1$) [15,47]. (The 1-2 transition comes into play for pathways that produce a population in the $v = 1$ level following the second pulse. The third pulse can then create a 1-2 coherence with subsequent vibrational echo emission at the 1-2 transition frequency, which differs from the 0-1 transition frequency by the anharmonic shift.) The response function can be separated into nonrephasing $\mathbf{R}^{(3)}_{NR}(t_3, t_2 t_1)$ and rephasing $\mathbf{R}^{(3)}_R(t_3, t_2, t_1)$ quantum pathways depending on the oscillation frequencies during the evolution and detection time periods. $\mathbf{R}^{(3)}_{NR}(t_3, t_2, t_1)$ pathways involve oscillations at the same frequency (same sign of the frequency) during both time periods, and thus they decay in time (nonrephasing). On the other hand, $\mathbf{R}^{(3)}_R(t_3, t_2, t_1)$ pathways can be "rephrased" because they involve oscillations during the detection time period with frequencies that are conjugate

or nearly conjugate (opposite sign) to those during the evolution time period. The nonrephasing and rephasing signals are measured in the same phase-matched direction $\mathbf{k}_{sig} = -\mathbf{k}_1 + \mathbf{k}_2 + \mathbf{k}_3$ by changing the sequence of three input electric fields $\mathbf{E}_1(\mathbf{k}_1)$, $\mathbf{E}_2(\mathbf{k}_2)$, and $\mathbf{E}_3(\mathbf{k}_3)$. In the beam geometry shown in Fig. 1b, the nonrephasing signal $\mathbf{E}_{sig}(\mathbf{k}_{sig} = +\mathbf{k}_2 - \mathbf{k}_1 + \mathbf{k}_3)$ is measured with pulse sequence of $\mathbf{E}_2(\mathbf{k}_2)$, $\mathbf{E}_1(\mathbf{k}_1)$, and $\mathbf{E}_3(\mathbf{k}_3)$ while the rephasing signal $\mathbf{E}_{sig}(\mathbf{k}_{sig} = -\mathbf{k}_1 + \mathbf{k}_2 + \mathbf{k}_3)$ is obtained with $\mathbf{E}_1(\mathbf{k}_1)$, $\mathbf{E}_2(\mathbf{k}_2)$, and $\mathbf{E}_3(\mathbf{k}_3)$ [14].

When it is heterodyne detected, the vibrational echo signal is written with three experimental time variables by

$$S(\mathbf{k}_{sig}; \tau, T_w, t) \propto \mathbf{E}^*_{LO}(\mathbf{k}_{sig}) \cdot \mathbf{E}_{sig}(\mathbf{k}_{sig}; \tau, T_w, t) , \tag{3}$$

where $\mathbf{E}_{LO}(\mathbf{k}_{sig})$ is a time-independent local oscillator. Without heterodyne detection, the signal is proportional to $|\mathbf{E}_{sig}|^2$, and it contains no phase information. With the heterodyne detection method the signal $\mathbf{E}_{sig}(\mathbf{k}_{sig}; \tau, T_w, t)$ is linear in the electric field and amplified, giving full information on its amplitude, phase, and frequency. $\mathbf{E}_{sig}(\mathbf{k}_{sig}; \tau, T_w, t)$ has both absorptive and dispersive contributions. To obtain either pure or almost pure absorptive 2D vibrational echo spectra, which considerably narrows the spectra, nonrephasing and rephasing signals are separately collected by using different pulse sequences. These signals are Fourier transformed along both t-axis and τ-axis and are added [14]

$$S(\mathbf{k}_{sig}; \omega_\tau, T_w, \omega_m) = \tag{4}$$

$$= S_{NR}(\mathbf{k}_{sig}; \omega_\tau, T_w, \omega_m) + S_R(\mathbf{k}_{sig}; \omega_\tau, T_w, \omega_m) ,$$

where $S_{NR}(\mathbf{k}_{sig}; \omega_\tau, T_w, \omega_m)$ and $S_R(\mathbf{k}_{sig}; \omega_\tau, T_w, \omega_m)$ are the nonrephasing and rephasing 2D spectra defined as

$$S_{NR}(\mathbf{k}_{sig}; \omega_\tau, T_w, \omega_m) = \tag{5}$$

$$= \mathrm{Re}\left[\int_0^\infty d\tau \int_0^\infty dt \exp[i\omega_m t + i\omega_\tau \tau] S_{NR}(\mathbf{k}_{sig}; \tau, T_w, t)\right] ,$$

$$S_R(\mathbf{k}_{sig}; \omega_\tau, T_w, \omega_m) = \tag{6}$$

$$= \mathrm{Re}\left[\int_0^\infty d\tau \int_0^\infty dt \exp[i\omega_m t - i\omega_\tau \tau] S_R(\mathbf{k}_{sig}; \tau, T_w, t)\right] .$$

In practice, the t-axis in $S(\mathbf{k}_{sig}; \tau, T_w, t)$ is Fourier transformed during data collection by the monochromator and the τ-axis in $S(\mathbf{k}_{sig}; \tau, T_w, t)$ is numerically Fourier transformed after data acquisition is finished as described in more detail in the next section.

3. Experimental procedures

3.1. Femtosecond laser system and 2D-IR spectrometer

Femtosecond IR pulses employed in the experiments are generated using a Ti:Sapphire regeneratively amplified laser/OPA system. The output of a modified Spectra

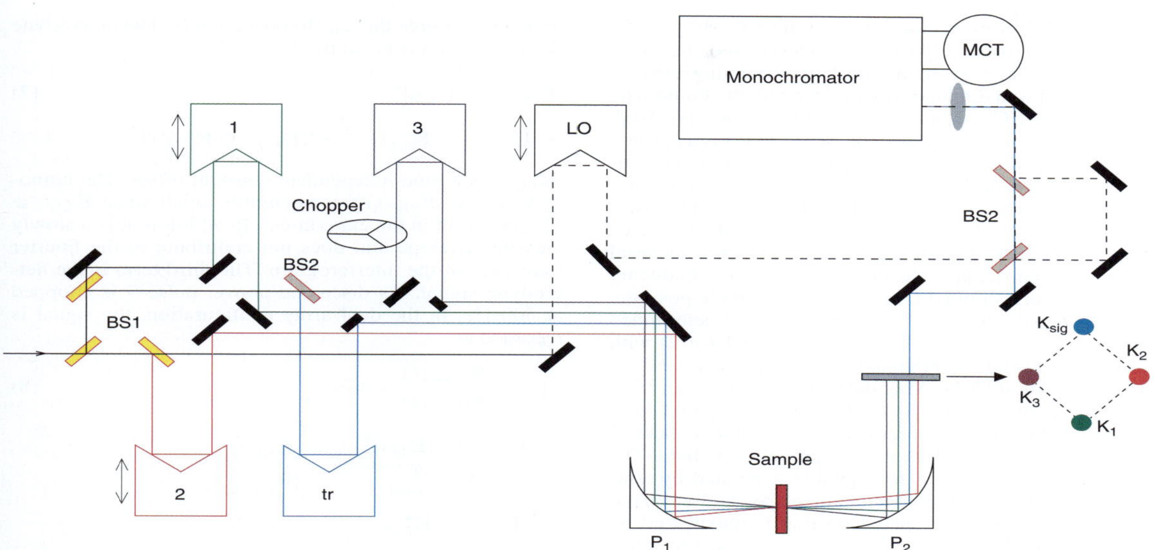

Figure 2 (online color at www.lphys.org) Schematic of 2D-IR spectrometer. Beam 1 (green), 2 (red), and 3 (purple) are the paths of the three excitation pulses. LO (black, dotted) and tr (blue) are a local oscillator and tracer, respectively. Following the sample, the vibrational echo pulse follows the same path as the tracer beam. Beam 1, 2, and LO are delayed by precision motorized translational stages (step size 10 nm) and their time delay positions are controlled by computer. BS1(yellow), 50:50 ZnSe beam splitters; BS2(light blue), ZnSe window; P1 and P2, 90° off-axis parabolic mirrors (3-inch in diameter and 15 cm in focal length); MCT, liquid N_2 cooled dual 2×32 element MCT array detector. The "diamond" at the right of the figure shows the non-collinear beam geometry and the phase-matched direction

Physics Regen is 30 fs transform-limited 0.6 mJ pulses centered at $\sim$ 800 nm with a 1 kHz repetition rate. These are used to pump an IR OPA to generate IR pulses μm in 0.5 mm thick $AgGaS_2$ crystal by difference frequency generation. For the experiments described below, the frequency is centered at 2510 cm^{-1}. The IR pulses are $\sim$ 55 fs transform-limited following compensation for chirp [16].

A schematic layout of the 2D-IR spectrometer is shown in Fig. 2. A HeNe laser beam (632 nm) is used to align all optics in the 2D-IR spectrometer. The IR beam is collimated allowing for propagation of several meters without significant divergence. A 2 mm thick Ge Brewster plate is used to overlap the IR beam with a mode-matched HeNe alignment beam. The IR beam passes through the plate and the HeNe reflects from it. 50:50 ZnSe beam splitters are used to split the IR beam into five beams. Three of them (denoted as 1 (red), 2 (green), and 3 (purple) in Fig. 2) are the excitation beams for the stimulated vibrational echo. The fourth beam (tr) is a tracer beam for alignment and is blocked during the 2D-IR experiment. All four beams are focused with a 3-inch diameter off-axis parabolic mirror (f.l. = 15 cm) in a diamond geometry as shown in Fig. 2 and are spatially overlapped at the

sample position by having them go through a 100 μm pinhole. A fifth beam (LO, black dotted) does not go through the sample and is further split into a local oscillator and a reference beam. These two beams are dispersed in a monochromator (iHR320, Horiba) and are sent to the upper and lower 32 element stripes of a liquid N_2 cooled dual MCT (HgCdTe) array detector (Infrared Associates Inc.) with high speed data acquisition system (Infrared System Development Corp.), respectively. The vibrational echo signal is generated in the phase matched direction to the fourth corner of the diamond pattern shown in the figure (the tracer beam direction) and is temporally and spatially overlapped with the local oscillator for heterodyne detection. Beam 3 is chopped at 500 Hz to remove scattered light from the other excitation beams. It is important to chop beam 3. As discussed below, beams 1 and 2 are scanned to produce interferograms between the vibrational echo signal and the local oscillator. Scattered light from these beams will also produce interferograms.

However, beam 3 is fixed. Although scattered light from beam 3 will be heterodyne detected, it only produces a signal offset, which does not contribute to the Fourier transformed spectrum.

Laser Phys. Lett. **4**, No. 10 (2007)

The chirp on excitation pulses is measured by a frequency resolved optically gating (FROG) method in a transient grating (TG) geometry [49]. By scanning pulse 2 with pulses 1 and 3 temporally overlapped, the homodyne TG signal of CCl_4 is measured and frequency-resolved. The frequency-resolved transients are fit to Gaussian functions to obtain the time zero position and width of the transient at each frequency. The plot of the time zero positions against frequency is a measure of chirp on the input pulses. CaF_2 plates with different thickness are used to balance the positive chirp introduced by other dielectric material in the setup, particularly the Ge Brewster plate, producing nearly transform-limited IR pulses at the sample position. The time zeros between excitation pulses are determined within $\pm$ 5 fs by iteratively scanning pulse 1 (or 2) with pulses 2 (or 1) and 3 overlapped and measuring a nonresonant TG signal of CCl_4. After the time zeros between excitation pulses are determined, the temporal and spatial overlap of the LO with the vibrational echo signal is checked by measuring a spectral interferogram between them. The nonresonant vibrational echo signal is generated by temporally overlapping three excitation pulses and is mixed with the LO pulse. The spatial overlap is maximized by increasing the depth of modulation in the spectral interferogram between the vibrational echo signal and the LO pulse by adjusting the mirrors steering the vibrational echo signal on top of the LO, which has previously be aligned through the monochromator onto the two array stripes. Inverse Fourier transformation of the spectral interferogram converts the fringe spacing into the corresponding temporal separation so that the LO pulse can be moved temporally and placed on top of pulse 3 within $\pm$ 5 fs.

Prior to the monochromator, the LO is split into two. One beam is combined with the vibrational echo, passed through the monochromator, and detected by the upper MCT array stripe. The other beam is used as a reference beam. It is passed through the monochromator and detected by the lower array stripe. 32 pixel frequencies in the MCT array were calibrated by using a spectral interferogram between the LO and tracer pulses. The LO and tracer pulses were spatially overlapped and shifted temporally by a few picoseconds to produce a spectral interferogram on the MCT array detector. The same spectral interferogram was measured through a slit at a different exit port of the monochromator with a single element MCT detector by scanning the monochromator. Corrections to the array detected spectrum were made to make it correspond to the scanned spectrum.

3.2. Data acquisition

In 2D-IR vibrational echo spectroscopy, the vibrational echo signal was measured as a function of one frequency variable, ω_m, and two time variables, τ and T_w. τ is defined by the time between the first and second IR pulses and T_w is the time between the second and third IR pulses. The monochromator performs an experimental Fourier transform to give the ω_m frequency axis. The heterodyne detected signal is given by

$$|\mathbf{E}_{sig}(t) + \mathbf{E}_{LO}|^2 = \tag{7}$$

$$= |\mathbf{E}_{LO}|^2 + |\mathbf{E}_{sig}(t)|^2 + 2\mathbf{E}_{sig}(t) \cdot \mathbf{E}_{LO}^* .$$

$|\mathbf{E}_{LO}|^2$ is a time-independent constant offset. The homodyne signal $|\mathbf{E}_{sig}(t)|^2$ is negligibly small since $\mathbf{E}_{LO} \gg \mathbf{E}_{sig}(t)$ is set in the experiment. In addition, it is a slowly varying envelope and does not contribute to the Fourier transform of the interferogram. The third term is the heterodyne signal. As discussed above, pulse 3 is chopped at 500 Hz. In the dual array configuration, the signal is recorded as

$$\mathbf{S}(t) = \frac{\mathbf{S}_{open}(t)}{\mathbf{S}_{close}(t)} - 1 = \tag{8}$$

$$= \frac{|\mathbf{E}_{sig}(t) + \mathbf{E}_{LO}|^2_{upper}}{|\mathbf{E}_{ref}|^2_{lower}} \cdot \frac{|\mathbf{E}_{ref}|^2_{lower}}{|\mathbf{E}_{LO}|^2_{upper}} - 1 \propto$$

$$\propto 2\,\mathbf{E}_{sig}(t) \cdot \mathbf{E}_{LO}^* .$$

$\mathbf{S}_{open}(t)$ and $\mathbf{S}_{close}(t)$ are the signals measured with a chopper open and closed, respectively. $\mathbf{E}_{ref}$ represents a reference beam sent to the lower stripe of the dual array. $\mathbf{E}_{ref} = \mathbf{E}_{LO}$ when the dual array is normalized.

In a dual MCT 2×32 array configuration with upper and lower stripes, the LO and reference beams are sent to the upper and lower stripes, respectively. First, 32 pixels in the upper stripe are normalized so that each pixel gives the same responsivity. 32 pixels in the lower stripe are normalized to the corresponding 32 pixels in the upper stripe. In this normalization scheme, the lower stripe works identically as the upper stripe if there is no pixel noise. However, the pixel noise is much smaller than shot-to-shot laser intensity fluctuations. As a result, the dual array configuration effectively reduces the shot-to-shot spectral noise arising from instability of the generated IR pulses and improves the signal-to-noise ratio. At each monochromator setting, the dual array detects 32 individual frequencies at the same time.

As τ is scanned at a fixed T_w, the phase of the vibrational echo signal field is scanned relative to the fixed LO field, resulting in an interferogram as a function of τ. The interferograms contain full information on the amplitude, sign, frequency, and phase of the vibrational echo signal field. The interferograms obtained at given monochromator frequencies are numerically Fourier transformed to the frequency domain to give the ω_τ axis.

The interferogram measured as a function of τ contains both the absorptive and dispersive parts of the vibrational echo signal. To obtain the pure or nearly pure absorptive part of the vibrational echo signal, a dual scan method is used [14]. As explained in Sec. 2, there are two sets of quantum pathways (nonrephasing and rephasing) depending on the phase relation of the signal between the evolution and detection period. Nonrephasing and rephrasing signals can be measured separately by changing the sequence of excitation pulses in the 2D-IR vibrational echo experiment [14]. In the beam geometry shown in Fig. 1, the nonrephasing signal is obtained if pulse 2 precedes pulse 1 whereas the rephasing signal is obtained if pulse 1 precedes pulse 2. The dispersive parts of the nonrephasing and rephrasing signals are 180 out of phase while the absorptive parts of them are in-phase. Adding the nonrephasing and rephrasing signals, the dispersive parts are cancelled, leaving only the absorptive part. Nonrephasing and rephasing interferograms at a monochromator frequency are shown in Fig. 3a for OD hydroxyl stretching vibration of dilute HOD in H_2O. The experimental results are discussed in detail below. The nonrephasing interferogram was obtained by scanning pulse 2 from a negative time to the time zero (where pulse 1 is located) while the rephasing interferogram was measured by scanning pulse 1 from the time zero to a positive time.

3.3. Phasing 2D vibrational echo spectra

As mentioned above, there is an experimental uncertainty in determining the time zeros between excitation pulses, compensating the chirp of the excitation pulses, and considering the chirp on the emitted vibrational echo pulse. These experimental factors distort the spectral phase in the absorptive 2D correlation spectra. To get a correct absorptive 2D spectrum, the spectral phase needs to be adjusted, i.e., a "phasing" procedure must be employed [16,43]. The projection slice theorem is employed to correct the spectral phase to obtain the absorptive 2D spectrum [14–17,22,23,43,50,51]. The projection of the 2D spectrum onto the ω_m axis is equal to the IR pump-probe spectrum measured at the same T_w, as long as all of the contributions to the vibrational echo are purely absorptive. Numerical adjustments are made that compensate for timing errors and chirp using the theoretical approach that has been described previously [16]. The spectral phase of the 2D correlation spectrum is very sensitive to small errors in the time origin, even on the order of 1 fs, and to chirp. The frequency-dependent phasing factor used to adjust the spectral phase in the 2D correlation spectra has the form

$$S_C(\omega_m, \omega_\tau) = S_{NR}(\omega_m, \omega_\tau) \exp\left[i\phi_{NR}(\omega_m, \omega_\tau)\right] + \quad (9)$$

$$+ S_R(\omega_m, \omega_\tau) \exp\left[i\phi_R(\omega_m, \omega_\tau)\right],$$

$$\phi_{NR}(\omega_m, \omega_\tau) = \quad (10)$$

$$= -\omega_\tau \Delta\tau_{1,2} + \omega_m \Delta\tau_{3,LO} + \omega_m\omega_\tau Q_1 + \omega_m^2\omega_\tau^2 Q_2,$$

$$\phi_R(\omega_m, \omega_\tau) = \quad (11)$$

$$= \omega_\tau \Delta\tau_{1,2} + \omega_m \Delta\tau_{3,LO} + \omega_m\omega_\tau Q_1 + \omega_m^2\omega_\tau^2 Q_2.$$

$S_C(\omega_m, \omega_\tau)$ is the phased 2D correlation spectrum. $S_{NR}(\omega_m, \omega_\tau)$ and $S_R(\omega_m, \omega_\tau)$ are the nonrephasing and

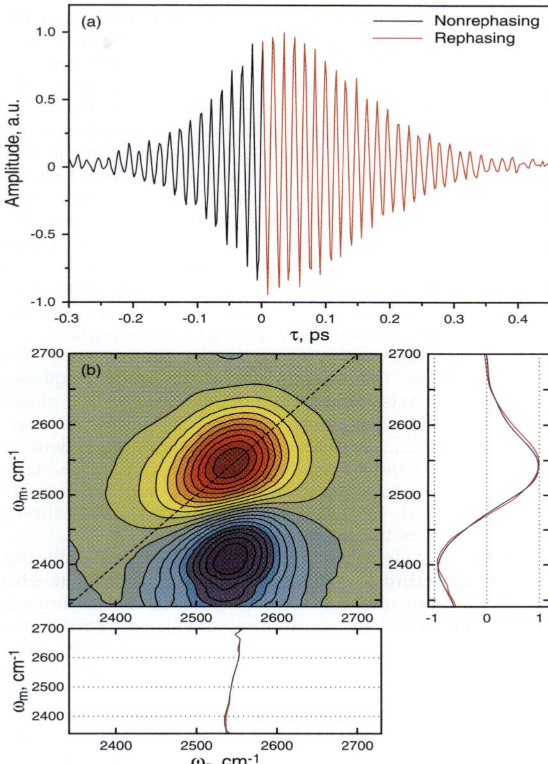

Figure 3 (online color at www.lphys.org) (a) – nonrephasing and rephasing interferograms at a single monochromator frequency measured by the dual phase scan method. (b) – absorptive 2D-IR vibrational echo spectrum obtained by using a phasing procedure with the projection slice theorem (right panel) along the ω_m axis and the center of gravity lines (lower panel) along the ω_τ axis. In the right panel, the blue line is the pump-probe spectrum and the red line is the projection of 2D spectrum onto the ω_m axis. In the lower panel, the two types of center of gravity lines. See the text for details. The dashed line is the diagonal

rephasing 2D spectra, respectively. $\Delta\tau_{1,2}$ is the uncertainty of time zero between the excitation pulses 1 and 2, while $\Delta\tau_{3,LO}$ accounts for the time zero error between the excitation pulse 3 and the LO. The third and fourth terms in $\phi_{NR}(\omega_m, \omega_\tau)$ and $\phi_R(\omega_m, \omega_\tau)$ are correction factors for chirp on excitation pulses and the vibrational echo pulse. Note that $\Delta\tau_{1,2}$ comes in with opposite sign for nonrephasing and rephasing 2D spectra. The four parameters, $\Delta\tau_{1,2}$, $\Delta\tau_{3,LO}$, Q_1, and Q_2, are adjusted until the projected 2D spectrum matches the pump-probe spectrum.

In most instances, it is possible to obtain essentially the correct 2D spectrum using only the projection theorem procedure. In many situations in which the peaks in the 2D spectra are relatively narrow, and particularly if the amplitudes and positions of the peaks are all that is required, the projection theorem is sufficient for phasing the 2D spectrum [14,15,22,23,50–52]. However, for the very broad water spectrum discussed below, in which very accurate T_w dependent shapes of the 2D spectra are required, to adjust the spectral phase unambiguously, it is preferable to have constraints on both ω_m axis and ω_τ axis. Therefore, an additional constraint is applied for the ω_τ axis in the phasing procedure. The absolute value nonrephasing $|S_{NR}(\omega_m,\omega_\tau)|$ and rephasing $|S_R(\omega_m,\omega_\tau)|$ spectra, which are independent of the phase factors $\phi_{NR}(\omega_m,\omega_\tau)$ and $\phi_R(\omega_m,\omega_\tau)$, are projected onto the ω_τ axis. The peaks of these two spectra are at the same frequency along the ω_τ axis. This should be also true for the purely absorptive 2D correlation spectra. The projected spectrum of $|S_C(\omega_m,\omega_\tau)|$ onto the ω_τ axis is dependent on $\phi_{NR}(\omega_m,\omega_\tau)$ and $\phi_R(\omega_m,\omega_\tau)$, because of the cross term in Eq. (9). Peak positions in the projected spectrum of $|S_{NR}(\omega_m,\omega_\tau)| + |S_R(\omega_m,\omega_\tau)|$ onto the ω_τ axis should be the same as the projected spectrum of $|S_C(\omega_m,\omega_\tau)|$ onto the ω_τ axis when $S_C(\omega_m,\omega_\tau)$ is properly phased. This peak position matching along the ω_τ axis is self-contained because the same nonrephasing and rephrasing spectra measured in 2D-IR experiments are used while the projection slice theorem requires pump-probe spectra measured separately. By putting these two constraints along both the ω_m axis and ω_τ axis and using a nonlinear fitting algorithm, the four parameters $\Delta\tau_{1,2}$, $\Delta\tau_{3,LO}$, Q_1, and Q_2 in Eqs. (10) and (11) are varied to obtain the correct absorptive 2D spectrum [16,25,43]. This method is quite general and can be applied to phase 2D correlation spectra having multiple peaks along the ω_τ axis.

When 2D vibrational echo spectrum has only one 0-1 peak and the corresponding 1-2 peak along the ω_τ axis, as is the case of water, a "center of gravity" line can be used as a very strong constraint for the ω_τ axis. The peak positions of slices along ω_τ at all ω_m frequencies of the $|S_{NR}(\omega_m,\omega_\tau)| + |S_R(\omega_m,\omega_\tau)|$ spectrum (absolute value spectrum) and the $|S_C(\omega_m,\omega_\tau)|$ spectrum (absolute value of the absorptive 2D spectrum) are determined. The plot of ω_τ peak frequencies vs. ω_m is the center of gravity line [16,17,43]. An example is shown in Fig. 3b. A 2D vibrational echo spectrum is of the OD stretching vibration of 5% HOD in H_2O in a high concentration NaBr solution at $T_w = 1.5$ ps. The dashed line is the diagonal. The peak on the diagonal is from the 0-1 transition. It is positive going. The off-diagonal peak is negative going. It results from vibrational echo emission at the 1-2 transition frequency. It is shifted along the ω_m axis by the anharmonicity of the OD stretching vibration. To the right of the 2D spectrum are the pump-probe spectrum (blue) and the projection of the 2D spectrum (red). Below the 2D spectrum are the center of gravity lines of the absolute value spectrum (blue) and the absolute value of the absorptive 2D

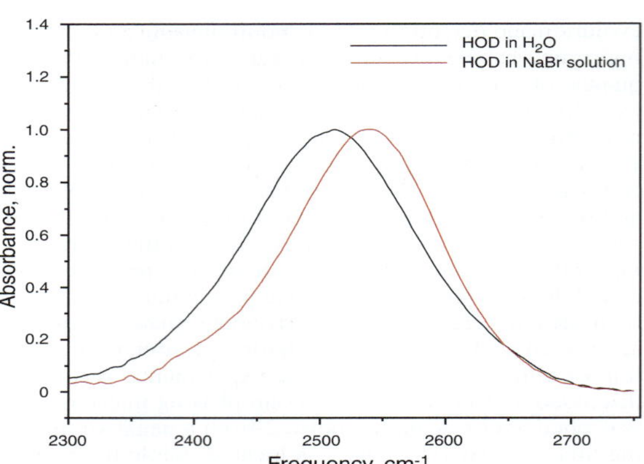

Figure 4 (online color at www.lphys.org) FT-IR absorption spectra of 5% HOD in H_2O (neat water) and in NaBr:H_2O=1:8 solution. The H_2O background is subtracted

spectrum (red). The projected and pump-probe spectra in the right panel and the center of gravity lines in the lower panel in Fig. 3b show excellent agreement, demonstrating that the 2D vibrational echo spectrum is properly phased.

3.4. Sample preparation and 2D-IR vibrational echo measurements

D_2O was purchased from Sigma-Aldrich and used as received. 5% of HOD in H_2O solution was prepared by adding 2.5 wt.% of D_2O to H_2O. A NaBr:H_2O = 1:8 solution was prepared by mixing 0.1 mol of NaBr salt with 0.8 mol of 5% of HOD in H_2O. Samples were placed in a cell with two 3 mm thick CaF_2 windows. The optical path length was set with a 6 μm thick Teflon spacer. Optical densities of the two samples at peaks at ~ 2500 cm^{-1} (OD stretching vibration) were less than 0.3 including the ~ 0.1 H_2O background absorption. Absorption spectra of the samples were measured before and after 2D-IR vibrational echo experiments. There were no changes in absorption spectra. Normalized absorption spectra of two samples are shown in Fig. 4. The IR pulses were centered at 2520 cm^{-1} with a bandwidth of ~ 250 cm^{-1} FWHM. Frequency resolution of each pixel in the MCT array was 5–7 cm^{-1} depending on the recorded frequency. Two array blocks (64 pixels) were used to scan the spectral range of 2340 to 2700 cm^{-1}. The IR portion of the 2D-IR setup was purged with dry air during the experiment to eliminate CO_2 and H_2O. The time zeros between excitation pulses were checked before and after measurements to make sure that there was no time zero drift. After the 2D-IR vibrational echo experiments were finished, the 2D-

IR setup was reconfigured to a pump-probe setup. Pump-probe spectra were measured at all T_w times and used for phasing of the 2D spectra.

4. The effect of concentrated NaBr salt on water dynamics

Water plays an important role in chemistry, biology, geology, and materials science. It is a solvent for a wide variety of molecules and compounds, and it is a particularly good solvent for ionic compounds because of its high dielectric constant. Water dissolves ions by ion-dipole interactions. Because of the large electronegativity of oxygen, the oxygen atom in water carries a partial negative charge while the hydrogens carry partial positive charges. Thus, schematically for a negative ion, HOH-Y$^-$, where, for example, Y = F, Cl, Br, or I, and for a positive ion, H$_2$O-X$^+$, with, for example, X = Na or K. The structure and dynamics of water molecules around ions are of fundamental importance in understanding ionic solvation and chemical reactions involving charged species in aqueous solutions. Vibrational population relaxation and orientational relaxation dynamics of water in bulk [53,54], nanoscopic environments [55–58], and ionic solutions [59–61] have been extensively studied by femtosecond IR pump-probe experiments. Vibrational echo experiments on hydroxyl stretching vibration have explicated the dynamics of hydrogen bond network in bulk water. Vibrational echo peak shift measurements [62–65] and 2D-IR vibrational echo spectroscopy experiments [17,22,23,43] combined with molecular dynamics simulations [66–75] have revealed detailed information on the nature of structural evolution of hydrogen bond network of bulk water. The results have shown that the dynamics of water responsible for spectral diffusion observed in the experiments occur over time scales ranging from sub-100 fs to picoseconds. The fastest dynamics are associated with very local hydrogen bond fluctuations. The longest time scale dynamics involve global structural rearrangement of hydrogen bond network that completes the randomization of hydrogen bond structures and the sampling of all frequencies in the absorption spectrum of hydroxyl stretching vibration (complete spectral diffusion).

Here the results of initial 2D-IR vibrational echo experiments used to study spectral diffusion dynamics of a concentrated aqueous ionic solution are presented. Previously, intensity level single frequency vibrational echo peak shift data for a concentrated NaCl solution showed that salt slows the spectral diffusion of the water's hydroxyl stretching vibration relative to that of pure water [56,58]. Spectral diffusion is related to how fast a molecular system undergoes structural fluctuations under a thermal equilibrium condition. The relationship between the vibrational echo observables and the microscopic dynamics of a system can be described by a frequency-frequency correlation function (FFCF). Spectral diffusion is observed

as a change in peak shape of the 2D-IR vibrational echo spectra as a function of T_w [46,76]. It has been observed that spectral diffusion dynamics of water are slower when the size of hydrogen bond network is reduced by encapsulating water molecules in nanoscopic environments [55,58] and hydrogen bond network is disrupted significantly by dissolved ions [59]. To investigate hydrogen bond dynamics of water around ions, 2D-IR vibrational echo spectroscopy was performed on OD stretching vibration in a high concentration of NaBr solution with 5% of HOD in H$_2$O. The concentration was set so that there were four water molecules per ion on average (see Sec. 3.3). At this very high concentration the hydrogen bond network is disrupted. There is essentially no bulk water in which there is an extended network of interconnected water molecules. Basically, all water molecules are hydrogen bonded to ions. The question of the dynamics of the structural evolution of hydrogen bonds under these extreme conditions is addressed with the 2D-IR vibrational echo experiments.

4.1. Linear FT-IR absorption spectra

FT-IR absorption spectra of 5% HOD in neat water and NaBr:H$_2$O = 1:8 solution are shown in Fig. 4. The peaks at 2510 and 2540 cm^{-1} in FT-IR spectra are OD stretching vibrational mode in the neat water and the NaBr salt solution, respectively. It is known for neat water that the broad spectrum is caused by different numbers and different strengths of the hydrogen bonds [77]. More hydrogen bonds and stronger hydrogen bonds red shift the absorption, while few hydrogen bonds and weaker hydrogen bonds blue shift the absorption [16,17,43]. FT-IR spectrum of the OD stretching vibration in the NaBr solution is blue-shifted by $\sim$ 30 m^{-1} compared with that of neat water. The width of the neat water band is also somewhat wider, $\sim$ 160 cm^{-1} vs. $\sim$ 140 cm^{-1} for the salt solution. The salt solution line is still very broad, indicating a wide distribution of OD-ion interactions. In neat water, a blue shift is associated with weaker and/or fewer hydrogen bonds. However, the spectrum of the salt solution arises from a very different type of interaction. The hydroxyl group is mainly interacting with an ion rather than with another water molecule. Therefore, it is not safe to assume that the blue shift should be associated with weaker hydrogen bonding. Water molecules are hydrogen-bonded to ions by three ion-dipole interactions in HOD-Br$^-$, DOH-Br$^-$, and HDO-Na$^+$. The last two configurations have less influence on the OD stretching vibration.

4.2. 2D-IR vibrational echo spectra

2D-IR vibrational echo spectra of the OD stretching vibration in the NaBr solution are shown in Fig. 5 for a series of T_ws. The nature of the spectral peaks was discussed in

Laser Phys. Lett. **4**, No. 10 (2007)

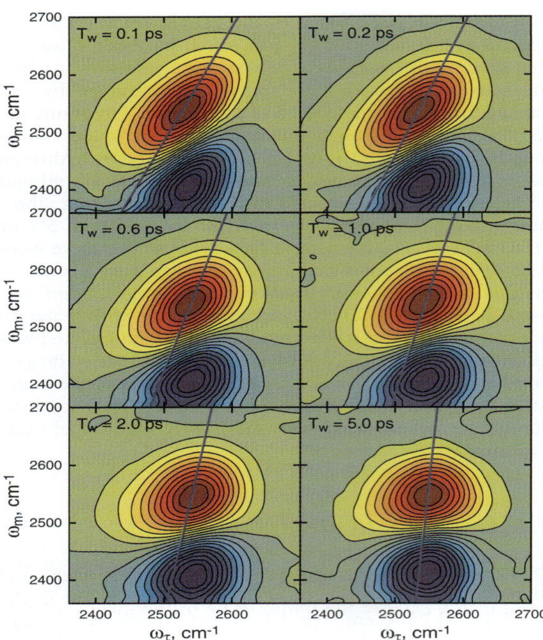

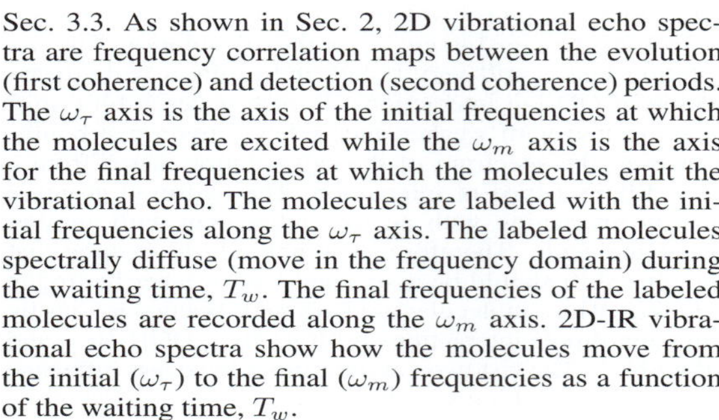

Figure 5 (online color at www.lphys.org) 2D-IR vibrational echo spectra of the OD stretching vibration obtained for the NaBr solution at increasing T_ws. Straight lines in the 2D spectra are the center lines defined in connection with the discussion of Fig. 6

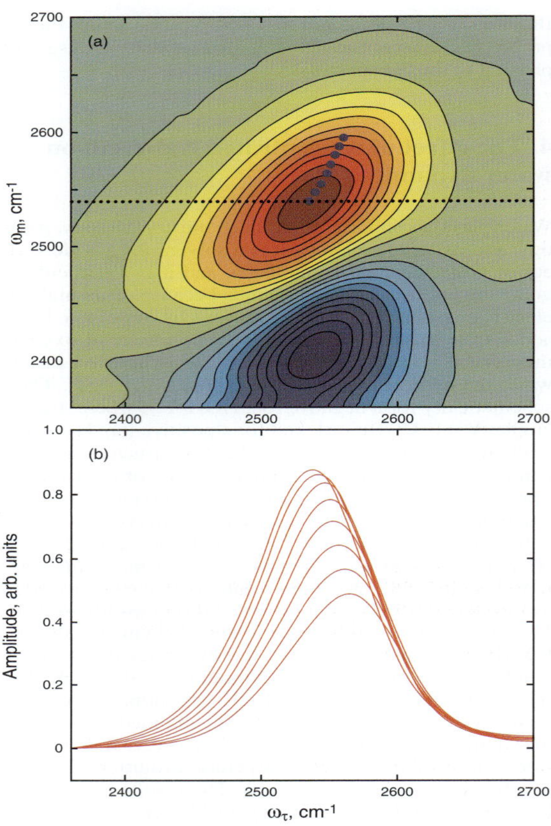

Figure 6 (online color at www.lphys.org) (a) – 2D-IR spectrum at $T_w = 0.2$ ps. A dotted horizontal line defines a frequency slice at $\omega_m = 2540$ cm^{-1} along the ω_τ axis. This slice gives the left most spectrum in (b). (b) – frequency slice spectra at different ω_m. The peak positions are obtained by fitting each to a Gaussian function. ω_m increases from left to right. Peak positions are marked with blue dots in (a). The center line slope (CLS) is obtained by a linear fit to the peak positions (blue dots) in (a)

Sec. 3.3. As shown in Sec. 2, 2D vibrational echo spectra are frequency correlation maps between the evolution (first coherence) and detection (second coherence) periods. The ω_τ axis is the axis of the initial frequencies at which the molecules are excited while the ω_m axis is the axis for the final frequencies at which the molecules emit the vibrational echo. The molecules are labeled with the initial frequencies along the ω_τ axis. The labeled molecules spectrally diffuse (move in the frequency domain) during the waiting time, T_w. The final frequencies of the labeled molecules are recorded along the ω_m axis. 2D-IR vibrational echo spectra show how the molecules move from the initial (ω_τ) to the final (ω_m) frequencies as a function of the waiting time, T_w.

Looking in particular at the fundamental transition (0-1 transition, red positive peak), a noticeable feature of 2D correlation spectra shown in Fig. 5 is a change of peak shape as T_w increases. The red peak is highly elongated along the diagonal at $T_w = 0.1$ ps. (The diagonal lines are not shown in Fig. 5. The blue lines are discussed below. See Fig. 3 for and example of the diagonal.) As T_w

increases the shape changes and the band becomes more symmetrical. This change is the signature of spectral diffusion. In the long time limit, all structural configurations are sampled and, therefore, all frequencies are sampled. The band would become completely symmetrical. However, because of the overlap with the negatively going band that arises from emission at the 1-2 transition frequency, the bottom of the 0-1 band is eaten away. The shape of the band also has a T_w independent contribution. As discussed in detail below, ultrafast fluctuations produce a motionally

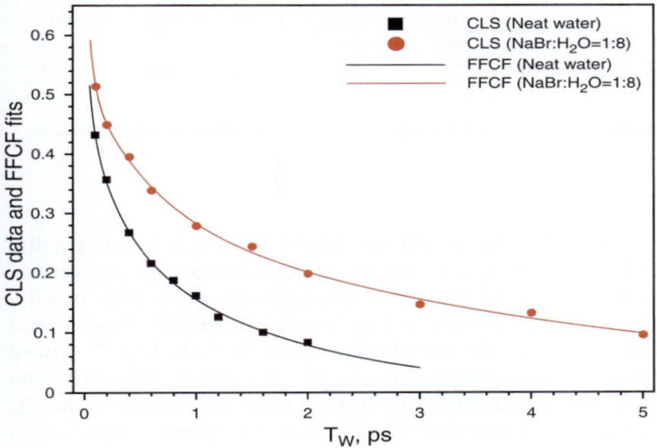

Figure 7 (online color at www.lphys.org) CLS plots (dots) vs. T_w for neat water and the NaBr solution. Solid lines are the fits to the CLS, which give the FFCFs for the two samples (see text)

narrowed (homogeneous) component of both 2D spectrum as well as the linear absorption spectrum. The motionally narrowed contribution produces a width along the ω_τ axis even at $T_w = 0$ ps. Although T_w independent, the motionally narrowed Lorentzian contribution to the 2D line shape can be determined by the procedure described in the next section.

4.3. The FFCF and the center line slope method

The frequency-frequency correlation function (FFCF) connects the experimental observables to the underlying dynamics. The FFCF is the probability that an oscillator with an initial frequency has the same frequency at time t later, averaged over all starting frequencies. It is generally written in terms of changes in frequency rather than the absolute frequency. Once the FFCF is known, all linear and nonlinear optical experimental observables can be calculated by time-dependent diagrammatic perturbation theory [46,76,78]. Conversely, the FFCF can be extracted from 2D vibrational echo spectra with input from the linear absorption. Generally, to determine the FFCF from 2D and linear spectra, full calculations of linear and nonlinear third-order response functions based on the time-dependent diagrammatic perturbation theory are performed iteratively until the calculation results converge to the experimental results. Here, a simpler method is introduced and used to determine the FFCF by circumventing the iterative fitting of the response functions. The inverse of the center line slope (CLS) of a 2D vibrational echo spectrum is defined as a new experimental observable to quantify spectral diffusion dynamics of water in bulk and in NaBr solution.

Fig. 6 illustrates how to define the CLS. A slice at a frequency ω_m (2540 cm^{-1}) along the ω_τ axis is marked as a dotted horizontal line in Fig. 6a. The projection of this slice onto the ω_τ axis is a spectrum. This particular slice is the left most spectrum in Fig. 6b. The peak position of the frequency slice can be obtained by fitting the frequency slice to a Gaussian or other appropriate function. The shape of the slice is not important. It is only necessary to determine its peak frequency. Frequency slices at different ω_m frequencies are shown in Fig. 6b, and their peak positions are marked with blue dots in Fig. 6a. The center line slope is obtained by plotting the ω_τ peak positions vs. ω_m. The "center lines" are shown as the blue lines in Fig. 5. The important feature of the center line is that as T_w increases, the slope becomes greater. In the limit of complete spectral diffusion, the long time limit, the 2D spectrum is symmetrical and the center line is vertical. In the other limit, $T_w = 0$ ps, and in the absence of a motionally narrowed component, the 2D spectrum is a thin line along the diagonal. The center line would be at 45°. As discussed below, the FFCF is related to the inverse of the center line slope. The inverse of the CLS has a maximum value of 1 at $T_w = 0$ ps and goes to 0 in the long time limit. The $T_w = 0$ ps maximum value of 1 can only occur in the absence of a motionally narrowed component. The larger the motionally narrowed contribution to the line shape is, the smaller the initial value of the inverse of the center line slope will be. The plot of the inverse of the CLS (simply referred to as the CLS for the remainder of this paper) vs. T_w is shown in Fig. 7 for both neat water and the NaBr solution. The decay of the CLS for the salt solution is considerably slower than that of bulk water.

In a subsequent publication, it will be shown theoretically that the CLS is the T_w dependent part of the FFCF to second order in time [79]. It will also be shown that the errors in determining the FFCF from the CLS arising from treating the problem to only second order are small. Furthermore, it is also possible to obtain the T_w independent part of the FFCF (the Lorentzian component) by combining the CLS and the linear absorption line shape [79].

The linear absorption spectrum $\sigma(\omega)$ can be calculated by a Fourier transformation of the linear response function $R^{(1)}(t)$,

$$\sigma(\omega) = \text{Re}\left[\int_0^\infty dt \exp(i\omega t) R^{(1)}(t)\right], \qquad (12)$$

$$R^{(1)}(t) = \qquad\qquad\qquad\qquad\qquad\qquad (13)$$

$$= |\mu_{01}|^2 \exp\left[-i\langle\omega_{01}\rangle t - g(t) - \frac{t}{2T_1} - \frac{t}{3T_{or}}\right],$$

where μ_{01} is the transition dipole moment of the fundamental transition (0-1), $\langle\omega_{10}\rangle$ is the ensemble-average of the fundamental transition frequency, and the population relaxation and orientational relaxation are included phenomenologically with time constants T_1 and T_{or}, respectively. The line shape function $g(t)$ is defined as

$$g(t) = \int\limits_0^t d\tau_2 \int\limits_0^{\tau_2} d\tau_1 \, \langle \delta\omega_{10}(\tau_1)\delta\omega_{10}(0)\rangle \,, \qquad (14)$$

where $\langle \delta\omega_{10}(t)\delta\omega_{10}(0)\rangle$ is the FFCF, $C(t)$, for the 0-1 transition, which can be written in terms of the Kubo model as [80]

$$C(t) = \langle \delta\omega_{10}(t)\delta\omega_{10}(0)\rangle = \sum_i \Delta_i^2 \exp(-t/\tau_i)\,, \qquad (15)$$

where Δ_i and τ_i are the amplitudes and correlation times, respectively, of the various contributions to the frequency fluctuations. In the motionally narrowed limit (fast modulation limit) $\Delta\tau \ll 1$. In this limit, Δ and τ cannot be determined independently, but $T_2^* = (\Delta^2\tau)^{-1}$ or $\Gamma^* = (\pi T_2^*)^{-1}$ is the single parameter describing the motionally narrowed Lorentzian contribution to the line shape. T_2^* is the pure dephasing that gives rise to the pure homogeneous linewidth, Γ^*. Including contributions from population relaxation and orientational relaxation, the total dephasing time T_2 is

$$\frac{1}{T_2} = \frac{1}{T_2^*} + \frac{1}{2T_1} + \frac{1}{3T_{or}}\,, \qquad (16)$$

where $\Gamma = (\pi T_2)^{-1}$ is the Lorentzian homogeneous line width. If there is only a single term in Eq. (15) and it is motionally narrowed, the absorption line will be a Lorentzian and its line width yields T_2. In general, there are additional terms in $C(t)$, and the absorption line is inhomogeneously broadened. In this case, the FFCF cannot be determined from the absorption spectrum, and requires the 2D-IR vibrational echo experiment to measure the spectral diffusion and to obtain the FFCF.

The FFCFs, $C(t)$, of the OD stretching vibration for neat water and the NaBr solution are determined by using the FT-IR linear spectra shown in Fig. 4 and the CLSs in Fig. 7 without resorting to a full calculation of nonlinear response functions based on time-dependent diagrammatic perturbation theory [79]. For neat water, $C(t)$ has been found previously to be the sum of three terms [17,43],

$$C(t) = \Delta_1^2 \exp(-t/\tau_1)+ \qquad (17)$$
$$+\Delta_2^2 \exp(-t/\tau_2) + \Delta_3^2 \exp(-t/\tau_3)\,.$$

By inserting Eq. (17) into Eq. (14), the corresponding line shape function is obtained as

$$g(t) = \Delta_1^2\tau_1^2[\exp(-t/\tau_1) + t/\tau_1 - 1]+ \qquad (18)$$
$$+\Delta_2^2\tau_2^2[\exp(-t/\tau_2) + t/\tau_2 - 1]+$$
$$+\Delta_3^2\tau_3^2[\exp(-t/\tau_3) + t/\tau_3 - 1]\,.$$

The linear absorption spectrum $\sigma(\omega)$ can be calculated from Eq. (12) with Eqs. (13) and (18).

sample	Γ^*, cm^{-1}	Δ_2, cm^{-1}	τ_2, ps	Δ_3, cm^{-1}	τ_3, ps
water	34	43	0.3	52	1.5
water/NaBr	29	37	0.6	44	4.4

Table 1 Frequency-frequency correlation function parameters

The full procedure for using the CLS to obtain the FFCF, including discussions of approximations and errors, will be described in detail subsequently [79]. Very briefly, the CLS data points in Fig. 7 were fit to a bi-exponential function. The fits are shown as the lines in Fig. 7. These fits determine parameters Δ_2, τ_2, Δ_3, and τ_3. The two time constants are accurate [79]. However, the values of the Δ_i are floors to the true values. For a τ_i greater than ~ 10 times the FID half width, the associated Δ_i has little error. This is the situation with water for the slowest component of the FFCF. Furthermore, it has been shown that the floor for Δ_i associated with faster τ_i is 70% of the true value. Then, the only unknown parameters are Δ_2 and T_2. Using the three known parameters, the absorption line shape is fit, using the 70% floor for Δ_2, which determines Δ_2 and T_2. (For a motionally narrowed component, $g(t)$ can be written in terms of the single parameter T_2 for that component.) In general, T_2 can be separated into the three contributions, pure dephasing T_2^*, population relaxation T_1, and orientational T_{or}, by using pump-probe experiments to measure T_1 and T_{or}. However, in water, T_2^* is ~ 100 times larger than the other two contributions, and T_1 and T_{or} are ignored here. They will be reported in a subsequent publication [79]. The FFCF parameters for neat water and the NaBr solution are given in Table 1. The neat water parameters are almost the same as those reported previously [17], and the experimental results used to determine them here are more extensive.

4.4. Hydrogen bond dynamics of neat water and NaBr solution

A water molecule can be a hydrogen bond donor and acceptor at the same time [77]. In liquid water, the water molecule can form up to four hydrogen bonds in a tetrahedral geometry with neighboring water molecules making an extended hydrogen bond network. The hydrogen bond network fluctuates and evolves continuously under a thermal equilibrium condition. Hydrogen bonds form and break. Strong (short) bonds become weak (long), and vice versa. In the NaBr solution there are only four water molecules per negative ion and four water molecules per positive ion. It seems reasonable to assume that all water molecules in the NaBr solution are in hydration shells around ions. Given the ratio of ions to water molecules, there is no bulk water. Water molecules in NaBr solution are bound (or hydrogen-bonded) to ions by ion-dipole interaction and the water-ion structure evolves in time as evidenced by the spectral diffusion. The linear absorption spectra shown in Fig. 4 shows that water in the salt solution is distinct from bulk water.

As shown in Table 1, the FFCF of OD stretching vibration is characterized by a motionally narrowed component, a subpicosecond intermediate component, and a slowest time component on a picosecond time scale. The motionally narrowed terms are reported as the pure homogeneous line width $\Gamma^* = \Delta_1^2 \tau_1 / \pi$. For neat water, molecular dynamics simulations and experiments show that the ultrafast motionally narrowed component arises from very local fluctuations in the hydrogen bond length with small contributions from angular changes [17,43,64,65,71,73]. The slowest, 1.5 ps component arises from global hydrogen bond structural rearrangements that randomize the hydrogen bond network. The two components (i.e. intermediate and slowest components) of the FFCF that are not motionally narrowed should be view as a model for a non-exponential decay. The intermediate component is a transition between the very local ultrafast single hydrogen bond fluctuations and the global complete randomization of the network.

In comparing the FFCF for the NaBr solution to that of neat water, the motionally narrowed components are similar. The amplitudes, Δ_2 and Δ_3 are also similar, although all are smaller for the NaBr solution. The large differences are in the time constants, τ_2 and τ_3, which are a factor of 2 and 3 longer respectively in the NaBr solution. This means that the time scale for complete randomization of the structure is much slower in the salt solution.

With only four water molecules per ion, the structural of water in the NaBr solution will have little resemblance to that of bulk water. Water molecules form hydration shells solvating ions in the NaBr solution by ion-dipole interaction and will have interactions like HOD-Br^-, DOH-Br^-, and HDO-Na^+. The dynamics will involve rearrangements of water molecules in relation to the ions rather than the water-water hydrogen bond rearrangements that occur in neat water. MD simulations performed on aqueous NaCl solutions at various concentrations of NaCl indicate that hydrogen bond structural relaxation slows down with increasing ion concentrations [81].

Recently, MD simulations were performed on a diluted I^-/HOD/D_2O system with a single iodide ion and 215 D_2O [82]. The results showed that the longest time component in the FFCF of OD stretching vibration was associated with the escape time of HOD from the first solvation shell of iodide ion to bulk. However, this simulation is not a model for a concentrated salt solution in which there is no bulk water. Furthermore, it does not give information on how water molecules solvating cations (Na^+) contribute to the overall hydrogen bond dynamics of water around ions, nor how the dynamics of water molecules that may bridge two anions or an anion and cation influence the spectral diffusion.

Bakker and coworkers performed two-color IR pump-probe experiments by pumping OH stretching vibration of DOH in D_2O salt solutions (6 M NaCl, NaBr, or NaI) near the center frequency of the OH absorption band and probing population relaxations at different frequencies [59,83]. It was observed that the decays of pump-probe signals got faster when the probing frequency moved away from the pumping frequency. The results were discussed in terms of a correlation time constant τ_c for spectral diffusion of the OH stretching vibration. A two Brownian oscillator model was necessary to extract the information from these indirect experiments and produced results that the correlations times were 20 to 30 times longer than that of neat water. The model used considered water bound to ions and water that was bulk like. These very long correlations times are in contrast to the measurements reported here in which the spectral diffusion components increase by factors of 2 and 3 compared to neat water.

Given the small number of water molecules per ion, it seems clear that the nature of water dynamics in concentrated salt solutions is fundamentally different from the dynamics of pure water. The inhomogeneous absorption spectrum must be composed of different configurations of a small number of water molecules associated with ions with few bulk-like water-water hydrogen bonds. The spectral diffusion is most likely associated with the structural rearrangements of water molecules strongly interacting with one or more ions and cannot involve exchange between a water-ion interaction and water in bulk. MD simulations will need to properly model the small numbers of waters per ion found in concentrated salt solutions.

5. Concluding remarks

The 2D-IR vibrational echo experimental method was described in considerable detail. 2D-IR vibrational echo spectroscopy has been developed over the last decade and has been utilized to study spectral diffusion, chemical exchange, isomerization, and the structure and dynamics of biomolecules. Investigations of such systems have just begun, and a great increase in the understanding of the fast structural dynamics of complex molecular systems is in the offing. In addition, 2D-IR vibrational echo spectroscopy has the potential to be applicable to many other microscopic molecular events occurring on subpicosecond and picosecond time scales.

In this paper, both the theoretical and experimental underpinnings of ultrafast 2D-IR vibrational echo spectroscopy were reviewed. In addition, a new method for extracting the frequency-frequency correlation function from T_w dependent 2D spectra was presented. The center line slope method makes it possible to obtain the FFCF without iterative fitting of the 2D spectra using time dependent perturbation theory techniques. As will be shown in a subsequent publication, the CLS method produces results that are not distorted by finite bandwidth of the excitation pulses or apodization procedures [79]. In addition, the first 2D-IR vibrational echo measurements of water spectral diffusion in a concentrated salt solution were presented. It was found that the structural randomization of the system

Laser Phys. Lett. **4**, No. 10 (2007)

is significantly slower than in pure water. Additional experiments on concentrated salt solutions will be reported in a subsequent publication. The dynamics of water in a variety of types of salt solutions over a range of concentrations are currently being studied.

Acknowledgements This work was supported by grants from the Air Force Office of Scientific Research (F49620-01-1-0018), the National Science Foundation (DMR-0332692), the Department of Energy (DE-FG03-84ER13251), and the National Institutes of Health (2 R01 GM-061137-05).

References

[1] T. Brixner, J. Stenger, H.M. Vaswani, M. Cho, and G.R. Fleming, Nature **34**, 625–628 (2005).

[2] J. Zheng, K. Kwak, J.B. Asbury, X. Chen, I. Piletic, and M.D. Fayer, Science **309**, 1338–1343 (2005).

[3] Y.S. Kim and R.M. Hochstrasser, Proc. Natl. Acad. Sci. USA **102**, 11185–11190 (2005).

[4] J. Zheng, K. Kwak, J. Xie, and M.D. Fayer, Science **313**, 1951–1955 (2006).

[5] R.R. Ernst, G. Bodenhausen, and A. Wokaun, Principles of Nuclear Magnetic Resonance in One and Two Dimensions (Oxford University Press, Oxford, 1987).

[6] D. Zimdars, A. Tokmakoff, S. Chen, S.R. Greenfield, M.D. Fayer, T.I. Smith, and H.A. Schwettman, Phys. Rev. Lett. **70**, 2718–2721 (1993).

[7] J. Zheng, K. Kwak, and M.D. Fayer, Acc. Chem. Res. **40**, 75–83 (2007).

[8] I.J. Finkelstein, J. Zheng, H. Ishikawa, S. Kim, K. Kwak, and M.D. Fayer, Phys. Chem. Chem. Phys. **9**, 1533–1549 (2007).

[9] M.C. Asplund, M.T. Zanni, and R.M. Hochstrasser, Proc. Natl. Acad. Sci. USA **97**, 8219–8224 (2000).

[10] M.T. Zanni, M.C. Asplund, and R.M. Hochstrasser, J. Chem. Phys. **114**, 4579–4590 (2001).

[11] M.T. Zanni and R.M. Hochstrasser, Curr. Opin. Struct. Biol. **11**, 516–522 (2001).

[12] J.B. Asbury, T. Steinel, C. Stromberg, K.J. Gaffney, I.R. Piletic, A. Goun, and M.D. Fayer, Chem. Phys. Lett. **374**, 362–371 (2003).

[13] J.B. Asbury, T. Steinel, C. Stromberg, K.J. Gaffney, I.R. Piletic, A. Goun, and M.D. Fayer, Phys. Rev. Lett. **91**, 237402 (2003).

[14] M. Khalil, N. Demirdöven, and A. Tokmakoff, Phys. Rev. Lett. **90**, 047401 (2003).

[15] M. Khalil, N. Demirdöven, and A. Tokmakoff, J. Phys. Chem. A **107**, 5258–5279 (2003).

[16] J.B. Asbury, T. Steinel, and M.D. Fayer, J. Lumin. **107**, 271–286 (2004).

[17] J.B. Asbury, T. Steinel, K. Kwak, S.A. Corcelli, C.P. Lawrence, J.L. Skinner, and M.D. Fayer, J. Chem. Phys. **121**, 12431–12446 (2004).

[18] N. Demirdöven, C.M. Cheatum, H.S. Chung, M. Khalil, J. Knoester, and A. Tokmakoff, J. Am. Chem. Soc. **126**, 7981–7990 (2004).

[19] Y.S. Kim, J. Wang, and R.M. Hochstrasser, J. Phys. Chem. B **109**, 7511–7521 (2005).

[20] J. Zheng, K. Kwak, T. Steinel, J.B. Asbury, X. Chen, J. Xie, and M.D. Fayer, J. Chem. Phys. **123**, 164301 (2005).

[21] K. Kwak, J. Zheng, H. Cang, and M.D. Fayer, J. Phys. Chem. B **110**, 19998–20013 (2006).

[22] J.J. Loparo, S.T. Roberts, and A. Tokmakoff, J. Chem. Phys **125**, 194521 (2006).

[23] J.J. Loparo, S.T. Roberts, and A. Tokmakoff, J. Chem. Phys **125**, 194522 (2006).

[24] J. Zheng, K. Kwak, X. Chen, J.B. Asbury, and M.D. Fayer, J. Am. Chem. Soc **128**, 2977–2987 (2006).

[25] I.J. Finkelstein, H. Ishikawa, S. Kim, A.M. Massari, and M.D. Fayer, Proc. Nat. Acad. Sci. USA **104**, 2637–2642 (2007).

[26] J.B. Asbury, T. Steinel, C. Stromberg, K.J. Gaffney, I.R. Piletic, and M.D. Fayer, J. Chem. Phys. **119**, 12981–12997 (2003).

[27] M. Khalil, N. Demirdöven, and A. Tokmakoff, J. Chem. Phys. **121**, 362–373 (2004).

[28] L.P. DeFlores, Z. Ganim, S.F. Ackley, H.S. Chung, and A. Tokmakoff, J. Phys. Chem. B **110**, 18973–18980 (2006).

[29] J.B. Asbury, T. Steinel, and M.D. Fayer, Chem. Phys. Lett. **381**, 139–146 (2003).

[30] P. Mukherjee, I. Kass, I.T. Arkin, and M.T. Zanni, Proc. Natl. Acad. Sci. USA **103**, 3528–3533 (2006).

[31] J. Wang, J. Chen, and R.M. Hochstrasser, J. Phys. Chem. B **110**, 7545–7555 (2006).

[32] H.S. Chung, M. Khalil, A.W. Smith, Z. Ganim, and A. Tokmakoff, Proc. Natl. Acad. Sci. USA **102**, 612–617 (2005).

[33] S. Woutersen, R. Pfister, P. Hamm, Y. Mu, D.S. Kosov, and G. Stock, J. Chem. Phys. **117**, 6833–6840 (2002).

[34] Y. Kim and R.M. Hochstrasser, J. Phys. Chem. B **109**, 6884–6891 (2005).

[35] C. Fang and R.M. Hochstrasser, J. Phys. Chem. B **109**, 18652–18663 (2005).

[36] P. Mukherjee, A.T. Krummel, E.C. Fulmer, I. Kass, I.T. Arkin, and M.T. Zanni, J. Chem. Phys. **120**, 10215–10224 (2004).

[37] M.F. DeCamp, L. DeFlores, J.M. McCracken, A. Tokmakoff, K. Kwac, and M. Cho, J. Phys. Chem. B. **109**, 11016–11026 (2005).

[38] S. Mukamel and D. Abramavicius, Chem. Rev. **104**, 2073–2098 (2004).

[39] J.H. Choi, H. Lee, K.K. Lee, S. Hahn, and M. Cho, J. Chem. Phys. **126**, 045102–045114 (2007).

[40] A.M. Massari, I.J. Finkelstein, and M.D. Fayer, J. Am. Chem. Soc. **128**, 3990–3997 (2006).

[41] I.J. Finkelstein, A.M. Massari, and M.D. Fayer, Biophys. J. **92**, 3652–3662 (2007).

[42] A.M. Massari, B.L. McClain, I.J. Finkelstein, A.P. Lee, H.L. Reynolds, K.L. Bren, and M.D. Fayer, J. Phys. Chem. B **110**, 18803–18810 (2006).

[43] J.B. Asbury, T. Steinel, C. Stromberg, S.A. Corcelli, C.P. Lawrence, J.L. Skinner, and M.D. Fayer, J. Phys. Chem. A **108**, 1107–1119 (2004).

[44] O. Golonzka, M. Khalil, N. Demirdöven, and A. Tokmakoff, Phys. Rev. Lett. **86**, 2154–2157 (2001).

[45] O. Golonzka, M. Khalil, N. Demirdöven, and A. Tokmakoff, J. Chem. Phys. **115**, 10814–10828 (2001).

[46] S. Mukamel, Principles of Nonlinear Optical Spectroscopy (Oxford University Press, New York, 1995).

[47] J. Sung and R.J. Silbey, J. Chem. Phys. **115**, 9266–9287 (2001).

[48] D.M. Jonas, Annu. Rev. Phys. Chem. **54**, 425–463 (2003).

[49] R. Trebino, K.W. DeLong, D.N. Fittinghoff, J.N. Sweetser, M.A. Krumbugel, B.A. Richman, and D.J. Kane, Rev. Sci. Instr. **69**, 3277–3295 (1997).

[50] J.D. Hybl, Y. Christophe, and D.M. Jonas, Chem. Phys. **266**, 295–309 (2001).

[51] J.D. Hybl, A.A. Ferro, and D.M. Jonas, J. Chem. Phys. **115**, 6606–6622 (2001).

[52] T. Brixner, T. Mancal, I.V. Stiopkin, and G.R. Fleming, J. Chem. Phys **121**, 4221–4236 (2004).

[53] S. Woutersen, U. Emmerichs, and H. J. Bakker, Science **278**, 658-660 (1997).

[54] T. Steinel, J.B. Asbury, J.R. Zheng, and M.D. Fayer, J. Phys. Chem. A **108**, 10957–10964 (2004).

[55] H.-S. Tan, I.R. Piletic, and M.D. Fayer, J. Chem. Phys. **122**, 174501–174509 (2005).

[56] H.-S. Tan, I.R. Piletic, R.E. Riter, N.E. Levinger, and M.D. Fayer, Phys. Rev. Lett. **94**, 057405 (2005).

[57] I.R. Piletic, H.-S. Tan, and M.D. Fayer, J. Phys. Chem. B **109**, 21273–21284 (2005).

[58] I. Piletic, D.E. Moilanen, D.B. Spry, N.E. Levinger, and M.D. Fayer, J. Phys. Chem. A **110**, 4985–4999 (2006).

[59] M.F. Kropman and H.J. Bakker, J. Chem. Phys. **115**, 8942–8948 (2001).

[60] M.F. Kropman and H.J. Bakker, Science **291**, 2118–2120 (2001).

[61] M.F. Kropman, H.-K. Nienhuys, S. Woutersen, and H.J. Bakker, J. Phys. Chem. A **105**, 4622–4626 (2001).

[62] J. Stenger, D. Madsen, J. Dreyer, P. Hamm, E.T.J. Nibbering, and T. Elsaesser, Chem. Phys. Lett. **354**, 256–263 (2002).

[63] J. Stenger, D. Madsen, P. Hamm, E.T.J. Nibbering, and T. Elsaesser, J. Phys. Chem. A **106**, 2341–2350 (2002).

[64] C.J. Fecko, J.D. Eaves, J.J. Loparo, A. Tokmakoff, and P.L. Geissler, Science **301**, 1698–1702 (2003).

[65] C.J. Fecko, J.J. Loparo, S.T. Roberts, and A. Tokmakoff, J. Chem. Phys. **122**, 054506–054518 (2005).

[66] R. Rey, K.B. Moller, and J.T. Hynes, J. Phys. Chem. A **106**, 11993–11996 (2002).

[67] C.P. Lawrence and J.L. Skinner, J. Chem. Phys. **117**, 8847–8854 (2002).

[68] A. Piryatinski and J.L. Skinner, J. Phys. Chem. B **106**, 8055–8063 (2002).

[69] A. Piryatinski, C.P. Lawrence, and J.L. Skinner, J. Chem. Phys. **118**, 9672–9679 (2003).

[70] C.P. Lawrence and J.L. Skinner, Chem. Phys. Lett. **369**, 472–477 (2003).

[71] C.P. Lawrence and J.L. Skinner, J. Chem. Phys. **118**, 264–272 (2003).

[72] S. Corcelli, C.P. Lawrence, and J.L. Skinner, J. Chem. Phys. **120**, 8107–8117 (2004).

[73] S.A. Corcelli, C.P. Lawrence, J.B. Asbury, T. Steinel, M.D. Fayer, and J.L. Skinner, J. Chem. Phys. **121**, 8897–8900 (2004).

[74] R. Rey, K.B. Moller, and J.T. Hynes, Chem. Rev. **104**, 1915–1928 (2004).

[75] K. Moller, R. Rey, and J. Hynes, J. Phys. Chem. A **108**, 1275–1289 (2004).

[76] S. Mukamel, Ann. Rev. Phys. Chem. **51**, 691–729 (2000).

[77] G.C. Pimentel and A.L. McClellan, The Hydrogen Bond (W.H. Freeman and Co., San Francisco, 1960).

[78] W.M. Zhang, V. Chernyak, and S. Mukamel, J. Chem. Phys. **110**, 5011–5028 (1999).

[79] K. Kwak, S. Park, I.J. Finkelstein, and M.D. Fayer, J. Chem. Phys., in preparation.

[80] R. Kubo, in: D. Ter Haar (ed.), Fluctuation, Relaxation, and Resonance in Magnetic Systems (Oliver and Boyd, London, 1961).

[81] A. Chandra, Phys. Rev. Lett. **85**, 768–771 (2000).

[82] B. Nigro, S. Re, D. Laage, R. Rey, and J.T. Hynes, J. Phys. Chem. A **110**, 11237–11243 (2006).

[83] S. Woutersen and H.J. Bakker, Phys. Rev. Lett. **83**, 2077–2080 (1999).